OXFORD
UNIVERSITY PRESS

Cambridge International AS & A Level

Complete
Physics

Third Edition

Jim Breithaupt

Hossam Attya
Camille Pervenche
Jaykishan Sharma

OXFORD
UNIVERSITY PRESS

OXFORD
UNIVERSITY PRESS

Great Clarendon Street, Oxford, OX2 6DP, United Kingdom

Oxford University Press is a department of the University of Oxford.
It furthers the University's objective of excellence in research, scholarship, and education by publishing worldwide. Oxford is a registered trade mark of Oxford University Press in the UK and in certain other countries

British Library Cataloguing in Publication Data
Data available

978-1-38-200539-5

10 9 8 7 6

Paper used in the production of this book is a natural, recyclable product made from wood grown in sustainable forests. The manufacturing process conforms to the environmental regulations of the country of origin.

Printed in India by Multivista Global Pvt. Ltd

Acknowledgements

The author would like to thank the following:

I would like to thank my family for their support in the preparation of the book, particularly my wife, Marie, for secretarial support and encouragement. I am also grateful to the publishing team at Oxford University Press, in particular Oliver Thornton, who initiated the production of the book, and Ben Rout and Emma Baxter, who coordinated the production of the book.

The publisher and authors would like to thank the following for permission to use photographs and other copyright material:

Cover: Karl Gaff/Science Photo Library

Artworks: Aptara

Photos: pxiv: Sheila Terry/Science Photo Library; pxv: Andrew Lambert Photography/Science Photo Library; pxviii: Giphotostock/Science Photo Library; pxix: Martyn F. Chillmaid/Science Photo Library; pxx: Andrew Lambert Photography/Science Photo Library; p2-3: Funny Solution Studio/Shutterstock; p8: Narvikk/E+/Getty Images; p10: Dariusz Gora/Shutterstock; p18: NASA/Science Photo Library; p21: Loren Winters/Visuals Unlimited, Inc./Science Photo Library; p32: Germanskydiver/Shutterstock; p33: foto2800/123RF; p39: kozmoat98/E+/Getty Images; p42 (L): Phant/Shutterstock; p42 (R): James King-Holmes/Science Photo Library; p43 (L): Naten/Shutterstock; p43 (R): Ashley Cooper/Visuals Unlimited/Science Photo Library; p63: Boogasheski/Dreamstime; p70: Martin Bond/Science Photo Library; p72: Caia Image/Science Photo Library; p78: Merkushev Vasiliy/Shutterstock; p80: Gerard Koudenburg/Shutterstock; p94: Charles D. Winters/Science Photo Library; p97: nicolas_/E+/Getty Images; p98: Jacek Kita/iStockphoto; p99: Scanrail/iStock/Getty Images; p120: Science photo/Shutterstock; p122 (T): Sciencephotos/Alamy Stock Photo; p122 (B): Sciencephotos/Alamy Stock Photo; p123 (L): Andrew Lambert Photography/Science Photo Library; p123 (R): Andrew Lambert Photography/Science Photo Library; p126: Berenice Abbott/Science Photo Library; p131: Pressmaster/Shutterstock; p135: Ducu59Us/Shutterstock; p138: Giphotostock/Science Photo Library; p140: Jim Breithaupt; p141: Jim Breithaupt; p143: Dept. Of Physics, Imperial College/Science Photo Library; p147: Omikron/Science Photo Library; p148: Drs A. Yazdani & D.J. Hornbaker/Science Photo Library; p150: Sputnik/Science Photo Library; p151 (L): Science Photo Library; p151 (R): C.T.R. Wilson/Science Photo Library; p152: Carl Anderson/Science Photo Library; p155: Public Health England/Science Photo Library; p157 (T): Prof. G. Piragino/Science Photo Library; p157 (B): James King-Holmes/Science Photo Library; p166-167: koya979/Shutterstock;p168: Digital Vision/Getty Images; p171: Enigma/Alamy Stock Photo; p191 (T): Bettmann/Getty Images; p196: Digital Vision/Photodisc/Getty Images; p198: European Space Agency/CNES/Arianespace/CSG/S. Martin/Science Photo Library; p201: ©Science Museum/Science & Society Picture Library; p206: David Ducros/Science Photo Library; p212: ©Science Museum/Science & Society Picture Library; p239: Brandon Blinkenberg/Shutterstoc; p244: Science Photo Library; p245: Martin F. Chillmaid/Science Photo Library; p253: Trevor Clifford Photography/Science Photo Library; p255: d13/Shutterstock; p270: Ivan Smuk/Shutterstock; p272: Thierry Berrod, Mona Lisa Production/Science Photo Library; p297: Library Of Congress/Science Photo Library; p301 (T): Giphotostock/Science Photo Library; p301 (C): Physics Dept., Imperial College/Science Photo Library; p301 (B): Physics Dept., Imperial College/Science Photo Library; p303: Andrew Lambert Photography/Science Photo Library; p314: Stock-media/iStockphoto; p322: Efda-Jet/Science Photo Library; p329: Suzanne Tucker/Shutterstock; p336: springsky/Shutterstock; p340: xfox01/Shutterstock; p348 (L): NASA; p348 (R): NASA/WMAP Science Team;

Contents

AS Level

A Level

Introduction

Complete Physics aims to make your study of physics successful and interesting.

It has been written specifically to meet the requirements of the Cambridge International AS & A Level Physics syllabus (9702).

The book is divided into 24 chapters:

- Chapters 1–10 cover AS level.
- Chapters 11–24 cover A level.

New ideas are presented in the book in a careful step-by-step manner to allow you to develop a firm understanding of concepts and ideas. Key concepts that occur throughout the course are identified and discussed in detail at appropriate intervals in the book. The questions at the end of each chapter and after each key concept interval will help you to develop your grasp of the key concepts. These concepts are essential ideas, theories, principles or mental tools that help you to develop a deep understanding of their subject and make links between the different topics.

The key concepts in physics cover the following ideas.

- The development of **models of physical systems** is central to physics. Physics is the science that seeks to understand the behaviour of the Universe. Models simplify, explain and predict how physical systems behave.
- Physical models are usually based on prior observations or experiments, and their **predictions** are tested to check that they are consistent with **evidence** from further observations and experiments.
- **Mathematics** is used to express physical principles and models. It is also a **tool** to analyse theoretical models, solve quantitative problems and make predictions.
- Everything in the Universe comprises **matter** and/or **energy**. **Waves** transfer energy and are essential to many modern applications of physics.
- Matter and energy interact through **forces** and **fields**. The behaviour of the Universe is governed by fundamental forces that act over different length scales and magnitudes, ranging from the very small (particle physics) to the very large (cosmology).

Physics at AS and A Level will require you to describe and explain facts and processes in detail and with accuracy. The course is also about developing skills so that you can apply what you have learned.

The more basic topics are covered in the early chapters of the book and each chapter contains related topics. However, the topics do not have to be studied in strict numerical order, but can be followed in any sequence which suits you or your teacher.

The layout of the book is designed to cover information in a clear way that is easy to access. Its features include:

- **Accessible language** to improve your comprehension and understanding.
- **Purple type** to highlight words that can be found in the glossary. This enables you to easily access a full explanation of important terms used in the text. The comprehensive glossary provides you with clear and concise definitions of over 250 terms used throughout the book.
- **Extension material** helps to widen your horizons and stimulate an interest in broader aspects of physics. This extension material is **not** required for the Cambridge Assessment International Examinations (CAIE), but will serve you well in your further studies.
- **Notes** in boxes and the text to give you useful advice about certain aspects of a topic and so aid your learning. These may be helpful information or explanations to make concepts clearer.
- **Summary tests** to provide a quick check on how well you have learnt and understood the factual content of each topic. Answers are provided at the back of the book, so that the accurately completed test gives you a concise summary of the information in each topic.
- **End of chapter questions** to effectively consolidate the information learned across all of the topics in a chapter. Answers are provided at the back of the book.
- A **syllabus mapping grid** (p.vii) is provided showing the topic in the book that covers each syllabus topic. In addition, these pages also include an outline of the CAIE A level physics examinations structure.

In the Exam-style and Practice Questions sections at the end of each chapter, you will find the following icon:

In the Enhanced Online Student Book, this icon will launch additional digital resources to support your learning further. This content includes:

- **Worksheets** containing additional questions and practice
- **Interactive online quizzes**, including **multiple-choice** question practice
- Further support to improve your **mathematical skills**
- **Animations** illustrating key concepts

Visit **www.oxfordsecondary.com/bookshelf** to redeem your token code and access the Enhanced Online Student Book.

Answers to questions in this book and the syllabus matching grid are also available on the support website: **www.oxfordsecondary.com/caie-al-complete-science**

Specification grid

Specification topic		Sub-topic	Book reference
1.1	Physical quantities	1,2	**Part 1: Skills for starting physics**
1.2	SI units	1,2,3	
1.3	Errors and uncertainties	1,2,3,4	**Part 2: Practical skills**
1.4	Scalars and vectors	1,2,3	1.1 Vectors and scalars
2.1	Equations of motion	1,4	1.2 Speed and velocity
		1,5,6	1.3 Acceleration
		5,6	1.4 Motion along a straight line at a constant acceleration
		7,8	1.5 Free fall
		3,4,5	1.6 Motion graphs
		7	1.7 More calculations on motion along a straight line
		9	1.8 Projectile motion 1
		9	1.9 Projectile motion 2
3.1	Momentum and Newton's laws of motion	1,2,5,6	2.1 Force and acceleration
		1,2,5	2.2 Using $F = ma$
		3,4,5	4.1 Momentum
3.2	Non-uniform motion	1,2	2.3 Terminal speed
3.3	Linear momentum and its conservation	2	4.2 Impact force
		1,2	4.3 Conservation of momentum
		1,2,3,4	4.4 Elastic and inelastic collisions
		1,2,4	4.5 Explosions
4.1	Turning effects of forces	1,2	3.2 The principle of moments
		3,4	3.3 More on moments
		1,2,3,4	3.6 Statics calculations
4.2	Equilibrium of forces	1	3.2 The principle of moments
		1	3.3 More on moments
		1,2,3,4	3.6 Statics calculations
		1,2,3	3.4 Stability
		1,2,3	3.5 Equilibrium rules
4.3	Density and pressure	1	5.1 Density
		2,3,4	5.2 Pressure
		5,6	5.3 Upthrust
5.1	Energy conservation	1,2	2.4 Work and energy
		5,6,7	2.6 Power
		3,4	2.7 Efficiency
		1–7	2.8 Renewable energy
5.2	Gravitational potential energy and kinetic energy	1–4	2.5 Kinetic energy and potential energy
		1,2,3,4	2.5 Kinetic energy and potential energy

	Specification topic	Sub-topic	Book reference
6.1	Stress and strain	3,4,	5.4 Springs
		1,2,5,6	5.5 Deformation of solids
		1,2,5,6	5.6 More about stress and strain
6.2	Elastic and plastic behaviour	3,4	5.4 Springs
		1	5.5 Deformation of solids
		1,2,3	5.6 More about stress and strain
7.1		2,3,4	8.2 Measuring waves
7.2	Transverse and longitudinal waves	1,2	8.1 Waves and vibrations
7.3		1,2	8.9 The Doppler effect
7.4	Electromagnetic spectrum	1,2,3	8.1 Waves and vibrations 8.4 Electromagnetic waves and polarisation
7.5	Polarisation	2	8.4 Electromagnetic waves and polarisation
8.1	Stationary waves	1,2,3	8.5 Superposition
		2,3	8.6 Stationary and progressive waves
		2,3,4	8.7 More about stationary waves on strings
		2,3,4	8.8 Stationary waves in pipes
8.2	Diffraction	1,2	8.3 Wave properties
8.3	Interference	1, 3,4	9.1 Interference of light
8.4		1,2,3,4	9.2 More about interference
8.5	The diffraction grating	1,2	9.3 The diffraction grating
9.1	Electric current	1,2	6.1 Electric charge
		3,4	6.2 Current and charge
9.2	Potential difference and power	1,2,3	6.3 Potential difference and power
9.3	Resistance and resistivity	1,2,3,5,6	6.4 Resistance
		3,4,7,8	6.5 Components and their characteristics
10.1	Practical circuits	1,2	6.3 Potential difference and power
		3,4,5	7.3 E.m.f. and internal resistance
10.2	Kirchoff's laws	1,2 3,4,5,6,	7.1 Circuit rules 7.2 More about resistance
10.2		4,6,7	7.4 More circuit calculations
10.3	Potential dividers	1,2,3,4	7.5 The potential divider
11.1	Atoms,nuclei and radiation	1,2	10.1 The discovery of the nucleus
		2,3,4,5	10.2 Inside the atom
		7,9	10.3 The properties of α, β and γ radiation
		7,8,9,10,11	10.4 More about α, β and γ radiation
11.2	Fundamental particles	1,2,3,4,5,6	10.6 Fundamental particles
12.1	Kinematics of uniform circular motion	1,2,3	11.1 Uniform circular motion
	Centripetal acceleration	1,2	11.2 Centripetal acceleration
		1,,3,4	11.3 On the road
		1, 3,4	11.4 At the fairground
13.1	Gravitational field	1,2	13.1 Gravitational field strength

Specification topic		Sub-topic	Book reference
13.2	Gravitational force between point masses	3,4	13.3 Newton's law of gravitation 13.5 Satellite motion
13.3	Gravitational field of a point mass	1,2,3	13.4 Planetary fields
13.4	Gravitational potential	1,2,3	13.2 Gravitational potential 13.4 Planetary fields
14.1 14.2	Thermal equilibrium Temperature scales	1,2 1,2,3,4	18.2 Temperature scales
14.3	Specific heat capacity and change of state	1 2	18.3 Specific heat capacity 18.4 Change of state
15.1	The mole	1,2	19.2 The ideal gas law
15.2	Equation of state	1,2,3	19.1 Experiments on gases 19.2 The ideal gas law
15.3	Kinetic theory of gases	1,2,3,	19.3 The kinetic theory of gases
16.1	Internal energy	1,2	18.1 Internal energy and temperature 19.4 Thermodynamics of ideal gases
16.2	The first law of thermodynamics	1,2	18.1 Internal energy and temperature 19.4 Thermodynamics of ideal gases
17.1	Simple harmonic oscillations	1	12.1 Measuring oscillations
		1,2,5	12.2 The principles of simple harmonic motion
		1,2,3,4,5	12.3 More about sine waves
		1,2,3,4,5	12.4 Applications of simple harmonic motion
17.2	Energy in simple harmonic motion	1,2	12.5 Energy and simple harmonic motion
17.3	Damped and forced oscillations, resonance	3	12.6 Forced oscillations and resonance
18.1 18.2	Electric field and field lines Uniform electric fields	1,2,3 1	14.1 Electric field strength
18.3	Electric force between two point charges	1,2	14.1 Coulomb's law
18.4	Electric field of a point charge	1	14.3 Electric field strength of point charges
18.5	Electric potential	1,2, 3,4	14.2 Electric potential
19.1	Capacitors and capacitance	1,2	15.1 Capacitance
		3,4	15.2 Capacitors in series and parallel
19.2	Energy stored in a capacitor	1,2	15.3 Energy stored in a charged capacitor
19.3	Discharging a capacitor	1,2,3	15.4 Capacitor discharge
20.1	Concept of a magnetic field	1,2	16.1 Magnetic field patterns
20.2	Force on a current carrying conductor	1,2,3	16.2 The motor effect 16.3 Magnetic flux density
20.3	Force on a moving charge	1,2,3,4,5,6	16.4 Moving charges in a magnetic field 16.5 Charged particles in circular orbits
20.4	Magnetic fields due to currents		16.1 Magnetic field patterns
20.5	Electromagnetic induction	1,2,3,4,5	17.1 Generating electricity 17.2 The laws of electromagnetic induction
21.1	Characteristics of alternating currents	1,2,3,4	19.3 Alternating current

	Specification topic	Sub-topic	Book reference
21.2	Rectification and smoothing	1,2,3,4	19.3 Alternating current
22.1	Energy and momentum of the photon	1,2,3	20.1 Photoelectricity
		5	20.4 Wave–particle duality
22.2	Photoelectric effect	4	20.2 More about photoelectricity
22.3	Wave-particle duality	1,2,3,4,5	20.4 Wave–particle duality
22.4	Energy levels and line spectra	1,2,3	20.3 Energy levels and spectra
23.1	Mass defect and binding energy	1,2	21.4 Energy from the nucleus
		3,4,5,6	21.5 Binding energy
		5,6,7	21.6 Fission and fusion
23.2	Radioactive decay	1,2,3	21.1 Radioactive decay
		3,4,5,6	21.2 The theory of radioactive decay
		3,4,5,6	21.3 Radioactive isotopes in use
24.1	Production and uses of ultraound	1,2,3,4,5,6	22.1 Ultrasonic imaging
24.2	Production and use of X-rays	1,2,3,4	22.2 X-rays
24.3	PET scanning	1,2,3,4,5,6,	22.3 PET scanning
25.1	Standard candles	1,2,3,4	23.1 Astronomical distances
25.2	Stellar radii	1,2,3	23.2 Stellar radii
25.3	Hubble's theory and the Big Bang theory	1,2,3,4	23.3 Hubble's Law and the Big Bang

Summary of the CAIE AS and A Level physics examination assessment structure

The assessment objectives (AOs) are:

- **AO1** Knowledge and understanding
- **AO2** Handling, applying and evaluating information
- **AO3** Experimental skills and investigations

The assessment objectives are examined through the following papers.

Paper 1: Multiple choice Written paper, 1 hour 15 minutes, 40 marks	Questions are based on the AS Level syllabus content.	40 multiple-choice items of the four-choice type, testing assessment objectives AO1 and AO2
Paper 2: AS Level structured questions Written paper, 1 hour 15 minutes, 60 marks	Questions are based on the AS Level syllabus.	Structured questions testing assessment objectives AO1 and AO2 content
Paper 3: Advanced practical skills Practical test, 2 hours, 40 marks	This paper tests assessment objective AO3 in a practical context. Question 1 involves carrying out an experiment in which measurements are made, recorded and used in calculations and/or to plot graphs to draw conclusions. Question 2 involves carrying out an inaccurate experimental method, evaluating it and suggesting improvements.	Two questions assess the practical skills listed in the practical assessment section of the syllabus. The content of the questions may be outside the syllabus content.
Paper 4: A Level structured questions Written paper, 2 hours, 100 marks	Structured questions testing assessment objectives AO1 and AO2. Questions are based on the A Level syllabus.	Knowledge of material from the AS Level syllabus content will be required.
Paper 5: Planning, analysis and evaluation Written paper, 1 hour 15 minutes, 30 marks	Two questions testing assessment objective AO3. Questions are based on the A Level practical skills of planning, analysis and evaluation.	Knowledge of practical skills from the AS Level syllabus may be required. The content of the questions may be outside of the syllabus content.

Skills for starting physics

1. Using a calculator

Practice makes perfect when it comes to using a calculator. For A level Physics, you need no more than a scientific calculator. You need to make sure you master the technicalities of using a scientific calculator as early as possible in your A level Physics course. At this stage, you should be able to use a calculator to add, subtract, multiply, divide, find squares and square roots and calculate sines, cosines and tangents of angles. Further important calculator functions are described in Chapter 24 Mathematical skills.

2. Making measurements

You should know at this stage how to make measurements using basic equipment such as metre rules, protractors, stopwatches, thermometers, measuring cylinders, balances (for weighing an object) and ammeters and voltmeters. During the course, you will also be expected to use micrometers, (vernier) calipers, centre-reading meters (galvanometers), oscilloscopes and data-loggers, and Hall probes for measuring magnetic fields (A2 only). The use of these items is described for reference in part 2 of this introduction, on p.xvi–xxii (except for the Hall probe: see p.245).

Some useful reminders about making measurements:

- Check the **zero reading** when you use an instrument to make a measurement. For example, a metre ruler worn away at one end might give a zero error or a digital meter or analogue meter (i.e. a pointer with a scale) might not read zero when the input to the meter is zero.
- When a multi-range instrument is used, start with the **highest range** and switch to a lower range if the reading is too small to measure accurately.
- Make sure you record all your measurements in a **logical order**, stating the correct unit of each measured quantity.
- Don't pack equipment away until you are sure you have enough measurements or you have checked unexpected readings. See p.xxiv for **anomalous** measurements.

How to obtain accurate results from an experiment

- Make your measurements as precisely as possible and repeat them if possible to ensure they are reliable. See p.xv–xvi for more about **precision** and **reliability**.

Learning outcomes

On these pages you will learn to:

- appreciate that all physical quantities consist of a numerical magnitude and a unit
- know the unit of each physical quantity listed in the syllabus
- make reasonable estimates of physical quantities in the syllabus
- recall the following SI base quantities and their units: mass (kg), length (m), time (s), current (A), temperature (K), amount of substance (mol) (Topic 19.2)
- express derived units in terms of SI base units
- use SI base units to check the homogeneity of physical equations
- use the following prefixes and their symbols to indicate decimal submultiples or multiples of both base and derived units: pico (p), nano (n), micro (µ), milli (m), centi (c), deci (d), kilo (k), mega (M), giga (G), tera (T)
- use the scientific conventions for labelling graph axes and table columns

- Reduce experimental errors in making measurements. Such errors are either:
  1. Random errors that cause repeat readings to differ at random and are caused by uncontrolled variables,
  2. Systematic errors such as zero errors that cause readings to be consistently higher or lower than they should be. See p.xvi for more about random and systematic errors.
- Where measurements are repeated, calculate and use the **mean values**. Assess the uncertainty in a measurement from the spread of the readings about the mean value. See p.xxiii for more about how to assess and use **uncertainties** in your measurements to find out how **accurate** your results are.

3. Using measurements in calculations

When you record a measurement, you must always note the correct unit as well as the numerical value of the measurement.

The scientific system of units is called the SI system. The SI system is described in more detail later in this introduction, on p.xiv. The base units of the SI system you need to remember are listed below. All other units are derived from the SI base units. The symbol for each unit is shown in brackets. Later in your course you will also meet the kelvin (K), which is the SI unit of temperature (see p.272), and the mole (mol) which is the SI unit for the amount of a substance (see p.287).

- The metre (m) is the SI unit of length. Note also that $1\,m = 100\,cm = 1000\,mm$.
- The kilogram (kg) is the SI unit of mass. Note that $1\,kg = 1000\,g$.
- The second (s) is the SI unit of time.
- The ampere (A) is the SI unit of current.

All other units are derived from the SI base units. For example,

- the SI unit of speed is the metre per second because speed = distance ÷ time. The unit is abbreviated as $m\,s^{-1}$ because it is the unit of distance (m) ÷ the unit of time (s)
- the SI unit of force is the newton (N). Because 'force = mass × acceleration' as explained in Topic 2.1, the newton expressed in terms of base units is the kilogram metre per second squared ($kg\,m\,s^{-2}$).

See p.xiii–xiv for more about derived units.

Powers of ten and numerical prefixes are used to avoid unwieldy numerical values. For example,

- $1\,000\,000 = 10^6$ which is 10 raised to the power 6 (usually stated as '10 to the 6')
- $0.000\,000\,1 = 10^{-7}$ which is 10 raised to the power −7 (usually stated as '10 to the −7').

Prefixes are used with units as abbreviations for powers of ten. For example, a distance of 1 kilometre may be written as $1000\,m$ or $10^3\,m$ or $1\,km$. The most common prefixes are shown in Table 1.

Table 1 Prefixes

Prefix	pico-	nano-	micro-	milli-	kilo-	mega-	giga-	tera-
Value	10^{-12}	10^{-9}	10^{-6}	10^{-3}	10^3	10^6	10^9	10^{12}
Prefix symbol	p	n	μ	m	k	M	G	T

Note that the prefix c for centi- stands for 10^{-2} or a hundredth (as in cm for centimetre) and prefix d for deci- stands for 10^{-1} or a tenth. For example, $1\,cm^3 = 1 \times (10^{-2})^3 = 10^{-6}\,m^3$.

Also, note that cubic centimetre (cm^3) and the gram (g) are in common use and are therefore allowed as exceptions to the prefixes in Table 1.

Standard form is usually used for numerical values smaller than 0.001 or larger than 1000.

- The numerical value is written as a number between 1 and 10 multiplied by the appropriate power of ten. For example:

$$64\,000\,m = 6.4 \times 10^4\,m$$
$$0.000\,005\,1\,s = 5.1 \times 10^{-6}\,s$$

- A prefix may be used instead of some or all of the powers of ten. For example,

$$35\,000\,m = 35 \times 10^3\,m = 35\,km$$
$$0.000\,000\,59\,m = 5.9 \times 10^{-7}\,m = 590\,nm$$

To convert a number to standard form, count how many places the decimal point must be moved to make the number between 1 and 10. The number of places moved is the power of ten that must accompany the number between 1 and 10. Moving the decimal place:

- to the left gives a positive power of ten (e.g. $64\,000 = 6.4 \times 10^4$)
- to the right gives a negative power of ten (e.g. $0.000\,005\,1 = 5.1 \times 10^{-6}$).

4. Using trigonometry

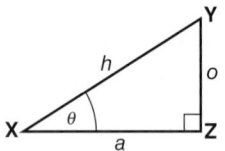

Figure 1 A right-angled triangle

The sine, cosine and tangent of an angle are defined from the right-angled triangle. Figure 1 shows a right-angled triangle XYZ, in which side XY is the **hypotenuse** (i.e. the side opposite the right angle), side YZ is **opposite** to angle θ and side XZ is **adjacent** to angle θ.

$$\sin\theta = \frac{YZ}{XY} = \frac{o}{h} \quad \text{where } o = YZ, \text{ the side opposite angle } \theta$$

$$\cos\theta = \frac{XZ}{XY} = \frac{a}{h} \quad h = XY, \text{ the hypotenuse}$$

$$\tan\theta = \frac{YZ}{XZ} = \frac{o}{a} \quad a = XZ, \text{ the side adjacent to angle } \theta$$

Notice that when:
i $\theta = 0$, YZ = 0 so $\sin\theta = 0$ and XZ = XY so $\cos\theta = 1$
ii $\theta = 90°$, YZ = XY so $\sin\theta = 1$ and XZ = 0 so $\cos\theta = 0$

Pythagoras' theorem states that for any right-angled triangle:

The square of the hypotenuse = the sum of the squares of the other two sides

Applying Pythagoras' theorem to the right-angled triangle XYZ gives:

$$\mathbf{XY^2 = XZ^2 + YZ^2}$$

5. Symbols in science

Symbols are used in equations and formulae to represent physical quantities. In your previous course, you may have used equations with words instead of symbols to represent physical quantities. For example , you will have met the equation 'distance moved = speed × time'. Perhaps you were introduced to the same equation in the symbolic form '$s = vt$', where s is the symbol used to represent distance , v is the symbol used to represent speed and t is the symbol used to represent time. The equation in symbolic form is easier to use because the rules of algebra are more easily applied to it than to a word equation.

If you used symbols in your GCSE course, you might have met the use of s for distance and I for current. Maybe you wondered why we don't use d for distance instead of s; or C for current instead of I. The answer is that physics discoveries have taken place in many countries. The first person to discover the key ideas about speed was Galileo, the great Italian scientist; he used the word '*scale*' from his own language for distance, and therefore assigned the symbol s to distance. Important discoveries about electricity were made by Ampère, the great French scientist; he wrote about the intensity of an electric current, so he used the symbol I for electric current. The symbols we now use are used in all countries in association with the **SI system** of units.

Table 2

Physical quantity	Symbol	Unit	Unit symbol
Distance	s	metre	m
Speed or velocity	v	metre per second	$m s^{-1}$
Acceleration	a	metre per second per second	$m s^{-2}$
Mass	m	kilogram	kg
Force	F	newton	N
Energy or work	E	joule	J
Power	P	watt	W
Density	ρ	kilogram per cubic metre	$kg m^{-3}$
Current	I	ampere	A
Potential difference or voltage	V	volt	V
Resistance	R	ohm	Ω

A full list of all the physical quantities, their symbols and units, that you are likely to meet in the AS or A level Physics course is given at the end of Topic 24.3 More about algebra.

6. Using equations

Equations often need to be rearranged. This can be confusing if you don't learn the following basic rules at an early stage:

- **Read an equation properly**. For example, the equation $v = 3t + 2$ is not the same as the equation $v = 3(t + 2)$. If you forget the brackets when you use the second equation to calculate v when $t = 1$, then you will get $v = 5$ instead of the correct answer $v = 9$. The first equation tells you to multiply t by 3 then add 2. The second equation tells you to add t and 2, then multiply the sum by 3.
- **Rearrange an equation properly**. In simple terms, always make the same change to both sides of an equation. For example, to make t the subject of the equation $v = 3t + 2$:

Step 1: Subtract 2 from both sides of the equation, so $v - 2 = 3t + 2 - 2 = 3t$

Step 2: The equation is now $v - 2 = 3t$ and can be written $3t = v - 2$

Step 3: Divide both sides of the equation by 3, so $\dfrac{3t}{3} = \dfrac{v - 2}{3}$

Step 4: Cancel 3 on the top and the bottom of the left-hand side, to finish with $t = \dfrac{v - 2}{3}$

To use an equation as part of a calculation:

- Start by making the quantity to be calculated the **subject** of the equation.
- Write the equation out with the **numerical values** in place of the symbols.
- Carry out the calculation and make sure you give the answer with the **correct unit**.

Unless the equation is simple (e.g. $V = IR$), don't insert numerical values then rearrange the equation. Rearrange, then insert the numerical values; you are less likely to make an error if the numbers are inserted later in the process.

Estimating the result of a measurement or calculation is a useful technique to ensure big errors have not been made. For example, in a density measurement of a metal object, you would expect the result to be several times the density of water. A result of $800\,kg\,m^{-3}$ compared with a value of $1000\,kg\,m^{-3}$ for water means an error must have been made. Before carrying out an exact calculation of a physical quantity, a mental estimate of the approximate result of the calculation gives an 'order of magnitude' value. Combined with an awareness of the magnitudes of physical quantities, errors can thus be spotted and corrected before carrying out an accurate calculation. See Chapter 24 Mathematical skills for more about signs, equations and formulae.

7. More about SI units

In the SI system of units, every derived unit can be expressed in terms of the **base units** of the SI system. Table 3 at the end of Topic 24.3 More about algebra, shows how the derived units you meet at A level are related to the SI base units and, in some cases, how they relate to each other. For example, in terms of base units, the coulomb (C) is the ampere second (A s). Such knowledge can prove useful. For example, if you can't quite recall the formula for:

- Density, but you know that its unit is the kilogram per cubic metre ($kg\,m^{-3}$), you should be able to see that $kg\,m^{-3}$ is the unit for mass divided by volume.
- Centripetal acceleration, remembering that the unit of acceleration (i.e. $m\,s^{-2}$) is the same as the unit of speed^2 divided by the unit of distance (i.e. $m^2\,s^{-2}/m$) can lead you to the formula v^2/r for centripetal acceleration. See Topic 11.2 if necessary.

Derived units written in terms of their base units can be used to check equations. The physical quantities on each side of an equation must match in terms of base units. If they don't match, the equation cannot be correct. For example, consider:

- the equation $v = \sqrt{2gR}$, which is used to calculate the escape speed v of an object from the surface of a planet of radius R and surface gravitational field strength g
  Left-hand side base units = $m\,s^{-1}$
  Right-hand side base units = $\sqrt{m\,s^{-2} \times m} = m\,s^{-1}$
  The equation has the same combination of base units on each side, so it is correct; we say it is **homogeneous** in terms of the base units.
- the equation $W = QV$ is used to calculate the work done W to move a charge Q through a potential difference V

Left-hand side base units (see Table 3) = $kg\,m^2\,s^{-2}$
Right-hand side base units = $(A\,s) \times (kg\,m^2\,s^{-3}\,A^{-1})$
= $kg\,m^2\,s^{-2}$
The equation has the same combination of base units on each side, so it is homogeneous. Note that for simple equations such as this, homogeneity can be checked more quickly by recalling basic relationships between physical quantities. In this example, one volt is one joule per coulomb, so the unit of QV is the joule per coulomb × the coulomb, which is the joule.

The links between different units do not need to be made through the SI base units. For example, the volt (V) is the joule per coulomb ($J\,C^{-1}$), which is a useful link to remember as it helps you to develop your understanding of potential difference.

There are some units in the A level specification that are not SI units but they are used in specific situations for convenience. Those listed below are in common use.

1 The **atmosphere** (atm) is a unit of pressure equal to the mean pressure of the atmosphere at sea level and is equal to $101\,kPa$, or $1.01 \times 10^5\,Pa$.
2 The **electron volt** (eV) is a unit of energy defined as the work done when an electron moves through a p.d. of 1 V.
  $1\,eV = 1.6 \times 10^{-19}\,J$.
  Note that $1\,MeV = 10^6\,eV = 1.6 \times 10^{-13}\,J$
3 The **kilowatt hour** (kW h) is a unit of energy equal to the energy supplied to a 1 kilowatt appliance in 1 hour which is $3.6\,MJ$.
4 The **light year** is the distance travelled in space by light in 1 year.
5 The **litre** is a unit of volume equal to $10^{-3}\,m^3$.

In the laboratory

The experimental skills you will develop during your course are part of the 'tools of the trade' of every physicist. Data loggers and computers are commonplace in modern physics laboratories, but awareness on the part of the user of precision, reliability, errors and accuracy are just as important as when measurements are made with much simpler equipment. The practical skills you develop in the laboratory will prepare you for the CIE Paper 3 practical tests and for the CIE Paper 5 written practical examination. Let's consider in more detail what you need to be aware of when you are working in the physics laboratory.

Safety and organisation

A science laboratory should be a safe place, but it can be dangerous if the safety rules and common sense are disregarded. Your teacher will give you a set of safety rules and should explain them to you. You must comply with them at all times. In addition, you must use your common sense and organise yourself so that you work safely. For example, if you set up an experiment with pulleys and weights, you need to ensure they are stable and will not topple over. Before you set up an electrical circuit, think about where to place the meters so you can read them easily.

Working with others

Most scientists work in teams, each person cooperating with the other team members to achieve specific objectives.

In your practical activities, you will often work in a small group of two or three people. In such group work, you need to cooperate with the others in the group so everyone understands the objectives of the practical activity and everyone participates in planning and carrying out the activity.

Planning

You will be asked to plan an experiment or investigation in Paper 5. The practical activities you carry out during your course should enable you to prepare a plan. Here are the key steps in drawing up a plan:

1 Decide in detail what you intend to investigate. Identify the variable you will alter (the **independent variable**) and the variable that will be altered as a result (the **dependent variable**). Note these variables and note the other variables that need to be controlled and kept constant. A **control variable** that can't be kept constant would cause the dependent variable to alter.
2 Select the equipment necessary for the measurements.

Specify the range of any electrical meters you need.
3 List the key stages in the method you intend to follow and make some preliminary measurements to check your initial plans. Consider safety issues before you do any preliminary tests. If necessary, modify your plans as a result of your preliminary tests.
4 If the aim of your investigation is to test a hypothesis or theory or use the measurements to determine a physical quantity (e.g. resistivity), you need to describe how you will use your measurements to achieve the objectives of your plan.

Carrying out instructions and recording your measurements

In some investigations, and in the CIE Paper 3 tests, you will be expected to follow instructions supplied to you, either verbally or on a worksheet. You should be able to follow a sequence of instructions without guidance. However, always remember safety first and, if the instructions are not clear, ask your teacher to clarify them.

When you record your measurements, tabulate them with a column for the independent variable and one or more columns for the dependent variable. Remember to allow for repeat readings and average values, if appropriate. Include columns for values of quantities calculated from your measurements. The table should have a clear heading for each of the measured variables, with the unit symbol shown after the heading (e.g. current/A or I/A), as below.

Single measurements of other variables (e.g. control variables) should be recorded together, immediately before or after the table. Make sure you record the unit of each measurement. In addition, you should record the precision (i.e. the least detectable reading) of each measurement (which should be the same for each measurement of the same measured variable). This information is important when you come to analyse and evaluate your measurements.

Table 1 *Tabulating the measurements from an investigation of p.d. against current for a wire*

potential difference / V	current /A			average current/A
	1st set	2nd set	3rd set	

length of wire/m = _____
diameter of wire/mm = _____, _____, _____
average diameter of wire/mm = _____

Making careful measurements

Learning outcomes

On these pages you will learn to:

- measure lengths using a ruler, vernier calipers and a micrometer
- weigh an object and determine its mass using a spring balance, a lever balance or a top-pan balance
- use a protractor to measure an angle and use a set square
- measure time intervals using clocks, stopwatches and the time base of an oscilloscope
- measure temperature using a thermometer
- use ammeters and voltmeters with appropriate scales
- use an oscilloscope.

In addition you should be able to:
- distinguish between systematic errors (including zero errors) and random errors
- understand what is meant by accuracy, sensitivity, linearity, reliability, precision and validity
- read analogue and digital displays

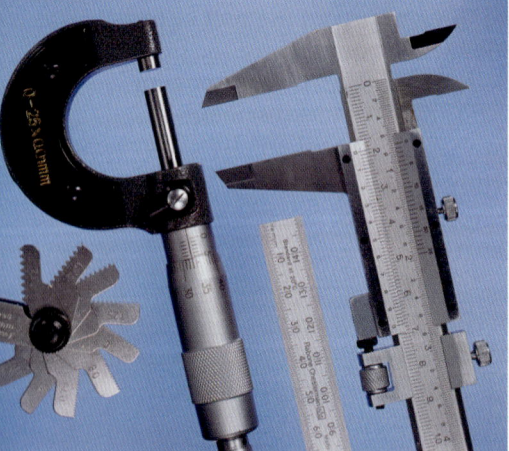

Figure 1 *Physics instruments*

Measurements and errors

Measurements play a key role in science, so they must be:

1 **Reliable** – which means that a consistent value should be obtained each time the same measurement is repeated. An unreliable weighing machine in a shop would cause the customers to go elsewhere. In science, you can't go elsewhere so the measurements must be reliable. Each time a given measurement is repeated, it should give the same value within acceptable limits.
2 **Valid** – which means the measurements are of the required data or can be used to give the required data. For example, a voltmeter connected across a variable resistor in series with a lamp and a battery would not measure the potential difference across the lamp.

Errors of measurement are important in finding out how accurate a measurement is. We need to consider errors in terms of differences from the mean value. Consider the example of measuring the diameter of a uniform wire using a micrometer. Suppose the following diameter readings are taken for different positions along the wire from one end to the other:

0.34 mm, 0.33 mm, 0.36 mm, 0.33 mm, 0.35 mm

- The **range** of the measurements is 0.03 mm. This is the difference between the lowest and the highest reading. We will see later we can use this to estimate the **uncertainty** of the measurement.
- The **mean value**, $<d>$, is 0.34 mm, calculated by adding the readings together and dividing by the number of readings. If the difference between each reading and $<d>$ changed regularly from one end of the wire to the other, it would be reasonable to conclude that the wire was non-uniform. Such differences are called **systematic errors**. If no such differences can be seen in the set of readings, in other words, there is no obvious trend in the differences, the differences are said to be **random errors**.

What causes random errors? In the case of the wire, vibrations in the machine used to make the wire might have caused random variations in its diameter along its length. The experimenter might not use or read the micrometer correctly consistently. This shouldn't happen to a skilled experimenter.

The range of the diameter readings is from 0.33 mm to 0.36 mm. The readings lie within 0.015 mm (i.e. half the range) of the mean value, which we will round up to 0.02 mm. The diameter can therefore be written as 0.34 ± 0.02 mm. The diameter is accurate to ±0.02 mm. The uncertainty in the mean value of the diameter is therefore ±0.02 mm.

Using instruments

Instruments used in the physics laboratory range from the very basic (e.g. a millimetre scale) to the highly sophisticated (e.g. a multichannel data recorder). Whatever type of instrument you use, you need to know what the following terms mean.

Zero error
Does the instrument read zero when it is supposed to? If not, the zero reading must be taken into account when measuring the gap width, otherwise there will be a systematic error in the measurements.

Uncertainty

The uncertainty in the mean value of a measurement may be taken as half the range, expressed as $a \pm$ value (e.g. $I = 2.6 \pm 0.2\,A$). If the readings are the same, the instrument precision should be used as the uncertainty.

Accuracy

An **accurate** measurement is one that has been obtained using accurately-calibrated instruments correctly and where there are no systematic errors. Accuracy is a measure of confidence in a measurement and is usually expressed as the uncertainty in the measurement.

Precision

The precision of a measurement is the degree of exactness of the measurement. The precision of an instrument is the smallest non-zero reading that can be measured using the instrument.

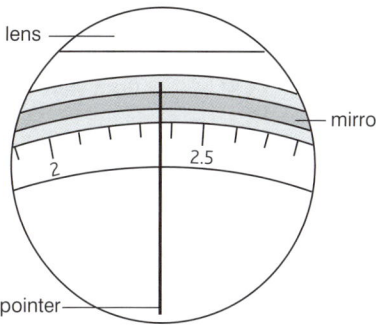

Figure 2 Magnifying a scale

- If the reading of an instrument fluctuates when it is being taken, take several readings and calculate the mean value and range of the measurements. The precision of the measurement is then given by the uncertainty of the readings (i.e. half the range of the readings).

- If the reading is constant, estimate the precision of a measurement directly from the instrument (or use the stated precision from the instrument specification).

Precise readings are not necessarily accurate readings, because systematic errors could make precise readings all lower or all higher than they ought to be.

Linearity

This is a design feature of many instruments: it means the reading is directly proportional to the magnitude of the variable that causes the reading to change. For example, if the scale of a moving coil meter is linear, the reading of the pointer against the scale should be proportional to the current.

A lens may be used to read a scale with greater precision. Provided the pointer is thin compared with a scale division, using a lens enables the scale to be read to within 0.2 of a division. In Figure 2, the scale can be read to $\pm 0.02\,A$ ($= 0.2$ of a scale division $\times 0.1\,A$).

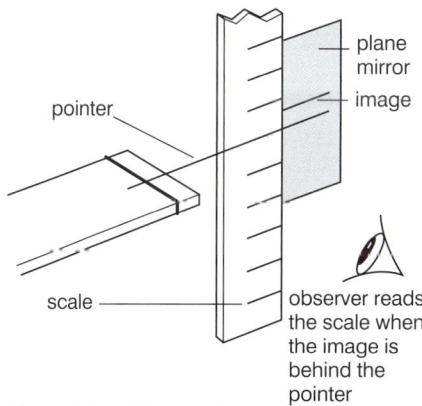

Figure 3 Reading a scale

Measurement errors, referred to as parallax errors, are caused in analogue instruments if the pointer on the scale is not observed correctly. The observer must be directly in front of the pointer when the reading is made. Figure 3 shows how a plane mirror is used for this purpose. The image of the pointer must be directly behind the pointer to ensure the observer views the scale directly in front of the pointer.

Instrument range

Multi-range instruments such as multimeters have a 'range' dial that needs to be set according to the maximum reading to be measured. For example, if the dial can be set at $0-0.10\,A$, $0-1.00\,A$ or $0-10.0\,A$, you would use the $0-1.00\,A$ range to measure the current through a 0.25 A torchbulb as the $0-0.10\,A$ range is too low and the $0-1.00\,A$ range is more sensitive than the $0-10.0\,A$ range.

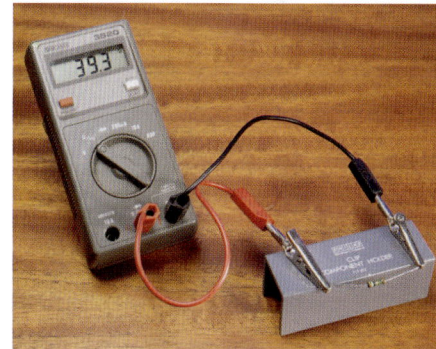

Figure 4 A multimeter

Everyday physics instruments

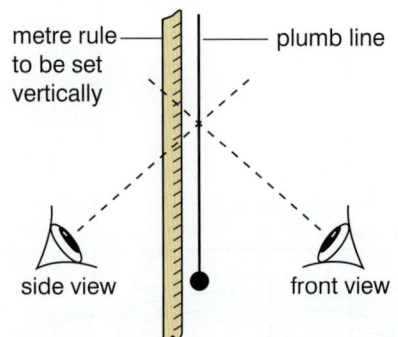

metre rule to be set vertically — plumb line

side view — front view

If the metre rule appears parallel to the plumb line from the front and the side, the ruler must be vertical.

Figure 1 *Finding the vertical*

Rules and scales

Metre rules are often used as vertical or horizontal scales in mechanics experiments.

To set a metre rule in a vertical position:

- use a set square perpendicular to the rule and the bench, if the bench is known to be horizontal, or
- use a plumb line (a small weight on a string) to see if the rule is vertical. You need to observe the rule next to the plumb line from two perpendicular directions. If the rule appears parallel to the plumb line from both directions, then it must be vertical. See Figure 1.

To ensure a metre rule is horizontal, use a set square to align the metre rule perpendicular to a vertical metre rule.

Micrometers and vernier calipers

Micrometers give readings to within 0.01 mm. A **digital** micrometer gives a read-out equal to the width of the micrometer gap. An **analogue** micrometer has a barrel on a screw thread with a pitch of 0.5 mm. For such a micrometer:

- the edge of the barrel is marked in 50 equal intervals, so each interval corresponds to changing the gap of the micrometer by 0.5/50 mm = 0.01 mm
- the stem of the micrometer is marked with a linear scale graduated in 0.5 mm marks
- the reading of a micrometer is where the linear scale intersects the scale on the barrel.

Figure 2 shows a reading of 4.06 mm. Note that the edge of the barrel is between the 4.0 mm and 4.5 mm marks on the linear scale. The linear scale intersects the 6th mark after the zero mark on the barrel scale. The reading is therefore 4.00 mm from the linear scale + 0.06 mm from the barrel scale.

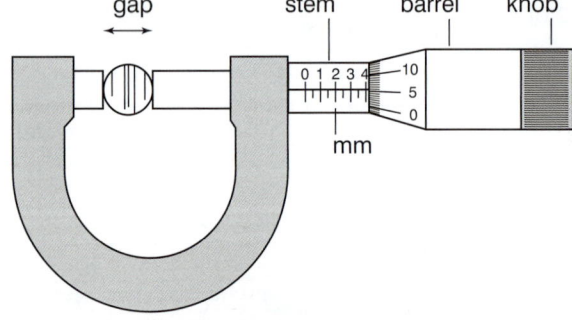

gap   stem   barrel   knob

0 1 2 3 4   10   5   0

mm

Figure 2 *Using a micrometer*

To use a micrometer correctly:

1 Check its zero reading and note the zero error if there is one.
2 Open the gap (by turning the barrel if analogue) then close the gap on the object to be measured. Turn the knob until it slips. Don't overtighten the barrel.
3 Take the reading and note the measurement after allowing, if necessary, for the zero error.
4 Note that the precision of the measurement is ±0.010 mm because the precision of the reading and the zero reading are both ±0.005 mm. So the difference between the two readings (i.e. the measurement) has a precision of 0.010 mm.

Vernier calipers are used for measurements of distances up to 100 mm or more. Readings can be made to within 0.1 mm. The sliding scale of an analogue vernier caliper has ten equal intervals covering a distance of exactly 9 mm, so each interval of this scale is 0.1 mm less than a 1 mm interval. To make a reading:

1 The zero mark on the sliding scale is used to read the main scale to the nearest millimetre. This reading is rounded down to the nearest millimetre.
2 The mark on the sliding scale closest to a mark on the millimetre scale is located and its number noted. Multiplying this number by 0.1 mm gives the distance to be added on to the rounded-down reading.

Figure 3 shows the idea. The zero mark on the sliding scale is between 39 and 40 mm on the mm scale. So the rounded-down reading is 39 mm. The 5th mark after the zero on the sliding scale is nearest to a mark on the millimetre scale. So the extra distance to be added on to 39 mm is 0.5 mm (= 5 × 0.1 mm). Therefore, the reading is 39.5 mm.

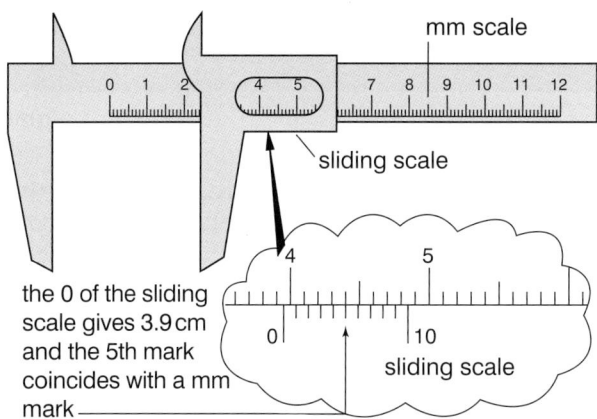

Figure 3 *Using vernier calipers*

Timers

Stopwatches used for interval timings are subject to human error because reaction time, about 0.2 s, is variable for any individual. With practice, the delays when starting and stopping a stopwatch can be reduced. Even so, the precision of a single timing is unlikely to be better than 0.1 s. Digital stopwatches usually have read-out displays with a resolution of 0.01 s but human variability makes such precision unrealistic and the precision of a single timing is the same as an analogue stopwatch.

Timing multiple oscillations requires timing for as many cycles as possible. The timing should be repeated several times to give an average (mean) value. Any timing that is significantly different to the other values is probably due to miscounting the number of oscillations so that timing should be rejected. For accurate timings, a fiducial mark (a pointer or marker in the field of view of the experimenter that can be used as a point of reference) is essential. The mark should be lined up with the centre of the oscillations so that it provides a reference position to count the number of cycles as the object swings past it each cycle.

Electronic timers use automatic switches or 'gates' to start and stop the timer. However, just as with a digital stopwatch, a timing should be repeated, if possible several times, to give an average value. Light gates may be connected via an interface unit to a computer. Interrupt signals from the light gates are timed by the computer's internal clock. A software program is used to provide a set of instructions to the computer.

Balances

A balance is used to measure the weight of an object. Spring balances are usually less precise than lever balances. Both types of balance are usually much less precise than an electronic top-pan balance. The scale or read-out of a balance may be calibrated for convenience in kilograms or grams. The accuracy of an electronic top-pan balance can easily be tested using accurately known masses.

More about measurements

Learning outcomes

On these pages you will learn to:

- use the instruments described in this section
- use a calibrated Hall probe (Topic 16.4)
- use a galvanometer (centre-reading meter) (Topic 7.5)
- use calibration curves where appropriate

During your course you are also expected to be able to use instruments that are more complex. Such instruments may include the oscilloscope, the data logger, sensors and light gates and the Geiger tube with a scaler counter. You should be able to describe how to use these instruments in paper 5. In addition, you should know how to time multiple oscillations and how to avoid parallax errors when reading a scale.

An oscilloscope

This is used to display waveforms and to measure p.d.s and time intervals. You will have probably used an oscilloscope when you studied alternating current. In topics such as capacitor discharge, you will use an oscilloscope to measure the p.d. across a discharging capacitor. When you use an oscilloscope, you should assume that the control dials for its time base and voltage gain are calibrated accurately. However:

- Always check if the oscilloscope has a variable control for either the time base or the voltage gain in addition to the fixed settings of each control dial. If so, you need to ensure that the variable control is at the correct setting (e.g. fully clockwise) for the calibration figure for each of the fixed settings to apply.

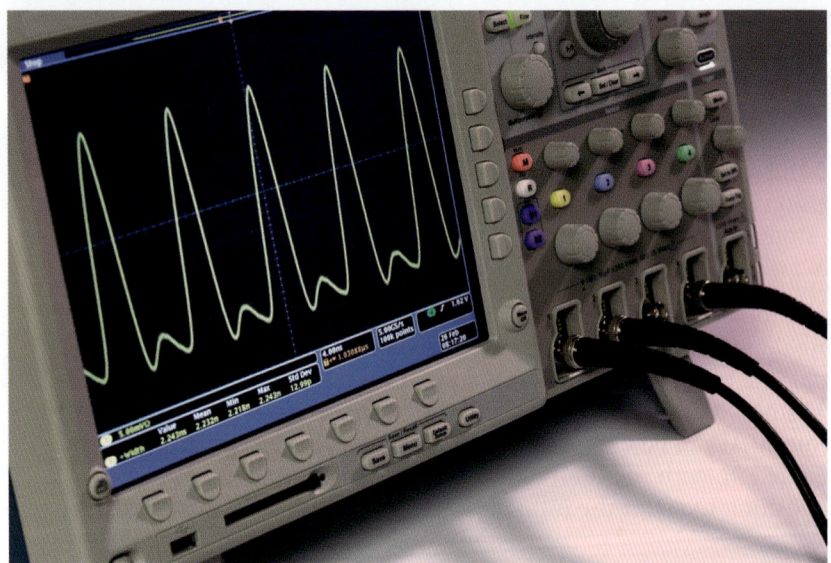

Figure 1 *The oscilloscope*

- Make sure if you are measuring direct p.d.s that the oscilloscope input is set for direct p.d. measurements rather than for a.c. measurements. Likewise, if you are measuring an a.c. waveform, you should check that the input is set for a.c. measurements rather than d.c. measurements.

In addition, when measuring:

- an a.c. waveform, ensure that the y-gain is adjusted so the vertical height of the waveform is as large as possible with the full waveform from top to bottom on the screen. When measuring a time period, ensure that several cycles are displayed across the screen and that you measure across as many cycles as possible to reduce experimental uncertainty.
- d.c. potentials, for example in a capacitor discharge experiment, ensure that the zero reading is correct for zero input p.d. and check that it has not drifted during the investigation.

A data logger

A data logger enables routine or remote measurements to be made as well as measurements over very long or short time scales. Electronic sensors connected to a data logger can be used to record the variation of a physical property such as temperature. Current and voltage sensors can be used to measure currents and potential differences.

Data loggers vary considerably in complexity and ease of use. Assuming that the data logger and sensors are set up, before using a data logger, you may need to choose:

- the most appropriate time scale for the recording
- the time interval between successive recordings (or the number of recordings per second/minute/hour)
- the most appropriate range of each sensor.

If a recording is too fast or too long or if the sensors are out of range, the recording should be repeated if possible.

Most data loggers will be linked to computers which are loaded with appropriate software for recording, processing and/or plotting graphs of the results. You may need to print any graphs out if you intend to use them to measure, for example, the gradient if it is a straight-line graph. However, the computer software may do such measurements for you.

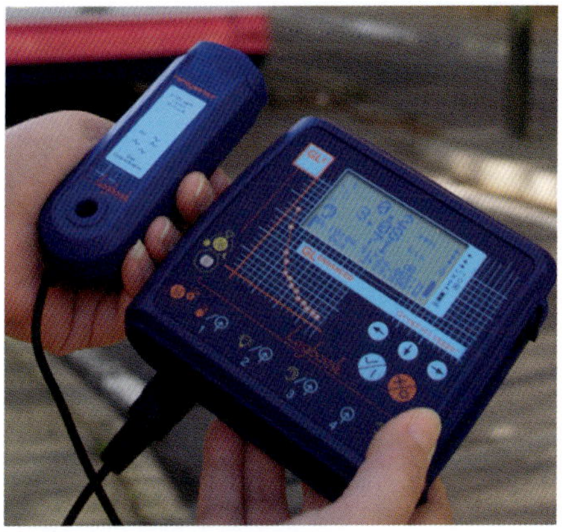

Figure 2 *Using a data logger*

Light gates

These are used with a computer or a data logger or timer to remove some of the random errors associated with personal judgements when a moving object passes a certain position, for example, if you have to time an object to move from rest through a certain distance, as shown in Figure 3.

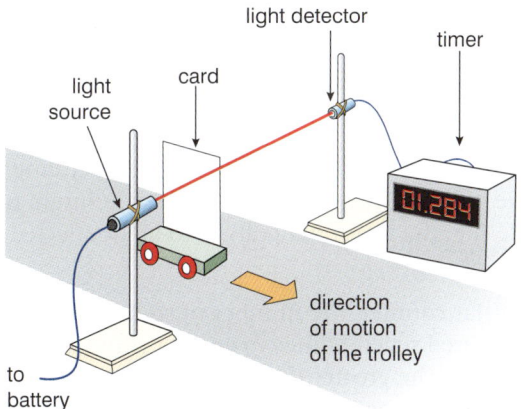

Figure 3 *Using a light gate*

The effect of using light gates should be to reduce the range of the readings for a given measurement. However, light gates may not be suitable for every experiment in which a moving object has to be timed. For example, the time period of an oscillating object that repeatedly moves backwards and forwards through a light gate could only be timed for one half cycle of the object's motion, corresponding to the object moving through the light gate in one direction to start the timing and then in the opposite direction to stop the light gate. The light gate would need to be exactly at the centre of the oscillations otherwise the timing would not be exactly one half cycle. Repeated measurements of one half cycle could be made to give a more reliable mean value and this might give better results than using a stopwatch if the oscillations are too fast to time manually.

The Geiger tube

This is used with a scaler counter which counts the number of ionising particles that enter the tube. The tube p.d. must be set at its operating p.d. which is normally in excess of 300 V. The number of counts in a certain time interval is measured by setting the counter to zero, then starting the counter and stopping it after a certain time.

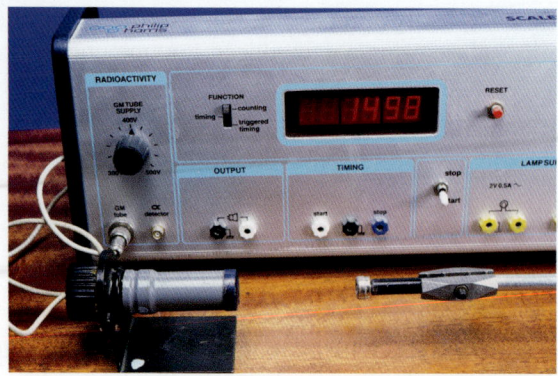

a

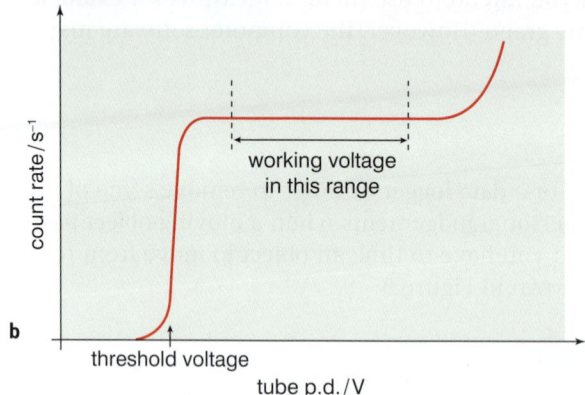

b

Figure 4 *Using a Geiger tube **a** a Geiger tube connected to a scalar counter **b** a graph of count rate v. tube p.d.*

Figure 4b shows how the count rate varies with the tube p.d. The operating p.d. corresponds to the plateau of the graph sufficiently far from the minimum p.d. necessary for the tube to operate (i.e. the threshold p.d.) as to be unaffected by random fluctuations in the tube p.d.

- When using the tube with a scalar counter, the number of counts in a given time (e.g. 100 s) should be measured several times to give a mean value of the count rate (i.e. counts per second). The greater the total number of counts, the smaller the uncertainty in the measurement. If the time interval is too short, random errors that may occur in starting and stopping the counter could be more significant than if a longer time interval were used. If the time interval is too long, it would be difficult to tell if the activity of the source is decreasing or if an error in starting or stopping the timer has occurred.
- To measure the count rate due to a radioactive source, remember to measure the background count rate (ie the count rate without the source present) and subtract it from the measurements made when the source is present.

AS Level

Data processing

For a single reading, the precision of the measuring instrument determines the precision of the reading. A micrometer with a precision of 0.01 mm gives readings that each have a precision of 0.01 mm.

For several readings, the number of significant figures of the mean value should be the same as the precision of each reading. For example, consider the following measurements of the diameter of a wire: 0.34 mm, 0.33 mm, 0.36 mm, 0.33 mm, 0.35 mm. The mean value of the diameter readings works out at 0.342 mm but the third significant figure cannot be justified, as the precision of each reading is 0.01 mm. Therefore the mean value is rounded down to 0.34 mm.

Estimating uncertainties

How confident can you be in your measurements and any results or conclusions you draw from your measurements? If you work out what each uncertainty is, as a percentage of the measurement (the percentage uncertainty), you can then see which measurement is least accurate. You can then think about how that measurement could be made more accurately.

> **Note**
>
> **Always** show your working and reasoning in calculations. At the end of a calculation, don't give the answer to as many significant figures as shown on the calculator display. Give your answer to the same number of significant figures as (or one more than) the data with the least number of significant figures.

Worked example

The mass and diameter of a ball bearing were measured and the uncertainty of each measurement was estimated.

The mass, m, of a ball bearing $= 4.85 \times 10^{-3} \pm 0.02 \times 10^{-3}\,\text{kg}$

The diameter, d, of the ball bearing $= 1.05 \times 10^{-2} \pm 0.01 \times 10^{-2}\,\text{m}$

Calculate and compare the percentage uncertainty of these two measurements.

Solution

The percentage uncertainty of the mass $m = 0.02/4.85 \times 100\% = 0.4\%$

The percentage uncertainty of the diameter $d = 0.01/1.05 \times 100\% = 1.0\%$

The uncertainty in the diameter measurement is therefore more than twice as much as the mass measurement.

When two measurements are added or subtracted, the uncertainty of the result is the sum of the uncertainties of the measurements. For example, the mass of a beaker is measured when it is empty and then when it contains water:

- The mass of an empty beaker $= 65.1 \pm 0.1\,\text{g}$,
- The mass of the beaker and water $= 125.6 \pm 0.1\,\text{g}$

Then the mass of the water could be as much as

$(125.6 + 0.1) - (65.1 - 0.1)\,\text{g} = 60.7\,\text{g}$, or as little as
$(125.6 - 0.1) - (65.1 + 0.1)\,\text{g} = 60.3\,\text{g}$.

The mass of water is therefore $60.5 \pm 0.2\,\text{g}$.

Tips on graphs

- Label graph axes using the same convention as for tables (see p.xv).

- Choose scales that ensure the plotted points cover at least half of each scale.

- Number the whole of each scale every 2 cm and choose scales that are easy to read.

- Plot points to an accuracy better than 1 mm using a fine pencil to make a fine cross or a circled dot.

- Draw a **line of best fit** using a fine pencil so that there is an even distribution of points either side of the line. The line should be no thicker than 1 mm and should be without breaks or kinks.

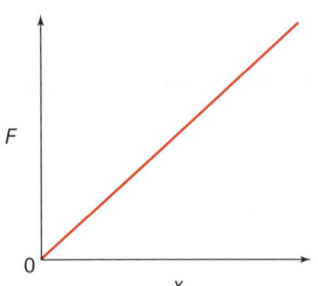

Figure 1 Graph of F = kx

Notes

1. To find the gradient of a straight-line graph, the two points chosen to determine the gradient should be separated by more than half the length of the line. In Figure 1, one of the two points could be the origin 0 of the graph.

2. If the y-intercept cannot be read directly off the graph, read off the coordinates of a point on the line and insert their values and the value of the gradient m into the equation $y = mx + c$ to calculate c.

3. Chapter 24 provides more helpful information about the maths skills you need to develop, for example in Topics 24.4 and 24.5 you will find useful and detailed notes about straight-line graphs.

Straight line graphs

Straight-line graphs are important because they are used to establish the relationship between two physical quantities. For example, consider a set of measurements of the distance fallen by an object released from rest and the time it takes. A graph of distance fallen against (time)^2 should be a straight line through the origin. If the line is straight, the theoretical equation $s = \frac{1}{2}gt^2$ (where s is the distance fallen and t is the time taken) is confirmed. The value of g can be calculated as the gradient of the graph (equal to $\frac{1}{2}g$). If the straight line does not pass through the origin, there is a systematic error in the distance measurement. Even so, the gradient is still $\frac{1}{2}g$.

A best-fit test

Suppose you have obtained your own measurements for an experiment and you use them to plot a graph that is predicted to be a straight line. The plotted points are unlikely to be exactly straight in line with each other. The next stage is to draw a straight line of best fit so that the points are on average as close as possible to the line. Some problems may occur at this stage:

1 There might be a point much further than any other point from the line of best fit. The point is referred to as an **anomaly**. Methods for dealing with an anomalous point are as follows:

- If possible, the measurements for that point should be repeated and used to replace the anomalous point, if the repeated measurement is much nearer the line.

- If the repeated measurement confirms the anomaly, there could be a fault in the equipment or the way it is used. For example, in an electrical experiment, it could be caused by a change of the range of a meter to make that measurement. If no fault is found, make more measurements near the anomaly to see if these measurements lead towards the anomaly. If they do, it is no longer an anomaly and the measurements are valid.

- If a repeat measurement is not possible, the anomalous point should be ignored and a comment made in your report (or on the graph) about it.

2 In physics, a straight-line graph indicates there is a linear relationship between the quantities plotted on each axis. The mathematical equation for a straight line graph is $y = mx + c$, where m is the gradient of the graph and c is the y-intercept (i.e. the point where the line intercepts the y-axis. At AS level you should be able to:

- Determine the gradient of a straight-line graph and the y-intercept, and relate them to a physics equation of the form $y = mx + c$. For example, the force F needed to produce an extension x of a spring is given by the equation $F = kx$ (where k is called the **spring constant**). So in an investigation to test a spring, measurements of the extension for different values of force applied to the spring should give a straight-line graph as in Figure 1. If F is plotted on the y-axis and e on the x-axis, then comparing $F = kx$ with $y = mx + c$ should tell you that the gradient m represents the spring constant k, and the line passes through the origin because $c = 0$.

- Draw a tangent at a point on a graph line which is curved and determine the gradient of the tangent. You will find more in Topic 24.5 about how to draw accurately a tangent to a curved line.

Evaluating your results

You should be able to form a conclusion from the results of an investigation. This might be a final calculation of a physical quantity or property (e.g. resistivity) or a statement of the relationship established between two variables. As explained earlier, the degree of accuracy of the measurements could be used as a guide to the number of significant figures in a 'final result' conclusion. Mathematical links established or verified between quantities should be stated in a 'relationship' conclusion.

You always need to evaluate the conclusion(s) of an experiment or investigation to establish its validity. This evaluation could start with a discussion of the strength of the experimental evidence used to draw the conclusions:

- Discuss the reliability of the data and suggest improvements, where appropriate, that would improve the reliability. You may need to consider the effect of the control variables, if the experimental evidence is not as reliable as it should be.
- Discuss the methods taken (or proposed) to eliminate or reduce any random or systematic errors. Describe the steps taken to deal with anomalous results.
- Evaluate the accuracy of the results by considering the percentage uncertainty in the measurements. These can be compared to identify the most significant sources of uncertainty in the measurements, which can then lead to a discussion of how to reduce the most significant sources of uncertainty.

A Level

The practical skills assessed in Paper 5 involve planning an experiment, carrying out the plan and evaluating the results. Although planning skills are in Paper 5, such skills can be developed throughout the course. For this reason, information about planning an experiment is given in Practical Skills, p.xv. The section below concentrates on analysis, conclusions and evaluation skills beyond the Paper 3 requirements.

More about uncertainties

When a measurement in a calculation is raised to a power n, the percentage uncertainty is increased n times. For example, suppose you need to calculate the area A of cross-section of a wire that has a diameter of 0.34 ± 0.01 mm. You will need to use the formula $A = \pi d^2/4$. The calculation should give an answer of 9.08×10^{-8} m^2. The percentage uncertainty of d is $0.01/0.34 \times 100\% = 2.9\%$. So the percentage uncertainty of A is 5.8% (= $2 \times 2.9\%$). The consequence of this rule is that in any calculation where a quantity is raised to a higher power, the uncertainty of that quantity becomes much more significant.

When two or more quantities are multiplied or divided by each other in a calculation, the overall percentage uncertainty in the result is the sum of the uncertainties of each quantity. For example, if the uncertainty in a resistance R is 5% and in a capacitance C is 4%, the uncertainty in RC is 9%.

Percentage uncertainties

To work out the percentage uncertainty of A, you could calculate:

- the area of cross-section A where $d = 0.34 - 0.01$ mm = 0.33 mm

  This should give an answer of 8.55×10^{-8} m^2.

- the area of cross-section A where $d = 0.34 + 0.01$ mm = 0.35 mm

  This should give an answer of 9.62×10^{-8} m^2.

Therefore, the area lies between 8.55×10^{-8} m^2 and 9.62×10^{-8} m^2.

In other words, the area is $(9.08 \pm 0.53) \times 10^{-8}$ m^2

  (as $9.08 - 0.53 = 8.55$ and $9.08 + 0.53 = 9.62$).

The percentage uncertainty of A is $0.53/9.08 \times 100\% = 5.8\%$.

This is twice the percentage uncertainty of d.

It can be shown as a general rule that for a measurement x:

  the percentage uncertainty in x^n is n times the percentage uncertainty in x.

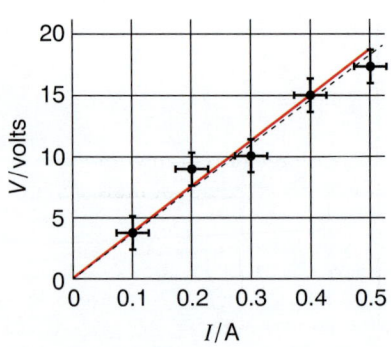

Figure 2 *Error bars*

Notes

1. An example of the worst acceptable line on a graph is shown in Figure 2 as the dashed line. In this case it still passes through the origin as the current was definitely zero when the voltage was zero. This might not always be the case. For example, in an investigation of a spring, there is an uncertainty in the unstretched length of a spring which gives an uncertainty at zero extension.

2. In the above example, check for yourself that the gradient of the dashed line is about 2.5% less than the gradient of the best-fit line. The gradient of the best-fit line is $37.5\,\Omega$ ($= 15.0\,V / 0.40\,A$). So the uncertainty in the gradient is $\pm 0.9\,\Omega$ (0.9 rounded down from 0.9375 as the gradient value is given to one decimal place.)

More about Graphs

Error bars

The uncertainty of each measurement can be used to give a small range or **error bar** for each measurement. Figure 2 shows the idea. **A straight line of best fit** should pass through all the error bars such that there is an even distribution of points either side of the line. To estimate the uncertainty in the gradient, you could draw the steepest straight line of (or least steep straight line) passing through all the error bars. The uncertainty of the gradient is the difference between the gradient of this worst acceptable line and the 'even distribution' best fit line.

For graphs in which the points are clearly not in a straight line:

- The points may lie along a straight line over most of the range, but the points curve away from the straight line further along. If so, a straight-line relationship between the plotted quantities is valid only over the range of measurements which produced the straight part of the line.
- Two or three points might seem to lie on a straight line (see Figure 3). In this case, it cannot be concluded that there is a **linear** relationship between the plotted quantities.

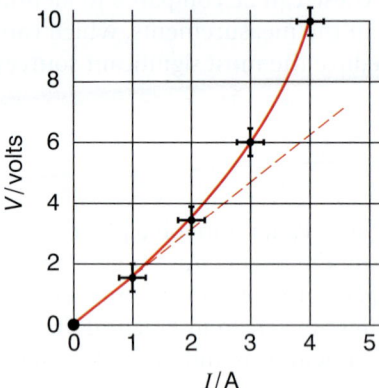

Figure 3 *Curves*

If there is a known mathematical equation between the two plotted quantities, this equation could be rearranged into a linear equation in order to produce a straight-line graph.

Some examples are:

1. In a falling ball experiment, the time t taken for a small ball to fall from rest through two light gates at different distances s apart is measured. The top light gate would need to be a fixed distance below the point of release of the ball. The kinematics equation $s = ut + \frac{1}{2}\,at^2$ would give a curved graph if distance s is plotted on the y-axis and time t is plotted on the x-axis. If both sides of the equation are divided by t, the equation becomes $s/t = u + \frac{1}{2}\,at$. This would give a straight-line graph of the form $y = mx + c$ if s/t is plotted on the y-axis and t on the x-axis. The gradient in this case is $\frac{1}{2}\,a$, where a is the acceleration, and the y-intercept is u, the speed of ball at the upper light gate.

2. In an investigation of the relationship between the current I and the p.d. V for a lamp, the relationship is non-linear as shown by Figure 2. The relationship could be of the form $V = kI^n$ where k and n are constants. This possible power relationship can be tested by taking base 10 logarithms (lg) of both sides to give $\lg V = \lg(kI^n) = \lg k + n \lg I$. Plotting $y = \lg V$ against $x = \lg I$ would give a straight line if the relationship is of the form $V = kI^n$. If so, the value of n can be determined as it is equal to the gradient of the line. You can learn more about logarithms in Topic 24.6.

3. In a capacitor discharge experiment, the capacitor is charged and then discharged through a resistor. The capacitor voltage V is measured at regular intervals as the capacitor discharged. A graph of V on the y-axis against time t (from the start of the discharge) on the x-axis gives an exponential decrease curve corresponding to the equation $V = V_0 \, e^{-t/RC}$ where RC is the time constant of the discharge. The time constant can be determined by taking natural logarithms (ln) of both sides to give $\ln V = \ln (V_0 \, e^{-t/RC}) = \ln V_0 - \left(\dfrac{1}{RC} \times t \right)$.

Comparing this with the straight-line equation $y = mx + c$ means that a graph of $\ln V$ against t should be a straight line with a negative gradient equal to $-\dfrac{1}{RC}$ and a y-intercept equal to $\ln V_0$. You will learn more about capacitor discharge and the exponential decrease in Topic 15.4 and in Topic 24.7.

More about evaluation

In Paper 5, your evaluation skills will be tested in a written paper in which you will be given an equation and some experimental data. From these you need to find the value of a constant to estimate the uncertainty in your answer. So, in addition to the practical skills you were tested on in Paper 3, apart from Paper 3 measurements, observation and data presentation skills, you need to be able to work out the uncertainty of a derived quantity from the uncertainties in the quantities within the derived quantity.

For example, in a magnetic field context, you may be given values and uncertainties for the force F on a current-carrying wire of length L in a uniform magnetic field, when the wire is at right angles to the field lines and there is a current I in the wire. You will be expected to calculate the magnetic flux density B of the field using an appropriate equation, in this case the equation $F = BIL$. To calculate B, you would be expected to rearrange the equation and use the rearranged equation $B = F/IL$ to calculate B. In addition, you would be expected to add together the percentage uncertainties in F, I and L to obtain the percentage uncertainty in B and then calculate the actual uncertainty in B.

If you are given data to plot a graph, you could be expected to estimate the uncertainty in the gradient of the line if the line is straight (or of a tangent to the line if it is curved) and use your estimate to find the uncertainty in a derived quantity. For example, in a capacitor discharge experiment, you could be expected to plot a straight-line graph of $\ln V$ against t in order to find the time constant and its uncertainty from the gradient of the line as explained on p.xxvi. You could then be asked to calculate the uncertainty of the capacitance C given the uncertainty in the resistance and your own calculated uncertainty in the time constant.

Uncertainty calculations are essential to give a proper evaluation of an investigation. By comparing percentage uncertainties in each measured quantity, you can pinpoint weaknesses in the method or procedure used, and you can also make suggestions about how to improve the methods or procedure. By combining percentage uncertainties, you can estimate the overall uncertainty in a derived quantity and thereby compare its value and uncertainty with a known value or with values obtained by other methods.

Notes

1. A data analysis software package on a computer could be used to test different possible relationships, but you need to learn about the above use of logarithms in case they occur in your exams. See Topic 24.6 for more about logarithms.

2. Where logarithms are applied to data, units should be shown in a bracket with the quantity whose logarithm is being taken (e.g. $\ln (I/\text{A})$ not $\ln I$). Also, for a logarithm of the value of a quantity, the number of decimal places should correspond to the number of significant figures in the value of the quantity.

AS Level

This section of the book contains the material that you will cover in the first year of the Cambridge International AS & A Level Physics course.

The content builds on the physics you have studied earlier and is a foundation for the second year of your A Level studies.

The material is divided into two parts:

- Forces: Chapters 1–5
- Electricity, waves and radioactivity : Chapters 6–10

Each chapter is matched to the syllabus and is followed by practice questions that will test your understanding and give you practice at tackling examination-style questions.

1.1 Vectors and scalars

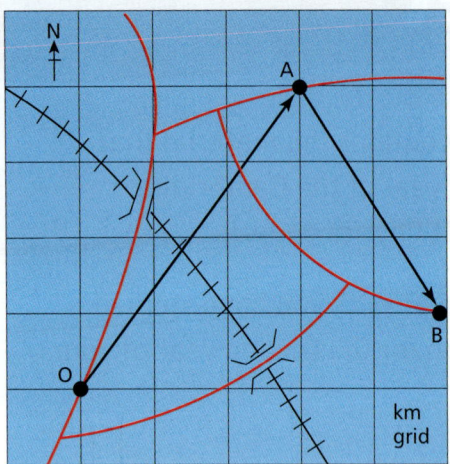

Figure 1 *Map of locality*

Imagine you are planning to cycle to a friend's home several kilometres away from your home. The **distance** you travel depends on your route. However, the direct distance from your home to your friend's home is the same whichever route you choose. Distance in a given direction or **displacement** is an example of a **vector** quantity because it has magnitude and direction.

> **A vector is any physical quantity that has a direction as well as a magnitude.**

Further examples of vectors include velocity, **acceleration**, force and **weight**.

> **A scalar is any physical quantity that has magnitude only and is not directional.**

For example, distance is a scalar because it takes no account of direction. Further examples of scalars include mass, density, volume and energy.

Representing vectors

Any vector can be represented as an arrow. The length of the arrow represents the **magnitude** of the vector quantity. The direction of the arrow gives the **direction** of the vector.

- **Displacement** is distance in a given direction. The displacement from one point to another can be represented on a map or a scale diagram as an arrow from the first point to the second point. The length of the arrow must be in proportion to the least distance between the two points.
- **Velocity** is **speed** in a given direction. The velocity of an object can be represented by an arrow with length in proportion to the speed pointing in the direction of motion of the object.
- **Force and acceleration** are both vector quantities and therefore can each be represented by an arrow in the appropriate direction and of length in proportion to the magnitude of the quantity.

On a journey

Cyclists and hill walkers should always take a map and compass to make sure they do not get lost. A compass tells the user which direction is north. A map tells the user how far he or she has gone. Consider the map shown in Figure 1. Suppose your home is at O and your friend's home is at A. Your route-plan is to cycle along the road heading north, over the railway bridge, then turn east at the next road junction.

- The **distance** to be cycled can be estimated by measuring the length of the route on the map in centimetres, then using the map scale.
- The **displacement** or direct distance from O to A is marked on the map as an arrow OA.
- Your **velocity** at any point on your journey changes, because you change direction and because your speed changes. Suppose your speed as you pass over the railway bridge is $2.0\,\mathrm{m\,s^{-1}}$. The direction in which you are travelling as you pass over the bridge is about 10° east of due north. You can check this on Figure 1 using a protractor. So your velocity at the railway bridge is $2.0\,\mathrm{m\,s^{-1}}$ in a direction which is 10° east of due north.

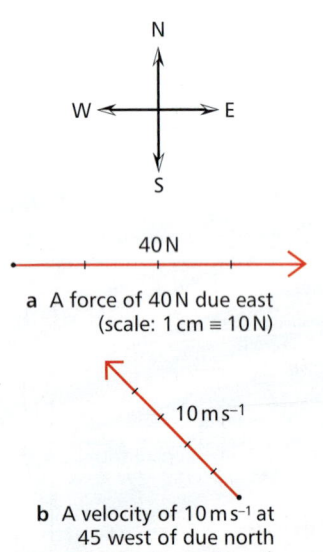

a A force of 40 N due east
(scale: 1 cm ≡ 10 N)

b A velocity of $10\,\mathrm{m\,s^{-1}}$ at 45 west of due north
(scale: 1 cm ≡ $4\,\mathrm{m\,s^{-1}}$)

Figure 2 *Representing a vector*

Vector components

The displacement vector **OA**, shown on Figure 1, can be represented:

- as an arrow of length in proportion to the direct distance of 5.0 km from O to A. The direction of the arrow would be 53° north of due east.

- as a map reference, one part stating how far A is east or west of O and the other stating how far A is north or south of O. The two parts of the map reference, referred to as **components**, may be written as (3.0 km, 4.0 km) where east/west is first. This is the same as writing the **coordinates** of a point on a graph as (x, y), where x is the distance from the origin along the x-axis and y is the distance from the origin along the y-axis.

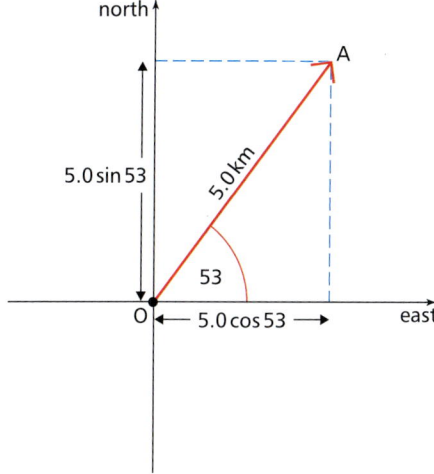

Resolving a vector into two perpendicular components

Resolving a vector is the process of working out the components of a vector in two perpendicular directions from the magnitude and direction of the vector. Figure 3 shows the displacement vector OA represented on a scale diagram that also shows lines due north and due east. The components of this vector along these two lines are 5.0 cos 53° km (= 3.0 km) along the line due east and 5.0 sin 53° km (= 4.0 km) along the line due north.

Figure 3 Resolving a vector

In general, to resolve any vector into two perpendicular components, draw a diagram showing the two perpendicular directions and an arrow to represent the vector. Figure 4 shows this diagram for a vector OP. The components are represented by the projection of the vector onto each line. If the angle θ between the vector OP and one of the lines is known:

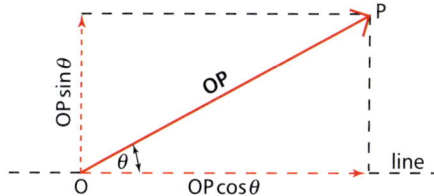

- the component along that line $= OP \cos\theta$, and
- the component perpendicular to that line (i.e. along the other line) $= OP \sin\theta$.

Figure 4 The general rule for resolving a vector

Worked example

An aircraft in level flight has a constant velocity of 50 m s^{-1} in a direction of 40° north of due east, as shown in Figure 5. The angle between the direction of its velocity and due east is therefore 40°.

The components of its velocity are, in m s^{-1},

- 50 cos 40° due east, and
- 50 sin 40° due north.

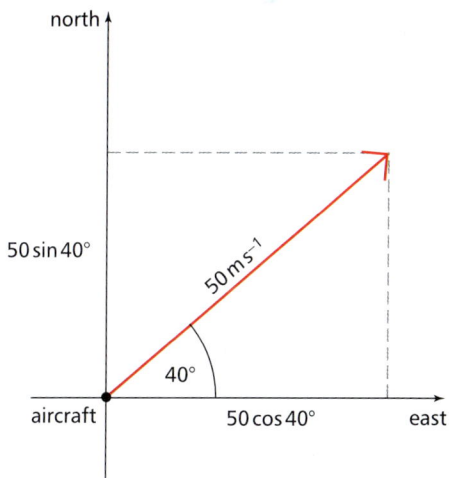

Addition of vectors

Using a scale diagram

Let's go back to the cycle journey in Figure 1. Suppose when you reach your friend's home at A, you then go on to another friend's home at B. Your journey is now a two-stage journey:

- **Stage 1, from O to A,** is represented by the displacement vector **OA**.

- **Stage 2, from A to B,** is represented by the displacement vector **AB**.

Figure 5 Resolving velocity

Figure 6 shows how the overall displacement from O to B, represented by vector **OB**, is the result of adding vector **AB** to vector **OA**. The **resultant** is the third side of a triangle, where OA and AB are the other two sides.

$$\mathbf{OB} = \mathbf{OA} + \mathbf{AB}$$

Use Figure 1 to show that the resultant displacement **OB** is 5.1 km in a direction 11° north of due east.

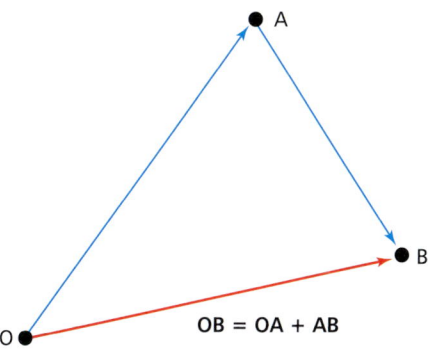

Figure 6 Displacement from O to B

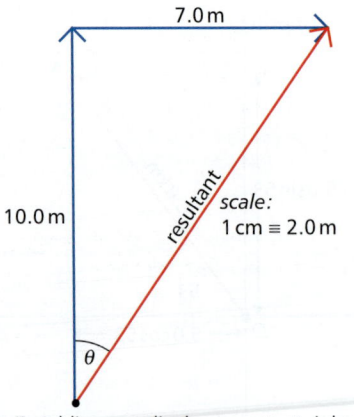

Figure 7 Adding two displacements at right angles to each other

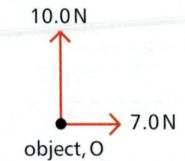

Figure 8 Two forces acting at right angles to each other

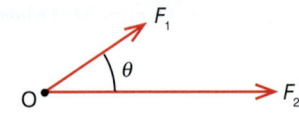

a Vector diagram for F_1 and F_2

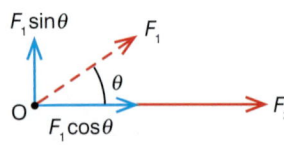

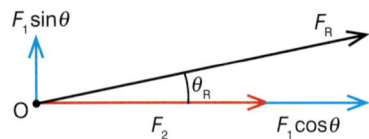

b Using the components to find the resulting force F_r

Figure 9 Using a calculator to find a resultant force

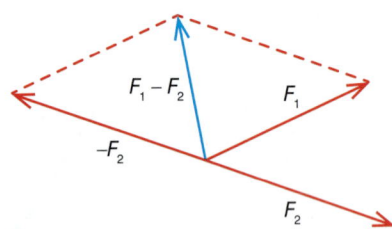

Figure 10 Vector subtraction

Using a calculator

1 Adding two perpendicular vectors

Suppose you walk 10.0 m forward then turn through exactly 90° and walk 7.0 m. At the end, how far will you be from your starting point? The vector diagram to add the two displacements is shown in Figure 7, drawn to a scale of 1 cm to 2.0 m. The two displacements form two sides of a right-angled triangle with the resultant as the hypotenuse. To find the resultant displacement:

- using Pythagoras' theorem gives 12.2 m (= $(10.0^2 + 7.0^2)^{½}$) for the magnitude of the resultant displacement (i.e. the distance), and
- using the trigonometry equation $\tan\theta = \dfrac{7.0}{10.0}$ gives 35° for the angle θ between the direction of the resultant displacement and the initial direction.

The method above can be applied to any two vectors at right angles to each other. For example:

Figure 8 shows an object, O, acted on by two forces, 7.0 N and 10.0 N perpendicular to each other. The vector diagram to add the two forces together would be as shown in Figure 7 if the labels were changed to 7.0 N and 10.0 N instead of 7.0 m and 10.0 m. As explained above, the resultant force is therefore 12.2 N at an angle of 35° to the 10 N force.

2 Adding two vectors that are at angle *Θ* to each other

Consider an object, O, acted on by forces F_1 and F_2 at angle θ to each other, as shown in Figure 9a. The magnitude and direction of the resultant force F_R can be found by resolving one of the forces into components that are parallel and perpendicular to the other force, as shown in Figure 9b.

- Resolving F_1 parallel and perpendicular to F_2 gives $F_1\cos\theta$ for the parallel component and $F_1\sin\theta$ for the perpendicular component.
- Adding the components in each direction therefore gives the parallel component of F_R as $F_1\cos\theta + F_2$ and the perpendicular component as $F_1\sin\theta$.

Using Pythagoras' theorem to find the magnitude of the resultant force gives

$$F_R = [(F_1\cos\theta + F_2)^2 + (F_1\sin\theta)^2]^{½}$$

Because $\sin^2\theta + \cos^2\theta = 1$ for all angles of θ, it can be shown that

$$F_R{}^2 = F_1{}^2 + F_2{}^2 + 2F_1F_2\cos\theta.$$

Using the trigonometry rule for $\tan\theta$ to find θ_R, the angle between the resultant force and F_2 gives

$$\tan\theta_R = \frac{F_1\sin\theta}{(F_1\cos\theta + F_2)}.$$

> **Note**
>
> The resultant of two vectors that act along the same line has a magnitude that is:
>
> - the **sum,** if the two vectors are in the *same* direction. For example, if an object is acted on by a force of 6.0 N and a force of 4.0 N, both acting in the same direction, the resultant force is 10.0 N.
>
> - the **difference,** if the two vectors are in *opposite* directions. For example, if an object is acted on by a 6.0 N force and a 4.0 N force in opposite directions, the resultant force is 2.0 N in the direction of the 6.0 N force.

To subtract one vector from another, the above example illustrates a general method which is to reverse one vector and then add the reversed vector to the other. To reverse a vector, point it in the opposite direction to its original direction so the sign of each component is changed from + to − or − to +.

Summary test 1.1

1 A ship leaves a port, A, and travels a distance of 60 km due north; it then changes direction at B and travels a distance of 20 km due east to a second port, C. Calculate:

   a the direct distance from A to C,

   b the angle, θ, between due north and the line from A to C.

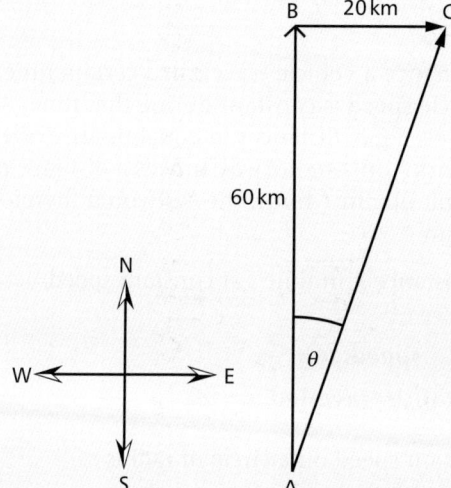

Figure 11

2 An aircraft travels on a straight flight path at a constant speed of 80 m s⁻¹, in a direction 60° north of due east.

   a Calculate the components of the aircraft's velocity due north and due east.

   b Calculate how far north the aircraft travels in 300 seconds.

3 A crane is used to raise one end of a steel girder off the ground, as shown in Figure 12. When the cable attached to the end of the girder is at 20° to the vertical, the force of the cable on the girder is 6.5 kN. Calculate the horizontal and vertical components of this force.

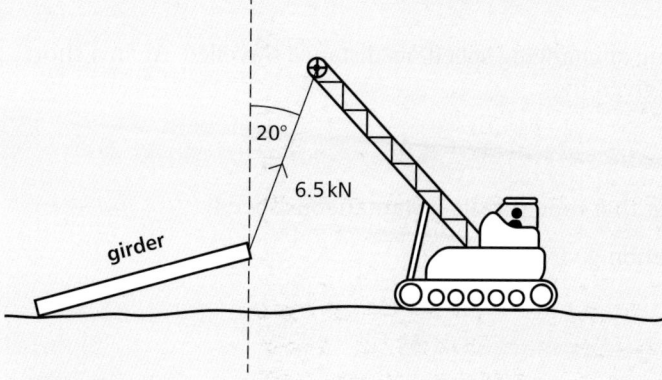

Figure 12

4 On an orienteering exercise, a team is told to go from a car park, C, to a meeting point, P, 12 km away in a direction which is 30° north of due east, as shown in Figure 13.

   a Calculate how far P is: **i** due east of C, **ii** due north of C.

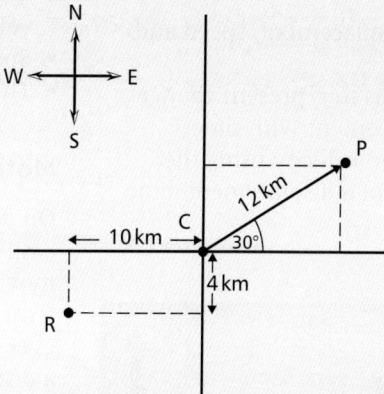

Figure 13

   b A railway station, R, is 10 km due west and 4 km due south of car park C.

   Calculate how far P is: **i** due east of R, **ii** due north of R.

   c Calculate the direct distance from P to R.

5 Figure 14 shows three situations, **a–c**, where an object is acted on by two known forces. For each situation, calculate the magnitude and direction of the resultant force.

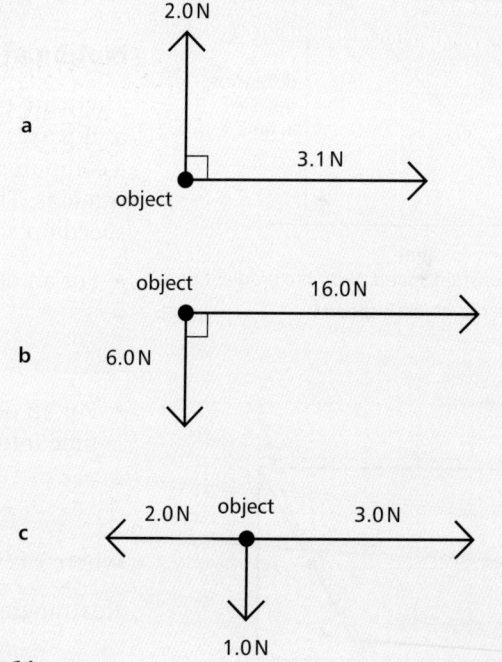

Figure 14

6 Calculate the magnitude and direction of the resultant force on an object that is acted on by a force of 4.0 N and a force of 10 N that are:

   a in the same direction   **c** at right angles to each other

   b in opposite directions   **d** at 35° to each other.

Speed and velocity

Learning outcomes

On these pages you will learn to:

- define displacement, speed and velocity
- use graphs to represent changes of displacement with time
- determine velocity using the gradient of a displacement–time graph

Figure 1 *A speeding vehicle*

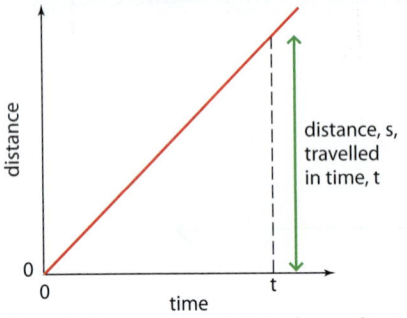

Figure 2 *Constant speed. Note the gradient of the line, $\frac{s}{t}$ = the object's speed v*

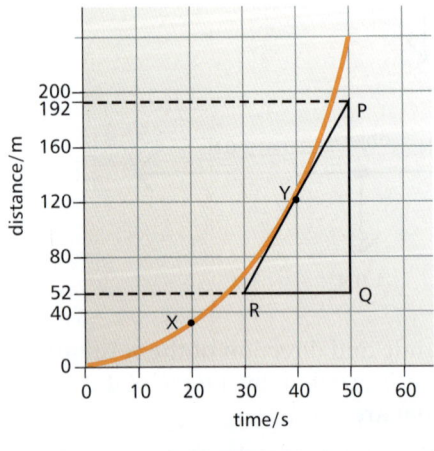

speed at Y = $\frac{PQ}{QR} = \frac{192-52}{20} = 7\,\text{m/s}$

Figure 3 *Changing speed*

Fact file

- **Speed** is defined as change of distance per unit time.
- **Velocity** is defined as change of displacement per unit time. In other words, velocity is speed in a given direction.
- Speed is a **scalar** quantity. Velocity is a **vector** quantity.
- The unit of speed and of velocity is the **metre per second (m s⁻¹)**.

Motion at constant speed

On a motorway journey, the distance a vehicle travels in a certain time can easily be worked out if the vehicle speed is constant during that time. An object moving at a constant speed travels equal distances in equal times. For example, a car travelling at a speed of 30 m s⁻¹ on a motorway travels a distance of 30 m every second or 1800 m every minute. In 1 hour, the car would therefore travel a distance of 108 000 m or 108 km.

- For an object which travels distance s, in time t at constant speed,

$$\text{speed, } v = \frac{s}{t}$$
$$\text{distance travelled, } s = vt$$

- For an object moving at constant speed on a circle of radius r,

$$\text{speed, } v = \frac{2\pi r}{T}$$

where T is the time to move round once and $2\pi r$ is the circumference of the circle.

Motion at changing speed

There are two types of speed cameras. One type measures the speed of a vehicle as it passes the camera. The other type is linked to a second speed camera and a computer, which works out the average speed of the vehicle between the two cameras. This will catch drivers who slow down for a speed camera and then speed up again!

- For an object that travels a distance s in time t,

$$\text{average speed, } v_{\text{av}} = \frac{s}{t}$$

- For an object moving at changing speed, its distance travelled, Δs, in a short time interval, Δt, is given by:

$$\Delta s = v\Delta t$$

where v is the speed at that time (i.e. its instantaneous speed).

Rearranging this equation gives:

$$v = \frac{\Delta s}{\Delta t}$$

Distance–time graphs

- For an object moving at **constant** speed, the graph is **a straight line with a constant gradient** (see Figure 2).

$$\text{Speed of the object} = \frac{\text{distance travelled}}{\text{time taken}} = \text{gradient of the line}$$

- For an object moving with **changing** speed, **the gradient of the line changes**. The gradient of the line at any point can be found by drawing a tangent to the line at that point and then measuring the gradient of the tangent. This is shown in Figure 3, where PR is the tangent at point Y on the line. Show for yourself that the speed at point X on the line is $2.7\,\mathrm{m\,s^{-1}}$.

Velocity

Take care when cycling or driving along a country lane. Such roads can be deceptive, because they wind round 'blind bends' where drivers are unable to see oncoming traffic. In such circumstances, a driver has to be ready to change speed, and change direction, very rapidly. The velocity of a car on a country road is very unlikely to be constant!

- An object moving at constant velocity moves at the same speed without changing its direction of motion. In other words, an object moving at constant velocity travels along a straight line, covering equal distances in equal times.

- If an object changes or reverses its direction of motion, its velocity changes. For example, the velocity of an object moving on a circular path at constant speed changes continuously, because its direction of motion changes continuously (Figure 4).

- An object travelling along a straight line has two possible directions of motion. To distinguish between the two directions, we need a direction code where + values are in one direction and − values in the opposite direction.

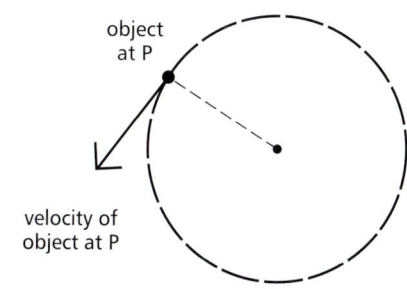

Figure 4 Circular motion

Summary test 1.2

1 kilometre (km) = 1000 metres (m)

1 A car travels a distance of 60 km in 45 minutes at constant speed. Calculate its speed in:
  a $\mathrm{km\,h^{-1}}$, **b** $\mathrm{m\,s^{-1}}$.

2 A satellite moves round the Earth at constant speed on a circular orbit of radius 8000 km, with a time period of 120 minutes. Calculate its orbital speed in:
  a $\mathrm{km\,h^{-1}}$, **b** $\mathrm{m\,s^{-1}}$.

3 A vehicle joins a motorway and travels at a steady speed of $25\,\mathrm{m\,s^{-1}}$ for 30 minutes, then it travels a further distance of 40 km in 20 minutes before leaving the motorway (Figure 5). Calculate:

  a the distance travelled in the first 30 minutes,
  b its average speed on the motorway.

4 a Explain the difference between speed and velocity.

  b A police car joins a straight motorway at Junction 4 and travels for 12 km at a constant speed for 400 s; then it leaves at Junction 5 and rejoins on the opposite side, and travels for 8 km at a constant speed for 320 s to reach an accident (Figure 6). Calculate:
    i the displacement from Junction 4 to the accident,
    ii the velocity of the car on each side of the motorway.

  c Sketch a displacement–time graph for the journey.

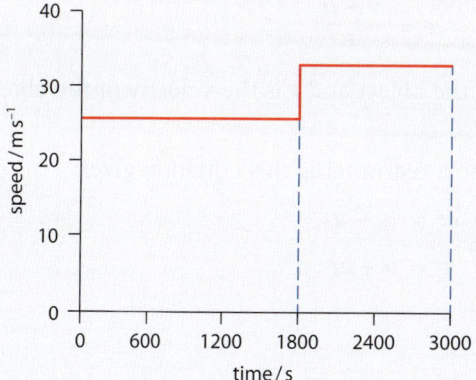

Figure 5

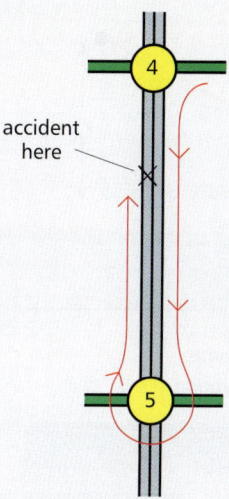

Figure 6

Learning outcomes

On these pages you will learn to:

- define acceleration
- calculate the acceleration of an object undergoing uniform acceleration
- use graphs to represent displacement, speed, velocity and acceleration
- determine acceleration using the gradient of a velocity–time graph

Figure 1

Note

The acceleration of the new model above is $5.0\,\mathrm{m\,s^{-2}}$, because its speed increased by $5.0\,\mathrm{m\,s^{-1}}$ every second. The acceleration of the old model above is $4.0\,\mathrm{m\,s^{-2}}$, because its speed increased by $4.0\,\mathrm{m\,s^{-1}}$ every second.

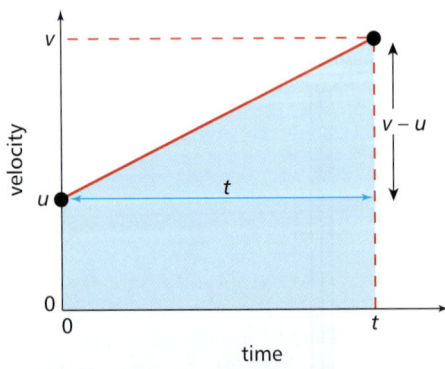

Figure 2 *Uniform acceleration*

Note

Notice that the gradient of the line represents the acceleration $\left(= \dfrac{v - u}{t}\right)$.

Performance tests

A car maker wants to compare the performance of a new model with the model being replaced. To do this, the velocity of each car is measured on a test track. Each vehicle accelerates as fast as possible to top velocity from a standing start. The results are listed in Table 1 below.

Table 1 *Performance of two models of a car*

Time from a standing start/s	0	2	4	6	8	10
Velocity of old model/$\mathrm{m\,s^{-1}}$	0	8	16	24	32	32
Velocity of new model/$\mathrm{m\,s^{-1}}$	0	10	20	30	30	30

Which car accelerates faster?

- The old model took $8\,\mathrm{s}$ to reach its top velocity of $32\,\mathrm{m\,s^{-1}}$. Its velocity must have increased by $4.0\,\mathrm{m\,s^{-1}}$ every second for 8 seconds to reach its top velocity.
- The new model took $6\,\mathrm{s}$ to reach its top velocity of $30\,\mathrm{m\,s^{-1}}$. Its velocity must have increased by $5.0\,\mathrm{m\,s^{-1}}$ every second for $6\,\mathrm{s}$ to reach its top velocity.

Clearly, the new model speeds up at a faster rate than the old model. In other words, its acceleration is greater.

> **Acceleration is defined as change of velocity per unit time.**

- The unit of acceleration is the **metre per second per second ($\mathrm{m\,s^{-2}}$)**.
- Acceleration is a **vector** quantity.
- **Deceleration** is in the opposite direction to velocity so its values are negative and signify that the velocity decreases with respect to time.

For a moving object that does not change direction, its acceleration at any point can be worked out from its rate of change of speed because there is no change of direction.

Uniform acceleration

Uniform acceleration is where the velocity of an object moving along a straight line changes at a constant rate. In other words, the acceleration is constant. Consider an object that accelerates uniformly from velocity, u, to velocity, v, in time, t, along a straight line. Figure 2 shows how its velocity changes with time.

$$\textbf{Acceleration, } a, \textbf{ of the object } = \frac{\textbf{change of velocity}}{\textbf{time taken}} = \frac{v - u}{t}$$

$$a = \frac{v - u}{t}$$

where u is the initial velocity of the object and v is the velocity of the object at time t.

To calculate the velocity v at time t, rearranging this equation gives:

$$at = (v - u)$$

$$\therefore \qquad v = u + at$$

Worked example

The driver of a vehicle travelling at $8\,\mathrm{m\,s^{-1}}$ applies the brakes for $30\,\mathrm{s}$ and reduces the velocity of the vehicle to $2\,\mathrm{m\,s^{-1}}$. Calculate the deceleration of the vehicle during this time.

Solution

$u = 8\,\mathrm{m\,s^{-1}}; v = 2\,\mathrm{m\,s^{-1}}; t = 30\,\mathrm{s}$

$\therefore a = \dfrac{v - u}{t} = \dfrac{(2 - 8)}{30} = \dfrac{-6}{30} = -0.2\,\mathrm{m\,s^{-2}}$

Non-uniform acceleration

Non-uniform acceleration is where the direction of motion of an object changes, or its speed changes, at a varying rate. Figure 3 shows how the velocity of an object increases for an object moving along a straight line with an increasing acceleration. This can be seen directly from the graph, because the gradient increases with time (i.e. the graph becomes steeper and steeper). The **gradient** represents the acceleration.

The acceleration at any point is the gradient of the tangent to the curve at that point. In Figure 3:

$$\text{Gradient of tangent at point P} = \frac{\text{height of gradient triangle}}{\text{base of gradient triangle}}$$

$$= \frac{4.0 - 1.0\,\mathrm{m\,s^{-1}}}{2.0\,\mathrm{s}} = 1.5\,\mathrm{m\,s^{-2}}$$

Therefore, the acceleration at P is $1.5\,\mathrm{m\,s^{-2}}$.

Acceleration = gradient of the line on the velocity–time graph

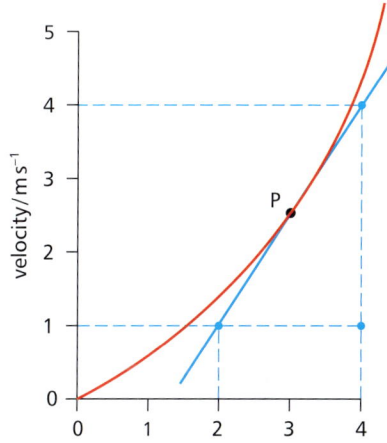

Figure 3 Non-uniform acceleration

Summary test 1.3

1 a An aeroplane taking off accelerates uniformly on a runway from a velocity of $4\,\mathrm{m\,s^{-1}}$ to a velocity of $64\,\mathrm{m\,s^{-1}}$ in $40\,\mathrm{s}$. Calculate its acceleration.

  b A car travelling at a velocity of $20\,\mathrm{m\,s^{-1}}$ brakes sharply to a standstill in 8.0 seconds. Calculate its deceleration, assuming its velocity decreases uniformly.

2 A cyclist accelerates uniformly from a velocity of $2.5\,\mathrm{m\,s^{-1}}$ to a velocity of $7.0\,\mathrm{m\,s^{-1}}$ in a time of $10\,\mathrm{s}$. Calculate:

  a the acceleration,

  b the velocity $2.0\,\mathrm{s}$ later, if she continued to accelerate at the same rate.

3 A train, on a straight journey between two stations, accelerates uniformly from rest for $20\,\mathrm{s}$ to a velocity of $12\,\mathrm{m\,s^{-1}}$. It then travels at a constant velocity for a further $40\,\mathrm{s}$, before decelerating uniformly to rest in $30\,\mathrm{s}$.

  a Sketch a velocity–time graph to represent its journey.

  b Calculate its acceleration in each part of the journey.

4 The velocity of an object released in water increased as shown in Figure 4.

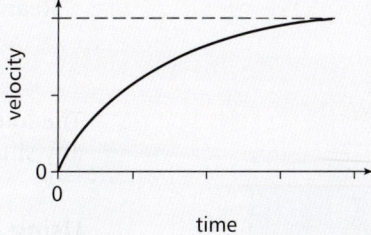

Figure 4

Describe how:

  a the velocity of the object changed with time,

  b the acceleration of the object changed with time.

1.4 Motion along a straight line at a constant acceleration

Learning outcomes

On these pages you will learn to:

- derive equations for uniformly accelerated motion in a straight line
- solve problems using equations for uniformly accelerated motion in a straight line
- calculate the distance travelled from the area under a speed–time graph

The dynamics equations for constant acceleration

Consider an object that accelerates uniformly from initial speed, u, to final speed v, in time t without change of direction. Figure 2, p.8, shows how its speed changes with time.

1 The acceleration $a = \dfrac{v - u}{t}$ (as explained in unit 1.3)

As before, rearranging this equation gives:

$$v = u + at \qquad \text{(Equation 1)}$$

2 The distance moved, s = average speed × time taken

Because the acceleration is uniform, the average speed $= \dfrac{(u + v)}{2}$

Therefore $s = \dfrac{(u + v)t}{2}$ (Equation 2)

3 By combining the two equations above, to eliminate v, a further useful equation is produced.

Substitute $u + at$ in place of v in Equation 2. This gives:

$$s = \frac{(u + (u + at))t}{2} = \frac{(u + u + at)t}{2} = \frac{2ut + at^2}{2}$$

Therefore $s = ut + \frac{1}{2}at^2$ (Equation 3)

4 A fourth useful equation is obtained by combining Equations 1 and 2 to eliminate t. This can be done by multiplying

$$a = \frac{v - u}{t} \quad \text{and} \quad s = \frac{(u + v)t}{2} \text{ together to give}$$

$$as = \frac{v - u}{t} \times \frac{(u + v)t}{2}$$

which simplifies to

$$as = \frac{(v - u)(v + u)}{2} = \frac{(v^2 - uv + uv - u^2)}{2} = \frac{(v^2 - u^2)}{2}$$

Therefore $2as = v^2 - u^2$

Rearranging this equation gives:

$$v^2 = u^2 + 2as \qquad \text{(Equation 4)}$$

The four equations, often referred to as the 'suvat' equations, are invaluable in any situation where the acceleration is constant.

Using a speed–time graph to find the distance moved

An object moving at constant speed

The distance moved in time t, s = speed × time taken = vt. This distance is represented on the graph by the **area under the line between the start and time t**. This is a rectangle of height corresponding to speed v and of base corresponding to the time t (Figure 1a).

Worked example

A driver of a vehicle travelling at a speed of $30\,\text{m s}^{-1}$ on a motorway brakes sharply to a standstill in a distance of $100\,\text{m}$. Calculate the deceleration of the vehicle.

$u = 30\,\text{m s}^{-1}$, $v = 0$, $s = 100\,\text{m}$, $a = ?$

Solution
To find a, use $v^2 = u^2 + 2as$

Therefore $0 = u^2 + 2as$, because $v = 0$

Rearranging this equation gives:

$2as = -u^2$

$a = \dfrac{-u^2}{2s} = \dfrac{-30^2}{2 \times 100} = -4.5\,\text{m s}^{-2}$

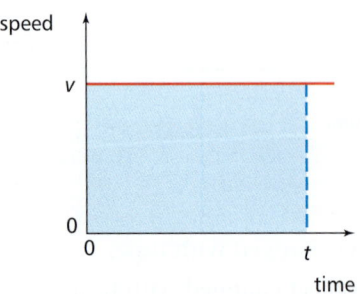

Figure 1a Constant speed

An object moving at constant acceleration

From Equation 2, the distance moved in time t, $s = \frac{(u + v)t}{2}$.

This distance is represented on the graph by the **area under the line between the start and time t**. This is a trapezium which has a base corresponding to time t and an average height corresponding to the average speed $\frac{(u + v)t}{2}$.

Therefore the area of the trapezium (= average height × base) corresponds to $\frac{(u + v)t}{2}$, which is the distance moved (Figure 1b).

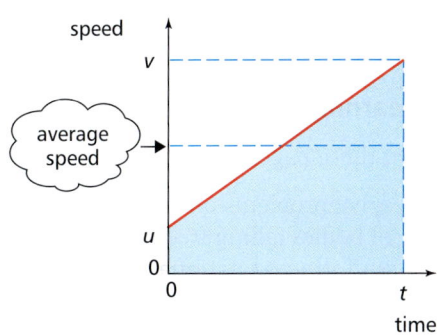

Figure 1b *Constant acceleration*

An object moving with a changing acceleration

Let v represent the speed at time t and $v + \delta v$ represent the speed a short time later at $t + \delta t$ (δ is pronounced 'delta' from Greek).

Because the speed change δv is small compared with the speed v (Figure 1c), the distance travelled δs in the short time interval δt is given by $\delta s = v\delta t$.

This is represented on the graph by the **area of the shaded strip under the line which has a base corresponding to δt and a height corresponding to v.** In other words, $\delta s = v\delta t$ is represented by the area of this strip.

By considering the whole area under the line in strips of similar width, the total distance travelled from the start to time t_1 is therefore represented by the sum of the areas of every strip: which is therefore the **total area** under the line.

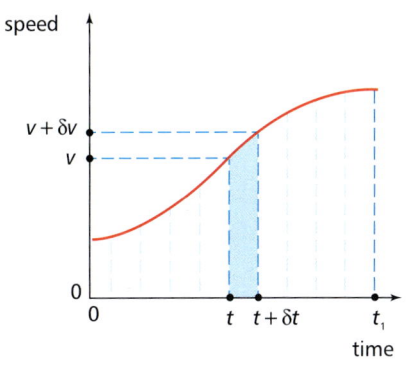

Figure 1c *Changing acceleration*

Whatever the shape of the line of a speed–time graph:

> **Distance travelled = area under the line of a speed–time graph**

Summary test 1.4

1 A vehicle accelerates uniformly along a straight road, increasing its speed from $4.0\,\mathrm{m\,s^{-1}}$ to $30.0\,\mathrm{m\,s^{-1}}$ in 13 s. Calculate:

a its acceleration,

b the distance it moves in this time.

2 An aircraft lands on a runway at a speed of $40\,\mathrm{m\,s^{-1}}$ and brakes to a halt in a distance of 860 m. Calculate:
a the braking time, **b** the deceleration of the aircraft.

3 A cyclist accelerates uniformly from rest to a speed of $6.0\,\mathrm{m\,s^{-1}}$ in 30 s, then brakes at uniform deceleration to a halt in a distance of 24 m.

a For the first part of the journey, calculate:
   i the acceleration, **ii** the distance travelled.

b For the second part of the journey, calculate:
   i the deceleration, **ii** the time taken.

c Sketch a speed–time graph for this journey,

d Use the graph to determine the average speed for the journey.

4 The speed of an athlete for the first 5 s of a sprint race is shown in Figure 2. Use the graph to determine:

a the initial acceleration of the athlete,

b the distance moved in the first 2 s,

c the distance moved in the next 2 s,

d the average speed over the first 4 s.

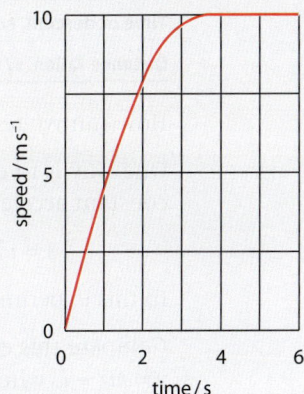

Figure 2

13

Learning outcomes

On these pages you will learn to:

- solve problems on the motion of bodies falling freely in a uniform gravitational field using equations for uniformly accelerated motion in a straight line
- describe an experiment to determine the acceleration of free fall, g, using a falling body

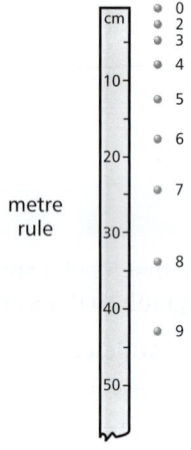

Figure 1 *Investigating free fall*

Experimental tests

Does a heavy object fall faster than a lighter object?

Release a stone and a small coin at the same time. Which one hits the ground first? Galileo first discovered the answer to this question about four centuries ago. He reasoned that because any number of identical objects must fall at the same rate, then any one such object must fall at the same rate as the rest put together. So he concluded that any two objects must fall at the same rate, regardless of their relative weights. He was reported to have demonstrated the correctness of his theory by releasing two different weights from the top of the leaning Tower of Pisa.

The inclined plane test

Galileo wanted to know if a falling object speeds up as it falls. Clocks and stopwatches were devices of the future. The simplest test he could think of was to time a ball as it rolled down a plank. He devised a 'dripping water' clock, counting the number of the drips as a measure of time. He measured how long the ball took to travel equal distances down the slope from rest. His measurements showed that the ball gained speed as it travelled down the slope. In other words, he showed that the ball accelerated as it rolled down the slope. He reasoned that the acceleration would be greater the steeper the slope. So he concluded that an object falling vertically accelerates.

Acceleration due to gravity

One way to investigate the free fall of a ball is to make a multi-flash photo or video clip of the ball's flight as it falls after being released from rest. To do this, a camera is used to record the ball's descent in a dark room with a flashing light switched on. The flashing light needs to flash at a known constant rate of about 20 flashes per second. A vertical metre rule can be used to provide a scale. Figure 1 shows a possible arrangement using a steel ball and a multi-flash photo taken with this arrangement.

For each image of the ball on the photograph, the time of descent of the ball and the distance fallen by the ball from rest can be measured directly. The photograph shows the ball speeds up as it falls, because it travels further between successive images. Measurements from this photograph are given in Table 1 below.

Table 1 *A free-falling ball*

Number of flashes after start	0	2	3	4	5	6	7	8	9
Time of descent, t / s	0	0.06	0.10	0.13	0.16	0.19	0.23	0.26	0.29
Distance fallen, s / m	0	0.02	0.04	0.07	0.12	0.17	0.24	0.33	0.42

How can we tell if the acceleration is constant from these results?

One way is to consider how the distance fallen, s, would vary with time, t, for constant acceleration. From unit 1.4, we know that:

$$s = ut + \tfrac{1}{2}at^2, \text{ where } u = \text{the initial speed and } a = \text{acceleration.}$$

In this experiment, $u = 0$ therefore $\boldsymbol{s = \tfrac{1}{2}at^2}$ for constant acceleration, a.

Compare this equation with the general equation for a straight-line graph $y = mx + c$, where m is the gradient and c is the y-intercept. If we let y represent s and let x represent t^2, then $m = \tfrac{1}{2}a$ and $c = 0$.

So a graph of s against t^2 should give a straight line through the origin. In addition, the gradient of the line $(= \tfrac{1}{2}a)$ can be measured and the acceleration $(= 2 \times \text{gradient})$ calculated (Figure 2).

As you can see, the graph is a straight line through the origin. We can therefore conclude that the equation $s = \frac{1}{2}at^2$ applies here, so the acceleration of a falling object is constant. Show for yourself that the gradient of the line is $5.0\,\text{m s}^{-2}$ ($\pm 0.2\,\text{m s}^{-2}$), giving an acceleration of $10\,\text{m s}^{-2}$. Because there are no external forces acting on the object, apart from its weight, this value of acceleration is known as the **acceleration of free fall** and is represented by the symbol, **g**. Accurate measurements give a value of $9.81\,\text{m s}^{-2}$ near the Earth's surface.

The 'suvat' equations on p.12 may be applied to any 'free fall' situation, where **air resistance** is negligible.

The equations can also be applied to situations where objects are thrown vertically upwards. As a general rule, apply the direction code + **for upwards and** − **for downwards** when values are inserted into the 'suvat' equations.

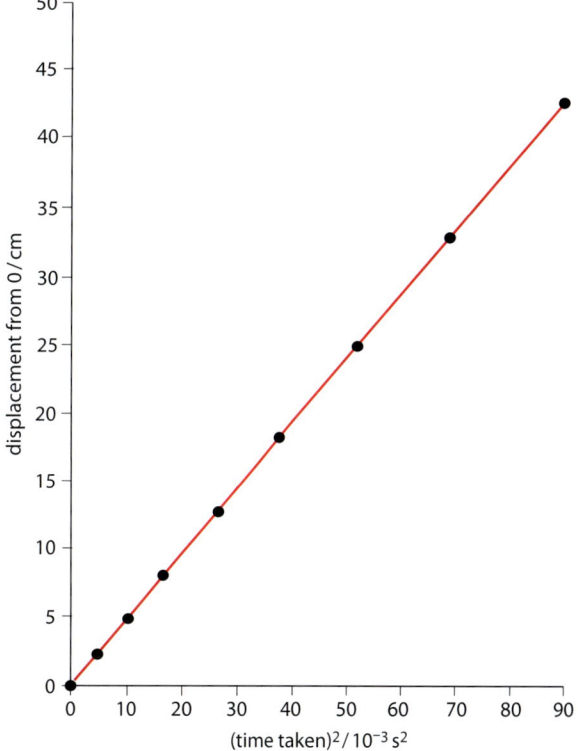

Figure 2 A graph of s against t^2

Worked example

$g = 9.81\,\text{m s}^{-2}$

1 A coin was released from rest at the top of a well. It took 1.6 s to hit the bottom of the well. Calculate: **a** the distance fallen by the coin, **b** its speed just before impact.

Solution

$u = 0$, $t = 1.6\,\text{s}$, $a = -9.81\,\text{m s}^{-2}$ (− as g acts downwards)

a To find s, use $s = \frac{1}{2}at^2$ as $u = 0$

Therefore, $s = \frac{1}{2} \times -9.81 \times 1.6^2 = -12.6\,\text{m}$ (− indicates 12.6 m downwards) so the distance fallen is 12.6 m

b To find v, use $v = u + at = 0 + (-9.81 \times 1.6) = -15.7\,\text{m s}^{-1}$ (− indicates downward velocity) so the speed is 15.7 m s^{-1}

Summary test 1.5

$g = 9.81\,\text{m s}^{-2}$

1 A pebble, released at rest from a canal bridge, took 0.9 s to hit the water.
   Calculate:

   a the distance it fell before hitting the water,

   b its speed just before hitting the water.

2 A spanner was dropped from a hot air balloon, when the balloon was at rest 50 m above the ground.
   Calculate:

   a the time taken for the spanner to hit the ground,

   b the speed of impact of the spanner on hitting the ground.

3 A bungee jumper jumped off a platform 75 m above a lake, releasing a small object at the instant she jumped off the platform.

   a Calculate: **i** the time taken by the object to fall to the lake, **ii** the speed of impact of the object on hitting the water, assuming air resistance is negligible.

   b Explain why the bungee jumper would take longer to descend than the time taken in part **a i**.

4 An astronaut on the Moon threw an object 4.0 m vertically upwards and caught it again 4.5 s later.
   Calculate:

   a the acceleration due to gravity on the Moon,

   b the speed of projection of the object,

   c how high the object would have risen on the Earth, for the same speed of projection.

Learning outcomes

On these pages you will learn to:

- determine displacement from the area under a velocity–time graph
- determine velocity using the gradient of a displacement–time graph
- determine acceleration using the gradient of a velocity–time graph

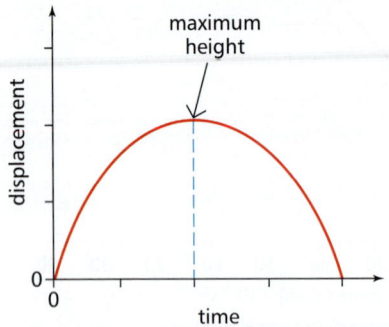

Figure 1 *Displacement against time for an object projected upwards*

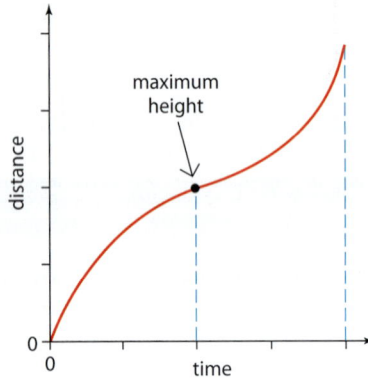

Figure 2 *Distance against time for an object projected upwards*

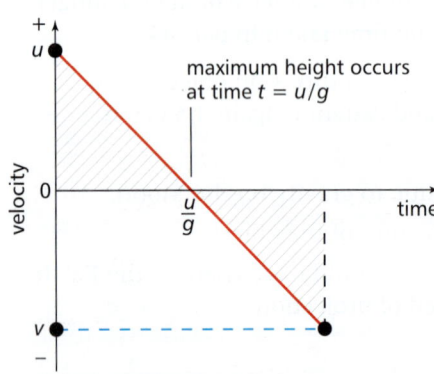

Figure 3 *Velocity–time graph*

The difference between a distance–time graph and a displacement–time graph

Displacement is distance in a given direction from a certain point. Consider a ball thrown directly upwards and caught when it returns. Its displacement from the instant it leaves the thrower:

- increases until it reaches maximum height;
- decreases to zero as it returns to the thrower from maximum height.

For example, if the ball rises to a maximum height of 2.0 m, on returning to the thrower its displacement from its initial position is zero. However, the distance it has travelled is 4.0 m.

The displacement of the object changes with time as shown by Figure 1. The line in this graph fits the equation $s = ut - \frac{1}{2}gt^2$, where s is the displacement, t is the time taken and u is the initial velocity of the object.

The **gradient** of the line represents the **velocity** of the object.

- Initially, the velocity is positive and large so the gradient is positive and large.
- As the ball rises, its velocity decreases so the gradient decreases.
- At maximum height, the velocity is zero so the gradient is zero.
- As the ball returns, the velocity becomes increasingly negative, corresponding to increasing speed in a downwards direction. So the gradient becomes increasingly negative.

The distance travelled by the object changes with time as shown by Figure 2. The **gradient** of this line represents the **speed**.

- From projection to maximum height, the shape is exactly the same as in Figure 1.
- After maximum height, the distance continues to increase so the line curves up, not down.

More about velocity–time graphs

Velocity is speed in a given direction. Consider how the velocity of an object thrown into the air changes with time. The object's velocity:

- decreases from its initial positive (i.e. upwards) value to zero at maximum height;
- increases in the negative (i.e. downwards) direction as it falls.

Figure 3 shows how the velocity of the object changes with time.

- **The gradient of the line is constant and negative, equal to the acceleration of free fall, g.** The acceleration of the object is the same when it descends as when it ascends, so the gradient of the line is always equal to $-9.81\,\text{m s}^{-2}$.
- **The area under the line represents the displacement of the object from its starting position.**
  a The area between the **positive** section of the line and the time axis represents the **displacement** during the **ascent**.
  b The area between the **negative** section of the line and the time axis represents the displacement during the **descent**.
     Taking the area for **a** as positive and the area for **b** as negative, the total area is zero. This corresponds to zero for the total displacement.

Worked example

$g = 9.81 \, \text{m s}^{-2}$

A ball released from a height of 1.20 m above a concrete floor rebounds to a height of 0.82 m.

a Calculate: **i** its time of descent, **ii** the speed of the ball immediately before it hits the floor.
b Calculate: the speed of the ball immediately after it leaves the floor.
c Sketch a velocity–time graph for the ball. Assume the contact time is negligible compared with the time of descent or ascent.

Solution

a $u = 0$, $a = -9.81 \, \text{m s}^{-2}$, $s = -1.2$ m

i To find t, use $s = ut + \frac{1}{2}at^2$

$$\therefore -1.2 = 0 + 0.5 \times -9.81 \times t^2$$
$$-1.2 = -4.905t^2$$
$$t^2 = \frac{-1.2}{-4.905} = 0.245$$
$$t = 0.49 \, \text{s}$$

ii To find v, use $v = u + at$
$$\therefore v = 0 + -9.81 \times 0.49 = -4.8 \, \text{m s}^{-1} \quad (-\text{ for downwards})$$
so the speed is $4.8 \, \text{m s}^{-1}$

b $v = 0$, $a = -9.81 \, \text{m s}^{-2}$, $s = +0.82$ m

To find u, use $v^2 = u^2 + 2as$

$$\therefore 0 = u^2 + 2 \times -9.81 \times 0.82$$
$$u^2 = 16.1 \, \text{m}^2 \, \text{s}^{-2}$$
$$u = +4.0 \, \text{m s}^{-1} \quad (+\text{ for upwards})$$

c See Figure 4. Note that the line has a constant gradient equal to the acceleration due to gravity, $-9.81 \, \text{m s}^{-2}$, except on impact.

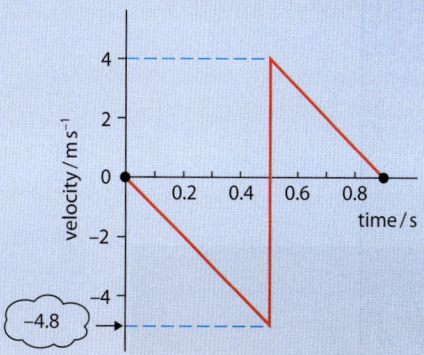

Figure 4

Summary test 1.6

$g = 9.81 \, \text{m s}^{-2}$

1 A swimmer swims 100 m from one end of a swimming pool to the other end at a constant speed of $1.2 \, \text{m s}^{-1}$; then swims back at constant speed, returning to the starting point 210 s after starting.

   a Calculate how long the swimmer takes to swim from:
   i the starting end to the other end,
   ii back to the start from the other end.
   b For the swim from start to finish, sketch:
   i a displacement–time graph,
   ii a distance–time graph.
   c Sketch a velocity–time graph for the swim.

2 A motorcyclist travelling along a straight road at a constant speed of $8.8 \, \text{m s}^{-1}$ passes a cyclist travelling in the same direction at a speed of $2.2 \, \text{m s}^{-1}$. After 200 s, the motorcyclist stops.

   a Calculate how long the motorcyclist has to wait before the cyclist catches up.
   b On the same axes, sketch a velocity–time graph for:
   i the motorcyclist, **ii** the cyclist.

3 The graph (Figure 5) shows the velocity of a train on a straight track for 50 min after it left a station.

   a **i** Describe how the displacement of the train from the station changed with time.
   ii Sketch a graph to show how the displacement in part **i** varied with time.

   b **i** Calculate how far from the station the train was after 50 min.
   ii Calculate the total distance travelled by the train in this time.

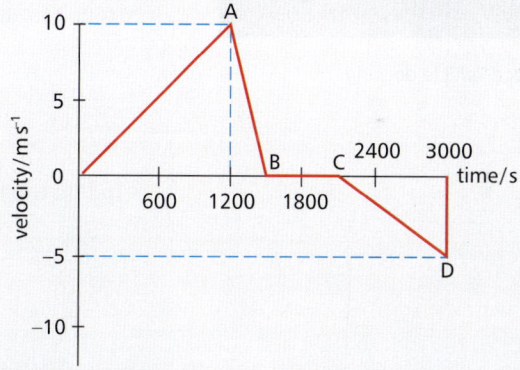

Figure 5

4 A ball is released from a height of 1.8 m above a level surface and rebounds to a height of 0.90 m.

   a Given $g = 9.81 \, \text{m s}^{-2}$, calculate: **i** the duration of its descent, **ii** its velocity just before impact, **iii** the duration of its ascent, **iv** its velocity just after impact.
   b Sketch a graph to show how its velocity changes with time from release to rebound at maximum height.
   c Sketch a further graph to show how the displacement of the object changes with time.

More calculations on motion along a straight line

Learning outcomes

On these pages you will learn to:

- solve problems including those with several stages using the equations for uniformly accelerated motion in a straight line

Motion along a straight line at constant acceleration

- The 'suvat' equations for motion at constant acceleration, a, are:

$$v = u + at \qquad \text{(Equation 1)}$$

$$s = \frac{(u + v)t}{2} \qquad \text{(Equation 2)}$$

$$s = ut + \tfrac{1}{2}at^2 \qquad \text{(Equation 3)}$$

$$v^2 = u^2 + 2as \qquad \text{(Equation 4)}$$

where s is the displacement in time t, u is the initial velocity and v is the final velocity.

- For motion along a straight line at constant acceleration, one direction along the line is 'positive' and the other direction is negative.

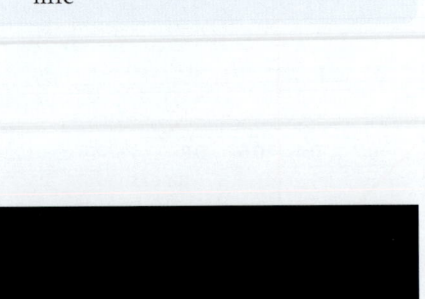

Figure 1 *A space vehicle docking*

Worked example

A space vehicle moving towards a docking station, at a speed of $2.5\,\text{m s}^{-1}$, is 26 m from the docking station when its reverse thrust motors are switched on (to slow it down and stop it when it reaches the station). The vehicle decelerates uniformly until it comes to rest at the docking station, when its motors are switched off.

Calculate: **a** its deceleration, **b** how long it takes to stop, **c** its velocity, if its motors remained on for 5.0 s longer than necessary.

Solution
Let the + direction represent motion towards the docking station and − away from the station.

a and **b** Initial velocity, $u = +2.5\,\text{m s}^{-1}$, final velocity, $v = 0$, displacement, $s = +26\,\text{m}$.

To find its deceleration, a, use

$$v^2 = u^2 + 2as.$$
$$0 = 2.5^2 + 2a \times 26$$
$$-52\,a = 2.5^2$$
$$a = -\frac{2.5^2}{52} = -0.12\,\text{m s}^{-2}$$

To find the time taken, use

$$v = u + at.$$
$$0 = 2.5 - 0.12t$$
$$0.12t = 2.5$$
$$t = \frac{2.5}{0.12} = 21\,\text{s}$$

c Initial velocity, $u = 2.5\,\text{m s}^{-1}$, acceleration, $a = -0.12\,\text{m s}^{-2}$, time taken $t = 26\,\text{s}$

To calculate its velocity v after 26 s, use $\quad v = u + at = 2.5 - 0.12 \times 26$

$$= -0.62\,\text{m s}^{-1}$$

The velocity of the space vehicle would be $0.62\,\text{m s}^{-1}$ away from the docking station after 26 s.

Two-stage problems

Consider an object released from rest, falling then hitting a bed of sand. The motion is in two stages:

1 **Falling** motion due to gravity; acceleration = g (downwards)
2 **Deceleration** in the sand; initial velocity = velocity of object just before impact.

The acceleration in each stage is **not** the same. The link between the two stages is that the velocity at the end of the first stage is the same as the velocity at the start of the second stage.

For example, consider a ball released from a height of 0.85 m above a bed of sand that creates an impression in the sand of depth 0.025 m (Figure 2). For directions, let + represent upwards and – represent downwards.

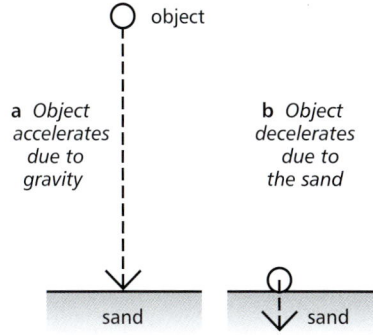

a Object accelerates due to gravity

b Object decelerates due to the sand

Figure 2 A two-stage problem

Stage 1
$u = 0$, $s = -0.85$ m, $a = -9.81$ m s^{-2}.

To calculate the speed of impact v, use $v^2 = u^2 + 2as$

$\therefore$ $\quad v^2 = 0^2 + 2 \times -9.81 \times -0.85 = 16.7$

$\therefore$ $\quad v = -4.1$ m s^{-1}

> **Note**
>
> $v^2 = 16.7$, so $v = -4.1$ or $+4.1$ m s^{-1}. The negative answer is chosen, as the ball is moving downwards.

Stage 2
$u = -4.1$ m s^{-1}, $v = 0$ (as the ball comes to rest in the sand), $s = -0.025$ m

To calculate the deceleration, a, use $v^2 = u^2 + 2as$.

$\therefore$ $\quad 0^2 = (-4.1)^2 + 2a \times -0.025$

$\therefore$ $\quad 2a \times 0.025 = 16.7$

$$a = \frac{16.7}{2 \times 0.025} = 334 \text{ m s}^{-2}$$

> **Note**
>
> $a > 0$ and therefore in the opposite direction to the direction of motion, which is downwards. Thus the ball slows down in the sand with a deceleration of 334 m s^{-2}.

Summary test 1.7

$g = 9.81$ m s^{-2}

1 A vehicle on a straight downhill road accelerated uniformly from a speed of 4.0 m s^{-1} to a speed of 29 m s^{-1} over a distance of 850 m, when the driver braked and stopped the vehicle in 28 s.

    a Calculate: **i** the time taken to reach 29 m s^{-1} from 4 m s^{-1}, **ii** its acceleration during this time.

    b Calculate: **i** the distance it travelled during deceleration, **ii** its average deceleration as it slowed down.

2 A rail wagon moving at a speed of 2.0 m s^{-1} on a level track reached a steady incline, which slowed it down in 15.0 s and caused it to reverse. Calculate:

    a the distance it moved up the incline,

    b its acceleration on the incline,

    c its velocity and position on the incline after 20.0 s.

3 A cyclist accelerated from rest at a constant acceleration of 0.4 m s^{-2} for 20 s, then stopped pedalling and slowed to a standstill at a constant deceleration over a distance of 260 m.

    a Calculate: **i** the distance travelled by the cyclist in the first 20 s, **ii** the speed of the cyclist at the end of this time.

b Calculate: **i** the time taken to cover the distance of 260 m after she stopped pedalling, **ii** her deceleration during this time.

4 A rocket was launched directly upwards from rest. Its motors operated for 30 s after it left the launch pad, providing it with a constant vertical acceleration of 6.0 m s^{-2} during this time. Its motors then switched off (Figure 3).

    a Calculate: **i** its initial velocity, **ii** its height above the launch pad when its motors switched off.

    b Calculate its maximum height gain after its motors switched off.

    c Calculate the velocity with which it would hit the ground, if it fell from maximum height without the support of a parachute.

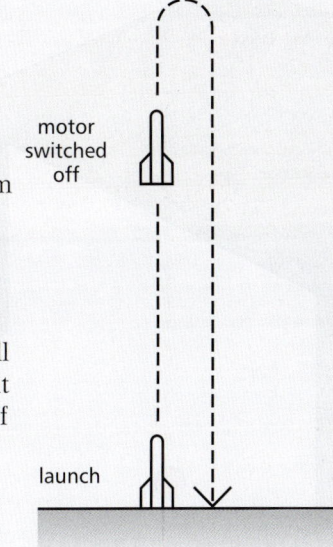

Figure 3

Projectile motion 1

On these pages you will learn to:

- describe and explain the motion due to a uniform velocity in a certain direction and a uniform acceleration in a perpendicular direction
- solve projectile problems where projection is horizontal using the equations for uniform acceleration in a vertical direction and for uniform velocity in a horizontal direction

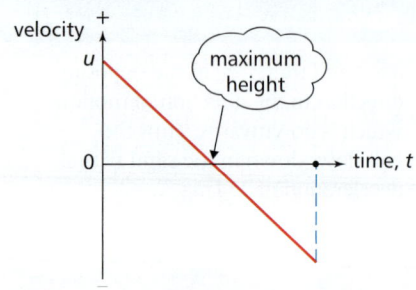

Figure 1 *Upward projection*

A **projectile** is any object acted upon only by the force of gravity. Air resistance is assumed to be negligible. Three key principles apply to all projectiles:

- The **acceleration** of the object is always equal to g and acts downwards, because the force of gravity acts downwards. The acceleration therefore only affects the vertical motion of the object.
- The **horizontal velocity** of the object is constant, because the acceleration of the object does not have a horizontal component.
- The motions in the horizontal and vertical directions are **independent** of each other.

Vertical projection

If an object is projected vertically, it moves vertically as it has no horizontal motion. Its acceleration is $9.81\,\text{m s}^{-2}$ downwards. Using the direction code '+ is upwards; – is downwards', its displacement, y, and velocity, v_y, after time t are given by:

$$v_y = u - gt$$

$$y = ut - \tfrac{1}{2}gt^2$$

where u is its initial velocity.

See Topic 1.7 for more about vertical projection.

Horizontal projection

A stone thrown from a cliff top follows a curved path downwards before it hits the water. If its initial projection was horizontal:

- Its path through the air becomes steeper and steeper as it drops.
- The faster it is projected, the further away it will fall into the sea.
- The time taken for it to fall into the sea does not depend on how fast it is projected.

Suppose two balls are released at the same time above a level floor, such that one ball drops vertically and the other is projected horizontally. Which one hits the floor first? In fact, they both hit the floor simultaneously. Try it! Why should the two balls hit the ground at the same time? They are both pulled to the ground by the force of gravity which gives each ball a downward acceleration, g. The ball that is projected horizontally experiences the same downward acceleration as the other ball. This downward acceleration does not affect the horizontal motion of the ball projected horizontally; only the vertical motion is affected.

Investigating horizontal projection

A stroboscope and a camera may be used to record the motion of a projectile. Figure 3 shows a multi-flash photograph of two balls, A and B, released at the same time. B was released from rest and dropped vertically; A was given an initial horizontal projection so it followed a curved path. The stroboscope flashed at a constant rate, so images of both balls were recorded at the same time.

- **The horizontal position** of A changes by equal distances between successive flashes. This shows that the horizontal component of A's velocity is constant.
- **The vertical position** of A and B changes at the same rate. At any instant, A is at the same level as B. This shows that A and B have the same vertical component of velocity at any instant.

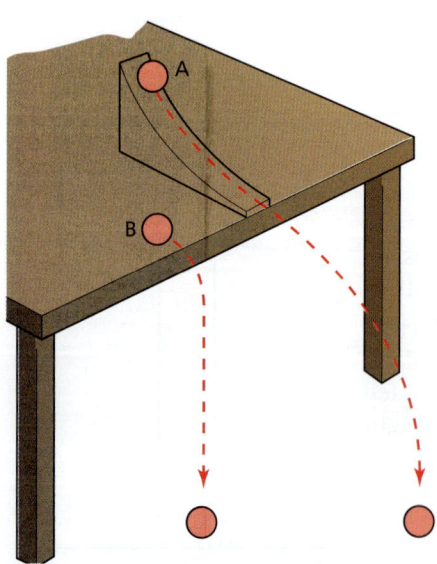

Figure 2 *Testing horizontal projection*

The projectile path of a ball projected horizontally
An object projected horizontally falls in an **arc** towards the ground. If its initial velocity is U, then at time t after projection (Figure 4):

- The **horizontal component** of its displacement,
$$x = Ut$$

(because it moves horizontally at a constant speed)

- The **vertical component** of its displacement,
$$y = \tfrac{1}{2}gt^2$$

(because it has no vertical component of its initial velocity)

- Its velocity has:

    a horizontal component $v_x = U$, and

    a vertical component $v_y = -gt$

Note Its speed at time $t = \sqrt{(v_x^2 + v_y^2)}$

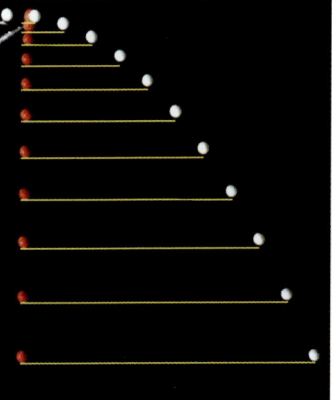

Figure 3 *Multi-flash photo of two falling balls*

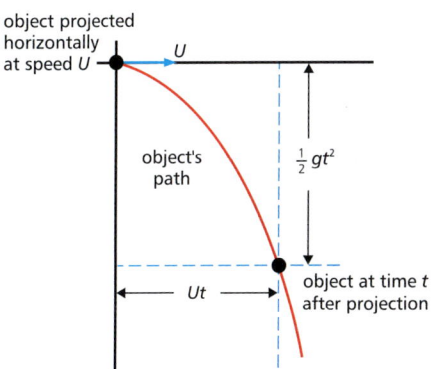

Figure 4 *Horizontal projection*

Worked example

$g = 9.81\,\mathrm{m\,s^{-2}}$

An object is projected horizontally at a speed of $15\,\mathrm{m\,s^{-1}}$ from the top of a tall tower (Figure 5) of height 35.0 m. Calculate:

a how long it takes to fall to the ground,

b how far it travels horizontally,

c its speed just before it hits the ground.

Solution
a $y = -35\,\mathrm{m}$, $a = -9.81\,\mathrm{m\,s^{-2}}$ (– for downwards)
$$y = \tfrac{1}{2}gt^2$$
$$\therefore t^2 = \frac{2y}{g} = \frac{2 \times -35}{-9.81} = 7.14\,\mathrm{s^2}$$
$$\therefore t = 2.67\,\mathrm{s}$$

b $U = 15\,\mathrm{m\,s^{-1}}$, $t = 2.67\,\mathrm{s}$
$x = Ut = 15 \times 2.67 = 40\,\mathrm{m}$

c Just before impact, $v_x = U = 15\,\mathrm{m\,s^{-1}}$ and $v_y = -gt = -9.81 \times 2.67 = 26.2\,\mathrm{m\,s^{-1}}$
$\therefore$ speed just before impact, $v = (v_x^2 + v_y^2)^{½} = (15^2 + 26.2^2)^{½} = 30.2\,\mathrm{m\,s^{-1}}$

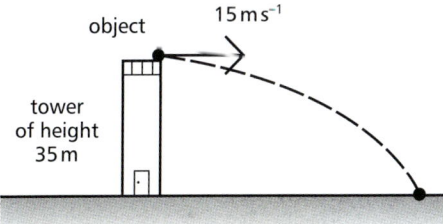

Figure 5

Summary test 1.8

$g = 9.81\,\mathrm{m\,s^{-2}}$

1 An object is released from a hot air balloon 50 m above the ground that is descending vertically at a speed of $4.0\,\mathrm{m\,s^{-1}}$. Calculate:

    a the velocity of the object at the ground,

    b the duration of descent of the object,

    c the height of the balloon above the ground when the object hits the ground.

2 An object is projected horizontally at a speed of $16\,\mathrm{m\,s^{-1}}$ into the sea from a cliff top of height 45.0 m. Calculate:

    a how long it takes to reach the sea,

    b how far it travels horizontally,

    c its impact velocity.

3 A dart is thrown horizontally along a line which passes through the centre of a dartboard 2.3 m away from the point at which the dart was released. The dart hits the dartboard at a point 0.19 m below the centre. Calculate:

    a the time of flight of the dart,

    b its horizontal speed of projection.

4 A parcel is released from an aircraft travelling horizontally at a speed of $120\,\mathrm{m\,s^{-1}}$ above level ground. The parcel hits the ground 8.5 s later. Calculate:

    a the height of the aircraft above the ground,

    b the horizontal distance travelled in this time by
       i the parcel, **ii** the aircraft,

    c the speed of impact of the parcel at the ground.

Projectile motion 2

Learning outcomes

On these pages you will learn to:

- compare projectile motion with the motion of electrons in a uniform electric field
- solve projectile problems where projection is not horizontal using the equations for uniform acceleration in a vertical direction and for uniform velocity in a horizontal direction

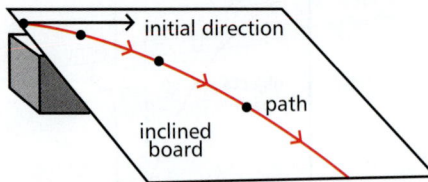

Figure 1 *Using an inclined board*

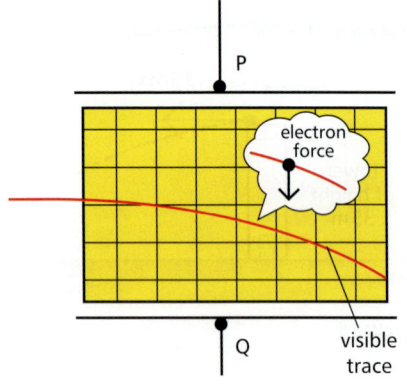

Figure 2 *An electron beam on a parabolic path*

(Vacuum tube used to contain apparatus not shown)

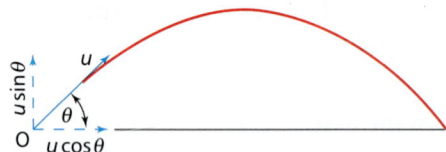

Figure 3 *Projectile motion*

Projectile-like motion

Any form of motion where an object experiences a constant acceleration in a different direction to its velocity will be like projectile motion. For example:

- The path of a ball rolling across an inclined board will be a projectile path. Figure 1 shows the idea. The object is projected across the top of the board from the side. Its path curves down the board and is **parabolic**. This is because the object is subjected to a constant acceleration acting down the board, and its initial velocity is across the board. The same equations as for projectile motion apply, with the motion down the board replacing the vertical motion.
- The path of a beam of **electrons** directed between two oppositely charged parallel plates is a **parabola**, as shown in Figure 2. Each electron in the beam is acted on by a constant force towards the positive plate, because the charge of an electron is negative. Therefore, each electron experiences a constant acceleration towards the positive plate. If its initial velocity is parallel to the plates, then its path is parabolic because its motion parallel to the plates is at **zero** acceleration; whereas its motion perpendicular to the plates is at **constant** (non-zero) acceleration.

Projection at angle θ and speed U above the horizontal

An object projected into the air follows a path that depends on its **speed** of projection and its **angle** of projection. In unit 1.8, we looked at the special cases of vertical projection and of horizontal projection. In this section, we will consider the general case of projection at angle θ to the horizontal. The analysis can be applied to the motion of any object acted on by a force that is constant in **magnitude** and **direction**. Air resistance is assumed to be negligible in this analysis.

For an object projected at initial speed U at an angle θ to the horizontal, as in Figure 3, its **initial velocity** has:

- a **horizontal** component $u_x = U\cos\theta$
- a **vertical** component $u_y = U\sin\theta$

Its **acceleration** has:

- a **horizontal** component $a_x = 0$
- a **vertical** component $a_y = -g$

For the **horizontal motion**, at time t after release:

- Its horizontal component of **velocity**, $v_x = U\cos\theta$ (unchanged), as its horizontal component of acceleration is zero.
- Its horizontal component of **displacement**, x = its horizontal component of velocity × time taken = $Ut\cos\theta$

For the vertical motion,

- Its vertical component of **velocity**, $v_y = U\sin\theta - gt$
  This is obtained by applying the equation '$v = u + at$' to the vertical motion with $U\sin\theta$ for the initial speed u and $-g$ for the acceleration a.
- Its vertical component of **displacement**, $y = Ut\sin\theta - \frac{1}{2}gt^2$
  This is obtained by applying the equation '$s = ut + \frac{1}{2}at^2$' to the vertical motion with $U\sin\theta$ for the initial speed u and $-g$ for the acceleration a.
- Its speed at time $t = \sqrt{(v_x^2 + v_y^2)}$, in accordance with Pythagoras' rule for adding the two perpendicular components of a vector.

A **positive value** of θ will signify projection at angle θ **above** the horizontal.

A **negative value** of θ will signify projection at angle θ **below** the horizontal.

Worked example

An arrow is fired at a speed of $48\,\text{m}\,\text{s}^{-1}$ at an angle of $50°$ above the horizontal. Calculate **a** how long it takes to reach maximum height; **b** its maximum height. $g = 9.81\,\text{m}\,\text{s}^{-2}$

Solution

$U = 48\,\text{m}\,\text{s}^{-1}$, $\theta = 50°$

a The vertical component of the initial velocity,
$u_y = U\sin\theta = 48\sin 50° = 36.7\,\text{m}\,\text{s}^{-1}$

The vertical component of velocity at maximum height,
$v_y = 0$

Using $v_y = U\sin\theta - gt$, $0 = 36.7 - 9.81t$

$$t = \frac{36.7}{9.81} = 3.74\,\text{s} = 3.7\,\text{s to 2 s.f.}$$

b The maximum height can be calculated by using
$y = Ut\sin\theta - \frac{1}{2}gt^2$ with $t = 3.74\,\text{s}$,
$\therefore$ maximum height $= 36.7 \times 3.74 - (0.5 \times 9.81 \times 3.74^2)$

$$= 68.6\,\text{m}$$

The effects of air resistance

A projectile moving through air experiences a force that drags on it because of the resistance of the air it passes through. This **drag force** is partly caused by friction between the layers of air near the projectile's surface where the air flows over its surface. The drag force:

- acts in the opposite direction to the direction of motion of the projectile, and it increases as the projectile's speed increases
- has a horizontal component that reduces its range
- reduces the maximum height of the projectile if its initial direction is above the horizontal and makes its descent steeper than its ascent.

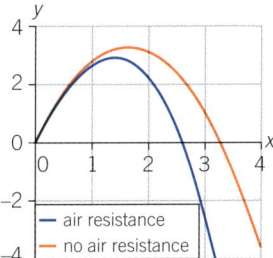

Summary test 1.9

$g = 9.81\,\text{m}\,\text{s}^{-2}$. For all questions, assume air resistance is negligible unless otherwise stated.

1 A ball was projected horizontally at a speed of $0.52\,\text{m}\,\text{s}^{-1}$ across the top of an inclined board of width 600 mm and length 1200 mm. It reached the bottom of the board 0.90 s later.

Calculate:

a the distance travelled by the ball across the board,

b its acceleration on the board,

c its speed at the bottom of the board.

2 An arrow is fired at a speed of $45\,\text{m}\,\text{s}^{-1}$ at an angle of $30°$ above the horizontal. Calculate:

a its maximum height,

b how long it takes to reach maximum height,

c how long it takes to return to the same horizontal level at which it started,

d the distance travelled horizontally to the point of return in **c**.

3 A cable car was travelling at a speed of $4.6\,\text{m}\,\text{s}^{-1}$ in a direction that was $40°$ above the horizontal when an object was released from the cable car (Figure 5).

a Calculate the horizontal and vertical components of velocity of the object at the instant it was released.

b The object took 5.8 s to fall to the ground below. Calculate: **i** the distance fallen by the object from the point of release, **ii** the horizontal distance travelled by the object from the point of release to where it hit the ground.

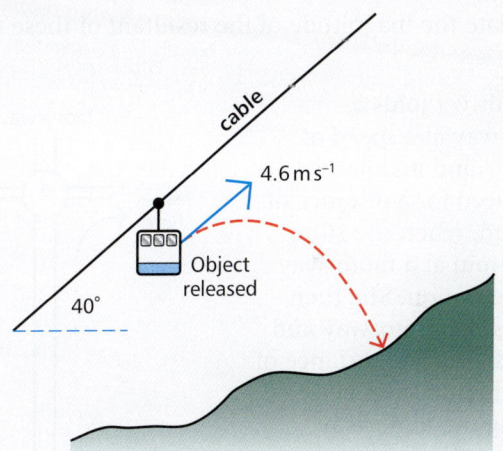

Figure 5

4 A cannon ball was fired from the top of a tower at an angle of elevation of $25°$. The ball hit the ground 2.7 s later at a distance of 58 m away from the foot of the tower.

a Calculate the horizontal and vertical components of the velocity of the cannon ball at the instant it was fired.

b Calculate the height of the tower above the ground.

c Calculate the speed of impact of the cannon ball at the ground.

d Discuss the effect on the range of the cannon ball if air resistance had not been negligible.

Launch additional digital resources for the chapter

$g = 9.81\,\mathrm{ms^{-2}}$

1 a i State the difference between a vector and a scalar quantity.
   ii Give one example of a vector quantity other than force.
   iii Give an example of a scalar quantity other than speed.

b A 6.0 N force and a 5.0 N force act at right angles to each other on an object of mass 3.0 kg.

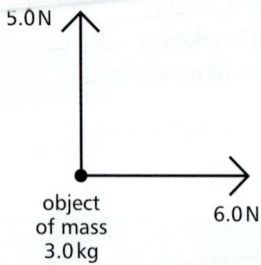

Figure 1.1

Calculate the magnitude of the resultant of these two forces.

2 A car driver joins a motorway at a speed of 22 m s^{-1} and maintains this speed for a distance of 32.5 km, when she stops for 20 min at a motorway service station. She then rejoins the motorway and travels a further distance of 47.5 km in 40 min before leaving the motorway.

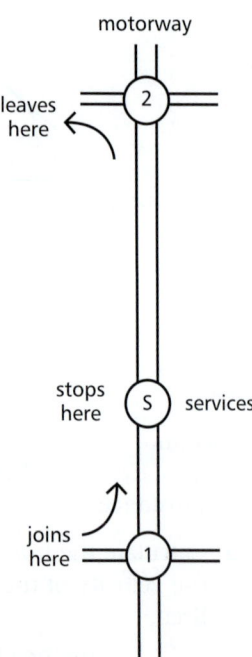

Figure 2.1

a Calculate the time taken for the first part of the journey.

b Sketch a speed–time graph for this journey on the motorway.

c Calculate the average speed of the car for the whole journey.

3 An aircraft taking off accelerated uniformly for 40 s from rest, before it left the ground after travelling a distance of 1600 m. Calculate:

a its average speed during take-off,

b its velocity at the instant it left the ground,

c its acceleration during take-off.

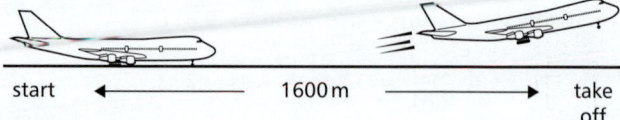

start ◄——— 1600 m ———► take off

Figure 3.1

4 A train accelerated uniformly from rest for 40 s until its speed reached 15 m s^{-1}. It then travelled at a constant speed of 15 m s^{-1} for a further 60 s, before slowing down at constant deceleration and stopping 20 s later.

a Sketch a speed–time graph to represent its motion.

b Calculate its acceleration and the distance it travelled in each of the three stages of its motion.

c Calculate its average speed.

5 An object released from rest at the side of a river bridge hit the water 2.2 s later. Calculate:

a the distance fallen in air by the object,

b the speed of the object just before it hit the water.

6 A stationary railway truck on an inclined track was struck by a shunting engine moving at a constant speed of 0.6 m s^{-1}. The impact caused the truck to move up the incline at an initial speed of 2.0 m s^{-1}. The truck slowed down and stopped 10 s later.

engine   truck

Figure 6.1

a Calculate: **i** the distance travelled by the truck before it stopped, **ii** the magnitude and direction of its acceleration.

b The truck stopped for an instant, then rolled back down the track and hit the engine again. The engine continued to move up the track at 0.6 m s^{-1}.
   i Sketch a velocity–time graph on the same axes for the truck and the engine.
   ii Use your graph to show that the truck hit the engine a second time 14 s after it was first hit.
   iii Calculate the velocity of the truck immediately before this second impact.

7 A ball thrown vertically into the air left the thrower's hand when it was 1.6 m above the ground and hit the ground on return 3.1 s later.

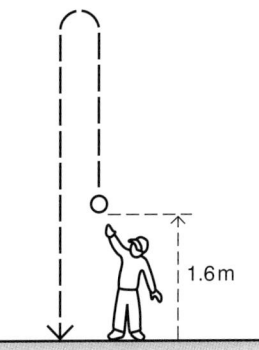

Figure 7.1

a Calculate: **i** the initial velocity of the ball, **ii** the maximum height of the ball above the ground.

b Calculate its velocity just before impact.

8 A cricketer on a level playing field threw a ball into the air at an angle of 30° above the horizontal. The ball was caught by another cricketer, 36 m away at the same level 1.8 s later.

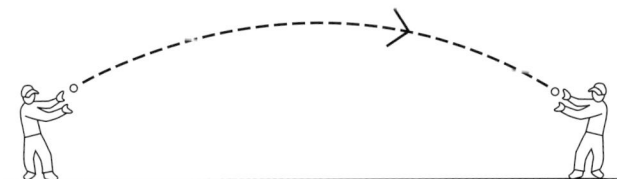

Figure 8.1

a Calculate the horizontal velocity of the ball.

b Show that the initial speed of the ball was 23 m s^{-1}.

c Calculate the maximum height reached by the ball.

9 A rocket launched vertically accelerated at a constant acceleration of 6.0 m s^{-2} for 25 s before its fuel supply ran out. It then rose to maximum height and fell back to the ground.

a i Calculate its velocity and its height when its fuel ran out.

   ii Show that it reached a maximum height of 3.0 km.

b i Calculate how long it took to fall from maximum height to the ground.

   ii Calculate its velocity immediately before impact with the ground.

   iii Sketch a velocity–time graph for the motion of the rocket from launch to impact.

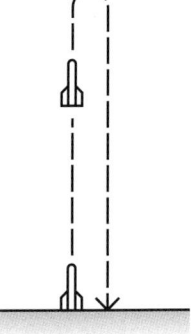

Figure 9.1

10 When a tennis player served a ball, the ball left the racquet horizontally at a speed of 23 m s^{-1} as shown in Figure 10.1.

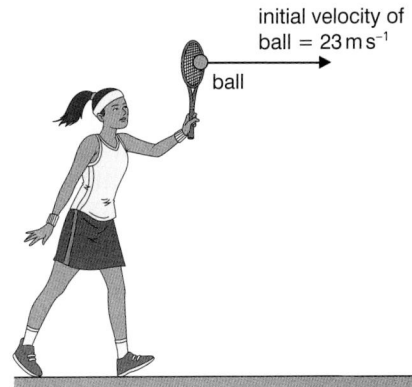

Figure 10.1

a After leaving the racquet, the ball just cleared the net after travelling a horizontal distance of 11.9 m to the net. Calculate the vertical displacement of the ball in this time. Assume air resistance is negligible.

b After clearing the net at a height of 0.92 m above the ground, the ball hit the ground, as shown in Figure 10.2.

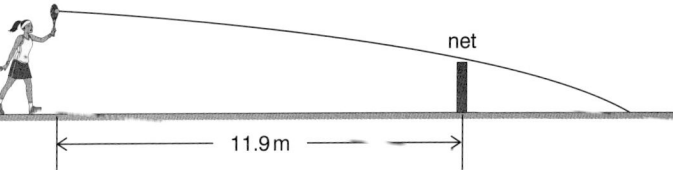

Figure 10.2

   i Calculate the horizontal distance from the net to the point where the ball hit the ground.

   ii Calculate the magnitude and direction of the ball's velocity immediately before it hit the ground.

c Discuss the effect on: **i** the magnitude, and **ii** the direction of the ball's velocity immediately before it hit the ground if the ball's initial direction had been above the horizontal.

2 Forces in action

2.1 Force and acceleration

Learning outcomes

On these pages you will learn to:

- show an understanding that mass is the property of a body that resists change in motion
- state Newton's first law of motion and give examples of where it applies
- recall and use Newton's second law of motion for constant mass in the form $F = ma$
- describe and use the concept of weight as the effect of a gravitational field on a mass
- recall that for a body of mass m, its weight $W = mg$, where g is the acceleration of free fall

Motion without force

Motorists on icy roads in winter need to be very careful, because the tyres of a car have little or no 'grip' on the ice. Moving from a standstill on ice is very difficult. Stopping on ice is almost impossible, as a car moving on ice will slide when the brakes are applied. **Friction** is a hidden force that we don't usually think about until it is absent!

If you have ever tried to push a heavy crate across a rough concrete floor, you will know all about friction. The push force is opposed by friction and as soon as you stop pushing, friction stops the crate moving. If the crate had been pushed onto a patch of ice, it would have moved across the ice without any further push needed.

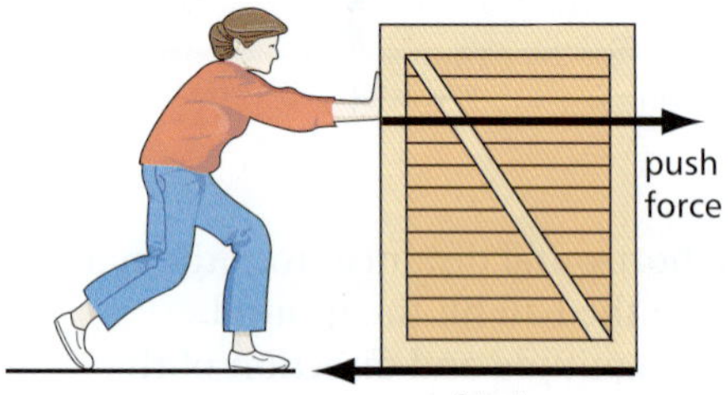

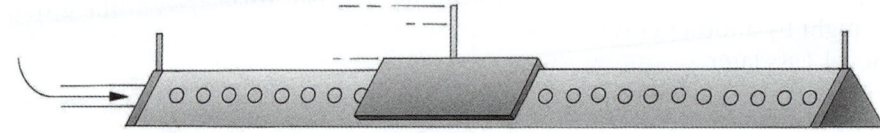

Figure 1 *Overcoming friction*

Figure 2 shows an air track which allows **motion** to be observed in the absence of friction. The glider on the air track floats on a cushion of air. Provided the track is level, the glider moves at a constant velocity along the track because friction is absent.

Figure 2 *The linear air track*

Newton's first law of motion

Objects either stay at rest, or remain in uniform motion (i.e. at constant velocity), unless acted on by a resultant force.

Sir Isaac Newton was the first person to realise that a moving object stays moving at constant velocity unless acted on by a force. He recognised that when an object is acted on by a **resultant** force, the result is to change the object's velocity. In other words, an object moving at **constant** velocity is either:

- acted on by **no forces**, or
- the forces acting on it are **balanced** (i.e. the **resultant force is zero**).

Investigating force and motion

How does the velocity of an object change if it is acted on by a constant force? Figure 3 shows how this can be investigated using a dynamics trolley and a ticker-tape timer. The timer prints dots on the tape at a constant rate of 50 dots per second; so the faster the tape is moving, the greater the spacing between adjacent dots. An electronic timer linked to a computer can be used in place of the ticker-tape timer.

The trolley is pulled along a sloped runway by means of one or more elastic bands stretched to the same length. The runway is sloped just enough to compensate for friction. To test for the correct slope, the trolley should move down the runway at constant speed after being given a brief push.

- If a ticker-tape timer is used, the length of each 'ten dot' section of the tape gives a measure of the speed as the middle of that section went through the ticker timer. Cutting the tape into 'ten dot' sections enables a speed–time chart to be made, as shown in Figure 4.

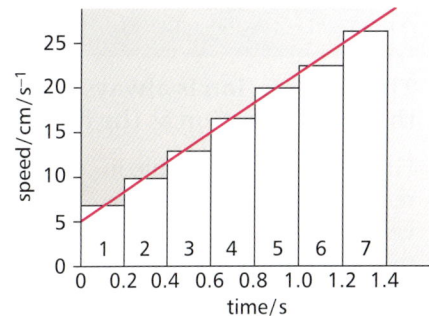

Figure 4 *Making a tape chart*

① Runway sloped just enough to compensate for friction

② Ticker timer prints 50 dots per second on the tape

③ Tape records the trolley's motion

④ Elastic bands stretched to the same length as the trolley, pull it down the runway with constant force

Figure 3 *Investigating force and motion*

• If an electronic timer linked to a computer is used, the measurements should be displayed directly as a speed–time graph (Figure 5).

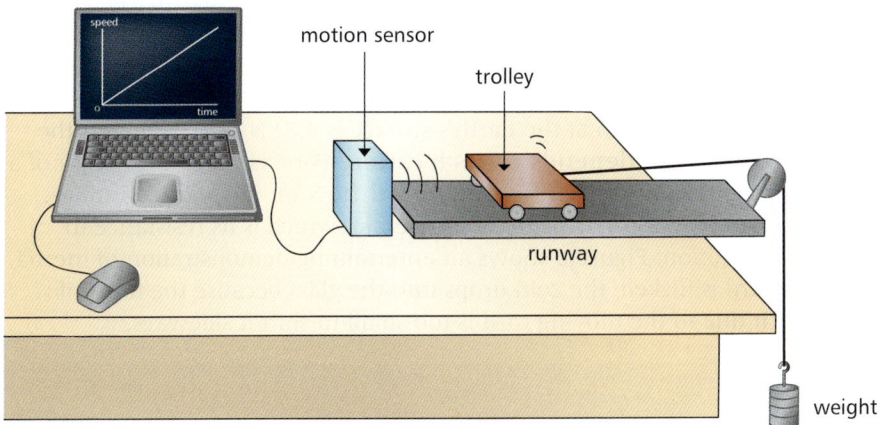

Figure 5 *Using a computer to measure acceleration*

The speed–time graph should show that the speed increased at a constant rate. The **acceleration** of the trolley is therefore constant, and can be measured from the speed–time graph. Table 1 shows typical measurements using different amounts of **force** (i.e. one, or two, or three elastic bands in 'parallel') and different amounts of **mass** (i.e. a single trolley, or a double trolley, or a triple trolley).

The results in Table 1 show that the force is proportional to the mass × the acceleration. In other words, if a force F acts on an object of mass m, the object undergoes an acceleration a such that:

$$F \text{ is proportional to } ma$$

Table 1 *Varying force and mass*

Force (i.e. no. of elastic bands)	1	2	3	1	2	3
Mass (i.e. no of trolleys)	1	1	1	2	2	2
Acceleration / m s⁻²	12	24	36	6	12	18
Mass × acceleration	12	24	36	12	24	36

Note

The acceleration is always in the same direction as the force

For example, a projectile in motion experiences a force vertically downwards due to gravity. Its acceleration is therefore vertically downwards, no matter what its direction of motion is.

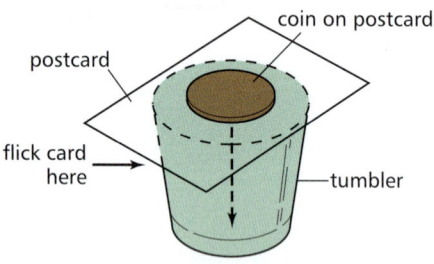

Figure 6 An 'inertia' trick

Newton's second law for constant mass

By defining the unit of force, the **newton (N)**, as the amount of force that will give an object of mass 1 kg an acceleration of $1\,\mathrm{m\,s^{-2}}$, the proportionality statement can be expressed as an equation:

$$F = ma$$

where F = force (in N), m = mass (in kg), a = acceleration (in $\mathrm{m\,s^{-2}}$).

This equation is known as **Newton's second law** for constant mass.

Worked example

A vehicle of mass 600 kg accelerates uniformly from rest to a speed of $8.0\,\mathrm{m\,s^{-2}}$ in 20 s. Calculate the force needed to produce this acceleration.

Solution

Acceleration, $a = \dfrac{v - u}{t} = \dfrac{8.0 - 0}{20} = 0.4\,\mathrm{m\,s^{-2}}$

Force, $F = ma = 600 \times 0.4 = 240\,\mathrm{N}$

Weight

- The acceleration of a **falling object** acted on by gravity only is g. Because the force of gravity on the object is the only force acting on it, its weight W (in newtons) is given by:

$$W = mg$$

where m = the mass of the object (in kg).

- When an object is **in equilibrium**, the support force on it is equal and opposite to its weight. Therefore, an object placed on a weighing balance (e.g. a spring balance or a top-pan balance) exerts a force on the balance equal to the weight of the object. Thus the balance measures the weight of the object.
- g is also referred to as the **gravitational field strength** at a given position, as it is the force of gravity per unit mass on a small object at that position. So the gravitational field strength at the Earth's surface is $9.81\,\mathrm{N\,kg^{-1}}$. Note that the weight of a fixed mass **depends on its location**. For example, the weight of a 1 kg object is 9.81 N on the Earth's surface and 1.6 N on the Moon's surface.
- The mass of an object is a measure of its **inertia**, which is its resistance to change of its motion. Figure 6 shows an entertaining demonstration of inertia. When the card is flicked, the coin drops into the glass because the force of friction on it due to the moving card is too small to shift it sideways.

Summary test 2.1

$g = 9.81\,\mathrm{N\,kg^{-1}}$

1 A car of mass 800 kg accelerates uniformly along a straight line from rest to a speed of $12\,\mathrm{m\,s^{-1}}$ in 50 s. Calculate:

a the acceleration of the car,

b the force on the car that produced this acceleration,

c the ratio of the accelerating force to the weight of the car.

2 An aeroplane, of mass 5000 kg, lands on a runway at a speed of $60\,\mathrm{m\,s^{-1}}$ and stops 25 s later. Calculate:

a the deceleration of the aeroplane,

b the braking force on the aeroplane.

3 a A vehicle, of mass 1200 kg, on a level road accelerates from rest to a speed of $6.0\,\mathrm{m\,s^{-1}}$ in 20 s, without change of direction. Calculate the force that accelerated the car.

b The vehicle in part **a** is fitted with a trailer of mass 200 kg. Calculate the time taken to reach a speed of $6.0\,\mathrm{m\,s^{-1}}$ from rest for the same force as in part **a**.

4 A bullet of mass 0.002 kg, travelling at a speed of $120\,\mathrm{m\,s^{-1}}$, hit a tree and penetrated a distance of 55 mm into the tree. Calculate:

a the deceleration of the bullet,

b the impact force of the bullet on the tree.

Using F = ma

Two forces in opposite directions

- When an object is acted on by two **unequal** forces acting in **opposite** directions, the object accelerates in the direction of the larger force. If the forces are F_1 and F_2 where $F_1 > F_2$:

$$\text{Resultant force, } F_1 - F_2 = ma$$

where m is the mass of the object and a is its acceleration.

- If the object is on a horizontal surface and F_1 and F_2 are **horizontal** and in opposite directions, the above equation still applies. The support force on the object is equal and opposite to its weight.

Learning outcomes

On these pages you will learn to:

- recall the relationship $F = ma$ and solve problems using it, including where more than one force acts on object
- appreciate that acceleration and force are always in the same direction

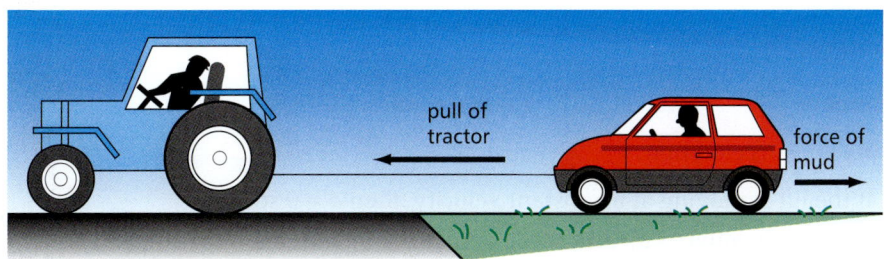

Figure 1 *Unbalanced forces*

Some examples are given below, where two forces act in different directions on an object.

Towing a trailer

Consider the example of a car of mass M fitted with a trailer of mass m on a level road. When the car and the trailer accelerate, the car pulls the trailer forward and the trailer holds the car back (Figure 2).

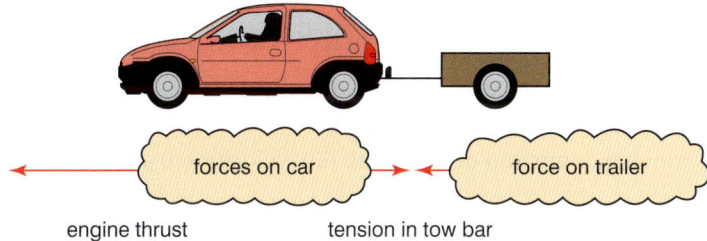

Figure 2 *Car and trailer*

- The car is subjected to a driving force F pushing it forwards (from its engine thrust) and the tension T in the tow bar holding it back. Therefore:

$$\text{Resultant force on car } = F - T = Ma$$

- The force on the trailer is due to the tension T in the tow bar pulling it forwards. Therefore:

$$T = ma$$

Combining the two equations gives:

$$F = Ma + ma = (M + m)a$$

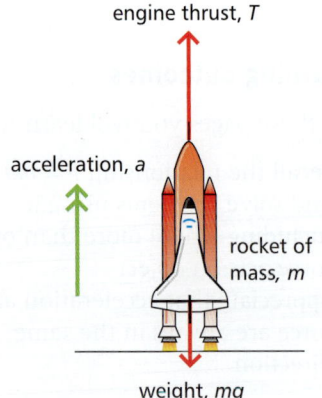

engine thrust, T

acceleration, a

rocket of mass, m

weight, mg

Figure 3 Rocket launch

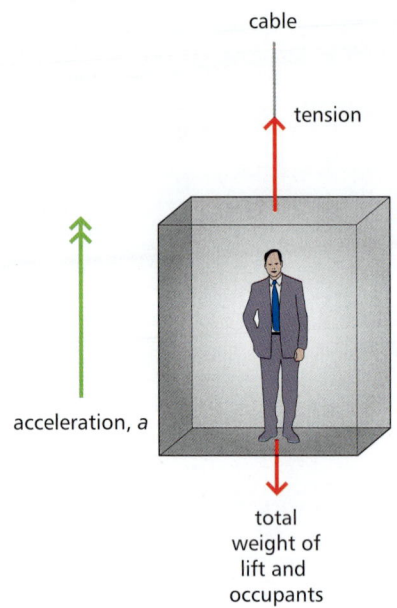

cable

tension

acceleration, a

total weight of lift and occupants

Figure 4 In a lift

Rocket problems

If T is the thrust of the rocket engine when its mass is m and the rocket is moving upwards, its acceleration a is given by:

$$T - mg = ma$$

Therefore:

$$\textbf{thrust, } T = mg + ma$$

The rocket thrust must therefore overcome the weight of the rocket for the rocket to take off (Figure 3).

Lift problems

Using 'upwards is positive' gives the resultant force on the lift as:

$$F = T - mg$$

where T is the tension in the lift cable and m is the total mass of the lift and occupants (Figure 4).

Therefore $$T - mg = ma$$

where a = acceleration.

- If the lift is moving at a constant velocity, then $a = 0$ so $T = mg$.
- If the lift is moving up and accelerating, then $a > 0$ so $T = mg + ma > mg$.
- If the lift is moving up and decelerating, then $a < 0$ so $T = mg + ma < mg$.
- If the lift is moving down and accelerating, then $a < 0$ (i.e. velocity and acceleration are both downwards and therefore negative) so $T = mg - ma < mg$.
- If the lift is moving down and decelerating, then $a > 0$ (i.e. velocity downwards and acceleration upwards and therefore positive) so $T = mg + ma > mg$.

Tension in the cable is less than the weight if:
- The lift is moving up and decelerating (i.e. velocity > 0 and acceleration < 0).
- The lift is moving down and accelerating (i.e. velocity < 0 and acceleration < 0).

Tension in the cable is greater than the weight if:
- The lift is moving up and accelerating (i.e. velocity > 0 and acceleration > 0).
- The lift is moving down and decelerating (i.e. velocity < 0 and acceleration > 0).

Worked example

$g = 9.81 \, \text{m s}^{-2}$

A lift of total mass 650 kg moving downwards decelerates at $1.5 \, \text{m s}^{-2}$ and stops. Calculate the tension in the lift cable during the deceleration.

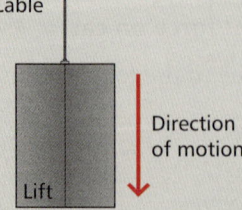

Cable

Direction of motion

Lift

Figure 5

Solution

The lift is moving down so its velocity $v < 0$. Since it is decelerating, its acceleration a is in the opposite direction to its velocity, so $a > 0$. Therefore use $a = +1.5 \, \text{m s}^{-2}$ in the equation

$$T - mg = ma$$
$$T = mg + ma$$
$$= 650 \times 9.81 + 650 \times 1.5 = 7400 \, \text{N}$$

Further *F = ma* problems

Pulley problems

Consider two masses M and m (where $M > m$) attached to a thread hung over a frictionless pulley, as in Figure 6. When released, mass M accelerates downwards and mass m accelerates upwards. If a is the acceleration and T is the tension in the thread, then:

- On mass M, the resultant force $= Mg - T = Ma$.
- On mass m, the resultant force $= T - mg = ma$.

Therefore, adding the two equations gives:

$$Mg - mg = (M + m)a$$

Sliding down a slope

Consider a block of mass m sliding down a slope (Figure 7). The component of the block's weight down the slope is $mg \sin \theta$. If the force of friction on the block is F_0, then:

$$\text{Resultant force on the block} = mg \sin \theta - F_0$$

Therefore: $\qquad mg \sin \theta - F_0 = ma$

where a is the acceleration of the block.

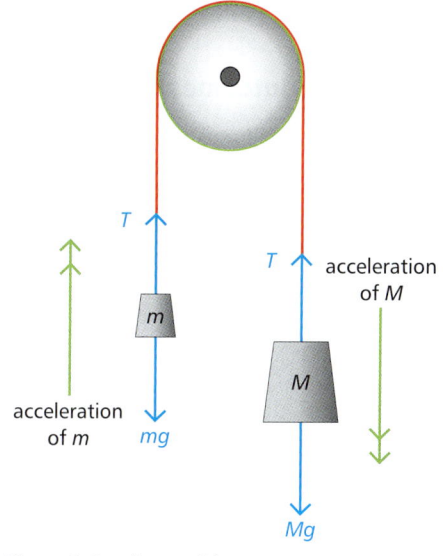

Figure 6 *A pulley problem*

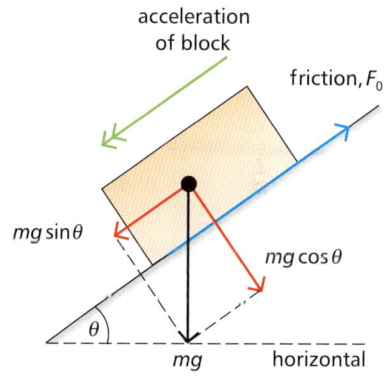

Figure 7

> ### Note
>
> With the addition of an engine force F_E, the above equation can be applied to a vehicle on a downhill slope of constant gradient. Thus:
>
> $$F_E + mg \sin \theta - F_0 = ma$$
>
> where F_0 is the combined sum of the force of friction and the braking force.

Summary test 2.2

$g = 9.81\,\text{m s}^{-2}$

1 A rocket of mass 550 kg blasts vertically from the launch pad at an acceleration of $4.2\,\text{m s}^{-2}$. Calculate:

  a the weight of the rocket,

  b the thrust of the rocket engines.

2 A car of mass 1400 kg, pulling a trailer of mass 400 kg, accelerates from rest to a speed of $9\,\text{m s}^{-1}$ in a time of 60 s on a level road. Assuming air resistance is negligible, calculate:

  a the tension in the tow bar,

  b the engine force.

3 A lift and its occupants have a total mass of 1200 kg. Calculate the tension in the lift cable when the lift is:

  a stationary,

  b ascending at constant speed,

  c ascending at a constant acceleration of $0.4\,\text{m s}^{-2}$,

  d descending at a constant deceleration of $0.4\,\text{m s}^{-2}$.

4 A brick, of mass 3.2 kg on a sloping flat roof, at 30° to the horizontal, slides at constant acceleration 2.0 m down the roof in 2.0 s from rest. Calculate:

  a the acceleration of the brick,

  b the frictional force on the brick due to the roof.

Learning outcomes

On these pages you will learn to:

- describe qualitatively the motion of bodies falling in a uniform gravitational field with air resistance
- explain what is meant by terminal velocity
- explain why bodies falling in a fluid and powered vehicles reach a terminal speed and why their acceleration decreases to zero

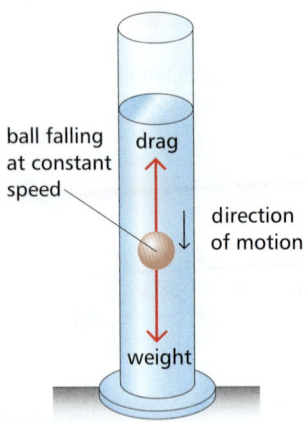

a Falling in a fluid

b Skydiving

Figure 1 At terminal velocity

Note

The acceleration at any instant is the gradient of the speed–time curve.

Drag forces

Any object moving through a fluid (something that can 'flow', i.e. a liquid or a gas) experiences a **drag force** due to the fluid. The drag force depends on:

- the shape and size of the object
- its speed
- the **viscosity** of the fluid (which is a measure of how easily the fluid flows past a surface).

The faster an object travels in a fluid, the greater the drag force on it.

Motion of an object falling in a fluid

- The speed of an object released from rest in a fluid increases as it falls, so the drag force on it due to the fluid increases. The resultant force on the object is the difference between the force of gravity on it (i.e. its weight) and the drag force. As the drag force increases, the resultant force decreases so the acceleration becomes less as it falls. If it continues falling, it attains **terminal velocity** when the drag force on it is equal and opposite to its weight. Its acceleration is then zero and its speed remains constant as it falls.
- Figure 2 shows how to investigate the motion of an object falling in a fluid. When the object is released, the thread attached to the object pulls a tape through a ticker timer which prints dots on the tape. The spacing between successive dots is a measure of the speed of the object, as the dots are printed on the tape at a constant rate. A tape chart can be made from the tape to show how the speed changes with time. The results show that:

The speed increases and reaches a constant value, which is the terminal velocity.

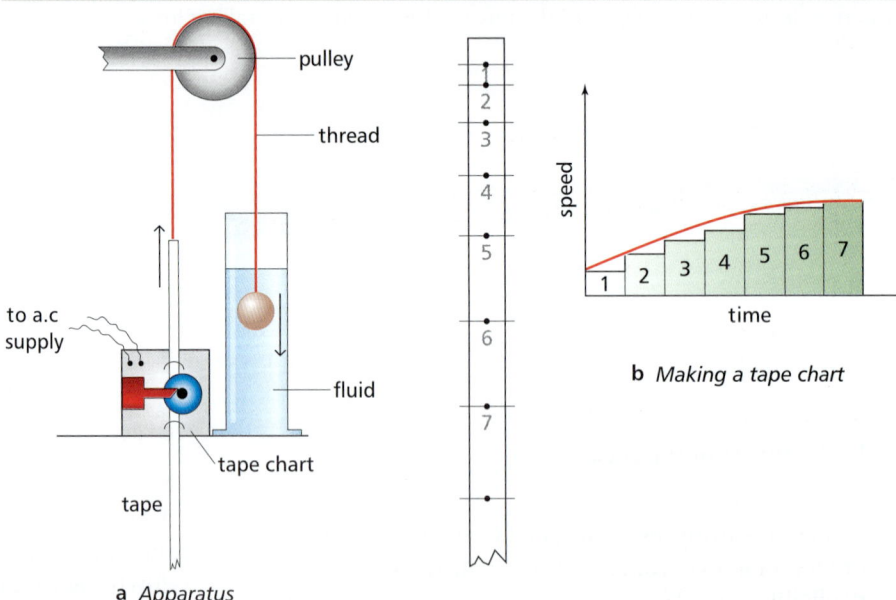

a Apparatus

Figure 2 Investigating the motion of an object falling in a fluid

- At any instant, the resultant force is $mg - F_D$, where m is the mass of the object and F_D is the drag force.

  Therefore: **Acceleration of the object** $= \dfrac{mg - F_D}{m} = g - \dfrac{F_D}{m}$

1 The **initial acceleration** is g, because the drag force is zero at the instant the object is released.
2 At the **terminal velocity**, the **potential energy** lost by the object as it falls is converted to **internal energy** of the fluid by the drag force.

Motion of a powered vehicle

The top speed of a road vehicle or an aircraft depends on its **engine power**, and its shape. A vehicle with a streamlined shape can reach a higher top speed than a vehicle with the same engine power that is not streamlined.

For a powered vehicle of mass m moving on a level surface, if F_E represents the **engine force** of the vehicle (i.e. the engine force driving the vehicle):

$$\text{Resultant force} = F_E - F_D$$

where F_D is the combined effect of friction and the drag force due to air resistance.

Therefore, its acceleration $a = \dfrac{(F_E - F_D)}{m}$

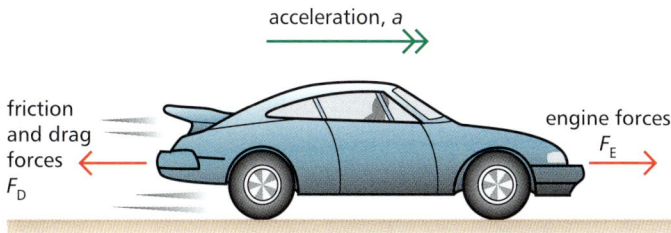

Figure 3

- Because the drag force increases with speed, the maximum speed (i.e. terminal velocity) of the vehicle, v_{max}, is attained when the drag force becomes equal and opposite to the motive force.

- At maximum speed, the **work done** by the engine is dissipated by the drag force and becomes internal energy of the surroundings. See Topic 2.6 for more about power.

Worked example

A car of mass 1200 kg has an engine which provides a motive force of 600 N. Calculate:

a its initial acceleration,

b its acceleration when the drag force is 400 N.

Solution

a The maximum acceleration is when the drag force is zero (i.e. when the car starts). The resultant force on the car is therefore 600 N at the start.

Therefore:

$$\text{initial acceleration} = \frac{\text{force}}{\text{mass}} = \frac{600\,\text{N}}{1200\,\text{kg}} = 0.5\,\text{m s}^{-2}$$

b When the drag force = 400 N,

$$\begin{aligned}\text{the resultant force} &= \text{motive force} - \text{drag force}\\ &= 600 - 400\,\text{N}\\ &= 200\,\text{N}\end{aligned}$$

$$\therefore \text{Acceleration} = \text{force/mass} = \frac{200\,\text{N}}{1200\,\text{kg}} = 0.16\,\text{m s}^{-2}$$

Hydrofoil physics

A hydrofoil boat travels much faster than an ordinary boat, because it has a powerful jet engine that enables it to 'ski' on its hydrofoils when the jet engine is switched on.

When the jet engine is switched on and takes over from the less-powerful propeller engine, the boat speeds up and the hydrofoils are extended. The boat rides on the hydrofoils so the drag force is reduced, as its hull is no longer in the water. At top speed, the motive force of the jet engine is equal to the drag force on the hydrofoils.

Figure 4 A hydrofoil ferry

Summary test 2.3

$g = 9.81\,\text{m s}^{-2}$.

1 a A steel ball of mass 0.15 kg released from rest in a liquid falls a distance of 0.20 m in 5.0 s. Assuming the ball reaches terminal velocity within a fraction of a second, calculate:
   i   its terminal velocity,
   ii the drag force on it when it falls at terminal velocity.

   b State and explain whether or not a smaller steel ball would fall at the same rate in the same liquid.

2 Explain why a cyclist can reach a higher top speed by crouching over the handlebars instead of sitting upright while pedalling.

3 A vehicle of mass 32 000 kg has an engine which has a maximum driving force of 4.4 kN and a top speed of 36 m s^{-1} on a level road.
   Calculate:

   a its maximum acceleration from rest,

   b the distance it would travel at maximum acceleration to reach a speed of 12 m s^{-1} from rest.

4 Explain why a vehicle has a higher top speed on a downhill stretch of road than on a level road.

Work and energy

Learning outcomes

On these pages you will learn to:

- give examples of energy in different forms, its conversion and conservation, and apply the principle of conservation of energy to simple examples
- define work as the product of a force and the displacement in the direction of the force
- calculate the work done in a number of situations including the work done by a gas that is expanding against a constant external pressure: $W = p\Delta V$

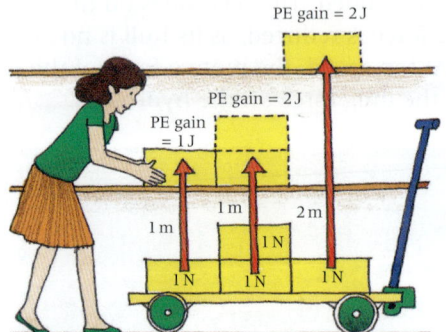

Figure 1 *Using joules*

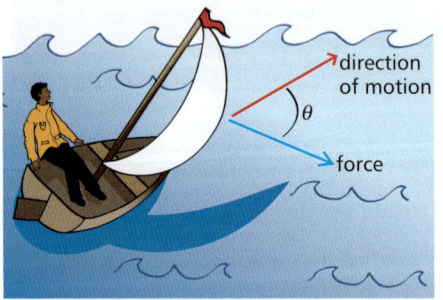

Figure 2 *Force and displacement*

Energy rules

Energy is needed to make stationary objects move, or to lift an object, or to change its shape, or to warm it up. When you lift an object, you **transfer energy** from your muscles to the object.

Objects can possess energy in **different forms**, including:
- **gravitational potential energy**, which is energy due to position in a gravitational field,
- **kinetic energy**, which is energy due to motion,
- **thermal energy**, which is energy due to the temperature of an object,
- **chemical or nuclear energy**, which is energy associated with chemical or nuclear reactions,
- **electrical potential energy**, which is energy of electrically charged objects,
- **elastic potential energy**, which is energy stored in an object when it is stretched or compressed.

Energy is measured in **joules (J)**. One joule is equal to the energy needed to raise a 1N weight through a vertical height of 1m.

Energy can be changed from one form into other forms. In any change, the total amount of energy after the change is always equal to the total amount of energy before the change. The total amount of energy is unchanged. In other words:

> **Energy cannot be created or destroyed.**

This statement is known as the **principle of conservation of energy.**

Work

Work is done on an object when a force acting on it due to another object makes it move. As a result, energy is transferred to the object by the force from the other object. The amount of work done depends on the force and the displacement of the object. The greater the force or the further the distance, the greater the work done.

> **Work done = force × displacement in the direction of the force**

The **unit of work** is the joule (J).

> **One joule is equal to the work done when a force of 1N moves its point of application by a displacement of 1m in the direction of the force.**

Energy is also measured in joules, because the energy transferred to an object when a force acts on it is equal to the work done on it by the force. For example, a force of 2N needs to be applied to an object of weight 2N to raise the object steadily. If the object is raised by 1.5m, the work done by the force is 3J (= 2N × 1.5m). Therefore, the gain of potential energy of the object is 3J.

Force and displacement

Imagine a yacht acted on by a wind force F at an angle θ to the direction in which the yacht moves, as in Figure 2. The wind force has a component $F\cos\theta$ in the direction of motion of the yacht, and a component $F\sin\theta$ at right angles to the direction of motion. If the yacht is moved a displacement s by the wind, the work done on it is equal to the component of force in the direction of motion × the displacement:

$$W = Fs\cos\theta$$

Force–distance graphs

- If a constant force F acts on an object and makes it move a displacement s in the direction of the force, the work done on the object $W = Fs$. Figure 3 shows a graph of force against distance in this situation. The area under the line is a rectangle, of height representing the force and of base length representing the distance moved. Therefore **the area represents the work done**.

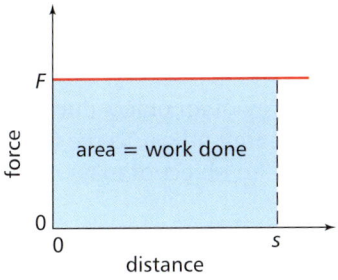

Figure 3 *A force–distance graph for a constant force*

- If a variable force F acts on an object and causes it to move in the direction of the force, for a small amount of distance Δs, the work done $\Delta W = F\Delta s$. This is represented on a graph of the force F against distance s by the area of a strip under the line of width Δs and height F. The total work done is therefore **the sum of the areas of all the strips** (i.e. the total area under the line). See Figure 4.

> **The area under the line of a force–distance graph represents the total work done.**

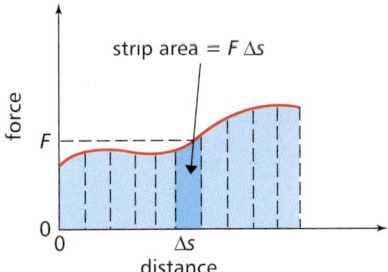

Figure 4 *A force–distance graph for a variable force*

Springs

For example, consider the force needed to stretch a spring. The greater the force, the more the spring is extended from its unstretched length. Figure 5 shows how the force needed to stretch a spring changes with the **extension** of the spring. The graph is a straight line through the origin. Therefore the force needed is proportional to the extension of the spring. This is known as **Hooke's law**. See Topic 5.3 for more about springs.

Figure 5 is a graph of force against distance; in this case, the distance the spring is extended. As explained above, the area under the line represents the work done to stretch the spring. Let F_0 represent the force needed to extend the spring to extension x_0. Therefore, because the area under the line from the origin to extension x_0 is a triangle of height F_0 and base length x_0:

$$\text{Area under line} = \tfrac{1}{2} \times \text{height} \times \text{base} = \tfrac{1}{2}F_0 x_0$$

$$\text{Work done to stretch spring to extension } e_0 = \tfrac{1}{2}F_0 x_0$$

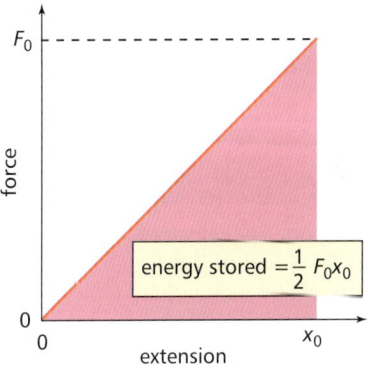

Figure 5 *Force against extension for a spring*

Summary test 2.4

1 Calculate the work done when:

   a a weight of 40 N is raised by a height of 5.0 m,

   b a spring is stretched to an extension of 0.45 m by a force of 20 N.

2 Calculate the energy transferred by a force of 12 N when it moves an object by a distance of 4.0 m:

   a in the direction of the force,

   b in a direction at 60° to the direction of the force,

   c in a direction at right angles to the direction of the force.

3 A luggage trolley of total weight 400 N is pushed at a steady speed 20 m up a slope by a force of 50 N acting in the same direction as the object moves in. At the end of this distance, the trolley is 1.5 m higher than at the start. Calculate:

   a the work done pushing the trolley up the slope,

   b the gain of potential energy of the trolley,

   c the energy wasted due to friction.

4 A spring that obeys Hooke's law requires a force of 1.2 N to extend it to an extension of 50 mm. Calculate:

   a the force needed to extend it to an extension of 100 mm,

   b the work done when the spring is stretched to an extension of 100 mm.

Learning outcomes

On these pages you will learn to:

- derive, from the equations of motion, the formula for kinetic energy $E_k = \frac{1}{2}mv^2$ and recall and apply the formula
- distinguish between gravitational potential energy and elastic potential energy
- show an understanding of and use the relationship between force and potential energy in a uniform field to solve problems
- derive, from the defining equation $W = Fs$, the formula $\Delta E_p = mg\Delta h$ for potential energy changes near the Earth's surface
- recall and use the formula $\Delta E_p = mg\Delta h$ for potential energy changes near the Earth's surface

Note

The formula does not hold at speeds **approaching the speed of light**. Einstein's theory of special relativity tells us that the mass of an object increases with speed and that the energy of an object can be worked out from the equation $E = mc^2$, where c is the speed of light in free space and m is the mass of the object.

Note

The formula does not hold unless the change of height Δh is much smaller than the **Earth's radius**. If height Δh is not insignificant compared with the Earth's radius, the value of g is not the same over height Δh. The force of gravity on an object decreases with increased distance from the Earth.

Kinetic energy

Kinetic energy is the energy of an object due to its motion. The faster an object moves, the more kinetic energy it has. To see the exact link between kinetic energy and speed, consider an object of mass m, initially at rest, acted on by a constant force F for a time t.

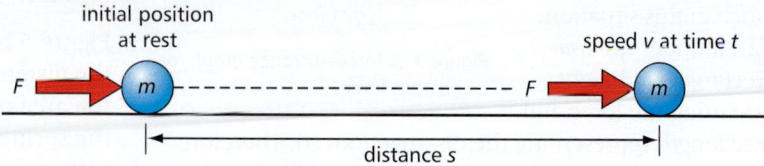

Figure 1 *Gaining kinetic energy*

Let the speed of the object at time t be v:

$$\therefore \qquad \text{Distance travelled, } s = \frac{1}{2}(u + v)t$$

$$= \frac{1}{2}vt \text{ because } u = 0$$

$$\text{Acceleration, } a = \frac{v - u}{t} = \frac{v}{t}$$

Using Newton's second law:

$$F = ma = \frac{mv}{t}$$

$\therefore$ the work done, by force F, to move the object through distance s:

$$W = Fs = \frac{mv}{t} \times \frac{vt}{2} = \frac{1}{2}mv^2$$

Because the gain of kinetic energy is due to the work done:

$$\textbf{Kinetic energy, } E_K = \tfrac{1}{2}mv^2$$

Potential energy

Potential energy is the energy of an object due to its position.

If an object of mass m is raised through a vertical height Δh at steady speed, the force needed to raise it is equal and opposite to its weight mg. Therefore:

$$\textbf{Work done to raise the object } = \textbf{force} \times \textbf{distance moved} = mg\Delta h$$

The work done on the object increases its **gravitational potential energy**:

$$\textbf{Change of gravitational potential energy, } \Delta E_P = mg\Delta h$$

At the Earth's surface, $g = 9.81\,\text{m s}^{-2}$.

Energy changes involving kinetic and potential energy

An object of mass m released above the ground

If air resistance is negligible, the object gains speed as it falls. Its potential energy therefore decreases and its kinetic energy increases.

After falling through a vertical height Δh, its kinetic energy is equal to its loss of potential energy.

In other words: $\qquad \frac{1}{2}mv^2 = mg\Delta h$

A pendulum bob

A pendulum bob is displaced from equilibrium and then released with the thread taut. The bob passes through the equilibrium position at maximum speed, and then slows down to reach maximum height on the other side of equilibrium. If its initial height above the equilibrium position is Δh_0, then whenever its height above the equilibrium position is Δh, its speed v at this height is such that:

Its kinetic energy = its loss of potential energy from maximum height

$$\tfrac{1}{2}mv^2 = mg\Delta h_0 - mg\Delta h$$

A fairground vehicle of mass m on a downward track

If a fairground vehicle was initially at rest at the top of the track, and its speed is v at the bottom of the track, then at the bottom of the track:

- its kinetic energy = $\tfrac{1}{2}mv^2$
- its loss of potential energy = $mg\Delta h$, where Δh is the vertical distance between the top and the bottom of the track
- the work done to overcome friction and air resistance = $mg\Delta h - \tfrac{1}{2}mv^2$.

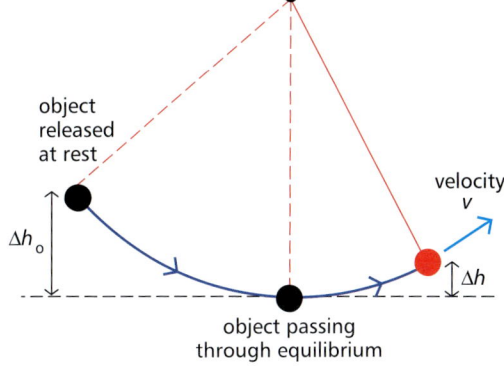

Figure 2 *A pendulum in motion*

Worked example

$g = 9.81\,\mathrm{m\,s^{-2}}$

On a fairground ride, the track descends by a vertical drop of 55 m over a distance of 120 m along the track. A train of mass 2500 kg on the track reaches a speed of $30\,\mathrm{m\,s^{-1}}$ at the bottom of the descent, after being at rest at the top. Calculate:

a the loss of potential energy of the train,

b its gain of kinetic energy,

c the average frictional force on the train during the descent.

Solution
a Loss of potential energy = $mg\Delta h = 2500 \times 9.81 \times 55 = 1.35 \times 10^6\,\mathrm{J}$

b Its gain of kinetic energy = $\tfrac{1}{2}mv^2 = 0.5 \times 2500 \times 30^2 = 1.13 \times 10^6\,\mathrm{J}$

c Work done to overcome friction = $mg\Delta h - \tfrac{1}{2}mv^2 = 1.35 \times 10^6 - 1.13 \times 10^6$
$$= 2.2 \times 10^5\,\mathrm{J}$$

Because the work done to overcome friction = frictional force × distance moved along track,

the frictional force $= \dfrac{\text{work done to overcome friction}}{\text{distance moved}} = \dfrac{2.2 \times 10^5\,\mathrm{J}}{120\,\mathrm{m}} = 1830\,\mathrm{N}$

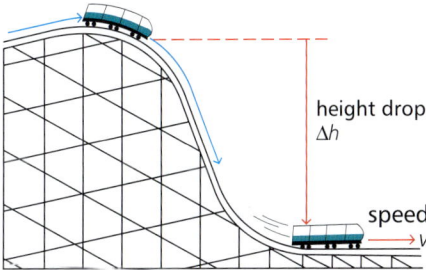

Figure 3

Summary test 2.5

1 A ball of mass 0.50 kg was thrown directly up at a speed of $6.0\,\mathrm{m\,s^{-1}}$. Calculate:

a its kinetic energy at $6\,\mathrm{m\,s^{-1}}$,

b its maximum gain of potential energy,

c its maximum height gain.

2 A ball of mass 0.20 kg, at a height of 1.5 m above a table, is released from rest and it rebounds to a height of 1.2 m above the table. Calculate:

a i the loss of potential energy on descent,
ii the gain of potential energy at maximum rebound height,

b the loss of energy due to the impact.

3 A cyclist of mass 80 kg (including the bicycle) freewheels from rest 500 m down a hill. The foot of the hill is 20 m lower than the cyclist's starting point, and the cyclist reaches a speed of $12\,\mathrm{m\,s^{-1}}$ at the foot of the hill.

Calculate:

a i the loss of potential energy,
ii the gain of kinetic energy of the cyclist and cycle,

b i the work done against friction and air resistance during the descent,
ii the average resistive force during the descent.

4 A fairground vehicle of total mass 1200 kg, moving at a speed of $2\,\mathrm{m\,s^{-1}}$, descends through a height of 50 m to reach a speed of $28\,\mathrm{m\,s^{-1}}$ after travelling a distance of 75 m along the track. Calculate:

a its loss of potential energy,

b its initial kinetic energy,

c its kinetic energy after the descent,

d the work done against friction,

e the average frictional force on it during the descent.

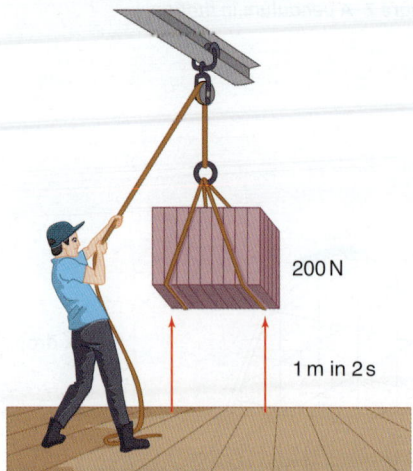

Figure 1 *A 100 watt worker*

Power and energy

Energy transfers

Energy can be **transferred** from one object to another by means of:

- **Work done** by a force due to one object making the other object move.

- **Heat transfer** from a hot object to a cold object. Heat transfer can be due to conduction, convection or radiation.

- In addition, **electricity**, **sound waves and electromagnetic radiation**, such as light or radio waves, transfer energy.

In any energy transfer process, the more energy transferred per second, the greater the **power** of the transfer process. For example, in a tall building where there are two elevators of the same total weight, the more powerful elevator is the one that can reach the top floor faster. In other words, its motor transfers energy from electricity at a faster rate than the motor of the other elevator. The energy transferred per second is the **power** of the motor:

> **Power is defined as the rate of transfer of energy.**

The unit of power is the watt (W), equal to an energy transfer rate of 1 joule per second.

If energy ΔE is transferred steadily in time t:

$$\text{Power, } P = \frac{\Delta E}{t}$$

Where energy is transferred by a force doing work, the energy transferred is equal to the work done by the force. Therefore, the **rate of transfer of energy** is equal to the work done per second. In other words, if the force does work W in time t:

$$\text{Power, } P = \frac{W}{t}$$

Power measurements

1 Muscle power

Test your own muscle power by timing how long it takes you to walk up a flight of steps. To calculate your muscle power, you will need to know your weight and the total height gain of the flight of steps:

- Your gain of potential energy = your weight × total height gain

- Your muscle power, $P = \dfrac{\text{energy transferred}}{\text{time taken}} = \dfrac{\text{weight} \times \text{height gain}}{\text{time taken}}$

A person of weight 480 N, who climbs a flight of stairs of height 10 m in 12 s, has leg muscles of power: $\dfrac{480\,\text{N} \times 10\,\text{m}}{12\,\text{s}} = 400\,\text{W}$

Since each leg does work while the other is being lifted, each leg would have an output power of 400 W.

2 Electrical power

The power of a 12 V light bulb can be measured using a joulemeter, as shown in Figure 2. The joulemeter is read before and after the light bulb is switched on. The difference between the readings is the energy supplied to the light bulb. If the light bulb is switched on for a measured time, the power of the light bulb can be calculated from the energy supplied to it / the time taken.

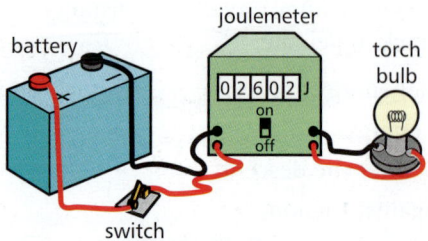

Figure 2 *Using a joulemeter*

Engine power

Vehicle engines, marine engines and aircraft engines are all designed to make objects move. The output power of an engine is called its **engine** power.

> **When a powered object moves at a constant velocity at a constant height, the resistive forces (e.g. friction, air resistance, drag) are equal and opposite to the engine force.**

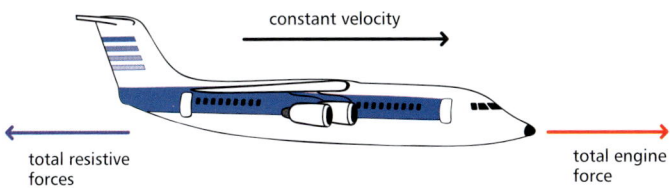

constant velocity

total resistive forces

total engine force

Figure 3 *Engine power*

The work done by the engine is converted to internal energy of the surroundings by the resistive forces.

For a powered vehicle driven by a constant force F moving at speed v:

Work done per second = force × distance moved per second

∴ Engine power of the engine, $P = Fv$

Worked example

An aircraft powered by engines that exert a force of 40 kN is in level flight at a constant velocity of 80 m s^{-1}. Calculate the engine power of the engine at this speed.

Solution
Power = force × velocity = 40 000 N × 80 m s^{-1} = 3.2 × 10^6 W

> **When a powered object gains speed, the engine force exceeds the resistive forces on it.**

Consider a vehicle that speeds up on a level road. The engine power of its engine is the work done by the engine per second. The work done by the engine increases the kinetic energy of the vehicle and enables the vehicle to overcome the resistive forces acting on it. Because the resistive forces increase the internal energy of the surroundings:

Engine power = energy per second wasted + gain of kinetic energy per second

Juggernaut physics

The maximum mass of a truck on UK roads must not exceed 44 tonnes, which corresponds to a total mass of 44 000 kg. This limit is set so as to prevent damage to roads and bridges. European Union regulations limit the engine power of a large truck to a maximum of 6 kW per tonne. Therefore, the maximum motive power of a 44 tonne truck is 264 kW. Prove for yourself that a truck with a engine power of 264 kW moving at a constant speed of 31 m s^{-1} (= 70 miles per hour) along a level road experiences a drag force of 8.5 kN.

Figure 4 *Heavy goods on the move*

Summary test 2.6

$g = 9.81$ m s^{-2}

1 A student of weight 450 N climbed 2.5 m up a rope in 18 s. Calculate:

    a the gain of potential energy of the student,

    b the energy transferred per second.

2 Calculate the power of the engines of an aircraft at a speed of 250 m s^{-1}, if the total engine thrust to maintain this speed is 2.0 MN.

3 A rocket of mass 5800 kg accelerates uniformly and vertically from rest to a speed of 220 m s^{-1} in 25 s. Calculate:

    a its gain of potential energy,

    b its gain of kinetic energy,

    c the power output of its engine, assuming no energy is wasted due to air resistance.

4 Calculate the height through which a 5 kg mass would need to be dropped to lose the same energy as a 100 W light bulb would use in 1 min.

Efficiency

Learning outcomes

On these pages you will learn to:

- recall and understand that the efficiency of a system is the ratio of useful work done by the system to the total energy input
- show an appreciation for the implications of energy losses in practical devices and use the concept of efficiency to solve problems

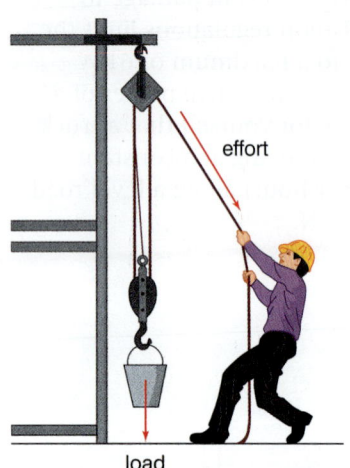

Figure 1 *Using pulleys*

Machine power

A machine that lifts or moves an object applies a force to the object to move it. If the machine exerts a force F on an object to make it move through a displacement s in the direction of the force, the work done W on the object by the machine can be calculated using the equation:

$$\textbf{Work done, } W = Fs$$

If the object moves at a constant velocity v due to this force being opposed by an equal and opposite force caused by friction, the object moves a displacement $s = vt$ in time t.

Therefore, the output power of the machine is given by

$$P_{OUT} = \frac{\text{work done by the machine}}{\text{time taken}} = \frac{Fvt}{t} = Fv$$

> **Output power, $P_{OUT} = Fv$, where F = 'output' force of the machine and v = speed of the object**

Examples

1 An electric motor operating a sliding door exerts a force of 125 N on the door, causing it to open at a constant speed of $0.40 \, \text{m s}^{-1}$. The output power is $125 \, \text{N} \times 0.40 \, \text{m s}^{-1} = 50 \, \text{W}$. The motor must therefore transfer 50 J every second to the sliding door while the door is being opened.
   Friction in the motor bearings and also electrical resistance of the motor wires means that some of the electrical energy supplied to the motor is wasted. For example, if the motor is supplied with electrical energy at a rate of $150 \, \text{J s}^{-1}$ and it transfers $50 \, \text{J s}^{-1}$ to the door, the difference of $100 \, \text{J s}^{-1}$ is wasted as a result of friction and electrical resistance in the motor.

2 A pulley system is used to raise a load of 80 N at a speed of $0.15 \, \text{m s}^{-1}$ by means of a constant '**effort**' of 30 N applied to the system. Figure 1 shows the arrangement. Note that for every metre the load rises, the effort needs to act over a distance of three metres because the load is supported by three sections of rope. The effort must therefore act at a speed of $0.45 \, \text{m s}^{-1}$ ($= 3 \times 0.15 \, \text{m s}^{-1}$).
   - The work done on the load each second = load × distance per second
     $= 80 \, \text{N} \times 0.15 \, \text{m s}^{-1} = 12 \, \text{J s}^{-1}$
   - The work done by the effort each second = effort × distance each second
     $= 30 \, \text{N} \times 0.45 \, \text{m s}^{-1} = 13.5 \, \text{J s}^{-1}$
   The difference of $1.5 \, \text{J s}^{-1}$ is the energy wasted each second in the pulley system. This is due to friction in the bearings and also because energy must be supplied to raise the lower pulley. For example, if the weight of the lower pulley is 6 N, the potential energy gain each second by the lower pulley would be $0.9 \, \text{J s}^{-1}$ ($= 6 \, \text{N} \times 0.15 \, \text{m s}^{-1}$) when the load is raised at a speed of $0.15 \, \text{m s}^{-1}$. Thus the energy wasted each second due to friction would be $0.6 \, \text{J s}^{-1}$ ($= 1.5 - 0.9 \, \text{J s}^{-1}$).

Efficiency measures

Useful energy is energy transferred for a purpose. In any machine, where friction is present, some of the energy transferred by the machine is wasted. In other words, not all the energy supplied to the machine is transferred for the intended purpose. For example, suppose a 500 W electric winch raises a weight of 150 N by 6.0 m in 10 s:

- The electrical energy supplied to the winch is $500 \, \text{W} \times 10 \, \text{s} = 5000 \, \text{J}$
- The useful energy transferred by the machine is the potential energy gain of the load $= 150 \, \text{N} \times 6 \, \text{m} = 900 \, \text{J}$.

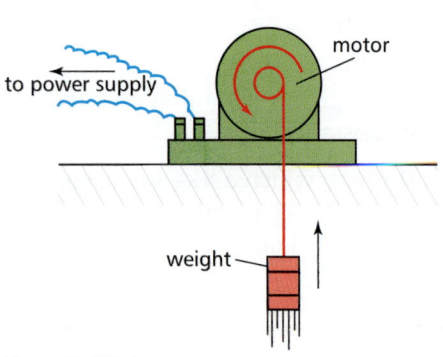

Figure 2 *Efficiency*

Therefore, in this example, 4100 J of energy is wasted.

$$\text{The efficiency of a machine} = \frac{\text{useful energy transferred by the machine}}{\text{energy supplied to the machine}}$$

$$= \frac{\text{work done by the machine}}{\text{energy supplied to the machine}}$$

Note

- **Percentage efficiency = efficiency × 100%**

In the above example, the efficiency of the machine is therefore 0.18 or 18%.

- Also: **Efficiency** $= \dfrac{\text{output power of a machine}}{\text{input power to the machine}}$

Wasting energy

In any process or device where energy is transferred for a purpose, the efficiency of the transfer process or the device is the fraction of the energy supplied which is used for the intended purpose. For example:

- A light bulb that is 10% efficient emits 10 J of energy as light for every 100 J of energy supplied to it by electricity. The rest of the energy is wasted as heat.

- An engine that is 45% efficient delivers 45 J of useful energy for every 100 J of energy supplied to it from its fuel. The rest of the energy is wasted as sound and heat.

Is it possible to stop energy being wasted as heat? In a power station, steam is used to drive turbines which turn the electricity generators. If the turbines were not kept cool, they would stop working because the **pressure** inside would build up and prevent steam entering. Stopping the heat transfer to the cooling water would stop the generators working. In general, energy tends to spread out when it is usefully used.

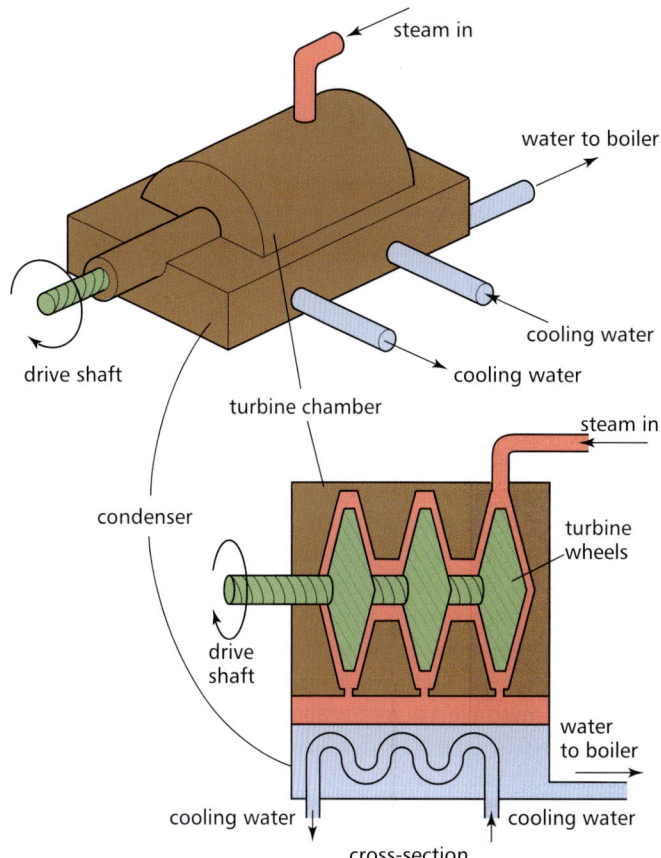

Figure 3 A power station generator

Summary test 2.7

1 In a test of muscle efficiency, an athlete on an exercise bicycle pedals against a braking force of 30 N at a speed of 15 m s^{-1}.

    a Calculate the useful energy supplied per second by the athlete's muscles.

    b If the efficiency of the muscles is 25%, calculate the energy per second supplied to the athlete's muscles.

2 A 60 W electric motor raises a weight of 20 N through a height of 2.5 m in 8.0 s. Calculate:

    a the electrical energy supplied to the motor,

    b the useful energy transferred by the motor,

    c the efficiency of the motor.

3 A power station has an overall efficiency of 35% and it produces 200 MW of electrical power. The fuel used in the power station releases 80 MJ kg^{-1} of fuel burned. Calculate:

    a the energy per second supplied by the fuel,

    b the mass of fuel burned per day.

4 A vehicle engine has a power output of 6.2 kW and uses fuel which releases 45 MJ kg^{-1} of energy when burned. At a speed of 30 m s^{-1} on a level road, the fuel usage of the vehicle is 18 km kg^{-1}. Calculate:

    a the time taken by the vehicle to travel 18 km at 30 m s^{-1},

    b the useful energy supplied by the engine in this time,

    c the overall efficiency of the engine.

Learning outcomes

On these pages you will learn to:

- recognise energy transfers in renewable energy sources
- use knowledge about power and energy to evaluate renewable energy sources

Renewable energy sources will contribute increasingly to the world's energy supplies in the future. Almost 80% of the energy we use at present is obtained from fossil fuels. Scientists think that the use of fossil fuels is causing increased global warming due to the increasing amount of carbon dioxide in the atmosphere. Increased global warming could have disastrous consequences (e.g. rising sea levels, changing weather patterns, etc.) in many regions. To cut carbon emissions, many countries are now planning to develop more renewable energy resources and to build new nuclear power stations. A nuclear power station can generate enough electricity for about 200 000 homes. In terms of physics, let's consider how much energy could be supplied from typical renewable energy resources.

Solar power

A solar panel of area $1\,m^2$ in space would absorb solar energy at a rate of about $1400\,J\,s^{-1}$ if it absorbed all the incident solar energy. At the Earth's surface, the incident energy would be less because some would be absorbed in the atmosphere. In addition, some of the Sun's energy would be reflected by the panel itself.

Figure 1 a *A solar heating panel*   *b* *A solar cell panel*

- **A solar heating panel** can heat water running through it to 70°C on a hot sunny day. In hot countries in summer, a $1\,m \times 3\,m$ solar heating panel can absorb up to $4000\,J\,s^{-1}$ of solar energy.
- **A solar electric panel** produces electricity directly. A potential difference is produced across each solar cell when light is incident on the cell. A large array of solar cells and plenty of sunshine is necessary to produce useful amounts of power. To make a significant contribution to national needs, millions of homes and buildings would need to be fitted with solar panels.

Wind power

A wind turbine is an electricity generator on a tall tower, driven by large blades pushed round by the force of the wind. A typical modern wind turbine on a suitable site can generate about $2\,MW$ of electrical power. Let's consider how this estimate is obtained.

- For a mass m of wind moving at speed v, its kinetic energy $= \frac{1}{2}mv^2$
  $\therefore$ the kinetic energy per unit volume of air from the wind at this speed $= \frac{1}{2}\rho v^2$, where ρ is the density (i.e. mass per unit volume) of air.
- Suppose the blades of a wind turbine sweep out an area A when they rotate. For wind at speed v, a cylinder of air of area A and length v passes every second through the area swept out by the blades. So the volume of air passing per second $= vA$.
  $\therefore$ the kinetic energy per second of the wind passing through a wind turbine $= \frac{1}{2}\rho v^2 vA = \frac{1}{2}\rho v^3 A$.

For a wind turbine with blades of length $20\,m$,
$A = \pi(20)^2 = 1300\,m^2$

The density of air, $\rho = 1.2\,kg\,m^{-3}$

$\therefore$ the power of the wind at $v = 15\,m\,s^{-1}$
$= \frac{1}{2} \times 1.2 \times 15^3 \times 1300$
$= 2.6 \times 10^6\,W = 2.6\,MW$

The calculation shows that the maximum power output of a large wind turbine at a windy site could not be more than about $2\,MW$. To generate the same power as a $5000\,MW$ power station, about 2500 wind turbines would need to be constructed and connected to the electricity network.

Water power

Hydroelectricity and tidal power stations both make use of the potential energy released by water when it runs to a lower level.

- A **tidal power station** covering an area of $100\,km^2$ could trap a depth of $6\,m$ of sea water twice per day. This would mean releasing a volume of $6 \times 10^8\,m^3$ of sea water over a few hours. The mass of such a volume is about $6 \times 10^{11}\,kg$ and if its centre of mass drops through an average height of about $1\,m$, the potential energy released would be about $6 \times 10^{12}\,J$. Prove for yourself that this would give an energy transfer rate of more than $250\,MW$ if released over about 6 hours.
- A **hydroelectric power station** releases less water per second than a tidal power station but the water drops through a much greater height. However, even

a large height drop of 500 m with rainfall over an area of 1000 km² at a depth of about 10 mm per day would transfer no more than about 500 MW.

Geothermal energy

Energy released by radioactive substances, deep within the Earth, heats the surrounding rock. As a result, heat or **geothermal energy** is transferred towards the Earth's surface. Geothermal power stations are located in volcanic areas or where there are hot rocks deep below the surface. Water gets pumped down to these rocks to produce steam. Then the steam produced drives electricity turbines at ground level. In some areas, buildings can be heated using geothermal energy directly. Heat flow from underground is called **ground heat**. It can be used to heat water which is then pumped round the building.

Figure 2 *A geothermal power station*

Biofuels and biomass

Methane from cows or animal manure and from sewage works, decaying rubbish and other sources can be used in small-scale gas-fired power stations. Methane is an example of a **biofuel** which is any fuel obtained from **biomass**, the word we use for biological material from living or recently

living organisms such as animal waste. Other biofuels include ethanol from sugar cane, straw, nutshells and woodchip. Ethanol may be used as a fuel for road vehicles.

Figure 3 *Using biofuel to generate electricity*

Renewable energy overview

A 600 MW power station can supply electricity to about 100 000 homes. Could renewable energy be used instead?

- 25 000 homes each with a 4 kW solar panel could produce 100 MW.
- 1 offshore wind farm with about 100 wind turbines could produce about 100 MW.
- 1 tidal power station or hydroelectric power station could produce about 100 MW.

The renewable energy resources available in any given area vary from one area to another. Also, renewables such as wind and wave power are unreliable. However, one or more of the renewable energy resources discussed above could make a significant contribution to energy supplies in many areas. For example, solar panels, biofuels and wind farms could reduce fossil fuel dependency in hot coastal regions.

Summary test 2.8

1 A solar cell panel of area 1 m² can produce 200 W of electrical power on a sunny day. Calculate the area of panels that would be needed to produce 2000 MW of electrical power.

2 The maximum power that can be obtained from a wind turbine is proportional to the cube of the wind speed. When the wind speed is 10 m s⁻¹, the power output of a certain wind turbine is 1.2 MW. Calculate the power output of this wind turbine when the wind speed is 15 m s⁻¹.

3 A hydroelectric power station produces electrical power at an overall efficiency of 25%. The power station is driven by water that has descended from

an upland reservoir 650 m above the power station. Calculate the volume of water passing through the power station per second when it produces 200 MW of electrical power.

The density of water = 1000 kg m⁻³.

4 At a tidal power station, water is trapped over an area of 200 km² when the tide is 3.0 m above the power station turbines. The trapped water is released gradually over a period of 6 hours. Calculate: **a** the mass of trapped water, **b** the average loss of potential energy per second of this trapped water when it is released over a period of 6 hours. The density of sea water = 1050 kg m⁻³.

2 Exam-style and Practice Questions

> 📑 Launch additional digital resources for the chapter

1 a Explain why a force is needed to make an object move at a steady speed across a rough floor.

b Explain why an object sliding across ice keeps moving even though no force is pushing it.

2 A vehicle of mass 450 kg accelerates from rest to a velocity of $9.5\,\text{m}\,\text{s}^{-1}$ in 38 s.

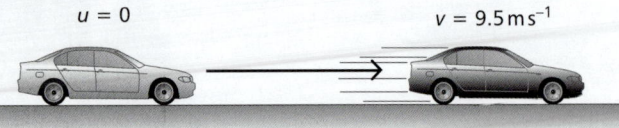

$u = 0$ $v = 9.5\,\text{m}\,\text{s}^{-1}$

Figure 2.1

Calculate:

a the acceleration of the vehicle,

b the motive force of the vehicle engine, assuming zero resistive force,

c the ratio of the motive force to the vehicle's weight.

3 A bullet of mass $2.4 \times 10^{-3}\,\text{kg}$ moving at a speed of $115\,\text{m}\,\text{s}^{-1}$ hits a tree and embeds itself in the tree to a depth of 85 mm.

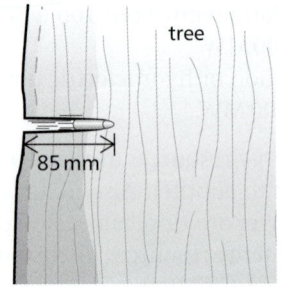

tree

85 mm

Figure 3.1

Calculate:

a the impact time,

b the deceleration of the bullet in the tree,

c the impact force on the bullet.

4 A heavy goods vehicle of total weight 320 kN has an engine which has a maximum motive force of 6 kN.

a Calculate:

i the mass of the vehicle,

ii the maximum acceleration of the vehicle,

iii the time taken for the vehicle to accelerate from rest to a speed of $20\,\text{m}\,\text{s}^{-1}$,

iv the distance moved by the vehicle in this time.

Figure 4.1

b The vehicle is used to pull a trailer of weight 60 kN. Calculate:

i the maximum acceleration of the vehicle when the trailer is attached,

ii the time taken by the vehicle to reach a speed of $20\,\text{m}\,\text{s}^{-1}$ from rest with the trailer attached.

5 A steel ball bearing was released in water and fell to the bottom of the water container. The time of descent was measured at different distances as it fell. The figure below shows the measurements plotted as a graph of distance fallen against time taken.

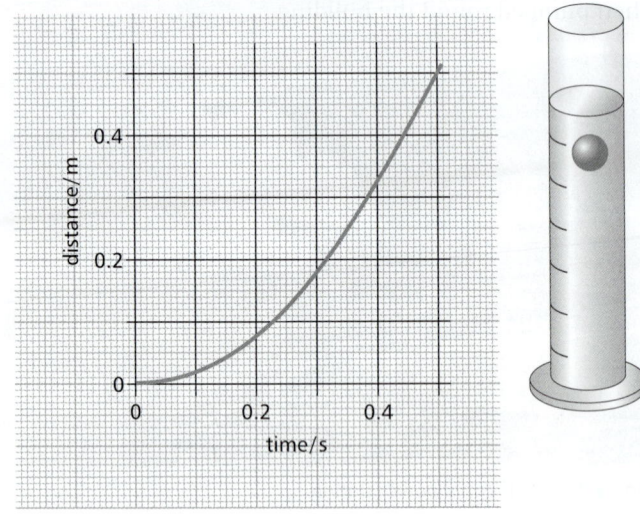

Figure 5.1

a i What feature of this graph represents the speed of the ball bearing at any instant?

ii Describe how the speed of the ball bearing changed as it fell.

iii Sketch a graph to show how the speed of the ball bearing changed with time during its descent.

b Explain, in terms of the forces acting on the ball, why the acceleration of the ball bearing decreased as it fell.

6 a Calculate the kinetic energy of a 96 kg rugby player running at a speed of $6.0\,\text{m}\,\text{s}^{-1}$.

b What speed would a 2.4 kg rugby ball need to be moving at to have the same kinetic energy as the rugby player in part **a**?

c What would be the height gain of the 2.4 kg rugby ball in part **b**, if all its kinetic energy was converted into potential energy?

7 A brick, of mass 2.5 kg, falls off the top of a wall and hits the ground 2.5 m below. Calculate:

a the loss of potential energy of the brick,

b the kinetic energy of the brick just before impact,

c the speed of the brick just before impact.

8 A ball, of mass 0.40 kg, released from a height of 1.2 m above a horizontal concrete floor hits the floor and rebounds to a maximum height of 0.80 m.

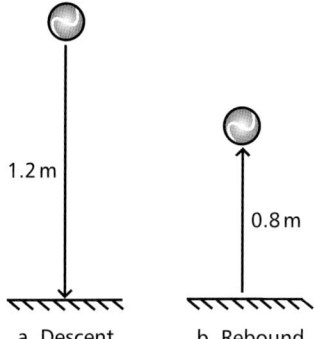

a. Descent   b. Rebound

Figure 8.1

a Calculate:
  i the loss of potential energy of the ball just before hitting the floor,
  ii the kinetic energy of the ball just before impact,
  iii the speed of the ball just before impact.

b Calculate:
  i the maximum gain of potential energy of the ball after impact,
  ii the kinetic energy of the ball just after impact,
  iii the speed of the ball just after impact.

9 An aircraft, of mass 2200 kg, takes off from an aircraft carrier and reaches a speed of 85 m s⁻¹ in 20 s at a height of 320 m above the aircraft carrier.

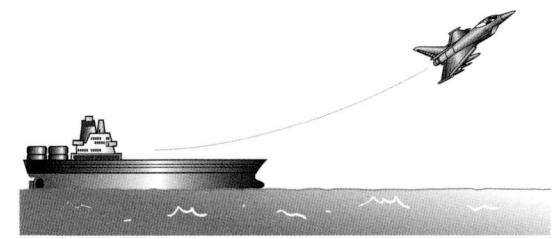

Figure 9.1

a Calculate:
  i the gain of potential energy of the aircraft,
  ii its gain of kinetic energy.

b **i** Estimate the average output power of the aircraft's engines in this time.
  ii If the aircraft's fuel has an energy value of 50 MJ kg⁻¹ and the engines have an efficiency of 30%, estimate the mass of fuel burned in the 20 s taken to reach a height of 320 m.

10 A car, of mass 1200 kg, accelerates at a steady rate of 1.5 m s⁻² along a straight level road.

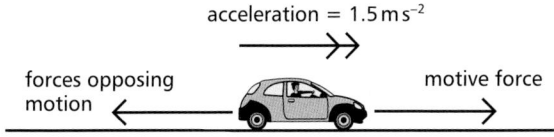

acceleration = 1.5 m s⁻²

forces opposing motion   motive force

Figure 10.1

a Calculate the resultant force acting on the car.

b The total force opposing the motion of the car is 500 N when the speed of the car is 7.0 m s⁻¹. Calculate the motive force of the engine at this speed when the acceleration is 1.5 m s⁻².

c Calculate:
  i the output power of the engine,
  ii the power wasted due to the forces opposing the motion of the car when the car's speed is 7.0 m s⁻¹.

d Account for the difference between your answers to parts **c i** and **ii**.

11 An object is dropped vertically from a stationary hot air balloon. The variation with time t of the object's speed, v, is shown in Figure 11.1.

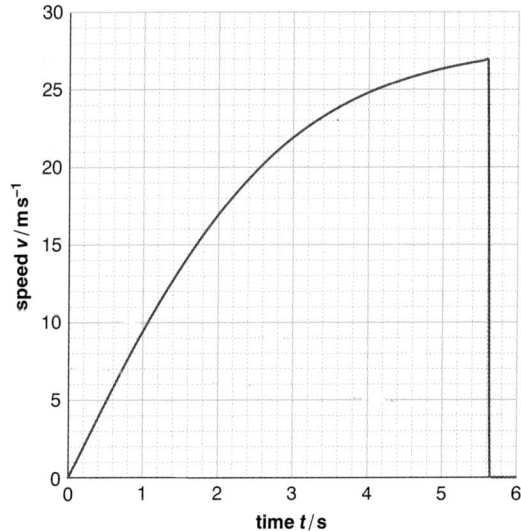

Figure 11.1

a i Explain why the object's initial acceleration is 9.81 m s⁻².
  ii Sketch a graph to show how the magnitude of the object's acceleration changes after the object is released.
  iii Explain, in terms of the forces acting on the object, why its acceleration changes.

b The object has a mass of 0.84 kg.
  i Calculate the weight of the object.
  ii Estimate the resultant force on the object when it is halfway between its point of release and the ground.

c i Calculate the kinetic energy of the object immediately before it hits the ground.
  ii Estimate the distance fallen and the change of potential energy of the object in the descent.
  iii Estimate the average resistive force on the object in the descent.

3.1 Balanced forces

Learning outcomes

On these pages you will learn to:

- recognise different forces and represent them on free-body force diagrams
- recognise the forces acting in simple equilibrium situations
- resolve a force into two perpendicular components

Force as a vector

At this moment, at least two forces are acting on you or on any object near you. In addition to the force of the Earth's gravitational field, one or more support forces will be acting on you preventing you from accelerating downwards. The support force on an object may be due to other solid objects under it and/or cables suspended above it. Other types of forces that can act on objects include:

- **electrostatic forces** which act on charged objects in an electric field
- **magnetic forces** which act on magnets and electromagnets in a magnetic field
- **friction forces** which oppose the motion of objects when they slide over each other
- **viscous forces** which oppose the motion of objects moving through fluids
- **upthrust forces** which help objects to float.

Whatever the forces acting on an object are, they can be represented by **vectors** in a vector diagram. A vector has **magnitude** and **direction**. The two key rules for dealing with vectors are explained in Topic 1.1.

- A force F can be resolved into two perpendicular components: $F\cos\theta$ and $F\sin\theta$ parallel and perpendicular to a line at angle θ to the line of action of the force.
- Two forces can be added together using a scale diagram or by using a calculator and Pythagoras' theorem after resolving one of the forces into components parallel and perpendicular to the other force. See Topic 1.1 if necessary.

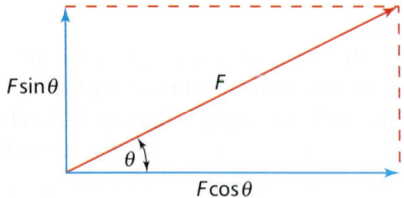

Figure 1 *Resolving a force*

Equilibrium of a point object

When **two forces** act on a point object, the object is in equilibrium (i.e. at rest or moving at constant velocity) only if the two forces are equal and opposite to each other. The resultant of the two forces is therefore zero. The two forces are said to be **balanced**. For example, an object resting on a surface is acted on by its weight W (i.e. the force of gravity on it) acting downwards and a support force S from the surface acting upwards. Hence S is equal and opposite to W, provided the object is at rest (or moving at constant velocity) i.e. **$S = W$**.

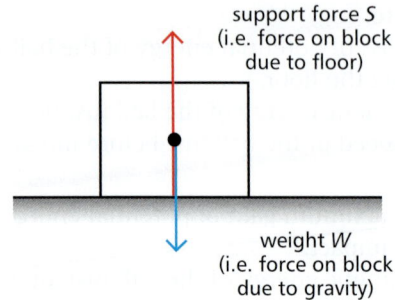

Figure 2 *Balanced forces*

When **three forces** act on a point object, their combined effect (i.e. resultant) is zero only if the resultant of any two of the forces is equal and opposite to the third force. One way of checking if the combined effect of the three forces is zero involves:

- resolving each force along the same parallel and perpendicular lines,
- balancing the components along each line.

Example 1

A child of weight W on a swing is at rest, due to the swing seat being pulled to the side by a horizontal force F_1. The rope is then at an angle θ to the vertical, as shown in Figure 4.

Assuming the swing seat is of negligible weight, the swing seat is acted on by three forces: the weight of the child W, the horizontal force F_1 and the tension T in the rope. Resolving the tension T vertically and horizontally gives $T\cos\theta$ for the vertical component of T (which is upwards) and $T\sin\theta$ for the horizontal component of T. Therefore, the balance of forces:

- Horizontally: $F_1 = T\sin\theta$
- Vertically: $W = T\cos\theta$

Because $\sin^2\theta + \cos^2\theta = 1$ (see p.6),

$$F_1{}^2 + W^2 = T^2\sin^2\theta + T^2\cos^2\theta = T^2$$

$$\therefore \qquad T^2 = F_1{}^2 + W^2$$

Also, because $\tan\theta = \dfrac{\sin\theta}{\cos\theta}$, then $\dfrac{F_1}{W} = \dfrac{T\sin\theta}{T\cos\theta} = \tan\theta$

$$\therefore \qquad \tan\theta = \frac{F_1}{W}$$

Example 2

An object of weight W at rest on a rough slope is acted on by a frictional force F, which prevents it sliding down the slope, and a support force S from the slope, perpendicular to the slope.

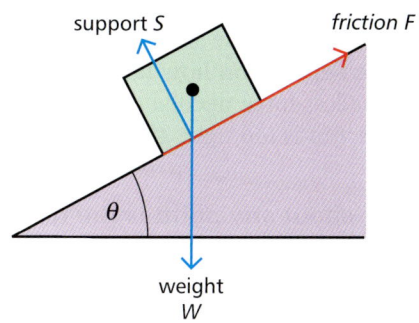

Figure 3

Resolving the three forces parallel and perpendicular to the slope gives:

- Parallel: $F = W\sin\theta$
- Perpendicular: $S = W\cos\theta$

Because $\sin^2\theta + \cos^2\theta = 1$, then $F^2 + S^2 = W^2\sin^2\theta + W^2\cos^2\theta = W^2$

$$\therefore \qquad W^2 = F^2 + S^2$$

Also, because $\tan\theta = \dfrac{\sin\theta}{\cos\theta}$, then $\dfrac{F}{S} = \dfrac{W\sin\theta}{W\cos\theta} = \tan\theta$

$$\therefore \qquad \tan\theta = \frac{F}{S}$$

a On a swing

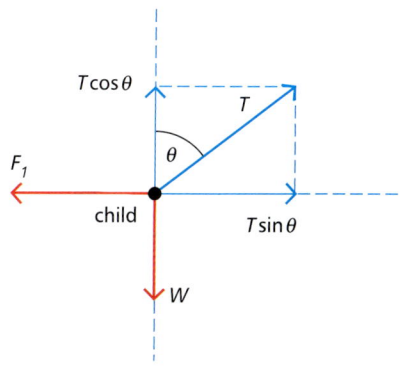

b Force diagram

Figure 4

Summary test 3.1

1 A point object of weight 6.2 N is acted on by a horizontal force of 3.8 N.

   a Calculate the resultant of these two forces.

   b Determine the magnitude and direction of a third force acting on the object for it to be in equilibrium.

2 A small object of weight 5.4 N is at rest on a rough slope which is at an angle of 30° to the horizontal.

   a Sketch a diagram and show the three forces acting on the object.

   b Calculate: **i** the frictional force on the object,
   ii the support force from the slope on the object.

3 An archer pulls a bow string back until the two halves of the string are at 140° to each other (Figure 5). The force needed to hold the string in this position is 95 N.

Calculate:

   a the tension in each part of the bow string in this position,

   b the resultant force on an arrow at the instant the bow string is released from this position.

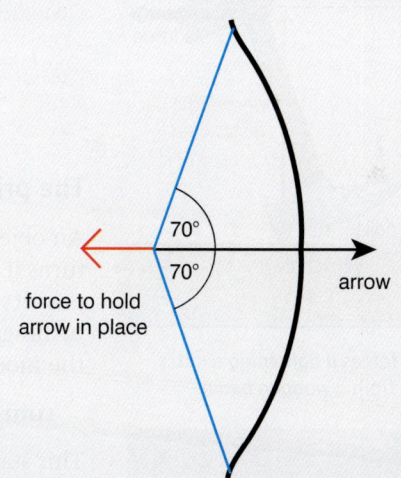

Figure 5

4 An elastic string is stretched horizontally between two fixed points, 0.80 m apart. An object of weight 4.0 N is suspended from the midpoint of the string, causing the midpoint to drop a distance of 0.12 m. Calculate:

   a the angle of each part of the string to the vertical,

   b the tension in each part of the string.

3.2 The principle of moments

Learning outcomes

On these pages you will learn to:

- define and apply the moment of a force
- know that the weight of a body may be taken as acting at a single point known as its centre of gravity
- state and apply the principle of moments to objects balanced on a single pivot

Turning effects

Whenever you use a lever or a spanner, you are using a force to turn an object about a **pivot**. For example, if you use a spanner to loosen a wheel nut on a bicycle, you need to apply a force to the spanner to make it turn about the wheel axle. The effect of the force depends on how far it is applied from the wheel axle. The longer the spanner, the less force needed to loosen the nut. However, if the spanner is too long and the nut is too tight, the spanner could snap if too much force is applied to it.

> The **moment of a force** about any point is the force × the perpendicular distance from the line of action of the force to the point. The unit of the moment of a force is the newton metre (N m).

For a force F acting along a line of action at perpendicular distance d from a certain point:

$$\text{moment of force} = Fd$$

> **Note**
>
> - The greater the distance d, the greater the moment.
> - The distance d is the **perpendicular** distance from the line of action of the force to the point (Figure 1).

> **Worked example**
>
> A spanner of length 0.24 m is used to tighten a wheel nut. The moment of the force must not exceed 60 N m, otherwise the wheel nut is damaged. Calculate the maximum force that should be applied to the spanner to tighten this nut.
>
> *Solution*
> Moment = force × distance
> $$\therefore F \times 0.24 = 60$$
> $$F = \frac{60}{0.24} = 250\,\text{N}$$

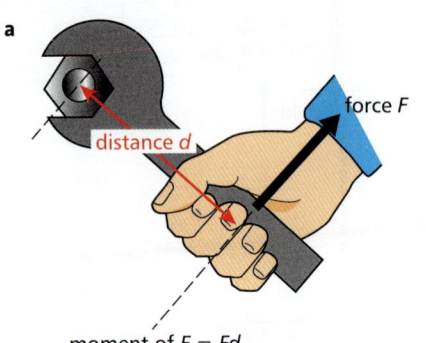

a

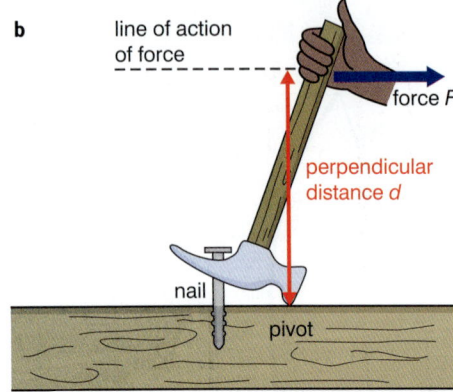

b

Figure 1 *Turning forces **a** tightening a nut **b** removing a nail from a wooden beam*

> **Note**
>
> In Figure 1a, the moment of F is Fd where d is the perpendicular distance to the line of action of the force from the centre of the nut, which is along the handle of the spanner.
>
> In Figure 1b, the moment of F is also Fd but the perpendicular distance d to the line of action of the force from the pivot is not along the handle of the hammer.

The principle of moments

An object that is not a point object is referred to as a **body**. Any such object turns if a force is applied to it anywhere other than through its **centre of gravity**. If a body is acted on by more than one force and it is in equilibrium, the turning effects of the forces must balance out. In more formal terms, considering the moments of the forces about any point, for equilibrium:

sum of the clockwise moments = sum of the anticlockwise moments

This statement is known as the **principle of moments**.

The balanced metre rule

1 Consider a uniform metre rule balanced on a pivot at its centre, supporting weights W_1 and W_2 suspended from the rule on either side of the pivot (Figure 2).

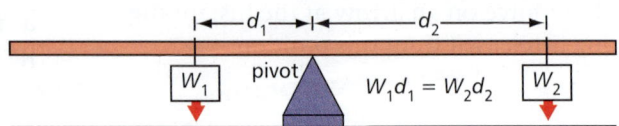

Figure 2 *The principle of moments*

- Weight W_1 provides an anticlockwise moment about the pivot of W_1d_1, where d_1 is the distance from the line of action of the weight to the pivot.
- Weight W_2 provides a clockwise moment about the pivot of W_2d_2, where d_2 is the distance from the line of action of the weight to the pivot.

For equilibrium, applying the principle of moments:

$$W_1d_1 = W_2d_2$$

2 If a third weight W_3 is suspended from the rule on the same side of the pivot as W_2 at distance d_3 from the pivot, then the rule can be rebalanced by increasing distance d_1.

At this new distance d_1' for W_1:

$$W_1d_1' = W_2d_2 + W_3d_3$$

Centre of gravity

A tightrope walker and a waiter serving drinks know just how important the centre of gravity of an object can be. One slight off-balance movement can be catastrophic.

> **The centre of gravity of a body is the point where the weight of the body may be considered to act.**

Centre of gravity tests
- Support a ruler at its centre on the end of your finger. The centre of gravity of the ruler is directly above the point of support. Tip the ruler too much and it falls off, because the centre of gravity is no longer above the point of support.
- Balance a postcard on the end of a pencil. The centre of gravity of the postcard is directly above the point of support when the card

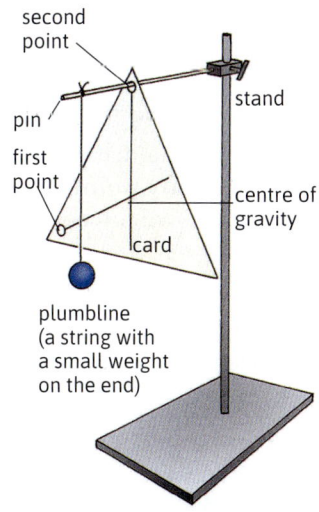

Figure 3 A centre of gravity test

is balanced. You should find that the centre of gravity of the card is at the midpoint of the card where the two diagonals cross.
- Find the centre of gravity of a triangular card. Figure 3 shows how to do this. It should be possible to balance the card at its centre of gravity on the end of a pencil.

Calculating the weight of a metre rule
- Locate the centre of gravity of a metre rule by balancing it horizontally on a horizontal knife edge. Note the position of the centre of gravity. The rule is said to be **uniform** if its centre of gravity is exactly at the middle of the rule.
- Balance the metre rule off-centre on a knife edge, using a known weight W_1, as shown in Figure 4. The position of the known weight needs to be adjusted gradually until the rule is exactly horizontal.

At this position:
- The known weight W_1 provides an anticlockwise moment about the pivot of W_1d_1, where d_1 is the perpendicular distance from the line of action of W_1 to the pivot.
- The weight of the rule W_0 provides a clockwise moment of W_0d_0, where d_0 is the distance from the centre of gravity of the rule to the pivot.
- Applying the principle of moments:

$$W_0d_0 = W_1d_1$$

By measuring distances d_0 and d_1, the weight W_0 of the rule can therefore be calculated.

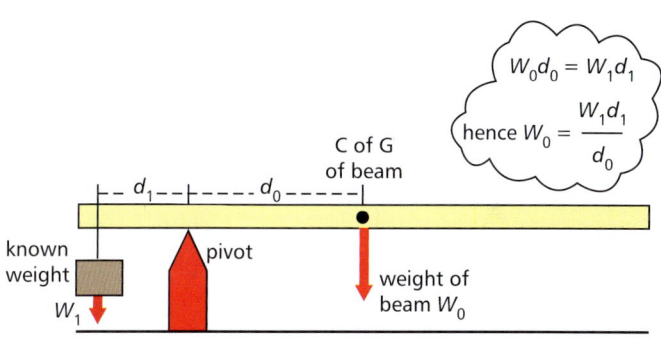

Figure 4 Finding the weight of a beam

Summary test 3.2

1 A child of weight 200 N sits on a seesaw at a distance of 1.2 m from the pivot at the centre. The seesaw is balanced by a second child sitting on it at a distance of 0.8 m from the centre. Calculate the weight of the second child.

2 A uniform metre rule pivoted at its centre of gravity supports a 3.0 N weight at its 5.0 cm mark, a 2.0 N weight at its 25 cm mark and a weight W at its 80 cm mark.

   a Sketch a diagram to represent this situation.

   b Calculate the weight W.

3 In question **2**, the 3.0 N weight and the 2.0 N weight are swapped with each other. Sketch the new arrangement and work out the new distance of weight W from the pivot.

4 A uniform metre rule supports a 4.5 N weight at its 100 mm mark. The rule is balanced horizontally on a horizontal knife edge at its 340 mm mark. Sketch the arrangement and calculate the weight of the rule.

Learning outcomes

On these pages you will learn to:

- state and apply the principle of moments to objects balanced on two supports
- show an understanding that a couple is a pair of forces that tends to produce rotation only
- define and apply the torque of a couple

Support forces

Single-support problems

When an object in equilibrium is supported at one point only, the support force on the object is equal and opposite to the total downward force acting on the object. For example, in Figure 1, a uniform rule is balanced on a knife edge pivot at its centre of gravity with two additional weights W_1 and W_2 attached to the rule. The support force S on the rule from the knife edge must be equal to the total downward weight. Therefore:

$$S = W_1 + W_2 + W_0$$

where W_0 is the weight of the rule.

As explained in Topic 3.2, taking moments about the knife edge gives:

$$W_1 d_1 = W_2 d_2$$

Taking moments about a different point

Let's consider moments about the point where W_1 is attached to the rule:

$$\text{Sum of the clockwise moments} = W_0 d_1 + W_2(d_1 + d_2)$$

$$\text{Sum of the anticlockwise moments} = S d_1 = (W_1 + W_2 + W_0) d_1$$

$$\therefore \qquad (W_1 + W_2 + W_0) d_1 = W_0 d_1 + W_2(d_1 + d_2)$$

Multiplying out the brackets gives:

$$W_1 d_1 + W_2 d_1 + W_0\, d_1 = W_0 d_1 + W_2 d_1 + W_2 d_2$$

which simplifies to become: $\qquad W_1 d_1 = W_2 d_2$

This is the same as the equation obtained by taking moments about the pivot. So moments can be taken about **any point**. It makes sense therefore to choose a point which one or more unknown forces act through, as such forces have **zero moment** about this point.

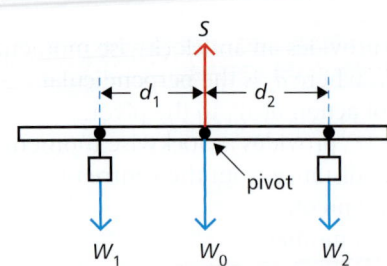

Figure 1 A single support force

Two-support problems

Consider a uniform beam supported on two pillars X and Y, which are at distance D apart. The weight of the beam is 'shared' between the two pillars according to how far the beam's centre of gravity is from each pillar. For example:

- If the centre of gravity of the beam is mid-way between the pillars, the weight of the beam is 'shared' equally between the two pillars. In other words, the support force on the beam from each pillar is equal to half the weight of the beam.
- If the centre of gravity of the beam is at distance d_X from pillar X and distance d_Y from pillar Y, as shown in Figure 2, then taking moments:

1 **Where X is in contact with the beam**:
   $S_Y D = W d_X$, where S_Y is the support force from pillar Y

$$\text{so } \mathbf{S_Y} = \frac{W d_X}{D}$$

2 **Where Y is in contact with the beam**:
   $S_X D = W d_Y$, where S_X is the support force from pillar X

$$\text{so } \mathbf{S_X} = \frac{W d_Y}{D}$$

Therefore, if the centre of gravity is closer to X than to Y, $d_X < d_Y$ so $S_Y < S_X$.

Figure 2 A two-support problem

A uniform beam of length 5.0 m and weight 120 N rests horizontally on the tops of two walls, X and Y (Figure 3), with its centre of gravity at a distance of 2.0 m from wall X and 1.5 m from wall Y. Calculate the support force on the beam from each wall.

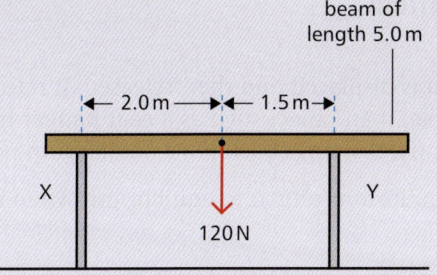

Figure 3

Solution

Let S_X and S_Y represent the support forces at X and Y.

$\therefore \qquad S_X + S_Y = 120\,\text{N}$

Taking moments about X gives:

Sum of the clockwise moments

$\quad$ = weight of beam × distance from centre of gravity to X

$\quad$ = 120 N × 2.0 m = 240 N m

Sum of the anticlockwise moments

$\quad$ = support force S_Y × distance from X to Y

$\quad = S_Y \times 3.5\,\text{m}$

$\therefore$ Applying the principle of moments gives:

$\qquad 3.5\,S_Y = 240$

$$S_Y = \frac{240}{3.5} = 69\,\text{N}$$

Therefore $\qquad S_X = W - S_Y = 120 - 69 = 51\,\text{N}$

Couples

A **couple** is a pair of equal and opposite forces acting on a body, but not along the same line. Figure 4 shows a couple acting on a coil. The couple turns, or tries to turn, the coil.

The moment of a couple is referred to as a **torque:**

> **Torque of a couple = one force × perpendicular distance between lines of action of the forces**

The total moment is the same, regardless of the point about which the moments are taken.

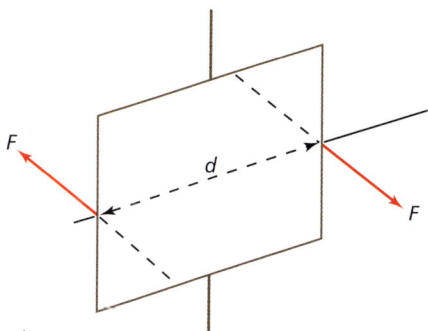

Figure 4 A couple

Summary test 3.3

1 A metre rule of weight 1.2 N rests horizontally on two knife edges at the 100 mm mark and the 800 mm mark. Sketch the arrangement and calculate the support force on the rule due to each knife edge.

2 A uniform beam of weight 230 N and of length 10 m rests horizontally on the tops of two brick walls 8.5 m apart, such that a length of 1.0 m projects beyond one wall and 0.5 m projects beyond the other wall (Figure 5).

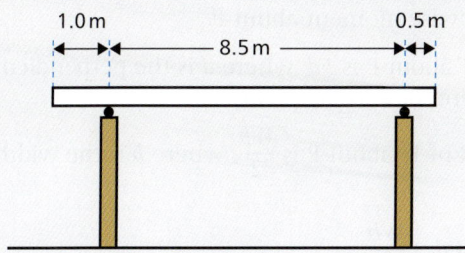

Figure 5

Calculate:

  a the support force of each wall on the beam,

  b the force of the beam on each wall.

3 A uniform bridge span of weight 1200 kN and of length 17.0 m rests on a support of width 1.0 m at either end. A stationary lorry of weight 60 kN is the only object on the bridge. Its centre of gravity is 3.0 m from the centre of the bridge (Figure 6). Calculate the support force on the bridge at each end.

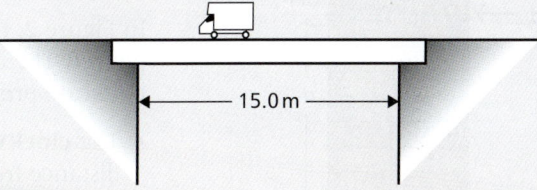

Figure 6

4 A uniform plank of weight 150 N and of length 4.0 m rests horizontally on two bricks. One of the bricks is at the end of the beam. The other brick is 1.0 m from the other end of the plank.

  a Sketch the arrangement and calculate the support force on the plank from each brick.

  b A child stands on the free end of the beam and just causes the other end to lift off its support.

  Sketch this arrangement and calculate the weight of the child.

Stability

Learning outcomes

On these pages you will learn to:

- recognise stable and unstable equilibrium situations
- recognise how the stability of an object at rest on a flat surface is affected when the object or the surface is tilted

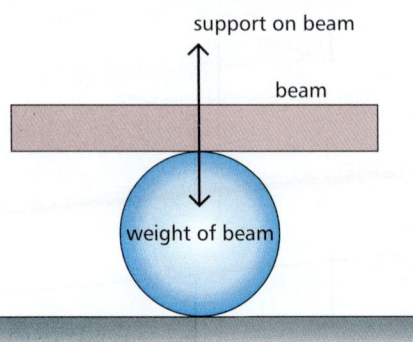

Figure 1 *Unstable equilibrium*

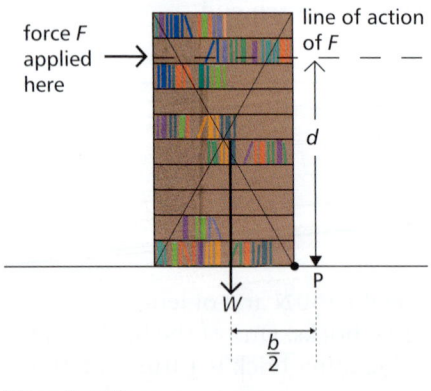

Figure 2 *Tilting*

Stable and unstable equilibrium

Stable equilibrium

If a body in **stable equilibrium** is displaced and then released, it returns to its equilbrium position. For example, if an object such as a coat hanger hanging from a support is displaced slightly, it swings back to its equilibrium position.

Why does an object in stable equilbrium return to equilibrium when it is displaced and then released?

- The reason is that the centre of gravity of the object is directly below the point of support when the object is at rest. The support force and the weight are directly equal and opposite to each other when the object is **in equilibrium**.
- However, when it is displaced and then released, at the instant of release, the line of action of the weight no longer passes through the point of support so the weight **returns the object to equilibrium**.

Unstable equilibrium

A plank balanced on a drum is in **unstable equilibrium**. If it is displaced slightly from equilibrium then released, the plank will roll off the drum.

- The reason is that the centre of gravity of the plank is directly above the point of support when it is **in equilibrium**. The support force is exactly equal and opposite to the weight.
- If the plank is displaced slightly, the centre of gravity is no longer above the point of support. The weight therefore acts to turn the plank **further from the equilibrium position**.

Tilting and toppling

Skittles at a bowling alley are easy to knock over because they are top-heavy. This means that the centre of gravity is too high and the base is too narrow. A slight nudge from a ball causes a skittle to tilt and then tip over.

Tilting

Tilting is where an object at rest on a surface is acted on by a force that raises it up on one side. For example, if a horizontal force is applied to the top of a tall free-standing bookcase, the force can make the bookcase tilt about its base along one edge.

In Figure 2, to make the bookcase tilt, the force must turn it clockwise about point P. The entire support from the floor acts at point P. The weight of the bookcase provides an anticlockwise moment about P.

- The **clockwise** moment of F about P is Fd, where d is the perpendicular distance from the line of action of F to the pivot.

- The **anticlockwise** moment of W about P is $\dfrac{Wb}{2}$, where b is the width of the base.

Therefore, for tilting to occur: $Fd > \dfrac{Wb}{2}$

Toppling

A tilted object will **topple** over if it is tilted too far. For example, a tractor on a hill could topple over sideways if the hill is too steep. If an object on a flat surface is tilted more and more, the line of action of its weight (which is through its centre of gravity) passes closer and closer to the 'pivot'. If the object is tilted so much that the line of action of its weight passes beyond the pivot, the object will topple over if allowed to. The position where the line of action of the weight passes through the 'pivot' is the furthest it can be tilted without toppling. Beyond this position, it topples over if it is released. See Figure 3.

On a slope

A tall object on a slope will topple over if the slope is too great. For example, a high-sided vehicle on a road with a sideways slope will tilt over. If the slope is too great, the vehicle will topple over. This will happen if the line of action of the weight (passing through the centre of gravity of the object) lies outside the **wheel base** of the vehicle. In Figure 4, the vehicle will not topple over because the line of action of the weight lies within the wheel base.

Consider the forces acting on the vehicle on a slope when it is at rest. The sideways friction F, the support forces S_X and S_Y and the force of gravity on the vehicle (i.e. its weight) act as shown in Figure 4. For equilibrium, resolving the forces parallel and perpendicular to the slope gives:

- **Parallel** to the slope:

$$F = W \sin \theta$$

- **Perpendicular** to the slope:

$$S_X + S_Y = W \cos \theta$$

Note that S_X is greater than S_Y because X is lower than Y.

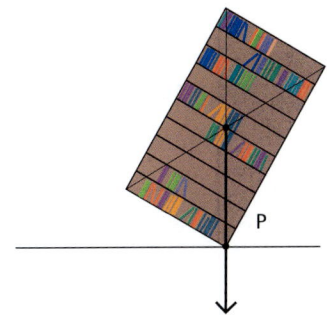

Figure 3 *Toppling over*

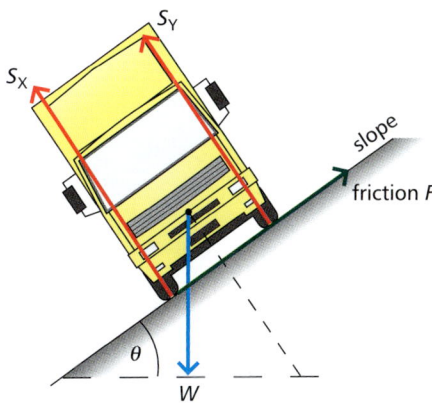

Figure 4 *On a slope*

Summary test 3.4

1 Explain why a bookcase with books on its top shelf only is less stable than if the books were on the bottom shelf.

2 An empty wardrobe of weight 400 N has a square base 0.8 m × 0.8 m and a height of 1.8 m. A horizontal force is applied to the top edge of the wardrobe to make it tilt. Calculate the force needed to lift the wardrobe base off the floor along one side (Figure 5).

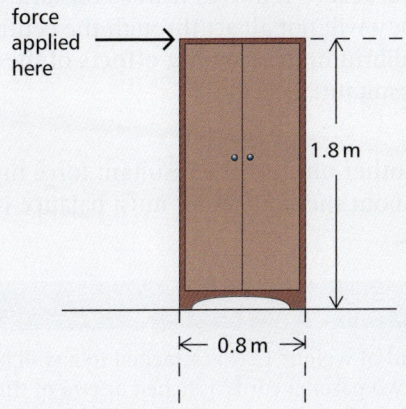

force applied here

1.8 m

0.8 m

Figure 5

3 A vehicle has a wheel base of 1.8 m and a centre of gravity, when unloaded, which is 0.8 m from the ground (Figure 6).

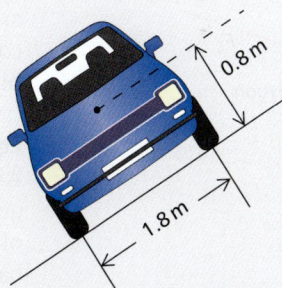

0.8 m

1.8 m

Figure 6

a The vehicle is tested for stability on an adjustable slope. Calculate the maximum angle of the slope to the horizontal if the vehicle is not to topple over.

b If the vehicle carries a full load of people, will it be more or less likely to topple over on a slope? Explain your answer.

4 Discuss whether or not a high-sided heavy goods lorry is more or less likely to be affected by strong side winds when it is fully loaded compared to when it is empty.

3.5 Equilibrium rules

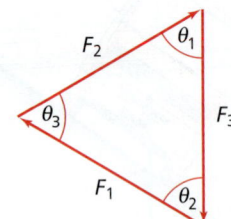

Figure 1 *Triangle of forces*

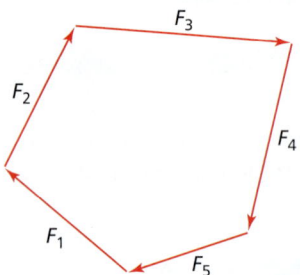

Figure 2 *The closed polygon*

The triangle of forces

For a point object acted on by **three forces** to be in equilibrium, the three forces must give an overall **resultant of zero**. The three forces as vectors should form a **triangle**. In other words, for three forces F_1, F_2 and F_3 to give zero resultant:

$$\textbf{Vector sum, } F_1 + F_2 + F_3 = 0$$

Any two of the forces gives a resultant which is represented by the third side of the triangle. Therefore, for equilibrium, the third force must be represented by the third side of the triangle. For example, the resultant of $F_1 + F_2$ is equal and opposite to F_3:

i.e. $$F_1 + F_2 = -F_3$$

The sine rule can be applied to the triangle of forces to find an unknown force or angle, given the other forces and angles in the triangle:

$$\frac{F_1}{\sin\theta_1} = \frac{F_2}{\sin\theta_2} = \frac{F_3}{\sin\theta_3}$$

where θ_1, θ_2 and θ_3 are the angles opposite sides F_1, F_2 and F_3 respectively.

The closed polygon

The triangle of forces rule can be extended for any number of forces acting on an object. If the object is in equilibrium, the force vectors drawn end-to-end must form a closed polygon. In other words, the tip of the last force vector must join the tail of the first force vector. Figure 2 shows the idea. Unfortunately, the sine rule can't be applied here; so the forces must be resolved in the same parallel and perpendicular directions to calculate an unknown force, given all the other forces.

The conditions for equilibrium of a body

Free-body force diagrams

When two objects interact, they always exert equal and opposite forces on one another. A diagram showing the forces acting on an object can become very complicated if it also shows the forces the object exerts on other objects. A **free-body force diagram** shows only the forces acting on the object.

Equilibrium

An object in equilbrium is either at rest or it moves with a constant velocity. In general, the forces acting on a body will not all act through the centre of gravity of the body. If the body is in equilibrium, the **turning effects** of the forces must balance out as well giving zero resultant.

For a body at rest:
- The **forces** must balance each other out (i.e. the resultant force must be zero).
- The **moments of the forces** about the same point must balance (i.e. the resultant torque must be zero).

Worked example

A uniform shelf of width 0.6 m and of weight 12 N is attached to a wall by hinges and is supported horizontally by two parallel cords attached at two of the corners of the shelf, as shown in Figure 3. The other end of each cord is fixed to the wall 0.4 m above the hinge. Calculate:

a the angle between each cord and the shelf,

b the tension in each cord.

Solution

a Let the angle between each cord and the shelf be θ.

From Figure 3, $\tan\theta = \dfrac{0.4}{0.6}$ so $\theta = 34°$

b Taking moments about the hinge gives:

- Sum of the clockwise moments
  = weight of shelf × distance from hinge to the centre of gravity of the shelf
  = $12 \times 0.3 = 3.6\,\text{N m}$

- Sum of the anticlockwise moments = $2\,Td$, where T is the tension in each cord and d is the perpendicular distance from the hinge to either cord.

  From Figure 3, it can be seen that $d = 0.6\sin\theta = 0.6\sin 34 = 0.34\,\text{m}$
  ∴ Applying the principle of moments gives:

$$2 \times 0.34 \times T = 3.6$$
$$T = 5.3\,\text{N}$$

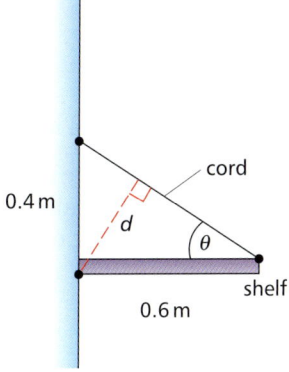

Figure 3

Summary test 3.5

1 A uniform plank of length 5.0 m rests horizontally on two bricks, which are 0.5 m from either end. A child of weight 200 N stands on one end of the plank and causes the other end to lift, so that it is no longer supported at that end.

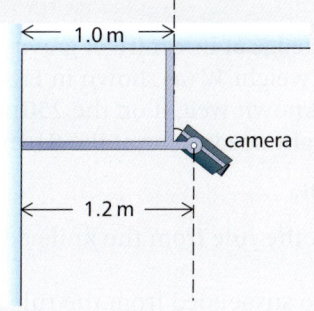

Figure 4

Calculate:

a the weight of the plank,

b the support force acting on the plank from the supporting brick.

2 A security camera is supported by a frame which is fixed to a wall and ceiling (Figure 5). The supporting structure must be strong enough to withstand the effect of a downward force of 1500 N acting on the camera, in case the camera is gripped by someone below it.

Figure 5

Calculate:

a the moment of a downward force of 1500 N on the camera about the point where the supporting structure is attached to the wall,

b the extra force of the vertical strut supporting the frame, when the camera is pulled with a downward force of 1500 N.

3 A crane is used to raise one end of a 15 kN girder of length 10.0 m off the ground. When the end of the girder is at rest 6.0 m off the ground, the crane cable is perpendicular to the girder (Figure 6). Calculate the tension in the cable.

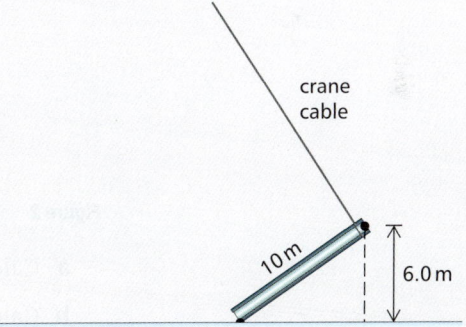

Figure 6

4 In question **3**, show that the support force on the girder from the ground has a horizontal component of 3.6 N and a vertical component of 10.2 kN. Hence calculate the magnitude of the support force.

1 Calculate the magnitude of the resultant of a 6.0 N force and a 9.0 N force acting on a point object when the two forces act:

   a in the same direction,

   b in opposite directions,

   c at 90° to each other.

2 A point object in equilbrium is acted on by a 3 N force, a 6 N force and a 7 N force. What is the resultant force on the object if the 7 N force is removed?

3 A point object of weight 5.4 N in equilibrium is acted on by a horizontal force of 4.2 N and a second force *F*.

   a By considering the triangle of forces rule, determine the magnitude of *F*.

   b Calculate the angle between the direction of *F* and the horizontal.

4 An object of weight 7.5 N hangs on the end of a cord which is attached to the midpoint of a wire, stretched between two points on the same horizontal level. Each half of the wire is at 12° to the horizontal. Calculate the tension in each half of the wire.

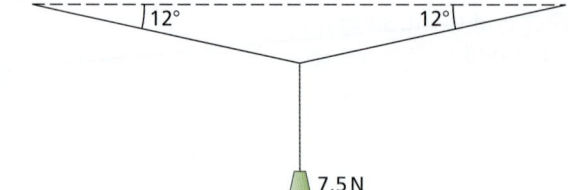

Figure 1

5 A ship is towed at constant speed by two tug boats, each pulling the ship with a force of 9 kN. The angle between the tug-boat cables is 40°.

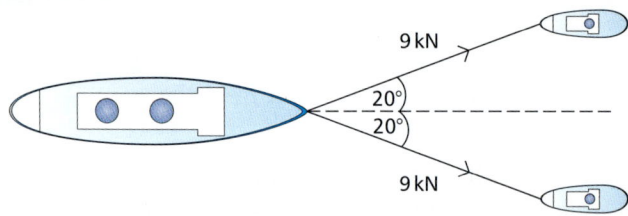

Figure 2

   a Calculate the resultant force on the ship due to the two cables.

   b Calculate the drag force on the ship.

6 A metre rule is pivoted on a knife edge at its centre of gravity, supporting a weight of 5.0 N and an unknown weight *W* (as shown in Figure 3). To balance the rule horizontally with the unknown weight on the 250 mm mark of the rule, the position of the 5.0 N weight needs to be at the 810 mm mark.

   a Calculate the unknown weight.

   b Calculate the support force on the rule from the knife edge.

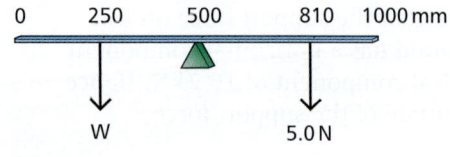

Figure 3

7 In Figure 3, a 2.5 N weight is also suspended from the rule at its 400 mm mark. What adjustment needs to be made to the position of the 5.0 N weight to rebalance the rule?

8 A uniform metre rule is balanced horizontally on a knife edge at its 350 mm mark by placing a 3.0 N weight on the rule at its 10 mm mark.

   a Sketch the arrangement and calculate the weight of the rule.

   b Calculate the support force on the rule from the knife edge.

9 A diving board has a length 4.0 m and a weight of 250 N. It is bolted to the ground at one end and projects by a length of 3.0 m beyond the edge of the swimming pool (Figure 4). A person of weight 650 N stands on the free end of the diving board. Calculate:

   a the force on the bolts,

   b the force on the edge of the swimming pool.

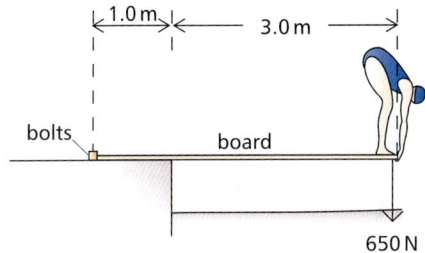

Figure 4

10 A uniform beam XY of weight 1200 N, and of length 5.0 m, is supported horizontally on a concrete pillar at each end. A person of weight 500 N sits on the beam, at a distance of 1.5 m from end X.

   a Sketch a free-body force diagram of the beam.

   b Calculate the support force on the beam from each pillar.

11 A bridge crane used at a freight depot consists of a horizontal span of length 12 m fixed at each end to a vertical pillar (Figure 5).

   a When the bridge crane supports a load of 380 kN at its centre, a force of 1600 kN is exerted on each pillar. Calculate the weight of the horizontal span.

   b The same load is moved across a distance of 2.0 m by the bridge crane. Sketch a free-body force diagram of the horizontal span and calculate the force exerted on each pillar.

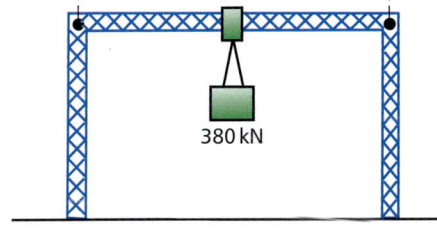

Figure 5

12 A curtain pole of weight 24 N and of length 3.2 m is supported horizontally by two wall-mounted supports X and Y, which are 0.8 m and 1.2 m from each end respectively.

   a Sketch the free-body force diagram for this arrangement, and calculate the force on each support when there are no curtains on the pole.

   b When the pole supports a pair of curtains of total weight 90 N drawn along the full length of the pole, what will be the force on each support?

13 A steel girder of weight 22 kN and of length 14 m is lifted off the ground at one end, by means of a crane. When the raised end is 2.0 m above the ground, the cable is vertical.

   a Sketch a free-body force diagram of the girder in this position.

   b Calculate the tension in the cable at this position and the force of the girder on the ground.

14 A rectangular picture of weight 24 N, hangs on a wall supported by a cord attached to the frame at each of the top corners (Figure 6). Each section of the cord makes an angle of 25° with the picture, which is horizontal along its width.

   a Copy the diagram and mark the forces acting on the picture.

   b Calculate the tension in each section of the cord.

Figure 6

3 Exam-style and Practice Questions

> Launch additional digital resources for the chapter

1 A tugboat is used to pull a barge along a river at constant velocity, as shown below. The tugboat pulls on the barge with a force of 4.5 kN. The tugboat's engine drives the tugboat forward with a force of 7.0 kN.

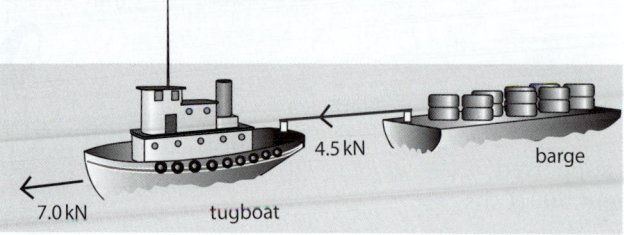

Figure 1.1

Calculate:

a the drag force on the tugboat due to the water,

b the drag force on the barge due to the water.

2 A yacht is moving due north as a result of a force, due to the wind, of 350 N in a horizontal direction of 40° east of due north, as shown below.

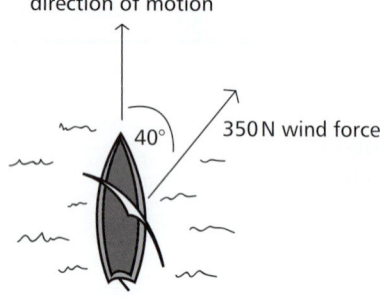

Figure 2.1

a Calculate the component of the force of the wind:
 i in the direction in which the yacht is moving,
 ii perpendicular to the direction in which the yacht is moving.

b Explain why the crew of the yacht need to lean out, as shown below, to prevent the yacht capsizing due to the force of the wind.

Figure 2.2

3 A car, of weight 6.2 kN, is towed at constant speed by a van on an uphill road, as shown below. The tow rope is at an angle of 15° above the horizontal, when the road is 5° above the horizontal. The tension in the tow rope pulls the car with a force of 580 N.

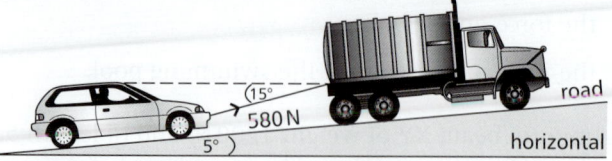

Figure 3.1

a Calculate the component of the tension in the tow rope:
 i parallel to the road,
 ii perpendicular to the road.

b Calculate the component of the car's weight:
 i parallel to the road,
 ii perpendicular to the road.

c By comparing your answers to parts **ai** and **bi**, calculate the magnitude and direction of the force of friction acting on the car.

d The van's weight is 12 kN. By considering the component of the van's weight parallel to the road and the component of the tension in the tow rope, show that the force of the van's engine must be at least 1.6 kN.

4 A uniform girder, of weight 14 kN and of length 8.0 m, rests horizontally on two wooden blocks, one at 1.0 m from one end of the girder and the other at 0.5 m from the other end (as shown below). Calculate the force of the girder on each block.

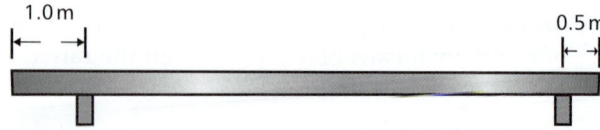

Figure 4.1

5 a i Explain what is meant by the centre of gravity of a body.
 ii Define the moment of a force about a point.

b A builder needs to measure the weight of a bag of sand using a set of scales that can measure weights up to 200 N. He discovers that the bag weighs more than 200 N, so he devises the arrangement shown on the next page to weigh the bag. This arrangement consists of a uniform plank, of length 4.5 m, resting horizontally with one end on a brick and the other end on the scales.

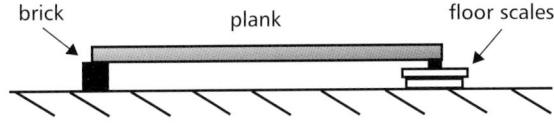

Figure 5.1 Without the bag of sand

i Without the bag of sand on the plank, the scales read 30 N. Calculate the weight of the plank.

ii With the bag of sand on the plank, the scales read 200 N when the bag is 1.5 m from the end of the plank on the scales. Show that the bag of sand has a weight of 255 N.

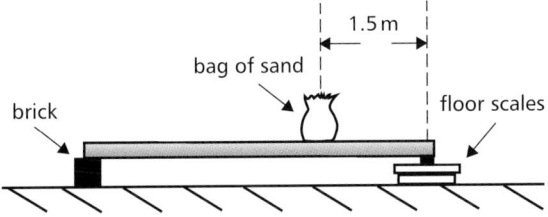

Figure 5.2 With the bag of sand

6 A flower basket, of weight 35 N, is suspended on the lower end of a rope which is tied at its upper end to a horizontal bar. The force of a horizontal wind blows the basket to one side, causing the rope to make an angle of 20° with the vertical, as shown below.

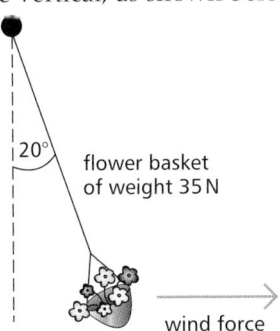

Figure 6.1

a Show that the tension in the rope in this position is 37 N.

b Calculate the force of the wind on the flower basket.

7 A paraglider, moving at constant velocity and constant height, is pulled by a cable which is at an angle of 35° to the horizontal. The force of the cable on the paraglider is 185 N.

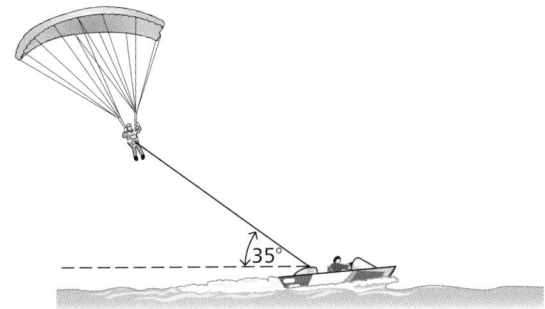

Figure 7.1

a Calculate the horizontal and vertical components of this force.

b With the aid of a force diagram, explain why the force of the parachute on the paraglider is not in the opposite direction to the force of the cable on the paraglider.

8 A bridge crane is used to transfer a container of weight 60 kN from a rail wagon onto a container lorry. The horizontal frame of the bridge crane has a length of 14 m and a total weight of 200 kN. It rests on two vertical steel pillars, as shown below.

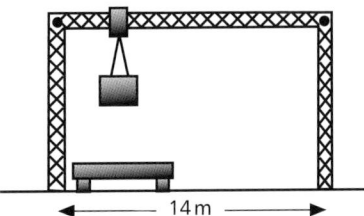

Figure 8.1

Calculate the force on each vertical pillar due to the weight of the horizontal frame and the container when:

a the container is suspended at the centre of the horizontal frame,

b the container is 4.0 m from the centre.

9 a i Describe in terms of forces what is meant by a *couple*.

   ii Define the *torque* of a couple.

b A *torque wrench* is a device used to tighten a nut and bolt to a pre-set torque. Figure 9.1 shows a torque wrench used to tighten a nut on a bolt.

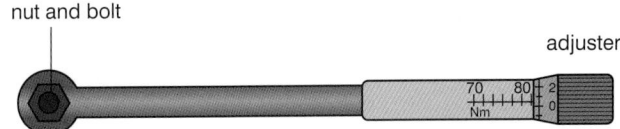

Figure 9.1

Before the wrench is used, the maximum torque it can apply is set by turning the adjuster on the handle to the required torque setting. When the wrench is used to tighten the nut, it clicks and stops turning the nut when the required torque setting is reached.

   i The wheel nuts on a certain type of vehicle are to be tightened to a torque of 84 N m. To turn the nut, the user applies a force F perpendicular to the wrench handle at a distance of 0.44 m from the axis of the bolt. Calculate the force F that needs to be applied to the handle to achieve the required torque.

   ii State and explain how the magnitude of the force needed to give the same torque would differ if its direction was not perpendicular to the handle.

Momentum

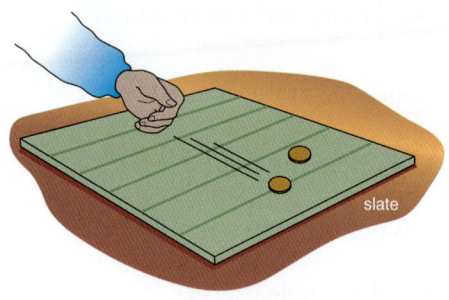

Figure 1 *A momentum game*

If you have ever run into someone on the sports field, you will know something about **momentum**. If the person you ran into is more massive than you, then you probably came off worse than the other person. When two bodies collide, the effect they have on each other depends not only on their initial velocities but also on the **mass** of each body. You can easily test the ideas using coins, as shown in Figure 1. You might already have developed your skill in this area! It's not too difficult to show that when a large coin and a small coin collide, the motion of the small coin is affected more.

Sir Isaac Newton was the first person to realise that a force was needed to change the velocity of an object. He realised that the effect of a force on an object depended on its mass as well as on the amount of force. He defined the momentum of a moving object as its mass multiplied by its velocity and showed how the momentum of an object changes when a force acts on it. In Topic 2.1, you learned that the force needed to give an object a certain acceleration can be calculated from the equation 'force = mass × acceleration'. In this topic, we consider the ideas that Newton established in full.

Although Newton put forward his ideas over 300 years ago, his laws continue to provide the essential mathematical rules for predicting the motion of objects in any situation except inside the atom (where the rules of quantum physics apply) or at speeds approaching the speed of light (where Einstein's theory of relativity applies). For example, the launch of a satellite into orbit is carefully planned using Newton's laws of motion.

The **momentum** of an object is defined as its **mass × its velocity**.

- The unit of momentum is $kg\,m\,s^{-1}$. The symbol for momentum is p.
- Momentum is a vector quantity. Its direction is the same as the direction of the object's velocity.
- For an object of mass m moving at velocity v, its momentum $\mathbf{p = mv}$.

For example, a ball of mass $2.0\,kg$ moving at a velocity of $10\,m\,s^{-1}$ has the same amount of momentum as a person of mass $50\,kg$ moving at a velocity of $0.4\,m\,s^{-1}$.

Momentum and Newton's laws of motion

Newton's first law of motion: An object remains at rest or in uniform motion unless acted on by a resultant force.

In effect, Newton's first law tells us that a force is needed to change the momentum of an object. If the momentum of an object is constant, there is no resultant force acting on it. Clearly, if the mass of an object is constant and the object has constant momentum, it follows that the velocity of the object is also constant. However, if a moving object with constant momentum gains or loses mass, its velocity would change to keep its momentum constant. For example, a cyclist in a race who collects a water bottle as he or she speeds past a 'service' point gains mass (i.e. the water bottle) and therefore loses velocity.

Newton's second law of motion: The rate of change of momentum of an object is proportional to the resultant force on it. In other words, the resultant force is proportional to the change of momentum per second.

Consider an object of constant mass m acted on by a constant force F. Its acceleration causes a change of its speed from initial speed u to speed v in time t without change of direction. See Figure 2.

- its initial momentum = mu, and its final momentum = mv
- change of momentum = its final momentum (mv) – its initial momentum (mu).

According to Newton's second law, the force is proportional to the change of momentum per second.

$$\text{Therefore, force } F \propto \frac{\text{change of momentum}}{\text{time taken}} = \frac{mv - mu}{t} = \frac{m(v - u)}{t} = ma$$

where $a = \dfrac{v - u}{t} = $ the acceleration of the object.

This proportionality relationship (i.e. $F \propto ma$) can be written as $F = kma$, where k is a constant of proportionality.

The value of k is made equal to 1 by defining the unit of force, the **newton**, as the amount of force that gives an object of mass 1 kg an acceleration of $1\,\text{m s}^{-2}$ (i.e. force $F = 1\,\text{N}$, mass $m = 1\,\text{kg}$, acceleration, $a = 1\,\text{m s}^{-2}$ so $k = 1$).

Therefore, with $k = 1$, the equation $F = ma$ follows from Newton's second law provided the mass of the object is constant.

In general, the change of momentum of an object may be written as $\Delta(mv)$, where the symbol Δ means 'change of'. Therefore, if the momentum of an object changes by $\Delta(mv)$ in time Δt, the force F on the object is given by the equation

$$F = \frac{\Delta(mv)}{\Delta t}$$

Note that $\dfrac{\Delta(mv)}{\Delta t}$ is the rate of change of momentum of an object of mass m. The equation above means that force may be defined as rate of change of momentum.

1 **If m is constant,** then $\Delta(mv) = m\Delta v$, where Δv is the change of velocity of the object.

$$\therefore \quad F = m\frac{\Delta v}{\Delta t} = ma \text{ where acceleration } a = \frac{\Delta v}{\Delta t}$$

2 **If m changes at a constant rate** as a result of mass being transferred at constant velocity, then $\Delta(mv) = v\Delta m$, where Δm is the change of mass of the object.

$$\therefore \quad F = v\frac{\Delta m}{\Delta t} \text{ where } \frac{\Delta m}{\Delta t} = \text{change of mass per second.}$$

This form of Newton's second law is used in any situation where an object gains or loses mass continuously. For example, if a hosepipe ejects water at speed v, the force F needed to eject the water is given by $F = v\dfrac{\Delta m}{\Delta t}$ where $\dfrac{\Delta m}{\Delta t} = $ mass of water lost per second. An equal and opposite reaction force acts on the hosepipe due to the water.

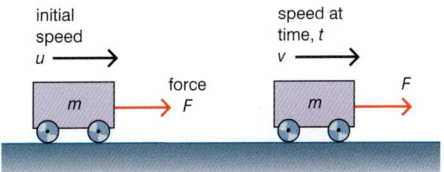

Figure 2 *Force and momentum*

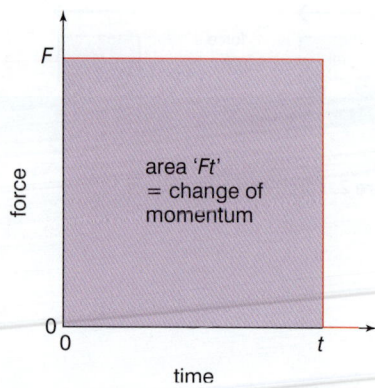

F

force

0

0

t

time

Figure 3 *Force v. time for constant force*

Note

The unit of momentum can be given as the newton second (Ns) or the kilogram metre per second ($kg\,m\,s^{-1}$).

area 'Ft'
= change of
momentum

Force v. time graphs

Suppose an object of constant mass m is acted on by a constant force F which changes its velocity from initial velocity u to velocity v in time t. As explained on the previous page, Newton's second law gives

$$F = \frac{mv - mu}{t}$$

Rearranging this equation gives $Ft = mv - mu$

Figure 3 is a graph of force v. time for this situation. Because force F is constant for time t, the area under the line represents Ft which is equal to $mv - mu$. In other words, the area under the line of a force v. time graph represents the change of momentum.

Worked example

A force of 10 N acts for 20 s on an object of mass 50 kg which is initially at rest. Calculate:

a the change of momentum of the object,

b the velocity of the object at 20 s.

Solution

a Change of momentum = Ft = 10 N × 20 s = 200 N s.

b Momentum at 20 s = 200 N s as the object was initially at rest.

$\therefore$ Velocity = $\dfrac{\text{momentum}}{\text{mass}}$ = $\dfrac{200\,\text{N s}}{50\,\text{kg}}$ = $4.0\,\text{m s}^{-1}$

Summary test 4.1

1 a Calculate the momentum of
   i an atom of mass 4.0×10^{-25} kg moving at a velocity of $3.0 \times 10^{6}\,\text{m s}^{-1}$,
   ii a pellet of mass 4.2×10^{-4} kg moving at a velocity of $120\,\text{m s}^{-1}$,
   iii a bird of mass 0.56 kg moving at a velocity of $25\,\text{m s}^{-1}$.

 b Calculate:
   i the mass of an object moving at a velocity of $16\,\text{m s}^{-1}$ with momentum of $96\,\text{kg m s}^{-1}$,
   ii the velocity of an object of mass 6.4 kg that has momentum of $128\,\text{kg m s}^{-1}$.

2 A train of mass 24 000 kg moving at a velocity of $15.0\,\text{m s}^{-1}$ is brought to rest by a braking force of 6000 N. Calculate:

 a the initial momentum of the train,

 b the time taken for the brakes to stop the train.

3 An aircraft of total mass 45 000 kg accelerates on a runway from rest to a velocity of $120\,\text{m s}^{-1}$ when it takes off. During this time, its engines provide a constant driving force of 120 kN.

Calculate:

 a the gain of momentum of the aircraft,

 b the 'take off' time.

4 The velocity of a vehicle of mass 600 kg was reduced from $15\,\text{m s}^{-1}$ by a constant force of 400 N which acted for 20 s then by a constant force of 20 N for a further 20 s.

 a Sketch the force v. time graph for this situation.

 b i Calculate the initial momentum of the vehicle.
   ii Use the force v. time graph to determine the total change of momentum.
   iii Hence calculate the final velocity of the vehicle.

Impact force

A sports person knows that the harder a ball is hit, the further it travels. The impact changes the momentum of the ball in a very short time when the object exerting the impact force is in contact with the ball.

Figure 1 *A golf ball impact*

Learning outcomes

On these pages you will learn to:

- recognise how the momentum of an object changes when it undergoes an impact
- calculate the force of an impact from the change of momentum of an object and the duration of the impact

- If the ball is initially stationary and the impact causes it to accelerate to speed v in time t, the gain of momentum of the ball due to the impact = mv, where m is the mass of the ball.

  Therefore, the force of the impact $F = \dfrac{\text{change of momentum}}{\text{contact time } t} = \dfrac{mv}{t}$

- If the ball is moving with an initial velocity, u, and the impact changes its velocity to v in time t, the change of momentum of the ball = $mv - mu$.

  Therefore, the force of impact $F = \dfrac{\text{change of momentum}}{\text{contact time } t} = \dfrac{mv - mu}{t}$

Worked example

A ball of mass 0.63 kg initially at rest was struck by a bat which gave it a velocity of 35 m s^{-1}. The contact time between the bat and ball was 25 ms. Calculate:

a the momentum gained by the ball,

b the average force of impact on the ball.

Solution

a Momentum gained = 0.63 kg × 35 m s^{-1} = 22 kg m s^{-1},

b Impact force = $\dfrac{\text{gain of momentum}}{\text{contact time } t} = \dfrac{22 \text{ kg m s}^{-1}}{0.025 \text{ s}} = 880 \text{ N}$

Vehicle safety reminders

Vehicle safety features such as crumple zones, seatbelts, collapsible steering wheels, and airbags are all designed to lessen the effect of an impact on people in the vehicle. The essential idea is to increase the time taken by an impact so the acceleration or deceleration is less and therefore the impact force is less. The result is the same using the idea of momentum; for a given change of momentum, the force is reduced if the impact time is increased. However, as explained in Topic 4.3, the ideas can be developed much further by using the concept of momentum.

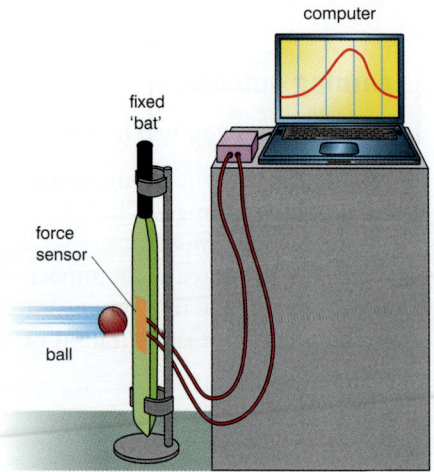

computer

fixed 'bat'

force sensor

ball

Figure 2 *Investigating an impact force on a ball*

Force v. time graphs for impacts

The variation of an impact force with time on a ball can be recorded using a force sensor connected via suitably long wires or a radio link to a computer. The force sensor is attached to the object (e.g. a bat) that causes the impact. Because equal and opposite forces act on the ball, the force on the ball due to the bat varies in exactly the same way as the force on the bat due to the ball. The variation of force with time is displayed on the computer screen.

Figure 3 shows a typical force v. time graph for an impact. The graph shows that the impact force increases then decreases during the impact. As explained on p.62, the area under the graph is equal to the change of momentum. The average force of impact can be worked out from the change of momentum divided by the contact time.

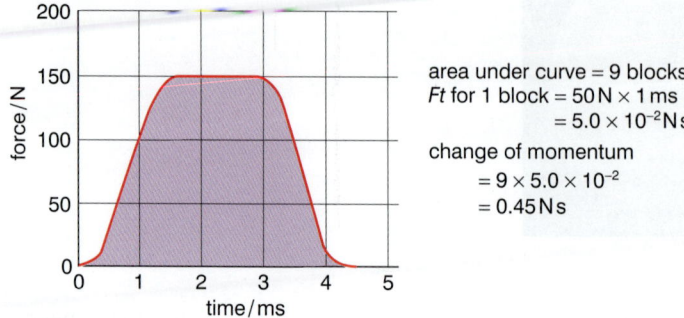

area under curve = 9 blocks
Ft for 1 block = 50 N × 1 ms
$\qquad\qquad\quad$ = 5.0×10^{-2} N s
change of momentum
$\qquad$ = $9 \times 5.0 \times 10^{-2}$
$\qquad$ = 0.45 N s

Figure 3 *Force v. time for an impact*

Rebound impacts

When a ball hits a wall and rebounds, its momentum changes direction due to the impact. If the ball hits the wall normally, it rebounds normally so the direction of its momentum is reversed. The velocity and, therefore, the momentum after the impact are in the opposite direction to the velocity before the impact and therefore have the opposite sign. Figure 4 shows the idea.

Suppose the ball hits the wall normally with an initial speed u and it rebounds at speed v in the opposite direction. Since its direction of motion reverses on impact, a sign convention is necessary to represent the two directions. Using + for 'towards the wall' and − for 'away from the wall', its initial momentum = $+mu$, and its final momentum = $-mv$.

Therefore, its change of momentum = final momentum − initial momentum
$$= (-mv) - (mu)$$

The impact force $\quad F = \dfrac{\text{change of momentum}}{\text{contact time } t} = \dfrac{(-mv) - (mu)}{t}$

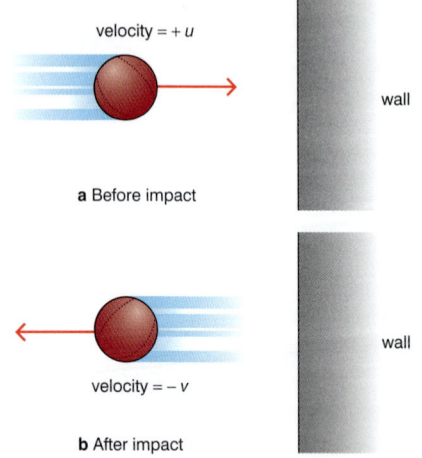

velocity = + u

wall

a Before impact

velocity = − v

wall

b After impact

Figure 4 *A rebound*

> ### Note
>
> If there is no loss of speed on impact, then $v = u$
>
> so the impact force $F = \dfrac{(-mu) - (mu)}{t} = -\dfrac{2mu}{t}$

Worked example

A tennis ball of mass $0.20\,kg$ moving at a speed of $18\,m\,s^{-1}$ was hit by a bat, causing the ball to go back in the direction it came from at a speed of $15\,m\,s^{-1}$. The contact time was $0.12\,s$. Calculate:

a the change of momentum of the ball,

b the impact force on the ball.

Solution

a Mass of ball $m = 0.20\,kg$, initial velocity $u = +18\,m\,s^{-1}$, final velocity $= -15\,m\,s^{-1}$,

Change of momentum $= mv - mu = (0.20 \times -15) - (0.20 \times 18)$

$$= -3.0 - 3.6 = -6.6\,kg\,m\,s^{-1}$$

b Impact force $= \dfrac{\text{change of momentum}}{\text{time taken}} = \dfrac{-6.6\,kg\,m\,s^{-1}}{0.12\,s} = -55\,N$

The minus sign indicates the force on the ball is in the same direction as the velocity after the impact.

Summary test 4.2

1 A $2000\,kg$ lorry reversing at a speed of $0.8\,m\,s^{-1}$ backs accidentally into a steel fence. The fence stops the lorry $0.5\,s$ after the lorry first makes contact with the fence. Calculate:

    a the initial momentum of the lorry,

    b the force of the impact.

2 A car of mass $600\,kg$ travelling at a speed of $3.0\,m\,s^{-1}$ is struck from behind by another vehicle. The impact lasts for $0.4\,s$ and causes the speed of the car to increase to $8.0\,m\,s^{-1}$. Calculate:

    a the change of momentum of the car due to the impact,

    b the impact force.

3 A molecule of mass $5.0 \times 10^{-26}\,kg$ moving at a speed of $420\,m\,s^{-1}$ hits a surface at right angles to the surface and rebounds at the same speed in the opposite direction in an impact lasting $0.22\,ns$. Calculate:

    a the change of momentum,

    b the force on the molecule.

4 Repeat the calculation in question **3** for a molecule of the same mass at the same speed which hits the surface at $60°$ to the normal and rebounds without loss of speed at $60°$ to the normal, as shown in Figure 5. You will need to work out the component of the molecule's velocity parallel to the normal before and after the impact. Assume the contact time is the same.

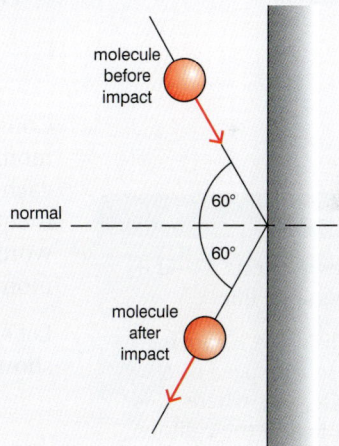

Figure 5

Learning outcomes

On these pages you will learn to:

- state Newton's third law of motion
- state the principle of conservation of momentum
- apply the principle of conservation of momentum to solve simple problems

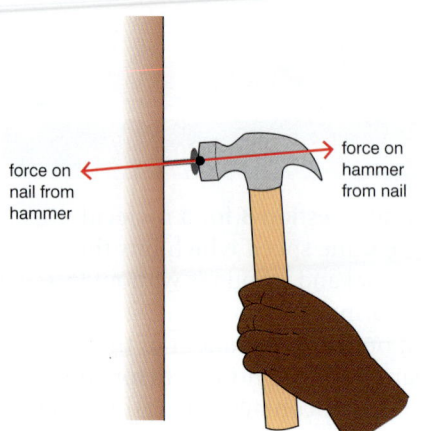

Figure 1 *Newton's third law*

 **Relative speed**

Figure 2 shows two objects A and B moving along the same line.

- Before the collision, A is moving faster than B, so the speed of A relative to B $= u_A - u_B$ where u_A is the speed of A before the collision and u_B is the speed of B before the collision.

- After the collision, B is moving faster than A, so the speed of B relative to A $= v_B - v_A$ where v_B is the speed of B after the collision and v_A is the speed of A after the collision.

The concept of relative speed is useful in Topic 4.4 when considering kinetic energy changes in straight line collisions.

Newton's third law of motion

> **When two objects interact, they exert equal and opposite forces on each other.**

In other words, if object A exerts a force on object B, there must be an equal and opposite force acting on object A due to object B.

For example,

- An object resting on a table exerts a force on the table which exerts an equal and opposite force on the object.
- A person leaning against a wall exerts a force on the wall which exerts an equal and opposite force on the person.
- A hammer hitting a nail exerts a force on the nail which exerts an equal and opposite force on the hammer.
- The Earth exerts a force due to gravity on an object which exerts an equal and opposite force on the Earth.
- A jet engine exerts a force on hot gas in the engine to expel the gas; the gas being expelled exerts an equal and opposite force on the engine.

The principle of conservation of momentum

When an object is acted on by a resultant force, its momentum changes. If there is no change of its momentum, there can be no resultant force on the object. Now consider several objects which interact with each other. If no external resultant force acts on the objects, the total momentum does not change. However, interactions between the objects can transfer momentum between them. But the total momentum does not change.

> **The principle of conservation of momentum states that for a system of interacting objects, the total momentum remains constant, provided no external resultant force acts on the system.**

Consider two objects that collide with each other then separate. As a result, the momentum of each object changes. They exert equal and opposite forces on each other when they are in contact. So the change of momentum of one object is equal and opposite to the change of momentum of the other object. In other words, if one object gains momentum, the other object loses an equal amount of momentum. So the total amount of momentum is unchanged.

Let's look in detail at the example of two snooker balls A and B in collision, as shown in Figure 2.

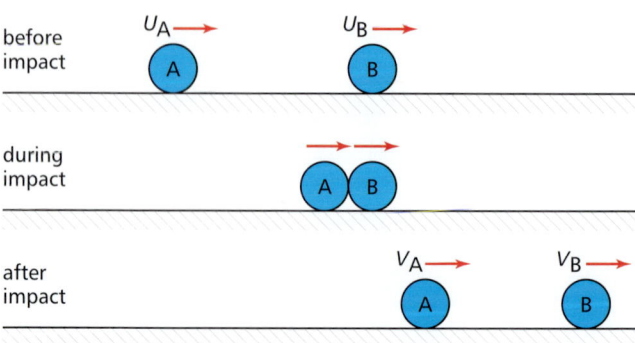

Figure 2 *Conservation of momentum*

- The impact force F_1 on ball A changes the velocity of A from u_A to v_A in time t.

Therefore, $\quad F_1 = \dfrac{m_A v_A - m_A u_A}{t}$,

where t = the time of contact between A and B,

and m_A = the mass of ball A.

- The impact force F_2 on ball B changes the velocity of B from u_B to v_B in time t.

Therefore, $\quad F_2 = \dfrac{m_B v_B - m_B u_B}{t}$,

where t = the time of contact between A and B,

and m_B = the mass of ball B.

Because the two forces are equal and opposite to each other, $F_2 = -F_1$

Therefore,

$$\frac{m_B v_B - m_B u_B}{t} = -\frac{m_A v_A - m_A u_A}{t}$$

Cancelling t on both sides gives

$$m_B v_B - m_B u_B = -m_A v_A + m_A u_A$$

Rearranging this equation gives

$$m_B v_B + m_A v_A = m_A u_A + m_B u_B$$

Therefore,

the total final momentum = the total initial momentum

In other words, the total momentum is unchanged by this collision.

> **Note**
>
> If the colliding objects stick together as a result of the collision, they have the same final velocity. The above equation may therefore be written
>
> $$(m_B + m_A) V = m_A u_A + m_B u_B$$

Testing conservation of momentum

Figure 3 shows an arrangement that can be used to test conservation of momentum using a motion sensor linked to a computer. The mass of each trolley is measured before the test. With trolley B at rest, trolley A is given a push so that it moves towards trolley B at constant velocity. The two trolleys stick together on impact. The computer records and displays the velocity of trolley A throughout this time. The computer display shows that the velocity of trolley A dropped suddenly when the impact took place. The velocity of trolley A immediately before the collision, u_A, and after the collision (V) can be measured. The measurements should show that the total momentum of both trolleys after the collision is equal to the momentum of trolley A before the collision. In other words,

$$(m_B + m_A) V = m_A u_A$$

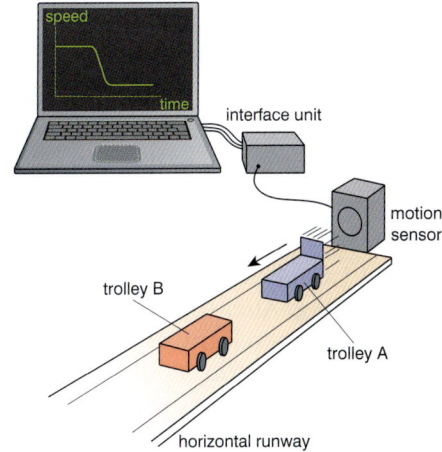

Figure 3 *Testing conservation of momentum*

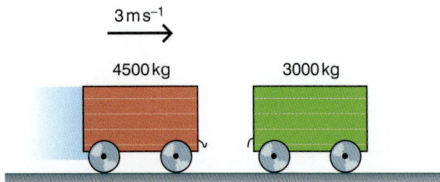

Figure 4 *Colliding wagons*

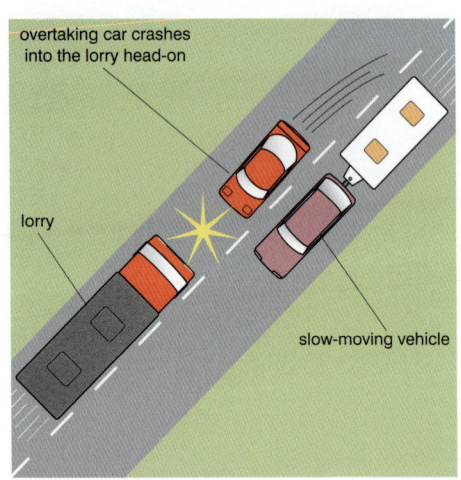

Figure 5 *A crash calculation*

A railway wagon of mass 4500 kg moving along a level track at a speed of 3.0 m s^{-1} collides with and couples to a second railway wagon of mass 3000 kg which is initially stationary. Calculate the speed of the two wagons immediately after the collision.

Solution

Total initial momentum

$$= \text{initial momentum of A} + \text{initial momentum of B}$$

$$= (4500 \times 3.0) + (3000 \times 0) = 13\,500\,\text{kg m s}^{-1}$$

Total final momentum

$$= \text{total mass of A and B} \times \text{velocity } V \text{ after the collision}$$

$$= (4500 + 3000)\,V = 7500V$$

Using the principle of conservation of momentum,

$$7500V = 13\,500$$

$$V = \frac{13\,500}{7500} = 1.8\,\text{m s}^{-1}$$

Head-on collisions

Consider two objects moving in opposite directions that collide with each other. Depending on the masses and initial velocities of the two objects, the collision could cause them both to stop. The momentum of the two objects after the collision would then be zero. This could only happen if the initial momentum of one object was exactly equal and opposite to that of the other object. In general, if two objects move in opposite directions before a collision, then the vector nature of momentum needs to be taken into account by assigning numerical values of velocity + or − according to the direction.

For example, if a car of mass 600 kg travelling at a velocity of 25 m s^{-1} collides head-on with a lorry of mass 2400 kg travelling at a velocity of 10 m s^{-1} in the opposite direction the total momentum before the collision is 9000 kg m s^{-1} in the direction in which the lorry was moving. As momentum is conserved in a collision, the total momentum after the collision is the same as the total momentum before the collision. Prove for yourself that if the two vehicles stick together after the collision, they must have a velocity of 3.0 m s^{-1} (= 9000 kg m s^{-1} / 3000 kg) immediately after the impact in the direction in which the lorry was moving.

1 A railway wagon of mass 3000 kg moving at a velocity of 1.2 m s^{-1} collides with a stationary wagon of mass 2000 kg. After the collision, the two wagons couple together. Calculate their speed immediately after the collision.

2 A railway wagon of mass 5000 kg moving at a velocity of 1.6 m s^{-1} collides with a stationary wagon of mass 3000 kg. After the collision, the 3000 kg wagon moves away at a velocity of 1.5 m s^{-1}. Calculate the speed and direction of the 5000 kg wagon after the collision.

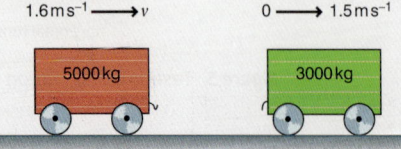

Figure 6

3 In a laboratory experiment, a trolley of mass 0.50 kg moving at a speed of 0.25 m s^{-1} collides with a trolley of mass 1.0 kg moving in the opposite direction at a speed of 0.20 m s^{-1}. The two trolleys couple together on collision. Calculate their speed and direction immediately after the collision.

4 A ball of mass 0.80 kg moving at a speed of 2.5 m s^{-1} along a straight line collides with a ball of mass 2.5 kg which is initially stationary. As a result of the collision, the 2.5 kg ball is given a velocity of 1.0 m s^{-1} along the same line. Calculate the speed and direction of the 0.80 kg ball immediately after the collision.

Elastic and inelastic collisions

Drop a bouncy rubber ball from a measured height onto a hard floor. The ball should bounce back almost to the same height. Try the same with a cricket ball and there will be very little bounce! A **perfectly elastic** ball would be one that bounces back to exactly the same height. Its kinetic energy just after impact must equal its kinetic energy just before impact. Otherwise, it cannot regain its initial height. There is no loss of kinetic energy in a perfectly elastic collision.

> **A perfectly elastic collision is one where there is no loss of kinetic energy.**

- A squash ball hitting a hard surface bounces off the surface with little or no loss of speed. If the ball is perfectly elastic, there is no loss of speed on impact and no loss of kinetic energy.
- A very low speed impact between two cars is almost perfectly elastic, provided no damage is done. Some of the initial kinetic energy of the two vehicles may be converted to sound. However, if the collision causes damage to the vehicles, the kinetic energy after the collision is less than before so the collision is not elastic.

> **A totally inelastic collision is one where the colliding objects stick together.**

- A railway wagon that collides with and couples to another wagon is an example of a totally inelastic collision. Some of the initial kinetic energy is converted to other forms of energy.
- A vehicle crash in which the colliding vehicles lock together is another example of a totally inelastic collision. The total kinetic energy after the collision is less than the total kinetic energy before the collision.

A **partially inelastic collision** is where the colliding objects:
1 move apart, and
2 have less kinetic energy after the collision than before.

To work out if a collision is elastic or inelastic, the kinetic energy of each object before and after the collision must be worked out.

For a ball of mass m falling in air from a measured height H and rebounding to a height h, as shown in Figure 3:

- the kinetic energy immediately before impact = loss of potential energy through height $H = mgH$,
- the kinetic energy immediately after impact = gain of potential energy through height $h = mgh$.

So the height ratio $\dfrac{h}{H}$ gives the fraction of the initial kinetic energy that is recovered as kinetic energy after the collision.

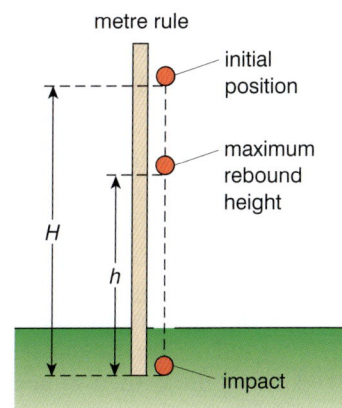

Figure 3 Testing an impact

Collisions between two objects

When two objects with known masses and initial velocities collide, if the final velocity of one of the objects is known, the final velocity of the other object can be determined using the principle of conservation of momentum. The kinetic energy of each object before or after the collision can then be calculated using the kinetic energy formula $E_k = \frac{1}{2}mv^2$. If the collision is elastic, the total kinetic energy of the objects before the collision is equal to their total kinetic energy after the collision.

Learning outcomes

On these pages you will learn to:

- distinguish between an elastic and an inelastic collision
- recognise that momentum is always conserved in any collision, whereas kinetic energy is only conserved in an elastic collision
- apply the principle of conservation of momentum to elastic and inelastic interactions between bodies in both one and two dimensions
- recognise that, for a perfectly elastic collision, the relative speed of approach is equal to the relative speed of separation

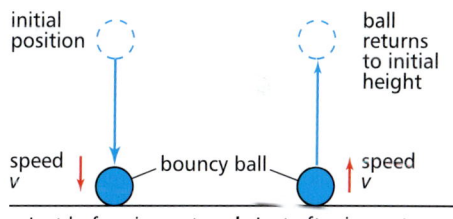

Figure 1 An elastic impact

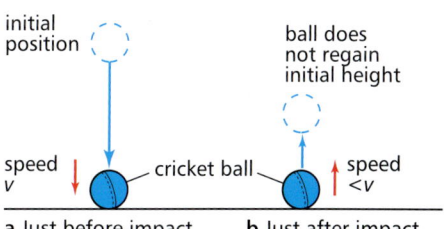

Figure 2 A partially elastic impact

For a perfectly elastic collison between two objects, using the principle of conservation of momentum and the rule that the total kinetic energy is unchanged in an elastic collision, it can be shown that **their relative speed of separation = their relative speed of approach.**

• **For a collision in which the objects move along the same straight line before and after the collision** (e.g. railway wagons on a straight track), velocities in one direction are assigned positive values and velocities in the opposite direction are assigned negative values, because velocity is a vector quantity.

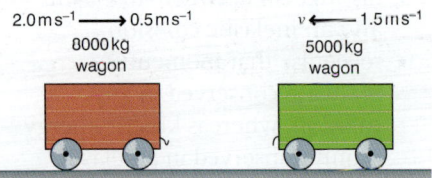

Figure 4

Figure 5 *This dramatic collision was used to demonstrate that special casks used to transport radioactive materials by rail could withstand high speed impacts. In this collision, a remotely controlled diesel locomotive was driven into a stationary cask. Even though the locomotive was destroyed, the cask itself was intact afterwards.*

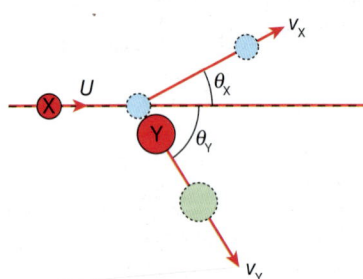

Figure 6 *An oblique collision*

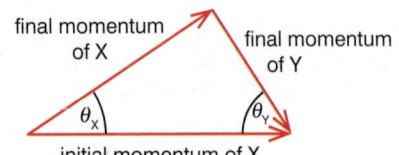

Figure 7 *Conservation of momentum*

> ### Worked example
>
> A railway wagon of mass 8000 kg moving at a speed of 2.0 m s^{-1} collides with a wagon of mass 5000 kg moving at a speed of 1.5 m s^{-1} in the opposite direction. The collision reduces the speed of the 8000 kg wagon to 0.5 m s^{-1} without changing its direction. Calculate the speed V and direction of the second wagon after the collision.
>
> *Solution*
> The total initial momentum $= (8000 \times 2.0) + (5000 \times -1.5)$
>
> $$= 16000 - 7500 = 8500 \, \text{kg m s}^{-1}$$
>
> (note the negative value of the initial velocity of the 5000 kg wagon)
>
> The total final momentum $= (8000 \times 0.5) + (5000 \times V)$
>
> Using the principle of conservation of momentum
>
> $$(8000 \times 0.5) + (5000 \times V) = 8500$$
>
> $$5000 \times V = 8500 - 4000 = 4500$$
>
> $$V = \frac{4500}{5000} = +0.9 \, \text{m s}^{-1}$$
>
> Note that the relative speed of separation (0.4 m s^{-1} (= 0.9 m s^{-1} − 0.5 m s^{-1})) is not equal to their relative speed of approach (3.5 m s^{-1} (= 2.0 m s^{-1} + 1.5 m s^{-1})). So the collision is **inelastic**. Prove for yourself that 18.6 kJ of kinetic energy is transferred to the surroundings (as heat and sound).

• **In an 'oblique' collision, two objects collide then move away in different directions from their initial directions.** Figure 6 shows an oblique collision between two spheres X and Y, where X is initially moving at velocity V towards Y, which is initially stationary. After the collision, the two objects move away at velocities v_X and v_Y in directions at angles θ_X and θ_Y to the initial direction of X. If the initial and final velocity of X is known and the masses of the two spheres are known, the momentum of Y after the collision can be determined using the principle of conservation of momentum:

  • **either by drawing a vector diagram**, as shown in Figure 7

  • **or by calculation,** as follows.

  1 Resolve the velocities v_X and v_Y parallel and perpendicular to the initial direction of X to give $v_X \cos\theta_X$ and $v_X \sin\theta_X$ for the parallel and perpendicular components of X and $v_Y \cos\theta_Y$ and $v_Y \sin\theta_Y$ for the parallel and perpendicular components of Y.

  2 Apply the principle of conservation of momentum in each direction:
  • In the parallel direction: $m_X U = m_X v_X \cos\theta_X + m_Y v_Y \cos\theta_Y$
  • In the perpendicular direction: $0 = m_X v_X \sin\theta_X + m_Y v_Y \sin\theta_Y$, where m_X and m_Y are the masses of X and Y, respectively.

  3 Insert the known values into the above equations to give a value of $v_Y \cos\theta_Y$ from the first equation and a value of $v_Y \sin\theta_Y$ from the second equation. Note that if the perpendicular component of X is assigned a positive value, then the perpendicular component of Y needs to be assigned a negative value (because they are in opposite directions).

  4 The values of v_Y and θ_Y can then be found using $v_Y^2 = (v_Y \cos\theta_Y)^2 + (v_Y \sin\theta_Y)^2$ and $\tan\theta_Y = v_Y \sin\theta_Y / v_Y \cos\theta_Y$.

Worked example

A sphere X of mass 0.10 kg moving at a speed of 2.0 m s^{-1} collides with a sphere Y of mass 0.20 kg, which is initially at rest. As a result, X is deflected by 30° at a speed of 1.6 m s^{-1}. Show that Y rebounds at a speed of 0.50 m s^{-1} in a direction at 53° to the initial direction of X.

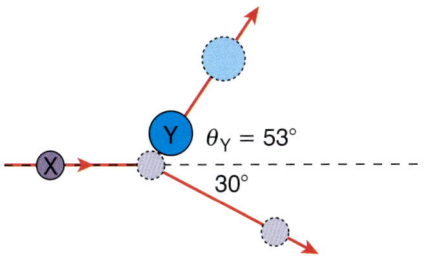

Solution 1: Vector diagram method

Initial momentum of X = 0.10 × 2.0 = 0.20 kg m s^{-1}

Final momentum of X = 0.10 × 1.6 = 0.16 kg m s^{-1} at 30° to the initial direction of X

Final momentum of Y = 0.20 × 0.50 = 0.10 kg m s^{-1} at cos 53° to the initial direction of X

Figure 9 shows the momentum vectors for the above three momentum values. The diagram shows that the vector sum of the final momentum vector for X and the final momentum vector for Y is equal to the initial momentum vector for X.

Figure 8

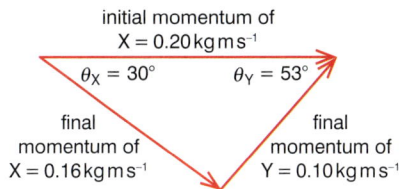

Figure 9

Solution 2: Calculation method

Apply the principle of conservation of momentum parallel and perpendicular to the initial direction of X.

In the parallel direction:

$$m_X U = m_X v_X \cos\theta_X + m_Y v_Y \cos\theta_Y$$

$$(0.10 \times 2.0) = (0.10 \times 1.6 \times \cos 30°) + (0.20 v_Y \times \cos\theta_Y)$$

$$(0.20 v_Y \times \cos\theta_Y) = (0.10 \times 2.0) - (0.10 \times 1.6 \times \cos 30°) = 0.20 - 0.139 = 0.061$$

$$v_Y \times \cos\theta_Y = 0.061 \div 0.20 = 0.305$$

In the perpendicular direction:

$$0 = m_X v_X \sin\theta_X + m_Y v_Y \sin\theta_Y$$

$$0 = (0.10 \times 1.6 \times \sin 30°) + (0.20\, v_Y \times \sin\theta_Y)$$

$$0.20 v_Y \times \sin\theta_Y = -0.10 \times 1.6 \times \sin 30° = -0.080$$

$$v_Y \times \sin\theta_Y = -0.080 \div 0.20 = -0.400$$

Therefore $v_Y = [(v_Y \cos\theta_Y)^2 +$

$(v_Y \sin\theta_Y)^2]^{0.5} = [0.305^2 + 0.400^2]^{0.5}$

$= 0.50$ m s^{-1} to two significant figures

$$\tan\theta_Y = \frac{v_Y \sin\theta_Y}{v_Y \cos\theta_Y}$$

$$= \frac{-0.400}{0.305}$$

$$= -1.31$$

Therefore, $\theta_Y = 53°$

Summary test 4.4

1 a A squash ball is released from rest above a flat surface. Describe how its energy changes if:
   i it rebounds to the same height,
   ii it rebounds to a lesser height.
   b In a ii, the ball is released from a height of 1.2 m above the surface and it rebounds to a height of 0.9 m above the surface. Show that 25% of its kinetic energy is lost in the impact.

2 A vehicle of mass 800 kg moving at a speed of 15.0 m s^{-1} collides with a vehicle of mass 1200 kg moving in the same direction at a speed of 5.0 m s^{-1}. The two vehicles lock together on impact. Calculate:

   a the velocity of the two vehicles immediately after impact,
   b the loss of kinetic energy due to the impact.

3 An ice puck of mass 1.5 kg moving at a speed of 4.2 m s^{-1} collides head on with a second ice puck of mass 1.0 kg moving in the opposite direction at a speed of 4.0 m s^{-1}. After the impact, the 1.5 kg ice puck continues in the same direction at a speed of 0.8 m s^{-1}. Calculate:

   a the speed and direction of the 1.0 kg ice puck after the collision,
   b the loss of kinetic energy due to the collision.

4 Bumper cars at fairgrounds are designed to withstand low-speed impacts without damage. A bumper car of mass 250 kg moving at a velocity of 0.9 m s^{-1} collides elastically with a stationary car of mass 200 kg. Immediately after impact, the 250 kg car has a velocity of 0.1 m s^{-1} in the same direction as it was initially moving.

   a i Calculate the velocity of the 200 kg car immediately after the impact.
      ii Show that the collision is elastic.
   b i What is the relative velocity of approach?
      ii What is the relative velocity of separation?

4.5 Explosions

When two objects fly apart after being initially at rest, they recoil from each other with equal and opposite amounts of momentum.

Figure 1 *One false move and the rider and the skateboard will fly apart*

Consider Figure 2 where a trolley of mass m_A and a trolley of mass m_B, initially at rest, move away at speeds v_A and v_B respectively.

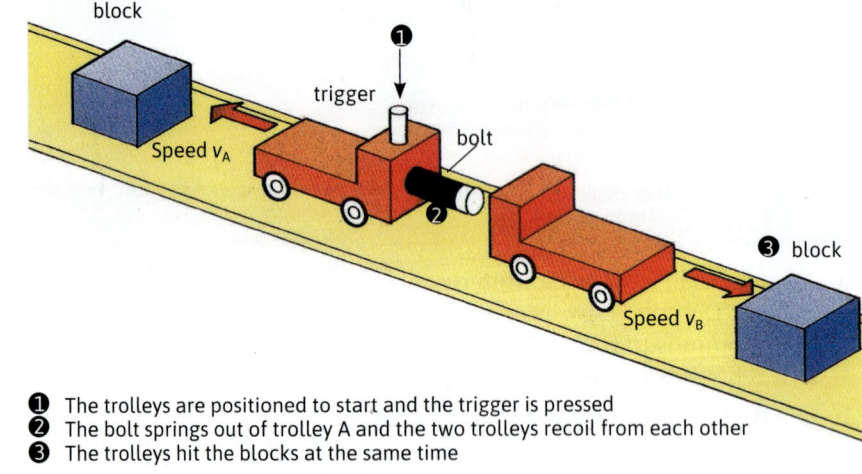

❶ The trolleys are positioned to start and the trigger is pressed
❷ The bolt springs out of trolley A and the two trolleys recoil from each other
❸ The trolleys hit the blocks at the same time

Figure 2 *Flying apart*

The total initial momentum = 0

The total momentum immediately after the explosion = momentum of A + momentum of B

$$= m_A v_A + m_B v_B$$

Using the principle of conservation of momentum, $m_A v_A + m_B v_B = 0$

$$\therefore m_B v_B = -m_A v_A$$

The minus sign means that the two masses move away from each other in opposite directions. For example, if $m_A = 1.0\,\text{kg}$, $v_A = 2\,\text{m s}^{-1}$ and $m_B = 0.5\,\text{kg}$, then $v_B = \dfrac{-m_A v_A}{m_B} = -4.0\,\text{m s}^{-1}$. So A and B move away at speeds of $2\,\text{m s}^{-1}$ and $4\,\text{m s}^{-1}$ in opposite directions.

Testing a model explosion

In Figure 2, when the spring is released from one of the trolleys, the two trolleys, A and B, push each other apart. The bricks are positioned so that the trolleys hit the bricks at the same moment. The distance travelled by each trolley to the point of impact with the brick is equal to its speed multiplied by the time taken to travel that distance. Because the time taken is the same for the two trolleys, the distance ratio is the same as the speed ratio. Because the trolleys have equal (and opposite) amounts of momentum, the ratio of their speeds is the inverse of the mass ratio. The distance ratio should therefore be equal to the inverse of the mass ratio. In other words, if trolley A travels twice as far as trolley B, then the mass of A must be half the mass of B (so that they carry away equal amounts of momentum).

Summary test 4.5

1 A shell of mass 2.0 kg is fired at a speed of 140 m s^{-1} from an artillery gun of mass 800 kg. Calculate the recoil velocity of the shell.

2 In a laboratory experiment to measure the mass of an object X, two identical trolleys A and B, each of mass 0.50 kg, were initially stationary on a track. Object X was fixed to trolley A. When a trigger was pressed, the two trolleys moved apart in opposite directions at speeds of 0.30 m s^{-1} and 0.25 m s^{-1}.

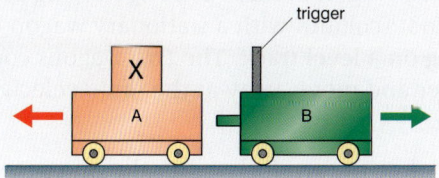
trigger

Figure 3

a Which of the two speeds given above was the speed of trolley A? Give a reason for your answer.

b Show that the mass of X must have been 0.10 kg.

3 Two trolleys, X of mass 1.2 kg and Y of mass 0.8 kg, are initially stationary on a level track.

a When a trigger is pressed on one of the trolleys, a spring pushes the two trolleys apart. Trolley Y moves away at a velocity of 0.15 m s^{-1}.
  i Calculate the velocity of X.
  ii Calculate the total kinetic energy of the two trolleys immediately after the impact.

b If the test had been carried out with trolley X held firmly, calculate the speed at which Y would have recoiled, assuming that the energy stored in the spring before release is equal to the total kinetic energy calculated in **a ii**.

4 A person in a stationary boat of total mass 150 kg throws a rock of mass 2.0 kg out of the boat. As a result, the boat recoils at a speed of 0.12 m s^{-1}. Calculate:

a the speed at which the rock was thrown from the boat,

b the kinetic energy gained by:
  i the boat, ii the rock.

Chapter summary

- Momentum = mass × velocity

- Force = $\dfrac{\text{change of momentum}}{\text{time taken}}$

- The principle of conservation of momentum states that when two or more bodies interact, the total momentum is unchanged, provided no external forces act on the bodies.

- An elastic collision is one in which the total kinetic energy after the collision is the same as before the collision. For such a collision between two objects, the relative speed of approach is equal to the relative speed of separation.

- A totally inelastic collision is one in which the colliding objects stick together.

- In an explosion where two objects fly apart, the two objects carry away equal and opposite momentum.

┌───┐
│ 🗐 Launch additional digital resources for the chapter │
└───┘

1 An object of mass 7.0 kg, initially at rest, is acted on by a force of 14 N for 10 s. Calculate:

   a the gain of momentum of the object,

   b the velocity of the object after 10 s.

2 A vehicle of mass 600 kg travelling at a velocity of 15 m s^{-1} is acted on by a braking force of 150 N. Calculate:

   a the momentum of the object at 15 m s^{-1},

   b the time taken to stop the object.

3 An object of mass 5.0 kg, initially at rest, is acted on by a force of 8.0 N for 12 s and is then brought to rest by a different force in 20 s.

   a Show that the change of momentum of the object due to the 8.0 N force is 96 N s.

   b Calculate the speed of the object after 12 s.

   c Calculate the force needed to bring the object to rest in 20 s.

4 A stream of identical atoms of mass 1.0×10^{-25} kg hit a vertical surface normally at a speed of 750 m s^{-1} at a rate of 2000 atoms per second. The atoms stick to the surface on impact.

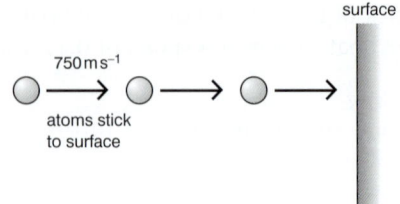

surface

750 m s^{-1}

atoms stick to surface

Figure 4.1

Calculate:

   a the loss of momentum of a single atom,

   b the average force of the atoms on the surface.

5 A molecule of mass 2.5×10^{-26} kg moving at a speed of 520 m s^{-1} collides normally with a wall and rebounds normally without loss of speed.

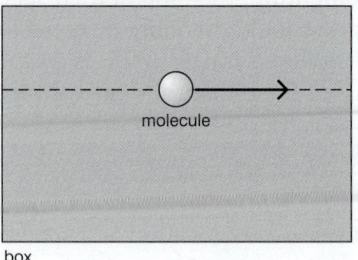

molecule

box

Figure 5.1

   a Calculate the change of momentum of the molecule.

   b The molecule is in a rectangular box and collides repeatedly with the same side of the box every 2.0 ms. Calculate the average force of impact of the molecule on the box.

6 A railway wagon of mass 2500 kg moving at a speed of 2.4 m s^{-1} collides with a stationary wagon of mass 1500 kg on a level track. The two wagons couple together and move away at the same velocity after the impact.

   a Calculate their velocity after the impact.

   b i Calculate the loss of kinetic energy due to the impact.
     ii Discuss the energy changes that take place as a result of the impact.

7 In a road accident, a van of mass 1500 kg moving at a speed of 28 m s^{-1} runs into the back of a car of mass 900 kg moving in the same direction at a speed of 11 m s^{-1}. As a result of the impact, the car is pushed forward at a speed of 18 m s^{-1}.

   a Calculate the velocity of the van immediately after the impact.

   b Calculate:
     i the loss of kinetic energy of the van,
     ii the gain of kinetic energy of the car,
     iii the total change of kinetic energy of the two vehicles.

   c Discuss the effect of the impact on a person in the car.

8 In a radioactive decay, a nucleus of mass 4.0×10^{-25} kg, initially at rest, emitted an α-particle of mass 6.7×10^{-27} kg with a velocity of 1.5×10^7 m s^{-1}.

a Calculate the velocity of recoil of the nucleus.

b Calculate the kinetic energy of:
i the recoil nucleus,
ii the α-particle.

9 A railway wagon P of mass 15 000 kg moving at a speed of 1.8 m s^{-1} collides on a level track with a 25 000 kg wagon Q, which then moves away from wagon P. Figure 9.1 shows how the velocity of Q changes during the impact.

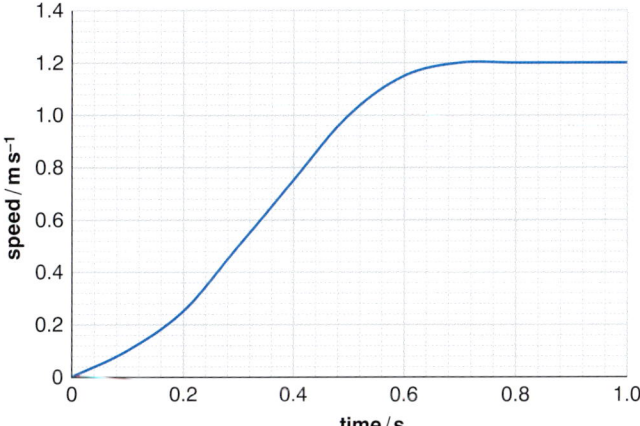

Figure 9.1

a Estimate the duration of the impact and hence calculate the force of the impact.

b i State the principle of conservation of momentum.
ii Determine the velocity of P immediately after the impact.

c i State what is meant by a *perfectly elastic collision*.
ii Calculate the change of kinetic energy of each truck due to the impact.
iii Discuss whether or not the collision is elastic.

10 A radioactive isotope emits α-particles which all have the same initial kinetic energy. When an α-particle is emitted by a stationary nucleus, the nucleus and the α-particle move away from each other in opposite directions at different speeds.

a Explain, in terms of momentum, why the nucleus and the α-particle move away from each other:
i in opposite directions,
ii at different speeds.

b The nuclei of the polonium isotope $^{210}_{94}$Po emits α-particles of kinetic energy 8.5×10^{-13} J. Each such nucleus changes into a nucleus of a lead isotope when it emits an α-particle. The process is represented in Figure 10.1.

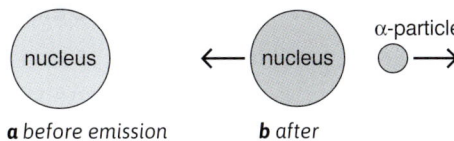

a *before emission*　　**b** *after*

Figure 10.1

i Calculate the speed of an α-particle with this kinetic energy.
The mass of an α-particle = 6.6×10^{-27} kg.
ii The mass of the lead nucleus is 3.4×10^{-25} kg. Calculate the speed and the kinetic energy of the nucleus immediately after the α-particle has been emitted.

c The α-particles from this polonium isotope have a range in air of 39 mm at atmospheric pressure.
i Calculate the time taken for an α-particle to travel this distance.
ii Estimate the average deceleration of an α-particle over this distance.

11 A rubber band is stretched between two frictionless trolleys A and B in fixed positions on a horizontal surface, as shown in Figure 11.1. The trolleys, which have different masses, are then released at the same time so that they move towards each other until they collide. The mass of A is greater than the mass of B.

Figure 11.1

a i Immediately before the trolleys collide, explain why the two trolleys have equal and opposite momentum.
ii Explain why the two trolleys have different speeds immediately before they collide.

b State and explain which trolley has the greater kinetic energy immediately before the collision.

5.1 Density

Learning outcomes

On these pages you will learn to:

- define density and use the density equation $\rho = m/V$ in calculations
- measure the density of liquids and regular and irregular solids

Table 1 Density

Substance	Density/kg m^{-3}
Air	1.2
Aluminium	2700
Copper	8900
Gold	19300
Hydrogen	0.083
Iron	7900
Lead	11300
Oxygen	1.3
Silver	10500
Water	1000

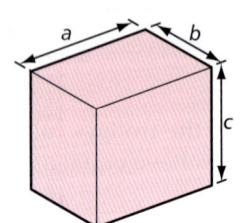

i Volume of cuboid = $a \times b \times c$

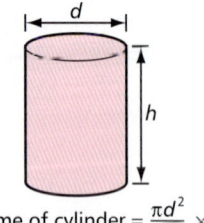

ii Volume of cylinder = $\dfrac{\pi d^2}{4} \times h$

Figure 1 Volume formulae

Density and its measurement

Lead is much more dense than aluminium. Sea water is more dense than tap water. To see how dense one substance is compared with another, we need to measure the mass of equal volumes of the two substances. The substance with the greater mass in the same volume is more dense. For example, a lead sphere of volume $1 \, \text{cm}^3$ has a mass of $11.3 \, \text{g}$; whereas an aluminium sphere of the same volume has a mass of $2.7 \, \text{g}$, so lead is more dense than aluminium.

The density of a substance is defined as its mass per unit volume.

For a certain amount of a substance of mass m and volume V, its density, ρ (pronounced 'rho'), may be calculated using the equation

$$\text{density, } \rho = \frac{m}{V}$$

- The unit of density is the **kilogram per metre^3 (kg m^{-3})**.
- Rearranging the above equation gives: $m = \rho V$

$$\text{or } V = \frac{m}{\rho}$$

More about units

Mass	$1 \, \text{kg} = 1000 \, \text{g}$
Length	$1 \, \text{m} = 100 \, \text{cm} = 1000 \, \text{mm}$
Volume	$1 \, \text{m}^3 = 10^6 \, \text{cm}^3$
Density	$1000 \, \text{kg m}^{-3} = \dfrac{10^6 \, \text{g}}{10^6 \, \text{cm}^3} = 1 \, \text{g cm}^{-3}$

Table 1 shows the density of some common substances in kg m^{-3}. You can see that gases are much less dense than solids or liquids.

Worked example

Using the data above, calculate:

a the mass, in kilograms, of a piece of aluminium of volume $3.6 \times 10^{-5} \, \text{m}^3$,

b the volume, in m^3, of a mass of $0.50 \, \text{kg}$ of iron.

Solution

a $\rho = 2700 \, \text{kg m}^{-3}$; mass $m = \rho V = 2700 \, \text{kg m}^{-3} \times 3.6 \times 10^{-5} \, \text{m}^3 = 9.7 \times 10^{-2} \, \text{kg}$

b $\rho = 7900 \, \text{kg m}^{-3}$; volume $= \dfrac{m}{\rho} = \dfrac{0.50 \, \text{kg}}{7900 \, \text{kg m}^{-3}} = 6.3 \times 10^{-5} \, \text{m}^3$

Density measurements

An unknown substance can often be identified if its density is measured and compared with the density of known substances. The following procedures may be used to measure the density of a substance.

- **A regular solid.** Measure its mass using a top-pan balance; measure its dimensions using Vernier calipers or a micrometer. Calculate its volume using the appropriate formula (e.g. volume of a sphere of radius r is $\frac{4}{3}\pi r^3$; volume of a cylinder of radius r and length L is $\pi r^2 L$). Calculate the density from $\dfrac{\text{mass}}{\text{volume}}$.

- **A liquid.** Measure the mass of an empty measuring cylinder. Fill the cylinder with the liquid and measure the volume of the liquid directly. Measure the mass of the cylinder and liquid to enable the mass of the liquid to be calculated. Calculate the density from $\dfrac{\text{mass of liquid}}{\text{volume}}$.

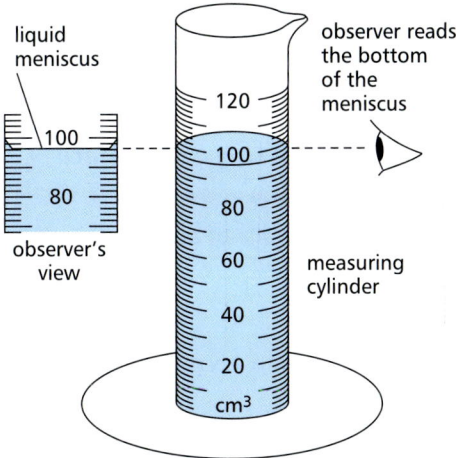

Figure 2 *Using a measuring cylinder*

- **An irregular solid.** Measure the mass of the object. Fill a displacement can with water up to the spout, as shown in Figure 3. Place a beaker of known mass under the spout. Immerse the object on a thread in the liquid and collect the overflow. Measure the mass of the beaker and overflow. Hence determine the mass of the overflow water and calculate its volume, given that the density of water is $1000 \, \text{kg m}^{-3}$.

Calculate the density of the object from

$$\dfrac{\text{its mass}}{\text{the overflow volume}}.$$

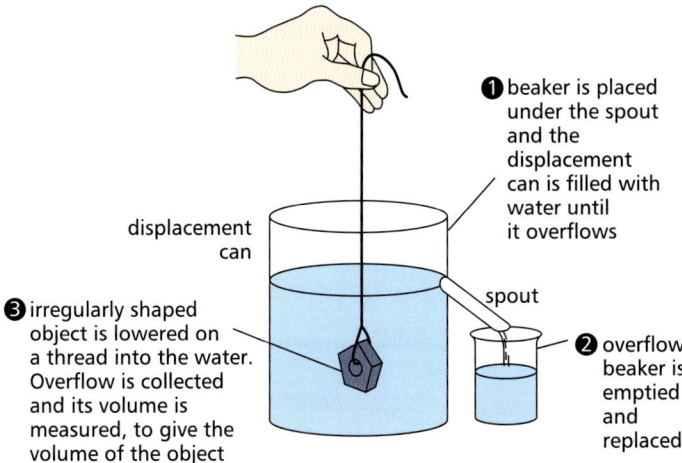

❶ beaker is placed under the spout and the displacement can is filled with water until it overflows

❷ overflow beaker is emptied and replaced

❸ irregularly shaped object is lowered on a thread into the water. Overflow is collected and its volume is measured, to give the volume of the object

Figure 3 *Measuring the volume of an irregularly shaped object*

Density of alloys

An **alloy** is a solid mixture of two or more metals. For example, **brass** is an alloy of copper and zinc which has good resistance to corrosion and wear.

If an alloy consists of two metals A and B, then for volume V of the alloy:

- If the volume of metal A is v_A, the mass of metal A is $\rho_A v_A$, where ρ_A is the density of metal A.
- If the volume of metal B is v_B, the mass of metal B = $\rho_B v_B$, where ρ_B is the density of metal B.

$$\therefore \qquad \textbf{Mass of the alloy,}\ m = \rho_A v_A + \rho_B v_B$$

Hence the density of the alloy, $\rho = \dfrac{m}{V} = \dfrac{\rho_A v_A + \rho_B v_B}{V}$

$$= \dfrac{\rho_A v_A}{V} + \dfrac{\rho_B v_B}{V}$$

Worked example

A brass object consists of $3.3 \times 10^{-5} \, \text{m}^3$ of copper and $1.7 \times 10^{-5} \, \text{m}^3$ of zinc. Calculate the mass and the density of this object. The density of copper is $8900 \, \text{kg m}^{-3}$; the density of zinc is $7100 \, \text{kg m}^{-3}$.

Solution

Mass of copper = density of copper × volume of copper
$$= 8900 \, \text{kg m}^{-3} \times 3.3 \times 10^{-5} \text{m}^3 = 0.29 \, \text{kg}$$

Mass of zinc = density of zinc × volume of zinc
$$= 7100 \, \text{kg m}^{-3} \times 1.7 \times 10^{-5} \text{m}^3 = 0.12 \, \text{kg}$$

Total mass, $m = 0.29 + 0.12 = 0.41 \, \text{kg}$

Total volume, $V = 5.0 \times 10^{-5} \text{m}^3$

Density of alloy, $\rho = \dfrac{m}{V} = \dfrac{0.41 \, \text{kg}}{5.0 \times 10^{-5} \text{m}^3} = 8200 \, \text{kg m}^{-3}$

Summary test 5.1

1 A rectangular brick of dimensions $5.0 \, \text{cm} \times 8.0 \, \text{cm} \times 20.0 \, \text{cm}$ has a mass of $2.5 \, \text{kg}$. Calculate:
   a its volume,
   b its density.

2 An empty paint tin of diameter $0.150 \, \text{m}$ and of height $0.120 \, \text{m}$ has a mass of $0.22 \, \text{kg}$. It is filled with paint to within $7 \, \text{mm}$ of the top. Its total mass is then $6.50 \, \text{kg}$. Calculate, for the paint in the tin:
   a the mass,
   b the volume,
   c the density.

3 A solid steel cylinder has a diameter of $12 \, \text{mm}$ and a length of $85 \, \text{mm}$. Calculate:
   a its volume (in m^3),
   b its mass (in kg); the density of steel is $7800 \, \text{kg m}^{-3}$.

4 An alloy tube of volume $1.8 \times 10^{-4} \text{m}^3$ consists, by volume, of 60% aluminium and 40% magnesium.
   a Calculate the mass, in the tube, of:
      i aluminium,
      ii magnesium.
   b Calculate the density of the alloy; the density of aluminium is $2700 \, \text{kg m}^{-3}$; the density of magnesium is $1700 \, \text{kg m}^{-3}$.

Learning outcomes

On these pages you will learn to:

- define and use the pressure equation $p = F/A$
- derive and use the equation $\Delta p = \rho g \Delta h$
- describe how to measure gas pressure
- use $\Delta p = \rho g \Delta h$ to compare the densities of two different liquids

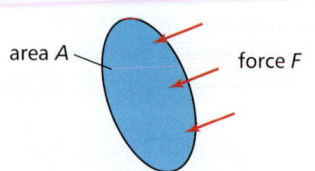

Figure 1 Pressure

Figure 2 A snowmobile

In climates where snow falls, caterpillar tracks on snow vehicles allow the vehicle to travel across snow without sinking into the snow. The tracks have a much greater contact area with the ground than ordinary tyres. Therefore, the pressure of the tracks on snow is much less than the pressure would be if the vehicle was fitted with tyres.

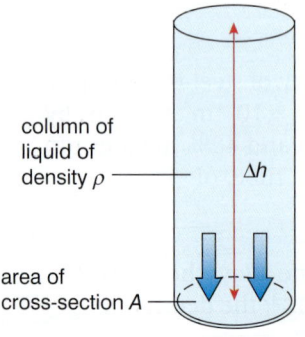

Figure 3 Calculating liquid pressure

Pressure and force

Lots of people need to measure **pressure**. For example, nurses measure blood pressure, motor vehicle technicians measure tyre pressure and gas engineers measure the pressure of the gas supply. In these examples, a liquid or a gas inside a container presses on the container surface wherever the liquid or gas is in contact with the surface. An example of pressure due to a solid is where a brick rests on a surface. The weight of the brick presses down on the surface.

Pressure is defined as the force per unit area acting on a surface perpendicular to the surface.

- The pressure of a force F acting at right angles to a surface of area A is given by the equation

$$\text{pressure, } p = \frac{F}{A}$$

- The unit of pressure is the **pascal (Pa)**, which is equal to $1\,\text{N m}^{-2}$.

Solid pressure

When a force acts on an object, the smaller the area over which the force acts, the greater the pressure of the force on a surface.

Liquid pressure

For any fluid at rest:

- the pressure at any point acts equally in all directions,
- the pressure increases with depth.

The **upthrust** on an object in a fluid is due to the increase of pressure with depth in the liquid. Imagine a ball in a liquid. The average force due to the liquid pressure on its lower half acts upwards. This is greater than the average force on its upper half which acts downwards. The ball therefore experiences an upward resultant force or upthrust due to the pressure of the liquid.

The greater the density of a liquid is, the greater its pressure is at any given depth. Consider the column of liquid in the container shown in Figure 3. The pressure caused by the liquid column on the bottom of the container is due to the weight of the liquid.

For a column of height Δh and area of cross-section A:

- the volume of liquid in the container $= A\Delta h$,
- the mass of liquid = its density × its volume $= \rho A\Delta h$ where ρ is the density of the liquid,
- the weight of the liquid = mass × g = $\rho A\Delta h g$ where the **gravitational field strength** $g = 9.81\,\text{N kg}^{-1}$.

The pressure Δp at the base of the liquid column $= \dfrac{\text{weight}}{\text{area of cross-section } A}$

$$\Delta p = \frac{\rho g A\Delta h}{A} = \rho g \Delta h$$

Worked example

$g = 9.81\,\text{N kg}^{-1}$

Calculate the pressure due to sea water of density $1050\,\text{kg m}^{-3}$ at a depth of $200\,\text{m}$.

Solution
pressure $\Delta p = \rho g \Delta h = 1050\,\text{kg m}^{-3} \times 9.81\,\text{N kg}^{-1} \times 200\,\text{m} = 2.1 \times 10^{6}\,\text{Pa}$

Gas pressure

The pressure of a gas is due to molecules repeatedly undergoing elastic collisions with the surface of the container. The pressure of the impacts of the gas molecules on the surface is constant, assuming that the temperature of the gas is constant, and that there are a large number of gas molecules in the container.

The pressure of a gas in a container can be measured using an electronic pressure gauge or a U-tube manometer, as shown in Figure 4. The pressure of the gas forces the liquid in the manometer up the open side of the U-tube until it is at rest.

Because atmospheric pressure acts on the liquid on the 'open' side of the manometer, the gas pressure = the pressure due to the difference in liquid levels ($h\rho g$) + atmospheric pressure, where ρ is the liquid density.

- The 'excess' pressure of the gas (i.e. its pressure above atmospheric pressure) is equal to the pressure due to the difference in the level of liquid on each side.

Therefore, the pressure of the gas relative to atmospheric pressure = $h\rho g$

Atmospheric pressure varies slightly from day to day, changing with the local weather conditions. The mean pressure of the atmosphere at sea level is 101 kPa. Atmospheric pressure decreases with height. The Earth's atmosphere extends more than 100 km into space.

Comparison of liquid densities

The pressure of a liquid depends on its density. The densities of two immiscible liquids, such as oil and water, can be compared using the U-tube arrangement shown in Figure 5. The arrangement is set up by pouring water into one side of the U-tube and then when the water has settled carefully pouring oil into the other side. When the liquids settle, the oil level is higher than the water level on the other side. This is because the pressure is equal at the same level in both columns. So the pressure at the bottom of the column of oil (X in Figure 5) is equal to the pressure at the same level of the water column (Y in Figure 5).

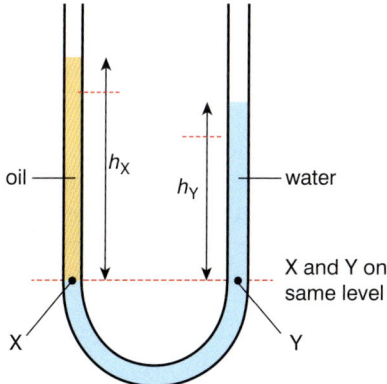

Figure 5 Comparing densities

To compare the densities of the two liquids, the length of each column is measured from the same level as the bottom of the oil column. Since the pressure of each measured column is the same, then for columns of height h_X for the oil and h_Y for the water,

$$\rho_X g h_X = \rho_Y g h_Y$$

Therefore:

$$\frac{\rho_X}{\rho_Y} = \frac{h_Y}{h_X}$$

Given that the density of water is $1000\,\text{kg}\,\text{m}^{-3}$ and knowing the measured values for h_X and h_Y, the density of the oil can be calculated.

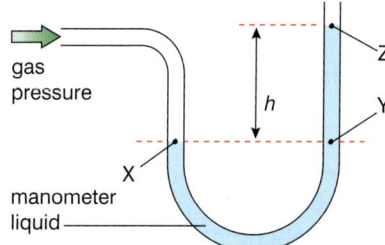

Figure 4 The U-tube manometer

Summary test 5.2

$g = 9.81\,\text{m}\,\text{s}^{-2}$

1 Calculate the pressure exerted by a paving stone of density $2500\,\text{kg}\,\text{m}^{-3}$ and of dimensions $0.80\,\text{m} \times 0.80\,\text{m} \times 0.05\,\text{m}$ when the stone rests flat on a smooth horizontal surface.

2 Calculate the force due to the atmosphere on the panes of a sealed double-glazed window of dimensions $1.5\,\text{m} \times 0.80\,\text{m}$ which has a vacuum between the two panes. The mean value of atmospheric pressure = 101 kPa.

3 The pressure in the tyres of a vehicle is 280 kPa. The contact area of each of the four tyres on the ground is $0.012\,\text{m}^2$. Calculate the weight of the vehicle.

4 The mean value of atmospheric pressure is 101 kPa.

   a The density of water is $1000\,\text{kg}\,\text{m}^{-3}$. Calculate the depth of water that will give a pressure of 101 kPa.

   b Domestic gas is supplied to homes at a pressure which is normally 2.5% above atmospheric pressure.

   A gas engineer used a water-filled U-tube manometer to measure the gas pressure at a house and observed a difference of 0.15 m between its levels.

   i Calculate the pressure of the gas supply in this house.

   ii State and explain whether or not the gas pressure at the house is normal.

5.3 Upthrust

Learning outcomes

On these pages you will learn to:

- describe and explain what is meant by an upthrust on an object in a fluid
- explain the cause of an upthrust
- Calculate the upthrust on an object in a fluid
- explain whether or not an object in a fluid sinks or floats

When you go swimming, have you noticed that you feel lighter in the water? People with mobility problems often find it much easier to move in water than in air. Water exerts an upward force on the body. This force is called an **upthrust**.

Investigating upthrust

Use a newton meter to weigh a metal object in air.

- Repeat the test by weighing the same object when it is completely in the water.

You should find that the newton meter reading is less when the object is in water. This is because when the metal object is in the water, it experiences an upthrust. The difference between the two newton meter readings is equal to the upthrust on the object.

Repeat the test with the same object only partly immersed in the water. You should find that the newton meter reading is in between the two earlier readings. This is because the upthrust is less when the object is only partly immersed in the water.

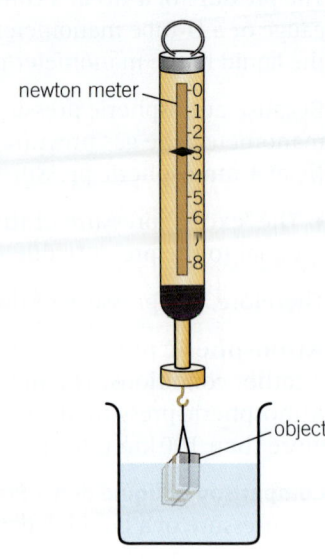

Figure 1 *Measuring an upthrust*

Explaining upthrust

The water level in a water container rises when an object is lowered into the water. This is because the object **displaces** some of the water.

- The more the object is lowered into the water, the bigger the volume of water displaced, and the bigger the upthrust.
- When the object is fully immersed, the volume of water displaced is equal to the volume of the object.

Figure 2 shows a cylinder fully immersed in water. Because pressure increases with depth, the pressure of the water at the bottom of the cylinder is greater than the pressure on the top of the cylinder. So the upward force of the water on the bottom of the cylinder is greater than the downward force of the water on the top of the cylinder. The upthrust is the resultant of these two forces.

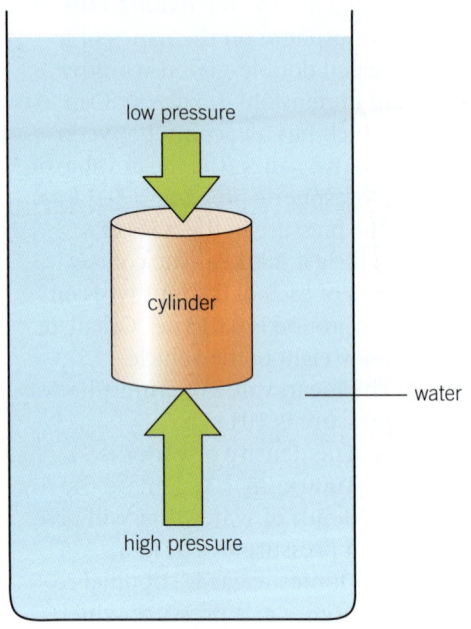

Figure 2 *Explaining upthrust*

Float or sink?

A ship being loaded with cargo will float lower and lower in the water as the load is increased. The ship displaces more water when the load increases, so the upthrust increases. At any instant, the upthrust on the boat is equal to the weight of the ship and its cargo. If the ship is loaded too much, it sinks because it has displaced as much water as it possibly can, and because the upthrust cannot support the total weight.

An object floats when its weight is equal to the upthrust.

An object sinks when its weight is greater than the upthrust.

Figure 3 *A loaded ship*

The upthrust equation

In Figure 2, because the cylinder is vertical, the pressure difference Δp between the top and the bottom of the cylinder $= \rho g \Delta h$, where Δh is equal to the length of the cylinder, ρ is the density of the liquid, and $g = 9.81\,\text{N kg}^{-1}$.

Therefore the upthrust on the cylinder is the pressure difference $\Delta p \times$ the cylinder's cross-sectional area:

$$(\rho g \Delta h) \times A = \rho g \Delta h A = \rho g V$$

where $V = \Delta h \times A =$ the cylinder volume.

For any object in a fluid:

the upthrust on the object $= \rho g V$

where ρ is the density of the fluid and V is the volume of fluid displaced by the object.

Notes

1 The upthrust equation applies to any object wholly or partially immersed in any fluid provided V is the volume of fluid displaced by the object.

2 Because mass $m =$ density $\rho \times$ volume V, the upthrust $= \rho V g = mg$, where m is the mass of liquid displaced by the cylinder. Hence **the upthrust is equal to the weight of liquid displaced by the cylinder**. This statement is known as **Archimedes' principle**.

Worked example

A raft in water floats with its base horizontal at a depth d of 0.22 m below the water line. The raft has a base area A of $1.5\,\text{m}^2$.

a Calculate the volume of water displaced by the raft.

b The density ρ of water is $1000\,\text{kg m}^{-3}$. Calculate the upthrust on the raft and hence determine its weight.

Solution

a Volume of water displaced $V = dA = 0.22\,\text{m} \times 1.5\,\text{m}^2 = 0.33\,\text{m}^3$

b Upthrust = weight of water displaced

$= \rho g V = 1000\,\text{kg m}^{-3} \times 9.81\,\text{N kg}^{-1} \times 0.33\,\text{m}^3 = 3240\,\text{N}$

Since the raft is floating, its weight is equal to the upthrust on it; hence its weight is 3230 N.

Density tests

Objects made of material such as cork or wood float in water, but metal objects sink. Objects that sink have a density greater than that of water. Objects that float have a density less than that of water. By observing if an object floats or sinks in water, you can tell if its density is less or greater than that of water.

An object that is more dense than water sinks because its weight is greater than the weight of water it displaces. So when it is fully immersed, it sinks because its weight is greater than the upthrust on it.

An object that is less dense than water floats because its weight is equal to the upthrust on it. This is because the density of the water is greater than that of the object so it displaces just enough water for the upthrust to be equal to the weight of the object.

Summary test 5.3

1 **a** Explain why it is difficult to hold an inflated plastic ball under water.

   b Explain why cork is a suitable material for filling a life belt.

2 When an object is weighed using a newton meter, the reading on the newton meter is 5.20 N when the object is in air, and 3.3 N when the object is totally immersed in water.

   a Explain why the reading on the newton meter is less when the object is in water.

   b i Calculate the upthrust on the object and hence determine its volume.

   ii Show that the density of the object is about $2700\,\text{kg m}^{-3}$.

3 Three blocks A, B, and C are released in a bowl of a water.

   Block A sinks to the bottom.

   Block B floats with its top half above the water.

   Block C floats with a small proportion above the water.

   a Which object has the greatest density? Give a reason for your answer.

   b Which object has the lowest density? Give a reason for your answer.

4 Figure 4 shows a weighted test tube floating vertically in water.

   a Make a reasoned prediction about how the length of tube above the water depends on the total weight of the tube.

   b Design an experiment to test your prediction, using the test tube and any other apparatus necessary.

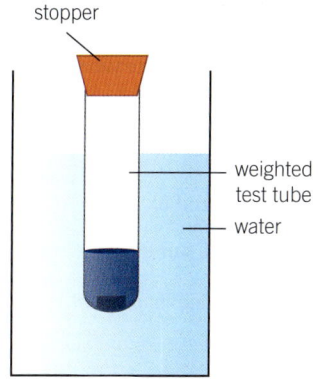

stopper

weighted test tube

water

Figure 4

Springs

Learning outcomes

On these pages you will learn to:

- state Hooke's law for springs and explain what is meant by the elastic limit and the spring constant of a spring
- use Hooke's law to solve problems involving springs, including springs in parallel and in series
- calculate the energy of a stretched spring

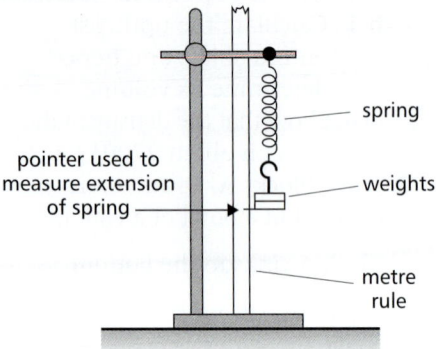

a Testing the extension of a spring

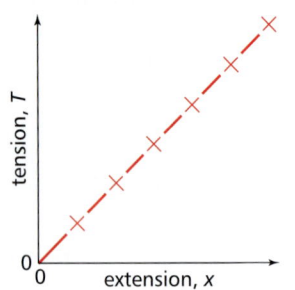

b Hooke's law

Figure 1

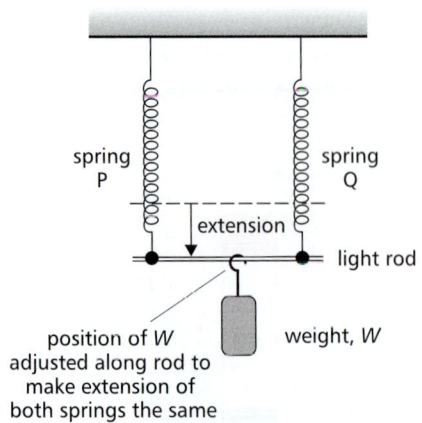

Figure 2 Two springs in parallel

Hooke's law

A stretched spring exerts a pull on the object holding each end of the spring. This pull, referred to as the **tension** in the spring, is equal and opposite to the force needed to stretch the spring. The more a spring is stretched, the greater the tension in it. Figure 1a shows a stretched spring at rest supporting a **load** consisting of some weights. This arrangement may be used to investigate how the tension in a spring depends on its extension from its unstretched length. The measurements may be plotted on a graph of tension v. extension (Figure 1b). The graph shows that the force needed to stretch a spring is proportional to the extension of the spring. This is known as **Hooke's law**, after Robert Hooke, a 17th-century scientist.

> **Hooke's law states that the force needed to stretch a spring is proportional to the extension of the spring from its natural length.**

Hooke's law may be written as:

$$\text{Force, } F = kx$$

where k is the spring constant (sometimes referred to as the stiffness constant) and x is the extension.

- The greater the value of k, the stiffer the spring is. The unit of k is N m^{-1}.
- The graph of F against x is a straight line of gradient k through the origin.
- If a spring is stretched beyond its **elastic limit**, it does not regain its initial length when the force applied to it is removed.
- A level maths students may meet Hooke's law in the form $F = \dfrac{\lambda x}{L}$ where L is the unstretched length of the spring and $\lambda \ (= kL)$ is the spring modulus. λ is not in the specification for A level physics.

Worked example

A vertical steel spring fixed at its upper end has an unstretched length of 300 mm. Its length is increased to 385 mm when a 5.0 N weight attached to the lower end is at rest. Calculate:

a the spring constant,

b the length of the spring when it supports an 8.0 N weight at rest.

Solution

a Use $F = kx$ with $F = 5.0\,\text{N}$ and $x = 385 - 300\,\text{mm} = 85\,\text{mm} = 0.085\,\text{m}$.

Therefore $k = \dfrac{F}{x} = \dfrac{5.0\,\text{N}}{0.085\,\text{m}} = 59\,\text{N m}^{-1}$

b Use $F = kx$ with $F = 8.0\,\text{N}$ and $k = 59\,\text{N m}^{-1}$ to calculate x:

$x = \dfrac{F}{k} = \dfrac{8.0\,\text{N}}{59\,\text{N m}^{-1}} = 0.136\,\text{m}$

Therefore the length of the spring $= 0.300\,\text{m} + 0.136\,\text{m} = 0.436\,\text{m}$

Springs in parallel

Figure 2 shows a weight W supported by means of two springs, P and Q, in parallel with each other. The extension, x, of each spring is the same. Therefore:

- The force needed to stretch P, $F_P = k_P x$
- The force needed to stretch Q, $F_Q = k_Q x$, where k_P and k_Q are the spring constants of P and Q respectively.

The weight W is supported by both springs, $W = F_P + F_Q = k_P x + k_Q x = k_{eff} x$ where the **effective spring constant, $k_{eff} = k_P + k_Q$**

Springs in series

Figure 3 shows a weight W supported by means of two springs joined end-on in 'series' with each other. The tension in each spring is the same and is equal to the weight W. Therefore:

- Extension of spring P, $x_P = \dfrac{W}{k_P}$

- Extension of spring Q, $x_Q = \dfrac{W}{k_Q}$, where k_P and k_Q are the spring constants of P and Q respectively.

$$\text{Total extension, } x = x_P + x_Q = \frac{W}{k_P} + \frac{W}{k_Q} = \frac{W}{k_{\text{eff}}}$$

where k_{eff}, the effective spring constant, is given by the equation $\dfrac{1}{k_{\text{eff}}} = \dfrac{1}{k_P} + \dfrac{1}{k_Q}$

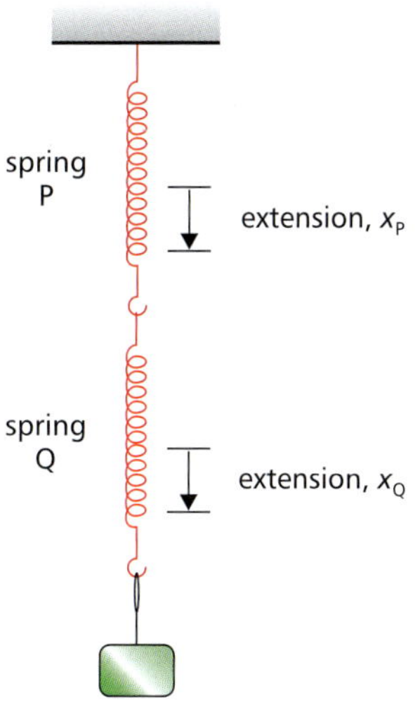

The energy stored in a stretched spring

Elastic potential energy is stored in a stretched spring. If the spring is suddenly released, the elastic energy stored in it is suddenly converted to kinetic energy of the spring. As explained in Topic 2.4, the work done to stretch a spring by extension x_0 from its unstretched length is $\frac{1}{2}F_0x_0$, where F_0 is the force needed to stretch the spring to extension x_0. The work done on the spring is stored as elastic potential energy. Therefore the elastic potential energy E_p in the spring is $\frac{1}{2}F_0x_0$. Also, since $F_0 = kx_0$, where k is the spring constant, $E_p = \frac{1}{2}kx_0^2$.

Elastic potential energy stored in a stretched spring, $E_p = \frac{1}{2}F_0x_0 = \frac{1}{2}kx_0^2$

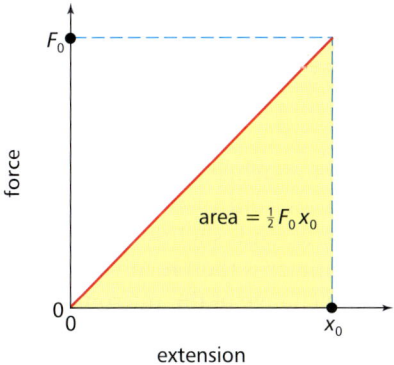

Figure 4 *Energy stored in a stretched spring*

Figure 3 *Two springs in series*

Summary test 5.4

$g = 9.81\,\text{m s}^{-2}$

1. A steel spring has a spring constant of $25\,\text{N m}^{-1}$. Calculate:
   a. the extension of the spring when the tension in it is equal to 10 N,
   b. the tension in the spring when it is extended by 0.50 m from its unstretched length.

2. Two identical steel springs of length 250 mm are suspended vertically side-by-side from a fixed point. A 40 N weight is attached to the ends of the two springs. The length of each spring is then 350 mm. Calculate:
   a. the tension in each spring,
   b. the extension of each spring,

   c. the spring constant of each spring.

3. Repeat questions **2a** and **b** for the two springs in 'series' and vertical.

4. An object of mass 0.150 kg is attached to the lower end of a vertical spring of unstretched length 300 mm, which is fixed at its upper end. With the object at rest, the length of the spring becomes 420 mm as a result. Calculate:
   a. the spring constant,
   b. the energy stored in the spring,
   c. the weight that needs to be added to extend the spring to 600 mm.

5.5 Deformation of solids

Learning outcomes

On these pages you will learn to:

- appreciate that deformation is caused by a force and it can be tensile or compressive
- define stress, strain and the Young modulus and carry out calculations involving these quantities
- describe an experiment to determine the Young modulus of a metal in the form of a wire
- use a stress–strain graph to calculate the Young modulus of a material
- distinguish between elastic and plastic deformation of a material
- compare the properties of different materials in terms of their stress–strain curves

Force and solid materials

Look around at different materials and think about the effect of force on each material. To stretch, twist, or compress the material, a pair of forces is needed. For example, stretching a rubber band requires the rubber band to be pulled by a force at either end. Some materials, such as rubber, bend or stretch easily. The **elasticity** of a solid material is its ability to regain its shape after it has been deformed or distorted, and the forces that deformed it have been removed. An object that regains its shape after being deformed is said to have undergone **elastic deformation**. Deformation that stretches an object is **tensile**, whereas deformation that compresses an object is **compressive**.

The arrangement shown in Figure 1a in Topic 5.4 may be used to test different materials to see how easily they stretch. In each case, the material is held at its upper end and loaded by hanging weights at its lower end. The position of the pointer is measured as the weight of the load is increased in steps then decreased to zero. The extension of the strip of material at each step is its increase in length from its unloaded length. The tension in the material is equal to the weight. The measurements may be plotted as a graph of extension v. weight, as shown in Figure 1.

- A steel spring gives a straight line, in accordance with Hooke's law (see Topic 5.4).
- A rubber band at first extends easily when it is stretched. However, it becomes 'fully stretched' and very difficult to stretch further when it has been lengthened considerably.
- A polythene strip 'gives' and stretches easily after its initial stiffness is overcome. However, after 'giving', it extends little and becomes difficult to stretch.

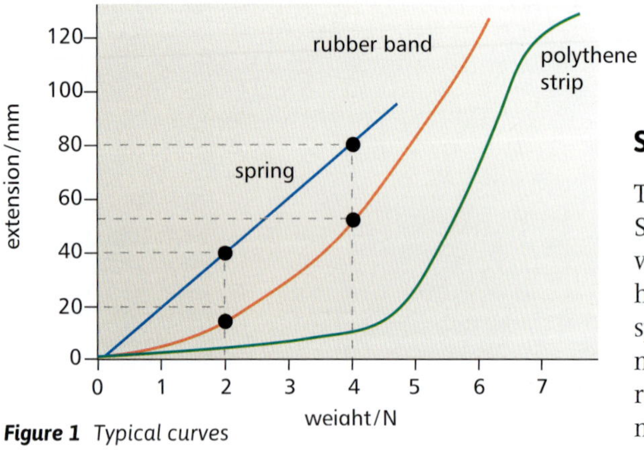

Stress and strain

The extension of a wire under tension may be measured using Searle's apparatus (Figure 2). A micrometer attached to the control wire is adjusted so the spirit level between the control and test wire is horizontal. When the test wire is loaded, it extends slightly causing the spirit level to drop on one side. The micrometer is then readjusted to make the spirit level horizontal again. The change of the micrometer reading is therefore equal to the extension. The extension may be measured for different values of tension by increasing the load (i.e. the test weights) in steps. At each step, the tension is equal to the total weight of the load.

For a wire of length L and area of cross-section A under tension:

- The **tensile stress** in the wire, $\sigma = \dfrac{F}{A}$, where F is the tension.

  The unit of stress is the **pascal (Pa)**, equal to $1\,\mathrm{N\,m^{-2}}$.

- The **tensile strain** in the wire, $\varepsilon = \dfrac{x}{L}$, where x is the extension of the wire (i.e. change of length, ΔL). Strain is a ratio and therefore has **no unit**.

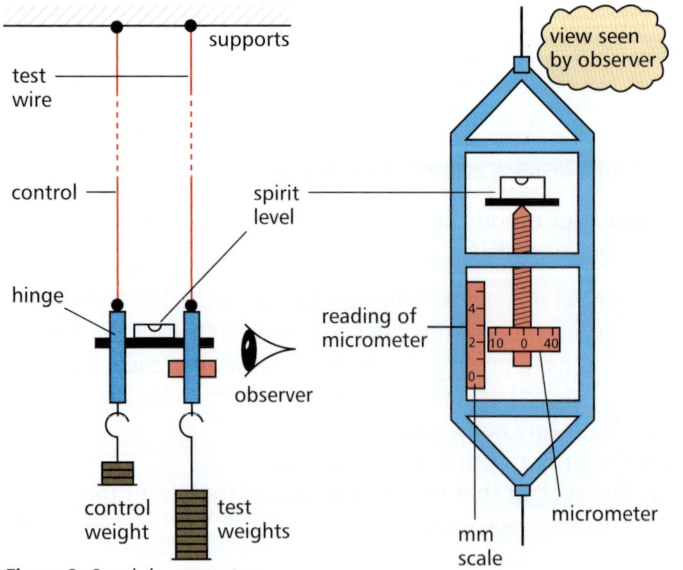

Figure 1 Typical curves

Figure 2 Searle's apparatus

> ### Note
>
> To calculate the stress in the wire, the diameter d of the wire should be measured then used in the formula $A = \frac{1}{4}\pi d^2$ to calculate A. The stress can then be calculated using $\sigma = F/A$.

Stress strain graphs

Figure 3 shows how the stress in a wire varies with strain.

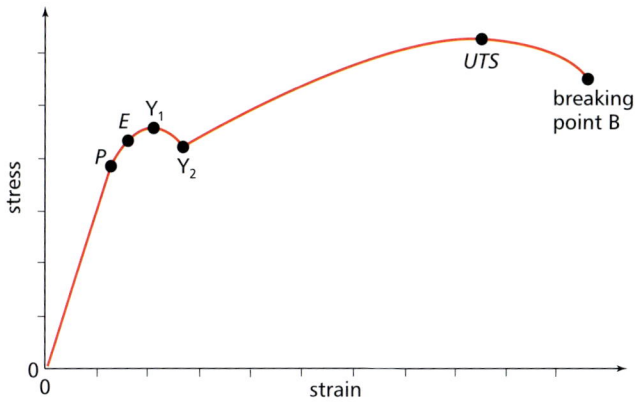

Figure 3 *Stress v. strain for a metal wire*

- **From 0 to the limit of proportionality P**, the stress is proportional to the strain.

The value of stress/strain is a constant, known as the **Young modulus** of the material:

$$\text{Young modulus, } E = \frac{\text{stress, } \sigma}{\text{strain, } \varepsilon} = \frac{\dfrac{F}{A}}{\dfrac{x}{L}} = \frac{FL}{Ax}$$

The unit of E is the pascal (Pa).

Because the line OP is straight and it passes through the origin, the gradient of this section is equal to stress ÷ strain. Hence the Young modulus is equal to the gradient of OP.

To determine the Young modulus of a metal in the form of a wire, the diameter and initial length of the wire are measured using a micrometer and metre ruler. The extension of the wire for different loads is then measured as described opposite. The corresponding stress and strain values can then be calculated and plotted as in Figure 3. The value of E is equal to the gradient of the initial straight section of the graph corresponding to section OP of Figure 3.

- **Beyond P**, the line curves and continues beyond the **elastic limit** E to the **yield point** Y_1, which is where the wire weakens temporarily. The elastic limit is the point beyond which the wire is permanently stretched and suffers **plastic deformation**.
- **Beyond Y_2**, a small increase in the stress causes a large increase in strain as the material of the wire undergoes **plastic flow**. Beyond maximum stress, or the **Ultimate Tensile Stress** (UTS), the wire loses its strength, extends and becomes narrower at its weakest point. Increase of stress occurs at this point due to the reduced area of cross-section, until the wire breaks at B. The stress at B is called the **breaking stress**.

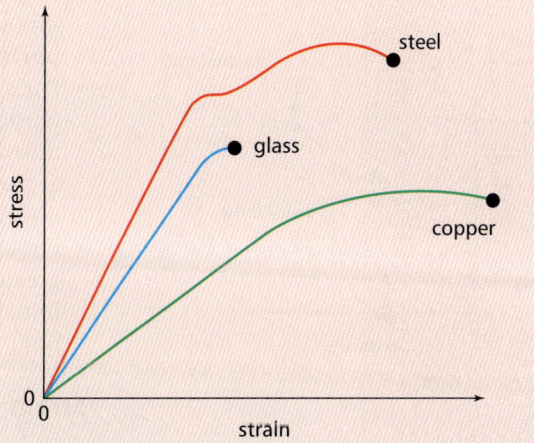
Summary test 5.5

1. Calculate the stress in a wire of diameter 0.25 mm when the tension in the wire is 50 N.
2. A metal wire of diameter 0.23 mm and of unstretched length 1.405 m is suspended vertically from a fixed point. When a 40 N weight is suspended from the lower end of the wire, the wire stretches by an extension of 10.5 mm. Calculate the Young modulus of the wire material.
3. A vertical steel wire of length 2.5 m and diameter 0.35 mm supports a weight of 90 N. Calculate:
   a the stress in the wire,
   b the extension of the wire;
   Young modulus of steel = 2.0×10^{11} Pa.
4. Use Figure 4 to decide:
   a if copper is stronger than glass,
   b if glass is stiffer than steel,
   c if copper is more ductile than glass. Give a reason for each answer.

5.6 More about stress and strain

Learning outcomes

On these pages you will learn to:

- compare loading and unloading curves for a material under stress
- deduce the strain energy in a deformed material from the area under the force–extension graph
- calculate the elastic potential energy stored per unit volume in a material when its elastic limit is not exceeded

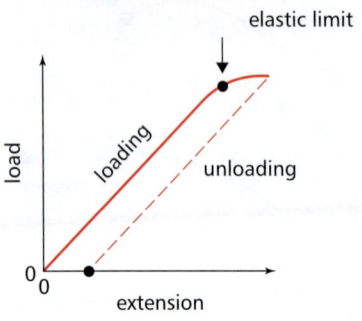

a *Metal wire*

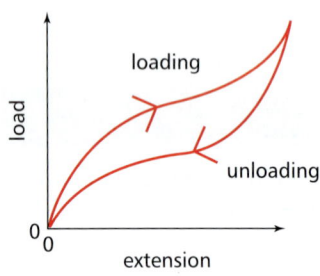

b *Rubber*

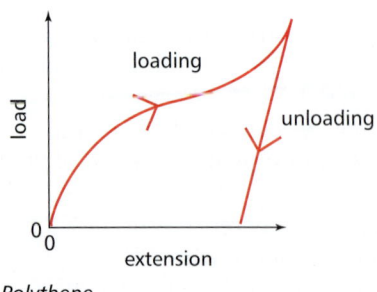

c *Polythene*

Figure 1 *Loading and unloading curves*

Investigating loading and unloading of different materials

How does the **stiffness** of a material change as a result of being stretched? To investigate this question, the tension in a strip of material is increased by increasing the weight it supports in steps. At each step, the extension of the material is measured. Typical results for different materials are shown in Figure 1. For each material, the loading curve and the subsequent unloading curve are shown.

- For a metal wire, its extension at any given tension when it is being unloaded is exactly the same as when it was being loaded (Figure 1a). The unloading curve and the loading curve are the same straight line. Provided its elastic limit is not exceeded, the wire returns to the same length when it has been completely unloaded as it had before it was loaded. If the wire is stretched beyond its elastic limit, the unloading line is parallel to the loading line. In this case, when it is completely unloaded, the wire is slightly longer so it has a **permanent extension**.
- For a rubber band (Figure 1b), the extension during unloading at any given tension is greater than during loading. The rubber band returns to the same unstretched length, but the unloading curve is below the loading curve except at zero extension and maximum extension. The rubber band remains elastic as it regains its initial length, but it has a **low limit of proportionality**.
- For a polythene strip (Figure 1c), the extension during unloading is also greater than during loading. However, the strip does not return to the same initial length when it is completely unloaded. The polythene strip has a low limit of proportionality and suffers **plastic deformation**.

Strain energy

The area under a graph of force v. extension is equal to the work done to stretch the wire. The work done to deform an object is referred to as **strain energy**. Consider the energy changes for each of the three materials in Figure 1 when each material is loaded then unloaded.

Metal wire (or spring)

Provided the limit of proportionality is not exceeded, the work done to stretch a wire to extension x is $\frac{1}{2}Fx$, where F is the tension in the wire at this extension. Because the elastic limit is not reached, the work done is stored as elastic potential energy in the wire. Therefore:

Elastic potential energy stored in a stretched wire $= \frac{1}{2}Fx$

Because the graph of tension against extension is the same for unloading as for loading, all the energy stored in the wire can be recovered when the wire is unloaded.

Since the volume of the wire $= AL$, where A is its cross-sectional area and L is its length:

Elastic potential energy stored per unit volume $= \dfrac{1}{2}\dfrac{Fx}{AL} = \frac{1}{2} \times$ **stress $\times$ strain**

Worked example

A steel wire of uniform diameter 0.35 mm and of length 810 mm is stretched to an extension of 2.5 mm. Calculate:

a the tension in the wire, **b** the elastic potential energy stored in the wire.

The Young modulus for steel $= 2.1 \times 10^{11}$ Pa

Solution

a Extension, $x = 2.5\,\text{mm} = 2.5 \times 10^{-3}\,\text{m}$

Area of cross-section of wire $= \dfrac{\pi(0.35 \times 10^{-3})^2}{4}$

$$= 9.6 \times 10^{-8}\,\text{m}^2$$

To find the tension, rearranging the Young modulus equation $E = \dfrac{FL}{Ax}$ gives

$$F = \dfrac{EAx}{L} = \dfrac{2.1 \times 10^{11} \times 9.6 \times 10^{-8} \times 2.5 \times 10^{-3}}{0.810} = 62\,\text{N}$$

b Elastic potential energy stored in the wire
$= \frac{1}{2}Fx = 0.5 \times 62 \times 2.5 \times 10^{-3} = 7.8 \times 10^{-2}\,\text{J}$

Rubber band

The work done to stretch the rubber band is represented in Figure 1b by the area under the loading curve. The work done by the rubber band when it is unloaded is represented by the area under the unloading curve. The area between the loading curve and the unloading curve therefore represents the difference between energy stored in the rubber band when it is stretched and useful energy recovered from it when it unstretches. The difference is because some of the energy stored in the rubber band becomes internal energy of the molecules when the rubber band unstretches. Figure 2 shows how the area between the loading and unloading curve can be used to determine the internal energy retained when the rubber band is stretched then released.

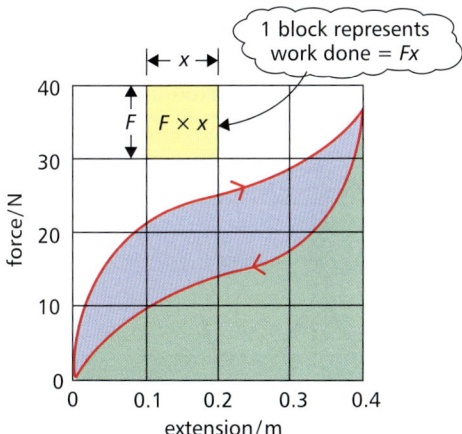

Figure 2 *Energy changes when loading and unloading rubber*

Polythene

As it does not regain its initial length, the area between the loading and unloading curve represents work done to deform the material permanently, as well as internal energy retained by the polythene when it unstretches.

The plastic behaviour of polythene is because polythene is a polymer, so each molecule is a long chain of atoms. Before being stretched, the molecules are tangled together or folded against each other. Weak bonds referred to as **cross-links** form between atoms where molecules are in contact with each other. When placed under tension, a thin sample of polythene easily stretches as the original weak cross-links break and the molecules align parallel to each other. New weak cross-links form in the stretched state and, when the tension is removed, the polythene strip remains stretched.

Extension

Notes Rubber is also a polymer, but its molecules are curled up and tangled together when it is unstretched. When placed under tension, its molecules are straightened out as its length increases more and more. When the tension is removed, its molecules curl up again and it regains its initial length.

A metal in the solid state is **crystalline** because it consists of tiny crystals or **grains** packed together. The atoms in each grain are arranged in layers in a regular pattern as in any crystal. When a metal is stretched beyond its elastic limit the layers of atoms in some of the grains slide past each other. When the distorting forces are removed, the displaced atoms do not return to their original places.

Solids such as glass are **amorphous**, which is neither crystalline nor polymeric. The atoms in an amorphous solid are randomly arranged without any regular pattern. Glass is brittle because stress causes any tiny cracks at its surface to propagate through the material.

Summary test 5.6

Young modulus for steel $= 2.1 \times 10^{11}\,\text{Pa}$; Young modulus for copper $= 1.3 \times 10^{11}\,\text{Pa}$

1 A vertical steel cable of diameter 24 mm and of length 18 m supports a weight of 1500 N attached to its lower end. Calculate:
   a the tensile stress in the cable,
   b the extension of the cable,
   c the elastic potential energy stored in the cable, assuming its elastic limit has not been reached.

2 A vertical steel wire of diameter 0.28 mm and of length 2.0 m is fixed at its upper end and has a weight of 15 N suspended from its lower end. Calculate:
   a the extension of the wire,
   b the elastic potential energy stored in the wire.

3 A steel bar of length 40 mm and cross-sectional area $4.5 \times 10^{-4}\,\text{m}^2$ is placed in a vice, and compressed by 0.20 mm when the vice is tightened. Calculate:
   a the compressive force exerted on the bar,
   b the work done to compress it.

4 Figure 2 shows a force v. extension curve for a strip of rubber. Use the graph to determine:
   a the work done to stretch the rubber to an extension of 0.40 m,
   b the internal energy retained by the rubber when it unstretches.

5 Exam-style and Practice Questions

$$\boxed{\text{🗐 Launch additional digital resources for the chapter}}$$

$g = 9.81\,\mathrm{m\,s^{-2}}$

1 A uniform copper wire of length 2.50 m has a diameter of 0.32 mm. Calculate:

   a the volume of the wire,

   b the mass of the wire,

   c the mass per unit length of the wire.
Density of copper = 8900 kg m^{-3}

2 A hydraulic lift is used in a garage to raise a vehicle to enable its underside to be repaired. Figure 2.1 below shows how such a lift works. The maximum pressure of the fluid in the hydraulic lift is 1.2 MPa.

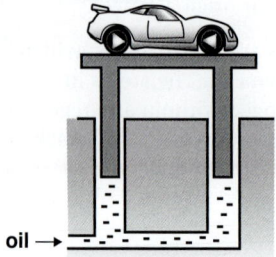

oil →

Figure 2.1

   a The area of cross-section of each of the four 'legs' of the lift is 9.0 × 10^{-3} m^2. Calculate the maximum safe load the lift can raise if the lift platform has a weight of 1.5 × 10^4 N.

   b The lift is used to raise a vehicle of mass 2200 kg through a height of 2.4 m.
     i Calculate the gain of potential energy of the vehicle.
     ii Calculate the work done by the hydraulic system to lift the vehicle by 2.4 m.
     iii Account for the difference between your answers to **i** and **ii**.

3 A steel spring of length 300 mm fixed at its upper end hangs vertically. When a 4.0 N weight is suspended from its lower end, the spring extends to an equilibrium length of 420 mm.

   a Calculate:
     i the spring constant of this spring,
     ii the length of the spring when it supports a weight of 6.0 N.

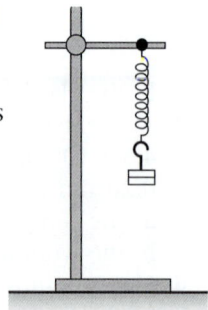

Figure 3.1

   b A second identical spring is suspended from the lower end of the first spring. Calculate the weight that would need to be suspended on the end of the lower spring to give a total extension for both springs equal to 90 mm.

4 **a** With the aid of a suitable example, explain what is meant by:
     i a brittle material,
     ii a ductile material.

   b A steel guitar wire of diameter 0.28 mm and of length 800 mm is tightened by turning its tension key by 2 turns, each turn increasing the length of the wire by 1.8 mm. Calculate the increase of tension in the wire as a result.
Young modulus of steel = 2.1 × 10^{11} N m^{-2}

5 **a** Explain what is meant by the following terms:
     i the elastic limit of a strip of material,
     ii plastic behaviour.

   b **i** Sketch a graph to show how the extension of a rubber band varies with the force used to stretch it.
     ii When a car is in motion, each part of its tyres that make contact with the road is squashed and stretched as it passes through the point of contact with the road. In energy terms, explain why this repeated squashing and stretching causes the tyre to become warm.

6 A crane is designed to lift a load with a maximum weight of 9000 N to a maximum height of 65 m. The crane cable has an area of cross-section of 4.5 × 10^{-4} m^2 and is attached to a pulley block and hook of weight 600 N, so that the load is supported by two lengths of the crane cable (as shown in the figure below).

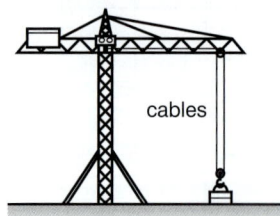

cables

Figure 6.1

   a Calculate the maximum weight of the suspended crane cable, pulley block and hook when the hook is just off the ground.

   b **i** Calculate the increase of stress in the crane cables when a load of 9000 N is lifted off the ground.
     ii Show that the extension of each of the two lengths of the crane cable when the load is raised off the ground is 3.3 mm.
Density of steel = 8000 kg m^{-3}
Young modulus of steel = 2.0 × 10^{11} N m^{-2}
     iii Calculate the elastic energy stored in the two crane cables due to this extension.

7 The figure below shows a graph of the force used to stretch a metal wire against the extension of the wire for a wire of length 1620 mm and of cross-sectional area 1.40×10^{-7} m².

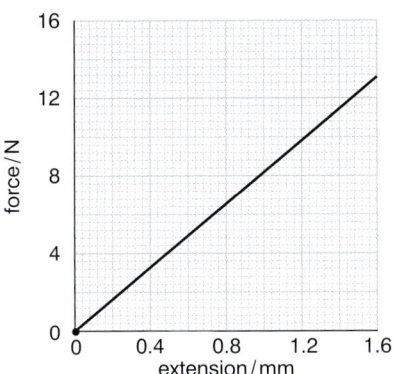

Figure 7.1

a Calculate the Young modulus of the wire.

b Calculate the energy stored in the wire when it extended from its unstretched length by 1.60 mm.

8 A spring which has a spring constant k hangs vertically from a fixed point. The spring supports a load of weight W attached to the lower end of the spring. When the load is at rest, the extension of the spring is within its elastic limit.

a i State what is meant by *elastic limit*.
ii State what is meant by *spring constant*.
iii Write down an expression in terms of k and W for the energy stored in the spring when the load is at rest.

b Figure 8.1 shows two arrangements P and Q of three identical springs used to support a load of weight W. The spring constant of each spring is k.

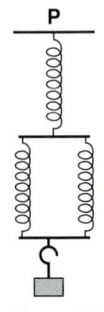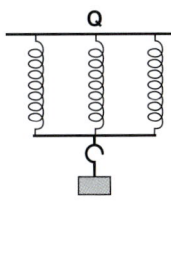

Figure 8.1

For each arrangement, determine:

i the extension in terms of e (the extension of a single spring for a load of weight W),
ii the effective spring constant in terms of k.

9 The variation of the tension T with extension e for a rubber band being stretched to an extension of 0.40 m is shown in Figure 9.1.

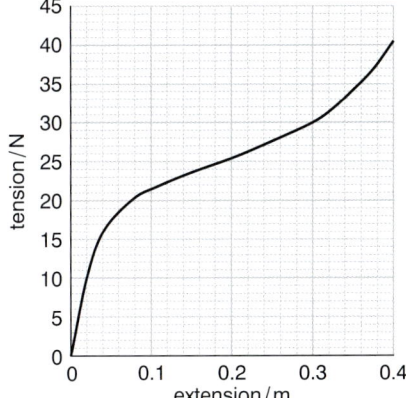

Figure 9.1

a Estimate the work done in extending the rubber band to an extension of 0.40 m.

b Describe how the stiffness of the rubber band changed as it was stretched.

10 In an investigation on upthrust, a metal cylinder of uniform cross-section was suspended vertically from a newton meter above a beaker of water, as shown in Figure 10.1.

Depth, d/mm	Newton meter reading/N	Upthrust, U/N
0	1.95	0
10	1.92	
21	1.88	
30	1.85	
39	1.82	
50	1.80	

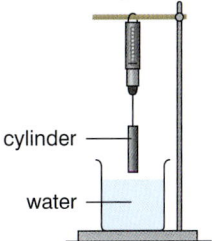

Figure 10.1

a The reading of the newton meter was 1.95 N. The cylinder was then partly lowered 50 mm into the water, causing the newton meter reading to decrease to 1.80 N.
i Explain why the newton meter reading decreased.
ii Calculate the upthrust on the cylinder.

b The cylinder was lowered to different measured depths in the water and the newton meter reading was recorded at each depth. See table above.
i Calculate the upthrust for the other depth measurements and copy and complete the table.
ii Plot a graph of the upthrust, U, against depth, d.
iii Determine the gradient of the graph.

c The length of the cylinder was 80 mm.
i Estimate the upthrust if the cylinder had been fully immersed.
ii Hence calculate the density of the cylinder. Density of water = 1000 kg m^{-3}

Key concepts: Forces

In studying chapters 1 to 5, you will have gained a wide appreciation of what forces do, as well as a deeper understanding of how physical principles relating to forces are combined with mathematical methods to predict and test their effects. The effects can then be compared with the predictions to find out if the mathematical methods give results consistent with observations. Such methods, developed by Galileo, Newton and other scientists, form the basis of many branches of physics and design engineering, from tunnels under the ground to robot vehicles on Mars. In addition, you should know now what we mean by 'conservation of energy', how we can use equations to calculate kinetic energy and potential energy changes and what we mean by 'efficiency' in relation to energy transfers. Momentum is another physical quantity you should now be familiar with – in particular, how you can use the relationship between force and momentum to calculate impact forces and what is meant by 'conservation of momentum' and how this important principle is used to work out what happens when objects collide or explode.

As you progressed through chapters 1–5, you will have developed your mathematical skills to a level where you can use equations and graphs to solve physical problems. In addition, you should now be able to carry out calculations accurately and express your answers in standard form in the correct units. You should also know that numerical answers should not be expressed to more significant figures than can be justified by the data provided.

Throughout the first five chapters, you will have had opportunities to develop your practical skills through measuring time intervals, distance, mass, weight and volume. Some practical experiments will have required you to use your measured data to calculate physical quantities such as density, enabling you to compare the accuracy of your results with accepted values. In addition, you should have become aware that each measurement you make has a degree of uncertainty and that uncertainties need to be estimated and used to give an overall uncertainty in a calculated quantity. Also, you will have plotted graphs, in particular straight-line graphs that can be linked to theoretical equations, from which the links can be tested. In addition, straight-line graphs can be used to determine physical quantities with less uncertainty than if a single set of measurements were used.

The following table provides an overview of the topics in chapters 1 to 5, showing where the key concepts have been developed.

Chapters 1–5: Key concepts

	Topic		Key concepts	
1.1	introduction to vectors	use of vectors to add displacement and to add forces	use of maths	problem-solving
1.2–1.7	motion along a straight line	use of dynamics equations and graphs to solve mechanics problems involving motion along a straight line	use of maths	problem-solving, including the physical significance of gradients and areas under graphs
1.5	free fall	Galileo's inclined plane test	testing	concept of acceleration
		measurement of g	forces and fields	graph of $s = \frac{1}{2}gt^2$ to determine g
1.8–1.9	projectile motion	investigating projectile paths	testing	concept of vertical and horizontal motion
		calculations of vertical and horizontal motion	use of maths	problem-solving
2.1–2.3	force and motion	investigating force and acceleration	testing	verification of $F = ma$
		calculations using $F = ma$	use of maths	problem-solving using relevant equations
		resistive forces and terminal speed	forces and fields	concept of terminal speed
2.4–2.6	energy and power	calculations using the equations for kinetic energy, potential energy, power and work	energy and matter	conservation of energy
2.7–2.8	efficiency	causes of inefficiency and energy losses		concept of dissipative forces
3.1–3.6	forces in equilibrium	principle of moments	forces and fields	problem-solving using key principles regarding equilibrium
		conditions for equilibrium	use of maths	
4.1 4.5	momentum	Newton's second law of motion	forces and fields	problem-solving using $Ft = \Delta mv$
		momentum, impact forces and conservation of momentum	use of maths	problem-solving using conservation of momentum
5.1–5.4	properties of materials	density	energy and matter	practical skills, density measurements
		pressure, upthrust, Hooke's law	testing	Hooke's law
		calculations and graphs involving density, pressure, upthrust and Hooke's law	use of maths	problem-solving
5.5–5.6	deformation of solids	stress, strain	energy and matter	practical skills
		Young modulus	testing	measurement of Young modulus
		calculations and graphs involving Young modulus	use of maths	problem-solving

Question

This question is about an investigation into the flight path of a ball projected horizontally from the edge of a table. It tests how we use models to make predictions about a physical system and how we then test these predictions with an experimental investigation.
*Notice also the use of mathematics to express the relationship between two perpendicular distances, **x** and **y**. In this case, the predicted relationship is of the form $y = kx^2$. So, by plotting a graph of **y** against x^2, we can tell if the relationship is valid by looking at whether or not the graph is linear (i.e. a straight-line graph) and passes through the origin. If the graph is linear, we can also determine the constant **k**, as it is equal to the gradient of the graph. You will notice as well that this investigation is about the behaviour of a particle in a field: in this case a uniform gravitational field. The path of such a particle is similar to that of a charged particle in a uniform electric field, because in both cases the acceleration of the particle is constant in magnitude and direction.*

Before answering the question, students should be given the opportunity to do this or a similar investigation.

1 A student wanted to measure the path of a small steel ball projected horizontally. Figure 1 shows the arrangement he used. The thin metal bar B was positioned at different horizontal distances from the edge of a flat laboratory bench. For each distance, the vertical position of the bar was adjusted until it was positioned so that the ball hit the bar after being released from rest inside the top of the cardboard tube T.

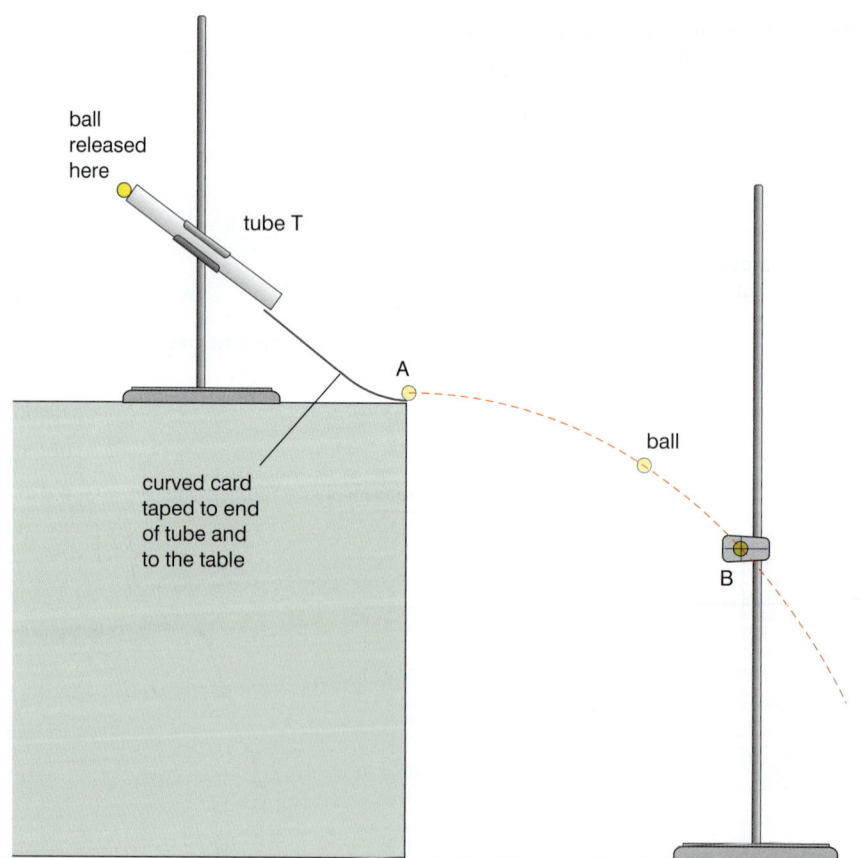

Figure 1

a i Why was it important to ensure that the ball was released at rest from the same position each time the test was carried out?

ii Give two reasons why a large steel ball would have been unsuitable.

(3 marks)

b For each horizontal distance x from B to the point A where the ball left the bench, x was measured and the corresponding vertical distance y was measured in three separate tests. The results obtained are shown below.

x/m	0	0.205	0.402	0.596	0.803	0.995	1.205
y/mm	0	19	78	155	272	432	643
	0	20	65	164	277	438	631
		22	72	158	286	429	639
mean value of y/mm							
x²/m²							

i Copy and complete the table of results above.

The student reckoned that at time t after the ball left the bench, $y = \frac{1}{2} g t^2$ and $x = Ut$, where U is the horizontal speed of projection of the ball.

ii Use these equations to show that $y = kx^2$, where k is a constant, and derive an expression for k in terms of U and g.

iii Plot a graph of y against x^2 to find out if the student's hypothesis is correct.

iv Use the graph to determine U, the horizontal speed of projection of the ball.

(13 marks)

c The student reckoned that the distances were measured to within 2 mm. Discuss this claim in terms of the above measurements.

(4 marks)

6 Electric current

6.1 Electric charge

Learning outcomes

On these pages you will learn to:

- explain how to create and detect static charge
- explain what is meant by an electric field and a field line

Static electricity

Most plastic materials can be charged quite easily by rubbing with a dry cloth. When charged, they attract small bits of paper. Do charged pieces of plastic material attract one another? Figure 1 shows an arrangement to test for attraction. A charged perspex ruler will attract a charged polythene comb, but two charged rods of the same material always repel one another.

There are just two types of charge which are referred to as positive and negative charge. A charged object will always:

- repel another charged object that has the same type of charge (i.e. like charge)
- attract another charged object that has the other type of charge (i.e. unlike charge).

> **Like charges repel; unlike charges attract**

Electrons are responsible for charging in most situations. An uncharged atom contains equal numbers of **protons** and electrons. An electron carries a tiny negative charge. A proton carries a tiny positive charge equal in magnitude to that of the electron.

- Add one or more electrons to an uncharged atom and it becomes negatively charged.
- Remove one or more electrons from an uncharged atom and it becomes positively charged.

An uncharged solid contains equal numbers of electrons and protons. When an uncharged perspex rod is rubbed with an uncharged dry cloth, electrons transfer from the rod to the cloth so the rod becomes positively charged and the cloth becomes negatively charged.

- Electrical conductors such as metals contain lots of free electrons. These are electrons which move about inside the metal and are not attached to any one atom. To charge a metal up, it must first be isolated from the Earth. Otherwise, any charge given to it is neutralised by electrons transferring between the conductor and the Earth. Then the isolated conductor can be charged by direct contact with any charged object. If an isolated conductor is charged positively then 'earthed', electrons transfer from the Earth to the conductor to neutralise or discharge it. See Figure 2.
- Insulating materials do not contain free electrons. All the electrons in an insulator are attached to individual atoms. Some insulators, such as perspex or polythene, are easy to charge because their surface atoms easily gain or lose electrons.

The gold leaf electroscope is used to detect charge. If a charged object is in contact with the metal cap of the electroscope, some of the charge on the object transfers to the electroscope. As a result, the gold leaf and the metal stem which is attached to the cap gain the same type of charge and the leaf rises because it is repelled by the stem.

If another object with the same type of charge is brought near the electroscope, the leaf rises further because the object forces some charge on the cap to transfer to the leaf and stem.

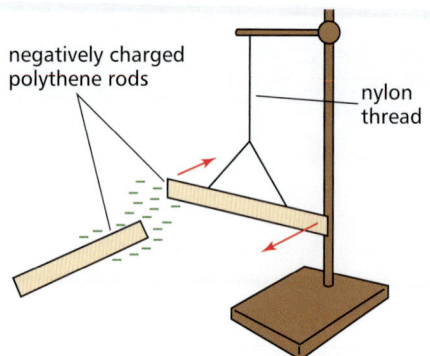

Figure 1 *Electrostatic forces*

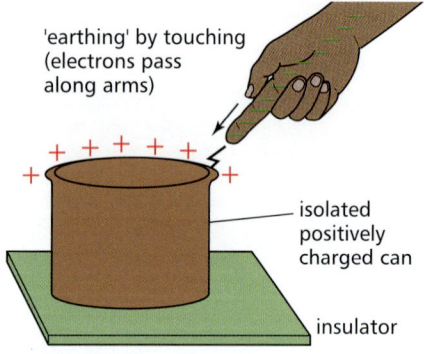

Figure 2 *Discharge to Earth*

Figure 3 *The electroscope*

Field lines and patterns

Any two charged objects exert equal and opposite forces on each other without being directly in contact. An **electric field** is said to surround each charge. Suppose a small positive charge is placed as a test charge near a body with a much bigger charge which is also positive. If the test charge is free to move, it will follow a path away from the body with the bigger charge. The path a free test charge follows is called a **line of force** or a **field line**.

The direction of an electric field line is the direction in which a positive test charge would move. Figure 4 shows the patterns of fields around different charged objects. Each pattern is produced by semolina grains sprinkled on oil. An electric field is set up across the surface of the oil by connecting two metal conductors in the oil to the output terminals of a high voltage supply unit. The grains line up along the field lines, like plotting compasses in a magnetic field.

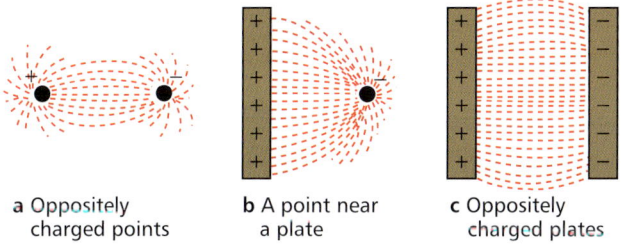

a Oppositely charged points   **b** A point near a plate   **c** Oppositely charged plates

Figure 4 *Electric field patterns*

- Oppositely charged point objects create a field as shown in Figure 4a. The field lines become concentrated at the points. A positive test charge released from an off-centre position would follow a curved path to the negative point charge.

- A point object near an oppositely charged flat plate produces a field as shown in Figure 4b. The field lines are concentrated at the point object but they meet the plate at right angles to it.
- Two oppositely charged plates create a field as shown in Figure 4c. The field lines run parallel from one plate to the other, meeting the plates at right angles. The field is **uniform** between the plates as the field lines are parallel to each other.

Devices that use electric fields directly include:

- electrostatic precipitators which extract ash, dust and other fine particles from gases released in power stations and factories. The gases pass through a grid of wires at high voltage between earthed metal plates fixed vertically, as shown in Figure 5. The particles become charged as they pass through the grid and are then attracted onto the plates by the electric field between the grid wires and the plates.
- electrostatic sprays such as airborne crop sprays which produce charged droplets that spread out as they repel each other and are attracted onto the ground. Industrial paint sprays are used to coat metal panels with an even film of paint.

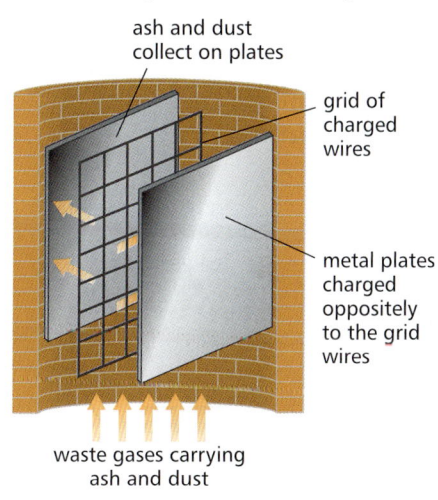

ash and dust collect on plates

grid of charged wires

metal plates charged oppositely to the grid wires

waste gases carrying ash and dust

Figure 5 *An electrostatic precipitator*

Summary test 6.1

1 Explain each of the following observations in terms of transfer of electrons:

   a An insulated metal can is given a positive charge by touching it with a positively charged rod.

   b A negatively charged metal sphere suspended on a thread is discharged by connecting it to the ground using a wire.

2 An insulated metal conductor is earthed before a negatively charged object is brought near to it.

   a Explain why the free electrons in the conductor move as far away from the charged object as they can.

   b The conductor is then briefly earthed. The charged object is then removed from the vicinity of the conductor. Explain why the conductor is left with an overall positive charge.

3 A positively charged point object is placed near an earthed metal plate, as shown in Figure 6.

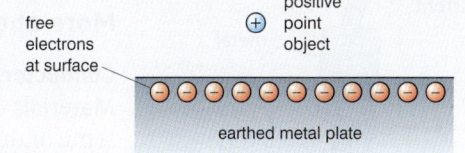

free electrons at surface

positive point object

earthed metal plate

Figure 6

   a Explain why electrons gather at the surface of the metal plate near the object.

   b Explain why there is a force of attraction between the object and the metal plate.

4 Sketch the pattern of the field lines of the electric field:

   a between two oppositely charged parallel plates,

   b between a positively charged point object and an earthed metal plate.

Current and charge

Learning outcomes

On these pages you will learn to:

- recognise that an electric current is a flow of charge due to the passage of charged particles
- know that the charge on a charge carrier is quantised and that the charge carriers in a metal are electrons
- distinguish between a conductor, an insulator and a semiconductor
- state and use the relationship between the current and the charge flow in a certain time
- derive and use $I = Anvq$, where n is the number density of charge carriers

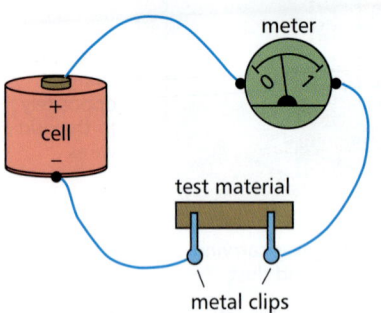

Figure 1 Testing for conduction

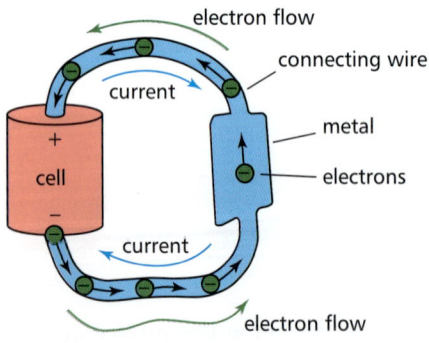

Figure 2 Convention for current

Electrical conduction

An electric current is a flow of charge due to the passage of charged particles. These charged particles are referred to as **charge carriers**.

- In metals, the charge carriers are **conduction electrons**. They move about inside the metal, repeatedly colliding with each other and the fixed positive ions in the metal.
- In comparison, when an electric current is passed through a salt solution, the charge is carried by ions which are charged atoms or molecules.
- Charge is quantised, which means it is only found in whole number multiples of 1.6×10^{-19} C. This basic amount is referred to as the 'quantum' of charge, denoted by the symbol e. For example, the charge on a calcium Ca^{2+} ion is equal to $+2e$.

A simple test for conduction of electricity is shown in Figure 1. If the test material is a metal, electrons pass round the circuit: they leave the battery at its negative terminal, pass though the metal, then re-enter the battery at its positive terminal.

The convention for the direction of current in a circuit is from **+ to −**, as shown in Figure 2. The convention was agreed long before the discovery of electrons. When it was set up, it was known that an electric current is the rate of flow of charge one way round a circuit. However, it was not known if the current was due to positive charge flowing round the circuit from + to −, or if it was due to negative charge flowing from − to +.

- The unit of current is the **ampere (A)**, which is defined in terms of the force between two parallel wires when they pass the same current. See Topic 16.3.
- The unit of charge is the **coulomb (C)**, equal to the charge flow in one second when the current is one ampere. The magnitude of the charge of the electron, e, is 1.6×10^{-19} C. This is sometimes referred to as the **elementary charge**.
- For a current I, the **charge flow Q** in time t is given by the equation

$$Q = It$$

- For charge flow Q in a time interval t, the current I is given by:

$$I = \frac{Q}{t}$$

The equation shows that a current of 1 A is due to a flow of charge of 1 coulomb per second. As the charge of the electron is 1.6×10^{-19} C, a current of 1 A along a wire must be due to 6.25×10^{18} electrons passing along the wire each second.

More about charge carriers

Conductors, insulators and semiconductors

Materials can be classified in electrical terms as conductors, insulators or semiconductors.

- In an **insulator**, each electron is attached to an atom and cannot move away from the atom. When a voltage is applied across an insulator, no current passes through the insulator because no electrons can move through the insulator.
- In a **metallic conductor**, most electrons are attached to atoms; but some are not and these are the **charge carriers** in the metal. When a voltage is applied across the metal, these **conduction electrons** move towards the positive terminal of the metal.
- In a **semiconductor**, the number of charge carriers increases with increase of temperature. The resistance of a semiconductor therefore decreases as its temperature is raised. A pure semiconducting material is referred to as an **intrinsic semiconductor**, because conduction is due to electrons that break free from the atoms of the semiconductor.

- In an **electrolyte**, the charge carriers are **positive and negative ions**. When a voltage is applied to the electrodes, the positive ions are attracted to the **cathode** (i.e. the negative electrode) and the negative ions are attracted to the **anode** (i.e. the positive electrode). The ions are discharged on reaching the relevant electrode. The process is known as **electrolysis**.

Drift velocity

The charge carriers in a conductor or a semiconductor move about at random inside the material. When a p.d. is applied across the material, in addition to their random motion, the charge carriers 'drift' to the positive end. Their mean **displacement** per unit time is their **drift velocity**. The current through a material depends on the drift velocity v of the charge carriers.

Consider Figure 3, which shows charge carriers moving through a section of wire of cross-sectional area A and length L. The average time taken, t, for each charge carrier to move through it $= L \div v$.

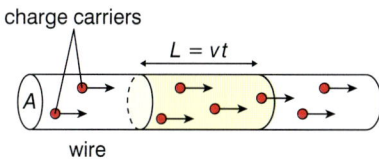

Figure 3 *Drift velocity*

If there are n charge carriers per unit volume in the wire,

- the number of charge carriers in this section $= n(A \times L)$
  since the section volume $= A \times L$
- their total charge Q = number of charge carriers × charge of each carrier (q)
  $$= n(A \times L)q$$

Therefore, the current through the section, $I = \dfrac{Q}{t} = \dfrac{nALq}{L/v}$

$$= Anvq$$

For electrons, $q = e$ gives $I = Anve$

Physics and the Human Genome Project

To map the human genome, fragments of DNA are tagged with a C, G, A, or T base containing a dye. Each tagged fragment carries a negative charge. A voltage is applied across a strip of gel containing a spot of liquid containing tagged fragments. The fragments are attracted to the positive electrode. The smaller the fragment, the faster it moves; so the fragments separate out according to size, as they move to the positive electrode. The fragments pass through a spot of laser light, which causes the dye attached to each tag to **fluoresce** as it passes through the laser spot. Light sensors linked to a computer detect the glow from each tag. The computer is programmed to work out and display the sequence of bases in the DNA fragments.

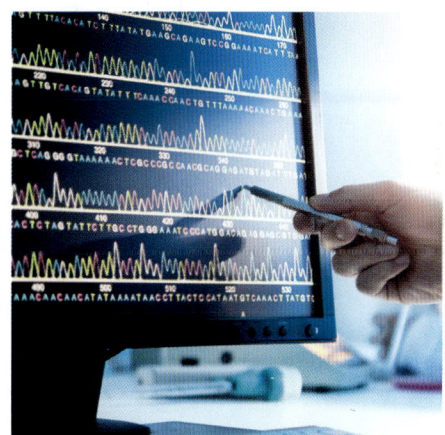

Figure 4 *Mapping the human genome*

Summary test 6.2

$e = 1.6 \times 10^{-19}$ C

1 a The current in a certain wire is 0.35 A. Calculate the charge passing a point in the wire:
  i in 10 s,   **ii** in 10 min.

b Calculate the average current in a wire through which a charge of 15 C passes in:
  i 5 s,   **ii** 100 s.

2 Calculate the number of electrons passing a point in the wire in 1 min when the current is:

  a 1 µA   **b** 5.0 A

3 A certain type of rechargeable battery is capable of delivering a current of 0.2 A for 4000 s, before its voltage drops and it needs to be recharged. Calculate:

a the total charge the battery can deliver before it needs to be recharged,

b the maximum time it could be used for without being recharged if the current through it were:
  i 0.5 A   **ii** 0.1 A

4 Estimate the drift velocity of the conduction electrons in a copper wire of diameter 0.4 mm when a current of 2.0 A passes through it. Assume copper contains 10^{29} electrons per m³.

Learning outcomes

On these pages you will learn to:

- define potential difference (p.d.) and the volt
- define the electromotive force (e.m.f.) of an electrical source
- distinguish between e.m.f. and p.d. in terms of energy considerations
- recall and use $V = W/Q$
- recall and use $P = VI$

Energy and potential difference

When a torch bulb is connected to a battery, electrons deliver energy from the battery to the torch bulb. Each electron which passes through the bulb takes a fixed amount of energy from the battery and delivers it to the bulb. After delivering energy to the bulb, each electron re-enters the battery via the positive terminal to be resupplied with more energy to deliver to the bulb.

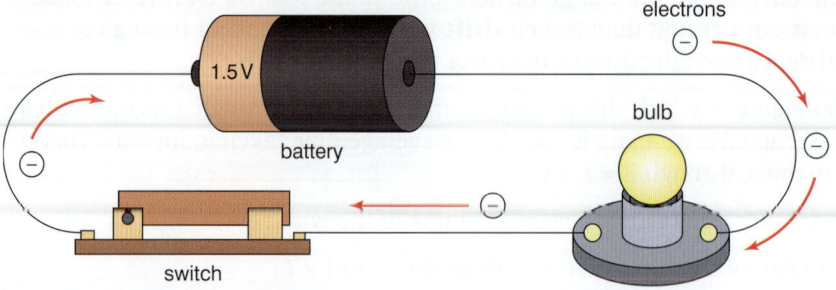

Figure 1 *Energy transfer by electrons*

Each electron in the battery has the **potential** to deliver energy, even if the battery is not part of a complete circuit. In other words, the battery supplies each electron with **electrical potential energy**. When the battery is in a circuit, each electron passing through the circuit component does work to pass through the component and therefore transfers some or all of its electrical potential energy to the component. The **work done** by an electron is equal to its loss of potential energy. The energy transfer (i.e. work done) per unit charge is defined as the **potential difference** (abbreviated as p.d.) or **voltage**, *V*, across the component.

> **The potential difference across a component is defined as the energy transferred per unit charge.**

The unit of p.d. is the **volt (V),** which is equal to 1 joule per coulomb.

If work *W* is done when charge *Q* flows through the component, the p.d. across the component, *V*, is given by:

$$V = \frac{W}{Q}$$

Rearranging this equation gives $W = QV$ for the work done or energy transfer when charge *Q* passes through a component which has a p.d. *V* across its terminals.

Examples

- If 30 J of work is done when 5 C of charge passes through a component, the p.d. across the component must be 6 V $\left(= \frac{30\,\text{J}}{5\,\text{C}}\right)$.

- If the p.d. across a component in a circuit is 12 V, then 3 C of charge passing through the component would transfer 36 J of energy from the battery to the component.

Energy transfer in different devices

An electric current has a **heating effect** in a component with **resistance**. It also has a **magnetic effect**, which is made use of in electric motors and loudspeakers.

- In a device that has resistance, such as an electrical heater, the work done on the device is transferred as **thermal energy**. This happens because the charge

Figure 2 *Electrical devices*

carriers repeatedly collide with atoms in the device and transfer energy to them, so the atoms **vibrate more** and the resistor becomes hotter.

- In an electric motor, the work done on the motor is transferred as **kinetic energy** of the motor. The charge carriers are electrons that need to be forced through the wires of the **spinning motor coil** against the opposing force on the electrons due to the motor's magnetic field.
- For a loudspeaker, the work done on the loudspeaker is transferred as **sound energy**. Electrons need to be forced through the wires of the **vibrating loudspeaker coil** against the force on them due to the loudspeaker magnet.

Electromotive force (e.m.f)

> **The e.m.f., E, of a source of electricity is defined as the electrical energy per unit charge produced inside the source.**

The unit of e.m.f. is the **volt (V)**, the same as the unit of p.d., which is equal to 1 joule per coulomb.

For a source of e.m.f., E, in a circuit, the electrical energy produced when charge Q passes through the source is QE. This energy is transferred to other parts of the circuit, and some may be dissipated in the source itself due to the source's **internal resistance** (see Topic 7.3 for internal resistance).

Figure 3 *Power supplies*

Electrical power and current

Consider a component, or device, which has a potential difference V across its terminals and a current I passing through it. In time t:

- The charge flowing through it, $Q = It$
- The work done by the charge carriers, $W = QV = (It)V = IVt$

Work done, $W = IVt$

The energy transfer ΔE in the component or device is equal to the work done W. Therefore, because power $= \dfrac{\text{energy}}{\text{time}}$, the electrical power P supplied to the device is given by:

$$P = \frac{IVt}{t} = IV$$

Electrical power, $P = IV$

Notes

- This equation can be rearranged to give: $I = \dfrac{P}{V}$ or $V = \dfrac{P}{I}$
- The unit of power is the **watt (W)**. Therefore one volt is equal to one watt per ampere. For example, if the p.d. across a component is 4 V, then the power delivered to the component is 4 W per ampere of current.
- Energy supplied by mains electricity is measured in **kilowatt hours (kW h)**, usually referred to as 'units'. One kilowatt hour is the energy transfer when 1 kW of power is supplied for exactly 1 h.
  Therefore 1 kW h = 3.6 MJ (= 1000 W × 3600 s).
- The correct value of a **fuse** for an electrical appliance can be worked out using the equation $I = \dfrac{P}{V}$ if the power and voltage of the appliance are known.
  For example, a 230 V, 2.5 kW electric kettle would take a current of 11 A $\left(= \dfrac{2500\,\text{W}}{230\,\text{V}}\right)$. Therefore, given the choice of a 5 A or a 13 A fuse, a 13 A fuse should be used.

Worked example

A 12 V, 48 W electric heater is connected to a 12 V battery. Calculate: **a** the heater current, **b** the energy transfer in 300 s.

Solution
a Rearrange $P = IV$ to give $I = \dfrac{P}{V} = \dfrac{48\,\text{W}}{12\,\text{V}} = 4\,\text{A}$
b $\Delta E = IVt = 4\,\text{A} \times 12\,\text{V} \times 300\,\text{s} = 14\,400\,\text{J}$

Summary test 6.3

1. Calculate the energy transfer in 1200 s in a component when the p.d. across it is 12 V and the current is:

   a 2 A,   **b** 0.05 A.

2. A 6 V, 12 W light bulb is connected to a 6 V battery. Calculate:

   a the current in the light bulb,

   b the energy transfer to the light bulb in 1800 s.

3. A 230 V electrical appliance has a power rating of 800 W.

   a Calculate:
      i the energy transfer in the appliance in 1 min,
      ii the current in the appliance.

   b Which of the following fuse values would be suitable for this appliance: 3 A, 5 A, or 13 A?

4. A battery has an e.m.f. of 9 V and a negligible internal resistance. It is capable of delivering a total charge of 1350 C. Calculate:

   a the maximum energy the battery could deliver,

   b the power it would deliver to the components of a circuit if the current in it was 0.5 A,

   c how long the battery would last if it supplies power at the rate calculated in part **b**.

Learning outcomes

On these pages you will learn to:

* recall the function and circuit symbols for commonly used components, including a diode, an LED, a thermistor and an LDR
* draw and interpret circuit diagrams with the above components in them
* describe how a potential divider is used to supply a variable p.d.
* describe how the resistance of a metal and a semiconductor thermistor changes with temperature

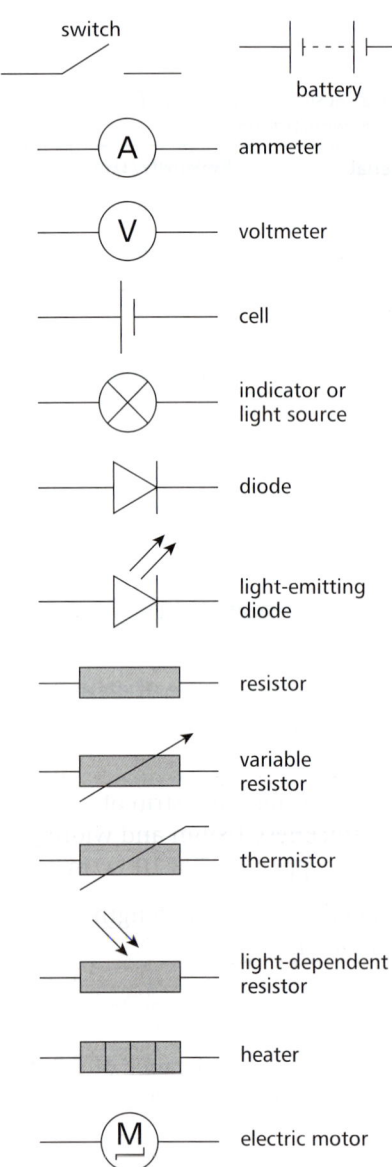

Figure 1 *Circuit components*

Circuit diagrams

Each type of component has its own **symbol**, which is used to represent the component in a **circuit diagram**. You need to recognise the symbols for different types of components to make progress – just like a motorist needs to know what different road signs mean. Note that, on a circuit diagram, the direction of the current is always shown from **+ to −** round the circuit.

The function of each of the components shown in Figure 1 is given in the list below:

* An **ammeter** measures the current in a circuit or a branch of a circuit.
* A **voltmeter** measures the p.d. between two points in a circuit.
* A **cell** is a source of electrical energy. Note that **a battery** is a combination of cells.
* The symbol for an **indicator** or any **light source** (including a filament lamp), except a light-emitting diode, is the same.
* A **diode** allows current to flow in one direction only. A **light-emitting diode** (or LED) emits light when it conducts. The direction in which the diode conducts is referred to as its 'forward' direction. The opposite direction is referred to as its 'reverse' direction.
* A **resistor** is a component designed to have a certain resistance.
* A **variable resistor** is used to control the current in a circuit.
* The resistance of a **thermistor** decreases with increase of temperature, if the thermistor is an intrinsic semiconductor such as silicon. Such a thermistor is described as a **negative temperature coefficient** (n.t.c.) thermistor.
* The resistance of a **light-dependent resistor** decreases with increase of light intensity.
* A **heater** is designed to transform electrical energy to heat.
* An **electric motor** is designed to transform electrical energy into kinetic energy.

Investigating the characteristics of different components

To measure the variation of current with p.d. for a component, use either:

* a **potential divider** to vary the p.d. from zero (Figure 2a), or
* a variable resistor to vary the current to a minimum (Figure 2b).

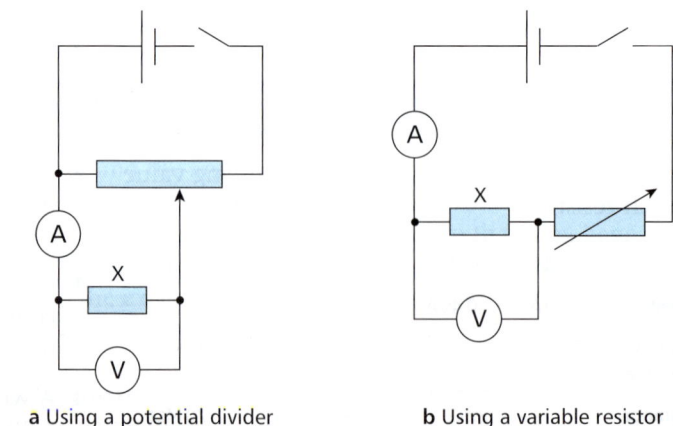

a Using a potential divider   **b** Using a variable resistor

Figure 2 *Investigating component characteristics*

The advantage of using a potential divider rather than a variable resistor is that the current in the component and the p.d. across it can be reduced to zero. However, in the variable resistor circuit, when the resistance of the variable resistor is at a maximum, the current cannot be reduced any further unless a further resistor (not shown) is connected in series with the variable resistor.

8 a Calculate the resistance of a 230 V, 100 W filament light bulb.

b The filament of the light bulb in part **a** consists of a coiled tungsten wire of diameter 0.05 mm. Calculate the total length of the wire of this filament. The resistivity of tungsten = $5.6 \times 10^{-8}\,\Omega\,\text{m}$

9 A light-dependent resistor has a resistance of 650 Ω in darkness. It is connected in series with a 4.5 V battery, a resistor, a milliammeter and a switch, as shown below.

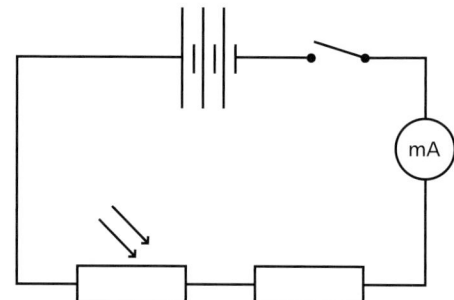

Figure 9.1

a The milliammeter reads 5.0 mA when the switch is closed and the LDR is in darkness. Calculate:
  i the p.d. across the LDR,
  ii the p.d. across the resistor,
  iii the resistance of the resistor.

b Describe how, and explain why, the milliammeter reading changes when the LDR is exposed to light.

10 The following measurements were made in an investigation to measure the resistivity of the material of a certain wire.

P.d. across the wire/V	0.00	2.0	4.0	6.0	8.0	10.0
Current through the wire/A	0.00	0.15	0.31	0.44	0.62	0.74

Length of wire = 1.60 m; Diameter of wire = 0.28 mm

a Plot a graph of the p.d. against the current.

b Use the graph to calculate the resistivity of the material of the wire.

11 A lamp X is connected in series with a variable resistor and an 18 V battery of negligible internal resistance. The variable resistor is adjusted until the lamp operates at its normal rating, which is 12 V and 24 W.

a Sketch the circuit diagram.

b Calculate the current in X and its resistance.

c Calculate the power supplied by the battery and the power dissipated in the variable resistor.

12 An overhead electric cable has 24 strands of aluminium wire of diameter 4.0 mm surrounding a core of 7 steel strands, each of diameter 3.0 mm (Figure 12.1).

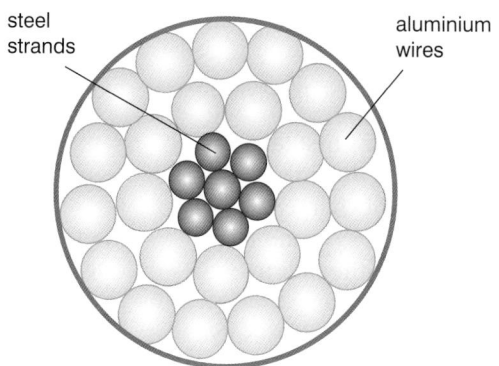

Figure 12.1

a i Calculate the resistance per metre of the aluminium strands if the resistivity of aluminium is $2.5 \times 10^{-8}\,\Omega\,\text{m}$.

  ii Calculate the resistance per metre of the steel strands if the resistivity of steel is $1.6 \times 10^{-7}\,\Omega\,\text{m}$.

b When a potential difference is applied across a 1000 m length of the cable, a current of 100 A passes through it.
  i Calculate the current in each of the two components of the cable.
  ii Calculate the potential difference across the cable and the power dissipated in each of the two components of the cable.

13 In an experiment, the current and potential difference for a 12 V filament lamp are measured at 2 V steps. The measurements are shown in the table below.

Potential difference/V	0	2.01	4.00	5.98	8.02	10.1	11.8
Current/A	0	1.36	2.07	2.64	3.13	3.58	4.05

a Plot a graph of the potential difference against the current.

b Calculate the resistance of the filament lamp at:
  i 1.00 A,   **ii** 2.00 A,   **iii** 3.00 A.

c The uncertainty in the current readings is ±0.02 A and the uncertainty in the potential difference readings is ±0.02 V. Estimate the uncertainty in your calculated value of resistance at 2.00 A.

d i Describe how the resistivity of the filament wire in the lamp changes as the wire becomes hotter.
  ii Discuss the effect on the *drift velocity* of the charge carriers in the wire when the current in the wire is increased.

7.1 Circuit rules

Learning outcomes

On these pages you will learn to:

- recall Kirchhoff's first law and appreciate the link to conservation of charge
- recall Kirchhoff's second law and appreciate the link to conservation of energy
- apply Kirchhoff's laws to solve simple circuit problems

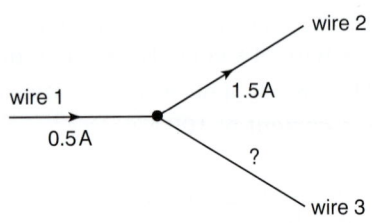

Figure 1 *Kirchhoff's first law*

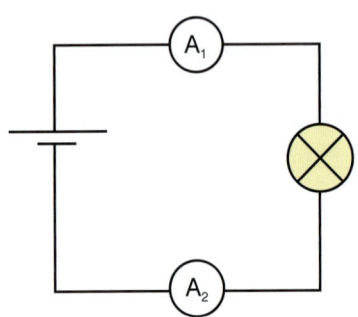

Figure 2 *Components in series*

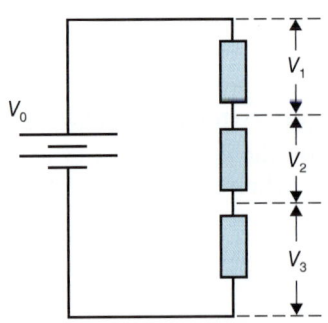

Figure 3 *Adding potential differences*

Currrent rules

Kirchhoff's first law

> **At any junction in a circuit, the total current leaving the junction is equal to the total current entering the junction.**

For example, Figure 1 shows a junction of three wires where the current in two of the wires (Wire 1 and Wire 2) is given. The current in Wire 3 must be 1.0 A into the junction, because the total current into the junction (1.0 A along Wire 3 + 0.5 A along Wire 1) is the same as the total current out of the junction (1.5 A along Wire 2).

Kirchhoff's first law follows from the **conservation of charge**, as charge flow in and charge flow out of a junction are always equal. The current along a wire is the charge flow per second. In Figure 1, the charge entering the junction each second is 0.5 C along Wire 1 and 1.0 C along Wire 3. The charge leaving the junction each second must therefore be 1.5 C, as the junction does not retain charge.

Components in series

- **The current entering a component is the same as the current leaving the component.** In other words, components do not use up current. At any instant, the charge entering a component each second is equal to the charge leaving it because the same number of charge carriers enter and leave the component each second. In Figure 2, A_1 and A_2 show the same reading because they are measuring the same current.
- **The current passing through two or more components in series is the same through each component.** This is because each charge carrier passes through every component and the same number of charge carriers pass through each component each second. At any instant, charge flows at the same rate through each component.

Potential difference rules

Energy and potential difference

> **The potential difference, or voltage, between any two points in a circuit is defined as the energy transfer per coulomb of charge that flows from one point to the other.**

If the charge carriers lose energy, the potential difference is a potential drop. If the charge carriers gain energy, which happens when they pass through a battery or cell, the potential difference is a potential rise equal to the e.m.f. of the battery or cell. The rules for potential differences are listed below with an explanation of each rule in energy terms.

- **For two or more components in series, the total p.d. across all the components is equal to the sum of the potential differences across each component.**

  Figure 3 shows a circuit consisting of a battery and three resistors in series. The p.d. across the battery terminals is equal to the sum of the potential differences across the three resistors. This is because each coulomb of charge from the battery delivers energy to each resistor as it flows round the circuit. The p.d. across each resistor is the energy delivered per coulomb of charge to that resistor. So the sum of the potential differences across the three resistors is the **total energy** delivered to the resistors per coulomb of charge that passes through them, which is the p.d. across the battery terminals.

- **The p.d. across components in parallel is the same.**
  In Figure 4, charge carriers can pass through either of the two resistors in parallel. The same amount of energy is delivered by a charge carrier, regardless of which of the two resistors it passes through.

  If the variable resistor is adjusted so that the p.d. across it is 4 V, and if the battery p.d. is 12 V, the p.d. across the two resistors in parallel is 8 V (12 V – 4 V). This is because each coulomb of charge leaves the battery with 12 J of electrical energy, and uses 4 J on passing through the variable resistor. Therefore, each coulomb of charge has 8 J of electrical energy to deliver to either of the two parallel resistors.

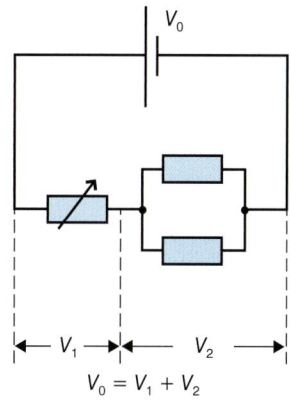

Figure 4 *Components in parallel*

- **For any complete loop of a circuit, the sum of the e.m.f.s round the loop is equal to the sum of the potential drops round the loop.**
  This statement is known as **Kirchhoff's second law**. It follows from the fact that the total e.m.f. in a loop is the total electrical energy per coulomb produced in the loop, and the sum of the potential drops is the electrical energy per coulomb delivered round the loop. The above statement follows, therefore, from the **conservation of energy**.

  For example, Figure 5 shows a 9 V battery connected to a 6 V light bulb in series with a variable resistor. If the variable resistor is adjusted so that the p.d. across the light bulb is 6 V, the p.d. across the variable resistor must be 3 V (9 V – 6 V). The only source of electrical energy in the circuit is the battery, so the sum of the e.m.f.s in the circuit is 9 V. This is equal to the sum of the p.ds round the circuit (3 V across the variable resistor + 6 V across the light bulb). In other words, the battery forces charge round the circuit. Every coulomb of charge leaves the battery with 9 J of electrical energy and supplies 3 J to the variable resistor and 6 J to the light bulb.

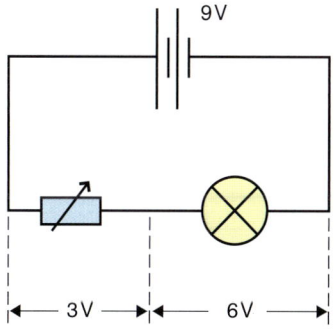

Figure 5 *Applying Kirchhoff's second law*

Summary test 7.1

1 A battery, which has an e.m.f. of 6 V and negligible internal resistance, is connected to a 6 V, 6 W light bulb in parallel with a 6 V, 24 W light bulb, as shown in Figure 6.

Calculate:

   a the current in each light bulb,
   b the current in the battery,
   c the power supplied by the battery.

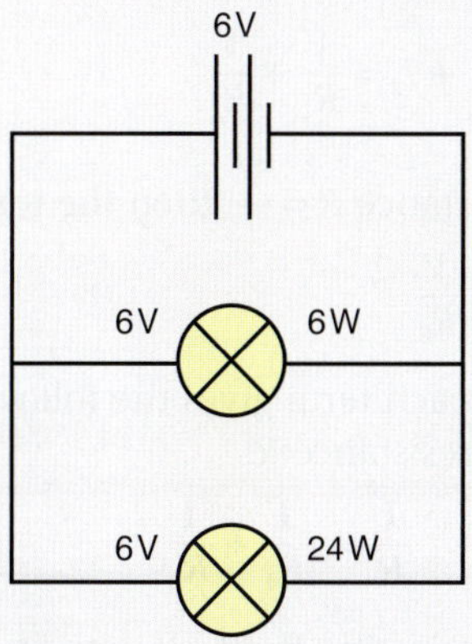

Figure 6

2 A 4.5 V battery is connected in series with a variable resistor and a 2.5 V, 0.5 W torch bulb.

   a Sketch the circuit diagram for this circuit.

   b The variable resistor is adjusted so that the p.d. across the torch bulb is 2.5 V. Calculate:
      i the p.d. across the variable resistor,
      ii the current in the torch bulb.

3 A 6.0 V battery is connected in series with an ammeter, a 20 Ω resistor and an unknown resistor R.

   a Sketch the circuit diagram.
   b The ammeter reads 0.20 A. Calculate:
      i the p.d. across the 20 Ω resistor,
      ii the p.d. across R,
      iii the resistance of R.

4 In question 3, when the unknown resistor is replaced with a torch bulb, the ammeter reads 0.12 A. Calculate:

   a the p.d. across the torch bulb,
   b the resistance of the torch bulb.

More about resistance

Learning outcomes

On these pages you will learn to:

- derive, using Kirchhoff's laws, the formula for the combined resistance of:
  - two or more resistors in series
  - two or more resistors in parallel
- solve problems using the formula for the combined resistance of:
  - two or more resistors in series
  - two or more resistors in parallel
- apply Kirchhoff's laws to solve simple circuit problems
- recall that rate of heat transfer in a resistor $= I^2R$

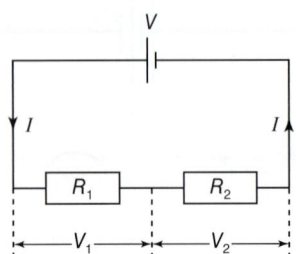

Figure 1 Resistors in series

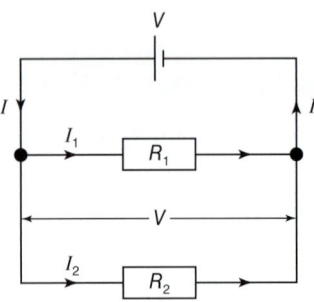

Figure 2 Resistors in parallel

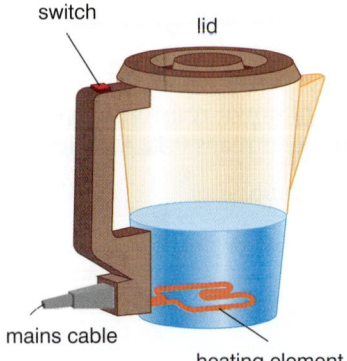

Figure 3 Heating element in a kettle

Resistor combination rules

Resistors in series

Resistors in series pass the same current, in accordance with Kirchhoff's first law. The total p.d. is equal to the sum of the individual potential differences.

- For two resistors R_1 and R_2 in series, as in Figure 1, when current I passes through the resistors:

  p.d. across R_1, $V_1 = IR_1$
  p.d. across R_2, $V_2 = IR_2$

  In accordance with Kirchhoff's second law, the cell p.d., V, is equal to the sum of the potential differences across the two resistors, $V = V_1 + V_2 = IR_1 + IR_2$

  therefore, the total resistance $R = \dfrac{V}{I} = \dfrac{IR_1 + IR_2}{I} = R_1 + R_2$

- For two or more resistors R_1, R_2, R_3, etc. in series, the theory can easily be extended to show that the **total resistance is equal to the sum of the individual resistances**:

$$R = R_1 + R_2 + R_3 + \ldots$$

Resistors in parallel

Resistors in parallel have the same p.d. The current through a parallel combination of resistors is equal to the sum of the individual currents.

- For two resistors R_1 and R_2 in parallel, as in Figure 2, the p.d. across the combination is V. Applying Kirchhoff's second law to the circuit in Figure 2 gives $V = I_1R_1$ for the loop containing the cell and resistor R_1 and $V = I_2R_2$ for the loop containing the cell and resistor R_2.

  the current through resistor R_1, $I_1 = \dfrac{V}{R_1}$

  the current through resistor R_2, $I_2 = \dfrac{V}{R_2}$

  In accordance with Kirchhoff's first law, the total current through the combination, I = the sum of the currents through the two resistors

$$= I_1 + I_2 = \frac{V}{R_1} + \frac{V}{R_2}$$

  Since the total resistance $R = \dfrac{V}{I}$, then the total current $I = \dfrac{V}{R}$

  therefore $\dfrac{V}{R} = \dfrac{V}{R_1} + \dfrac{V}{R_2}$

  Cancelling V from each term gives the following equation, which is used to calculate the total resistance R:

$$\frac{1}{R} = \frac{1}{R_1} + \frac{1}{R_2}$$

- For three or more resistors R_1, R_2, R_3, etc. in parallel, the theory can easily be extended to show that the total resistance R is given by:

$$\frac{1}{R} = \frac{1}{R_1} + \frac{1}{R_2} + \frac{1}{R_3} + \ldots$$

Resistance heating

The heating effect of an electric current in any component is due to the resistance of the component. As explained in Topic 6.3, the charge carriers repeatedly collide with the positive ions of the conducting material. There is a **net transfer of energy** from the charge carriers to the positive ions as a result of these collisions.

After a charge carrier loses kinetic energy in such a collision, the force due to the p.d. across the material accelerates it until it collides with another positive ion. For a component of resistance R, when current I passes through it:

$$\text{p.d. across the component, } V = IR$$

Therefore the power supplied to the component, $P = IV = I^2R = \dfrac{V^2}{R}$

Hence the **energy per second** transferred to the component as thermal energy $= I^2R$

- If the component is at constant temperature, heat transfer to the surroundings takes place at the same rate. Therefore:

$$\textbf{Rate of heat transfer} = I^2R = \dfrac{V^2}{R}$$

- If the component heats up, its temperature rise depends on the power supplied to it (i.e. I^2R), the rate of heat transfer to the surroundings and the heat capacity of the component.
- The energy transfer per second to the component (i.e. the power supplied to it) does not depend on the direction of the current. For example, the heating effect of an alternating current at a given instant depends **only on the magnitude of the current** not on the direction of the current.

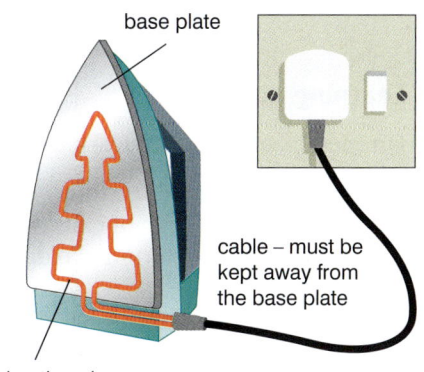

Figure 4 *Heating element in an iron*

base plate

cable – must be kept away from the base plate

heating element – on inside of base plate

Summary test 7.2

1 Calculate the total resistance of each of the resistor combinations in Figure 5.

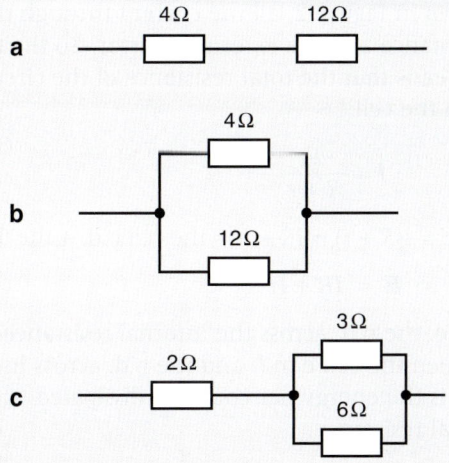

a 4Ω 12Ω

b 4Ω 12Ω

c 2Ω 3Ω 6Ω

Figure 5

2 A 3 Ω resistor and a 6 Ω resistor are connected in parallel. The parallel combination is connected in series with a 6 V battery of negligible internal resistance and a 4 Ω resistor, as shown in Figure 6.

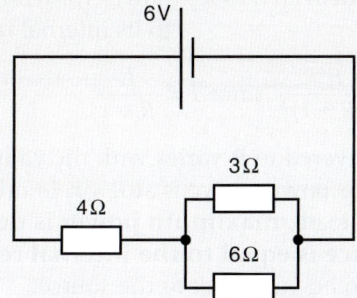

6V 4Ω 3Ω 6Ω

Figure 6

Calculate:

a the combined resistance of the 3 Ω resistor and the 6 Ω resistor in parallel,

b the total resistance of the circuit,

c the battery current,

d the power supplied to the 4 Ω resistor.

3 A 2 Ω resistor and a 4 Ω resistor are connected in series. The series combination is connected in parallel with a 9 Ω resistor and a 3 V battery of negligible internal resistance, as shown in Figure 7.

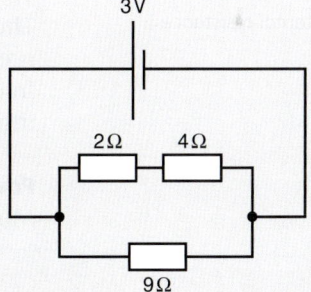

3V 2Ω 4Ω 9Ω

Figure 7

Calculate:

a the total resistance of the circuit,

b the battery current,

c the power supplied to each resistor,

d the power supplied by the battery.

4 Calculate:

a the power supplied to a 10 Ω resistor when the p.d. across it is 12 V,

b the resistance of a heating element designed to operate at 60 W and 12 V.

Learning outcomes

On these pages you will learn to:

- explain the effect of the internal resistance of a source of e.m.f. in a circuit on the terminal p.d. of the source
- solve circuit problems in which circuits include sources of e.m.f. with internal resistance

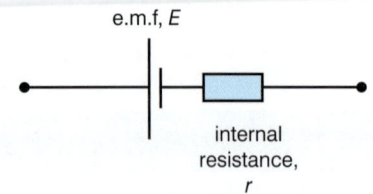

Figure 1 *Internal resistance*

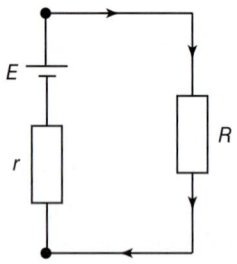

Figure 2 *E.m.f. and internal resistance*

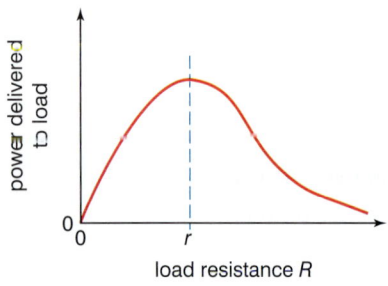

Figure 3 *Power delivered to a load v. load resistance*

Internal resistance

The **internal resistance** of a source of electricity is due to opposition to the flow of charge through the source. This causes electrical energy produced by the source to be dissipated inside the source when charge flows through it.

The **electromotive force (e.m.f.)** of the source is the electrical energy per unit charge produced by the source.

The **p.d. across the terminals** of the source is the electrical energy per unit charge that can be delivered by the source when it is in a circuit.

The **terminal p.d.** is less than the e.m.f. whenever current passes through the source. The difference is due to the internal resistance of the source.

> **The internal resistance of a source is the loss of potential difference per unit current in the source, when current passes through the source.**

In circuit diagrams, the internal resistance of a source may be shown as a resistor (labelled 'internal resistance') in series with the usual symbol for a cell or battery, as in Figure 1. If no internal resistance is shown, the symbol for the cell or battery should be labelled with symbols for its e.m.f. and its internal resistance.

When a cell of e.m.f. E and internal resistance r is connected to an external resistor of resistance R, as shown in Figure 2, all the current through the cell passes through its internal resistance and the external resistor. So the two resistors are in series, which means that the total resistance of the circuit is $r + R$. Therefore, the current through the cell:

$$I = \frac{E}{R + r}$$

In other words, the cell e.m.f., $E = I(R + r) = IR + Ir$ = the cell p.d. + the 'lost' p.d.

$$E = IR + Ir$$

The 'lost' p.d. inside the cell (i.e. the p.d. across the internal resistance of the cell) is equal to the difference between the cell e.m.f. and the p.d. across its terminals. In energy terms, the 'lost' p.d. is the energy per coulomb dissipated or wasted inside the cell due to its internal resistance.

Power

Multiplying each term of the above equation by the cell current I gives:

Power supplied by the cell, $IE = I^2R + I^2r$

In other words,

Power supplied by the cell = power delivered to R + power wasted in the cell due to its internal resistance

The power delivered to R = $I^2R = \frac{E^2R}{(R + r)^2}$ since $I = \frac{E}{R + r}$

Figure 3 shows how the power delivered to R varies with the value of R. It can be shown that the peak of this power curve is at $R = r$. In other words, when a source delivers power to a 'load', **maximum power is delivered to the load when the load resistance is equal to the internal resistance of the source.** The load is then said to be 'matched' to the source.

Measurement of internal resistance

The potential difference across the terminals of a cell when the cell is in a circuit can be measured by connecting a high-resistance voltmeter directly across the terminals of the cell. Figure 4 shows how the cell p.d. can be measured for different values of current. The current is changed by adjusting the variable resistor. The lamp limits the maximum current that can pass through the cell. The ammeter is used to measure the cell current.

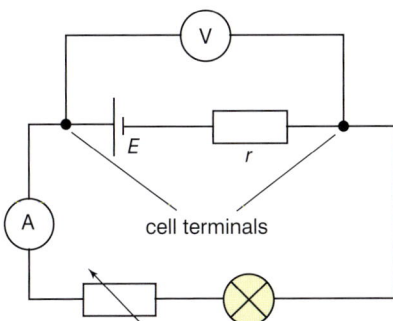

Figure 4 *Measuring internal resistance*

Graph of p.d. v. current

The measurements of cell p.d. and current for a given cell may be plotted on a graph, as shown in Figure 5.

The cell p.d. decreases as the current increases. This is because the 'lost' p.d. increases as the current increases.

- **The cell p.d. is equal to the cell e.m.f. at zero current**. This is because the 'lost' p.d. is zero at zero current.
- **The graph is a straight line with a negative gradient**. This can be seen by rearranging the equation

$$E = IR + Ir \text{ to become } IR = E - Ir$$

Because IR represents the cell p.d. V, then $V = E - Ir$. By comparison with the standard equation for a straight line $y = mx + c$, a graph of V on the y-axis against I on the x-axis gives a straight line with a gradient $-r$ and a y-intercept E. See Chapter 24, Mathematical skills.

Figure 5 shows the gradient triangle ABC, in which AB represents the lost p.d. and BC represents the current.

So the gradient $\dfrac{AB}{BC} = \dfrac{\text{lost voltage}}{\text{current}}$ = internal resistance r.

Note The internal resistance and the e.m.f. of a cell can be calculated, if the cell p.d. is measured for two different values of current. A pair of simultaneous equations can therefore be written, as follows:

- For current I_1, the cell p.d. $V_1 = E - I_1r$
- For current I_2, the cell p.d. $V_2 = E - I_2r$

Subtracting the first equation from the second gives:

$$V_1 - V_2 = (E - I_1r) - (E - I_2r) = I_2r - I_1r = (I_2 - I_1)r$$

Therefore, $r = \dfrac{(V_1 - V_2)}{(I_1 - I_2)}$

So r can be calculated from the above equation and then substituted into either equation for the cell p.d. to enable E to be calculated.

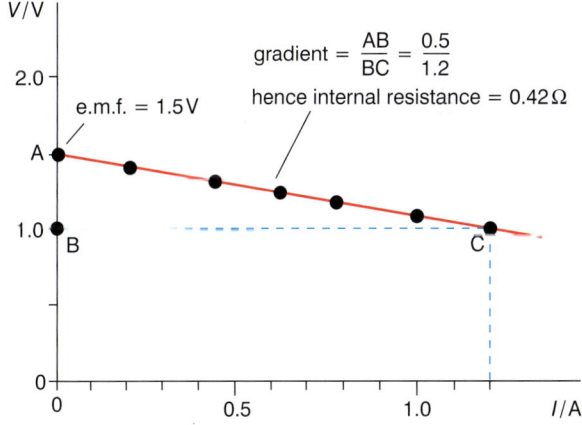

Figure 5 *A graph of cell p.d. v. current*

Summary test 7.3

1 A battery of e.m.f. 12 V and internal resistance 1.5 Ω is connected to a 4.5 Ω resistor. Calculate:

  a the total resistance of the circuit,

  b the current through the battery,

  c the lost p.d.,

  d the p.d. across the cell terminals.

2 A cell of e.m.f. 1.5 V and internal resistance 0.5 Ω is connected to a 2.5 Ω resistor. Calculate:

  a the current,

  b the terminal p.d.,

  c the power delivered to the 2.5 Ω resistor,

  d the power wasted in the cell.

3 The p.d. across the terminals of a cell is 1.1 V when the current from the cell is 0.20 A and 1.3 V when the current is 0.10 A. Calculate:

  a the internal resistance of the cell,

  b the cell's e.m.f.

4 A battery of unknown e.m.f., E, and internal resistance, r, is connected in series with an ammeter and a resistance box, R. The current was 2.0 A when $R = 4.0$ Ω, and 1.5 A when $R = 6.0$ Ω. Calculate E and r.

Learning outcomes

On these pages you will learn to:

- solve circuit problems in which circuits include two or more sources of e.m.f. in series or in parallel
- solve circuit problems in which circuits include diodes

Circuits with a single cell and one or more resistors

Here are some rules:

- Sketch the **circuit diagram** if it is not drawn.
- To calculate the **current** passing through the cell, calculate the total circuit resistance using the resistor combination rules. Don't forget to add on the internal resistance of the cell, if that is given:

$$\text{Cell current} = \frac{\text{cell e.m.f.}}{\text{total circuit resistance}}$$

- To work out the **current and p.d.** for each resistor, start with the **resistors in series** with the cell which pass the same current as the cell current:

$$\begin{array}{c}\text{p.d. across each resistor} \\ \text{in series with cell}\end{array} = \begin{array}{c}\text{current} \times \text{the resistance} \\ \text{of each resistor}\end{array}$$

- To work out the p.d. across the **parallel resistors**, find the difference between the cell e.m.f. and the sum of the p.d.s across the series resistors and the internal resistance.

The current through any parallel resistor is then given by dividing the p.d. across the parallel resistors by the resistance of that resistor.

Circuits with two or more cells in series

The same rules as above apply, except the current through the cells is calculated by dividing the **overall (i.e. net) e.m.f.** by the total resistance.

- If the cells are connected in the **same** direction in the circuit (Figure 1a), the net e.m.f. is the **sum** of the individual e.m.fs. For example, in Figure 1a the net e.m.f. is 3.5 V.
- If the cells are connected in **opposite** directions to each other in the circuit (Figure 1b), the net e.m.f. is the **difference** between the e.m.fs in each direction. For example, in Figure 1b, the net e.m.f. is 0.5 V in the direction of the 2.0 V cell.

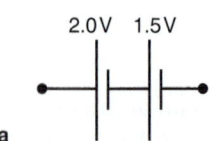 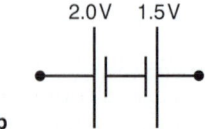

Figure 1 Cells in series

- The **total internal resistance** is the **sum** of the individual internal resistances. This is because the cells, and therefore the internal resistances, are in series.

Circuits with cells in parallel

Kirchhoff's laws are used to analyse circuits where cells and batteries are in parallel. For example, consider the circuit in Figure 2.

To determine the current through the 2.0 Ω resistor, let x and y represent the current through the 2.0 V and 1.5 V cells, which have internal resistances of 6.0 Ω and 4.0 Ω, respectively. The current through the 2.0 Ω resistor is therefore $x + y$.

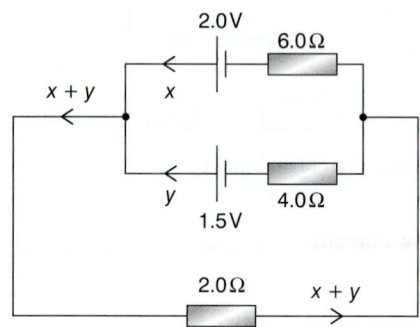

Figure 2

Applying Kirchhoff's second law

- to the outer loop (consisting of the 2.0 V cell with its internal resistance of 6.0 Ω and the 2.0 Ω resistor) gives
  $$2.0 = 6.0x + 2.0(x + y) \text{ or } 2.0 = 8.0x + 2.0y$$
- to the middle loop (consisting of the 1.5 V cell with its internal resistance of 4.0 Ω and the 2.0 Ω resistor) gives
  $$1.5 = 4.0y + 2.0(x + y) \text{ or } 1.5 = 2.0x + 6.0y$$

Solving these two simultaneous equations (see Chapter 24, Mathematical skills) gives $x = 0.205$ A and $y = 0.182$ A. Hence the current through the 2.0 Ω resistor $= x + y = 0.387$ A.

Note

In such calculations, a negative value for a current would mean the current is in the opposite direction to that assumed initially.

Diodes in circuits

Assume that a semiconductor diode has:

- **zero resistance in the forward direction** when the p.d. across it is 0.6 V or greater,
- **infinite resistance in the reverse direction** or at p.ds less than 0.6 V in the forward direction.

Therefore, in a circuit with one or more diodes as above:

- a p.d. of 0.6 V exists across a diode that is forward-biased and passing a current,
- a diode that is reverse-biased has infinite resistance.

For example, suppose a diode is connected in its forward direction in series with a 1.5 V cell and a 1.5 kΩ resistor, as in Figure 3.

The p.d. across the diode is 0.6 V, because it is forward-biased. Therefore, the p.d. across the resistor is 0.9 V (= 1.5 V − 0.6 V). The current through the resistor is therefore 6.0×10^{-4} A $\left(= \dfrac{0.9\,\text{V}}{1500\,\Omega}\right)$.

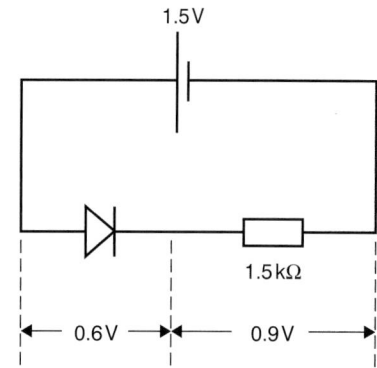

Figure 3 *Using a diode*

Summary test 7.4

1 A cell of e.m.f. 3.0 V, and negligible internal resistance, is connected to a 4.0 Ω resistor in series with a parallel combination of a 24.0 Ω resistor and a 12.0 Ω resistor (Figure 4).

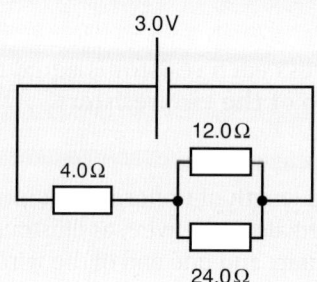

Figure 4

Calculate:

a the total resistance of the circuit,

b the cell current,

c the current and p.d. for each resistor.

2 A battery of e.m.f. 12.0 V, with an internal resistance of 3.0 Ω, is connected in series with a 15.0 Ω resistor and a battery of e.m.f. 9.0 V, which has an internal resistance of 2.0 Ω (Figure 5).

a Calculate:

i the total resistance of the circuit,

ii the battery current,

iii the current and p.d. across the 15 Ω resistor.

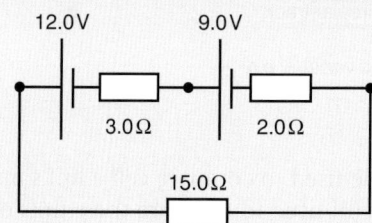

Figure 5

b A 1.0 Ω resistor is connected in parallel with the 12.0 V battery and the 3.0 Ω resistor. Calculate the current through: **i** the 1.0 Ω resistor, **ii** the 12.0 V battery.

3 a Two 8 Ω resistors and a battery of e.m.f. 12.0 V and internal resistance 8 Ω, are connected in series. Sketch the circuit diagram and calculate:

i the power delivered to each external resistor,

ii the power wasted due to internal resistance.

b The two 8 Ω resistors in part **a** are reconnected in parallel and are then connected to the same battery. Sketch the circuit diagram and calculate:

i the power delivered to each external resistor,

ii the power wasted due to internal resistance.

4 a For the circuit shown in Figure 6, calculate the p.d. and current for each resistor and diode.

b In the circuit (Figure 6), both diodes are reversed. Sketch the new circuit and calculate the p.d. and current for each resistor and diode for this new arrangement.

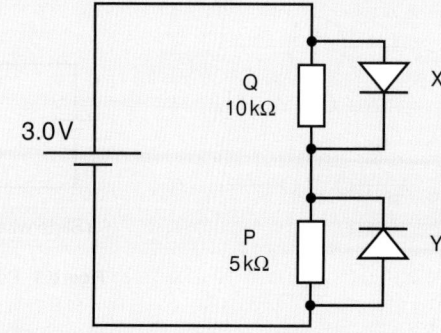

Figure 6

113

7.5 The potential divider

Learning outcomes

On these pages you will learn to:

- explain the principle of a potential divider circuit as a source of fixed p.d. or of variable p.d.
- recall and solve problems using the principle of the potentiometer as a means of comparing potential differences
- explain the operation of an electronic sensor consisting of a sensing device and a circuit that provides a voltage output
- explain the use of thermistors and light-dependent resistors in potential dividers

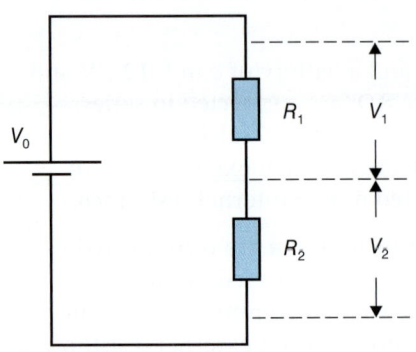

Figure 1 A potential divider

galvanometer (zero reading is at the centre)

power supply

potentiometer

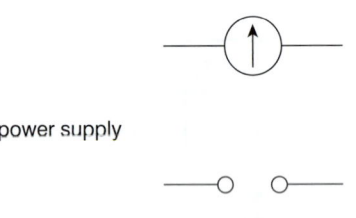

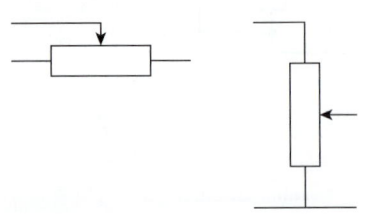

Figure 2 More circuit components

The theory of the potential divider

A **potential divider** consists of two or more resistances in series connected to a source of fixed potential difference. The source p.d. is divided between the resistors, as they are in series with each other. A potential divider can be used to supply a p.d. of any value between zero and the source p.d.

Figure 1 shows a potential divider consisting of two resistors R_1 and R_2, in series connected to a source of fixed p.d. V_0.

Total resistance of the combination = $R_1 + R_2$

Therefore, current through the resistors, $I = \dfrac{\text{p.d. across the resistors}}{\text{total resistance}} = \dfrac{V_0}{R_1 + R_2}$

so the p.d. across resistor R_1, $V_1 = IR_1 = \dfrac{V_0 R_1}{R_1 + R_2}$

and the p.d. across resistor R_2, $V_2 = IR_2 = \dfrac{V_0 R_2}{R_1 + R_2}$

Dividing the equation for V_1 by the equation for V_2 gives:

$$\frac{V_1}{V_2} = \frac{R_1}{R_2}$$

This equation shows that:

> **The ratio of the p.ds across each resistor is equal to the resistance ratio of the two resistors.**

To supply a variable p.d.

The source p.d. is connected to a fixed length of uniform resistance wire. A sliding contact on the wire can then be moved along the wire, as illustrated in Figure 3, giving a variable p.d. between the contact and one end of the wire. A uniform track of a suitable material may be used instead of resistance wire. The track may be linear or circular (Figure 3a,b). The circuit symbol for a variable potential divider which is also known as a **potentiometer** is shown in Figure 3c.

The variable potential divider in Figure 3c can be used to vary the brightness of a bulb by connecting the bulb between C and B. In contrast with using a variable resistor in series with the light bulb and the source p.d., the use of a potential divider enables the current through the light bulb to be reduced to zero. With a variable resistor at maximum resistance, there is a current through the light bulb.

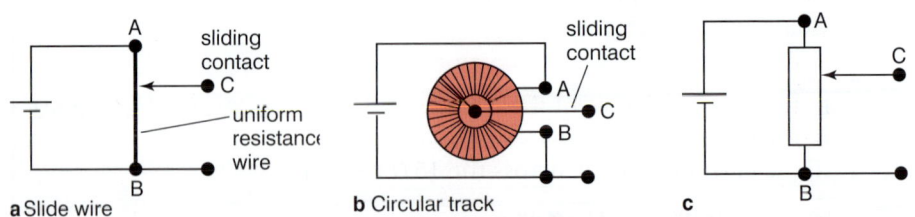

a Slide wire   **b** Circular track   **c**

Figure 3 Potential dividers used to supply a variable p.d.

To compare potential differences

The slide-wire potentiometer can be used to compare cell e.m.f.s and potential differences. Figure 4 shows the circuit used to compare the e.m.fs of two cells, X and Y. One of the cells (X) is shown connected in series with galvanometer G (ie a centre-reading meter) between end B of the wire and the sliding contact C on the wire. The driver cell provides a constant p.d. across the wire.

The position of the sliding contact on the wire is adjusted until the galvanometer G reads zero (ie there is a null reading). The cell e.m.f. E_x is then opposed equally (or 'balanced') by the p.d. between B and C. The length L_X of the wire from B to C is then measured.

The procedure is repeated using the other cell (Y). The ratio of the cell e.m.fs, E_X / E_Y is equal to the ratio of their balance lengths, L_X / L_Y.

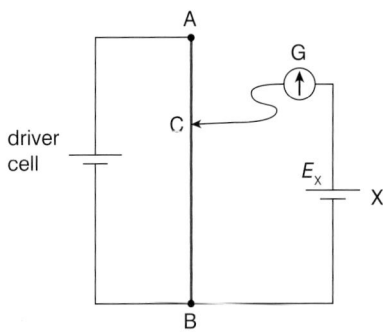

Figure 4 *Comparing cell e.m.f.s*

Sensor circuits

A **sensor circuit** produces an output p.d. which changes as a result of a change of a physical variable, such as temperature or pressure.

A temperature sensor

This consists of a potential divider made using a **thermistor** and a variable resistor (Figure 5).

With the temperature of the thermistor constant, the source p.d. is divided between the thermistor and the variable resistor. By adjusting the variable resistor, the p.d. across the thermistor can then be set at any desired value. When the temperature of the thermistor changes, its resistance changes so the p.d. across it changes. For example, suppose the variable resistor is adjusted so that the p.d. across the thermistor at 20 °C is exactly half the source p.d.; if the temperature of the thermistor is then raised, its resistance falls. As a result, the current increases, which means that the other resistor gets a bigger share of the source p.d.. So the thermistor p.d. therefore falls.

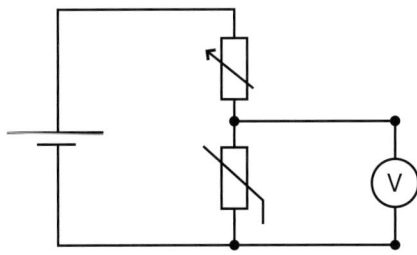

Figure 5 *A temperature sensor*

A light sensor

The circuit is similar to Figure 5 except a light-dependent resistor (LDR) is used instead of a thermistor. The p.d. across the LDR changes when the incident light intensity on the LDR changes. If the light intensity increases, the resistance of the LDR falls and the p.d. across the LDR falls.

Summary test 7.5

1 A potential divider consists of a 1.0 kΩ resistor in series with a 5.0 kΩ resistor and a battery of e.m.f. 4.5 V and negligible internal resistance.

   a Sketch the circuit and calculate the p.d. across each resistor.

   b A second 5.0 kΩ resistor is connected in the above circuit in parallel with the first 5.0 kΩ resistor. Calculate the p.d. across each resistor in this new circuit.

2 A 12 V battery, of negligible internal resistance, is connected to the fixed terminals of a variable potential divider (which has a maximum resistance of 50 Ω). A 12 V light bulb is connected between the sliding contact and the negative terminal of the potential divider. Sketch the circuit diagram and describe how the brightness of the light bulb changes when the sliding contact is moved from the negative to the positive terminal of the potential divider.

3 a A potential divider consists of an 8.0 Ω resistor in series with a 4.0 Ω resistor and a 6.0 V battery. Calculate:
    i the current,
    ii the p.d. across each resistor.

   b In the circuit in part a, the 4 Ω resistor is replaced by a thermistor with a resistance of 8 Ω at 20 °C and a resistance of 4 Ω at 100 °C. Calculate the p.d. across the fixed resistor at: i 20 °C, ii 100 °C.

4 A light sensor consists of a 5.0 V cell, an LDR and a 5.0 kΩ resistor in series with each other. A voltmeter is connected in parallel with the resistor. When the LDR is in darkness, the voltmeter reads 2.2 V.

   a Calculate:
    i the p.d. across the LDR,
    ii the resistance of the LDR when the voltmeter reads 2.2 V.

   b Describe and explain how the voltmeter reading would change if the LDR was exposed to daylight.

Launch additional digital resources for the chapter

1 Two resistors of resistances $6.0\,\Omega$ and $12.0\,\Omega$, in parallel, are connected to a $2.0\,\Omega$ resistor and a $6.0\,V$ battery. Figure 1.1 shows the circuit diagram.

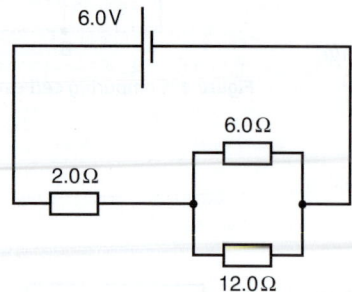

Figure 1.1

a Calculate the combined resistance of the three resistors.

b For each resistor, calculate:
  i the current,
  ii the p.d.,
  iii the power dissipated.

2 A light bulb is connected in series with a resistor and a battery, of negligible internal resistance, as shown below. A variable resistor is connected in parallel with the light bulb.

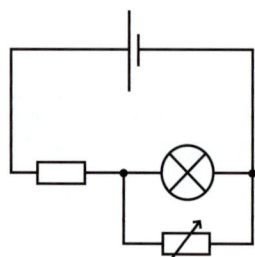

Figure 2.1

a Describe how the brightness of the light bulb changes as the resistance of the variable resistance is increased from zero.

b Explain why it is possible to reduce the bulb current to zero by adjusting the variable resistor.

3 A $15.0\,\Omega$ resistor and a $3.0\,\Omega$ resistor are connected in series with a $3.0\,V$ battery, which has an internal resistance of $2.0\,\Omega$, as shown in Figure 3.1.

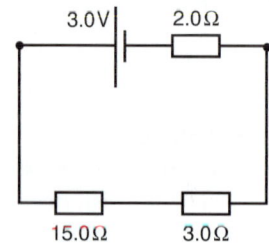

Figure 3.1

a Calculate:
  i the total resistance of the circuit,
  ii the p.d. across each of the external resistors,
  iii the p.d. across the battery terminals.

b Calculate the power dissipated:
  i in each resistor,
  ii in the battery due to its internal resistance.

4 The circuit diagram below is for the rear windscreen heater of a car. It consists of 4 heating elements, each of resistance $6.0\,\Omega$, connected to a $12.0\,V$ battery of internal resistance $2.0\,\Omega$.

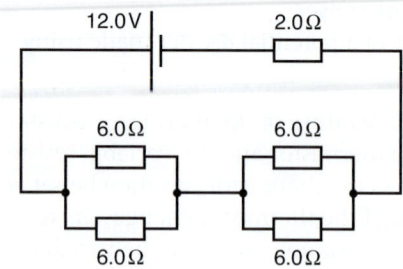

Figure 4.1

a Calculate:
  i the current,
  ii the p.d. across each heating element.

b Calculate the power dissipated:
  i in each heating element,
  ii in the battery due to its internal resistance.

c How would the operation of the heater differ if the heating elements were in parallel with each other, with the battery connected across the parallel combination?

5 A $6.0\,V$ battery of unknown internal resistance is connected in series with a switch, a resistor of resistance $4.0\,\Omega$ and an ammeter. When the switch is closed, the ammeter reads $1.0\,A$.

a Calculate the internal resistance of the battery.

b A second $4.0\,\Omega$ resistor was connected in parallel with the first resistor. Calculate the ammeter reading with this second resistor in the circuit.

6 A hand-held hair dryer has two $570\,\Omega$ identical heating elements, each in series with a switch. The heating elements are connected in parallel with each other and a $230\,V$ mains electricity supply. A $12\,W$, $230\,V$ electric fan in the hair dryer is used to blow air over the heating elements.

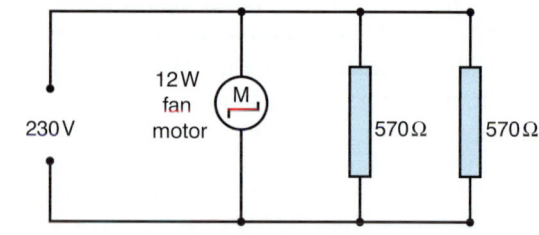

Figure 6.1

a When both heaters are being used, and the electric fan is on, calculate:
  i the current passing through each heating element,
  ii the power supplied by the electricity supply,
  iii the total current passing through the hair dryer.

b When one heater only is on, calculate:
  i the total current passing through the heater,
  ii the power supplied by the electricity supply.

7 The diagram shows a potential divider circuit consisting of a $10\,k\Omega$ resistor R connected in series with an n.t.c. thermistor and a $6.0\,V$ battery. A high resistance voltmeter is connected across the thermistor.

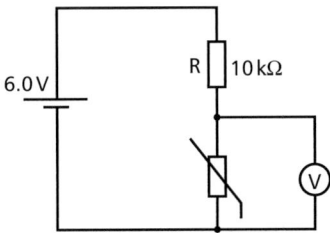

Figure 7.1

a The voltmeter reads $2.0\,V$ when the thermistor's temperature is $20\,°C$. Calculate the resistance of the thermistor at this temperature.

b A second $10\,k\Omega$ resistor was connected in parallel with R. Calculate the voltmeter reading with this second resistor in the circuit when the thermistor is at the same temperature.

c Describe and explain how the voltmeter reading would change if the the temperature of the thermistor were increased.

8 The diagram shows a circuit in which two $5.0\,k\Omega$ resistors are connected in series with each other and a $5.0\,V$ battery, of negligible internal resistance. A diode, is connected in parallel with each resistor, as shown.

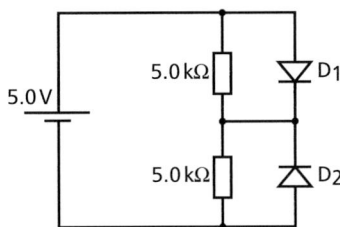

Figure 8.1

a Calculate:
  i the p.d.,
  ii the current through each resistor.

b Calculate the current through each resistor if the $5.0\,V$ battery was replaced with a $9.0\,V$ battery, also of negligible internal resistance.

9 In a potentiometer experiment using a slide-wire potentiometer as shown in Topic 7.5 Figure 3, a cell X is compared with a standard cell S of e.m.f. $1.50 \pm 0.01\,V$. The balance length for S was $735 \pm 3\,mm$ and for X was $530 \pm 3\,mm$.

a Calculate the e.m.f. of cell X.

b Calculate the percentage uncertainty in:
  i the e.m.f. of S,   **ii** each balance length.

c Hence calculate the uncertainty in the e.m.f. of X.

10 A slide wire potentiometer, as in Topic 7.5, Figure 3a, is used to compare the e.m.f. of a cell Y with the p.d. across the terminals of the same cell when a resistor S is connected across its terminals. The following measurements are made.
Without resistor S across Y, balance length $l_1 = 741$ mm
With resistor S across Y, balance length $l_2 = 625$ mm

a **i** Show that the ratio $\dfrac{l_1}{l_2} = \dfrac{R+r}{R}$
    where r is the internal resistance of cell Y and R is the resistance of S.
  ii The resistance of S is $5.00\,\Omega$. Calculate the internal resistance r of cell Y.

b Immediately after the second measurement, the first measurement is repeated in case the e.m.f. of the driver cell changes during the measurements. The repeat measurement is the same as the first measurement. If the repeat measurement had been:
i less than, **ii** greater than the first measurement, state one conclusion you would have drawn in each case.

11a A cell of e.m.f., E_X, and negligible internal resistance, X, is connected to a network of identical resistors A, B, C, D, E and F as shown in Figure 11.1. Each resistor has resistance R.

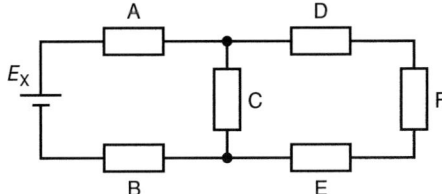

Figure 11.1

  i Calculate the total resistance of the network in terms of R.
  ii Determine the current in resistor F in terms of E_X and R.

b Resistor F is replaced by cell Y which is identical to X, as shown in Figure 11.2.

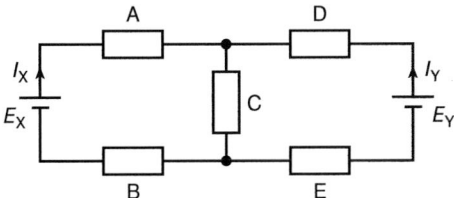

Figure 11.2

  i Use Kirchhoff's laws to show that $E_X = (3I_X + I_Y)\,R$, where I_X and I_Y are the currents in cells X and Y respectively.
  ii Derive a similar expression for E_Y, the e.m.f. of cell Y.
  iii If each cell has an e.m.f. of $6.0\,V$ and $R = 2.0\,\Omega$, calculate the potential difference across resistor C.

8 Waves

8.1 Waves and vibrations

Table 1 *The electromagnetic spectrum*

Type	Wavelength range
radio	>0.1 m
microwave	0.1 m to 1 mm
infrared	1 mm to about 650 nm
visible	about 650 nm to 350 nm
ultraviolet	about 350 nm to 1 nm
X-rays	<1 nm
gamma rays	<1 nm

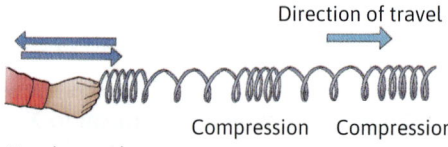

Hand moved backwards and forwards along the line of the slinky

Figure 2 *Longitudinal waves on a slinky*

Types of wave

Waves transfer energy from the source that creates them, to the objects that absorb them. Waves that pass through a substance are **vibrations** of the particles of the substance. Sound waves, seismic waves and waves on strings are examples of waves that pass through a substance. These types of wave are often referred to as **mechanical waves**. When waves progress through a substance, the particles of the substance vibrate in a certain way which makes other particles vibrate in the same way, and so on.

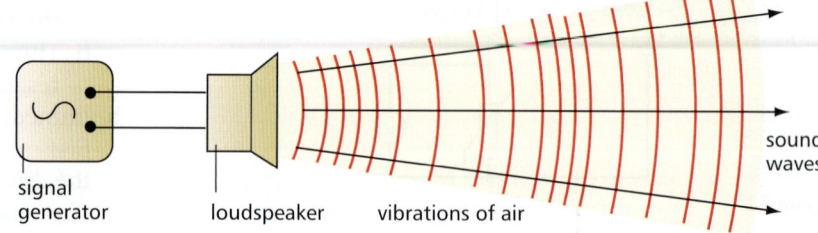

Figure 1 *Creating sound waves in air*

- **Electromagnetic waves** are vibrating **electric** and **magnetic fields** that progress through space, without the need for a substance. The vibrating electric field generates a vibrating magnetic field, which generates a vibrating electric field further away, and so on. Electromagnetic waves include radio waves, microwaves, infrared radiation, light, ultraviolet radiation, X-rays and gamma radiation. All electromagnetic waves travel with the same speed through free space. The full spectrum of electromagnetic waves is listed in Table 1. All types of waves transfer energy when they travel through a substance but only electromagnetic waves can transfer energy through a vacuum.

Longitudinal and transverse waves

Longitudinal waves

Longitudinal waves are waves in which the direction of vibration of the particles is **parallel** to (ie along) the direction in which energy is transferred by the waves. Sound waves, primary seismic waves and compression waves on a slinky are all longitudinal waves. Figure 2 shows how to send longitudinal waves along a 'slinky'. When one end of the slinky is moved to and fro repeatedly, each 'forward' movement causes a compression wave to pass along the slinky as the coils push into each other. Each 'reverse' movement causes the coils to move apart, so an 'expansion' wave passes along the slinky.

- **Sound waves** in air are created when a surface in contact with air vibrates. When the vibrating surface pushes on the air, the air molecules near the surface are pushed away from the surface, pushing on adjacent molecules which also push on adjacent molecules. This creates a wave of 'high density' air (i.e. a compression) that passes through the air. When the vibrating surface 'retreats', the air molecules near the surface move back into the space vacated by the surface, allowing adjacent air molecules to fall back and so on. This creates a wave of 'low density' air (referred to as a 'rarefaction') that passes through the air behind the compression wave. The vibrating surface therefore creates a series of **compressions** and **rarefactions** that pass through the air. The air molecules are therefore repeatedly pushed to and fro along the direction in which the sound travels.

Transverse waves

Transverse waves are waves in which the direction of vibration is **perpendicular** to the direction in which energy is transferred by the waves. Electromagnetic waves, secondary seismic waves and waves on a string or a wire are all transverse waves. Figure 3 shows transverse waves travelling along a rope. When one end of the rope is moved from side to side repeatedly, these sideways movements

travel along the rope: each unaffected part of the rope is pulled sideways when the part next to it moves sideways.

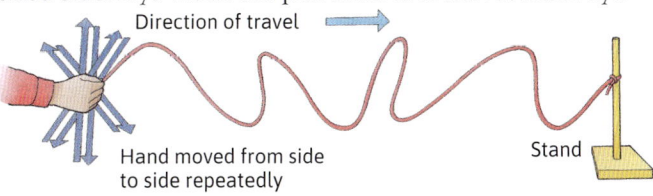

Figure 3 *Making rope waves*

Representations of transverse and longitudinal waves

Imagine a rope marked along its length into very short 'elements' of equal length. When a wave travels along the rope, each element is 'disturbed' by the wave and vibrates about its equilibrium (ie undisturbed) position along a line perpendicular to the direction of travel of the wave. Figure 4a shows a snapshot of a transverse wave on a rope. Figure 4b shows how the displacement of the rope at a certain time varies with distance from P along the rope. Figure 4c shows how the displacement of P varies with time.

The graph is easy to interpret because the waves on the rope are transverse waves and the graph has the same shape as a 'snapshot' of the wave. For example, two elements either side of an element at zero displacement are displaced in opposite directions. Each point on the rope oscillates about its undisturbed position along a line perpendicular to the direction of travel of the waves. For example, point P in Figure 4a is at zero displacement in the snapshot and it then moves up to maximum positive displacement then back to zero displacement in the next half cycle. Figure 4c shows how the displacement of point P changes with time in the next two cycles. Note that in Figure 4c the symbol T represents the time for one cycle.

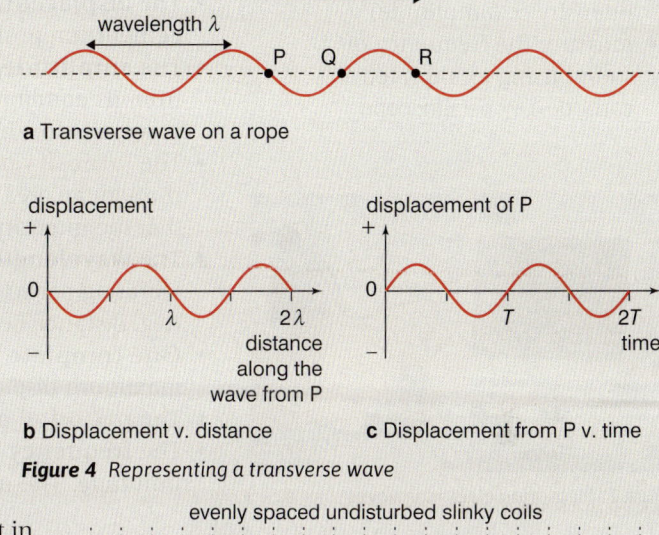

a Transverse wave on a rope

b Displacement v. distance   **c** Displacement from P v. time

Figure 4 *Representing a transverse wave*

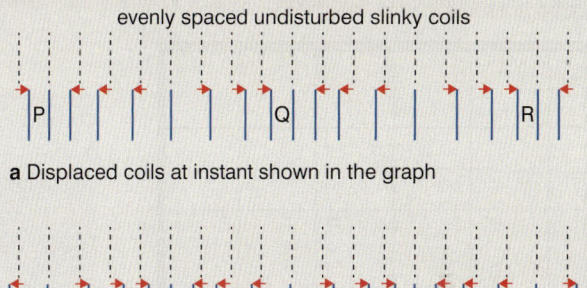

a Displaced coils at instant shown in the graph

b Displaced coils half a cycle later

Figure 5 *Longitudinal waves on a slinky*

Compare the waves on a rope with longitudinal waves on a 'slinky'. Figure 5 shows a representation of a longitudinal wave travelling along the slinky. The compressions are where the coils are closest and the rarefactions are where they are furthest apart. A displacement against distance graph for Figure 5a would be similar to Figure 4b.

- The coil at the centre of each 'compression' has zero displacement (for example at P in Figure 5a) and the coils on either side near the centre are displaced towards the centre of the compression. On a displacement v. distance graph, this would be where the displacement changes from + to – along the wave.
- The coil at the centre of each 'rarefaction' has zero displacement (for example midway between P and Q) and the coils on either side near the centre are displaced away from the centre of the rarefaction. On a displacement v. distance graph, this would be where the displacement changes from – to + along the wave.

Summary test 8.1

1 Classify the following types of wave as either longitudinal or transverse:
   radio waves   sound waves
   secondary seismic waves   microwaves
2 Sketch a snapshot of a longitudinal wave travelling on a slinky coil, indicating the direction in which the waves are travelling and areas of compression and expansion.
3 Sketch a snapshot of a transverse wave travelling along a rope, indicating the direction in which the waves are travelling and the direction of motion of the particles at the peaks and troughs.
4 **a** Describe how the position of the coil marked Q in Figure 5a varies with time over the next cycle.
   b Sketch a graph to illustrate your description in **a**.

On these pages you will learn to:

- explain and use the terms displacement, amplitude, phase difference, period, frequency, wavelength and speed
- deduce, recall and use the wave equation $v = f\lambda$
- recall and use the relationship intensity $\propto$ (amplitude)^2
- measure the frequency of sound using a calibrated cathode-ray oscilloscope

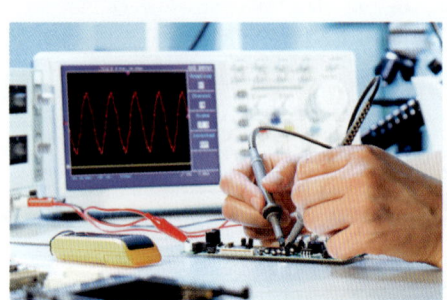

Figure 1 *Measuring electrical waves*

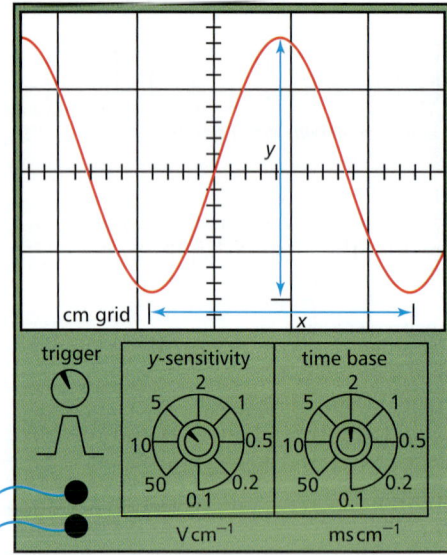

a Trace height $y = 32$ mm
∴ trace amplitude = 16 mm = 1.6 cm
given y-sensitivity = 5 V cm^{-1}
voltage amplitude = 5 × 1.6 = 8.0 volts

b x-distance from peak to peak = 32 mm = 3.2 cm
given time base 2 ms cm^{-1}
time period = 2 × 3.2 = 6.4 ms

$$\text{frequency } f = \frac{1}{\text{time period}} = \frac{1}{6.4 \times 10^{-3}\,\text{s}}$$

$f = 1.56 \times 10^2$ Hz = 156 Hz

Figure 3 *Measuring a waveform*

When an intercontinental phone call is made, sound waves are converted to electrical waves. These waves are carried by electromagnetic waves from ground transmitters to satellites in space and back to receivers on Earth, where they are converted back to electrical waves, then back to sound waves. The engineers who design and maintain communications systems need to measure the different types of wave at different stages to make sure the waves are not distorted.

Key terms

The following terms, illustrated in Figure 2, are used to describe waves:

- The **displacement** of a vibrating particle is its distance and direction from its equilibrium position.
- The **amplitude** of a wave is the maximum displacement of a vibrating particle from its equilibrium position. For a transverse wave, this is the height of a wave crest or the depth of a wave trough from the middle.
- The **intensity of a wave** is the **power per unit area** the waves would transfer through an area perpendicular to the direction of the waves. It can be shown that the intensity of a wave is proportional to the square of its amplitude.
- The **wavelength** of a wave is the least distance between two adjacent vibrating particles with the same displacement and velocity at the same time (e.g. distance between adjacent crests).
- **One complete cycle** of a wave is from a maximum displacement to the next maximum displacement (e.g. from one wave peak to the next).
- The **period of a wave** is the time for one complete wave to pass a fixed point.
- The **frequency** of a wave is the number of cycles of vibration of a particle per unit time. The unit of frequency is the **hertz (Hz)**. For waves of frequency f:

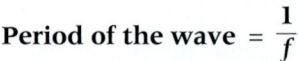

$$\textbf{Period of the wave} = \frac{1}{f}$$

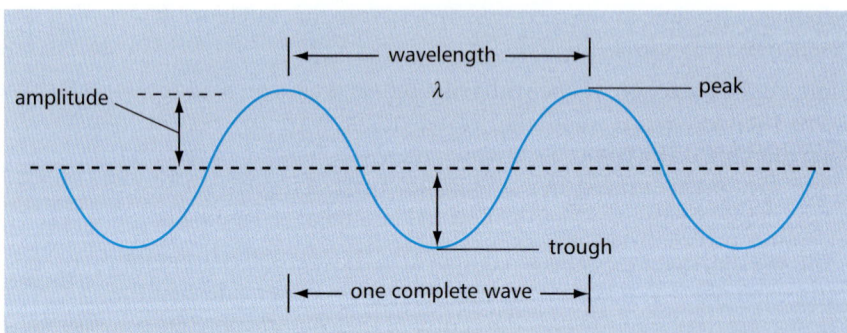

Figure 2 *Parts of a wave*

Using an oscilloscope to measure sound waves

The oscilloscope shown in Figure 3 is connected to a **microphone**, which detects sound waves emitted by a loudspeaker connected to a signal generator. The trace on the oscilloscope screen is the **waveform** of the sound waves produced by the loudspeaker. Figure 3 shows how to measure the amplitude and the frequency of this waveform using the oscilloscope controls. Note that the waveform **amplitude** is the distance **from the middle to the top**.

Wave speed

The higher the frequency of a wave, the shorter its wavelength. For example, if waves are sent along a rope, the higher the frequency at which they are produced, the closer together their wave peaks. The same effect can be seen in a ripple tank, when straight waves are produced at a constant frequency. If the frequency is raised to a higher value, the waves are closer together. Consider

Figure 4, which represents the crests of straight waves in a ripple tank travelling at constant speed.

- Each wave crest travels a distance equal to one wavelength (λ) in the time for one cycle.
- The time taken for one cycle = $\frac{1}{f}$, where f is the frequency of the waves.

Therefore, the speed of the waves:

$$v = \frac{\text{distance travelled in one cycle}}{\text{time taken for one cycle}} = \frac{\lambda}{1/f} = f\lambda$$

For waves of frequency f and wavelength λ:

Wave speed, $v = f\lambda$

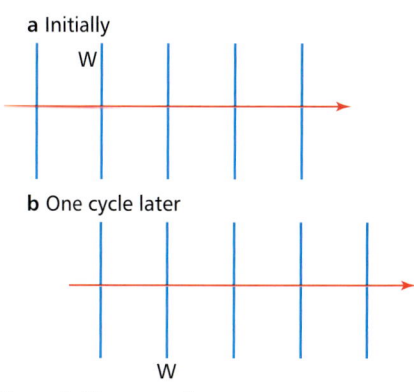

Figure 4 *Wave speed*

Phase difference

The **phase difference** between two vibrating particles is the fraction of a cycle between the vibrations of the two particles, measured either in **degrees or radians** where:

$$1 \text{ cycle} = 360° = 2\pi \text{ radians}$$

For two points at distance d apart along a wave of wavelength λ,

Phase difference, in radians $= \dfrac{2\pi d}{\lambda}$

> **Note**
>
> The symbol c is used for the speed of electromagnetic waves.

Figure 5 shows three successive snapshots of the particles of a transverse wave that progresses from left to right across the diagram. Particles O, P, Q, R and S are spaced approximately $\frac{1}{4}$ of a wavelength apart. Table 1 shows the phase difference between O and each of the other particles.

Table 1 *Phase differences*

	P	Q	R	S
Distance from O	$\frac{1}{4}\lambda$	$\frac{1}{2}\lambda$	$\frac{3}{4}\lambda$	λ
Phase difference relative to O / radians	$\frac{1}{2}\pi$	π	$\frac{3}{2}\pi$	2π

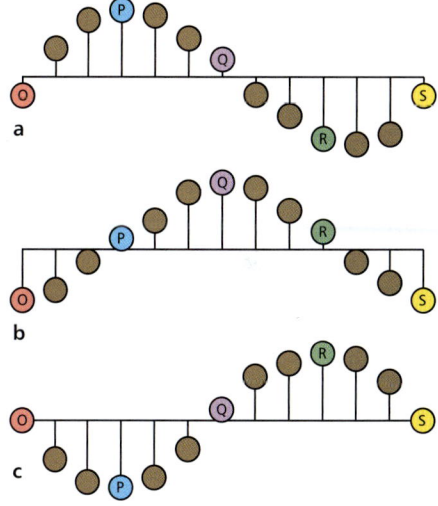

Figure 5 *Progressive waves*

Summary test 8.2

1 Sound waves, in air, travel at a speed of $340\,\text{m s}^{-1}$ at 20 °C. Calculate the wavelength of sound waves, in air, which have a frequency of:
   a 3400 Hz   b 18 000 Hz.

2 Electromagnetic waves, in air, travel at a speed of $3.0 \times 10^8\,\text{m s}^{-1}$. Calculate the frequency of electromagnetic waves of wavelength:
   a 0.030 m   b 600 nm.

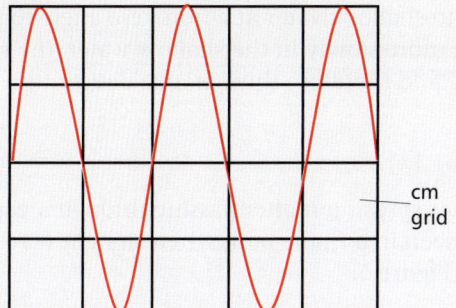

Figure 6

3 Figure 6 shows a waveform on an oscilloscope screen, when the y-sensitivity of the oscilloscope was $0.50\,\text{V cm}^{-1}$ and the time base was set at $0.5\,\text{ms cm}^{-1}$. Determine the amplitude and the frequency of this waveform.

4 For the waves in Figure 5:
   a determine:
      i the amplitude if the wavelength is 55 mm,
      ii the phase difference between P and R,
      iii the phase difference between P and S.
   b What would be the displacement and direction of motion of Q three-quarters of a cycle after the last snapshot?

Wave properties

Learning outcomes

On these pages you will learn to:

- know what is meant by reflection and refraction of waves
- describe examples of reflection and refraction
- explain what is meant by diffraction
- show an understanding of experiments that demonstrate diffraction, including the diffraction of water waves in a ripple tank with both a wide gap and a narrow gap
- describe wave motion in ripple tanks

Wave properties, such as reflection, refraction and diffraction, occur with many different types of wave. A **ripple tank** may be used to study these wave properties. The tank is a shallow transparent tray of water with sloping sides. The slopes prevent waves reflecting off the sides of tank. If they did reflect, it would be difficult to see the waves to be observed.

- The waves observed in a ripple tank are referred to as '**wave fronts**', which are lines of constant phase (e.g. crests).
- The direction in which a wave travels is at right angles to the wave front.

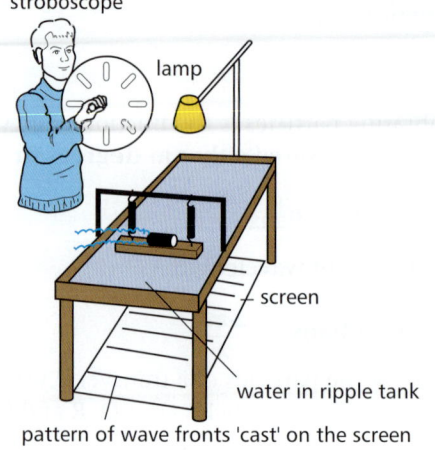
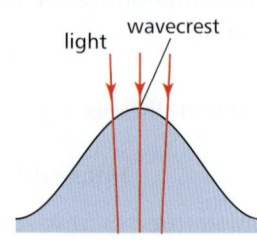

Each wave crest acts like a convex lens and concentrates the light onto the screen. So the pattern on the screen shows the wave crests.

Figure 1 The ripple tank

Reflection

- **Straight waves** directed at a certain angle to a hard flat surface (the '**reflector**') reflect off at the same angle, as shown in Figure 2. The angle between the reflected wave front and the surface is the **same** as the angle between the incident wave front and the surface. Therefore the direction of the reflected wave is at the same angle to the reflector as the direction of the incident wave. This same effect is observed when a light ray is directed at a plane mirror. The angle between the incident ray and the mirror is the **same** as the angle between the reflected ray and the mirror.

Refraction

When waves pass across a boundary at which the wave speed changes, the wavelength also changes. If the wave fronts are at a non-zero angle to the boundary, they change direction as well as changing speed. This effect is known as **refraction**.

Figure 4 shows the refraction of water waves in a ripple tank when they pass across a boundary, from deep to shallow water at a non-zero angle to the boundary. Because they move more slowly in the shallow water, the wavelength is smaller in the shallow water and therefore they change direction.

Diffraction

Diffraction occurs when waves spread out after passing through a gap, or round an obstacle. The effect can be seen in a ripple tank when straight waves are directed at a gap, as shown in Figure 5.

- The narrower the gap, the more the waves spread out.
- The longer the wavelength, the more the waves spread out.

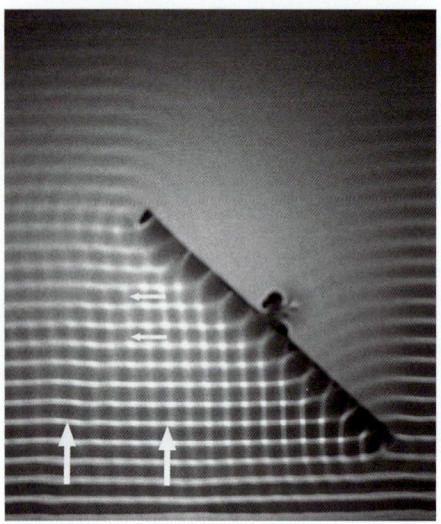

Figure 2 Reflection of plane waves

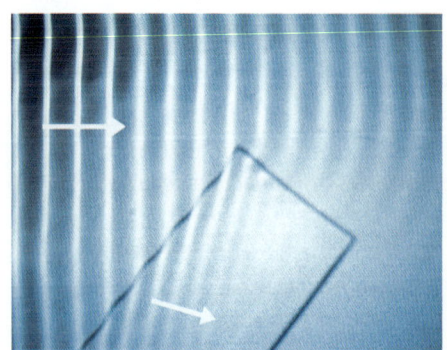

Figure 3 Refraction

To explain why the waves are diffracted on passing through the gap, consider each point on a wave front as a secondary emitter of 'wavelets'. The wavelets from the points along a wave front travel only in the direction in which the wave is travelling, not in the reverse direction, and they combine to form a new wave front spreading beyond the gap.

Investigating diffraction
Use a ripple tank to direct plane waves continuously at a gap between two metal barriers.

1 Change the gap spacing and observe the effect on the diffraction of the waves that pass through the gap. You should find that the diffraction of the waves increases as the gap is made narrower, as shown in Figure 4.
2 Keep the gap spacing constant and change the wavelength of the waves by altering the frequency of the vibrating beam. Observe the effect on the diffraction of the waves. You should find that the smaller the wavelength of the waves, the less they are diffracted.

a) at a wide gap

b) at a narrow gap

Figure 4 The effect of the gap width

Summary test 8.3

1 Copy and complete Figure 5, by showing the wave front after it has been reflected from the straight reflector. Also, show the direction of the reflected wave front.

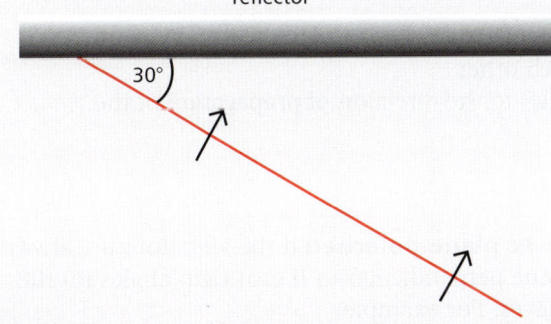

reflector

30°

Figure 5

2 Copy and complete Figure 6, by showing the wave fronts after they have passed across the boundary and have been refracted. Also, show the direction of the refracted waves.

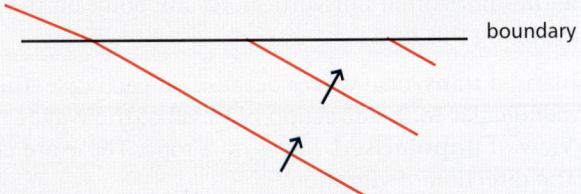

boundary

Figure 6

3 Water waves are diffracted on passing through a gap. How is the amount of diffraction changed as a result of:
 a widening the gap without changing the wavelength,
 b increasing the wavelength of the water waves without changing the gap width,

c increasing the wavelength of the water waves and reducing the gap width,
d widening the gap and increasing the wavelength of the waves?

4 Microwaves are reflected by a metal plate. In Figure 7, microwaves from a transmitter are directed at a gap between two metal plates. The detector placed on the other side of the plates receives a signal from the transmitter.
 a Explain why the detector at the position shown receives a signal even though it is not directly in line with the transmitter.
 b With the detector in the same position, describe how the signal would change if the gap was made wider.

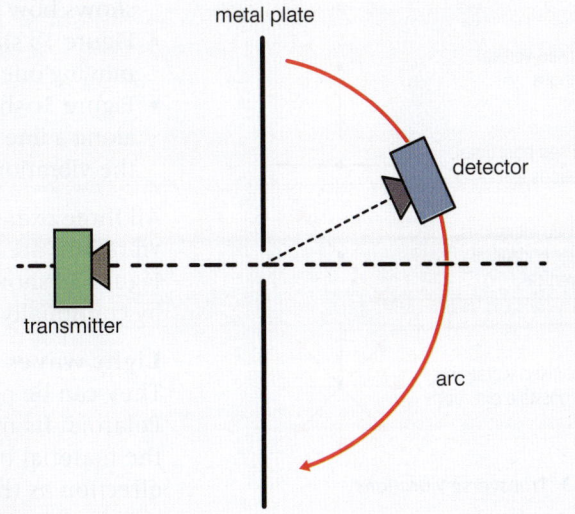

metal plate

detector

transmitter

arc

Figure 7

123

Electromagnetic waves and polarisation

Learning outcomes

On these pages you will learn to:

- recall that electromagnetic waves are transverse waves
- describe what an electromagnetic wave is
- explain what is meant by polarisation
- recall and use Malus's law

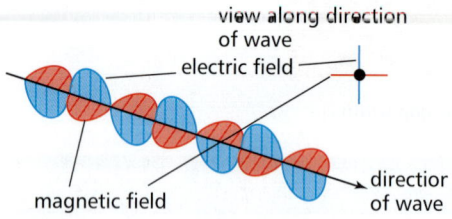

Figure 1 *Electromagnetic waves*

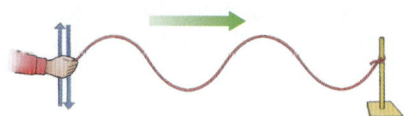

Figure 2 *Transverse waves*

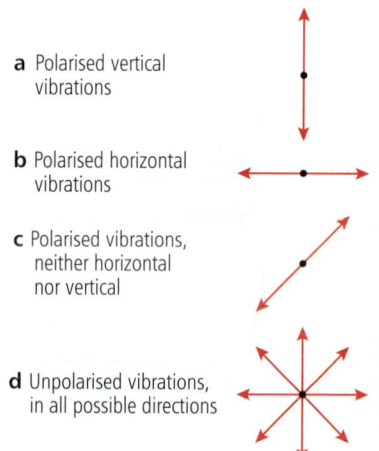

a Polarised vertical vibrations

b Polarised horizontal vibrations

c Polarised vibrations, neither horizontal nor vertical

d Unpolarised vibrations, in all possible directions

Figure 3 *Transverse vibrations*

The nature of electromagnetic waves

Electromagnetic waves were predicted by James Clerk Maxwell in 1862. Maxwell knew that a magnetic field is created round a wire when an electric current passes along the wire. He knew about Michael Faraday's discovery that a changing magnetic field in a coil of wire induces a voltage in the coil. He wondered if the two effects could be linked, and he used his mathematical skills to discover the link. In effect, Maxwell showed that the changing magnetic field created by an alternating current in a wire creates an alternating electric field, which creates an alternating magnetic field further away, which creates an alternating electric field, and so on.

Maxwell showed that the result is an **electromagnetic wave** consisting of an alternating magnetic field in phase with an alternating electric field. He worked out that the speed of an electromagnetic wave should be 300 000 km s^{-1}, the same as the speed of light in a vacuum! Maxwell realised, from his theory, that light must be an electromagnetic wave and that electromagnetic waves must exist beyond the visible spectrum. He knew that infrared and ultraviolet radiation are outside the visible spectrum, and he correctly predicted electromagnetic waves beyond these two invisible forms of radiation.

An electromagnetic wave consists of an **alternating magnetic field** (the magnetic wave) and an **alternating electric field** (the electric wave), in which the magnetic wave and the electric wave:

- are at right angles to each other
- vibrate in phase with each other
- both vibrate at right angles to the direction of propagation of the electromagnetic wave.

Polarisation

Tranverse waves are said to be **plane-polarised** if the vibrations are always along the same line in a plane perpendicular to (i.e. at right angles to) the direction of travel of the waves. For example:

- In Figure 2, each point on the rope vibrates in a vertical line only. Figure 3a shows how the rope would look if it was viewed end-on.
- Figure 3b shows an end-view of a rope made to vibrate horizontally by moving one end from side to side along a horizontal line.
- Figure 3c shows an end-view of the rope if one end is moved from side to side along a line which is neither horizontal nor vertical. At any point on the rope, the vibrations are always along this line.

All three examples are polarised transverse waves because, in each case, the vibrations are always perpendicular to the direction of travel of the waves. Figure 3d shows an end-view of **unpolarised** waves on a rope. These are created by continually changing the direction of vibration.

Light waves from a lamp bulb are unpolarised transverse waves. They can be polarised by passing them through special material called Polaroid. Its molecules are all lined up in the same direction. As a result, the material only allows light waves through that vibrate in the same direction as the molecules (see Figure 4).

By convention, the plane of polarisation of an electromagnetic wave is defined as the plane in which the electric field oscillates.

Investigating polarisation

If unpolarised light is passed through two Polaroid filters as in Figure 5, the transmitted light intensity changes if one polaroid is rotated relative to the other one. The filters are said to be 'crossed' when the transmitted intensity is a minimum. At this position, the polarised light from the first filter (the 'polariser') cannot pass through the second filter (the analyser), as the alignment of molecules in the second filter is at 90° to the alignment in the first filter. This is like passing rope waves through two 'letter boxes' at right angles to each other, as shown in Figure 5.

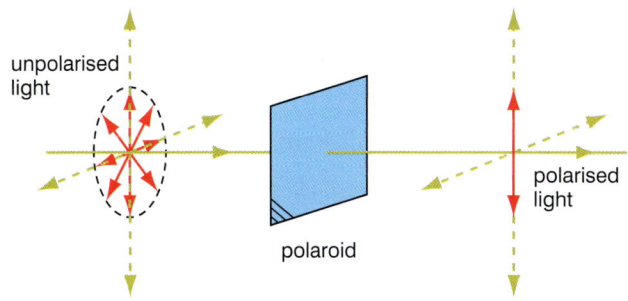

Figure 4 *Polarising a light beam using a Polaroid filter*

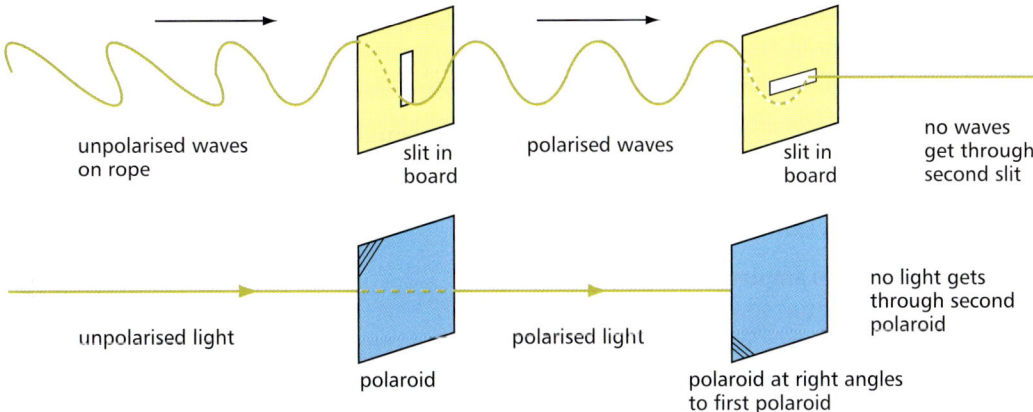

Figure 5 *Investigating polarisation*

The intensity of the polarised light transmitted through the analyser changes as the analyser is rotated about the transmitted beam. If the analyser is rotated from the position of maximum transmitted intensity I_0, the intensity I for an angle of rotation θ is given by the equation

$$I = I_0 \cos^2 \theta$$

This equation is known as **Malus's law**. The equation follows from the fact that the amplitude A of the wave at an angle of rotation θ is equal to $A_0 \cos \theta$. Since the intensity is proportional to the amplitude A, it follows that the intensity is proportional to $\cos^2 \theta$. Figure 6 shows how the intensity of the transmitted beam changes as the angle of rotation increases from 0 to 360°.

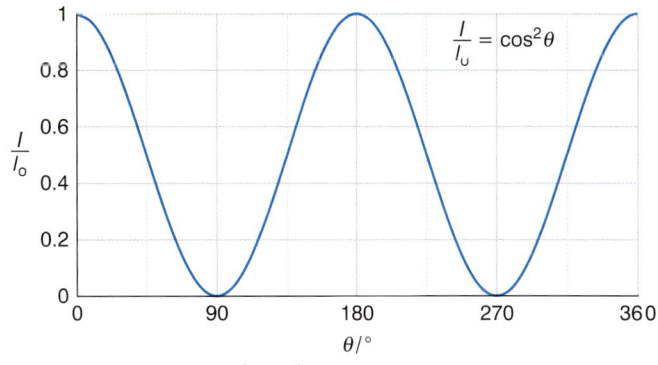

Figure 6 *Graph of* $\dfrac{intensity,\ I}{maximum\ intensity,\ I_0}$ *against angle of rotation,* θ

Summary test 8.4

1 a Describe an electromagnetic wave.

 b Explain what is meant by electromagnetic waves that are:

  i polarised,

  ii unpolarised.

2 A light source is observed through two Polaroid filters. Describe and explain what you would expect to observe when one of the filters is rotated through 180° about the line of sight.

3 When the aerial of a portable radio is turned away from a vertical position, the radio signal becomes weaker. Explain this effect.

4 a State Malus's law.

 b A polarised beam of light is directed through a Polaroid filter. Calculate the percentage reduction in the intensity of the transmitted light when the filter is rotated about the beam by an angle of 10° from the position of maximum intensity.

Learning outcomes

On these pages you will learn to:

- explain the principle of superposition in simple applications
- use the principle of superposition in simple applications, including interference and stationary waves
- describe and explain experiments that demonstrate two-source interference using water ripples, sound waves and microwaves

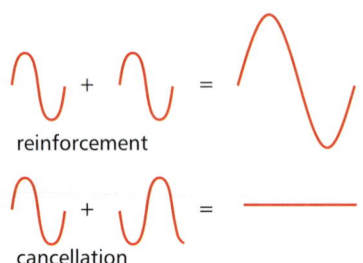

Figure 1 *Superposition*

Figure 2 *Making stationary waves*

Figure 3 *Interference of water waves*

The principle of superposition

When waves meet, they pass through each other. At the point where they meet, they combine for an instant before they move apart. This combining effect is known as **superposition**. Imagine a boat hit by two wave crests at the same time from different directions. Anyone on the boat would know it had been hit by a **supercrest**, the combined effect of two wave crests.

> The principle of superposition states that when two waves meet, the total displacement at a point is equal to the sum of the individual displacements at that point.

- Where a crest meets a crest, a **supercrest** is created; the two waves reinforce each other.
- Where a trough meets a trough, a **supertrough** is created; the two waves reinforce each other.
- Where a crest meets a trough, the resultant displacement is **zero**; the two waves cancel each other out.

Further examples of superposition

1 Stationary waves on a rope

Stationary waves are formed on a rope if two people send waves continuously along a rope from either end, as shown in Figure 2. The two sets of waves are referred to as **progressive waves**, to distinguish them from stationary waves. They combine at fixed points along the rope to form points of no displacement, or **nodes**, along the rope. At each node, the two sets of waves are always 180° out of phase so they cancel each other out. Stationary waves are described in more detail in Topic 8.6.

2 Water waves in a ripple tank

A vibrating dipper on a water surface sends out circular waves. Figure 3 shows a snapshot of two sets of circular waves produced in this way in a ripple tank. The waves pass through each other continuously:

- Points of **cancellation** are created where a crest from one dipper meets a trough from the other dipper. These points of cancellation are seen as gaps in the wave fronts.
- Points of **reinforcement** are created where a crest from one dipper meets a crest from the other dipper, or where a trough from one dipper meets a trough from the other dipper.

As the waves are continuously passing through each other at constant frequency and at a constant phase difference, cancellation and reinforcement occur at fixed positions. This effect is known as **interference**. The two dippers are said to be **coherent** emitters of waves, because they vibrate with a constant phase difference. If the phase difference changed at random, the points of cancellation and reinforcement would move about at random and no interference pattern would be seen. Interference of light is described in more detail in Topic 9.1.

> **Note**
>
> The points of cancellation and reinforcement would be further apart if the wavelength were increased or the dippers were closer together.

Tests using microwaves

A microwave transmitter and receiver can be used to demonstrate reflection, refraction, diffraction, interference and polarisation of microwaves. The transmitter produces microwaves of wavelength 3.0 cm. The receiver can be connected to a suitable meter, which gives a measure of the intensity of the microwaves at the receiver.

- Place the receiver in the path of the microwave beam from the transmitter. Move the receiver gradually away from the transmitter and note that the receiver signal decreases with distance from the transmitter. This shows that the microwaves become weaker as they travel away from the receiver.
- Place a metal plate between the transmitter and the receiver to show that microwaves cannot pass through metal.
- Use two metal plates to make a narrow slit and show that the receiver detects microwaves that have been diffracted as they pass through the slit. Show that if the slit is made wider, less diffraction occurs.
- Use a narrow metal plate with the two plates from above to make a pair of slits, as in Figure 4. Direct the transmitter at the slits and use the receiver to find points

of cancellation and reinforcement where the microwaves from the two slits overlap.

Interference of sound waves

To demonstrate interference of sound, a signal generator connected to two loudspeakers in series may be used. The two loudspeakers in series are coherent emitters of sound waves because their phase difference is constant. With the speakers operating at about 2 kHz about a metre apart, a careful listener walking slowly in front of the speakers ought to be able to locate points of cancellation and reinforcement. If the speakers are moved further apart, the points of cancellation and reinforcement will be closer together.

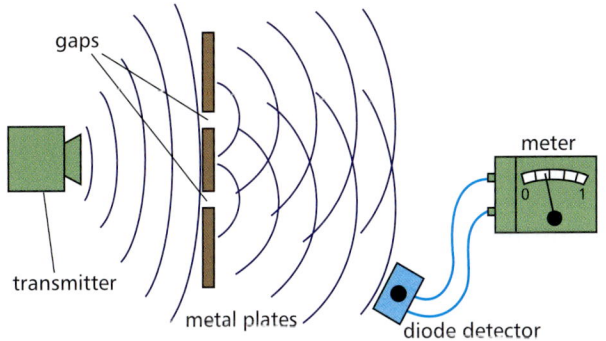

Figure 4 *Interference of microwaves*

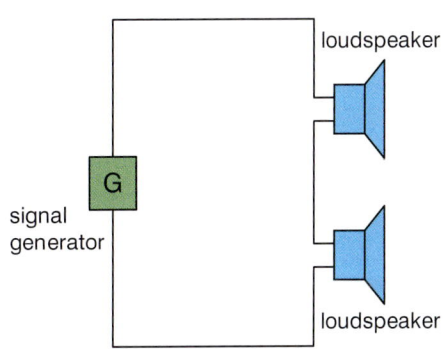

Figure 5

Summary test 8.5

1 Figure 6 shows two wave pulses on a rope travelling towards each other. Sketch a snapshot of the rope:

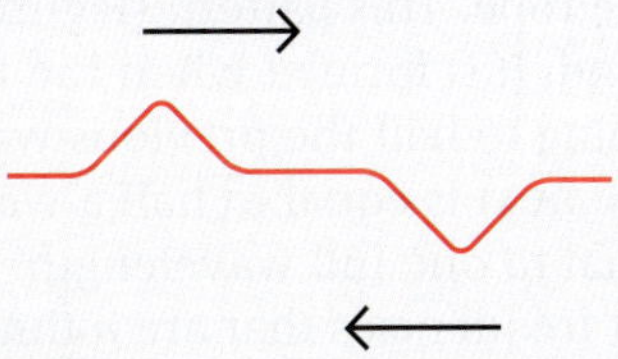

   Figure 6

   a when the two waves are passing through each other,

   b when the two waves have passed through each other.

2 How would you expect the interference pattern in Figure 3 to change if:

   a the two dippers are moved further apart,

   b the frequency of the waves produced by the dippers is reduced?

3 Microwaves from a transmitter are directed at a narrow slit between two metal plates. A receiver is placed in the path of the diffracted microwaves.

   How would you expect the receiver signal to change if:

   a the receiver is moved directly away from the slit,

   b the slit is then made narrower?

4 Microwaves, from a transmitter, are directed at two parallel slits in a metal plate. A receiver is placed on the other side of the metal plate. When the receiver is moved a short distance along a line AB parallel to the plate, the receiver signal decreases then increases again.

   a Explain why the signal decreased and then increased when it was moved from A along the line AB.

   b Explain why the signal increased as it moved towards B.

Learning outcomes

On these pages you will learn to:

- compare progressive and stationary waves in terms of particle vibrations
- explain the formation of a stationary wave on a string and identify nodes and antinodes
- describe and explain experiments that demonstrate stationary waves using microwaves and stretched strings

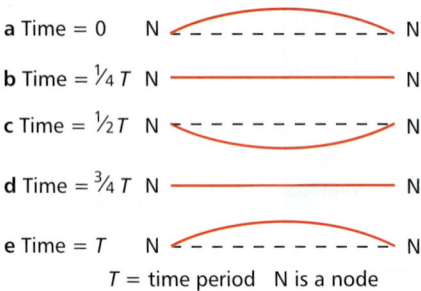

a Time = 0   N ⌒ ⌒ N
b Time = ¼ T   N —— N
c Time = ½T   N ⌣ ⌣ N
d Time = ¾ T   N —— N
e Time = T   N ⌒ ⌒ N

T = time period   N is a node

Figure 1 *Fundamental or first harmonic vibrations*

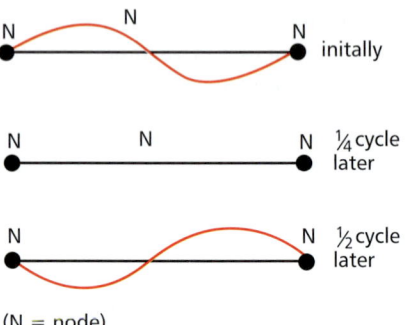

initially

¼ cycle later

½ cycle later

(N = node)

Figure 2 *A stationary wave of two loops*

Note

Stationary waves do not transfer energy. The amplitude of vibration is zero at the nodes, so there is no energy at the nodes. The amplitude of vibration is a maximum at the antinodes, so there is maximum energy at the antinodes. Because the nodes and antinodes are at fixed positions, no energy is transferred in a stationary wave pattern.

Formation of stationary waves

When a guitar string is plucked, the sound produced depends on the way in which the string vibrates. If the string is plucked gently at its centre, a stationary wave of constant frequency is set up on the string. The sound produced therefore has a constant frequency. If the guitar string is plucked harshly, the string vibrates in a more complicated way and the note produced contains other frequencies as well as the frequency produced when it is plucked gently.

As explained in Topic 8.5, a stationary wave is formed when two progressive waves pass through each other. This can be achieved on a string in tension by:

- Sending progressive waves along the string from either end.
- Fixing one end of the string and sending progressive waves along it from the other end. The waves reflect at the fixed end and pass through progressive waves moving towards the fixed end.
- Fixing both ends and making the middle part vibrate, so progressive waves travel towards each end, reflect at the ends, and then pass through each other.

The simplest stationary wave pattern on a string is shown in Figure 1. This is the **fundamental** or **first harmonic** mode of vibration of the string. It consists of a single loop that has a **node** (i.e. a point of no displacement) at either end. The string vibrates with maximum amplitude mid-way between the nodes. This position is referred to as an **antinode**. In effect, the string vibrates from side-to-side repeatedly. For this pattern to occur, the distance between the nodes at either end (i.e. the length of the string) must be equal to one half-wavelength of the waves on the string:

$$\text{Distance between adjacent nodes} = \tfrac{1}{2}\lambda$$

If the frequency of the waves sent along the rope from either end is raised steadily, the pattern in Figure 1 disappears: a new pattern is observed with two equal loops along the rope. This pattern (Figure 2) has a node at the centre as well as at either end. It is formed when the frequency is twice as high as in Figure 1, corresponding to half the previous wavelength. Because the distance from one node to the next is equal to half a wavelength, the length of the rope is therefore equal to one full wavelength. As explained in the next topic, the patterns occur at frequencies that are n times the fundamental frequency where n is a whole number, each pattern corresponding to n loops.

Explanation of stationary waves

Consider a snapshot of two progressive waves passing through each other:

- When they are in phase, they reinforce each other to produce a large wave (Figure 3a).
- A quarter of a cycle later, the two waves have each moved one quarter of a wavelength in opposite directions. They are now in **antiphase**, so they cancel each other (Figure 3b).
- After a further quarter cycle, the two waves are back in phase. The resultant is again a large wave as in Figure 3a, except reversed.

The points where there is no displacement (i.e. the nodes) are fixed in position throughout. Between these points, the stationary wave oscillates between the nodes.

In general, in any stationary wave pattern:

The amplitude of a vibrating particle in a stationary wave pattern varies with position from:

- zero at a node,
- to maximum amplitude at an antinode.

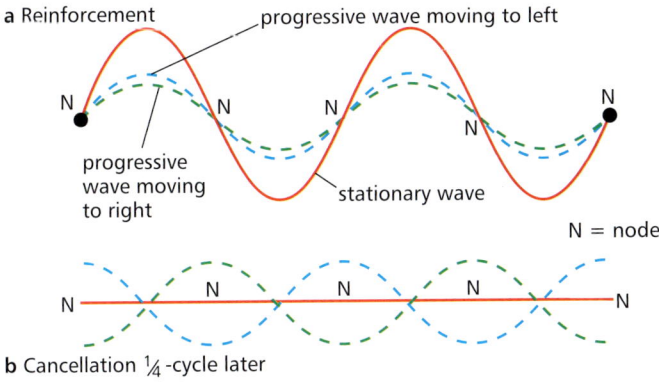

a Reinforcement

b Cancellation ¼-cycle later

Figure 3 *Explaining stationary waves*

The phase difference between two vibrating particles:

- is zero if the two particles are between adjacent nodes or separated by an even number of nodes,
- is 180° if the two particles are separated by an odd number of nodes.

Table 1 *Comparison between stationary waves and progressive waves in terms of particle vibrations*

	Stationary waves	Progressive waves
Frequency	All particles, except at the nodes, vibrate at the the same frequency.	All particles vibrate at same frequency.
Amplitude	The amplitude varies from zero at the nodes to a maximum at the antinodes.	The amplitude is the same for all particles.
Phase difference between two particles	$m\pi$, where m is the number of nodes between the two particles.	$2\pi x/\lambda$, where x = distance apart and λ is the wavelength.

More examples of stationary waves

Sound in a pipe

Sound resonates at certain frequencies in an air-filled tube or pipe. In a pipe closed at one end, these resonant frequencies occur when there is an antinode at the open end and a node at the other end. (See Topic 8.8.)

Using microwaves

Microwaves from a transmitter are directed normally at a metal plate, which reflects the microwaves back towards the transmitter. When a detector is moved along the line between the transmitter and the metal plate, the detector signal is found to be zero at equally spaced positions along the line. The reflected waves and the waves from the transmitter form a stationary wave pattern. The positions where no signal is detected are where nodes occur. They are spaced at intervals of one half of a wavelength.

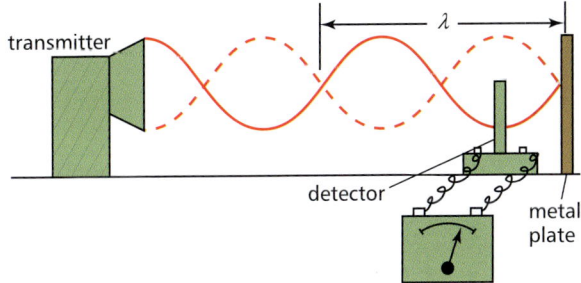

Figure 4 *Using microwaves*

1 a Sketch the stationary wave pattern seen on a rope, when there is a node at either end and an antinode at its centre.

   b If the rope in part **a** is 4.0 m in length, calculate the wavelength of the waves on the rope for the pattern you have drawn.

2 The stationary wave pattern shown in Figure 5 is set up on a rope of length 3.0 m.

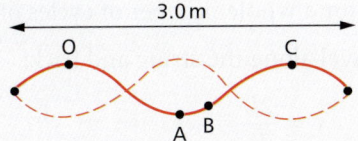

   Figure 5

   a Calculate the wavelength of these waves.

   b State the phase difference between the particle vibrating at O and the particle vibrating at:

   i A

   ii B

   iii C

3 State two differences between a stationary wave and a progressive wave in terms of the vibrations of the particles.

4 Microwaves from a transmitter are reflected back to the transmitter using a flat metal plate. A detector is moved along the line between the transmitter and the metal plate. The detector signal is zero at positions 15 mm apart.

   a Explain why the signal is zero at certain positions.

   b Calculate the wavelength of the microwaves.

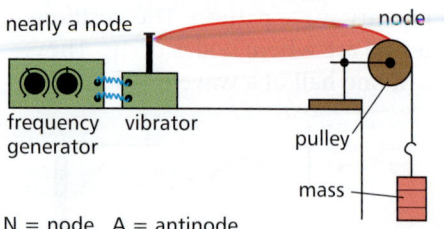

string at maximum displacement

nearly a node · node

frequency · vibrator · pulley
generator

mass

N = node   A = antinode
(dotted line shows string half a cycle earlier)

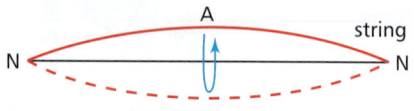

a Fundamental or first harmonic

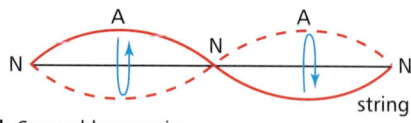

b Second harmonic

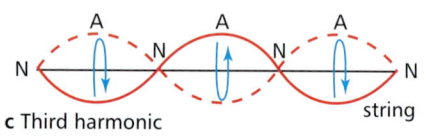

c Third harmonic

Figure 1 *Stationary waves on a string*

Stationary waves on a vibrating string

A controlled arrangement for producing stationary waves is shown in Figure 1. A string or wire is tied at one end to a mechanical vibrator, connected to a frequency generator. The other end of the string passes over a pulley and supports a mass, which keeps the tension in the string constant. As the frequency of the generator is increased from a very low value, different stationary wave patterns are seen on the string. In every case, the length of string between the pulley and the vibrator has a node at either end.

- The **fundamental** or **first harmonic pattern** of vibration is seen at the lowest possible frequency. This has an antinode at the middle, as well as a node at either end. Because the length L of the vibrating section of the string is between adjacent nodes, the wavelength of the waves that form this pattern, the fundamental wavelength λ_1, is equal to $2L$. Therefore, the fundamental or first harmonic frequency:

$$f_1 = \frac{v}{2L}$$

where v is the speed of the progressive waves on the wire.

- The next stationary wave pattern, the second harmonic, is where there is a node at the middle so the string is in two loops. The wavelength of the waves that form this pattern λ_2 is L, because each loop has a length of half a wavelength. Therefore, the frequency of the second harmonic vibrations:

$$f_2 = \frac{v}{\lambda_2} = \frac{v}{L} = 2f_1$$

- The next stationary wave pattern, the **third harmonic**, is where there are nodes at a distance of $\frac{1}{3}L$ from either end and an antinode at the middle. The wavelength of the waves that form this pattern λ_3 is $\frac{2}{3}L$, because each loop has a length of half a wavelength. Therefore, the frequency of the third harmonic vibrations:

$$f_3 = \frac{v}{\lambda_3} = \frac{3v}{2L} = 3f_1$$

In general, stationary wave patterns occur at frequencies f_1, $2f_1$, $3f_1$, $4f_1$, etc., where f_1 is the frequency of the first harmonic vibrations. This is the case in any vibrating linear system that has a node at either end.

Explanation of the stationary wave patterns on a vibrating string

In the arrangement shown in Figure 1, consider what happens to a wave sent out by the vibrator. The crest reverses its phase when it reflects at the fixed end, and travels back along the string as a trough. When it reaches the vibrator, it reflects and reverses phase again, travelling away from the vibrator once more as a crest. If this crest is reinforced by a crest created by the vibrator, the amplitude of the wave is increased. This is how a stationary wave is formed. The key condition is that the time taken for a wave to travel along the string and back should be equal to the time taken for a whole number of cycles of the vibrator:

- The time taken for a wave to travel along the string and back:

$$t = \frac{2L}{v}$$

where v is the speed of the waves on the string.

- The time taken for the vibrator to pass through a whole number of cycles:

$$t = \frac{m}{f}$$

where f is the frequency of the vibrator and m is a whole number (i.e. $m = 1$ or 2 or 3, etc.).

Therefore the key condition may be expressed as $\dfrac{2L}{v} = \dfrac{m}{f}$

Rearranging this equation gives:

$$f = \frac{mv}{2L} = mf_1 \text{ and } \lambda = \frac{v}{f} = \frac{2L}{m}$$

In other words:

- Stationary waves are formed at frequencies f_1, $2f_1$, $3f_1$, etc.
- The length of the vibrating section of the string $L = \dfrac{m\lambda}{2}$ = whole number of half wavelengths.

Making music

A guitar produces sound when its strings vibrate as a result of being plucked. In an electronic guitar, the vibrations of the string are detected by a microphone which produces electrical waves. These are amplified and converted back to sound waves in a loudspeaker. In an acoustic guitar, the string vibrations make the guitar surfaces vibrate and send out sound waves.

When a stretched string or wire vibrates, its **pattern of vibration** is a mix of its fundamental (i.e. first harmonic) mode of vibration and the harmonics of higher frequencies. The sound produced is the same mix of frequencies which change with time as the pattern of vibration changes. A **spectrum analyser** can be used to show how the intensity of a sound varies with frequency and with time. Combined with a **sound synthesiser**, the original sound can be altered by amplifying or suppressing different frequency ranges.

Figure 2 A sound synthesiser

The **pitch** of a note produced by a stretched string can be altered by changing the **tension** of the string, or by altering its **length**. The pitch is raised by raising the tension or shortening the length. Lowering the tension or increasing the length lowers the pitch. By changing the length or altering the tension, a vibrating string or wire can be tuned to the same pitch as a vibrating tuning fork. However, the sound from a vibrating string includes the frequencies of higher harmonics as well as the first harmonic (i.e. fundamental) frequency, whereas a tuning fork vibrates only at a single frequency. The wire is tuned when its first harmonic frequency is the same as the tuning fork frequency. A simple visual check is to balance a small piece of paper on the wire at its centre, and place the base of the vibrating tuning fork on one end of the wire. If the wire is tuned to the tuning fork, it will vibrate in its first harmonic mode and the piece of paper will fall off.

Summary test 8.7

1 A stretched wire of length 0.80 m vibrates in its first harmonic mode at a frequency of 256 Hz. Calculate:

  a the wavelength of the progressive waves on the wire,

  b the speed of the progressive waves on the wire.

2 The first harmonic frequency of vibration of a stretched wire is inversely proportional to the length of the wire. For the wire in question **1**, calculate the length of the wire needed to produce a frequency of:

  a 512 Hz

  b 384 Hz

  Assume that the tension of the wire is not changed.

3 Describe how you would expect the note from a vibrating guitar string to change if the string is:

  a shortened,

  b tightened.

4 The speed, v, of the progressive waves on a stretched wire varies with the tension T in the wire, in accordance with the equation $v = \left(\dfrac{T}{\mu}\right)^{1/2}$, where μ (Greek *mu*) is the mass per unit length of the wire. Use this formula to explain why a nylon wire and a steel wire, of the same length, diameter and tension, produce notes of different pitch. State, with a reason, which wire would produce the higher pitch.

Stationary waves in pipes

Learning outcomes

On these pages you will learn to:

- explain the formation of a stationary wave in a pipe closed at one end
- explain the formation of a stationary wave in a pipe open at both ends
- identify nodes and antinodes
- determine the wavelength of sound using stationary waves in a pipe

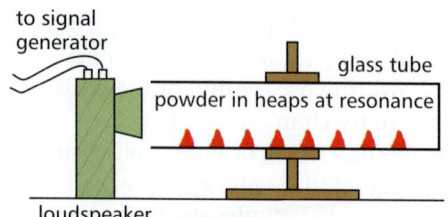

to signal generator

glass tube

powder in heaps at resonance

loudspeaker

Figure 1 *Acoustic resonance*

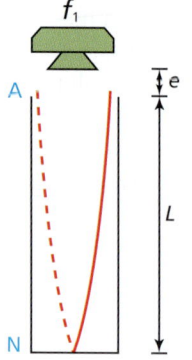

a Fundamental or first harmonic

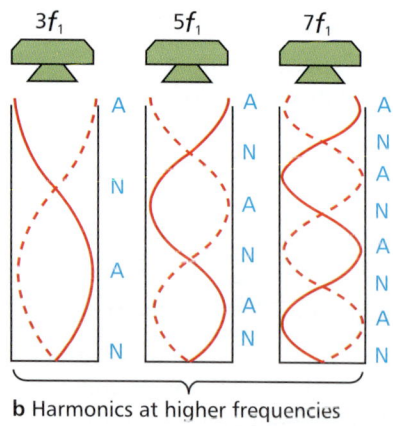

b Harmonics at higher frequencies

——— at $t = 0$ A = antinode

- - - - after half a cycle   N = node

Figure 2 *Resonances in a pipe closed at one end*

Stationary waves in a pipe closed at one end

If a small loudspeaker connected to a signal generator is placed near the open end of a pipe, the pipe resonates with sound at certain frequencies as the signal frequency is changed. Each **resonance** is due to stationary sound waves in the pipe. Sound waves from the transmitter at the open end are reflected at the other end of the pipe. The reflected waves, and the waves directly from the transmitter, pass along the pipe in opposite directions and therefore form a stationary wave pattern at certain frequencies. Figure 1 shows how this effect can be seen, as well as heard. The powder forms small heaps at the nodes, where there are no vibrations. Antinodes occur mid-way between the nodes and just beyond the open end. For greatest effect, the position of the loudspeaker needs to be at the antinode just beyond the open end.

- **At the fundamental frequency** or **first harmonic**, the lowest frequency at which the pipe resonates with sound, there is a node at the closed end and an antinode at the open end with no nodes or antinodes between. As shown in Figure 2, the pipe length $L = \frac{1}{4}\lambda_1$ (ignoring the small end-correction, e, at the open end). Therefore, the first harmonic wavelength $\lambda_1 = 4L$, the first harmonic frequency, $f_1 = \frac{v}{4L}$, where v is the speed of sound in the pipe.

- **Harmonics at higher $3f_1$, $5f_1$, $7f_1$, etc.** In each case, there is a node at the closed end and an antinode at the open end, with one or more equally spaced nodes or antinodes between. As shown in Figure 2:

1 At the **third harmonic**, the pipe length $L = \frac{3}{4}\lambda_1$ (ignoring the small end-correction at the open end).

Therefore, the third harmonic wavelength $\lambda_2 = \frac{4L}{3}$, where L is the pipe length, the third harmonic frequency, $f_2 = \frac{3v}{4L} = 3f_1$

2 At the **fifth harmonic**, the pipe length $L = \frac{5}{4}\lambda_3$ (ignoring the small end-correction).

Therefore, the fifth harmonic wavelength $\lambda_3 = \frac{4L}{5}$, where L is the pipe length, and the fifth harmonic frequency, $f_3 = \frac{5v}{4L} = 5f_1$

Further resonances occur at $7f_1$, $9f_1$, etc., corresponding to an odd number of quarter wavelengths equal to the pipe length. Note there are no even harmonics for a pipe closed at one end.

To determine the wavelength of sound in a pipe closed at one end, a vertical glass tube is used as the pipe, as shown in Figure 3. The length of the air column in the pipe is varied by using the valves to alter the level of water in the pipe. A small loudspeaker (or a tuning fork) is used to direct sound waves of constant frequency into the open end of the pipe.

- The water level in the pipe is gradually lowered from near the top of the pipe until the pipe resonates at its shortest length. The length, L_1, of the air column at this resonance is then measured. This length plus the small end-correction, e, is equal to one-quarter of a wavelength.

- The water level is then lowered further until the pipe resonates with sound at the third harmonic. The length, L_2, of the air column at this resonance is then measured. This length plus the small end-correction is equal to three-quarters of a wavelength.

The difference between the two length measurements, $L_2 - L_1$, is equal to half of the wavelength of the sound waves in the pipe. Hence the wavelength λ ($= 2(L_2 - L_1)$) can be calculated.

Stationary waves in a pipe open at both ends

Sound waves travelling along the pipe partially reflect at the open end because the speed changes at the exit. An antinode is formed at either end. Therefore the pipe resonates with sound at any frequency corresponding to the pipe length L equal to a whole number of half wavelengths (the distance between two adjacent antinodes). Figure 4 shows the stationary wave patterns in this situation.

- **At the fundamental frequency** or **first harmonic**, there is an antinode at either end of the pipe with a node mid-way. As shown in Figure 4, the pipe length $L = \frac{1}{2}\lambda_1$ (ignoring the small end-corrections at each end). Therefore, the first harmonic wavelength $\lambda_1 = 2L$, where L is the pipe length,

$$\text{the first harmonic frequency, } f_1 = \frac{v}{2L},$$

where v is the speed of sound in the pipe.

- **Harmonics at higher frequencies** occur at frequencies $2f_1$, $3f_1$, $4f_1$, etc. In each case, there is an antinode at either end, with one or more equally spaced nodes and antinodes between, as shown in Figure 4,

1 At the **second harmonic**, the wavelength λ_2 = the pipe length L (ignoring the small end-corrections).

Therefore, the first second harmonic frequency, $f_2 = \frac{v}{L} = 2f_1$

2 At the **third harmonic**, the wavelength λ_3 is such that the pipe length $L = \frac{3}{2}\lambda_3$ (ignoring the small end-corrections).

Therefore, the third harmonic wavelength $\lambda_3 = \frac{2L}{3}$, where L is the pipe length,

the third harmonic frequency, $f_3 = \frac{3v}{2L} = 3f_1$

Further resonances occur at $4f_1$, $5f_1$, etc., corresponding to a whole number of half-wavelengths equal to the pipe length.

Summary test 8.8

1 A pipe, of length 0.60 m, is closed at one end and open at the other end. The speed of sound in the pipe is 340 m s^{-1}. Estimate:

   a its first harmonic wavelength and frequency,

   b the frequency of its third harmonic.

2 A pipe of length 1.50 m, closed at one end and open at the other end, resonates at a frequency of 170 Hz. The speed of sound in the pipe is 340 m s^{-1}. Calculate:

   a the wavelength of the sound waves in the pipe,

   b the first harmonic frequency of this pipe.

3 A pipe, of length 2.40 m, is open at both ends. The speed of sound in the pipe is 340 m s^{-1}. Calculate:

   a its first harmonic frequency,

   b its second harmonic frequency.

4 A wind organ has pipes open at both ends, which are of lengths from 0.25 m to 2.50 m. Calculate the range of first harmonic frequencies from these pipes. The speed of sound in the pipes is 340 m s^{-1}.

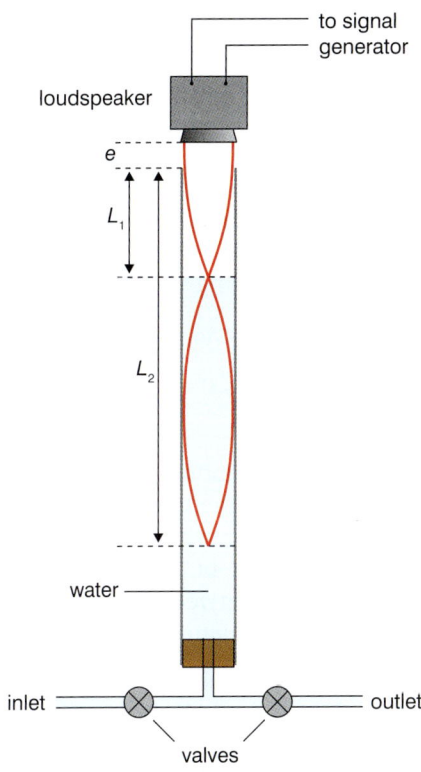

Figure 3 *Using stationary waves to measure the wavelength of sound*

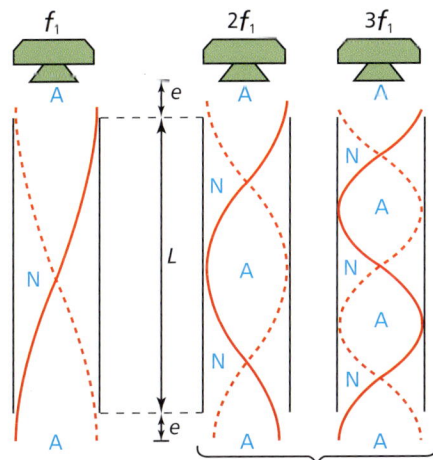

a First harmonic   b Second and third harmonics

— at time $t = 0$ A = antinode

- - - after half a cycle   N = node

Figure 4 *Stationary waves in an open pipe*

8.9 The Doppler effect

Learning outcomes

On these pages you will learn to:

- show an understanding of the Doppler effect and recall that the change of frequency is called a Doppler shift
- appreciate that Doppler shift is observed with all waves, including sound and light
- use the expression $f_0 = \dfrac{f_s v}{v \pm v_s}$ for a source of sound waves moving relative to a stationary observer

When a vehicle sounding its horn speeds past, the note heard by a bystander changes pitch. When approaching, the pitch is higher; when moving away, the pitch is lower. This is an example of the Doppler effect: when there is relative motion between a source of sound and an observer, the observed frequency differs from the emitted frequency.

The Doppler effect occurs with any form of wave motion, provided there is relative motion between the observer and the source of the waves along the line between them. To understand the cause of the Doppler effect, consider a source of waves moving at constant speed along a straight line, as in Figure 1, emitting waves of a constant frequency f_s.

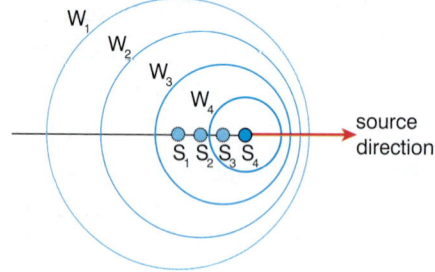

Figure 1 *The Doppler effect*

In Figure 1, the source has moved from position S_1 to S_4, where it emits wavefront W_4. It emitted wavefront W_1 when it was at S_1, wavefront W_2 when it was at S_2 and wavefront W_3 when it was at S_3. Because the source is moving in one direction, the wavefronts ahead of the source are bunched together and the wavefronts behind it are further apart.

More generally, if the wave speed is v and the speed of the source is v_s, then in one second the source emits f_s wavefronts and moves through a distance v_s, the leading wavefront having moved through a distance v.

Ahead of the source, the waves are bunched up because the source is moving in the same direction as the waves. So the distance from the leading wavefront to the source is equal to $v - v_s$ after the source has emitted f_s wavefronts. As this distance is equal to f_s wavelengths, the wavelength λ of the waves moving in the same direction as the source is given by:

$$\lambda = \frac{v - v_s}{f_s}$$

A stationary observer ahead of the source would therefore detect waves of frequency f_0, where:

$$f_0 = \frac{\text{wave speed}}{\text{wavelength}} = \frac{v}{\left(\dfrac{v - v_s}{f_s}\right)} = \frac{v f_s}{(v - v_s)}$$

Note that the observed frequency f_0 is greater than the source frequency f_s because:

$$\frac{v}{v - v_s} > 1$$

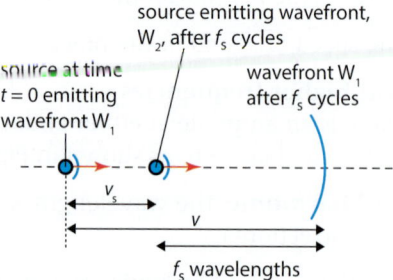

Figure 2 *Ahead of the source*

Behind the source, the waves are stretched out because the waves are moving in the opposite direction to the source direction. So the distance from the leading wavefront to the source is equal to $v + v_s$ after the source has emitted f_s wavefronts. As this distance is equal to f_s wavelengths, the wavelength λ of the waves moving in the opposite direction to the source is given by:

$$\lambda = \frac{v + v_s}{f_s}$$

A stationary observer behind the source would therefore detect waves of frequency f, where:

$$f_0 = \frac{\text{wave speed}}{\text{wavelength}} = \frac{v}{\left(\dfrac{v + v_s}{f_s}\right)} = \frac{v f_s}{(v + v_s)}$$

The equation shows that the observed frequency f_0 is less than the source frequency f_s because:

$$\frac{v}{(v + v_s)} < 1$$

Notes

1 The above equations may be summarised as a single equation:

$$f_0 = \frac{vf_s}{(v \pm v_s)}$$

where the − sign applies ahead of the source and the + sign applies behind it.

2 The change of frequency:

$$\Delta f = f_0 - f_s = \frac{vf_s}{(v \pm v_s)} - f_s$$
$$= \frac{vf_s}{(v \pm v_s)} - \frac{(v \pm v_s)f_s}{(v \pm v_s)} = \frac{(v_s f_s)}{(v \pm v_s)}$$

where the − sign applies when the observer is ahead of the source (i.e. the source is moving towards the observer) and the + sign applies when the observer is behind the source (i.e. the source is moving away from the observer). Notice the difference in the top line of this equation and the equation in Note 1.

3 For electromagnetic waves, where the speed of the source is much less than the speed of the waves, it can be shown that the change of frequency (or **Doppler shift**)

$$\Delta f = \frac{v_s f_s}{c},$$

where c is the speed of electromagnetic waves. You will meet this in Chapter 23 Astrophysics and cosmology'.

Worked example

A train sounds its horn as it travels at a speed of $30\,\mathrm{m\,s^{-1}}$ towards a level crossing. The sound is emitted from the horn at a frequency of $840\,\mathrm{Hz}$. Calculate:

a the wavelength of the sound waves travelling towards the level crossing

b the observed frequency of the sound waves at the level crossing.

The speed of sound in air = $340\,\mathrm{m\,s^{-1}}$.

a $\quad \lambda = \dfrac{v - v_s}{f_s} = \dfrac{340 - 30}{840} = 0.37\,\mathrm{m}$

b $\quad f_0 = \dfrac{v}{\lambda} = \dfrac{340}{0.37} = 920\,\mathrm{Hz}$

Speed measurement using the Doppler effect

A radar speed camera emits pulses of microwave radiation and detects any pulses reflected from a vehicle back to the camera. If the vehicle is moving towards or away from the camera, the frequency of such a reflected pulse differs from the emitted pulse. The difference in frequency is proportional to the speed of the vehicle and so can be used to measure the vehicle speed.

The same principle is used in:

• the Doppler anemometer, which uses pulses of laser light to measure wind speed by detecting pulses reflected by small particles in the wind

• Doppler echocardiography, which uses ultrasound pulses to create an image of the heart showing the speed and direction of blood flow inside it.

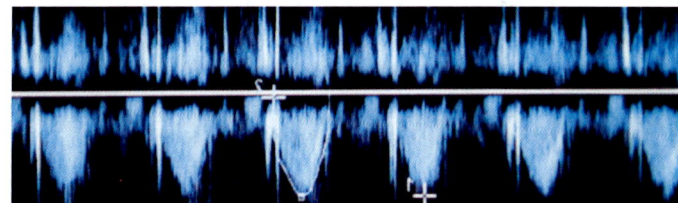

Figure 4 An echocardiogram

Summary test 8.9

The speed of sound in air = $340\,\mathrm{m\,s^{-1}}$.

The speed of electromagnetic waves in a vacuum = $3.0 \times 10^8\,\mathrm{m\,s^{-1}}$.

1 A car on a motorway sounds its horn as it approaches an overhead bridge at a speed of $28\,\mathrm{m\,s^{-1}}$. The horn emits sound waves at a frequency of $1100\,\mathrm{Hz}$.

  a Calculate the frequency of the sound heard by an observer on the bridge as the car approaches.

  b State and explain what the frequency of the sound heard by the observer would be when the car was directly below the bridge.

2 A train is travelling at a speed of $40\,\mathrm{m\,s^{-1}}$ on a straight section of track between two overhead bridges, X and Y, when it sounds its horn. An observer on bridge X hears the sound from the horn at a lower frequency than does an observer on bridge Y.

  a State whether the train is moving towards X or towards Y and explain why.

  b The observer at X hears sound of frequency $950\,\mathrm{Hz}$ from the horn. Calculate the frequency of the sound emitted by the horn.

3 The Sun has an equatorial diameter of $1.4 \times 10^9\,\mathrm{m}$ and spins at its equator at a steady rate of one revolution every 25 days.

  a Calculate the speed of rotation of an atom at the Sun's equator.

  b Calculate the change in frequency observed in light of frequency $5.0 \times 10^{14}\,\mathrm{Hz}$ emitted by atoms on the Sun's equator when they are moving away from the Earth at the speed calculated in **a**.

4 The Doppler anemometer uses pulses of laser light to measure wind speed by detecting pulses reflected by small particles in the wind. Explain why: **a** ultrasound, and **b** microwaves would be unsuitable for use instead of laser pulses.

⟦📄 Launch additional digital resources for the chapter ⟧

1 a i Explain the difference between longitudinal waves and transverse waves.
   ii State one example of a longitudinal wave.
   iii State one example of a transverse wave.

b With the aid of a diagram, explain what is meant by:
   i diffraction of a wave,
   ii refraction of a wave.

2 a Explain what is meant by a 'polarised' transverse wave.

b Explain why sound waves cannot be polarised, whereas light waves can.

3 An oscilloscope is used to display the signal from a microphone when sound waves of constant frequency are directed at the microphone (see graph below).

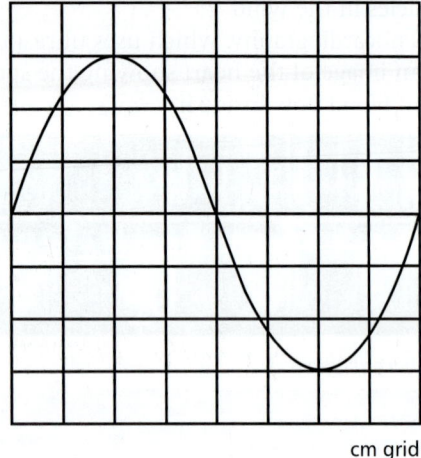

cm grid

Figure 3.1

a The oscilloscope time base is set at $0.5\,\text{ms}\,\text{cm}^{-1}$. Calculate the frequency of the sound waves.

b Sketch the display you would observe on the oscilloscope if the frequency of the sound waves was doubled with no change of loudness.

4 a Explain what is meant by the 'superposition' of waves.

b Two small loudspeakers, 1.2 m apart, are connected to a signal generator, as shown opposite. The signal generator is adjusted to produce an alternating voltage of constant amplitude, at a frequency of 1.0 kHz.
   i An observer moves along the line XY at a perpendicular distance of 2.0 m from the loudspeakers, as shown. The intensity of the sound rises and falls as she moves along the line. Explain this effect.
   ii In part **i**, the observer notes that successive minima are further apart along the line XY when the test is repeated at a lower frequency. Explain this observation.

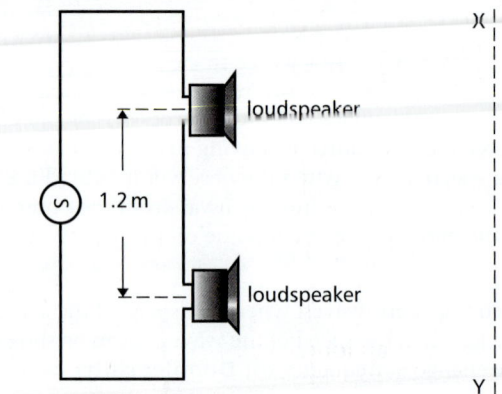

Figure 4.1

5 Microwaves from a transmitter were directed at two narrow gaps between three metal plates. A detector was placed on the other side of the plates at P, as shown below.

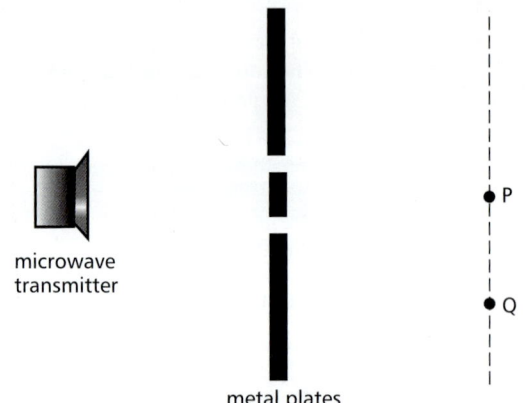

Figure 5.1

a The detector was moved along a line parallel to the plates to Q. The signal from the detector decreased from a maximum at P to a minimum mid-way between P and Q, then to a maximum at Q.
   i Explain why there was a maximum at P and at Q.
   ii Explain why there was a minimum mid-way between P and Q.

b Explain how the detector signal would change when the detector is mid-way between P and Q and a metal plate is placed over one of the gaps.

6 A stationary wave pattern of a wire, of length 0.60 m, vibrating at a frequency of 300 Hz is shown below.

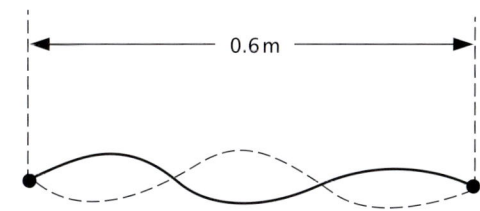

Figure 6.1

a In terms of amplitude and phase difference, compare the motion of the wire 0.10 m from the left-hand end with:
  i its motion at the midpoint,
  ii 0.10 m from the other end.

b **i** Calculate the wavelength of the waves on the wire.
  ii The frequency of vibrations is increased to 400 Hz, and a different stationary wave pattern is produced. Sketch the pattern you would expect to observe at 400 Hz.

7 Sound from a small loudspeaker connected to a signal generator was directed into a vertical glass tube containing some water, as shown below. The length of the air column in the tube could be varied by altering the level of water in the tube.

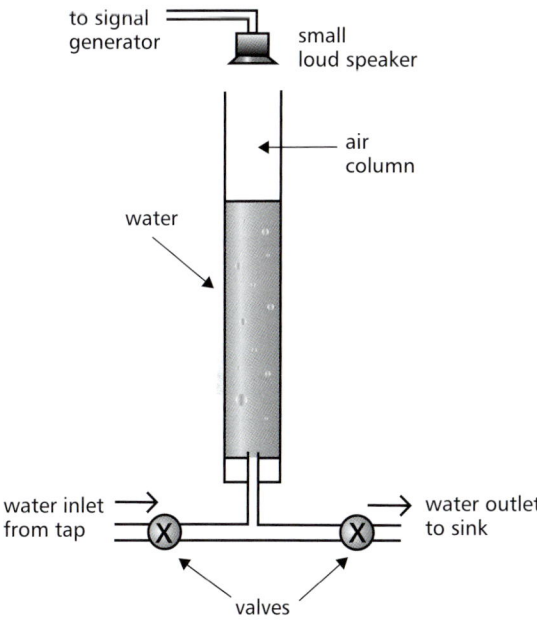

Figure 7.1

a The length of the air column was gradually increased from a few centimetres, by lowering the level of water in the tube slowly. The tube resonated with sound at certain positions of the water level. Explain why this effect happened only at certain lengths of the air column.

b The tube resonated with sound of frequency 300 Hz when the length of the air column was 270 mm and 820 mm.

i Diagram **a** shows how the amplitude of the sound waves in the tube varied with position along the tube when the air column length was 270 mm. Using a copy of diagram **b**, sketch the corresponding pattern when the air column length was 820 mm.

ii Show that the wavelength of the sound in the tube was 1.10 m, and hence calculate the speed of sound in the tube.

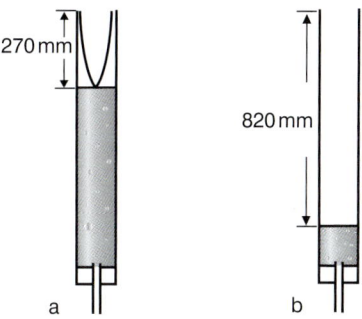

Figure 7.2

8 An open-ended pipe, of length 0.80 m, resonates with sound at a frequency of 200 Hz. The speed of sound in the pipe is 330 m s^{-1}.

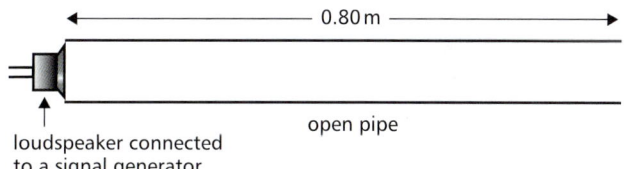

Figure 8.1

a **i** Calculate the wavelength of sound waves of frequency 200 Hz in the pipe.
  ii Sketch the stationary wave pattern in the pipe when it resonates at 200 Hz.

b **i** Calculate the next highest frequency at which the pipe would resonate if the frequency was gradually increased.
  ii Explain why the pipe resonates at certain frequencies only.

9 **a** State two differences between the vibrations of a stationary wave and those of a progressive wave.

b **i** Figure 9.1 shows a progressive wave travelling from left to right. Describe how the vibrations of the particle at point P compare with the vibrations of the particle at Q, and state the phase difference between their vibrations.

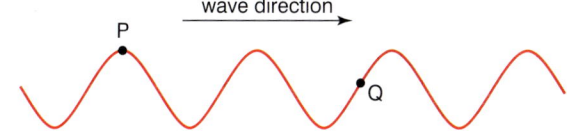

Figure 9.1

  ii Describe two essential conditions for the formation of a stationary wave pattern from two progressive waves.

137

9.1 Interference of light

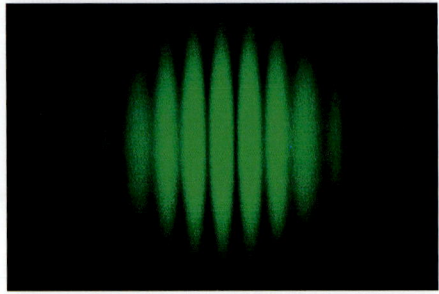

Figure 1 *Double-slit interference pattern as seen on a screen*

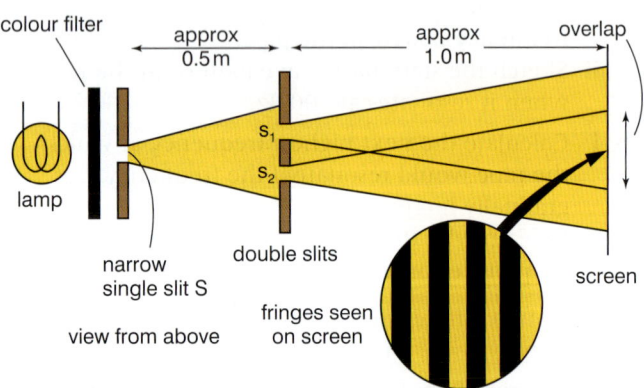

Figure 2 *Young's double-slit experiment*

The wave nature of light

Young's double-slit experiment

The wave nature of light was first suggested by Christiaan Huygens in the seventeenth century, but it was rejected at the time in favour of Sir Isaac Newton's corpuscular theory of light. Newton considered that light was composed of tiny particles he referred to as corpuscles and he was able to explain reflection and refraction using his theory. Huygens was also able to explain reflection and refraction using his wave theory. However, the two theories differed about whether or not light in a transparent substance travels faster (as predicted by Newton's theory) or slower (as predicted by Huygens' theory) than in air. Because of Newton's much stronger scientific reputation, Newton's theory of light remained unchallenged for over a century, until 1803, when Thomas Young at the Royal Institution first demonstrated interference of light. Even so, Newton's theory of light was not rejected in favour of Huygens' wave theory until several decades later when the speed of light in water was measured and found to be slower than in air.

An arrangement like the one used by Thomas Young to observe interference is shown in Figure 2. Young would have used a candle instead of a light bulb to illuminate a narrow single slit. A pair of narrow slits, referred to as the 'double slits', is illuminated by light from the single slit. Alternate bright and dark fringes can be seen using a microscope, or on a white screen placed where the diffracted light from the double slits overlaps. The fringes are evenly spaced and parallel to the double slits.

The fringes are formed due to **interference of light** from the two slits:

- Where a **bright fringe** is formed, the light from one slit reinforces the light from the other slit. In other words, the light waves from each slit arrive **in phase** with each other.
- Where a **dark fringe** is formed, the light from one slit cancels the light from the other slit. In other words, the light waves from the two slits arrive **180° out of phase**.

The distance from the centre of a bright fringe to the centre of the next bright fringe is called the **fringe separation, x**. This depends on the slit spacing a and the distance D from the slits to the screen, in accordance with the equation

$$\text{fringe separation, } x = \frac{\lambda D}{a}$$

where λ is the wavelength of light.

The equation shows that the fringes become more widely spaced if:

- the distance D from the slits to the screen is increased,
- the wavelength λ of the light used is increased,
- the slit spacing, a, is reduced.

The theory of the double-slit equation

Consider the two slits S_1 and S_2 shown in Figure 3. At a point P on the screen where the fringes are observed, light emitted from S_1 arrives later than light from S_2 emitted at the same time. This is because the distance S_1P is greater than the distance S_2P. The difference between distances S_1P and S_2P is referred to as the **path difference**.

- **For reinforcement at P**, the path difference $S_1P - S_2P = m\lambda$, where $m = 0, 1, 2$, etc.

  Therefore, light emitted simultaneously from S_1 and S_2 arrives in phase at P, if reinforcement occurs at P.

- **For cancellation at P**, the path difference $S_1P - S_2P = (m + \frac{1}{2})\lambda$, where $m = 0, 1, 2$, etc.

  Therefore, light emitted simultaneously from S_1 and S_2 arrives at P out of phase by 180°, if cancellation occurs at P.

In Figure 3, a point Q along line S_1P has been marked such that $QP = S_2P$. Therefore, the path difference $S_1P - S_2P$ is represented by the distance S_1Q.

Because triangles S_1S_2Q and MOP are very nearly similar in shape, where M is the midpoint between the two slits and O is the midpoint of the central bright fringe of the pattern, then:

$$\frac{S_1Q}{S_1S_2} = \frac{OP}{OM}$$

If P is the mth bright fringe from the centre (where $m = 0$, 1, 2, etc.), then $S_1Q = m\lambda$ and $OP = mx$, where x is the distance between centres of adjacent bright fringes.

Also, OM = distance D and S_1S_2 = slit spacing a.

Therefore,

$$\frac{m\lambda}{a} = \frac{mx}{D}$$

Rearranging this equation gives: $\boldsymbol{\lambda = \dfrac{ax}{D}}$

By measuring the slit spacing, a, the fringe separation, x, and the slit–screen distance D, the wavelength λ of the light used can be calculated. The formula is valid only if the fringe separation, a, is much less than the distance D from the slits to the screen. This condition is to ensure the triangles S_1S_2Q and MOP are very nearly similar in shape.

Note

Light wavelengths are usually expressed in **nanometres (nm)**, where $1\,\text{nm} = 10^{-9}\,\text{m}$.

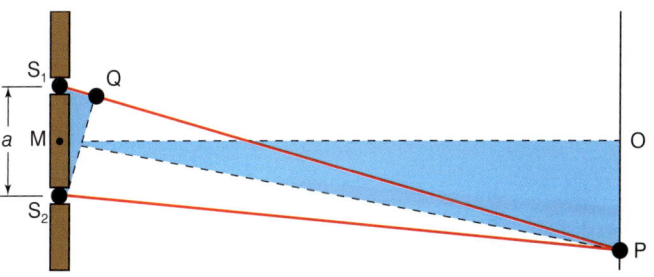

Figure 3 The theory of the double-slit experiment

Summary test 9.1

1 In a double-slit experiment using red light, a fringe pattern is observed on a screen at a fixed distance from the double slits. How would the fringe pattern change if:

   a the screen was moved closer to the slits,

   b one of the double slits was blocked completely?

2 The following measurements were made in a double-slit experiment:

   Slit spacing, $a = 0.4\,\text{mm}$; fringe separation, $x = 1.1\,\text{mm}$; slit–screen distance, $D = 0.80\,\text{m}$.

   Calculate the wavelength of light used.

3 In question **2**, the double slits were replaced by a pair of slits with a slit spacing of 0.5 mm. Calculate the fringe separation for the same slit–screen distance and wavelength.

4 The following measurements were made in a double-slit experiment:

   Slit spacing, $a = 0.4\,\text{mm}$; fringe separation, $x = 1.1\,\text{mm}$; wavelength of light used, $\lambda = 590\,\text{nm}$.

   Calculate the distance from the slits to the screen.

Learning outcomes

On these pages you will learn to:

- explain the terms monochromatic light and coherence
- show an understanding of the conditions required to observe two-source interference
- describe two-source interference fringes using monochromatic light and white light

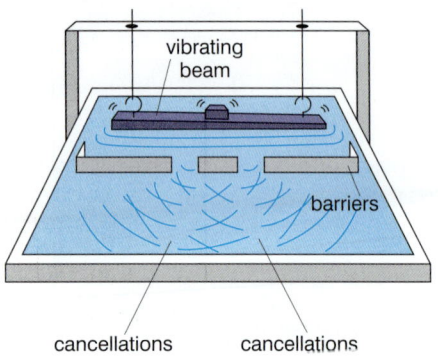

cancellations   cancellations

Figure 1 *Interference of water waves*

Coherence

The double slits are described as **coherent sources,** because they emit light waves with a constant phase difference. This is because each wave crest, or wave trough, from the single slit always passes through one of the double slits a fixed time after it passes through the other slit. The double slits therefore emit wave fronts with a constant phase difference.

The arrangement is like the ripple tank demonstration in Figure 1. Straight waves from the beam vibrating on the water surface diffract after passing through the two gaps in the barrier, and produce an interference pattern where the diffracted waves overlap. If one gap is closer to the beam than the other, each wave front from the beam passes through the nearer gap first. However, the time interval between the same wave front passing through the two gaps is always the same, so the waves emerge from the gaps with a constant phase difference.

Light from two separate light bulbs could not form an interference pattern, because the two light sources emit light waves at random. The points of cancellation and reinforcement would change at random, so no interference pattern is possible.

Wavelength and colour

In the double-slit experiment, the fringe separation depends on the colour of light used. White light is composed of a continuous spectrum of colours, corresponding to a continuous range of wavelengths from about 350 nm for violet light to about 650 nm for red light. Each colour of light has its own wavelength, as shown in Figure 2.

The fringe patterns shown in Figure 3 show that the fringe separation is greater for red light than for blue light. This is because red light has a longer wavelength than blue light. The fringe spacing, x, depends on the wavelength, λ, of the light according to the formula $\frac{x}{D} = \frac{\lambda}{a}$, as explained in Topic 9.1.

Rearranging this formula gives $x = \frac{\lambda D}{a}$. Thus the longer the wavelength of the light used, the greater the fringe separation is.

Light sources

- **Vapour lamps and discharge tubes** produce light with a dominant colour. For example, the sodium vapour lamp produces a yellow/orange glow which is due to light of wavelength 590 nm. Other wavelengths of light are also emitted from a sodium vapour lamp, but the colour due to light of wavelength 590 nm is much more intense than any other colour. A sodium vapour lamp is, in effect, a **monochromatic** light source because its spectrum is dominated by light of a certain colour.

- **Light from a filament lamp or from the Sun** is composed of the colours of the spectrum, and therefore covers a continuous range of wavelengths from about 350 nm to about 650 nm.

- **Light from a laser** is of a specific wavelength, and therefore laser light is highly monochromatic. For example, a helium–neon laser produces red light

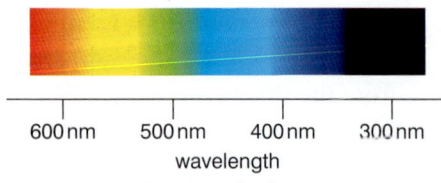

600 nm   500 nm   400 nm   300 nm
wavelength

Figure 2 *Wavelength and colour*

Figure 3 *The double slits fringe pattern*

of wavelength 635 nm only. Because a laser beam is almost perfectly parallel and monochromatic, a convex lens can focus it to a very fine spot. The beam power is then concentrated in a very small area. This is why a laser beam is very dangerous if it enters the eye. The **eye lens** would focus the beam on a tiny spot on the **retina** and the intense concentration of light at that spot would destroy the retina.

> **Always wear safety goggles in the presence of a laser beam. Never look along a laser beam, even after reflection.**

Observing Young's fringes

Provided the light source is **not** a laser, a microscope may be used to observe the fringe pattern and to measure the fringe spacing. If so, the plane of viewing of the fringe pattern must be located to measure the distance D from the slits to the fringes accurately. Figure 4 shows how this can be done.

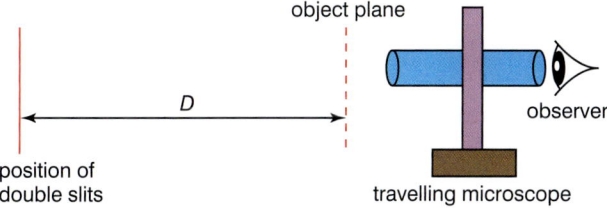

Figure 4 *Measuring the fringes using a microscope*

The contrast between the bright and dark fringes can be improved by narrowing the single slit (see Topic 9.1 Figure 2). If this slit is too wide, each part of it produces a fringe pattern which is displaced slightly from the pattern due to adjacent parts of the single slit. As a result, the dark fringes of the double slit pattern become narrower than the bright fringes, and contrast is lost between the dark and the bright fringes.

If a laser is used as the light source, a screen **must** be used on which to observe the fringe pattern produced by a laser. Never look along a laser beam, even after reflection. Safety goggles must always be worn in the presence of a laser beam.

White light fringes

Figure 3 shows the fringe patterns observed with blue light and with red light. As explained above, the blue light fringes are closer together than the red light fringes. The fringe pattern produced by white light is shown in Figure 5. Each component colour of white light produces its own fringe pattern, each pattern centred on the screen at the same position. As a result:

- The central fringe is **white**, because every colour contributes at the centre of the pattern.
- The inner fringes are tinged with **blue on the inner side and red on the outer side**. This is because the red fringes are more spaced out than the blue fringes, and the two fringe patterns do not overlap exactly.
- The outer fringes merge into an indistinct **background of white light**. This is because, where the fringes merge, different colours reinforce and therefore overlap.

Figure 5 *White light fringes*

Summary test 9.2

1 **a** Sketch an arrangement that may be used to observe the fringe pattern produced when light from a narrow slit, illuminated by a sodium vapour lamp, is passed through a pair of double slits.

  b Describe the fringe pattern you would expect to observe in part **a**.

2 In question **1**, describe how the fringe pattern would change if:

  a one of the double slits is blocked,

  b the narrow single slit is replaced by a wider slit.

3 Double slit interference fringes are observed using light of wavelength 590 nm and a pair of double slits of slit spacing 0.50 mm. The fringes are observed on a screen at a distance of 0.90 m from the double slits. Calculate the fringe separation of these fringes.

4 Describe and explain the fringe pattern that would be observed in question **3**, if the light source were replaced by a white light source.

The diffraction grating

Learning outcomes

On these pages you will learn to:

- explain how a diffraction grating works
- recall and solve problems using the formula $d\sin\theta = n\lambda$
- describe the use of a diffraction grating to determine the wavelength of light

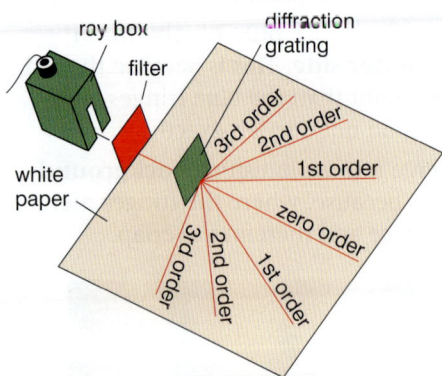

Figure 1 *The diffraction grating*

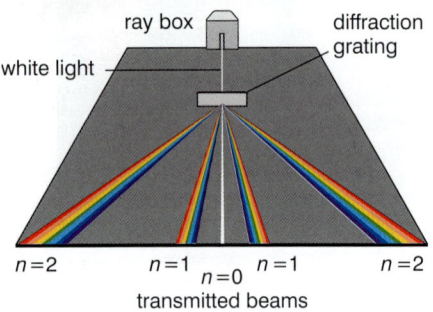

Figure 2 *Using white light*

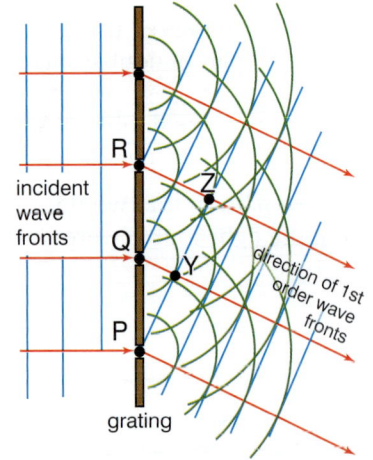

Figure 3 *Formation of the first order wavefronts*

Investigating the diffraction grating

A **diffraction grating** consists of a transparent plate with many closely spaced parallel slits ruled on it. When a parallel beam of monochromatic light is incident normally on a diffraction grating, as shown in Figure 1, light is transmitted by the grating in certain directions only. This is because:

- the light passing through each slit is diffracted
- the diffracted light waves from adjacent slits reinforce each other in certain directions only, including the incident light direction, and cancel out in all other directions.

The central beam, referred to as the 'zero order beam', is in the same direction as the incident beam. The other transmitted beams are numbered outwards from the zero order beam and are referred to as 'first order', 'second order', etc.

The **angle of diffraction**, θ, between each transmitted beam and the central beam increases if:

- light of a longer wavelength is used (e.g. by replacing a blue filter with a red filter)
- a grating with closer slits is used.

> **Note**
>
> Figure 2 shows what happens if white light instead of monochromatic light is directed at the grating. The central transmitted beam is white and each non-central transmitted beam is spread out as a spectrum of colour from violet to red. This is because white light consists of a continuous range of wavelengths from about 350 nm (violet) to about 650 nm (red). Therefore, in each order, red light is diffracted more than violet light as it has a longer wavelength than violet light.

The diffraction grating equation

The wavelength of the light in each transmitted beam can be calculated by measuring the angle of diffraction, θ, of the beam and using the **diffraction grating equation**

$$d\sin\theta = n\lambda$$

where n is the order number of the beam and d is the slit spacing of the grating.

To prove the diffraction grating equation, consider a magnified view of part of a diffraction grating, as shown in Figure 3. Each slit diffracts the light waves that pass through it. As each diffracted wavefront emerges from a slit, it reinforces a wavefront from each of the adjacent slits. For example, in Figure 3, the wavefront emerging at P reinforces the wavefront emitted from Q one cycle earlier which reinforces the wavefront emitted from R one cycle earlier, etc. The effect is to form a new wavefront PYZ which travels in a certain direction and contributes to the first order diffracted beam.

Figure 4 shows the formation of a wavefront of the nth order beam. The wavefront emerging from slit P reinforces a wavefront emitted n cycles earlier by the adjacent slit Q. This earlier wavefront therefore must have travelled a distance of n wavelengths from the slit. Thus the perpendicular distance QY from the slit to the wavefront is equal to $n\lambda$, where λ is the wavelength of the light waves.

Since the angle of diffraction of the beam, θ, is equal to the angle between the wavefront and the plane of the slits, it follows that $\sin\theta = \dfrac{\text{QY}}{\text{QP}}$, where QP is the grating spacing (i.e. the centre-to-centre distance d between adjacent slits).

Substituting d for QP and $n\lambda$ for QY therefore gives $\sin\theta = n\lambda/d$. Rearranging this equation gives the diffraction grating equation above.

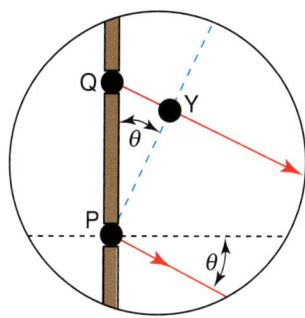

Figure 4 *The nth order wavefront*

Notes

1 The number of slits per metre on the grating, $N = 1/d$ where d is the grating spacing.
2 For a given order and wavelength, the smaller the value of d, the greater the angle of diffraction is. In other words, the larger the number of slits per metre, the bigger the angle of diffraction.
3 Fractions of a degree are usually expressed either as a decimal or in minutes (abbreviated '), where $1° = 60'$.
4 To find the maximum number of orders produced, substitute $\theta = 90°$ ($\sin 90° = 1$) into the diffraction grating equation and calculate n using $n = d/\lambda$.

The maximum number of orders is given by the value of d/λ rounded down to the nearest whole number.

Worked example

For a diffraction grating with 600 slits per millimetre, calculate the maximum order number for light of wavelength 580 nm.

Solution

$d = 1/N = 1/600\,\text{mm} = 1.67 \times 10^{-3}\,\text{mm}$

$\dfrac{d}{\lambda} = \dfrac{1.67 \times 10^{-3}\,\text{mm}}{580\,\text{nm}} = \dfrac{1.67 \times 10^{-6}\,\text{m}}{580 \times 10^{-9}\,\text{m}} = 2.88$

Therefore the maximum order number = 2 (as 2.88 rounded down to the nearest whole number = 2).

Note There would be a first order beam and a second order beam either side of the zero order beam.

Diffraction gratings in action

We can use a diffraction grating in a **spectrometer** to study the spectrum of light from any light source and to measure light wavelengths very accurately.

- A filament lamp produces a continuous spectrum of colour from deep violet at wavelengths of about 350 nm to deep red at about 650 nm, as shown in Topic 9.2 Figure 2.
- A glowing gas in a vapour lamp or a discharge tube produces a spectrum consisting of narrow vertical lines of different colours as shown in Figure 5. The wavelengths of the lines are characteristic of the chemical elements that produce the light.

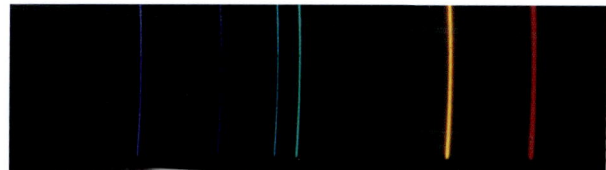

Figure 5 *A line spectrum*

A spectrometer is designed to measure angles to within 1 arc minute which is a sixtieth of a degree. The angle of diffraction of a diffracted beam can be measured very accurately. Using light of a known wavelength, the grating spacing of a diffraction grating can therefore be measured very accurately with a spectrometer. The spectrometer and the grating can then be used to measure light of any wavelength.

Summary test 9.3

1 A laser beam of wavelength 630 nm is directed normally at a diffraction grating which has 300 lines per millimetre. Calculate:

a the angle of diffraction of each of the first two orders,

b the number of diffracted orders produced.

2 Light directed normally at a diffraction grating contains wavelengths of 580 nm and 586 nm only. The grating has 600 lines per mm.

a How many diffracted orders are observed in the transmitted light?

b For the highest order, calculate the angle between the two diffracted beams.

3 Light of wavelength 430 nm is directed normally at a diffraction grating. The first order transmitted beams are at 28° to the zero order beam. Calculate:

a the number of slits per millimetre on the grating,

b the angle of diffraction for each of the other diffracted orders of the transmitted light.

4 A diffraction grating is used in a spectrometer to view light of wavelength 430 nm. The first order diffracted beam is observed at an angle of 14° 55' to the zero order beam.

a Calculate the number of lines per mm on the grating.

b i Determine the order number of the highest order beam.

   ii Calculate the angle of diffraction of this order.

(Reminder: 1 degree = 60 minutes of arc)

Launch additional digital resources for the chapter

1 An interference pattern of bright and dark fringes is observed when two closely spaced slits are illuminated by a parallel beam of monochromatic light, as shown below:

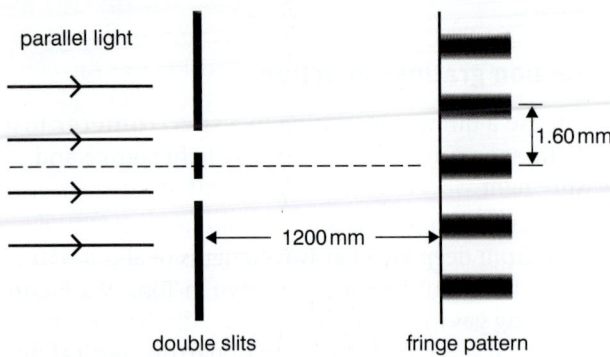

Figure 1.1

a Explain the formation of:
   i   a bright fringe,
   ii a dark fringe.

b The two slits act as coherent emitters of light.
   i   Explain what is meant by this statement.
   ii Explain why the fringe pattern could not be observed from two closely spaced light sources, such as two nearby filaments.

c The two slits were at a distance of 0.4 mm apart. The spacing between two adjacent bright fringes was 1.60 mm, when the fringes were observed at a distance of 1200 mm from the slits. Calculate the wavelength of the monochromatic light.

2 a Plane waves are incident on a slit as shown in Figure 3.1.

   i   Complete Figure 3.1 to show the two preceding wavefronts that have passed through the gap.

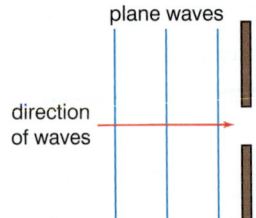

Figure 3.1

   ii Describe how the two preceding wavefronts would differ if the slit was significantly wider.

b A narrow parallel beam of light of wavelength 630 nm is directed normally at a diffraction grating which has 600 lines per millimetre.

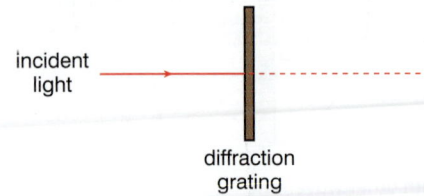

Figure 3.2

   i   Determine the number of orders of diffracted light that can be observed each side of the zero order.
   ii Calculate the angle of diffraction of the highest order.

3 In a double-slit experiment, when light of a certain wavelength was directed normally at the slits, an interference pattern was observed on a screen placed as shown in Figure 4.1.

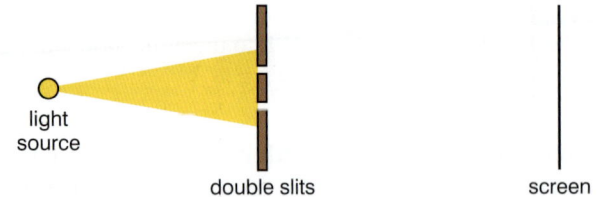

Figure 4.1

a i   The following measurements were made:
       Distance across five fringes = 4.8 mm
       Distance between the centres of each slit (i.e. the slit spacing) = 0.40 mm
       Distance from the slits to the screen = 810 mm
       Calculate the wavelength of the light.
   ii The slit–screen distance was estimated to have an uncertainty of ± 5 mm. The uncertainty in the spacing across the five fringes was estimated to have an uncertainty of ± 0.5 mm. The uncertainty in the slit spacing was estimated at ± 0.02 mm. Calculate the percentage uncertainty in each measurement and state and explain which of the three measurements is the least precise.

b Describe and explain how the fringe pattern in **a** would have differed if a beam of white light had been used instead.

4 A narrow beam of white light was directed normally, as shown in Figure 5.1, at a diffraction grating having 600 lines per millimetre. Two diffracted orders of the white light spectrum were observed on each side of a zero order beam.

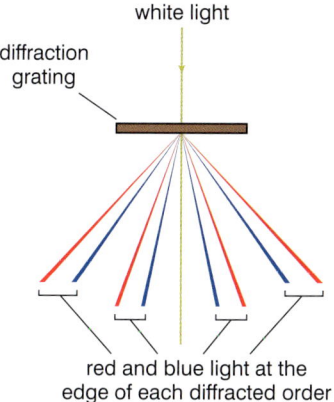

white light

diffraction grating

red and blue light at the edge of each diffracted order

Figure 5.1

a Describe the nature of white light.

b The second diffracted order on each side produced a white light spectrum with angles of diffraction of 48° for the red and 31° for the blue light. Use this data to calculate the wavelength of:
   i the red light,
   ii the blue light.

c Part of a third order spectrum was observed on each side of the zero order beam. Explain why only part of the third order spectrum was observed.

d A semi-circular glass block was placed against the diffraction grating, as shown in Figure 5.2.

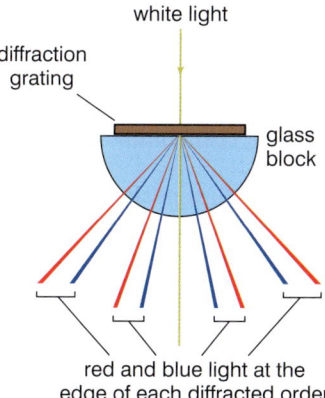

white light

diffraction grating

glass block

red and blue light at the edge of each diffracted order

Figure 5.2

The angle of diffraction of the second order red light was 30°, and the angle of diffraction of the second order blue light was 20°. The reduction in the angle of diffraction due to the glass block is because light in glass has a smaller wavelength than in air.

   i Use the information above to calculate the wavelength of each colour in glass.
   ii For each light colour, calculate the ratio:

$$\frac{\text{the wavelength in air}}{\text{the wavelength in glass}}$$

   iii The uncertainty in each measured angle was ± 1°. Discuss whether or not it is reasonable to conclude that the above ratio for red light differs from the ratio for blue light.

5 Light of wavelength 560 nm is directed normally at four parallel narrow slits spaced 0.12 mm apart. The two outer slits are blocked off and a screen is placed 1800 mm from the slits perpendicular to the direction of the incident beam.

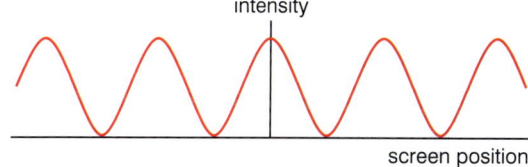

intensity

screen position

Figure 6.1 *The variation of intensity with position across the screen.*

a Calculate the fringe spacing.

b Copy the diagram and sketch on your copy the variation of intensity across the screen if the two outer slits had been used instead of the two inner slits. Give an explanation for this second fringe pattern.

c Discuss one aspect of the fringe pattern that would differ from either of the previous patterns if all four slits had been used.

6 A diffraction grating with 500 lines per millimetre was used to measure the wavelengths of a certain line emission spectrum.

a A fourth order blue line was measured at an angle of diffraction of 59°25′ and a prominent orange line was measured nearby at an angle of 62°15′.

   i Calculate the wavelength of the line at 59°25′.
   ii State the order number of the other line and calculate its wavelength.

b Describe the difference between a line emission spectrum and the spectrum of white light.

10.1 The discovery of the nucleus

Learning outcomes

On these pages you will learn to:

- describe the basic principles of Rutherford's α-particle scattering experiment
- describe what Rutherford discovered about the scattering of the alpha particles
- state and explain the key conclusions that Rutherford came to about the structure of the atom

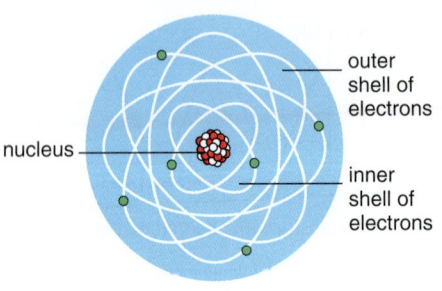

Figure 1 *The structure of an atom*

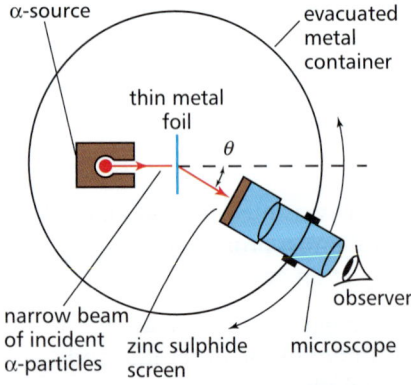

Figure 2 *Rutherford's α-scattering apparatus*

Rutherford's alpha scattering experiment

The nucleus was discovered by Ernest Rutherford in 1914. He knew from the work of J.J. Thomson that every atom contains one or more electrons. Thomson had shown that the electron is a negatively charged particle inside every atom but no one knew, until Rutherford's discovery, how the positive charge in the atom was distributed.

Rutherford knew that the atoms of certain elements were unstable and emitted radiation. It had been shown that there were three types of such radiation, referred to as **alpha radiation** (symbol α), **beta radiation** (symbol β) and gamma radiation (symbol γ). Rutherford knew that α-radiation consisted of fast-moving positively charged particles. He used this type of radiation to probe the atom. He reckoned that a beam of the particles directed at a thin metal foil might be scattered slightly by the atoms of the foil if the positive charge was spread out throughout each atom. He was astonished when he discovered that some of the particles bounced back from the foil – in his own words 'as incredible as if you fired a 15-inch naval shell at tissue paper and it came back'.

Rutherford's alpha scattering experiment in more detail

Rutherford used a narrow beam of alpha particles, all of the same kinetic energy, in an evacuated container to probe the structure of the atom. The diagram shows an outline of the arrangement he used. A thin metal foil was placed in the path of the beam. Alpha particles scattered by the metal foil were detected by a detector which could be moved round at a constant distance from the point of impact of the beam on the metal foil. See Figure 2.

Rutherford used a microscope to observe the pinpoints of light emitted by alpha particles hitting a fluorescent screen. He measured the number of alpha particles reaching the detector per minute for different angles of deflection from zero to almost 180°. His measurements showed that:

1 most alpha particles pass straight through the foil with little or no deflection
2 a small percentage of alpha particles deflect through angles of more than 90°.

Imagine throwing tennis balls at a row of vertical posts separated by wide gaps. Most of the balls would pass between the posts and therefore would not be deflected much. However, some would rebound as a result of hitting a post. Rutherford realised that the alpha scattering measurements could be explained in a similar way by assuming every atom has a 'hard centre' much smaller than the atom. His interpretation of each result was that:

1 most of the atom's mass is concentrated in a small region, the **nucleus**, at the centre of the atom,
2 the nucleus is positively charged because it repels alpha particles (which carry positive charge) that approach it too closely.

Figure 3 shows the paths of some alpha particles which pass near a fixed nucleus. The closer an alpha track deflection passes to a nucleus, the greater its deflection is because the alpha particle and the nucleus repel each other as they carry the same type of charge. Using **Coulomb's law of force** (i.e. the law of force between charged objects) and Newton's laws of motion, Rutherford used his nuclear model to explain the exact pattern of the results. By testing foils of different metal elements, he also showed that the magnitude of the charge of a nucleus is +Ze, where e is the charge of the electron and Z is the **atomic number** of the element.

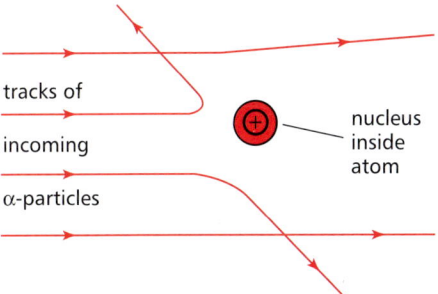

Figure 3 α-scattering paths

Notes

1 The alpha particles must have the same speed otherwise slow alpha particles would be deflected more than faster alpha particles on the same initial path.

2 The tube must be evacuated or the alpha particles would be stopped by air molecules.

3 The activity of the source (i.e. the number of alpha particles emitted each second) decreases because more and more nuclei become stable and the number of radioactive nuclei in the source decreases. The source of the alpha particles must therefore have a long half-life, otherwise later readings would be lower than earlier readings due to radioactive decay of the source nuclei.

Ernest Rutherford 1871–1937

Ernest Rutherford arrived in Britain from New Zealand in 1895. By the age of 28, he was a professor. He made important discoveries about radioactivity and was awarded the Nobel Prize for chemistry in 1908 for his investigations into the disintegration of radioactive substances. He worked in the universities of Montreal, Manchester and Cambridge. He put forward the nuclear model of the atom and proved it experimentally using α-scattering experiments. He was knighted in 1914 and made Lord Rutherford of Nelson in 1931. His co-worker Otto Hahn described him as a 'very jolly man'. In 1915, he expressed the hope that 'no one discovers how to release the intrinsic energy of radium until man has learned to live at peace with his neighbour'. After his death in 1937, his ashes were placed close to Newton's tomb in Westminster Abbey.

Figure 4 Rutherford's Cavendish laboratory, 1920s

Summary test 10.1

1 a In the Rutherford α-particle scattering experiment, most of the alpha particles passed straight through the metal foil. What did Rutherford deduce about the atom from this discovery?

   b A small fraction of the alpha particles were deflected through large angles. What did Rutherford deduce about the atom from this discovery?

2 In Rutherford's α-particle scattering experiment, why was it essential that:

   a the apparatus was in an evacuated chamber,

   b the foil was very thin,

   c the α-particles in the beam all had the same speed,

   d the beam was narrow?

3 An alpha particle collided with a nucleus and was deflected by it, as shown in Figure 5.

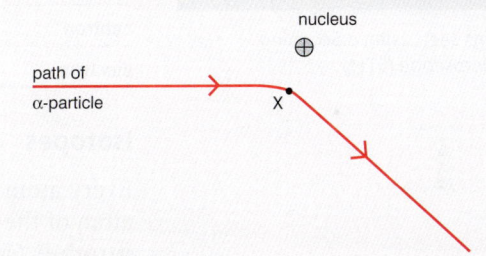

Figure 5

   a Copy the diagram and show on it the direction of the force on the alpha particle when it was at the position marked X.

   b Describe how:

      i the kinetic energy of the alpha particle, and
      ii the potential energy of the alpha particle changed during this interaction.

4 Explain why the radioactive source in Rutherford's alpha scattering experiment needed to have:

   a a very long half-life,

   b to emit alpha particles with the same kinetic energy.

10.2 Inside the atom

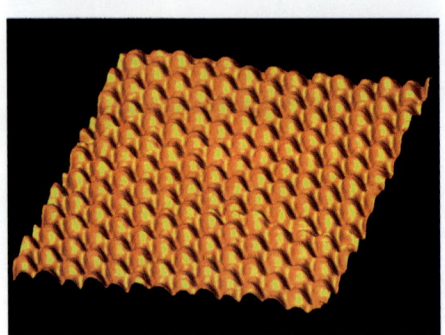

Figure 1 Atoms seen using a Scanning Tunnelling Microscope (STM)

The structure of the atom

Atoms are so small, less than a millionth of a millimetre in diameter, that we can only see images of them using the latest electron microscopes. We can't see inside them yet we know, from Rutherford's alpha scattering investigations, that every atom contains a positively charged nucleus composed of:

- **protons** and **neutrons**,
- **electrons** that surround the nucleus.

Each electron has a negative charge and is held in the atom by the electrostatic force of attraction between it and the nucleus because the nucleus is positively charged. Rutherford's investigations showed that the nucleus contains most of the mass of the atom and its diameter is of the order of 0.00001 times the diameter of a typical atom.

Table 1 shows the charge and the mass of the proton, the neutron and the electron in SI units (i.e. coulombs for charge and kilograms for mass) and relative to the charge and mass of the proton. Notice that:

1 the electron has a much smaller mass than the proton and the neutron,

2 the proton and the neutron have almost equal mass,

3 the electron has equal and opposite charge to the proton. The neutron is uncharged.

Table 1 Inside the atom

	Charge		Mass	
	/C	/charge of the proton	/kg	/mass of the proton
proton	$+1.60 \times 10^{-19}$	1	1.67×10^{-27}	1
neutron	0	0	1.67×10^{-27}	1
electron	-1.60×10^{-19}	-1	9.11×10^{-31}	0.0005

Isotopes

Every atom of a given element has the same number of protons as any other atom of the same element. The proton number is usually called the **atomic number** (symbol Z) of the element. For example:

- $Z = 6$ for carbon because every carbon atom has 6 protons in its nucleus,
- $Z = 92$ for uranium because every uranium atom has 92 protons in its nucleus.

The atoms of an element can have different numbers of neutrons. Atoms of the same element with different numbers of neutrons are called **isotopes**. For example, the most abundant isotope of natural uranium contains 146 neutrons and the next most abundant contains 143 neutrons.

> **Isotopes are forms of the same element with different numbers of neutrons and the same number of protons in their nuclei.**

A proton or a neutron in the nucleus is referred to as a **nucleon**. The total number of protons and neutrons in an atom is called the **nucleon number** (symbol A) or sometimes the **mass number** of the atom. This is because it is almost numerically equal to the mass of the atom in relative units (where the mass of a proton or neutron is approximately 1).

The isotopes of an element are labelled according to their atomic number Z, their mass number A and the chemical symbol of the element. Figure 2 shows how we do this. Notice that:

- Z is at the bottom left of the element symbol and gives the number of protons in the nucleus,
- A is at the top left of the element symbol and gives the number of protons and neutrons in the nucleus,
- The number of neutrons in the nucleus = $A - Z$.

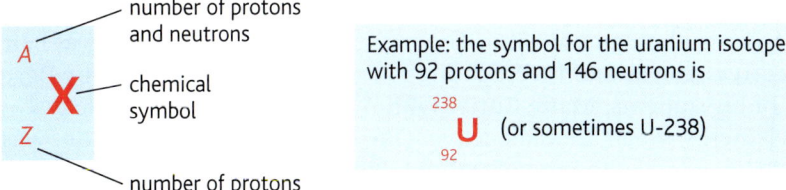

Figure 2 *Isotope notation*

Each type of nucleus is called a **nuclide** and is labelled using the isotope notation. For example, a nuclide of the carbon isotope $^{12}_{6}C$ has two fewer neutrons and two fewer protons than a nuclide of the oxygen isotope $^{16}_{8}O$.

About neutrons

From his α-scattering results, Rutherford deduced that the charge of a nucleus is $+Ze$, where Z is the atomic number of the element and the mass of the nucleus is $A m_u$, where A is the mass number of the nucleus. The $^{1}_{1}H$ nucleus is the smallest known nucleus, and physicists concluded that it is a single particle, which became known as the proton. They knew that Z and A are integers and that for all nuclei larger than the $^{1}_{1}H$ nucleus, A is greater than Z. So they put forward the hypothesis that the difference between A and Z could be due to the existence of an uncharged particle (which therefore did not contribute to Z) with about the same mass as the proton (as the difference is always an integer). They called this neutral particle the neutron, even though there was no direct evidence for it at the time. Such evidence was eventually found by Sir James Chadwick, who was one of Rutherford's former students.

Summary test 10.2

You will need to use data from Table 1 to answer some of the questions below.

1 State the number of protons and the number of neutrons in a nucleus of:

    a $^{12}_{6}C$ **b** $^{16}_{8}O$ **c** $^{235}_{92}U$ **d** $^{24}_{11}Na$ **e** $^{63}_{29}Cu$

2 Name the particle in an atom which:

    a has zero charge,

    b has the largest charge per unit mass,

    c when removed leaves a different isotope of the element.

3 a A $^{63}_{29}Cu$ atom loses two electrons. For the ion formed:

    i calculate its charge in C,

    ii state the number of nucleons it contains.

An ion has a mass of 2.67×10^{-26} kg and a negative charge of 3.2×10^{-19} C.

    b The ion has eight protons in its nucleus. How many neutrons and how many electrons does it have?

4 A block of wax contains hydrogen and carbon atoms. When a beam of neutrons is directed at a block of wax, the neutrons knock protons out of the block. Explain why protons are knocked out of the block rather than carbon nuclei.

Learning outcomes

On these pages you will learn to:

- show an understanding of the nature and properties of α, β and γ radiations

Figure 1 *Marie Curie (1867–1934)*

Marie Curie established the nature of radioactive materials. She showed how radioactive compounds could be separated and identified. She and her husband Pierre won the 1903 Nobel Prize for their discovery of two new elements, polonium and radium. After Pierre's death in 1906 she continued her painstaking research and was awarded a second Nobel Prize in 1911 – an unprecedented honour.

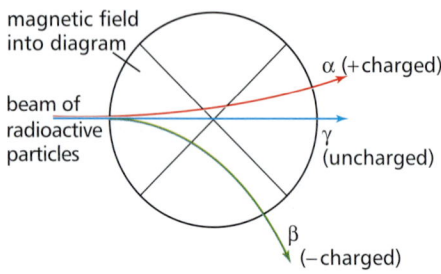

Figure 2 *Deflection by a magnetic field*

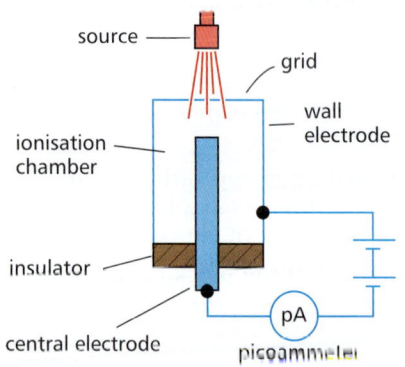

Figure 3 *Investigating ionisation*

The discovery of radioactivity

In 1896, Henri Becquerel was investigating materials that glow when placed in an X-ray beam. He wanted to find out if strong sunlight could make uranium salts glow. He prepared a sample and placed it in a drawer on a wrapped photographic plate, ready to test the salts on the next sunny day. When he developed the plate, he was amazed to see the image of a key. He had put the key on the plate in the drawer and then put the uranium salts on top of the key. He realised that uranium salts emit radiation which can penetrate paper and blacken a photographic film. The uranium salts were described as being **radioactive**. The task of investigating radioactivity was passed on by Becquerel to one of his students, Marie Curie. Within a few years, Marie Curie discovered other elements which are radioactive. One of these elements, radium, was found to be over a million times more radioactive than uranium.

Rutherford's investigations into radioactivity

Rutherford wanted to find out what the radiation emitted by radioactive substances was and what caused it. He found that the radiation:

- Ionised air, making it conduct electricity. He made a detector which could measure the radiation from its ionising effect.
- Was of two types. One type which he called **alpha** (α) radiation was easily absorbed. The other type which he called **beta** (β) radiation was more penetrating. A third type of radiation, called **gamma** (γ) radiation, even more penetrating than β radiation, was discovered a year later.

Further tests showed that a magnetic field deflects α and β radiation in opposite directions and has no effect on γ radiation. From the deflection direction, it was concluded that α radiation consists of positively charged particles and β radiation consists of negatively charged particles. γ radiation was later shown to consist of high-energy **photons**.

1 Ionisation

The ionising effect of each type of radiation can be investigated using an ionisation chamber and a picoammeter, as shown in Figure 3. The chamber contains air at atmospheric pressure. Ions created in the chamber are attracted to the oppositely charged electrode where they are discharged. Electrons pass through the picoammeter as a result of ionisation in the chamber. The current is proportional to the number of ions per second created in the chamber.

- α radiation causes strong ionisation. However, if the source is moved away from the top of the chamber, ionisation ceases beyond a certain distance. This is because α radiation has a range in air of no more than a few centimetres.
- β radiation has a much weaker ionising effect than air. Its range in air varies up to a metre or more. A β-particle, therefore, produces fewer ions per millimetre along its path than an α-particle does.
- γ radiation has a much weaker ionising effect than either α or β radiation. This is because photons carry no charge so they have less effect than α- or β-particles do.

2 Cloud chamber observations

A cloud chamber contains air saturated with a vapour at a very low temperature. Due to ionisation of the air, an α- or a β-particle passing through the cloud chamber leaves a visible track of minute condensed vapour droplets. This is because the air space is supersaturated. When an ionising particle passes through the supersaturated vapour, the ions produced trigger the formation of droplets.

- α-particles produce straight tracks that radiate from the source and are easily visible. The tracks from a given isotope are all approximately the same length, indicating that the α-particles have the same range.

- β-particles produce wispy tracks that are easily deflected as a result of collisions with air molecules. The tracks are not as easy to see as α-particle tracks because β-particles are less ionising than α-particles.

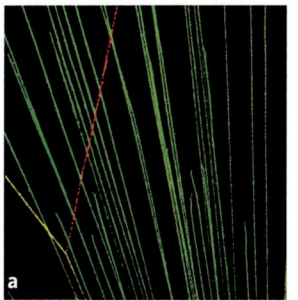 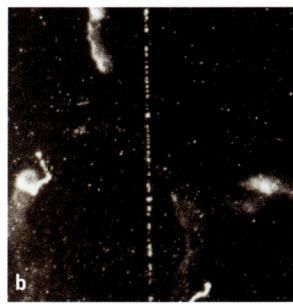

Figure 4 *Cloud chamber photographs* **a** *α-particle tracks* **b** *β-particle tracks*

3 Absorption tests

Figure 5 shows how a Geiger tube and a counter may be used to investigate absorption by different materials. Each particle of radiation that enters the tube is registered by the counter as a single count or 'click'. The clicks occur randomly which indicates that radioactive emission is a **random** process. The number of counts in a given time is measured and used to work out the **count rate** which is the number of counts divided by the time taken. Before the source is tested, the count rate due to **background radioactivity** must be measured. This is the count rate without the source present.

- The count rate is then measured with the source at a fixed distance from the tube without any absorber present. The background count rate is then subtracted from the count rate with the source present to give the **corrected (i.e. true) count rate** from the source.

- The count rate is then measured with the absorber in a fixed position between the source and the tube. The corrected count rates with and without the absorber present can then be compared.

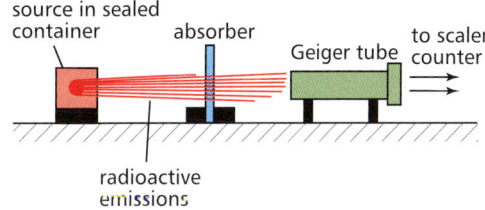

Figure 5 *Investigating absorption*

By using absorbers of different thicknesses of the same material, the effect of the absorber thickness can be investigated. Figure 6 shows a typical set of measurements for the absorption of β radiation by aluminium.

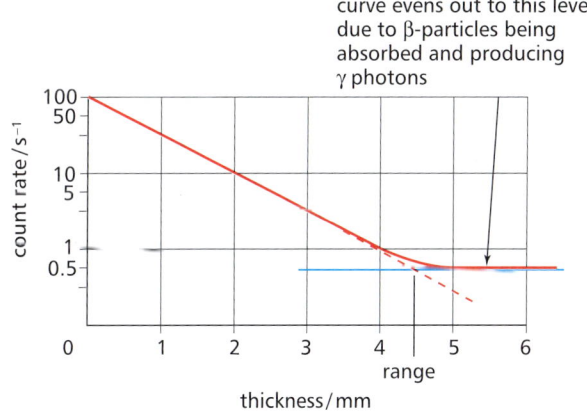

Figure 6 *Count rate v. absorber thickness*

Summary test 10.3

1 A beam of radiation from a radioactive substance passes through paper and is then stopped by an aluminium plate of thickness 5 mm.

  a What type of particles are in this beam?

  b Describe a further test you could do to check your answer in **a**.

2 a What type of radioactivity was responsible for the image of the key seen by Becquerel in the effect described on p.150?

  b Explain why an image of the key was produced on the photographic plate.

3 a Which type of radiation from a radioactive source is:
    i least ionising,   **ii** most ionising?

  b When an α-emitting source above an ionisation chamber grid was moved gradually away from the grid, the ionisation current suddenly dropped to zero. Explain why the current suddenly dropped to zero.

4 In an absorption test, as shown in Figure 5 using a β-emitting source and a Geiger counter, a count rate of 8.2 counts per second was obtained without the absorber present and a count rate of 3.7 counts per second was obtained with the absorber present. The background count rate was 0.4 counts per second. What percentage of the β-particles hitting the absorber:

  a pass through it,

  b are stopped by the absorber?

Learning outcomes

On these pages you will learn to:

- state that (electron) antineutrinos and (electron) neutrinos are produced during β⁻ and β⁺ decay
- use equations to represent the nuclear change when:
  - an α-particle is emitted
  - a β⁻ particle is emitted
  - a β⁺ particle is emitted
- recognise that a γ-photon is emitted by nuclei that have excess energy after emitting an α-particle or a β⁻ particle or a β⁺ particle

The range of α, β and γ radiation in air

The arrangement in Topic 10.3 Figure 5 without the absorbers may be used to investigate the range of each type of radiation in air. The corrected count rate is measured for different distances between the source and the tube, starting with the source close to the tube.

- α radiation has a range of several centimetres in air. The count rate decreases sharply once the tube is beyond the range of the α-particles. This can be seen in Topic 10.3 Figure 4 as the α-particle tracks are the same length indicating that the particles from the source have the same range and, therefore, the same initial kinetic energy. The range differs from one source to another indicating that the initial kinetic energy differs from one source to another.
- β radiation has a range in air of up to a metre or so. The count rate gradually decreases with increasing distance until it is the same as the background count rate at a distance of about 1 metre. The reason for the gradual decrease of count rate as the distance increases is that the β-particles from any given source have a range of initial kinetic energies up to a maximum. Faster β-particles travel further in air than slower β-particles as they have greater initial kinetic energy.
- γ radiation has an unlimited range in air. The count rate gradually decreases with increasing distance because the radiation spreads out in all directions so the proportion of the γ-photons from the source entering the tube decreases.

The nature of α, β and γ radiation

Alpha radiation consists of positively charged particles. Each α-particle is composed of two protons and two neutrons, the same as the nucleus of a helium atom.

Some years before his discovery that every atom contains a nucleus, Rutherford discovered that neutralised α-particles are the same as helium atoms. After he established the nuclear model of the atom, it was realised that the nucleus of the hydrogen atom, the lightest known atom, was a single positively charged particle which became known as

the **proton**. Rutherford realised that other nuclei contain protons and he predicted the existence of neutral particles of similar mass, **neutrons**, in the nucleus. For example, the helium nucleus carries twice the charge of the hydrogen nucleus and therefore contains two protons. However, its mass is four times the mass of the hydrogen nucleus so Rutherford predicted that it contained two neutrons as well as two protons.

β-radiation is emitted by a nucleus with either too many neutrons or too many protons.

A nucleus with too many neutrons emits a negative β-particle (symbol β⁻) as a result of one of its neutrons changing into a proton. In addition, an antineutrino (symbol $\bar{v}$) is emitted. The β⁻ particles were shown to be fast-moving electrons by deflecting them in electric and magnetic fields in order to measure their specific charge (their charge/mass value). This was found to be the same as the specific charge of the electron.

A nucleus with too many protons emits a positive β-particle (symbol β⁺), also called a positron, as a result of one of its protons changing into a neutron. In addition, a neutrino (symbol v) is emitted. See Figure 1 for more about the discovery of the positron.

Particles and antiparticles

For every known type of particle, there is a corresponding antiparticle with equal and opposite charge. The theory of antiparticles was predicted in 1928 by English physicist Paul Dirac. In his theory, he also predicted:

- **pair production**, whereby a photon of sufficient energy could produce a particle and its corresponding antiparticle
- **annihilation**, when a particle and its corresponding antiparticle collide, annihilate each other and produce photons.

Antiparticles were discovered in 1932 by American physicist Carl Anderson, who used a cloud chamber and a camera to photograph the trails produced by cosmic rays. He found trails that were curved by a magnetic field in the opposite direction to the trails produced by β⁻ particles, so they must have been produced by positively charge particles. He measured the specific charge of these particles and found it was the same as that of the β⁻ particles. So he concluded that each such positively charged particle was a **positron**, the antiparticle of the electron.

Figure 1 The discovery of the positron

An elusive particle

When the energy spectrum of beta particles was first measured, scientists were puzzled when they discovered the kinetic energy of β-particles varied up to a maximum even though the nucleus always lost a certain amount of energy in the process of emitting a β-particle. Either energy was not conserved in the change or some of it was carried away by unknown uncharged particles and antiparticles, which they called **neutrinos** and **antineutrinos**. This latter hypothesis was eventually shown to be correct when antineutrinos from a nuclear reactor were detected as a result of their interaction with cadmium nuclei in water. Now we know that billions of these elusive particles from the Sun sweep though our bodies every second without interacting.

γ radiation consists of photons with a wavelength of the order of a fraction of a nanometre or less. This discovery was made by using a crystal to diffract a beam of γ radiation in a similar way to the diffraction of light by a diffraction grating.

The equations for radioactive change

A nuclide $_{Z}^{A}X$ contains Z protons and $A - Z$ neutrons.

- its charge = $+Ze$, where e is the magnitude of the charge of an electron
- its mass in **atomic mass units, u,** = A approximately.

Note

The mass of a proton is $1.00728\,u$ and the mass of a neutron is $1.00866\,u$, where $1\,u = \frac{1}{12}$ of the mass of a $_{6}^{12}C$ atom.

1 α-emission

An α-particle is represented by the symbol $_{2}^{4}\alpha$. Its charge = $+2e$ so $Z = 2$ and it consists of 2 neutrons and 2 protons so $A = 4$.

When a nucleus $_{Z}^{A}X$ emits an α-particle, it loses two protons and two neutrons. Therefore, its proton number (Z) is reduced by 2 and its mass number (A) is reduced by 4.

$$_{Z}^{A}X \rightarrow {}_{2}^{4}\alpha + {}_{Z-2}^{A-4}Y$$

2 β-emission

β⁻ and β⁺ particles are represented in equations by the symbols $_{-1}^{0}\beta$ and $_{+1}^{0}\beta$, respectively, as their mass is much less than $1\,u$ and their charge is $-e$ and $+e$, respectively. Antineutrinos and neutrinos are uncharged and have very little mass compared with β-particles.

When a nucleus $_{Z}^{A}X$ emits a β⁻ particle (i.e. an electron), a neutron in the nucleus changes into a proton and an antineutrino, $\bar{\nu}$, is also emitted. Therefore, the proton number of the nucleus increases by 1 and the mass number is unchanged.

$$_{Z}^{A}X \longrightarrow {}_{Z+1}^{A}Y + {}_{-1}^{0}\beta + \bar{\nu}$$

When a nucleus $_{Z}^{A}X$ emits a β⁺ particle (i.e. a positron), a proton in the nucleus changes into a neutron and a neutrino (ν) is also emitted. Therefore, the proton number of the nucleus decreases by 1 and the mass number is unchanged.

$$_{Z}^{A}X \longrightarrow {}_{Z-1}^{A}Y + {}_{+1}^{0}\beta + \nu$$

3 γ-emission

A γ-photon is emitted if a nucleus has excess energy after it has emitted an α- or a β-particle.

No change occurs in the number of protons or neutrons of a nucleus when it emits a γ-photon.

Note

That in all the above changes, the nucleon number A and the charge are both conserved. In other words, the total number of protons and neutrons after the change is the same as before the change and the total charge after the change is the same as before the change.

Summary test 10.4

1 Copy and complete each of the following equations representing α-emission.

   a $_{92}^{238}U \rightarrow {}_{90}Th +$ **b** $_{90}Th \rightarrow {}_{88}^{224}Ra +$

2 Copy and complete each of the following equations representing β-emission.

   a $_{29}^{64}Cu \rightarrow {}_{30}Zn + {}^{0}\beta$ **b** $_{15}P \rightarrow {}^{32}S + {}_{-1}^{0}\beta$

3 The bismuth isotope $_{83}^{213}Bi$ decays by emitting a β-particle to form an unstable isotope of polonium (Po) which then decays by emitting an α-particle to form an unstable isotope of lead (Pb). This isotope then decays by emitting a β-particle to form a stable isotope of bismuth.

   a Write down the symbol for each of the three product nuclides in this sequence.

   b Write down the number of protons and the number of neutrons in a nucleus of:
      i the bismuth isotope $_{83}^{213}Bi$,
      ii the stable bismuth isotope.

4 The polonium isotope, $_{84}^{205}Po$, emits α radiation and decays into nuclei of an isotope of a lead (Pb) isotope which is unstable and decays into thallium, emitting β radiation in the form of positrons in the process.

   a Write down equations for the change when:
      i a nucleus of the polonium isotope emits an α particle, **ii** the lead nucleus formed in **i** emits a positron when it changes into a thallium nucleus.

   b State two differences in the physical properties of a positron and a β⁻ particle.

153

10.5 The dangers of radioactivity

Learning outcomes

On these pages you will learn to:

- explain what is meant by background radioactivity
- recall different types of ionising radiation
- recognise the hazards of ionising radiation and how it may be monitored
- describe procedures used to ensure safe use of radioactive materials

Figure 1 *A radioactive warning sign*

Table 1 *Sources of background radioactivity*

Source	Typical annual dose*
Natural radioactivity in the air	800
Ground and buildings	380
Food and drink	370
Cosmic rays	310
Nuclear weapons testing	10
Air travel	8
Nuclear power	3

* in microsieverts, a measure of the effect of radioactivity on cells.

The hazards of ionising radiation

Ionising radiation is hazardous because it damages living cells. **Ionising radiation** is any form of radiation that creates ions in substances it passes through. Such radiation includes **X-rays**, protons and neutrons as well as α, β and γ radiation. Ionising radiation affects living cells because:

- it can destroy cell membranes which causes cells to die, or
- it can damage vital molecules such as DNA directly or indirectly by creating 'free radical' ions which react with vital molecules. Normal cell division is affected and nuclei become damaged. Damaged DNA can cause cells to divide and grow uncontrollably, causing a tumour which may be cancerous. Damaged DNA in a sex cell (i.e. an egg or a sperm) can cause a mutation which might be passed on to future generations.

As a result of exposure to ionising radiation, living cells die or grow uncontrollably or mutate, affecting the health of the affected person (somatic effects) and possibly affecting future generations (genetic effects). High doses of ionising radiation kill living cells. Cell mutation and cancerous growth occur at low doses as well as at high doses. There is no evidence of the existence of a threshold level of ionising radiation below which living cells would not be damaged.

Background radioactivity occurs naturally due to cosmic radiation and from radioactive materials in rocks, soil and in the air. Everyone is exposed to background radioactivity which varies with location due to local geological features. In addition, radon gas, which is radioactive, can accumulate in poorly ventilated areas of buildings in such locations. Table 1 shows the many sources of background radioactivity that occur in most populated locations.

Radiation monitoring

The Geiger tube

The Geiger tube is a sealed metal tube that contains argon gas at low pressure. The thin mica window at the end of the tube allows α- and β-particles to enter the tube. γ-photons can enter the tube through the tube wall as well. A metal rod down the middle of the tube is at a positive potential as shown in Figure 2. The tube wall is connected to the negative terminal of the power supply and is earthed.

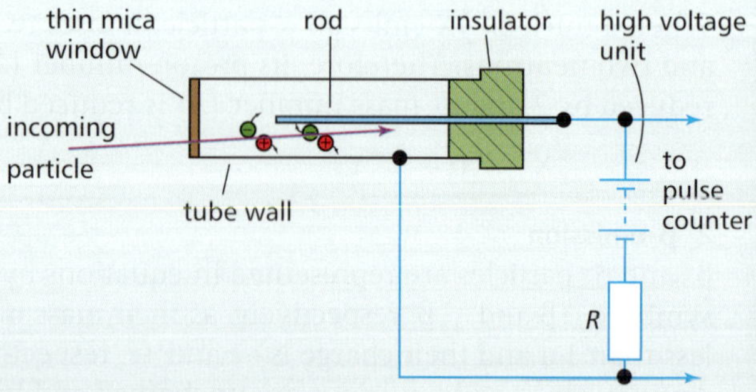

When a particle of ionising radiation enters the tube, the particle ionises the gas atoms along its track. The negative ions are attracted to the rod and the positive ions to the wall. The ions accelerate and collide with other gas atoms, producing more ions which produce further ions in the same way. Within a very short time, many ions are created and discharged at the electrodes. A pulse of charge passes round the circuit through resistor *R*, causing a voltage pulse across *R* which is recorded as a single count by the pulse counter.

Figure 2 *A Geiger tube*

The **dead time** of the tube, the time taken to regain its non-conducting state after an ionising particle enters it, is typically of the order of 0.2 ms. Another particle that enters the tube in this time will not cause a voltage pulse. Therefore, the count rate should be no greater than about $5000\,s^{-1}$ $(= \frac{1}{0.2\,ms})$.

The film badge

Anyone using equipment that produces ionising radiation must wear a film badge to monitor his or her exposure to ionising radiation. The badge contains a strip of photographic film in a light-proof wrapper. Different areas of the wrapper are covered by absorbers of different materials and different thicknesses. When the film is developed, the amount of exposure to each form of ionising radiation can be estimated from the blackening of the film. If the badge is overexposed, the wearer is not allowed to continue working with the equipment.

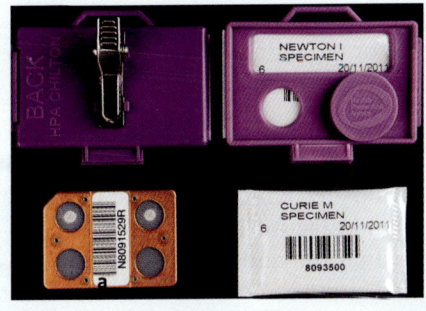

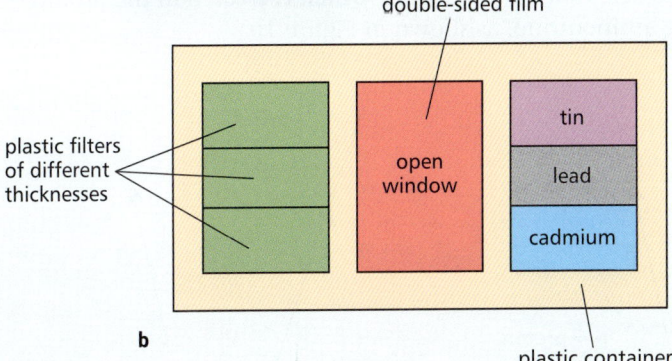

Figure 3 **a** A film badge **b** Inside a film badge

Safe use of radioactive materials

Because radioactive materials produce ionising radiation, they must be stored and used with care. In addition, disposal of a radioactive substance must be done in accordance with specific regulations. Only approved institutions are allowed to use radioactive materials. Approval is subject to regular checks and approved institutions are categorised according to purpose.

1 **Storage of radioactive materials** should be in lead-lined containers. Most radioactive sources produce γ radiation as well as α or β radiation so the lead lining of a container must be thick enough to absorb all the γ radiation from the sources in the container. In addition, regulations require that the containers are under 'lock and key' and a record of the sources is kept.

2 **When using radioactive materials**, established rules and regulations must be followed. No source should be allowed to come into contact with the skin.

   • Solid sources should be transferred using handling tools such as tongs or a glove-box or using robots. The handling tools ensure the material is as far from the user as practicable so the intensity of the γ radiation from the source at the user is as low as possible and the user is beyond the range of α or β radiation from the source.
   • Liquid and gas sources and solids in powder form should be in sealed containers. This is to ensure that radioactive gas cannot be breathed in and radioactive liquid cannot be splashed on the skin or drunk.

3 **Disposal** requires long-term storage until the radioactive material is no longer radioactive. The key aim is to ensure that people and the environment are not put at risk. As explained in the next section, radioactive half-lives differ according to the isotope. The half-life of a radioactive isotope is the time it takes for the activity of the isotope to decrease to half. For example, an isotope with a half-life of 5 years would decay to 6.25% of its initial activity after 20 years.

Summary test 10.5

1 **a** What is meant by ionisation?

  b Explain why a source of α radiation is not as dangerous as a source of β radiation provided the sources are outside the body.

2 **a** Discuss the reasons why ionising radiation is hazardous to a person exposed to the radiation.

  b **i** What is the purpose of a film badge worn by a radiation worker?

   ii With the aid of a diagram, describe what is in a film badge and how the film badge is tested.

3 Explain why a radioactive source should be:

   i kept in a lead-lined storage box when not in use,

   ii transferred using a pair of tongs with long handles.

4 Discuss the precautions you would take when carrying out an experiment using a source of γ radiation.

Learning outcomes

On these pages you will learn to:

- appreciate that protons and neutrons are not fundamental particles since they contain quarks
- describe a simple quark model of hadrons in terms of up, down and strange quarks and their respective antiquarks
- describe protons and neutrons in terms of a simple quark model
- appreciate that there is a weak interaction between quarks, giving rise to β decay
- describe β⁻ and β⁺ decay in terms of a simple quark model
- appreciate that electrons and neutrinos are leptons

📖 The fundamental forces of nature

In your physics course, you will meet four fundamental forces:

- the **electromagnetic force**, which acts between charged particles. This force is due to the charged particles creating and exchanging 'virtual photons', and its range is unlimited.
- the **strong nuclear force**, which holds the nucleus together and acts between neutrons and protons. Its range is of the order of 2–3 femtometres.
- the **weak nuclear force**, which causes a neutron to change into a proton or a proton to change into a neutron. Its range is of the order of a fraction of a femtometre.
- the **gravitational force**, which acts between any two matter particles. Its range is unlimited.

Physicists now know that at very high energies, the electromagnetic force and the weak nuclear force are unified as the 'electroweak' force. At even higher energies, the electroweak force and the strong nuclear force may possibly be unified. But the nature of gravity, the commonest force because we experience it all the time, may be beyond unification!

Forces and interactions

A stable isotope has nuclei that do not disintegrate, so there must be a force holding them together. We call this force the **strong nuclear force** or the 'strong interaction', because it overcomes the electrostatic force of repulsion between the protons in the nucleus and keeps the protons and neutrons together, except in unstable nuclei. Its range is no more than about 3–4 femtometres (fm), where 1 fm = 10^{-15} m. In comparison, the electrostatic force between two charged particles has an infinite range.

The strong nuclear force holds the neutrons and protons in a nucleus together. However, it doesn't cause a neutron to change into a proton in β⁻ decay or a proton to change into a neutron in β⁺ decay. These changes cannot be due to the electromagnetic force, as the neutron is uncharged. There must be a different force at work in the nucleus causing these changes. It must be weaker than the strong nuclear force, otherwise it would affect stable nuclei; hence we refer to it as the **weak nuclear force** or the 'weak interaction'. When a nucleus emits a beta particle, it also emits a neutrino or an antineutrino. These particles hardly interact with other particles but they sometimes do. For example:

- a neutrino can interact with a neutron and make it change into a proton. A β⁻ particle is created and emitted at the same time. The interaction is due to the exchange of a particle called a **W⁻ boson**, which is created in the neutron and absorbed by the neutrino, as shown in Figure 1a.
- an antineutrino can interact with a proton and make it change into a neutron. A β⁺ particle is created and emitted at the same time. The interaction is due to the exchange of a particle called a **W⁺ boson**, which is created in the proton and absorbed by the antineutrino, as shown in Figure 1b.

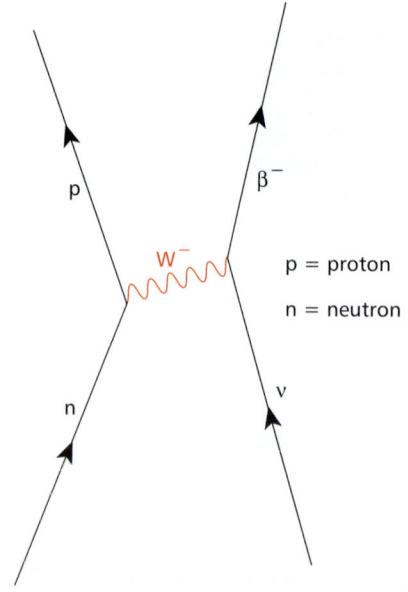

p = proton
n = neutron

a *A neutron–neutrino interaction*

b *A proton–antineutrino interaction*

Figure 1 *Weak interactions*

Collisions and colliders

Cosmic rays are high-energy particles that travel through space from the stars, including the Sun. When a cosmic particle enters the Earth's atmosphere, it interacts with the nuclei of atoms in the atmosphere and creates new short-lived particles and antiparticles. When they were first discovered, most physicists thought cosmic rays were from terrestial radioactive substances. This theory was disproved when physicist and amateur balloonist Victor Hess found that the ionising effect of the rays was significantly greater at 5000 m than at the ground.

Further investigations showed that most cosmic rays are fast-moving protons or small nuclei. They collide with the nuclei of gas atoms in the atmosphere, making them unstable and creating showers of particles and antiparticles that can be detected at ground level. By using cloud chambers and other detectors, new types of short-lived particles and antiparticles were discovered, including:

Figure 2 *Creation and decay of a π meson*

- the **muon** or 'heavy electron' (symbol μ), a negatively charged particle with a rest mass over 200 times the rest mass of the electron. Muons and antimuons decay into electrons and antineutrinos or positrons and neutrinos, respectively.
- the **pion** or 'π meson', a particle that can be positively charged (π^+), negatively charged (π^-) or neutral (π^0) and has a rest mass greater than a muon but less than a proton. Charged π mesons decay into muons and antineutrinos or antimuons and neutrinos.
- the **kaon** or 'K meson', which also can be positively charged (K^+), negatively charged (K^-) or neutral (K^0) and has a rest mass greater than a pion but still less than a proton. Like π mesons, kaons are produced through the strong interaction when protons moving at high speed crash into nuclei. Because they are produced in pairs through the strong interaction and decay through the weak interaction, they are called **strange** particles.

More strange and non-strange particles were discovered using high-energy accelerators such as:

- the Stanford Linear Accelerator in California. It accelerates electrons over a distance of 3 km through a potential difference (p.d.) of 50 000 million volts. The energy of an electron accelerated through this p.d. is 50 000 MeV (= 50 GeV). When the electrons collide with a target, they can create lots of particle–antiparticle pairs.
- the Large Hadron Collider at CERN near Geneva. It is designed to accelerate charged particles to energies of more than 7000 GeV. Unlike a linear accelerator, this accelerator is a 28 km diameter ring constructed in a circular tunnel below the ground. It is used by physicists from many countries to find out more about the fundamental nature of matter and radiation.

Figure 3 *The Large Hadron Collider*

Hadrons and leptons

Protons, neutrons, π mesons and K mesons and their antiparticles are examples of particles and antiparticles that we call **hadrons**, because they experience the strong interaction. Electrons, muons and neutrinos and their antiparticles are examples of particles and antiparticles that we call **leptons**, because they experience the weak interaction.

All leptons are fundamental particles in that they are elementary. Except for the proton, which is stable, hadrons decay in one or more stages into leptons and/or protons. Hadrons that include a proton in their decay products are referred to as **baryons**. Thus, a hadron is either a baryon or a **meson**, according to whether or not its decay includes a proton.

- Leptons do not interact through the strong interaction. They interact through the weak interaction and, if charged, through the electromagnetic interaction.

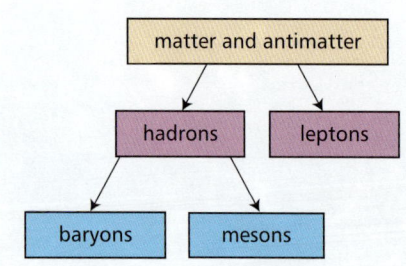

Figure 4 *Hadrons and leptons*

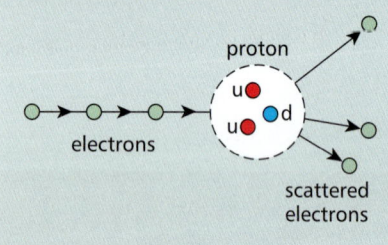
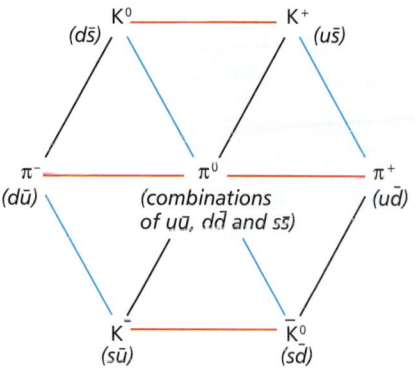

Figure 6 *Quark combinations for the mesons*

- Hadrons interact through the strong interaction and, if charged, through the electromagnetic interaction. Apart from the proton, which is stable, hadrons decay through the weak interaction.

Quarks

The properties of hadrons – can be explained by assuming they are composed of smaller particles known as **quarks** and **antiquarks**. Six different types of quarks and their corresponding antiquarks are necessary in the standard **quark model**. These six types are referred to as up, down, strange, charm, top and bottom quarks and are denoted by the symbols u, d, s, c, t, and b respectively. The six types are collectively called quark **flavours**. The equivalent antiquarks are denoted by relevant quark symbol with a bar over the top, for example $\bar{u}$ (pronounced 'u bar' denotes the antiparticle of the up quark. The properties of these six quarks are shown in Table 1.

Table 1 *Quark properties*

Type	Quarks						Antiquarks					
	u	d	c	s	t	b	$\bar{u}$	$\bar{d}$	$\bar{c}$	$\bar{s}$	$\bar{t}$	$\bar{b}$
Charge	$+\frac{2}{3}$	$-\frac{1}{3}$	$+\frac{2}{3}$	$-\frac{1}{3}$	$+\frac{2}{3}$	$-\frac{1}{3}$	$-\frac{2}{3}$	$+\frac{1}{3}$	$-\frac{2}{3}$	$+\frac{1}{3}$	$-\frac{2}{3}$	$+\frac{1}{3}$

The rules for combining quarks and antiquarks to form baryons and mesons are astonishingly simple.

Mesons: each meson is composed of a quark and an antiquark (e.g. a π^- meson is the $\bar{u}$d combination).

Baryons: each baryon is composed of three quarks (e.g. the proton is the uud combination).

Antibaryons: each antibaryon is composed of three antiquarks (e.g. the antiproton is the $\bar{u}\bar{u}\bar{d}$ combination).

Quark combinations

Mesons are hadrons that each consist of a quark and an antiquark. Figure 6 shows all nine different quark–antiquark combinations and the meson each combination forms in each case. Notice that:

- A π^0 meson can be any quark-corresponding antiquark combination.
- Each pair of charged mesons is a particle–different antiparticle pair.
- There are two uncharged K mesons, the K^0 meson and the $\overline{K^0}$ meson.
- A strange meson contains a strange quark or antiquark. A strange baryon contains one or more strange quarks. A strange antibaryon contains one or more strange antiquarks.

The antiparticle of any meson is a quark–antiquark pair and is therefore a meson.

Baryons and antibaryons are hadrons that consist of three quarks or three antiquarks:

- a proton is the uud combination
- a neutron is the udd combination.

The proton is the only stable baryon. A free neutron decays into a proton.

Quarks and beta decay

In β⁻ decay, a neutron in a neutron-rich nucleus changes into a proton and an electron and an antineutrino are released. In quark terms, a down quark changes to an up quark. In the process, a W⁻ boson is created and emitted from the down quark. The W⁻ boson then decays into a β⁻ particle and an antineutrino. A simple diagram for this change is shown in Figure 7(a).

In β⁺ decay, a proton in a proton-rich nucleus changes into a neutron and a positron and an neutrino are released. In quark terms, an up quark changes to a down quark. In the process, a W⁺ boson is created and emitted from the up quark. The W⁺ boson then decays into a β⁺ particle and a neutrino. A simple diagram for this change is shown in Figure 7(b).

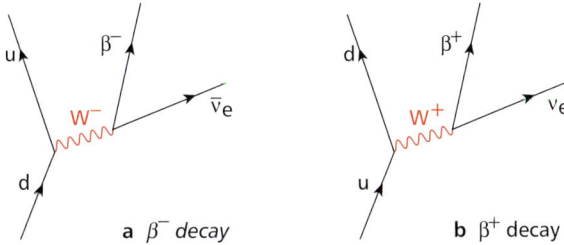

Figure 7 *Quark changes in beta decay*

Summary test 10.6

1 a State two differences between a W boson and a virtual photon.

b State the approximate range of a W boson and estimate its lifetime, given it cannot travel faster than the speed of light, which is $3.0 \times 10^8 \, \text{m s}^{-1}$.

2 a State the difference between a lepton and a hadron in terms of their interactions.

b Give an example of: **i** a baryon, **ii** a non-strange meson, **iii** a strange meson.

3 a State the quark composition of: **i** a proton, **ii** a neutron.

b With the aid of a diagram, explain in terms of quarks the changes that take place when a proton changes into a neutron in a nucleus.

4 The omega minus (Ω^-) consists of three strange quarks.

a Determine its charge relative to the proton.

b It decays into a π^- meson and a baryon X, which contains two strange quarks and another quark.
  i State the quark composition of the π^- meson.
  ii Determine the quark composition of X. Give a reason for your answer.

Launch additional digital resources for the chapter

$c = 3.00 \times 10^8 \, \text{m s}^{-1}$, $e = 1.60 \times 10^{-19} \, \text{C}$,
$h = 6.63 \times 10^{-34} \, \text{J s}$

1   A beam of α-particles was directed at normal incidence towards a thin metal foil. A detector was used to measure the number of α-particles per second scattered by different angles.

a   Explain why:
   i   most α-particles passed through the metal foil with little or no deflection,
   ii some α-particles were scattered through very large angles.

b   The figure below shows the path of two α-particles moving at the same speed in the same direction as they approached a nucleus of an atom. Copy the diagram and complete the paths of these particles.

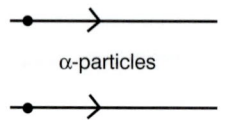 

α-particles   nucleus

Figure 1.1

2   An α-particle collided head-on with a nucleus, as shown below.

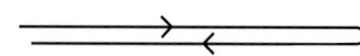    nucleus

Figure 2.1

a   Describe and explain how the kinetic energy of the α-particle changed as it approached then moved away from the nucleus.

b   i   Discuss the factors that determine how closely the α-particle can approach the nucleus.
   ii Explain why an α-particle with more kinetic energy on such a track might not rebound from the nucleus.

3   In a nuclear reaction, a neutron collided with a nucleus of the lithium isotope ^6_3Li. A tritium nucleus ^3_1H and another nucleus X was formed as a result.

Write down an equation that represents this reaction and identify the nucleus X.

4   A narrow beam of ionising radiation from a radioactive substance was directed at a detector at a distance of 30 cm from the source. The detector reading was unchanged when an aluminium plate 5 mm thick was placed between the source and the detector.

a   What type of radiation was emitted by this source?

b   With the source at a constant distance from the detector, discuss the effect on the detector reading of:
   i   placing a second identical aluminium plate in the path of the beam,
   ii placing a lead plate of thickness 10 mm in the path of the beam.

5   a   A Geiger tube in a room recorded a background count in 5 minutes of 130 counts. When a point source of γ radiation was placed at a distance of 0.25 m from the Geiger tube, the Geiger counter recorded 2450 counts in exactly 5 minutes. Calculate:
   i   the count rate in counts per second due to the source at this distance,
   ii the number of counts in 5 minutes with the source at 0.50 m from the tube.

b   When an experiment is carried out using a radioactive source in a school laboratory, explain why:
   i   the source should be kept in a lead-lined container when it is not in use,
   ii the user should keep as far from the source as possible and keep the time of use as short as possible.

6   a   A $^{238}_{92}\text{U}$ nucleus emits an α-particle. The nucleus then emits a β^--particle. Calculate the number of protons and neutrons in the nucleus formed immediately after the emission of:
   i   the α-particle,
   ii the β^--particle.

b   A copper disc is placed in the core of a nuclear reactor where the copper nuclei are bombarded by neutrons. As a result, the sample becomes radioactive and emits β^--particles.

Copy and complete the following equations representing these changes:

$$^{63}_{29}\text{Cu} + ^1_0\text{n} \rightarrow \text{X}$$

$$\text{X} \rightarrow \text{Zn} + ^{\ 0}_{-1}\beta$$

7 A radioactive nucleus decays with the emission of a beta particle and a gamma-ray photon.

a Describe the changes that occur in the proton number and the nucleon number of the nucleus.

b Comment on the relative penetrating powers of the two types of ionising radiation.

c Gamma rays from a point source are travelling towards a detector. The distance from the source to the detector is changed from 1.0 m to 4.0 m. Calculate:

$$\frac{\text{intensity of radiation at 4.0 m}}{\text{intensity of radiation at 1.0 m}}$$

8 In α-scattering experiments, α-particles with the same kinetic energy in a narrow beam are directed normally at a very thin metal foil.

Figure 8.1 shows the path of two α-particles, X and Y, that are deflected by the nucleus of an atom.

a Explain why the deflection of X is greater than the deflection of Y.

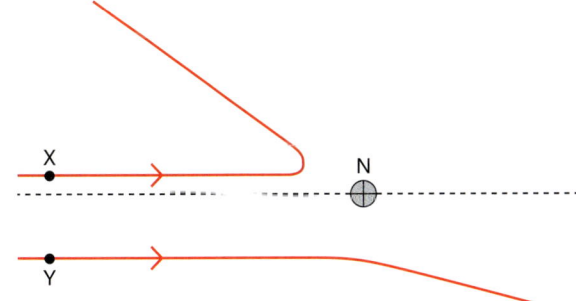

Figure 8.1

b A small proportion of the incident α-particles undergo very large deflections. What conclusions about the nature of atoms can be drawn from this observation?

c i State one reason why the metal foil needs to be very thin.

ii Explain why the α-particles in the beam need to have the same initial kinetic energy.

9 The bismuth isotope $^{209}_{83}$Bi has an atomic mass of 209 u and a radius of 7.1 fm. (See Topic 10.6 for explanation of fm.)

a State the number of: **i** protons, **ii** neutrons in this nucleus.

b Calculate the density of this nucleus.
$1\,\text{u} = 1.661 \times 10^{-27}\,\text{kg}$

c Bismuth has a density of 9800 kg m^{-3}. Explain why this is much smaller than the density of a bismuth nucleus.

10 The thorium isotope $^{232}_{90}$Th is radioactive and undergoes a series of α– and β–decays which terminate in the formation of the stable lead isotope $^{208}_{82}$Pb.

a The first two decays are an α–emission followed by a β-emission. The following equations represent these two changes. Determine the missing values *a* to *e* in these equations.

i $^{232}_{90}\text{Th} \rightarrow {}^{4}_{a}\alpha + {}^{b}_{c}\text{Ra}$ **ii** $^{b}_{c}\text{Ra} \rightarrow {}^{0}_{d}\beta + {}^{228}_{e}\text{Ac}$

b After the first two decays, the nucleus in **a** may undergo eight further α- and β-emissions of which five are α-emissions. Calculate the number of β-emissions in the series.

Key concepts: Electricity, waves and radioactivity

In studying electricity in Chapters 6 and 7, you will have gained a clear understanding of how the 'conduction electron' model is used to explain why metals conduct and insulators do not and why the resistivity of an intrinsic semiconductor decreases as its temperature is increased. Electrons are negatively charged particles, and their movement round an electric circuit causes energy to transfer from the battery or power supply to the components in the circuit. Conservation of charge and of energy are important concepts in electric circuit theory because these two principles underpin Kirchoff's laws and other circuit rules. In order to apply these rules to electric circuits, the characteristics of the components in a circuit need to be known before the currents and potential differences within the circuit can be determined. Such component characteristics include knowledge of its resistance and its variation, if any, with temperature and whether or not the component is an ohmic conductor. Internal resistance is an important characteristic of a battery or power source, as internal resistance transfers energy by heating from the power source to the surroundings and is therefore a source of inefficiency in the transfer of energy in an electric circuit.

You should also have developed your mathematical skills in studying electricity as a result of using circuit rules to calculate potential differences, currents and power supplied to the components in a circuit. Conservation of energy applied to a circuit means that the power supplied by a power source less the power dissipated to its internal resistance should be equal to the sum of the powers supplied to all the components.

Electricity is a practical subject, and you should now know how to measure currents and potential differences accurately and to use variable resistors and potential dividers to control currents and potential differences. The correct use of a potential divider in a sensor circuit requires a very clear grasp of the concepts of potential difference and resistance in order to predict how a sensor works.

Waves transfer energy and information. After studying chapters 8 and 9, you should understand the main differences between longitudinal waves and transverse waves and the difference between mechanical waves and electromagnetic waves. You should also know what is meant by wave characteristics such as wavelength, frequency and phase difference. You should also be able to describe wave properties such as reflection, refraction, diffraction and interference and be able to explain two-source interference. You should know how a diffraction grating works and how it is used to measure light wavelengths. Also, you should be able to describe stationary wave patterns and be able to explain their formation. A clear grasp of the principle of superposition is important in gaining a good understanding of interference and also of the formation of stationary waves.

The final chapter on particle physics develops previous knowledge from earlier chapters about the structure of the atom and about conservation of charge. In studying the nuclear model of the atom, the role of experimentation should be recognised as being crucial to our understanding of matter. Rutherford used his experimental skills to obtain measurements of the deflection of alpha particles and then successfully applied his theoretical knowledge of mechanics and electric fields to his nuclear model of the atom. Using the nuclear model of the atom, you should have a clear understanding of what is meant by an isotope and how equations based on conservation principles are used to represent nuclear changes. The nuclear model of the atom enables us to understand the nature and properties of alpha, beta and gamma radiation as well as gaining knowledge and awareness of the practical methods and devices used in determining such properties. In such practical work you should have become aware of the hazards of ionising radiation and of the importance of adhering to vital safety rules when radioactive materials are used.

The topic on fundamental particles provides awareness of how experimental discoveries have led to our present knowledge of the structure of matter, including matter and antimatter, the quark model and the weak nuclear force. Conservation rules on nuclear equations developed earlier provide the basic principles that underpin such knowledge.

The following table provides an overview of the topics in chapters 6 to 10, showing where key concepts have been developed.

Chapters 6–10: Key concepts

		Topic		Key concepts
6.1–6.2	conductors and insulators	electrons as charge carriers	models	drift velocity
		calculations	use of maths	problem-solving
	electric fields	nature of an electric field	forces and fields	action at a distance
6.3–6.5	potential difference and resistance	potential difference	energy and matter	energy transfer per unit charge
		I–V characteristics		
		resistance and resistivity		practical skills using an ammeter and a voltmeter and graph work
		I–V graphs		
		calculations involving resistance and resistivity		problem-solving
7.1–7.2	Kirchhoff's laws	currents at a junction	models	conservation of charge and of energy applied to electric circuits
		p.d. round a circuit	energy and matter	
		resistor rules		
		circuit calculations	use of maths	problem-solving
7.3–7.4	e.m.f. and internal resistance	measurement of e.m.f. and internal resistance	testing	practical skills involving electrical measurements and graph work
		conservation of energy	energy and matter	
		maximum power		
		circuit calculations	use of maths	application of Kirchoff's laws
7.5	the potential divider	theory and use of the potential divider	testing	practical skills involving measurements of p.d.
		sensor circuits	models	predicting output changes in sensor circuits
8.1–8.4	wave properties	longitudinal and transverse waves	energy and waves	amplitude, phase difference
		polarisation of transverse waves		
		reflection, refraction, diffraction, interference	testing	practical skills involved in measuring amplitudes and frequencies
8.5–8.8	stationary waves	stationary wave formation, nodes	energy and waves	principle of superposition
8.9	Doppler effect	Doppler shift	use of maths	problem-solving
		calculations using Doppler effect equations		

continued on p164

Chapters 6–10: Key concepts (*continued*)

Topic		Key concepts	
9.1–9.2	interference of light	theory of two-slit interference	models	principle of superposition
			energy and waves	path difference
				coherence
		calculations involving use of the two-slit interference equation	use of maths	problem-solving
9.3	diffraction grating	theory of the diffraction grating	energy and waves	superposition of diffracted wavefronts
		observation and measurement of spectra	testing	practical skills
		calculation of wavelength	use of maths	problem-solving
10.1–10.2	structure of the atom	Rutherford's alpha scattering experiment	models	nuclear model of the atom
		isotopes		
		calculation of nuclear size and nuclear density	use of maths	problem-solving
10.3–10.5	α, β and γ radiation	properties	testing	investigation of range, absorption, ionisation
		ionisation		
		background radioactivity		
		nuclear equations		conservation rules
10.6	fundamental particles	matter and antimatter	energy and matter	
		strong and weak nuclear forces	forces and fields	
		classification of particles		
		quarks	models	the quark model

Question

This question is about an investigation into the heating effect of an electric current and the use of physics equations to predict and test the results. It tests how we use models to make predictions about a physical system and how we then test these predictions with an experimental investigation. It also tests practical skills, when electrical and temperature measurements are used to plot a graph. Accuracy in making temperature measurements is often more difficult to achieve than in electrical or mechanical measurements such as distance or time intervals and simple practical techniques such as stirring the water being heated are important.

*Note the use of physics theory and mathematics to predict the relationship between the current I and the temperature change, ΔT, of the water being heated. In this case, the predicted relationship is of the form $\Delta T = kI^2$. So by plotting a graph of ΔT against I^2, we can tell if the relationship is valid from whether or not the graph is linear (i.e. a straight-line graph). If the graph is linear, we can also determine the constant **k**, as it is equal to the gradient of the graph.*

*This investigation involves a predicted relationship of the same form as that in the Key concepts 'projectile motion' question on p.92. In both situations, although the measurements give a straight-line graph, the spacing between the plotted points increases along the line. This uneven spacing causes unnecessary difficulty in judging the best-fit line. In practice, the values of **I** (or **x** in the previous question) should be selected to give approximately even intervals between successive values of I^2 rather than between successive values of **I** (or x^2 in the previous question).*

Before answering the question, students should be given the opportunity to do this or a similar investigation.

1 A student carried out an investigation to find out how the heating effect of an electric current changes when the current is increased. A 12 V heater in series with an ammeter, a variable resistor, a switch and a 12 V battery was used to heat some water in an insulated beaker. A thermometer was used to measure the temperature change of the water. The test was repeated several times with a different current each time. In each test, the same amount of water was heated for 600 seconds.

  a i Sketch the circuit diagram.

    ii In each test, the current was kept constant. Describe how this was achieved using the circuit you sketched. *(2 marks)*

  b i Outline a test you would carry out to ensure that the beaker insulation was effective.

    ii Why was it important to heat the same volume of water for the same length of time in each test? *(4 marks)*

  c The table below shows the results of these tests.

current I / A	1.2	2.0	3.1	3.7	4.5
temperature increase ΔT/°C	3.0	6.5	16.5	23.5	35.0
I^2 / A²					

The student predicted that the temperature rise ΔT should be directly proportional to the square of the current.

  i Give a justification for this prediction.

  ii Copy the table and complete the third row.

  iii Plot a suitable graph to find out if the prediction is valid. *(9 marks)*

  d The student noticed that in each test the resistance of the variable resistor had to be decreased to keep the current constant during the time the water was heated. Give an explanation of this observation. *(2 marks)*

A Level

This section of the book contains the material that you will cover in the second year of the Cambridge International A Level Physics course.

The content builds on the physics you have studied in the first year of your course and is a foundation for further studies in physics and physics-related courses.

The material is divided into two parts:
- Fields: Chapters 11–17
- Thermal and nuclear physics, medical physics and astrophysics : Chapters 18–23

Each chapter is matched to the syllabus and is followed by practice questions that will test your understanding and give you practice at tackling examination-style questions.

11 Motion in a circle

11.1 Uniform circular motion

Learning outcomes

On these pages you will learn to:

- explain what is meant by angular displacement and express angular displacement in radians
- explain what is meant by angular speed and angular velocity
- use the concept of angular speed to solve problems
- recall and use $v = r\omega$ to solve problems

In a cycle race, the cyclists pedal furiously at top speed. The speed of the perimeter of each wheel is the same as the cyclist's speed, provided the wheels do not slip on the ground. If the cyclist's speed is constant, the wheels must turn at a steady rate. An object rotating at a steady rate is said to be in **uniform circular motion**.

Consider a point on the perimeter of a wheel of radius r rotating at a steady speed.

- The circumference of the wheel = $2\pi r$
- The frequency of rotation $f = \dfrac{1}{T}$, where T is the time for one rotation.

The speed of a point on the perimeter $= \dfrac{\text{circumference}}{\text{time for 1 rotation}} = \dfrac{2\pi r}{T} = 2\pi rf$

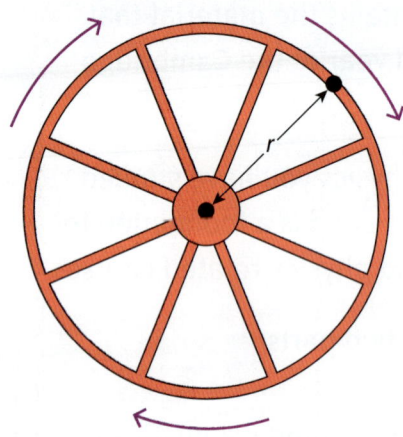

Figure 1 In uniform circular motion

Figure 2 The London Eye

> ### Worked example
>
> A cyclist is travelling at a speed of $13\,\text{m s}^{-1}$ on a bicycle which has wheels of radius 390 mm. Calculate:
>
> **a** the time for one rotation of the wheel,
>
> **b i** the frequency of rotation of the wheel.
>
>   **ii** the number of rotations of the wheel in 1 minute.
>
> **Solution**
>
> **a** Rearranging speed $v = \dfrac{2\pi r}{T}$ gives the time for 1 rotation, $T = \dfrac{2\pi r}{v}$
>
> Therefore, $T = \dfrac{2\pi \times 0.39\,\text{m}}{13\,\text{m s}^{-1}} = 0.19\,\text{s}$
>
> **b i** Frequency $f = \dfrac{1}{T} = \dfrac{1}{0.19\,\text{s}} = 5.3\,\text{Hz}$
>
>   **ii** Number of rotations in 1 minute = $60 \times 5.3 = 318$

Angular displacement

In the UK, the London Eye is a very popular tourist attraction. The wheel has a diameter of 130 m and takes passengers high above the surrounding buildings, giving a glorious view on a clear day. Each full rotation of the wheel takes 30 minutes. Each capsule therefore takes its passengers through an angle of $0.2°$ each second or $= \dfrac{0.2\pi}{180}$ radians each second as $360° = 2\pi$ radians.

For any object in uniform circular motion, the object turns through an angle of $\dfrac{2\pi}{T}$ radians each second, where T is the time taken for 1 complete rotation.

In other words, the angular displacement of the object each second is $\dfrac{2\pi}{T}$.

Therefore, the **angular displacement** of the object in time t is given by:

$$\theta \text{ (in radians)} = \frac{2\pi t}{T} = 2\pi ft$$

Angular velocity

The angular velocity of an object moving in uniform circular motion is defined as its angular displacement per second.

The unit of angular velocity is the radian per second (rad s^{-1}).

Using the previous equation $\theta = \dfrac{2\pi t}{T}$ therefore gives:

$$\text{angular velocity, } \omega = \frac{\text{angular displacement, } \theta}{\text{time taken, } t} = \frac{2\pi}{T}$$

The object travels once round the circle in its **time period**. For a circle of radius r, its circumference is equal to $2\pi r$. Hence its speed v is given by

$$v = \frac{\text{circumference of the circle}}{\text{time period}} = \frac{2\pi r}{T}$$

Substituting $\omega = \dfrac{2\pi}{T}$ gives:

$$v = \omega r$$

Figure 3 shows the angular displacement of the object in time t which is less than the time period T.

1 The angular displacement of the object, $\theta = \omega t$.
2 The object travels along an arc of the circle of length $s = vt = \dfrac{2\pi rt}{T} = r\theta$

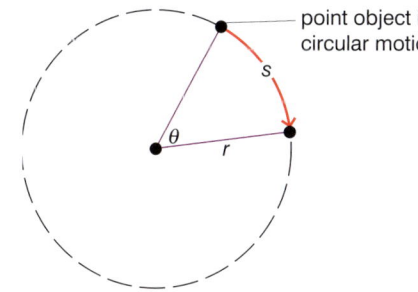
— point object in uniform circular motion

Figure 3 *Arcs and angles*

A cyclist travels at a speed of $12\,\text{m s}^{-1}$ on a bicycle which has wheels of radius $0.40\,\text{m}$. Calculate:

a the frequency of rotation of each wheel,

b the angular velocity of each wheel,

c the angle the wheel turns through in $0.10\,\text{s}$:

  i in radians,   **ii** in degrees.

Solution

a Circumference of wheel $= 2\pi r = 2\pi \times 0.4 = 2.5\,\text{m}$

  Time for 1 wheel rotation, $T = \dfrac{\text{circumference}}{\text{speed}} = \dfrac{2.5}{12} = 0.21\,\text{s}$

  Frequency $= \dfrac{1}{T} = \dfrac{1}{0.21} = 4.8\,\text{Hz}$

b Angular velocity $= \dfrac{2\pi}{T} = 30\,\text{rad s}^{-1}$

c i Angle the wheel turns through in $0.10\,\text{s}$, $\theta, = \dfrac{2\pi t}{T} = \dfrac{2\pi \times 0.10}{0.21}$
   $= 3.0\,\text{radians}$

  ii $\theta = 3.0 \times \dfrac{360}{2\pi} = 172°$

One radian is defined as the angle subtended by a circular arc of length equal to the radius of the circle. It is based on a scale of 2π radians being equal to $360°$. This is because the formula for the circumference of a circle $2\pi r$ (where r is the circle radius) then leads to the following useful formula for the length, s, of an arc of a circle:

$$s = r\theta$$

where θ is the angle in radians subtended by the arc at the centre of the circle. See Figure 3.

1 Calculate the angular displacement in radians of the tip of the minute hand of a clock in:

  a 1 second,

  b 1 minute,

  c 1 hour.

2 An electric motor turns at a frequency of $50\,\text{Hz}$. Calculate:

  a its time period,

  b the angle it turns through in radians in:
   i $1\,\text{ms}$,
   ii 1 second.

3 The Earth takes exactly 24 hours for 1 full rotation. Calculate:

  a the speed of rotation of a point on the Equator,

  b the angle the Earth turns through in 1 second in:
   i degrees,   **ii** radians.

  The radius of the Earth $= 6400\,\text{km}$.

4 A satellite in a circular orbit of radius $8000\,\text{km}$ takes 120 minutes per orbit. Calculate:

  a its speed,

  b its angular displacement in $1.0\,\text{s}$ in:
   i degrees,   **ii** radians.

Learning outcomes

On these pages you will learn to:

- describe uniform circular motion
- understand centripetal acceleration in the case of uniform circular motion
- recall and use centripetal acceleration equations $a = r\omega^2$ and $a = v^2/r$
- recall and use centripetal force equations $F = mr\omega^2$ and $F = mv^2/r$

The **velocity** of an object moving round a circle at constant speed continually changes direction. Because its velocity changes, the object therefore accelerates. If this seems odd because the speed is constant, remember that acceleration is change of velocity per second. Passengers on the London Eye might not notice they are being accelerated but if the wheel was made to rotate much faster, they undoubtedly would notice.

The velocity of an object in uniform circular motion at any point is along the tangent to the circle at that point. The direction of the velocity changes continuously as the object moves round on its circular path. The change in the direction of the velocity is towards the centre of the circle. So its acceleration is towards the centre of the circle and is referred to as **centripetal acceleration**. Centripetal means 'towards the centre of the circle'.

For an object moving at constant speed v in a circle of radius r, it can be shown that:

$$\text{centripetal acceleration, } a = \frac{v^2}{r}$$

Proof of this equation is not required for the specification. However, a proof is given to provide a better understanding of the idea of centripetal acceleration.

Proof of $a = \dfrac{v^2}{r}$

- Consider an object in uniform circular motion at speed v moving in a short time interval δt from position A to position B along its path. Therefore the distance AB along the circle, $\delta s = v\delta t$. Figure 2 shows the idea.
- The line from the object to the centre of the circle at C turns through angle θ when the object moves from A to B. The velocity direction of the object turns through the same angle θ, as shown in Figure 2.
- The change of velocity, δv = velocity at B − velocity at A, is shown in the velocity vector triangle in Figure 2.
- The triangles ABC and the velocity vector triangle have the same shape because they both have two sides of equal length with the same angle θ between the two sides.

Provided θ is small, then $\dfrac{\delta v}{v} = \dfrac{\delta s}{r}$

Because $\delta s = v\,\delta t$, then $\dfrac{\delta v}{v} = \dfrac{v\,\delta t}{r}$

Therefore, acceleration, $\quad a = \dfrac{\text{change of velocity } \delta v}{\text{time taken } \delta t}$

$\qquad\qquad\qquad\qquad = \dfrac{v^2}{r}$ towards the centre.

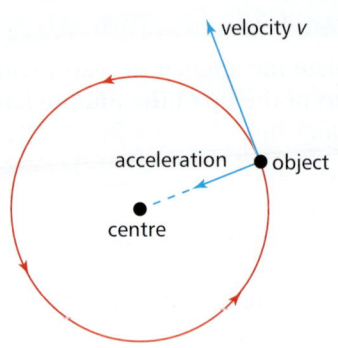

Figure 1 Centripetal acceleration

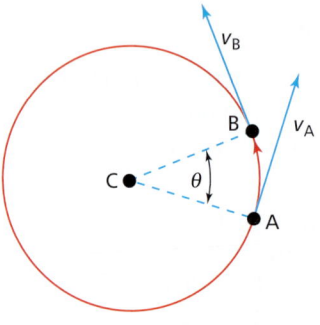

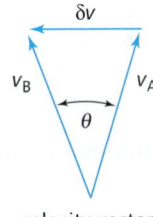

velocity vector triangle

Figure 2 Proving $a = \dfrac{v^2}{r}$

> **Note**
>
> Since $v = \omega r$, then $a = \dfrac{v^2}{r} = \dfrac{(\omega r)^2}{r} = \omega^2 r$

Centripetal force

To make an object move round on a circular path, it must be acted on by a resultant force which changes its direction of motion. Figure 3 shows a 'hammer' being whirled round in a circle. The tension in the cable pulls on the ball and changes its direction continuously. When the cable is released, the ball flies off at a tangent.

The resultant force on an object moving round a circle at constant speed is referred to as the **centripetal force** because it acts towards the centre of the circle.

- For an object whirling round on the end of a string, the tension in the string is the centripetal force.
- For a satellite moving round the Earth, the force of gravity between the satellite and the Earth is the centripetal force. See Figure 4.
- For a planet moving round the Sun, the force of gravity between the planet and the Sun is the centripetal force.
- For a capsule on the London Eye, the centripetal force is the resultant of the support force on the capsule and the force of gravity on it.
- In Chapter 16, you will meet the use of a magnetic field to bend a beam of charged particles (e.g. electrons) in a circular path. The magnetic force on the moving charged particles is the centripetal force.

Figure 3 *Centripetal force in action*

Any object that moves in circular motion is acted on by a resultant force which always acts towards the centre of the circle. The resultant force is the centripetal force and therefore causes a centripetal acceleration.

Notes

1 If the object is acted on by a single force only (e.g. a satellite in orbit round the Earth), that force is the centripetal force and causes the centripetal acceleration.

2 The centripetal force is at right angles to the direction of the object's velocity. Therefore, no work is done by the centripetal force on the object because there is no displacement in the direction of the force. The kinetic energy of the object is therefore constant so its speed is unchanged.

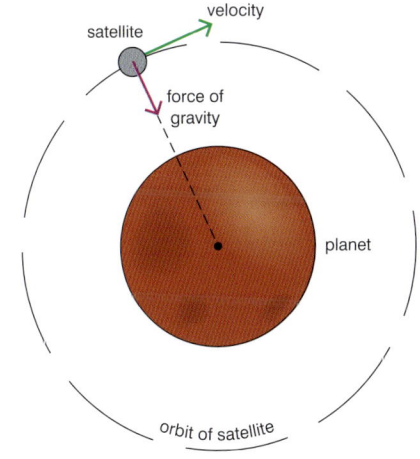

Figure 4 *A satellite in uniform circular motion*

Equation for centripetal force

For an object moving at constant speed v along a circular path of radius r, its centripetal acceleration $a = \dfrac{v^2}{r} \ (= \omega^2 r)$

Therefore, applying Newton's second law for constant mass in the form '$F = ma$' gives

$$\text{centripetal force } F = \frac{mv^2}{r} = m\omega^2 r$$

Summary test 11.2

1 The wheel of the London Eye has a diameter of 130 m and takes 30 minutes for a full rotation. Calculate:
   a the speed of a capsule,
   b i the centripetal acceleration of a capsule,
     ii the centripetal force on a person of mass 65 kg in a capsule.

2 An object of mass 0.15 kg moves round a circular path of radius 0.42 m at a steady rate once every 5.0 seconds. Calculate:
   a the speed and acceleration of the object,
   b the centripetal force on the object.

3 a The Earth moves round the Sun on a circular orbit of radius 1.5×10^{11} m, taking $365\frac{1}{4}$ days for each complete orbit. Calculate:
     i the speed, ii the centripetal acceleration of the Earth on its orbit round the Sun.

   b A satellite is in orbit just above the surface of a spherical planet which has the same radius as the Earth and the same acceleration of free fall at its surface. Calculate:
     i the speed,
     ii the time for 1 complete orbit of this satellite.
   The radius of the Earth = 6400 km
   Acceleration of free fall = 9.81 m s^{-2}

4 A hammer thrower whirls a 2.0 kg hammer on the end of a rope in a circle of radius 0.8 m. The hammer took 0.6 s to make one full rotation just before it was released. Calculate:
   a the speed of the hammer just before it was released,
   b its centripetal acceleration,
   c the centripetal force on the hammer just before it was released.

On the road

Learning outcomes

On these pages you will learn to:

- recall the equations for angular speed, centripetal acceleration and centripetal force
- solve problems on uniform circular motion in the context of road vehicles and aircraft

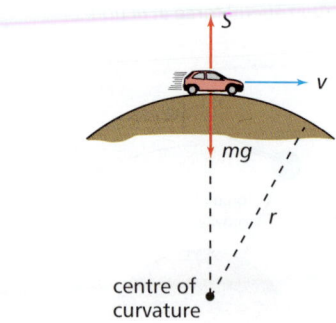

Figure 1 *Over the top*

centre of
curvature

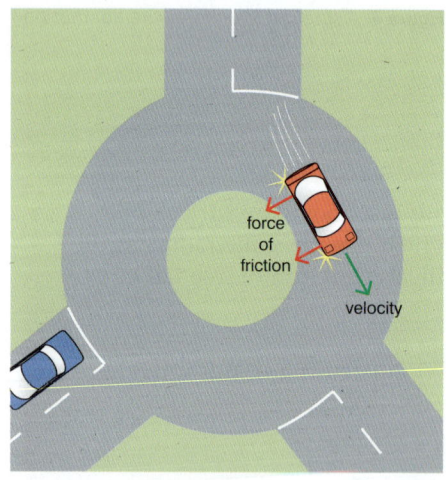

Figure 2 *On a roundabout*

Even on a very short journey, the effects of circular motion can be important. For example, a vehicle that turns a corner too fast could skid or topple over. A vehicle that goes over a curved bridge too fast might even lose contact briefly with the road surface. To make any object move on a circular path, the object must be acted on by a resultant force which is always towards the centre of curvature of its path.

Examples

1 Over the top of a hill

Consider a vehicle of mass m moving at speed v along a road that passes over the top of a hill or over the top of a curved bridge.

At the top of the hill, the support force S from the road on the vehicle is directly upwards in the opposite direction to its weight, mg. The resultant force on the vehicle is the difference between the weight and the support force. This difference acts towards the centre of curvature of the hill as the centripetal force. In other words,

$$mg - S = \frac{mv^2}{r}, \text{ where } r \text{ is the radius of curvature of the hill}$$

The vehicle would lose contact with the road if its speed is equal to or greater than a certain speed, v_0. If this happens, then $S = 0$ so $mg = \frac{mv_0^2}{r}$

Therefore, the vehicle speed should not exceed v_0, where $v_0^2 = gr$, otherwise the vehicle will lose contact with the road surface at the top of the hill. Prove for yourself that a vehicle that travels over a curved bridge of radius of curvature 5 m would lose contact with the road surface if its speed exceeded 7 m s⁻¹.

2 On a roundabout

Consider a vehicle of mass m moving at speed v in a circle of radius r as it moves round a roundabout on a level road. The centripetal force is provided by the force of friction between the vehicle's tyres and the road surface. In other words,

$$\text{force of friction } F = \frac{mv^2}{r}$$

For no skidding or slipping, the force of friction between the tyres and the road surface must be less than a limiting value F_0 which is proportional to the vehicle's weight.

Therefore, for no slipping, the speed of the vehicle must be less than a certain value v_0 which is given by the equation,

$$\text{limiting force of friction } F_0 = \frac{mv_0^2}{r}$$

3 On a banked track

A race track is often banked where it curves. Motorway slip roads in cities often bend in a tight curve. Such a road is usually banked to enable vehicles to drive round without any sideways friction on the tyres. Rail tracks on curves are usually banked to enable trains to move round the curve without slowing down too much.

- Without any banking, the centripetal force is provided only by sideways friction between the vehicle wheels and the road surface. As explained in the previous example, the vehicle on a bend slips outwards if its speed is too high.

- On a banked track, the speed can be higher. To understand why, consider Figure 3 which represents the front-view of a racing car of mass m on a banked track, where θ = the angle of the track to the horizontal. For there to be no sideways friction on the tyres due to the road, the horizontal component of the support forces N_1 and N_2 must act as the centripetal force.

Resolving these forces into horizontal components (= $(N_1 + N_2)\sin\theta$) and vertical components (= $(N_1 + N_2)\cos\theta$),

- because $(N_1 + N_2)\sin\theta$ acts as the centripetal force, then $(N_1 + N_2)\sin\theta = \dfrac{mv^2}{r}$

- because $(N_1 + N_2)\cos\theta$ balances the weight (mg), then $(N_1 + N_2)\cos\theta = mg$.

Therefore
$$\tan\theta = \frac{(N_1 + N_2)\sin\theta}{(N_1 + N_2)\cos\theta} = \frac{mv^2}{mgr}$$

Simplifying this equation gives the condition for no sideways friction; $\tan\theta = \dfrac{v^2}{gr}$

In other words, there is no sideways friction if the speed v is such that $v^2 = gr\tan\theta$.

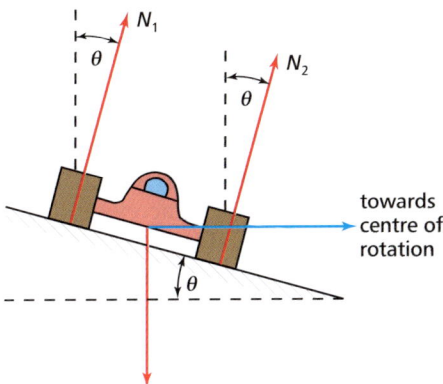

Figure 3 A racing car taking a bend

Notes

1 An aircraft banks when it turns in a circle. The horizontal component of the lift force acts as the centripetal force. Figure 4 shows the idea. The radius of the circle, r, depends on the speed and the angle of banking in accordance with the equation $v^2 = gr\tan\theta$.

2 The **conical pendulum** in Figure 5 shows an object in uniform circular motion due to the horizontal component of a force, in this case the tension in the supporting thread. Resolving the thread tension T into horizontal and vertical components gives.

- vertical component $T\cos\theta = mg$

- horizontal component $T\sin\theta = mv^2/r$, where the circle radius $r = L\sin\theta$

Hence $\tan\theta = \dfrac{\sin\theta}{\cos\theta} = \dfrac{mv^2/r}{mg} = \dfrac{v^2}{gr}$

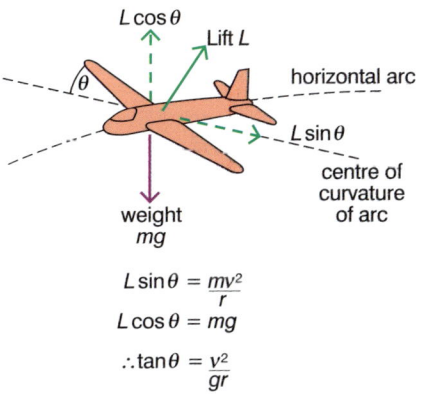

$L\sin\theta = \dfrac{mv^2}{r}$

$L\cos\theta = mg$

$\therefore \tan\theta = \dfrac{v^2}{gr}$

Figure 4 An aircraft circling

Summary test 11.3

$g = 9.81\,\text{m s}^{-2}$

1 A vehicle of mass 1200 kg passes over a bridge of radius of curvature 15 m at a speed of $10\,\text{m s}^{-1}$. Calculate:

  a the centripetal acceleration of the vehicle on the bridge,

  b the support force on the vehicle when it is at the top.

2 The maximum speed for no skidding of a vehicle of mass 750 kg on a roundabout of radius 20 m is $9.0\,\text{m s}^{-1}$. Calculate, for this speed:

  a the centripetal acceleration,

  b the centripetal force on the vehicle.

3 Explain why a circular athletics track that is banked is suitable for sprinters but not for marathon runners.

4 At a racing car circuit, the track is banked at an angle of 25° to the horizontal on a bend which has a radius of curvature of 350 m.

  a Calculate the speed of a vehicle on the bend if there is to be no sideways friction on its tyres.

  b Discuss and explain what could happen to a vehicle that took the bend too fast.

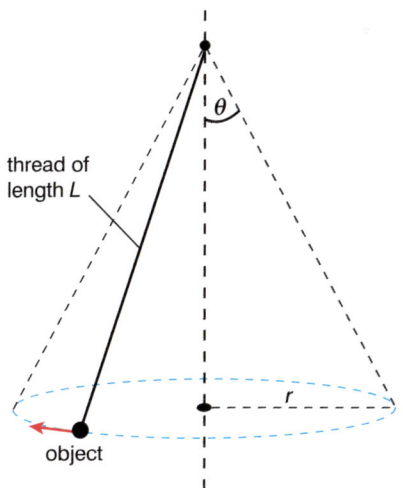

Figure 5 The conical pendulum

173

At the fairground

Many of the rides at a fairground take people round in circles. Some examples are analysed below. It is worth remembering that centripetal acceleration values of more than 2–$3g$ can be dangerous to the average person.

Examples

1 The Big Dipper

A ride that takes you at high speed through a big dip pushes you into your seat as you pass through the dip. The difference between the support force on you (acting upwards) and your weight acts as the centripetal force.

At the bottom of the dip, the support force S on you is vertically upwards, as shown in Figure 1.

Therefore, for a speed v at the bottom of a dip of radius of curvature r,

$$S - mg = \frac{mv^2}{r}$$

So the support force $S = mg + \frac{mv^2}{r}$

The extra force you experience due to your motion is therefore $\frac{mv^2}{r}$.

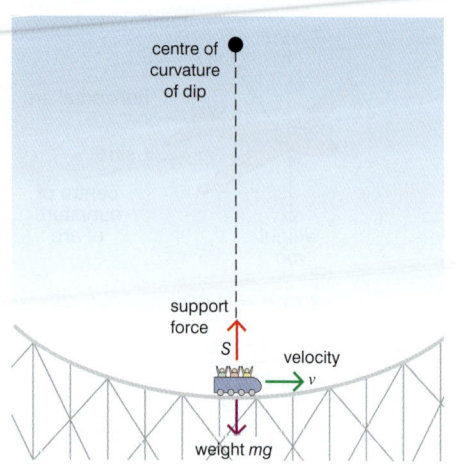

Figure 1 *In a dip*

2 The very long swing

Consider a person of mass m on a very long swing of length L, released from height h above the equilibrium position. The maximum speed is when the swing passes through the lowest point. This can be worked out by equating the loss of potential energy to the gain of kinetic energy.

$$\tfrac{1}{2}mv^2 = mgh$$

where v is its speed as it passes through the lowest point.

$$\text{Therefore } v^2 = 2gh$$

The person on the swing is on a circular path of radius L. At the lowest point, the support force S on the person due to the rope is in the opposite direction to the person's weight, mg. The difference, $S - mg$, acts towards the centre of the circular path and provides the centripetal force. Therefore

$$S - mg = \frac{mv^2}{L}$$

Because $v^2 = 2gh$, $S - mg = \dfrac{2mgh}{L}$

In other words, $\dfrac{2mgh}{L}$ represents the extra support force the person experiences due to circular motion. Prove for yourself that for $h = L$ (i.e. a 90° swing), the extra support force is equal to twice the person's weight.

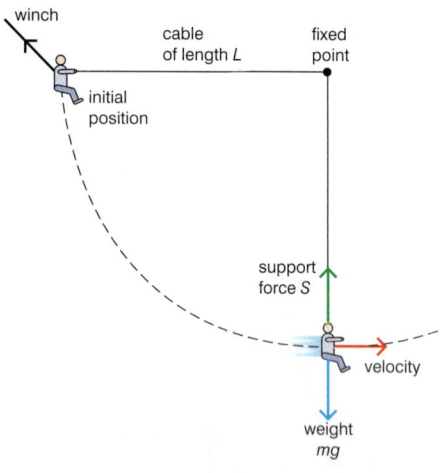

Figure 2 *The very long swing*

3 The Big Wheel

This ride takes its passengers round in a vertical circle on the inside of the circumference of a very large wheel. The wheel turns fast enough to stop the passengers falling out as they pass through the highest position.

At maximum height, the reaction R from the wheel on each person acts downwards. Therefore, the resultant force at this position $= mg + R$. This reaction force and the weight provide the centripetal force. Therefore, at the highest position when the wheel speed is v,

$$mg + R = \frac{mv^2}{r}, \qquad \text{where } r \text{ is the radius of the wheel}$$

$$\therefore \quad R = \frac{mv^2}{r} - mg$$

At a certain speed v_0 such that $v_0^2 = gr$, then $R = 0$ so there would be no force on the person due to the wheel.

A person in a Big Wheel with capsules (e.g. the London Eye) would be unsupported at speed v_0. Such a wheel must turn more slowly otherwise passengers would lose contact with the capsule floor.

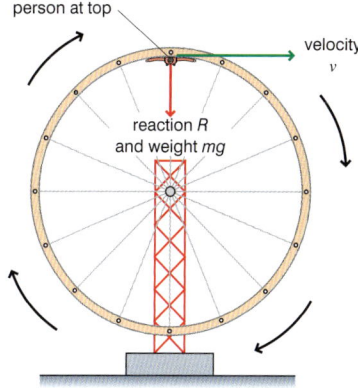

Figure 3 *The Big Wheel*

Summary test 11.4

$g = 9.81\,\mathrm{m\,s^{-2}}$

1 A train on a fairground ride is initially stationary before it descends through a height of 45 m into a dip which has a radius of curvature of 78 m, as shown in Figure 4.

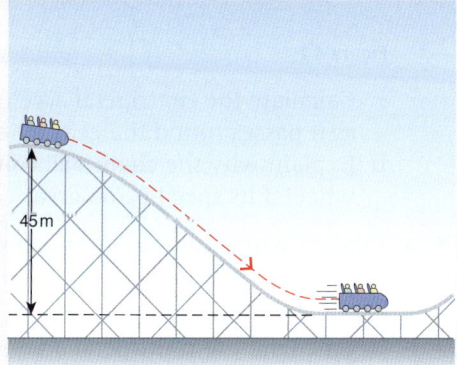

Figure 4

  a Calculate the speed of the train at the bottom of the dip, assuming air resistance and friction are negligible.

  b Calculate:

    i the centripetal acceleration of the train at the bottom of the dip,

    ii the extra support force on a person of weight 600 N in the train.

2 A very long swing at a fairground is 32 m in length. A person of mass 69 kg on the swing descends from a position when the swing is horizontal. Calculate:

  a the speed of the person at the lowest point,

  b the centripetal acceleration at the lowest point,

  c the support force on the person at the lowest point.

3 The Big Wheel at a fairground has a radius of 12.0 m and rotates in a vertical plane once every 6.0 seconds. Calculate:

  a the speed of rotation of the perimeter of the wheel,

  b the centripetal acceleration of a person on the perimeter,

  c the support force on a person of mass 72 kg at the highest point.

4 The wheel of the London Eye has a diameter of 130 m and takes 30 minutes to complete one revolution. Calculate the change, due to rotation of the wheel, of the support force on a person of weight 500 N in a capsule at the top of the wheel.

Chapter Summary

1 For an object of mass m in uniform circular motion on a circle of radius r:

- its **speed** $v = \dfrac{2\pi r}{T}$, where T is the time for one rotation

- its **frequency of rotation** $f = \dfrac{1}{T}$

- its **angular displacement**, in radians, in time $t = 2\pi f t = \dfrac{2\pi t}{T}$

- its **centripetal acceleration**, $a = \dfrac{v^2}{r}$ towards the centre of the circle

- the **centripetal force** on the object $= \dfrac{mv^2}{r}$

2 Examples:
- Car of mass m going over the top of a hill: support force $S = mg - \dfrac{mv^2}{r}$

- Object of mass m going through the bottom of a dip: support force $S = mg + \dfrac{mv^2}{r}$

- Person on the inside of a vertical fairground wheel at the top: reaction force $R = \dfrac{mv^2}{r} - mg$

- Banked track at angle θ to horizontal: $v^2 = gr\tan\theta$ for no sideways friction.

⊡ **Launch additional digital resources for the chapter**

$g = 9.81\,\mathrm{m\,s^{-2}}$

1 An object moves on a circular path at constant speed. Explain:

  a why the object's velocity continually changes even though its speed is constant,

  b the object accelerates even though its speed is constant.

2 The Moon moves round the Earth once every $27\frac{1}{3}$ days on a circular orbit of radius 380 000 km. Calculate:

  a the speed of the Moon,

  b its centripetal acceleration.

3 A pulley wheel of diameter 24 mm fitted to an electric motor in a machine rotates at a frequency of 30 Hz when the machine is in normal operation. A belt fitted to the wheel is used to drive a drum in the machine, as shown in Figure 3.1.

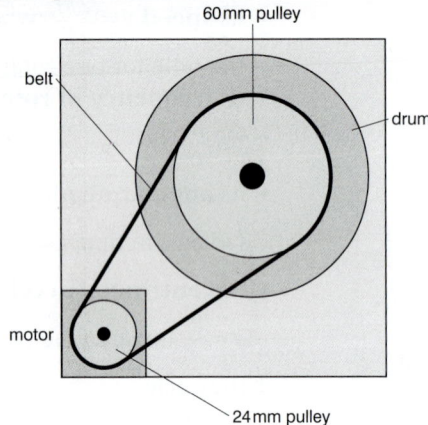

Figure 3.1

  a Calculate:

    i the speed of the belt on the wheel,

    ii the centripetal acceleration of the belt attached to the wheel as it moves round the pulley wheel.

  b The belt drives the drum via a second pulley wheel of diameter 60 mm attached to the drum axle. Calculate:

    i the frequency of rotation of the second pulley wheel,

    ii the centripetal acceleration of the belt as it passes round the second pulley wheel.

4 A cyclist travels at a speed of $15\,\mathrm{m\,s^{-1}}$ on a bicycle fitted with wheels of diameter 850 mm.

  a Calculate:

    i the frequency of rotation of each wheel,

    ii the centripetal acceleration of the tyre on each wheel.

b The rear wheel of the bicycle is driven by a chain which passes round a gear wheel of diameter 55 mm attached to the axle of the rear wheel.

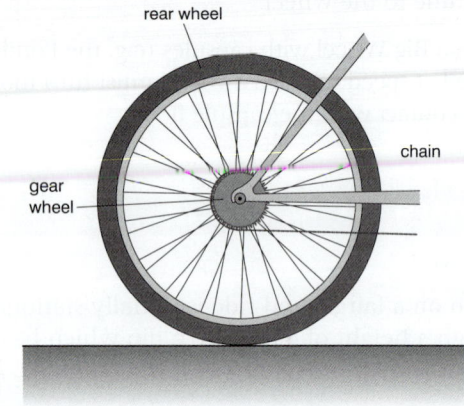

Figure 4.1

  i Calculate the centripetal acceleration of the chain as it passes round the gear wheel.

  ii Explain why the chain will come off the gear wheel if its speed is too great.

5 a Explain why a car on a roundabout will slide to the outside of the roundabout if it travels too fast on the roundabout.

  b i Calculate the centripetal acceleration of a vehicle travelling round a roundabout of diameter 40 m when the speed of the vehicle is $8\,\mathrm{m\,s^{-1}}$.

  ii For no slippage on the roundabout, the frictional force on the vehicle tyres must be less than $0.6 \times$ the vehicle weight. Calculate the maximum speed of the vehicle on the roundabout for no slippage.

6 On a fairground ride, a train of mass 1500 kg moving at a speed of $1.5\,\mathrm{m\,s^{-1}}$ descends from the highest point of the track through a height of 42 m to the bottom of a dip and then passes over a 'hill' which is 15 m higher than the bottom of the dip.

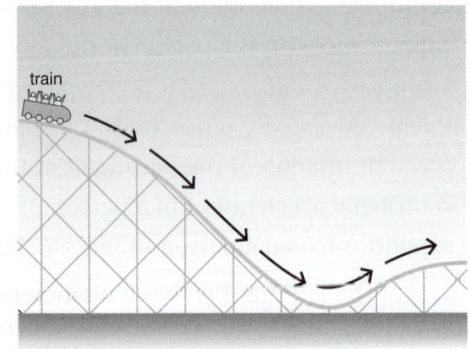

Figure 6.1

a i Calculate the speed of the train at the bottom of the dip.

ii Show that the speed of the train as it passes over the top of the hill is $23\,\mathrm{m\,s^{-1}}$. (Assume air resistance and friction are negligible.)

b i The dip has a radius of curvature of $65\,\mathrm{m}$. Show that a person of mass $80\,\mathrm{kg}$ in the train experiences an extra support force of $1020\,\mathrm{N}$ on passing through the bottom of the dip.

ii The passengers in the train momentarily leave their seats when the train passes over the hill. Show that the radius of curvature of the hill is $54\,\mathrm{m}$.

7 a Explain why a train can travel round a horizontal curve at a higher speed if the track is banked rather than flat.

b Discuss why the train would leave the track if it travelled round the curve too fast.

8 Figure 8.1 below shows a cross-section of an automatic brake fitted to a rotating shaft. The brake pads are held on the shaft by springs.

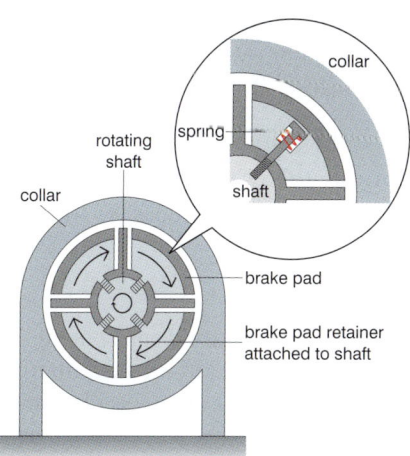

Figure 8.1

a Explain why the brake pads press against the inner surface of the stationary collar if the shaft rotates too fast.

b Each brake pad and its retainer has a mass of $0.30\,\mathrm{kg}$ and its centre of mass is $60\,\mathrm{mm}$ from the centre of the shaft. The tension in the spring attached to each pad is $250\,\mathrm{N}$. Calculate the maximum frequency of rotation of the shaft for no braking.

9 a Define *angular velocity* for an object moving in uniform circular motion.

b A ball B of mass $0.16\,\mathrm{kg}$ and diameter $24\,\mathrm{mm}$ is attached to a string of length $670\,\mathrm{mm}$. The ball is whirled round about a fixed point P in a vertical circle as shown in Figure 9.1.

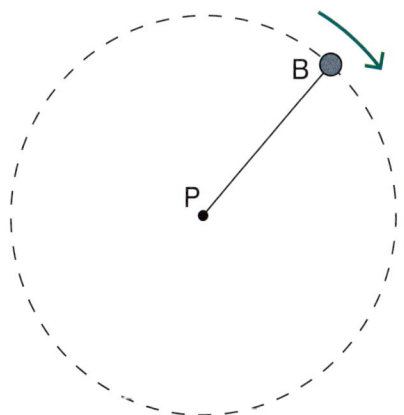

Figure 9.1

The ball rotates 24 times in $10.0\,\mathrm{s}$ at a constant frequency.

i Calculate the angular frequency of rotation of the ball.

ii Calculate the speed of the ball.

iii Explain why the tension in the string is greatest when the ball is at its lowest position and calculate the tension at this position.

iv The string would break if the tension in it exceeded $37\,\mathrm{N}$. If the frequency of rotation was gradually increased, calculate the maximum frequency of rotation of the ball.

12 Oscillations

12.1 Measuring oscillations

There are many examples of oscillations in everyday life. A car that travels over a bump bounces up and down for a short time afterwards. Every microcomputer has an electronic oscillator to drive its internal clock. A child on a swing moves forwards then backwards repeatedly. In this simple example, one full cycle of motion is from maximum height at one side to maximum height on the other side then back again. The lowest point is referred to as the **equilibrium** position as it is where the child eventually comes to a standstill. The child in motion is said to **oscillate** about equilibrium.

Further examples of oscillating motion include:

- an object on a spring moving up and down repeatedly,
- a pendulum moving to-and-fro repeatedly,
- a ball bearing rolling from side-to-side,
- a small boat rocking from side to side.

Displacement v. time for an oscillating object

An oscillating object moves repeatedly one way then in the opposite direction through its equilibrium position. The **displacement** of the object (i.e. magnitude and direction) from equilibrium continually changes during the motion. In one full cycle after passing through equilibrium, the displacement of the object:

- increases as it moves away from equilibrium, then
- decreases as it returns to equilibrium, then
- reverses and increases as it moves away from equilibrium in the opposite direction, then
- decreases as it returns to equilibrium.

The **amplitude** of the oscillations is the maximum displacement of the oscillating object from equilibrium. If the amplitude is constant and no frictional forces are present, the oscillations are described as **free** oscillations. See Topic 12.5.

The **time period** of the oscillating motion is the time for one complete cycle of oscillations. One full cycle after passing through any position, the object passes through that same position in the same direction.

The **frequency** of oscillations is the number of cycles per second made by an oscillating object.

The unit of frequency is the hertz (Hz) which is 1 cycle per second.

For oscillations of frequency f, the time period $T_P = \dfrac{1}{f}$

The **angular frequency** ω of the oscillating motion is defined as $\dfrac{2\pi}{T_P}$ $(= 2\pi f)$.

The unit of ω is the radian per second (rad s^{-1}).

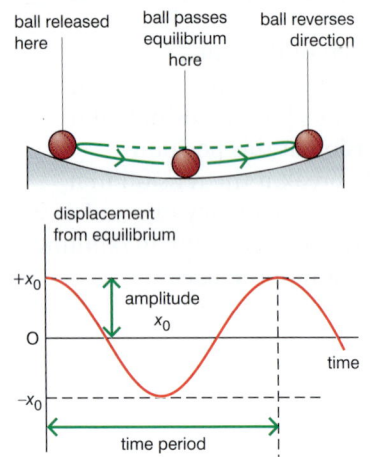

ball released here   ball passes equilibrium here   ball reverses direction

displacement from equilibrium

$+x_0$

O

amplitude x_0

time

$-x_0$

time period

Figure 1 *Oscillating motion*

Phase difference

Imagine two children on adjacent identical swings. The time period, T_P of the oscillating motion is the same, as the swings are identical. If one child reaches maximum displacement on one side a certain time, Δt, later than the other child, they oscillate out of phase. Their **phase difference** stays the same as they oscillate, always corresponding to a fraction of a cycle equal to $\dfrac{\Delta t}{T_P}$. For example,

if the time period is 2.0 s and one child reaches maximum displacement on one side 0.5 seconds later than the other child, the later child will always be a quarter

of a cycle $\left(=\dfrac{0.5\,\text{s}}{2.0\,\text{s}}\right)$ behind the other child. Their phase difference, in radians, is

therefore $0.5\,\pi\left(=2\pi\dfrac{\Delta t}{T_P}\right)$.

In general, for two objects oscillating at the same frequency,

$$\textbf{their phase difference, in radians, } = 2\pi\frac{\Delta t}{T_P}$$

where Δt is the time between successive instants when the two objects are at maximum displacement in the same direction.

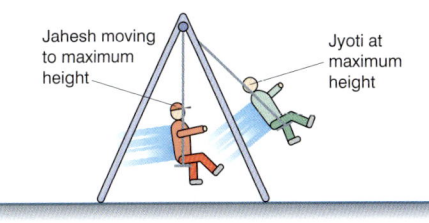

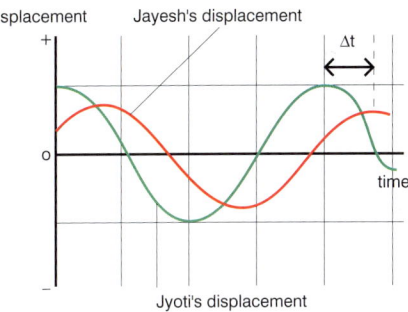

Figure 2 *Phase difference*

Notes

1 2π radians $= 360°$ so the phase difference in degrees is $360\dfrac{\Delta t}{T_P}$.

2 The two objects oscillate in phase if $\Delta t = T_P$. The phase difference of 2π is therefore equivalent to zero.

3 Table of phase differences

Δt	0	$0.25\,T_P$	$0.50\,T_P$	$0.75\,T_P$	T_P
Phase difference in radians	0	$\dfrac{\pi}{2}$	π	$\dfrac{3\pi}{2}$	2π
Phase difference in degrees	0	90	180	270	360

Summary test 12.1

1 Describe how the velocity of a bungee jumper changes from the moment he jumps off the starting platform to the moment he next returns to the platform.

2 a What is meant by *free oscillations*?

   b A metre rule is clamped to a table so that part of its length projects at right angles from the edge of the table, as shown in Figure 3. A 100 g mass is attached to the free end of the rule. When the free end of the rule is depressed downwards then released, the mass oscillates. Describe how you would find out if the oscillations of the mass are free oscillations.

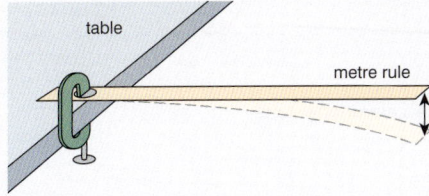

Figure 3

3 An object suspended from the lower end of a vertical spring is displaced downwards from equilibrium by a distance of 20 mm then released. It takes 9.6 s to undergo 20 complete cycles of oscillation. Calculate:

   a its time period,

   b its frequency of oscillation.

4 Two identical pendulums X and Y each consist of a small metal sphere attached to a thread of a certain length. Each pendulum makes 20 complete cycles of oscillation in 16 s. State the phase difference, in radians, between the motion of X and that of Y if:

   a X passes through equilibrium 0.2 s after Y passes through equilibrium in the same direction,

   b X reaches maximum displacement at the same time as Y reaches maximum displacement in the opposite direction.

179

12.2　The principles of simple harmonic motion

Learning outcomes

On these pages you will learn to:

- draw and describe graphs to show the changes in displacement, velocity and acceleration during simple harmonic motion
- define simple harmonic motion

An oscillating object speeds up as it returns to equilibrium and it slows down as it moves away from equilibrium. Figure 1 shows one way to record the displacement of an oscillating pendulum.

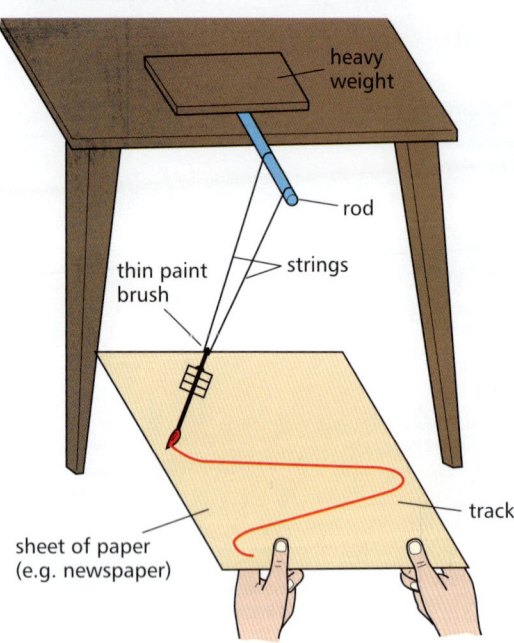

Figure 1 *Investigating oscillations*

The variation of displacement with time is shown in Figure 2 (i). Provided friction is negligible, the amplitude of the oscillations is constant.

The variation of velocity with time is given by the gradient of the displacement v. time graph, as shown by Figure 2 (ii).

- The velocity is greatest where the gradient of the displacement v. time graph is greatest (i.e. at zero displacement when the object passes through equilibrium).
- The velocity is zero where the gradient of the displacement v. time graph is zero (i.e. at maximum displacement).

The variation of acceleration with time is given by the gradient of the velocity v. time graph, as shown by Figure 2 (iii).

- The acceleration is greatest where the gradient of the velocity v. time graph is greatest. This is when the velocity is zero and occurs at maximum displacement.
- The acceleration is zero where the gradient of the velocity v. time graph is zero. This is when the displacement is zero.

By comparing Figure 2 (i) and (iii) directly, it can be seen that:

the acceleration is always in the opposite direction to the displacement.

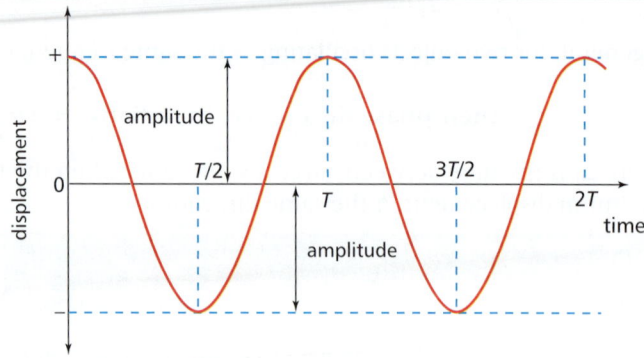

Figure 2 (i) *Displacement v. time*

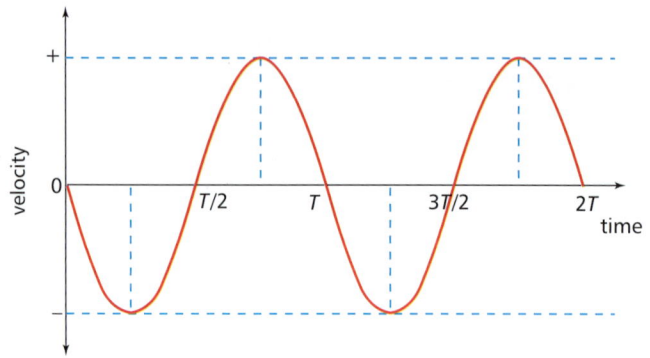

Figure 2 (ii) *Velocity v. time*

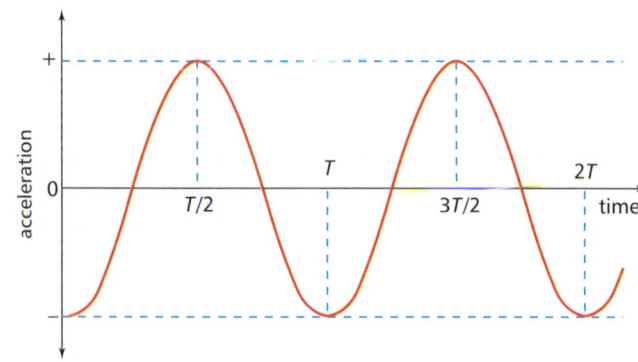

Figure 2 (iii) *Acceleration v. time*

In other words, if one direction is referred to as the positive direction and the other as the negative direction, the acceleration direction is always the opposite sign to the displacement direction.

Simple harmonic motion is defined as oscillating motion in which the acceleration is:

1 proportional to the displacement,
2 always in the opposite direction to the displacement.

Acceleration, $a = -$ constant $\times$ displacement x

The constant depends on the time period T_P of the oscillations. The shorter the time period, the faster the oscillations which means the larger the acceleration at any given displacement. So the constant is greater the shorter the time period. As shown in Topic 12.3, the constant in this equation is ω^2, where the angular frequency,

$$\omega = \frac{2\pi}{T_P} \ (= 2\pi f).$$

Therefore the defining equation for simple harmonic motion is

acceleration, $\quad a = -\omega^2 x,$ $\quad$ where x = displacement

$$\omega = \text{angular frequency} = \frac{2\pi}{T_P}$$

This equation may also be written as

$$a = -(2\pi f)^2 x, \quad \text{where } f = \text{frequency}$$

Summary test 12.2

1 A small object attached to the end of a vertical spring (Figure 3) oscillates with an amplitude of 25 mm and a time period of 2.0 s. The object passes through equilibrium moving upwards at time $t = 0$. What is the displacement and direction of motion of the object:

a $\frac{1}{4}$ cycle later, b $\frac{1}{2}$ cycle later,
c $\frac{3}{4}$ cycle later, d 1 cycle later?

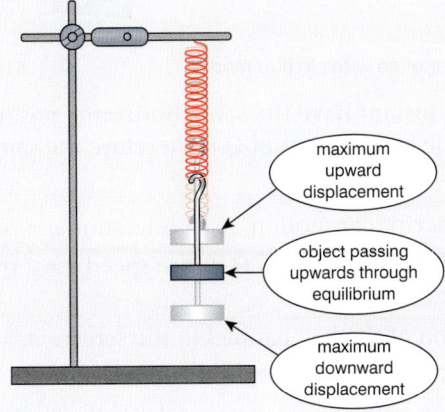

Figure 3

2 For the oscillations in **1**, calculate:
a the frequency,
b the acceleration of the object when its displacement is:
i +25 mm, ii 0, iii −25 mm.

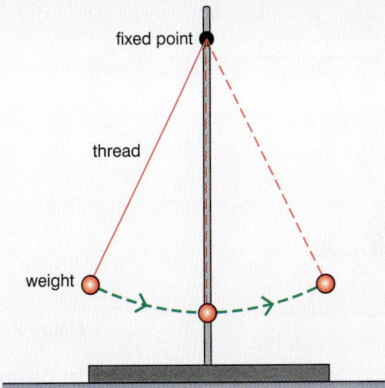

Figure 4

3 A simple pendulum consists of a small weight on the end of a thread. The weight is displaced from equilibrium and released. It oscillates with an amplitude of 32 mm, taking 20 s to execute 10 oscillations. Calculate:

a its frequency,
b its initial acceleration.

4 For the oscillations in **3**, the object is released at time $t = 0$. State the displacement and calculate the acceleration when:

a $t = 1.0$ s,
b $t = 1.5$ s.

More about sine waves

Learning outcomes

On these pages you will learn to:

- recall and use the equation $a = -\omega^2 x$ to solve simple harmonic motion problems
- recall and use $x = x_0 \sin \omega t$ as a solution to the equation $a = -\omega^2 x$

Circles and waves

Consider a small object P in uniform circular motion, as shown in Figure 1. Measured from the centre of the circle at O, the coordinates of P are therefore $x = r \cos \theta$ and $y = r \sin \theta$, where θ is the angle between the x-axis and the radial line OP. Figure 1 shows how the x-coordinate changes as angle θ changes. The shape of the curve is called a **sinusoidal wave**. It has the same shape as the simple harmonic motion curves in Topic 12.2 Figure 2.

Oscillating shadows

To see the link between simple harmonic motion and sine curves, consider the motion of the ball and the pendulum bob in Figure 2. A projector is used to cast a shadow of the ball (P) in uniform circular motion on to a screen alongside the shadow of the bob (Q) of a simple pendulum oscillating above the circle. The two shadows keep up with each other exactly when their time periods are matched.

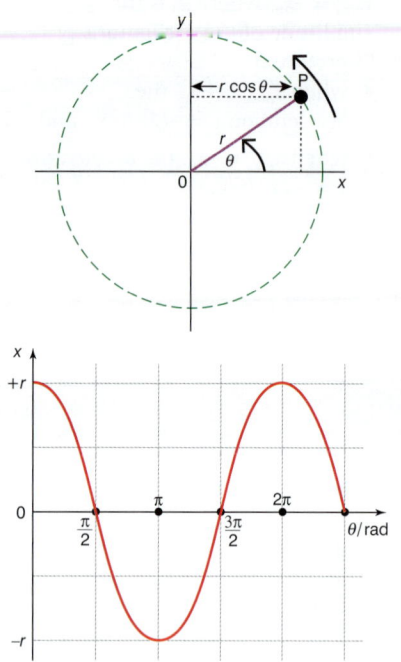

Figure 1 *Circles and waves*

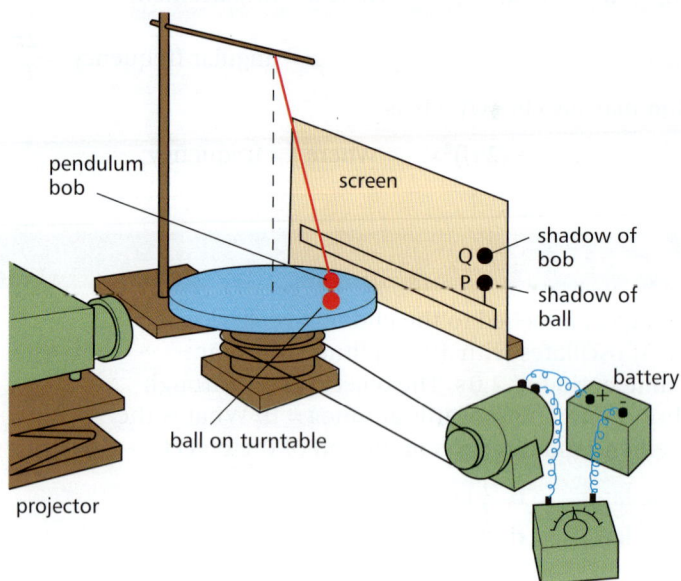

Figure 2 *Comparing simple harmonic motion with circular motion*

In other words, P and Q at any instant have the same horizontal position and the same horizontal motion. The acceleration of Q is therefore the same as the acceleration of P.

- Because the ball is in uniform circular motion, its acceleration $a = -\dfrac{v^2}{r}$, where the minus sign indicates its direction towards O. Since speed $v = 2\pi r f$ (see Topic 11.1), $a = -(2\pi f)^2 r$.

- The component of acceleration of the ball parallel to the screen, $a_x = a \cos \theta$, $\therefore$ the acceleration of P, $a_x = -(2\pi f)^2 r \cos \theta = -(2\pi f)^2 x$

Because the bob's motion is the same as the motion of the ball's shadow,

the acceleration of the bob, $a_x = -(2\pi f)^2 x$

This is the defining equation for simple harmonic motion and it shows why the constant of proportionality is $(2\pi f)^2$ or ω^2 where $\omega = 2\pi f$.

Note

The bob oscillates along the x-axis between $x = -r$ and $x = +r$. Its amplitude of oscillation, $x_0 = r$.

Sine wave solutions

For any object oscillating at frequency f in simple harmonic motion, its acceleration a at displacement x is given by

$$a = -(2\pi f)^2 x = -\omega^2 x$$

The variation of displacement with time depends on the initial displacement and the initial velocity (i.e. the displacement and velocity at time $t = 0$). For example:

1 If $x = 0$ when $t = 0$ and the object is moving to maximum displacement $+x_0$, then:

- Its displacement at time t is given by $x = x_0 \sin \omega t$. This is how the displacement of the bob's shadow Q varies with time if $t = 0$ is when the ball is nearest the screen.
- Its velocity at time t is given by $v = v_0 \cos \omega t$ where its maximum velocity $v_0 = \omega x_0$. This is how the velocity of the bob's shadow varies with time if time $t = 0$ when the ball is nearest the screen.

2 If $x - +x_0$ when $t = 0$ and the object has zero velocity at that instant, then its displacement at time t later is given by $x = x_0 \cos \omega t$. This is how the displacement of Q varies with time if $t = 0$ had been at the instant when Q was at maximum positive displacement.

Notes

1 In the CIE exams, you are expected to be able to use the equations $x = x_0 \sin \omega t$ and $v = v_0 \cos \omega t$, which give the displacement and velocity at time t after zero initial displacement.

2 The general solution is $x = x_0 \sin (\omega t + \phi)$, where ϕ is the phase difference between the instants when $t = 0$ and when $x = 0$. The general solution is not required in this physics specification.

3 The time period T_P does not depend on the amplitude of the oscillating motion. For example, the time period of an object oscillating on a spring is the same, regardless of whether the amplitude is large or small.

4 The symbol A is sometimes used for amplitude instead of x_0.

Worked example

A small object on a spring oscillates with a time period of 0.48 s and an amplitude of 15 mm. Its displacement, x, from equilibrium at time t is given by $x = x_0 \cos \omega t$. Calculate:

a its frequency,

b its displacement and acceleration at $t = 0.20$ s.

Solution

a $f = \dfrac{1}{T_P} = \dfrac{1}{0.48} = 2.08$ Hz

b $x_0 = 15$ mm $= 0.015$ m,
$\omega = 2\pi f = 2\pi \times 2.08 = 13.1$ rad s^{-1}

$x = 0.015 \cos \omega t = 0.015$
$\cos (13.1 \times 0.20) = -1.30 \times 10^{-2}$ m

$a = -\omega^2 x = -(13.1^2 \times -1.30 \times 10^{-2})$
$= 2.23$ m s^{-2}

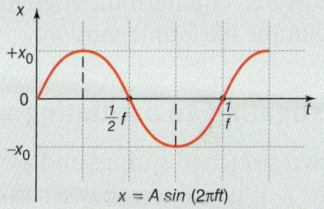
$x = A \sin (2\pi f t)$

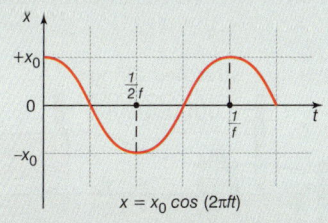
$x = x_0 \cos (2\pi f t)$

Figure 3 *Graphical solutions*

Summary test 12.3

1 An object oscillates in simple harmonic motion with a time period of 3.0 s and an amplitude of 58 mm. Calculate:

a its frequency,

b its maximum acceleration.

2 The displacement of an object oscillating in simple harmonic motion varies with time in accordance with the equation $x / \text{mm} = 12 \sin 10t$, where t is the time in seconds after the object's displacement was at its maximum positive value.

a Determine:
  i the amplitude, **ii** the time period.

b Calculate the displacement of the object at $t = 0.1$ s.

3 An object on a spring oscillates with a time period of 0.48 s and a maximum acceleration of 9.81 m s^{-2}. Calculate:

a its frequency,

b its amplitude.

4 An object oscillates in simple harmonic motion with an amplitude of 12 mm and a time period of 0.27 s. Calculate:

a its frequency,

b its displacement and its velocity:
  i at 0.10 s,
  ii at 0.20 s after its displacement was zero and it was moving towards the maximum positive displacement.

Applications of simple harmonic motion

Learning outcomes

On these pages you will learn to:

- describe how to measure the period of the oscillations of a loaded spring and a simple pendulum
- use the equations for the period of the oscillations of a loaded spring and a simple pendulum to solve problems

For any oscillating object, the resultant force acting on the object acts towards the equilibrium position. The resultant force is described as a restoring force as it always acts towards equilibrium. Provided the restoring force is proportional to the displacement from equilibrium, the acceleration is proportional to the displacement and always acts towards equilibrium. Therefore, the object oscillates with simple harmonic motion.

Investigating an oscillating system

Use two stretched springs and a trolley, as shown in Figure 1. When the trolley is displaced and then released, it oscillates backwards and forwards.

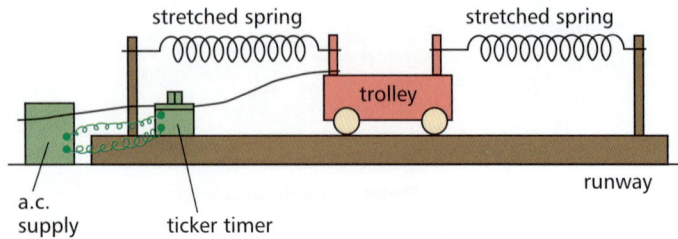

Figure 1 *Investigating oscillations*

- The first half-cycle of the trolley's motion can be recorded using a length of ticker tape attached at one end to the trolley. When the trolley is released, the tape is pulled through a ticker timer that prints dots on the tape at a rate of 50 dots per second.
  A graph of displacement v. time for the first half-cycle can be drawn using the tape, as shown in Figure 2. The graph can be used to measure the time period which can be checked as explained on the next page if the trolley mass m and the combined spring constant k are known.
- A motion sensor linked to a computer can be used to record the oscillating motion of the trolley. (See p.27.)

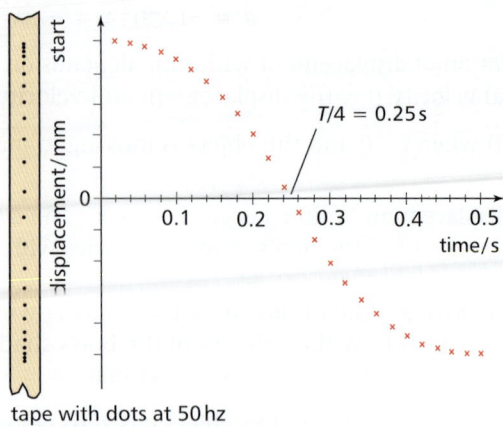

Figure 2 *Displacement–time curve from a tickertape*

What determines the frequency of oscillation of a loaded spring?

In the above investigation, the frequency of oscillation of the trolley can be changed by loading the trolley with extra mass or by replacing the springs with springs of different stiffness. The frequency is reduced by:

1 **Adding extra mass:** This is because the extra mass increases the inertia of the system. At a given displacement, the trolley would therefore be slower than if the extra mass had not been added. Each cycle of oscillation would therefore take longer.
2 **Using weaker springs:** The restoring force on the trolley at any given displacement would be less, as a weaker spring has a smaller spring constant. So the trolley's acceleration and speed at any given displacement would be less than it would have been if the weaker springs had not been added. Each cycle of oscillation would therefore take longer.

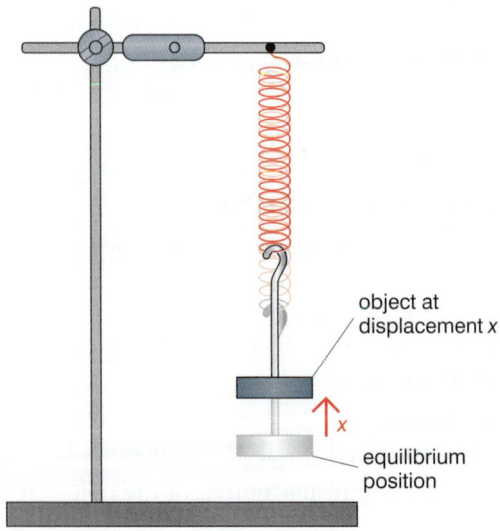

Figure 3 *The oscillations of a loaded spring*

To see exactly how the mass and the spring constant affect the frequency, consider a small object of mass m attached to a spring (see Figure 3).

- Assuming the spring obeys **Hooke's law**, the tension T in the spring is in proportion to its extension e from its unstretched length. This relationship can be expressed by means of the equation $T = ke$, where k is the spring constant.

- When the object is oscillating and is at displacement x from its equilibrium position, the change of tension in the spring provides the restoring force on the object. Using the equation $T = ke$, the change of tension ΔT from equilibrium is therefore given by $\Delta T = -kx$, where the minus sign represents the fact that the change of tension always tries to restore the object to its equilibrium position. Hence, the restoring force on the object $= -kx$.

- Therefore, the acceleration,

$$a = \frac{\text{restoring force}}{\text{mass}} = -\frac{kx}{m} = -\omega^2 x$$

where $\omega^2 = \dfrac{k}{m}$

The object therefore oscillates in simple harmonic motion with a time period

$$T_P = \frac{2\pi}{\omega} = 2\pi\sqrt{\frac{m}{k}}$$

Notes

1 The formula for the time period shows that T_P is increased by increasing m (i.e. using a larger mass) or by reducing k (i.e. using a weaker spring).

2 The time period does not depend on g. A mass–spring system on the Moon would have the same time period as it would on Earth.

3 The tension T in the spring varies from $mg + kx_0$ to $mg - kx_0$, where $x_0 =$ amplitude.

Maximum tension is when the spring is stretched as much as possible (i.e. $x = -x_0$)

Minimum tension is when the spring is stretched as little as possible (i.e. $x = +x_0$).

Worked example

$g = 9.81\,\text{m s}^{-2}$

A spring of natural length 300 mm hangs vertically with its upper end attached to a fixed point. When a small object of mass 0.20 kg is suspended from the lower end of the spring in equilibrium, the spring is stretched to a length of 379 mm. Calculate:

a i the extension of the spring at equilibrium, **ii** the spring constant,

b the time period of oscillations that the mass on the spring would have if the mass was to be displaced downwards slightly then released.

Solution

a i Extension of spring at equilibrium,

$e_0 = 79\,\text{mm} = 0.0790\,\text{m}$

ii Spring constant $k = mg / e_0 = 0.20 \times 9.81 / 0.079$
$= 24.8\,\text{N m}^{-1}$

b $T_P = 2\pi\sqrt{\dfrac{0.20}{24.8}} = 0.56\,\text{s}$

Investigating the oscillations of a loaded spring

To verify the formula $T_P = 2\pi\sqrt{\dfrac{m}{k}}$ for a loaded spring:

- measure the time period for at least six different known masses,

- time 20 complete cycles of oscillation three times for each mass to give an average value of the time period for each mass.

A graph of T_P^2 on the vertical axis against m on the horizontal axis should give a straight line through the origin. This is because $T_P^2 = \dfrac{4\pi^2}{k}m$, which means that a graph of T_P^2 against m should be a straight line through the origin with a gradient equal to $\dfrac{4\pi^2}{k}$.

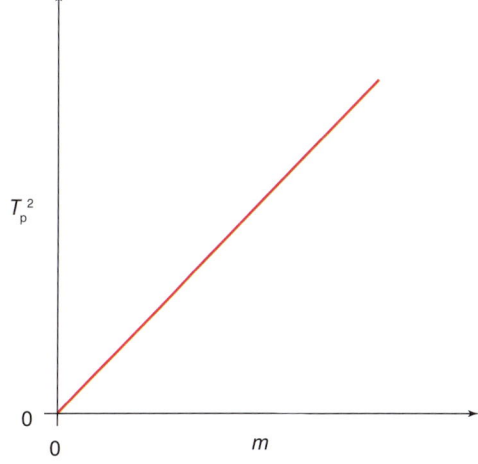

Figure 4 T_P^2 against m for a loaded spring

Extension

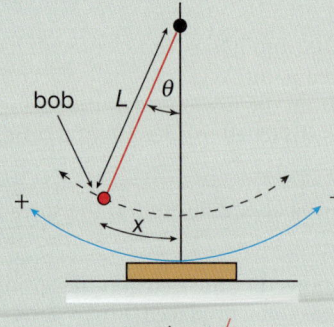

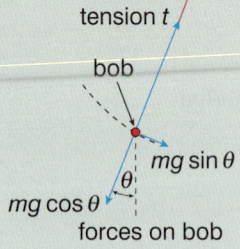

forces on bob

Figure 5 *The simple pendulum*

The theory of the simple pendulum

Consider a simple pendulum consisting of a bob of mass m attached to a thread of length L, as shown in Figure 5. If the bob is displaced from equilibrium then released, it oscillates about the lowest point. At displacement x, from the lowest point, when the thread is at angle θ to the vertical, the weight, mg, has components:

- $mg \cos \theta$ perpendicular to the path of the bob, and
- $mg \sin \theta$ along the path towards the equilibrium position.

The restoring force $F = -mg \sin \theta$, so the acceleration, $a = \dfrac{F}{m} = -\dfrac{mg \sin \theta}{m}$

$$= -g \sin \theta$$

Provided θ does not exceed about 10°, then $\sin \theta = \dfrac{x}{L}$,

therefore the acceleration $a = -\dfrac{g}{L}x = -\omega^2 x$, where $\omega^2 = \dfrac{g}{L}$

The object therefore oscillates in simple harmonic motion with a time period

$$T_P = \frac{2\pi}{\omega} = 2\pi \sqrt{\frac{L}{g}}$$

Notes

1 The length of the pendulum is the distance from the point of support to the centre of the bob.

2 As the bob passes through equilibrium, the tension T acts directly upwards. At this instant, the difference between the tension in the thread and the weight of the bob is equal to the centripetal force on the bob. In other words, $T - mg = mv^2/L$, where v is the speed as it passes through equilibrium.

Investigating the simple pendulum

To verify the formula $T_P = 2\pi \sqrt{\dfrac{L}{g}}$ for a simple pendulum:

- Measure the time period for at least six different measured lengths.
- Time 20 complete cycles of oscillation three times for each length to give an average value of the time period for each length. Ensure the angle between the thread and the vertical does not exceed 10°.
- Measure the length from the point of suspension of the thread to the centre of the spherical bob.

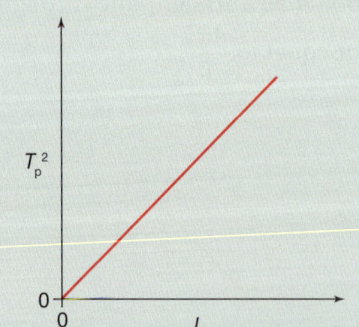

Figure 6 $T_P{}^2$ *against L for a simple pendulum*

A graph of $T_P{}^2$ on the vertical axis against L on the horizontal axis should give a straight line through the origin. This is because $T_P{}^2 = \dfrac{4\pi^2}{g}L$, which means that a graph of $T_P{}^2$ against L should be a straight line through the origin with a gradient $\dfrac{4\pi^2}{g}$.

Therefore g can be calculated from $g = \dfrac{4\pi^2}{\text{gradient}}$

Summary test 12.4

$g = 9.81\,\mathrm{m\,s^{-2}}$

1 An object was suspended from the end of a vertical spring and set into oscillating motion along a vertical line. The amplitude of its oscillations was 20 mm and it took 6.5 s to perform 20 oscillations. Calculate:

a i its time period,
   ii its frequency,

b its acceleration when its displacement was:
   i 0 mm,   ii 10 mm,   iii 20 mm.

2 In the arrangement described in 1, the object was replaced by an object of different mass.

When the second object was oscillating vertically, its acceleration, a, at displacement x was given by $a = -360x$.

a Calculate:
   i the frequency,
   ii the time period of the oscillations.

b By comparing the frequency of oscillation of the second object with that of the first object, discuss whether or not the mass of the second object is greater than or less than the mass of the first object.

3 The upper end of a vertical spring of natural length 250 mm is attached to a fixed point. When a small object of mass 0.15 kg is attached to the lower end of the spring, the spring stretches to an equilibrium length of 320 mm.

a Calculate:
   i the extension of the spring at equilibrium,
   ii the spring constant.

b When the object oscillates vertically, its acceleration, a, at displacement x is given by $a = -\left(\dfrac{k}{m}\right)x$, where m is the mass of the object and k is the spring constant of the spring.
   i Show that the object oscillates at a frequency of 1.9 Hz.
   ii Calculate the time period of the oscillations.

4 An object of mass 0.50 kg is attached to the lower end of a vertical spring which has a spring constant of 25 N m^{-1}. The mass is displaced downwards from equilibrium by a distance of 50 mm then released. Calculate:

a i the force on the object at a displacement of 50 mm,
   ii the acceleration of the object at the instant it was released.

b i Show that the acceleration, a, at displacement x is given by $a = -50x$.
   ii Calculate the frequency of the oscillations and the displacement of the mass 0.50 s after it was released.

5 Calculate the time period of a simple pendulum

a of length: i 1.0 m,   ii 0.25 m.

b Calculate the time period of a simple pendulum of length 1.0 m if it was on the surface of the Moon where $g = 1.6\,\mathrm{m\,s^{-2}}$.

6 A simple pendulum and a mass suspended on a vertical spring have equal time periods on the Earth. Discuss whether or not they would have the same time periods on the surface of the Moon where $g = 1.6\,\mathrm{m\,s^{-2}}$.

Energy and simple harmonic motion

Learning outcomes

On these pages you will learn to:

- describe the interchange between kinetic and potential energy during simple harmonic motion
- recognise and use the equations $v = v_0 \cos \omega t$ and $v = \pm \omega \sqrt{(x_0^2 - x^2)}$ for free oscillations
- explain what is meant by damped oscillations and give practical examples of different degrees of damping including critical damping

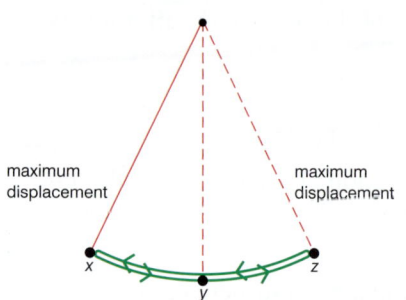

Position	E_p	E_k
x	E_{TOTAL}	0
y	0	E_{TOTAL}
z	E_{TOTAL}	0

Figure 1 *The energy changes of a simple pendulum*

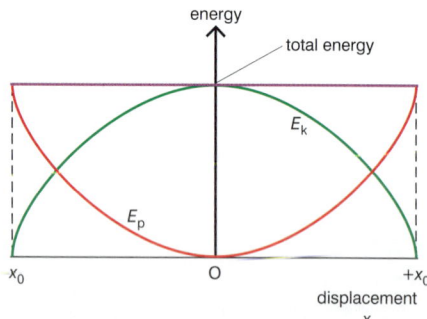

Figure 2 *Energy curves*

Free oscillations

A freely oscillating object oscillates with a constant amplitude because there is no friction acting on it. The only forces acting on it combine to provide the restoring force. If friction was present, the amplitude of oscillations would gradually decrease and the oscillations would eventually cease.

Observe the oscillations of a simple pendulum over many cycles and you should find that the decrease of amplitude from one cycle to the next is scarcely measurable. Nevertheless, over many cycles the amplitude does decrease noticeably. So, friction is present, even if its effect is insignificant over a single cycle.

Consider the example of a mass oscillating on a spring. The energy of the system changes from kinetic energy to potential energy and back again every half-cycle after passing though equilibrium. Provided friction is absent, the total energy of the system is constant and is equal to its maximum potential energy.

1 The potential energy changes with displacement x, in accordance with the equation below. See Topic 5.4 if necessary.

$$E_p = \tfrac{1}{2}kx^2$$

2 The total energy, E_{TOTAL}, of the system is, therefore, $\tfrac{1}{2}kx_0^2$, where x_0 is the amplitude of the oscillations. Since $k = m\omega^2$, **the total energy $E = \tfrac{1}{2}m\omega^2 x_0^2$**

3 The kinetic energy of the oscillating mass, $E_k = \tfrac{1}{2}k(x_0^2 - x^2)$ ($= E_{TOTAL} - E_p$)

The SHM speed equation

Using $E_k = \tfrac{1}{2}mv^2$ gives $\tfrac{1}{2}mv^2 = \tfrac{1}{2}k(x_0^2 - x^2)$, where v is the speed of the object at displacement x.

As $\omega^2 = \dfrac{k}{m}$, the above equation can be written as $v^2 = \omega^2(x_0^2 - x^2)$.

Hence $\qquad\qquad v = \pm\omega\sqrt{(x_0^2 - x^2)}, \qquad$ where $\omega = 2\pi f$.

Note that making $x = 0$ in this equation gives the maximum speed $v_{max} = 2\pi f x_0$.

Energy v. displacement graphs

- The potential energy curve is parabolic in shape, given by $E_p = \tfrac{1}{2}kx^2$.
- The kinetic energy curve is an inverted parabola, given by $E_k = \tfrac{1}{2}k(x_0^2 - x^2)$.
- The sum of the kinetic energy and the potential energy is always equal to $\tfrac{1}{2}kx_0^2$ which is the potential energy at maximum displacement. This is the same as the kinetic energy at zero displacement. So the two curves add together to give a flat line for the total energy.

Damped oscillations

The oscillations of a simple pendulum gradually die away because air resistance gradually reduces the total energy of the system. In any oscillating system where friction or air resistance is present, the amplitude decreases. The forces causing the amplitude to decrease are described as **dissipative forces** because they dissipate the energy of the system to the surroundings as thermal energy. The motion is said to be **damped** if dissipative forces are present. Figure 3 shows how the displacement of a lightly-damped oscillating system decreases with time. Note that:

- The amplitude gradually decreases, reducing by the same fraction each cycle.

- The time period is independent of the amplitude so each cycle takes the same length of time as the oscillations die away.

The greater the damping, the faster the amplitude decreases. For example, the amplitude of a mass on a spring oscillating in air decreases gradually because the damping is very light. However, the same mass oscillating on the same spring in oil would be subjected to much greater dissipative forces. The amplitude of its oscillations would decrease much more rapidly than in air.

Where the damping is strong enough, an object displaced from equilibrium and released returns to equilibrium without oscillating. The damping is said to be **critical** if the object returns to equilibrium in the shortest possible time without oscillating. If the damping is greater than critical, the time taken to return to equilibrium is greater.

Practical examples

1 **The suspension system of a car** consists of a coiled spring near each wheel between the wheel axle and the car chassis. When the wheel is jolted, for example on a bumpy road, the spring smoothes out the force of the jolts. An oil damper fitted with each spring prevents the chassis from bouncing up and down too much.

Without oil dampers, the occupants of the car would continue to be thrown up and down until the oscillations died away. The flow of oil through valves in the piston of each damper provides a frictional force which damps the oscillating motion of the chassis. The dampers are designed to ensure the chassis returns to its 'equilibrium' position in the shortest possible time after each jolt with little or no oscillations. The system is therefore at or near critical damping.

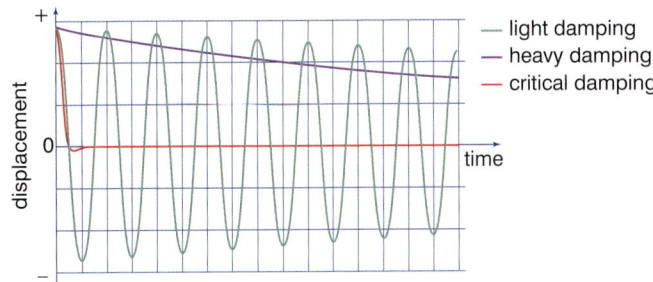

Figure 3 Damping

2 **Swing doors** need to be damped to prevent them swinging back on people. Imagine walking through a swing door without dampers; if you walk too slowly through it, it might bounce back and hit you as you are walking away from it. A damper fitted to the door slows its motion and stops it from swinging back. The damper needs to provide heavier than critical damping so the door closes slowly.

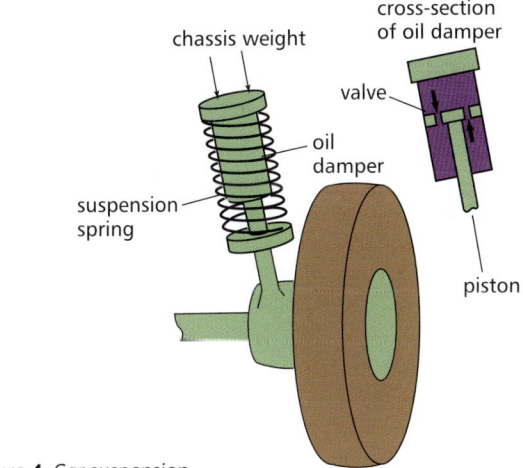

Figure 4 Car suspension

Summary test 12.5

1 a Describe the energy changes of a simple pendulum oscillating in air during one cycle of oscillation after it passes through equilibrium.

   b Sketch graphs on the same axes to show how the potential energy and the kinetic energy of a freely oscillating object vary with its displacement from equilibrium.

2 A glider of mass 0.45 kg on a frictionless air track is attached to two stretched springs at either end, as shown in Figure 5. A force of 3.0 N is needed to displace the glider from equilibrium and hold it at a displacement of 50 mm. The glider is then released and it oscillates freely on the air track. Calculate:

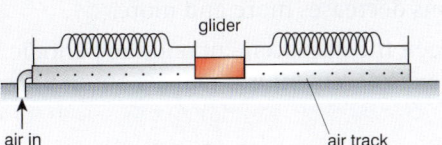

Figure 5

   a the spring constant k for the system,

   b i the initial potential energy of the system when the glider is held at a displacement of 50 mm,

   ii the maximum kinetic energy of the glider.

3 a In each of the following examples, describe the energy changes after the instant of release.

   i A child on a swing displaced from equilibrium then released.

   ii Water in a U-shaped tube displaced from equilibrium then released.

   b Discuss how effective a car suspension damper would be if the oil in the damper was replaced by oil that was much more viscous.

4 The amplitude of an oscillating mass on a spring decreases by 4% each cycle from an initial amplitude of 100 mm. Calculate the amplitude after:

   a 5 cycles of oscillation,

   b 20 cycles of oscillation.

Forced oscillations and resonance

Learning outcomes

On these pages you will learn to:

- describe practical examples of forced oscillations and resonance
- draw and describe graphs to show how the amplitude of a forced oscillation changes with frequency near to the natural frequency of the system
- show an understanding of the factors that determine the frequency response and sharpness of the resonance
- describe examples of circumstances in which resonance should be avoided

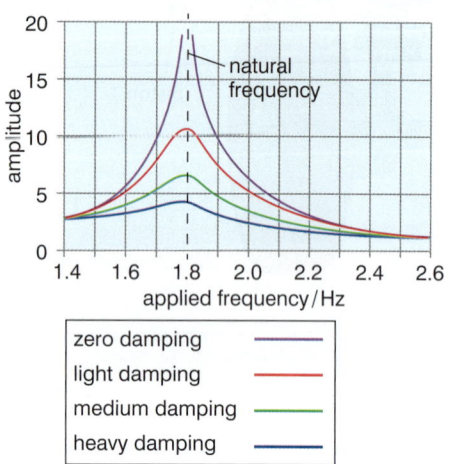

Figure 2 *Resonance curves*

Note

For each curve, the frequency of the periodic force at the peak amplitude is the **resonant frequency**. As the damping becomes less and less:

- the frequency at which resonance occurs tends towards the natural frequency (which is at 1.80 Hz in Figure 2 above)
- the curve becomes sharper in that the system becomes more responsive to change of frequency near the peak.

Imagine pushing someone on a swing at regular intervals. If each push is timed suitably, the swing goes higher and higher. The pushes are a simple example of a **periodic force** which is a force applied at regular intervals.

- When the system oscillates without a periodic force being applied to it, its frequency is referred to as its **natural frequency**.
- When a periodic force is applied to an oscillating system, the response depends on the frequency of the periodic force. The system undergoes **forced oscillations** when a periodic force is applied to it.

Investigating forced oscillations

Figure 1 shows how a periodic force can be applied to an oscillating system consisting of a mass attached to two stretched springs.

The bottom end of the lower spring is attached to a mechanical oscillator which is connected to a signal generator. The top end of the upper spring is fixed. The mechanical oscillator pulls repeatedly on the lower spring at a frequency that can be adjusted by adjusting the signal generator. The frequency of the oscillator is referred to as the **applied frequency**. The response of the system is measured from the amplitude of oscillations of the mass.

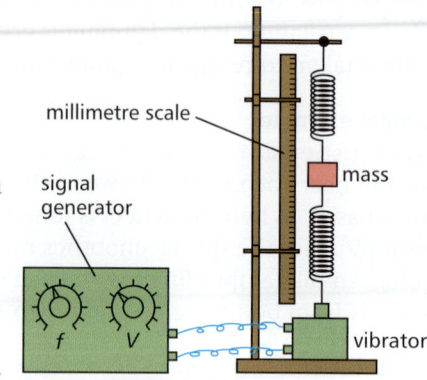

Figure 1 *Forced oscillations*

The variation of the response with the applied frequency is shown in Figure 2.

Consider the effect of increasing the applied frequency from zero:

1 As the applied frequency approaches the natural frequency of the mass–spring system:

- the amplitude of oscillations of the object increases more and more,
- the phase difference between the displacement and the periodic force increases from zero to $\frac{1}{2}\pi$ at the natural frequency.

2 When the applied frequency is equal to the natural frequency of the mass–spring system:

- the amplitude of oscillations becomes very large. The lighter the damping in the system, the larger the amplitude becomes. The system is said to be in **resonance** when the applied frequency is equal to the natural frequency.
- the phase difference between the displacement and the periodic force is $\frac{1}{2}\pi$ at resonance. The periodic force is then exactly in phase with the velocity of the oscillating object.

3 As the applied frequency becomes increasingly larger than the natural frequency of the mass–spring system:

- the amplitude of oscillations decreases more and more,
- the phase difference between the displacement and the periodic force increases from $\frac{1}{2}\pi$ until the displacement is π radians out of phase with the periodic force.

The amplitude of oscillations is greatest when the applied frequency is equal to the natural frequency, provided the damping is light.

For an oscillating system with little or no damping, at resonance,

the applied frequency of the periodic force **=** the natural frequency of the system

1 At resonance, the periodic force acts on the system at the same point in each cycle, causing the amplitude to increase to a maximum value limited only by damping. At maximum amplitude, energy supplied by the periodic force is lost at the same rate because of the effects of damping. So the system oscillates with a constant amplitude at resonance.

2 The applied frequency at resonance, the resonant frequency, is equal to the natural frequency only when there is little or no damping. Resonance occurs at a lower frequency than the natural frequency if the damping is not light. The lighter the damping, the closer the resonant frequency is to the natural frequency.

More examples of resonance

Barton's pendulums

Figure 3 shows five simple pendulums P, Q, R, S and T, of different lengths hanging from a supporting thread which is stretched between two fixed points. A single 'driver' pendulum, D, of the same length as one of the other pendulums is also tied to the thread.

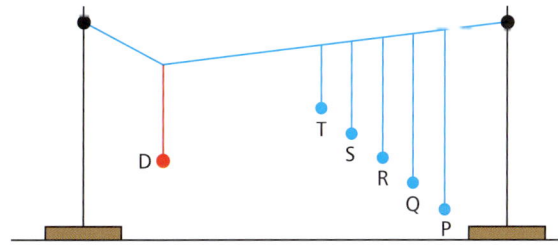

Figure 3 *Barton's pendulums*

The driver pendulum, D, is displaced and released so it oscillates in a plane perpendicular to the plane of the pendulums at rest. The effect of the oscillating motion of D is transmitted along the support thread, subjecting each of the other pendulums to forced oscillations. Pendulum R responds much more than any other pendulum. This is because it has the same length and, therefore, the same time period as D. So its natural frequency is the same as D's natural frequency. Therefore, R oscillates in resonance with D because it is subjected to forced oscillations of the same frequency as its own natural frequency of oscillation. The response of each of the other pendulums depends on how close its length is to the length of D.

Bridge oscillations

A bridge span can oscillate due to its 'springiness' and its mass. If a bridge span is not fitted with 'dampers', it can be made to oscillate at resonance if subjected to a suitable periodic force.

1 A cross-wind can cause a periodic force on a bridge span because of eddy currents created by the wind along the bridge span. If the wind speed is such that the frequency of the periodic force is equal to the natural frequency of the bridge span, resonance can occur in the absence of damping. The dramatic collapse of the Tacoma Narrows Bridge in the United States in 1940 was due to such resonance.

Figure 4 *The collapse of the Tacoma Narrows Bridge*

2 A steady trail of people in step with each other walking across a footbridge can cause resonant oscillations of the bridge span if there is insufficient damping. Soldiers marching in columns are taught to break step when crossing a footbridge to avoid causing resonance. Shortly after it was opened, the Millennium Bridge in London had to be closed and fitted with a suitable damping system because it swayed in resonance when people first walked across it.

Figure 5 *The Millennium Bridge, London*

Summary test 12.6

1 A mass suspended on a vertical spring is made to resonate by applying a periodic force of frequency, f_0.

  a Explain what is meant by *resonance*.

  b Explain why the frequency of the periodic force needs to be f_0 to cause resonance.

2 With reference to the mass–spring system shown in Figure 1, p.190, *state* and explain what the effect would be on the resonant frequency of:

  a increasing the mass,

  b replacing the springs with stiffer springs.

3 The panel of a certain washing machine vibrates loudly when the drum rotates at a certain frequency. Explain why this happens only when the drum rotates at this frequency.

4 A vehicle has a suspension system that is lightly damped. When it is driven without passengers at a certain speed over speed bumps spaced 15 m apart, the vehicle chassis bounces up and down violently.

  a Explain why this effect happens.

  b Discuss whether or not the effect would happen if:

  i the vehicle had been travelling at a different speed,

  ii the vehicle had been carrying passengers.

Chapter Summary

Measurements

1 For an oscillating object:

  • its **amplitude** is its maximum displacement from equilibrium,

  • its **time period** is the time taken for one complete cycle of oscillations,

  • its **frequency** is the number of cycles of oscillation per unit time.

2 For two objects oscillating with the same time period T_p,

$$\text{their phase difference, in radians,} = \frac{2\pi\Delta t}{T_P}$$

where Δt is the time between successive instants when the two objects are at maximum displacement in the same direction.

Simple harmonic motion

Definition:

An object oscillates in simple harmonic motion if its acceleration is:

1 proportional to the displacement of the object from equilibrium,

2 always directed towards equilibrium.

Equation:

Acceleration, $a = -\omega^2 x$,   where x = displacement from equilibrium, and ω = angular frequency of oscillations $= \frac{2\pi}{T_P}$

Solutions:

1 $x = x_0 \sin \omega t$ if $x = 0$ and the object is moving in the + direction at time $t = 0$.

2 $x = x_0 \cos \omega t$ and the object is at maximum displacement, $+ x_0$, at time $t = 0$.

3 Speed at displacement x, $v = \omega\sqrt{(x_0{}^2 - x^2)}$.

Applications

1 Time period of an oscillating mass m on the end of a vertical spring $= 2\pi\sqrt{\frac{m}{k}}$, where k is the spring constant.

2 Time period of a simple pendulum of length $L = 2\pi\sqrt{\frac{L}{g}}$.

Resonance condition

Frequency of the periodic force = natural frequency of the system, provided the damping is negligible.

📖 Launch additional digital resources for the chapter

1 In a test of an undamped suspension spring of a car, the chassis above the wheel is pushed down by 80 mm then released. It undergoes 5 complete cycles of oscillation in 7.0 seconds.

  a Calculate:
   i   the frequency of oscillation,
   ii the acceleration of the chassis when its displacement is +80 mm.

  b Discuss the effect on the oscillating motion if the test had been carried out with the damper fitted.

2 a Describe the energy changes of an oscillating simple pendulum that is lightly damped.

  b A simple pendulum is displaced 50 mm from equilibrium then released.
     It takes 35 s to complete 20 oscillations. Calculate:
   i   the frequency of oscillation of the pendulum,
   ii its initial acceleration.

3 Two identical simple pendulums X and Y with a time period of 2.0 s oscillate with an amplitude of 60 mm in the same vertical plane in which X is to the left of Y. When X is at maximum displacement, Y is at zero displacement and moving towards X.

  a State the phase difference between X and Y.

  b Sketch graphs to show how the displacement:
   i   of X, and
   ii of Y changes during the next 4.5 seconds.

4 Figure 4.1 shows a vehicle accelerometer that consists of a steel ring attached to springs on a horizontal slider.

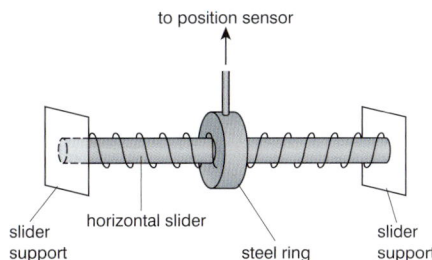

to position sensor

slider support   horizontal slider   steel ring   slider support

Figure 4.1

  a i   Explain why the ring is displaced from equilibrium when the vehicle accelerates.
     ii Describe how the displacement of the ring from equilibrium depends on the magnitude of the acceleration.

  b Describe the motion of the ring when the vehicle brakes sharply and stops.

  c What would be the effect on the accelerometer if:
   i   the spring was replaced by a weaker spring,
   ii the ring was replaced by a ring with more mass?

5 An object of mass 0.15 kg is suspended from the lower end of a spring, as shown below.

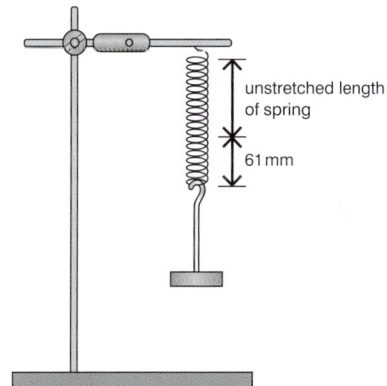

unstretched length of spring

61 mm

Figure 5.1

  a When the object is in equilibrium, the spring is extended by 61 mm from its unstretched length. Calculate the spring constant of the spring.

  b When the object is displaced vertically downwards and then released, it oscillates in simple harmonic motion. Its acceleration, a, at displacement x is given by $a = -\left(\dfrac{k}{m}\right)x$, where m is the mass of the object and k is the spring constant of the spring. Calculate:
   i   the frequency,
   ii the time period of these oscillations.

6 The pendulum bob of a simple pendulum is displaced from equilibrium by a distance of 36 mm with the thread taut then released. It performs 20 oscillations in 70 s. Its displacement at time t after passing through equilibrium from – to + displacement may be represented by the equation $x = x_0 \sin \omega t$.

   a What do x_0 and ω represent in this equation?

   b State the values of x_0 and ω.

   c Calculate the displacement of the object 2.0 s after it was released.

   d Calculate the maximum acceleration of the pendulum bob.

7 When a new car was tested, one of the metal panels of the car body vibrated loudly when the engine ran at a certain speed. When the engine speed was increased or decreased, the vibrations stopped.

   a Explain why the panel vibrated loudly at a certain engine speed.

   b A design engineer suggested the panel could be replaced by a thicker panel to eliminate the vibrations. Discuss whether or not this is a reasonable suggestion.

8 A floor-mounted diesel engine is used to drive an alternating current generator in a power station. To prevent vibrations from the engine shaking the building, the engine was mounted on a spring suspension system fitted with dampers, as shown below.

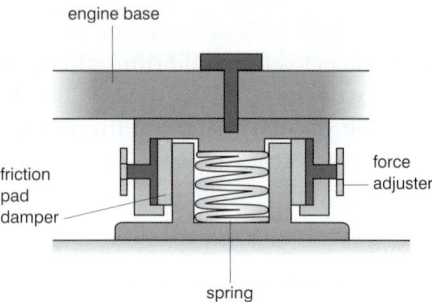

Figure 8.1

   a **i** Describe how the vibrations of the engine are absorbed in this system.

      ii Explain why the friction pads become warm when the engine is running.

   b At a certain engine speed, vibrations reached certain parts of the building until the force of the friction pads in the suspension system was increased. Explain why this stopped the vibrations.

9 A simple pendulum P consists of a metal ball on the end of a string which is attached to a fixed point. The mass of the ball is 3.30×10^{-3} kg, its diameter is 20 mm and the length of the string is 1200 mm.

   a The ball is displaced by 8.0° from its equilibrium position, as in Figure 9.1, and is then released.

      i Calculate the time period and frequency of the subsequent oscillations.

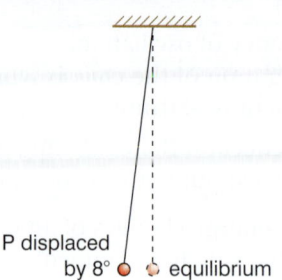

Figure 9.1

      ii Calculate the maximum kinetic energy of the ball in the first cycle.

      iii Describe measurements that could be made to investigate if the energy loss per cycle of the pendulum is significant.

   b The procedure in **a** is repeated with an identical pendulum Q, initially at rest and in contact with P, as shown in Figure 9.2.

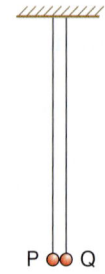

Figure 9.2

After P is displaced by 8.0° and released, it collides with Q such that P stops and Q is set in motion in the same vertical plane as P was in. Discuss whether or not the collision is elastic.

10 An investigation into the oscillations of a mass–spring system was carried out using the apparatus shown in Figure 10.1. The load L has a mass of 0.200 kg and is attached to two identical springs. A variable frequency oscillator is attached to the lower end of the lower spring.

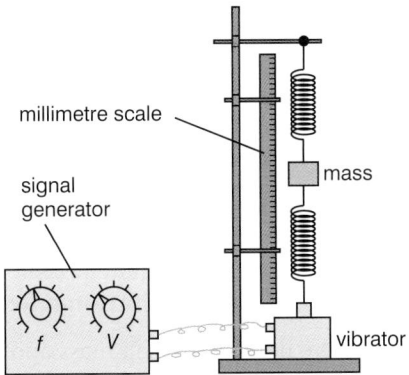

millimetre scale

signal generator

mass

f V

vibrator

Figure 10.1

The oscillator is operated at different measured frequencies. At each frequency, f, the load oscillates vertically with a constant amplitude, A.

a The table below gives the measurements from this investigation.

Frequency f/Hz	1.4	1.6	1.7	1.8	1.9	2.0	2.2	2.4	2.6
Amplitude A/mm	4	6	11	18	10	6	3	2	2

  i Plot a graph to show how the amplitude varies with the frequency.

  ii Estimate the maximum amplitude and the frequency at this amplitude.

  iii Calculate the maximum speed and kinetic energy of the load when it oscillates at maximum amplitude.

b i In a further test with the same load, a light card is attached horizontally to the load. On your graph, draw a curve to show how the oscillations are affected by the presence of the card.

  ii Describe the overall energy transfers when the arrangement with the card present is oscillating at maximum amplitude.

c In a separate test without the card, the load is doubled to 0.400 kg. Calculate the resonant frequency for the system.

13.1 Gravitational field strength

'What goes up must come down', or must it? Throw a ball into the air and it returns to you because of the Earth's gravity. The force of gravity on the ball pulls it back to Earth. The force of attraction between the ball and the Earth is an example of gravitational attraction which exists between any two masses. It isn't obvious that there is a force of attraction between you and any object near you, but it is true. Any two masses exert a gravitational pull on one another. But the force is usually too weak to notice unless at least one of the masses is very large.

The mass of an object creates a force field around itself. Any other mass placed in the field is attracted towards the object. The second mass also has a force field around itself and this pulls on the first object with an equal force in the opposite direction. The force field around a mass is called a **gravitational field**.

If a small mass is placed close to a massive body, the small mass and the body attract each other with equal and opposite forces. However, this force is too small to move the massive body noticeably. The small mass, assuming it is free to move, is pulled by the force towards the massive body. The path which the smaller mass would follow is called a **field line** or, sometimes, **a line of force**. Figure 2 shows the field lines near a planet. The lines are directed to the centre of the planet as a small object released near the planet would fall towards its centre.

> **The strength of a gravitational field, *g*, is the force per unit mass on a small test mass placed in the field.**

The test mass needs to be small, otherwise it might pull so much on the other object that it changes its position and alters the field. In general, the force on a small mass in a gravitational field varies from one position to another. If a small test mass, *m*, is at a certain position in a gravitational field where it is acted on by a gravitational force *F*, the gravitational field strength at that position is given by

$$g = \frac{F}{m}$$

The unit of gravitational field strength is the newton per kilogram ($N\,kg^{-1}$). For example, the gravitational field strength of the Earth at the surface of the Earth is $9.81\,N\,kg^{-1}$.

Figure 1 *Earth in space*

Free fall in a gravitational field

The weight of an object is the force of gravity on it. If an object of mass *m* is in a gravitational field, the gravitational force on the object $F = mg$, where *g* is the gravitational field strength at the object's position. If the object is not acted on by any other force, it accelerates with an acceleration given by

$$\text{acceleration } a = \frac{\text{force}}{\text{mass}} = \frac{mg}{m} = g$$

The object therefore falls freely with acceleration, *g*. Thus, *g* is also described as the acceleration of a freely falling object. See Topic 1.5 for more about the measurement of *g*.

An object that falls freely is unsupported. Although the object in this situation is commonly described as being weightless, it is better to describe it as 'unsupported' as it is acted on by the force of gravity alone.

Field patterns

1 **A radial field** is where the field lines are like the spokes of a wheel, which for a gravitational field are always directed to the centre. Figure 2 shows an example of a radial field. The force of gravity on a small mass near a much larger spherical mass is always directed to the centre of the larger mass. For example, the force on a small object near a spherical planet always acts towards the centre of the planet, regardless of the position of the object. The magnitude of g in a radial field decreases with increased distance from the massive body.

2 **A uniform field** is where the gravitational field strength is the same in magnitude and direction throughout the field. The field lines are therefore parallel to one another.

Is the Earth's gravitational field uniform or radial? The force of gravity due to the Earth on a small mass decreases with distance from the Earth so the gravitational field strength of the Earth falls with increasing distance from the Earth. The field is therefore radial. However, over small distances which are much less than the Earth's radius, the change of gravitational field strength is insignificant. In other words, over such small distances, the acceleration of free fall, g, (i.e. gravitational field strength) is constant. For example, the measured value of g has the same magnitude (= $9.81 \, \text{N kg}^{-1}$) and direction (downwards) 100 m above the Earth as it has on the surface. In theory, g is smaller higher up, but the difference is too small to be noticeable – provided we don't go too high! Only over distances which are small compared with the Earth's radius can the Earth's field be considered uniform.

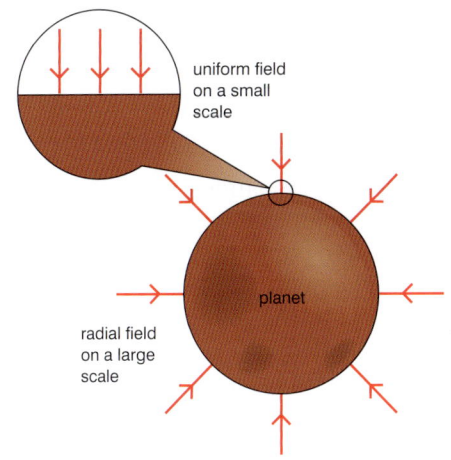

Figure 2 *Field patterns*

Summary test 13.1

1 **a** What is meant by a *field line* or a *line of force* of a gravitational field?

 b With the aid of a diagram in each case, explain what is meant by:
  i a radial field, **ii** a uniform field.

2 **a** Calculate the gravitational force on:
  i an object of mass 3.5 kg in a gravitational field at a position where $g = 9.5 \, \text{N kg}^{-1}$,
  ii an object of mass 100 kg in a gravitational field at a position where $g = 1.6 \, \text{N kg}^{-1}$.

 b Calculate the gravitational field strength at a position in a gravitational field where:
  i an object of mass 2.5 kg experiences a force of 40 N,
  ii an object of mass 18 kg experiences a force of 72 N.

3 Show that the acceleration of an object falling freely in a gravitational field is equal to g, where g is the gravitational field strength at that position.

4 Figure 3 represents a small part of the Earth's surface. Sketch the lines of force near this part of the Earth's surface:

 a if the density of the Earth in this part is uniform,

 b if there is a large mass of dense matter under this part of the surface.

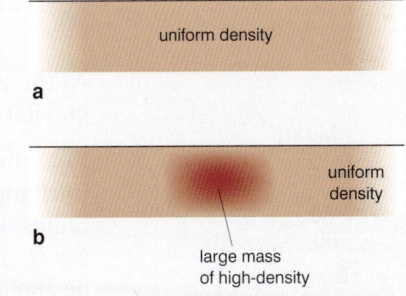

Figure 3

Learning outcomes

On these pages you will learn to:

- define gravitational potential at a point as the work done in bringing unit mass from infinity to the point
- explain what is meant by an equipotential
- solve problems using the equation $\Delta\phi = \Delta W/m$

Figure 1 Into space

Working in a gravitational field

Gravitational potential energy is the energy of an object due to its position in a gravitational field. The position for zero gravitational potential energy is at infinity – in other words, the object would be so far away that the gravitational force on it is negligible. A rocket climbing out of a planet's gravitational field needs to increase its gravitational potential energy to zero to escape completely. At the surface, its gravitational potential energy was negative so it needs to do work to escape from the field completely.

The gravitational potential at a point is defined as the work done per unit mass to move a small object from infinity to that point. In this book, the symbol for gravitational potential is ϕ (pronounced 'phi'). Note that ϕ is a scalar quantity and its unit is the joule per kilogram (J kg^{-1}).

For a small object of mass m at a position where the gravitational potential is ϕ, the work W that must be done on the object to enable it to escape completely is given by $W = m\phi$. Rearranging this gives

$$\phi = \frac{W}{m}$$

Suppose a rocket has a 'payload' mass of 1000 kg and the gravitational potential at the surface of the planet is −100 MJ kg^{-1}. Assume the fuel is used quickly to boost the rocket to high speed. For the rocket to escape completely, the gravitational potential energy of the 1000 kg payload must increase from −100 × 1000 MJ to zero. So the work done on the payload must be at least 100 000 MJ to escape. If the rocket payload is only given 40 000 MJ of kinetic energy from the fuel, then it can only increase its gravitational potential energy by 40 000 MJ. So it can only reach a position in the field where the gravitational potential is −60 MJ kg^{-1}.

> **Note**
>
> In general, if a small object of mass m is moved from gravitational potential ϕ_1 to gravitational potential ϕ_2, its change of gravitational potential energy $\Delta E_P = m(\phi_2 - \phi_1) = m\Delta\phi$ where $\Delta\phi = (\phi_2 - \phi_1)$.

As the work done ΔW to move it from ϕ_1 to ϕ_2 is equal to the change of its gravitational potential energy, then $\Delta W = m\Delta\phi$.

Near the Earth's surface, we can use $\Delta E_P = mg\Delta h$ because g is effectively unchanged from its surface value provided Δh is much smaller than the Earth's radius. Remember that $\Delta E_P = mg\Delta h$ can only be applied for values of Δh which are very small compared with the Earth's radius, whereas $\Delta E_P = m\Delta\phi$ can always be applied.

Equipotentials

Equipotentials are lines of constant potential. Hillwalkers ought to know all about equipotentials, since a map contour line is a line of constant potential. A contour line joins points of equal height above sea level. So a hillwalker following a contour line has constant potential energy. Sensible hillwalkers take great care where the contour lines are very close to one another. One slip and their gravitational potential energy might fall dramatically!

The equipotentials near a spherical planet are circles as shown in Figure 2. At increasing distance from the surface, the gravitational field becomes weaker, so

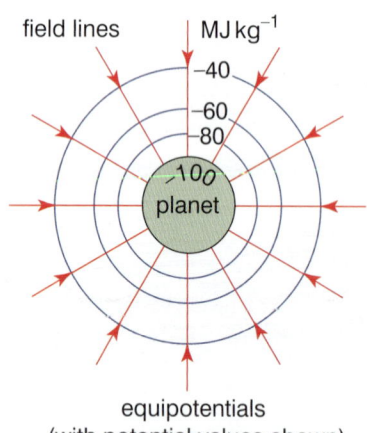

equipotentials
(with potential values shown)

Figure 2 Equipotentials near a planet

the gain of gravitational potential energy per metre of height gain becomes less. In other words, away from the surface, the equipotentials for equal increases of potential are spaced further apart.

However, near the surface over a small region, the equipotentials are horizontal (i.e. parallel to the ground) as shown in Figure 3. This is because the gravitational field over a small region is uniform. A 1 kg mass raised from the surface of the Earth by 1 m gains 9.81 J of gravitational potential energy; if it is raised another 1 m, it gains another 9.81 J. So its gravitational potential energy rises by 9.81 J for every metre of height it gains above the surface.

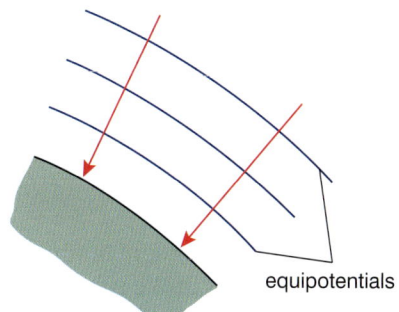

Figure 3 *Equipotentials near a surface*

Summary test 13.2

$g = 9.81 \, \text{N kg}^{-1}$

1 **a** Calculate the gain of gravitational potential energy of an object of mass 12 kg when its centre of mass is raised through a height of 2.0 m.

   b Show that the gravitational potential difference between the Earth's surface and a point 2.0 m above the surface is 19.6 J kg^{-1}.

2 A rocket of mass 35 kg launched from the Earth's surface gains 70 MJ of gravitational potential energy when it reaches its maximum height.

   a Calculate the gravitational potential difference between the Earth's surface and the highest point reached by the rocket.

   b The gravitational potential of the Earth's gravitational field at the surface of the Earth is −63 MJ kg^{-1}. Calculate: **i** the gravitational potential at the highest point reached by the rocket, **ii** the work that would need to have been done by the rocket to escape from the Earth's gravitational field.

3 Figure 4 shows the equipotentials near a non-spherical object.

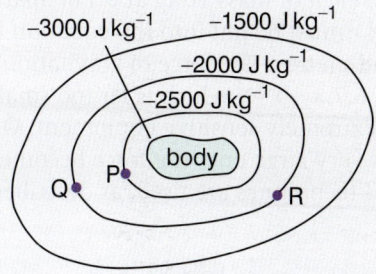

Figure 4

   a Calculate the gravitational potential energy of a 0.1 kg object at: **i** P, **ii** Q, **iii** R.

   b How much work must be done on the object to move it from: **i** P to Q, **ii** Q to R?

4 Figure 5 shows equipotentials at a spacing of 1.0 km near a planet. The point labelled X is on the −500 kJ kg^{-1} equipotential.

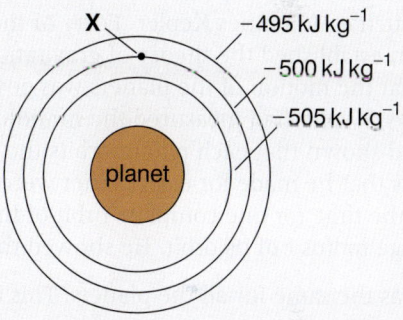

Figure 5

   a Show that the work done to move a 1 kg object from X to a position 10 m higher is 50 J.

   b Hence calculate the gravitational field strength at X.

   c Calculate the work that would need to be done to remove an object of mass 50 kg from X to infinity.

13.3 Newton's law of gravitation

Learning outcomes

On these pages you will learn to:

- recall and use Newton's law of gravitation in the form $F = Gm_1m_2/r^2$
- understand that, for a point outside a uniform sphere, the mass of the sphere may be considered to be a point mass at its centre

We owe our understanding of gravitation to Isaac Newton. 'The notion of gravity was occasioned by the fall of an apple!', said Newton when asked what made him develop the idea of gravity. Newton's theory of gravitation was an enormous leap forward because it explains events from the 'down-to-earth' falling apple to the motion of the planets. Like any good theory, it can be used to make predictions; for example, the return of a comet and its exact path can be calculated using the law of gravitation.

Newton realised that gravity is universal. Any two masses exert a force of attraction on each other. He knew about the careful measurements of planetary motion made by astronomers such as Johannes Kepler. Forty or more years before Newton established the theory of gravitation, Kepler had shown that the motion of the planets was governed by a set of laws. Kepler had measured the motion of each planet and had shown that each planet orbits the Sun. The measurements that he made for each planet were its time period T (i.e. the time for one complete orbit of the Sun) and the average radius r of its orbit. He showed that the value of $\frac{r^3}{T^2}$ was the same for all the planets. This is known as **Kepler's third law**.

	Mercury	Venus	Earth	Mars	Jupiter	Saturn
Average radius r of orbit/10^{10} m	6	11	15	23	78	143
Time T for one orbit/10^7 s	0.8	1.95	3.2	5.9	37.4	93.0
r^3/T^2/ 10^{16} m^3 s^{-2}	337	350	330	349	340	338

Figure 1 *Kepler's third law*

To explain Kepler's third law, Newton started by assuming that the planets and the Sun were point masses. A scale model of the Solar System with the Sun represented by a marble would put the Earth about a metre away, represented by a grain of sand! Newton assumed that the force of gravitation between a planet and the Sun varied inversely with the square of their distance apart.

In other words, if the force is F at distance d apart, then:

- at distance $2d$ apart, the force is $\frac{F}{4}$,
- at distance $3d$ apart, the force is $\frac{F}{9}$,
- at distance $4d$ apart, the force is $\frac{F}{16}$.

Using this **inverse-square law of force**, Newton was able to prove that $\frac{r^3}{T^2}$ was the same for all the planets. The actual proof is outlined in Topic 13.5. Newton then went on to use the inverse-square law of force to explain and make predictions for many other events involving gravity.

Newton's law of gravitation assumes that the gravitational force between any two point objects is:

- always an attractive force,
- proportional to the mass of each object,
- proportional to $\frac{1}{r^2}$, where r is their distance apart.

These three requirements can be summarised as

$$\textbf{gravitational force } F = -\frac{Gm_1m_2}{r^2}$$

where m_1 and m_2 = masses of the two objects. The minus sign in the equation is because the force is always an attractive force.

The constant of proportionality, G, in the above equation, is called the **universal constant of gravitation**. The unit of G can be worked out from $F = -\frac{Gm_1m_2}{r^2}$; rearranged, the equation gives $G = -\frac{Fr^2}{m_1m_2}$. So G can be given units of N m^2 kg^{-2}.

The value of G is 6.67×10^{-11} N m^2 kg^{-2}.

Work out for yourself the gravitational force between two point masses, each of mass 10 kg at 0.1 m apart. The values of m_1, m_2 and r must be put into the equation in units of kilograms and metres. The force of gravitational attraction works out at 6.7×10^{-7} N which is far too small to notice except with extremely sensitive equipment. Only if one of the masses is very large does the force become noticeable, unless special techniques are used, as described later.

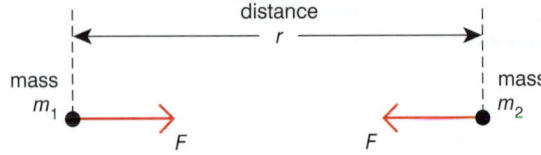

Figure 2 *Newton's law of gravitation*

Worked example

The distance from the centre of the Sun to the centre of the Earth is 1.5×10^{11} m. The mass of the Sun is 2.0×10^{30} kg and the mass of the Earth is 6.0×10^{24} kg.

a The Earth has a diameter of 1.3×10^7 m. The Sun has a diameter of about 1.4×10^9 m. Why is it reasonable to consider the Sun and the Earth at a distance of 1.5×10^{11} m apart as point masses on this distance scale?

b Calculate the force of gravitational attraction between the Sun and the Earth.

$G = 6.67 \times 10^{-11}$ N m^2 kg^{-2}

Solution

a On a scale model where the centre of the Sun was 1 m away from the centre of the Earth, the Sun would be a sphere of diameter about 1 cm and the Earth would be a sphere of diameter about 0.1 mm, no larger than a dot. The distance from the Earth to any part of the Sun is, therefore, the same to within 1%.

b $F = -\dfrac{6.67 \times 10^{-11} \times 2.0 \times 10^{30} \times 6.0 \times 10^{24}}{(1.5 \times 10^{11})^2}$

$= -3.6 \times 10^{22}$ N

Cavendish's measurement of G

The first accurate measurement of G was made by Henry Cavendish in 1798. He devised a torsion balance made of two small lead balls at either end of a rod. The rod was suspended horizontally by a torsion wire, as in Figure 3. The wire was calibrated by measuring the couple required to twist it per degree. Then, with the rod at rest in equilibrium, two massive lead balls were brought near the torsion balance to make the wire twist. By measuring the angle through which it twisted, the force of attraction between each massive lead ball and the small ball nearest to it was calculated. The distance between the centres of the small and large masses was also measured. Then G was calculated using the equation for the law of gravitation.

Figure 3 *Cavendish's measurement of G*

Summary test 13.3

$G = 6.67 \times 10^{-11}$ N m^2 kg^{-2}

1 a Calculate the force of gravitational attraction between two 'point objects' of masses 60 kg and 80 kg at a distance of 0.5 m apart.

b Calculate the distance between two identical point objects, each of mass 0.20 kg, that exert a force of 9.0×10^{-8} N on each other.

2 a Calculate the force of gravitational attraction between the Earth and an object of mass 80 kg on the surface of the Earth, where $g = 9.81$ N kg^{-1}.

b Use the result of your calculation in **a** to estimate the mass of the Earth. Assume that the mass of the Earth is concentrated at its centre.

The radius of the Earth $= 6.4 \times 10^6$ m

3 The Sun exerted a force of 6.0 N on a 1000 kg comet when it was at a distance of 1.5×10^{11} m from the Sun.

Calculate the force due to the Sun on the comet when it was at a distance of:

a 0.5×10^{11} m from the Sun,

b 7.5×10^{11} m from the Sun.

4 A space rocket of mass 1500 kg travelled from the Earth to the Moon, a distance of 3.8×10^8 m.

a When the space rocket was mid-way between the Earth and the Moon, calculate the force of gravitational attraction on it:

i due to the Earth,

ii due to the Moon.

b Calculate the magnitude and direction of the force of gravity of the Earth and the Moon on the space rocket when it was mid-way between the Earth and the Moon.

The mass of the Earth $= 6.0 \times 10^{24}$ kg

The mass of the Moon $= 7.4 \times 10^{22}$ kg

13.4 Planetary fields

Learning outcomes

On these pages you will learn to:

- derive the equation $g = \dfrac{GM}{r^2}$ for the gravitational field strength of a point mass
- recall and solve problems using the above equation
- solve problems using the equation $\phi = -\dfrac{GM}{r}$ for the potential in the field of a point mass

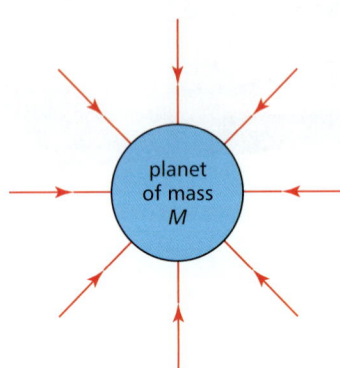

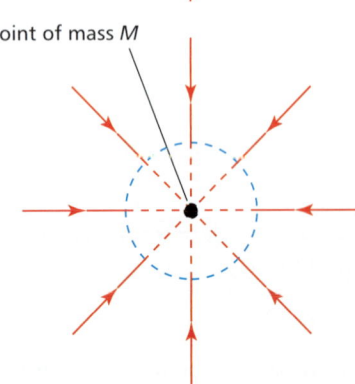

point of mass M

Figure 1 Comparing fields

Gravitational field strength near a spherical planet

The law of gravitation can be used to determine the gravitational field strength at any point in the field of a planet or any other spherical mass. Newton showed that the field of a spherical mass is the same as if it were a 'point mass' with all its mass M concentrated at its centre. The field lines of a spherical mass are always directed towards the centre, so the field pattern is just the same as for a point mass, as shown in Figure 1.

- For a point mass M, the force of attraction on a 'test mass' m (where $m \ll M$) at distance r from M is given by Newton's law of gravitation $F = \dfrac{GMm}{r^2}$.
  Therefore, the gravitational field strength at distance r is given by $g = \dfrac{F}{m} = \dfrac{GM}{r^2}$.
- For a spherical mass M of radius R, the force of attraction on a 'test mass' m at distance r **from the centre of M** is the same as if mass M was concentrated at its centre. Therefore,
  the force of attraction between m and M, $F = \dfrac{GMm}{r^2}$.

Therefore, the gravitational field strength at distance r is given by $g = \dfrac{F}{m} = \dfrac{GM}{r^2}$,

provided distance r is greater than or equal to the radius R of the sphere.

$$\text{Gravitational field strength, } g = \frac{GM}{r^2}$$

at distance r from a point object or the centre of a sphere of mass M.

Worked example

$G = 6.67 \times 10^{-11}\,\mathrm{N\,m^2\,kg^{-2}}$

The gravitational field strength at the surface of the Earth is $9.81\,\mathrm{N\,kg^{-1}}$. Calculate:

a the mass of the Earth,

b the gravitational field strength of the Earth at a height of 1000 km above the surface.

The radius of the Earth = 6400 km

Solution

a Rearranging $g_S = \dfrac{GM}{R^2}$ gives $M = \dfrac{g_S R^2}{G} = \dfrac{9.81 \times (6400 \times 10^3)^2}{6.67 \times 10^{-11}} = 6.0 \times 10^{24}\,\mathrm{kg}$

b At height $h = 1000\,\mathrm{km}$, $r = R + h = 7400\,\mathrm{km}$

$$\therefore g = \frac{GM}{r^2} = \frac{6.67 \times 10^{-11} \times 6.0 \times 10^{24}}{(7400 \times 10^3)^2} = 7.3\,\mathrm{N\,kg^{-1}}$$

The variation of g with distance from the centre of a spherical planet (or star)

1 At and beyond the surface of a spherical planet of mass M and radius R,

$$g = \frac{GM}{r^2},$$

where r is the distance from the centre of the sphere.

Because the surface gravitational field strength, $g_S = \dfrac{GM}{R^2}$, then $GM = g_S R^2$.

Therefore, $g = g_S \dfrac{R^2}{r^2}$.

The equation and Figure 2 shows how g changes with increase of distance r. The shape of the curve beyond $r = R$ is an **inverse-square law** curve.

2 Inside the planet

From the equation $g = \dfrac{GM}{r^2}$, you might think that g inside the Earth becomes ever larger and larger as r becomes smaller and smaller. However, inside the planet, only the mass in the sphere of radius r contributes to g. The rest of the mass outside r up to the surface gives no resultant force. So, as r becomes smaller, the mass M which contributes to g becomes smaller too. At the centre, the mass that contributes to g is zero. So g is zero at the centre. Figure 2 also shows how g varies with distance from the centre inside the planet, assuming its density is uniform.

Gravitational potential near a spherical planet

At or beyond the surface of a spherical planet, the gravitational potential ϕ at distance r from the centre of the planet of mass M is given by:

$$\phi = -\frac{GM}{r}$$

Applying this formula to the surface of the Earth with $M = 6.0 \times 10^{24}$ kg and $r = 6.4 \times 10^7$ m gives a value of -63 MJ kg^{-1}. This means that 63 MJ of work needs to be done to remove a 1 kg mass from the surface of the Earth to infinity.

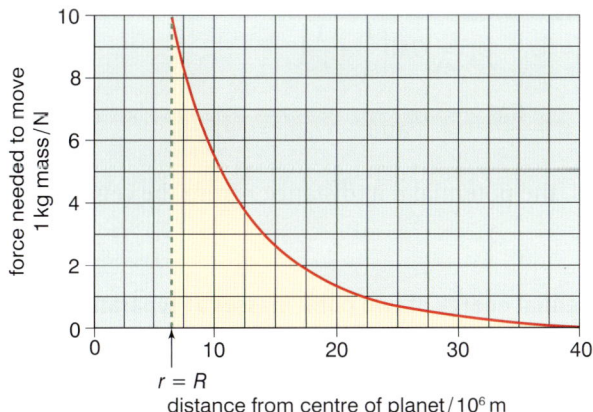

Figure 3 Work done

Figure 3 shows how the force of gravity on a 1 kg mass varies with distance r from the centre of the Earth. As explained previously, the mathematical equation for this curve is $g = g_S R^2 / r^2$. The area under the curve represents the work done to move the 1 kg mass from infinity to the surface.

- Each grid square in Figure 3 represents a 1 N force acting for a distance of 2.5×10^6 m, and therefore represents 2.5 MJ ($= 1$ N $\times 2.5 \times 10^6$ m) of work done.
- The work done to move the 1 kg mass from infinity to the surface can therefore be estimated by counting the number of grid squares under the curve and multiplying this number by 2.5 MJ.

By counting part-filled squares that are half-filled or more as wholly-filled squares and neglecting part-filled squares that are less than half-filled, show for yourself that this method gives an estimate for the work done which is very close to the value of 63 MJ determined above.

In effect, the area method used above is an application of 'work done = force $\times$ distance moved' with a variable force $F = GMm/r^2$ and $m = 1$ kg.

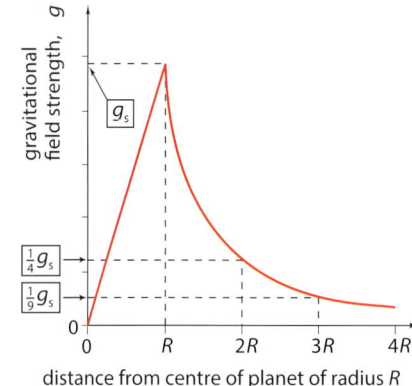

Figure 2 Gravitational field strength

> **Note**
>
> The equation $g = \dfrac{g_S R^2}{r^2}$ shows how g changes with increase of distance r.
>
> - At distance $r = 2R$,
> $$g = \frac{g_S R^2}{(2R)^2} = \frac{g_S}{4}$$
> - At distance $r = 3R$,
> $$g = \frac{g_S R^2}{(3R)^2} = \frac{g_S}{9}$$
> - At distance $r = 4R$,
> $$g = \frac{g_S R^2}{(4R)^2} = \frac{g_S}{16}$$

Consider one small step in moving the 1 kg mass from infinity to the surface. Suppose the distance from the centre of the Earth changes from r_1 to r_2 in making this small step of distance Δr.

The work done ΔW to make this small step is given by

$$\Delta W = F\Delta r = \frac{GMm\Delta r}{r^2}, \qquad \text{where } \Delta r = r_2 - r_1.$$

In Figure 3, the work done ΔW in making each small step $\Delta r = 2.5 \times 10^6$ m is represented by each column of grid squares under the curve. Thus the total work done to move from infinity to the surface is given by the total area under the curve.

Proof of the formula for gravitational potential $\phi = -GM/r$

Although the 'proof' below is not required in the specification, it is provided to give some further insight into the use of maths in physics.

The change of potential $\Delta\phi$ for the small step (from $r_1 = r$ to $r_2 = r - \Delta r$) $= \phi_2 - \phi_1$, where ϕ_2 is the potential at r_2 and ϕ_1 is the potential at r_1.

$$\Delta\phi = \frac{\Delta W}{m} = \frac{F\Delta r}{m} = -\frac{GM\Delta r}{r^2},$$

where the minus sign indicates a decrease in r.

As $\qquad \dfrac{1}{r_1} - \dfrac{1}{r_2} = \dfrac{r_2 - r_1}{r_1 r_2} = \dfrac{(r - \Delta r) - r}{r(r - \Delta r)} = -\dfrac{\Delta r}{r^2} \qquad$ provided $\Delta r \ll r$

then $\qquad \Delta\phi = \phi_2 - \phi_1 = GM\left(\dfrac{1}{r_1} - \dfrac{1}{r_2}\right)$

Hence $\qquad \phi_2 = -\dfrac{GM}{r_2} \qquad$ and $\qquad \phi_1 = -\dfrac{GM}{r_1}$

Therefore, in general, the potential ϕ at distance r from the centre of a planet is given by $\phi = -\dfrac{GM}{r}$

The gravitational potential energy of two point masses at separation r
When two point masses M and m are at distance r apart, their gravitational potential energy E_P is given by the following equation:

$$E_P = -\frac{GMm}{r}$$

To prove this equation, the gravitational potential ϕ of M at distance r from M is $-GMm/r$ and is zero at infinity. So if m is moved from infinity, the work done by their mutual force of gravitational attraction changes their potential energy from zero to $m\phi$ or $m \times (-GM/r)$. So at separation r their gravitational potential energy $= -GMm/r$. The same equation is obtained if M is moved from infinity to a distance r from m.

The variation of gravitational potential with distance from the centre of a spherical planet
The gravitational potential ϕ near a spherical planet is inversely proportional to the distance r from the centre of the planet, as given by the equation $\phi = -\dfrac{GM}{r}$, where M is the mass of the planet. Figure 4 shows how the gravitational potential of the Earth varies with distance. Note that the potential curve is a $1/r$ curve not an inverse square (i.e. $1/r^2$) curve like the field strength curve in Figure 3. So the potential:

• at distance $2R$ from the centre is $0.50 \times$ the potential at distance R from the centre,
• at distance $3R$ from the centre is $0.33 \times$ the potential at distance R from the centre,

- at distance 4R from the centre is 0.25 × the potential at distance R from the centre, etc.

The gradient of the potential curve at any point is equal to –g, where g is the gravitational field strength at the point.

Consider moving a small mass m a distance Δr away from the planet.

The work done on m = mΔφ where Δφ is the change of potential.

The force applied to m = work done ÷ distance moved in the direction of the force

$$= m\frac{\Delta\phi}{\Delta r}$$

As the gravitational force F is equal and opposite to the applied force, $F = -m\frac{\Delta\phi}{\Delta r}$.

Hence $$g = \frac{F}{m} = -\frac{\Delta\phi}{\Delta r} = \text{–gradient of the potential curve.}$$

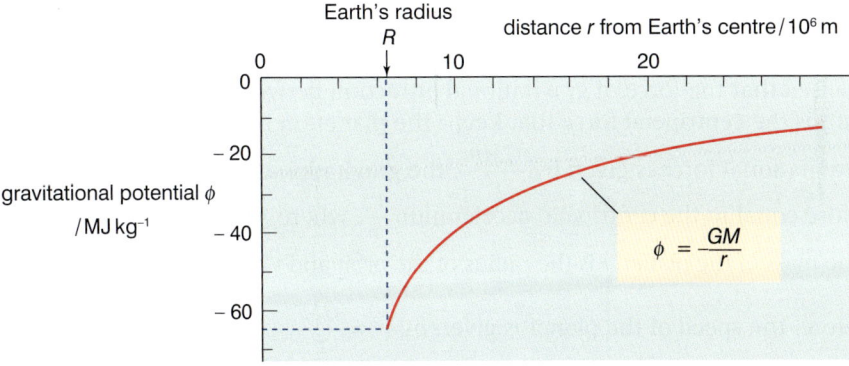

Figure 4 *Gravitational potential near the Earth*

Summary test 13.4

$G = 6.67 \times 10^{-11}\,N\,m^2\,kg^{-2}$

1 The Moon has a radius of 1740 km and its surface gravitational field strength is 1.62 N kg⁻¹ to three significant figures.

  a Calculate the mass of the Moon.

  b The Moon's gravitational pull on the Earth causes the ocean tides. Show that the gravitational pull of the Moon on the Earth's oceans is approximately three-millionths of the gravitational pull of the Earth on its oceans. The distance from the Earth to the Moon is 3.8×10^8 m.

2 The Sun has a mass of 2.0×10^{30} kg and a mean radius of 1.4×10^9 m. Calculate:

  a its gravitational field strength at:
    i its surface,
    ii the Earth's orbit which is at a distance of 1.5×10^{11} m from the Sun.

  b The Earth has a mass of 6.0×10^{24} kg. Show that the gravitational field strength of the Earth is equal and

opposite to the gravitational field strength of the Sun at a distance of 260 000 km from the centre of the Earth.

3 The tip of the tallest mountain on the Earth, Mount Everest, is 9 km above sea level. The mean radius of the Earth to the nearest kilometre is 6378 km.

  a Calculate the difference between the gravitational field strength of the Earth at sea level and the top of Mount Everest.

  b Discuss if it is reasonable to assume that the Earth's gravitational field is uniform between the surface and a height of 10 km above the surface.

  c Calculate the gain of potential energy of a mountaineer of mass 80 kg who travels to the top of the mountain from sea level.

4 Use the data in **1** to calculate the gravitational potential at the surface of the Moon and hence calculate the work done to launch a 500 kg rocket from the surface so that it escapes from the Moon's gravitational field.

13.5 Satellite motion

Learning outcomes

On these pages you will learn to:

- analyse circular orbits in inverse square law fields by relating the gravitational force to the centripetal acceleration it causes
- use the equation $r^3/T^2 = GM/4\pi^2$ for circular orbits
- explain what is meant by a geostationary orbit

Figure 1 *Space station in orbit*

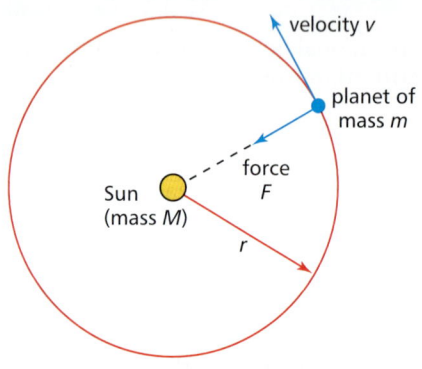

Figure 2 *Explanation of planetary motion*

On any clear night you ought to be able to see satellites passing overhead in the night sky. Although they are pinpoints of light, they are noticeable because they move steadily through the constellations. Websites supply information to enable you to identify some of them from their directions. However, **satellite motion** is not confined to artificial satellites orbiting the Earth. Any small mass which orbits a larger mass is a satellite. The Moon is the Earth's only natural satellite. Mars has two moons, *Phobos* and *Deimos*. Jupiter has more than 60 satellites including the four innermost satellites, *Io*, *Callisto*, *Ganymede* and *Europa*, first observed by Galileo four centuries ago.

From Kepler to Newton

Newton knew that the time period, T, of a planet orbiting the Sun depends on the mean radius, r, of the orbit in accordance with Kepler's third law:

$$\frac{r^3}{T^2} = \text{constant}$$

He realised that the force of gravitational attraction between each planet and the Sun is the centripetal force that keeps the planet on its orbit. By assuming the gravitational force is given by $\frac{GMm}{r^2}$, the gravitational field strength, $\frac{GM}{r^2}$, is therefore equal to the centripetal acceleration $\frac{v^2}{r}$, where M is the mass of the Sun, m is the mass of the planet, r is the radius of the orbit and v is the speed of the planet.

Therefore, the speed of the planet is given by $v^2 = \frac{GM}{r}$

Because speed, $v = \dfrac{\text{circumference of the orbit}}{\text{time period}} = \dfrac{2\pi r}{T}$

then $\dfrac{(2\pi r)^2}{T^2} = \dfrac{GM}{r}$

Rearranging this equation gives $\dfrac{r^3}{T^2} = \dfrac{GM}{4\pi^2}$

As $\dfrac{GM}{4\pi^2}$ is the same for all the planets, $\dfrac{r^3}{T^2}$ is the same for all the planets.

So, by assuming the force of attraction F varies with distance according to the inverse-square law (i.e. $F \propto \frac{1}{r^2}$), Newton was able to prove Kepler's third law.

Newton's theory not only explains Kepler's laws, but it also allows the mass M to be calculated if the value of G is known. The Earth orbits the Sun once per year on a circular orbit of radius 1.5×10^{11} m, so you can prove for yourself that the value of r^3/T^2 for any planet is 3.4×10^{18} m^{3}s^{-2}. Given $G = 6.67 \times 10^{-11}$ N m^{2}kg^{-2}, show that the mass of the Sun is 2.0×10^{30} kg.

Geostationary satellites

A geostationary satellite orbits the Earth from west to east directly above the Equator and has a time period of exactly 24 hours. It therefore remains in a fixed position above the Equator because it has exactly the same time period as the Earth's rotation.

The radius of orbit of a geostationary satellite can be calculated as follows using the equation $\dfrac{r^3}{T^2} = \dfrac{GM}{4\pi^2}$

$$T = 24 \text{ hours} = 24 \times 3600\,\text{s} = 86\,400\,\text{s}$$

$$r^3 = \frac{GM}{4\pi^2}T^2 = \frac{6.67 \times 10^{-11} \times 6.0 \times 10^{24} \times (86\,400)^2}{4\pi^2} = 7.6 \times 10^{22}\,\text{m}^3$$

$$\therefore \qquad r = 4.23 \times 10^7\,\text{m} = 42\,300\,\text{km}$$

The radius of the Earth is 6400 km. Therefore, the height of a geostationary satellite above the Earth is 36 000 km (= 42 300 − 6400 km to two significant figures).

Chapter Summary

Definitions
Gravitational field strength, g, is the force per unit mass on a small test mass placed in the field.

Gravitational potential, ϕ, at a point, is the work done per unit mass to move a small object from infinity to that point.

A line of force or **a field line** is the line followed by a small mass acted on by no other forces than the force of gravity.

A uniform field exists in a region where g is the same in magnitude and direction everywhere in the region.

Equations
1 $g = \dfrac{F}{m}$, where F is the gravitational force on a small mass m.

$\phi = \dfrac{W}{m}$ where W is the work done to move a small mass m from infinity.

2 Newton's law of gravitation; the gravitational force F between two point masses m_1 and m_2 at distance r apart is

given by $\qquad F = \dfrac{Gm_1m_2}{r^2}$

3 At distance r from a point mass M,

$g = \dfrac{GM}{r^2} \qquad\qquad \phi = -\dfrac{GM}{r}$

4 At or beyond the surface of a sphere of mass M,

$g = \dfrac{GM}{r^2}$

$\phi = -\dfrac{GM}{r}$

where r is the distance to the centre.

5 At the surface of a sphere of mass M and radius R,

$g_s = \dfrac{GM}{R^2}$

6 For a satellite in a circular orbit, its centripetal acceleration $\dfrac{v^2}{r} = g$

7 For point masses m and M at separation r,

$E_P = -\dfrac{GMm}{r}$

Summary test 13.5

$G = 6.67 \times 10^{-11}\,\text{N}\,\text{m}^2\,\text{kg}^{-2}$

The radius of the Earth = 6400 km,
$g = 9.81\,\text{N}\,\text{kg}^{-1}$ at the Earth's surface.

1 a Two satellites X and Y are seen from the ground crossing the night sky at the same time. Satellite X crosses the sky faster than Y. State with a reason which satellite is higher.

   b Explain why satellite TV dishes must be aligned carefully so they always point to the same position above the equator.

2 A space probe moving at a speed of $3.2\,\text{km}\,\text{s}^{-1}$ is in a circular orbit about a planet of mass M. The time period of the satellite is 110 minutes. Calculate:

   a the radius of the orbit,

   b the centripetal acceleration of the satellite,

   c the mass of the planet.

3 a A satellite moves at speed v in a circular orbit of radius r.
   i Write down an expression for the centripetal acceleration of the satellite.
   ii Show that the speed of the satellite is given by the equation $v^2 = gr$, where g is the gravitational field strength at the orbit.

   b A satellite orbits the Earth in a circular orbit at a height of 100 km. Calculate:
   i the gravitational field strength of the Earth at this distance,
   ii the speed of the satellite,
   iii the time period of the satellite.

4 a Show that the speed, v, of a satellite in a circular orbit of radius r about a planet of mass M is given by the equation $v^2 = \dfrac{GM}{r}$.

   b A weather satellite is in a polar orbit at a height of 1600 km.
   i Show that its speed is $7.1\,\text{km}\,\text{s}^{-1}$.
   ii Calculate its time period.
   iii Explain why such a satellite can survey global weather patterns every day.

13 Exam-style and Practice Questions

📑 **Launch additional digital resources for the chapter**

$G = 6.67 \times 10^{-11}\,\text{N}\,\text{m}^2\,\text{kg}^{-2}$

The Earth: radius = 6400 km, mass = 6.0×10^{24} kg, $g = 9.81\,\text{N}\,\text{kg}^{-1}$ at the surface.

1 a Sketch the pattern of lines of force of the gravitational field surrounding a uniform sphere.

b On your diagram:

i mark two points X and Y where the gravitational field strength has the same magnitude but is in opposite directions,

ii mark a point Z where the gravitational field strength is 0.25 times the gravitational field strength at X and in the same direction.

2 Jupiter's mass is 318 times the mass of the Earth.

a Calculate the gravitational force between the Earth and Jupiter when Jupiter is 6.3×10^{11} m from the Earth.

b Calculate the gravitational field strength of Jupiter at a distance of 6.3×10^{11} m from its centre.

3 a Sketch a graph to show how the gravitational field strength of the Earth varies with height from the surface to a height of 13 000 km above its surface.

b Calculate the gravitational field strength of the Earth at a distance of 5000 km above the surface.

4 a The Moon is 380 000 km from the Earth. Calculate the gravitational field strength of the Earth at the Moon.

b Use your answer to **a** to calculate:

i the centripetal acceleration of the Moon,

ii the speed of the Moon,

iii the time period of the Moon.

5 The mean radius of Jupiter's orbit round the Sun is 7.8×10^{11} m. Jupiter has a mass of 1.9×10^{27} kg. The Sun has a mass of 2.0×10^{30} kg.

a Calculate the distance from Jupiter to the point along the line between the Sun and Jupiter at which the gravitational field strength of Jupiter is equal and opposite to the gravitational field strength of the Sun.

b The asteroid belt consists of minor planets and smaller bodies in orbits round the Sun between the Sun and Jupiter. Discuss the view that the asteroids could be the remains of a larger body which was pulled apart by the gravitational force of Jupiter and of the Sun.

6 a Calculate the gravitational force of the Earth on an astronaut who weighs 690 N on the Earth's surface when the astronaut is in a spacecraft orbiting the Earth at a height of 100 km.

b Discuss whether or not an astronaut in an orbiting spacecraft is weightless.

7 A spy satellite orbits the Earth in a polar orbit at a height of 300 km.

a Calculate the gravitational field strength of the Earth at this height.

b i Show that the satellite has a speed of $7.7\,\text{km}\,\text{s}^{-1}$.

ii Calculate the time period of the satellite at this height.

8 a i What is meant by a geostationary orbit?

ii A communications satellite is in a geostationary orbit directly above the equator. Explain why it is advantageous for such a satellite to be in a geostationary orbit.

b A communications network company proposes to place 12 satellites in the same orbit, equally spaced along the orbit. The orbit is to be 500 km above the Earth.

i Calculate the gravitational field strength of the Earth at this height.

ii Show that the time period of a satellite in this orbit is 1 hour and 35 minutes.

iii Explain why a mobile phone at ground level using this system would need to switch to a different satellite every 8 minutes.

9 a Define *gravitational field strength*.

b Io is one of the four moons of the planet Jupiter that were first observed by Galileo. Io orbits Jupiter once every 1.77 days at an average orbital radius r of 4.22×10^5 km.

i Calculate the average speed, v, of Io about Jupiter.

ii Show that $v = \sqrt{\dfrac{GM}{r}}$, where M is the mass of Jupiter, and hence calculate the mass of Jupiter.

iii The radius of Jupiter is 6.99×10^4 km. Calculate the mean density of Jupiter.

c In 1979, Io was discovered to be volcanic. Scientists think it is continually being distorted by the varying gravitational forces due to Jupiter and its other moons. Europa, the nearest moon to Io, has a mass of 4.92×10^{22} kg and orbits Jupiter once every 3.8 days at an average orbital radius of 6.71×10^5 km.

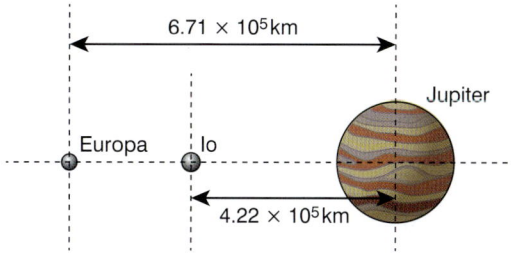

Figure 9.1

Use this data to discuss whether or not the gravitational force on Io due to Europa has a significant effect on Io compared with the gravitational force on Io due to Jupiter. Assume that the two moons orbit Jupiter in the same plane.

10 a i State Newton's law of gravitation.

ii Use Newton's law of gravitation to show that the gravitational field strength g at height h above the surface of a spherical planet is equal to $\dfrac{g_S}{(1 + \frac{h}{R})^2}$, where g_S is the gravitational field strength at the surface of the planet and R is the radius of the planet.

iii Draw a graph on a copy of the axes below to show how the Earth's gravitational field strength, g, varies with height, h, above the surface.

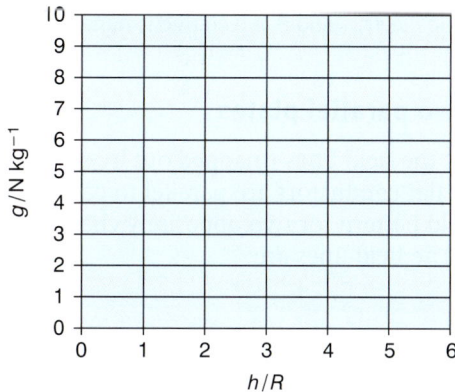

Figure 10.1

b The Earth's mass is 5.974×10^{24} kg. Its shape is not quite spherical. Its polar diameter is 12 714 km and its equatorial diameter is 12 756 km. Its gravitational field strength g at its poles is 9.83 N kg⁻¹ and 9.78 N kg⁻¹ at the Equator.

i The difference between the polar and equatorial values of g is partly due to the rotation of the Earth causing any object at the Equator to experience a centripetal acceleration. Calculate the centripetal acceleration of an object at the Equator and discuss whether or not the rotation of the Earth is a significant factor in explaining the difference between the polar and equatorial values of g.

ii Another factor that may contribute to the difference is that the Earth's equatorial diameter is 42 km greater than its polar diameter. Discuss whether or not this difference is a significant factor.

11 a Define *gravitational potential* at a point.

b A spherical planet has a radius R and a mass M.

i Write down expressions for the gravitational potential, ϕ, above the surface at distance r from the centre of the planet and for the surface gravitational potential, ϕ_S.

ii Draw a graph on a copy of the axes below to show how the ratio ϕ/ϕ_S at and above the surface varies with distance r from the centre of the planet.

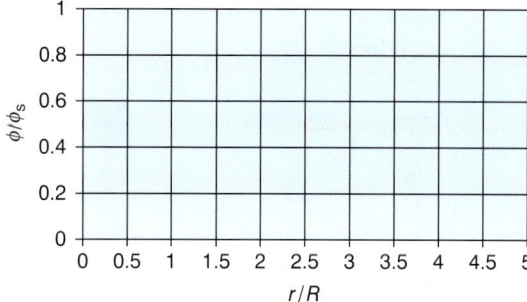

Figure 11.1

iii Show that the gravitational potential ϕ at distance r from the centre of the planet is given by the equation $\phi = -gr$, where g is the magnitude of the gravitational field strength at distance r from the centre of the planet.

c i The surface gravitational field strength of the Earth's moon is 1.62 N kg⁻¹ and its diameter is 1740 km. Calculate the surface gravitational potential due to the Moon at the surface of the Moon.

ii The escape speed from the Moon's surface is 2.38 km s⁻¹. Use this value to calculate the gravitational potential needed to escape from the Moon.

iii State one reason why the answers to **i** and **ii** differ.

14.1 Electric field strength

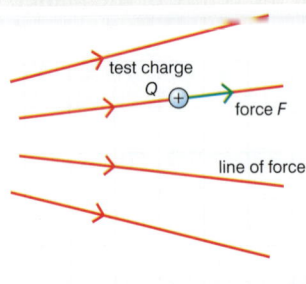

Figure 1 *Electric field strength*

Note

The unit of E may be written as the newton per coulomb ($\mathrm{N\,C^{-1}}$) or the volt per metre ($\mathrm{V\,m^{-1}}$).

The link between the two can be seen because $F = QE = \dfrac{Q\Delta V}{\Delta d}$.

Rearranging this equation gives $\dfrac{F}{Q} = \dfrac{\Delta V}{\Delta d}$ ($= E$). Therefore, the newton per coulomb and the volt per metre are both acceptable as the unit of E.

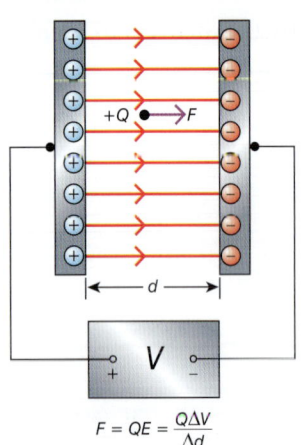

Figure 2 *The electric field strength between two parallel plates*

210

Inside an electric field

A charged object in an electric field experiences a force due to the field. Provided the object's size and charge are both sufficiently small, the object may be used as a 'test' charge to measure the strength of the field at any position in the field.

> **The electric field strength, E, at a point in the field is defined as the force per unit charge on a positive test charge placed at that point.**

The unit of E is the newton per coulomb ($\mathrm{N\,C^{-1}}$).

In Figure 1, a positive test charge Q at a certain point in an electric field is acted on by force F due to the electric field. The electric field strength, E, at that point is given by the equation

$$E = \frac{F}{Q}$$

Notes

1 Rearranging this equation gives $\boldsymbol{F = QE}$ for the force F on a test charge Q at a point in the electric field where the electric field strength is E.

2 Electric field strength is a vector which is in the same direction as the force on a positive test charge. In other words, the direction of a field line at any point is the direction of the electric field strength at that point. The force on a small charge in an electric field is:

- in the same direction as the electric field if the charge is positive,
- in the opposite direction to the electric field if the charge is negative.

The electric field between two parallel plates

Figure 4c in Topic 6.1 shows that the field lines (mapped out by semolina grains) between two oppositely charged flat conductors are parallel to each other and at right angles to the plates. The field pattern for two oppositely charged flat plates is similar, as shown in Figure 2. The field lines are:

- parallel to each other,
- at right angles to the plates,
- from the positive plate to the negative plate.

The field between the plates is **uniform**. This is because the electric field strength has the same magnitude and direction everywhere between the plates. The electric field strength E can be calculated from the potential difference ΔV between the plates and their separation Δd using the equation

$$E = \frac{\Delta V}{\Delta d}$$

To prove this equation, consider a small charge Q between the plates, as in Figure 2.

1 The force F on a small charge Q in the field is given by $F = QE$, where E is the electric field strength between the plates.

2 If the charge is moved from the positive to the negative plate, the work done W by the field on Q is given by $W =$ force $F \times$ distance moved $= QE\Delta d$.

3 By definition, the potential difference between the plates, ΔV is the work done per unit charge when a small charge is moved through potential difference ΔV.

Therefore, $\Delta V = \dfrac{W}{Q} = \dfrac{QE\Delta d}{Q} = Ed$, so rearranging $\Delta V = E\Delta d$ gives $E = \dfrac{\Delta V}{\Delta d}$.

Deflection of a beam of charged particles in a uniform electric field

A charged particle moving across a uniform electric field experiences a constant acceleration because the force on it is constant. Its motion is similar to that of a small object in a small region of the Earth's gravitational field, where g is constant.

Figure 3 shows an electron beam deflected by an electric field produced by applying a p.d. between the metal deflecting plates P and Q. The beam is produced by attracting electrons from a heated filament towards the positively charged anode; it is deflected towards the positive

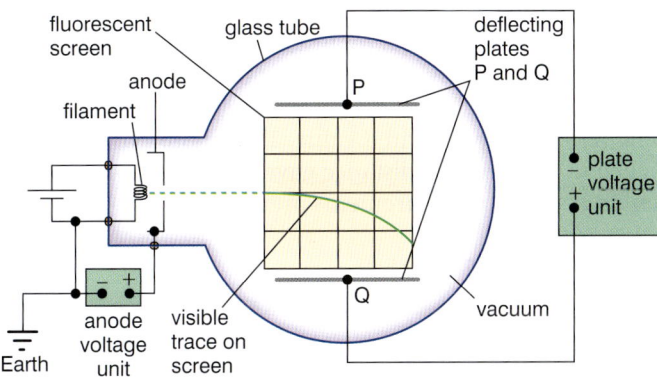

Figure 3 *Deflection of an electron beam; the hole in the anode lets some electrons through and the electron beam can be seen as it passes along the screen*

plate Q. By adjusting the strength of the field, the extent of the deflection can be controlled.

The beam curves in a parabolic path just as a projectile projected horizontally does. The projectile equations from Topic 1.8 can be used to determine its deflection, provided g is replaced by acceleration $a = \dfrac{eE}{m}$ where m is the mass of the electron and e is the charge of the electron.

Notes

1 If the p.d. between plates P and Q is ΔV_P, each electron experiences a force $F = eE = e\dfrac{\Delta V_P}{\Delta d}$, where Δd is the perpendicular distance between plates P and Q.

2 The acceleration of each electron towards the positive plate, $a = \dfrac{F}{m} = \dfrac{e}{m} \times \dfrac{\Delta V_P}{\Delta d}$, where m is the mass of the electron.

3 The time taken, t, by each electron to cross the field $= \dfrac{L}{v}$, where L is the length of each plate and v is the initial speed of the electron on entry to the field.

4 The deflection, y, of the electron on leaving the field is given by the equation $y = \frac{1}{2}at^2$.

Using the above equations, it can be shown that y is directly proportional to the plate p.d. ΔV_P.

Summary test 14.1

$e = 1.6 \times 10^{-19}\,C$

1 A +40 nC point charge Q_1 is placed in an electric field.

 a Calculate the magnitude of the force on Q_1 if the electric field strength where Q_1 is placed is $3.5 \times 10^4\,V\,m^{-1}$.

 b Q_1 is moved to a different position in the electric field. The force on Q_1 at this position is $1.6 \times 10^{-3}\,N$. Calculate the magnitude of the electric field strength at this position.

2 Figure 4 shows the path of a charged dust particle in an electric field.

 a The electric field strength at X is $65\,kV\,m^{-1}$. The force due to the field on the particle when it is at X is $8.2 \times 10^{-3}\,N$ towards the metal surface.

  i What type of charge does X carry?

  ii Calculate the charge carried by the particle.

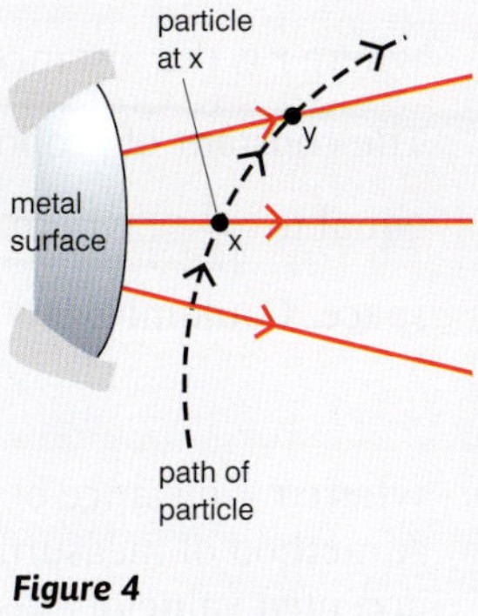

Figure 4

 b i Calculate the magnitude of the force on the particle when it is at Y where the electric field strength is $58\,kV\,m^{-1}$.

  ii State the direction of the force on the particle when it is at Y.

3 A high voltage supply unit is connected across a pair of parallel plates which are at a separation of 50 mm.

 a The voltage is adjusted to 4.5 kV. Calculate:

  i the electric field strength between the plates,

  ii the electric force on a droplet in the field that carries a charge of $8.0 \times 10^{-19}\,C$.

 b The separation between the plates is altered without changing the p.d. between the plates. The droplet in a is now acted on by a force of $4.5 \times 10^{-14}\,N$. Calculate the new separation between the plates.

4 A beam of electrons moving horizontally at speed v enters a vertical uniform electric field of width x and strength E. In terms of the charge e and the mass m of the electrons, show that the deflection y of the electrons on leaving the field is given by an equation of the form $y = kx^2$ and determine an expression for k.

Learning outcomes

On these pages you will learn to:

- recall and use Coulomb's law in the form $F = Q_1Q_2/4\pi\varepsilon_0r^2$ for the force between two point charges in free space or air
- understand that, for any point outside a spherical conductor, the charge on the sphere may be considered to act as a point charge at its centre

Like charges repel and unlike charges attract. The force between two charged objects depends on how close they are to each other. The exact link was first established by Charles Coulomb in France in 1784. He devised a very

Figure 1 Coulomb's torsion balance

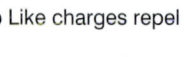

a Unlike charges attract

$$F = \frac{1}{4\pi\varepsilon_0}\frac{Q_1Q_2}{r^2}$$

b Like charges repel

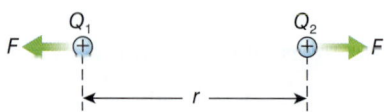

Figure 2 Coulomb's law

sensitive torsion balance to measure the force between charged pith balls. Figure 1 shows the arrangement. A needle with a ball made of pith (a substance obtained from plants) at one end and a counterweight at the other end was suspended horizontally by a vertical wire. Another pith ball on the end of a thin vertical rod could be placed in contact with the first ball.

The pith balls were small enough to be considered as point objects. The ball on the rod was charged and then placed in contact with the other ball on the needle. The contact between them charged the second ball which was then repelled by the ball fixed on the rod. This caused the wire to twist until the electrical repulsion was balanced by the twist built up in the wire. By turning the torsion head at the top of the wire, the distance between the two balls could be set at any required value. The amount of turning needed to achieve that distance gave the force. Some of Coulomb's many measurements are below.

Table 1 Some of Coulomb's results

Distance, r	36	18	8.5
Force, F	36	144	567

Measurements for both variables were actually made in degrees, so the above values are in relative units. Can you make out a pattern for these measurements? Halving the distance from 36 to 18 makes the force increase by a factor of 4. Halving the distance from 18 to 8.5 (near enough 9) increases the force again by a factor of about 4. The measurements fit the link that the force, F, is proportional to $\frac{1}{r^2}$. All the other measurements made by Coulomb fitted the same link.

Because the force is also proportional to the charge on each ball, Coulomb deduced the following equation, known as Coulomb's law, for the force, F, between two 'point charges', Q_1 and Q_2:

$$F = k\frac{Q_1Q_2}{r^2},$$

where r is the distance between the charges.

The constant of proportionality, k, can be shown to be equal to $\frac{1}{4\pi\varepsilon_0}$, where ε_0 is the absolute permittivity of free space. Coulomb's law is therefore written as

$$F = \frac{1}{4\pi\varepsilon_0}\frac{Q_1Q_2}{r^2},$$

where r = distance between two point charges Q_1 and Q_2 A method of measuring ε_0 is outlined in Topic 15.2. The accepted value of ε_0 is $8.85 \times 10^{-12}\,\mathrm{F\,m^{-1}}$ so

$$\frac{1}{4\pi\varepsilon_0} = 9.0 \times 10^9\,\mathrm{m\,F^{-1}}$$

Worked example

$e = 1.6 \times 10^{-19}\,\text{C}$, $\dfrac{1}{4\pi\varepsilon_0} = 9.0 \times 10^9\,\text{m}\,\text{F}^{-1}$

Calculate the force between a proton and an electron at a separation of $3.0 \times 10^{-10}\,\text{m}$.

Solution

$$F = \frac{1}{4\pi\varepsilon_0}\frac{Q_1 Q_2}{r^2} = \frac{9.0 \times 10^9 \times 1.6 \times 10^{-19} \times 1.6 \times 10^{-19}}{(3.0 \times 10^{-10})^2}$$
$$= 2.6 \times 10^{-9}\,\text{N}$$

Note on $k = \dfrac{1}{4\pi\varepsilon_0}$

If Coulomb's law in the form $F = k\dfrac{Q_1 Q_2}{r^2}$ is applied to the force on a 'test' charge q at distance r from a point charge Q, the force on the test charge $F = \dfrac{kQq}{r^2}$, so the electric field strength at distance r, $E = \dfrac{F}{q} = \dfrac{kQ}{r^2}$.

By introducing $\dfrac{1}{4\pi\varepsilon_0}$ as k, the equation $E = \dfrac{kQ}{r^2}$ may be written as $\dfrac{Q}{4\pi r^2} = \varepsilon_0 E$ or $\dfrac{Q}{A} = \varepsilon_0 E$. In this form, E is the electric field strength at the surface of a sphere of radius r and surface area $A(= 4\pi r^2)$ which has a charge Q

evenly distributed on its surface. The equation $\dfrac{Q}{A} = \varepsilon_0 E$ gives the surface charge density $\dfrac{Q}{A}$ needed to produce an electric field of strength E at the surface. Thus ε_0 represents the charge per unit area on a surface in a vacuum that produces an electric field of strength 1 volt per metre above the surface. The equation applies to any arrangement, for example, as in Figure 3. The explanation here is not part of this specification and is provided to give a better understanding of Coulomb's law.

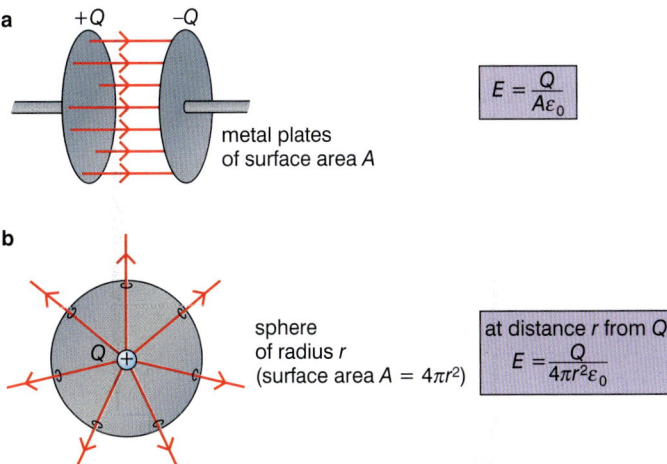

Figure 3 *Comparison of surface electric field strengths*

Summary test 14.2

$\varepsilon_0 = 8.85 \times 10^{-12}\,\text{F}\,\text{m}^{-1}$, $\dfrac{1}{4\pi\varepsilon_0} = 9.0 \times 10^9\,\text{m}\,\text{F}^{-1}$, $e = 1.6 \times 10^{-19}\,\text{C}$

1 Calculate the force between an electron and:

  a a proton at a distance of $2.5 \times 10^{-9}\,\text{m}$,

  b a nucleus of a nitrogen atom (charge $+7e$) at a distance of $2.5 \times 10^{-9}\,\text{m}$.

2 **a** Two point charges $Q_1 = +6.3\,\text{nC}$ and $Q_2 = -2.7\,\text{nC}$ exert a force of $3.2 \times 10^{-5}\,\text{N}$ on each other when they are at a certain distance, d, apart. Calculate:

    i the distance, d, between the two charges,

    ii the force between the two charges if they are moved to distance $3d$ apart.

  b A charge of $+4.0\,\text{nC}$ is added to each charge in **a**. Calculate the force between Q_1 and Q_2 when they are at separation d.

3 A $+30\,\text{nC}$ point charge is at a fixed distance of $6.2\,\text{mm}$ from a point charge Q. The charges attract each other with a force of $4.3 \times 10^{-2}\,\text{N}$.

  a Calculate the magnitude of charge Q and state whether Q is a positive or a negative charge.

  b The two charges are moved $2.5\,\text{mm}$ further apart. Calculate the force between them in this new position.

4 Two point objects, X and Y, carry equal and opposite amounts of charge at a fixed separation of $3.6 \times 10^{-2}\,\text{m}$. The two objects exert a force on each other of $5.1 \times 10^{-5}\,\text{N}$.

  a Calculate the magnitude, Q, of each charge and state whether the charges attract or repel each other.

  b The charge of each object is increased by adding a positive charge of $+2Q$ to each object. Calculate the separation at which the two objects would exert a force of $5.1 \times 10^{-5}\,\text{N}$ on each other and state whether the objects attract or repel each other.

Electric potential

Learning outcomes

On these pages you will learn to:

- define electric potential at a point in terms of the work done in bringing unit positive charge from infinity to the point
- recall the definition of electric field strength and the use of the equation $F = QE$
- state that the electric field strength of the field at a point is equal to the negative of potential gradient at that point

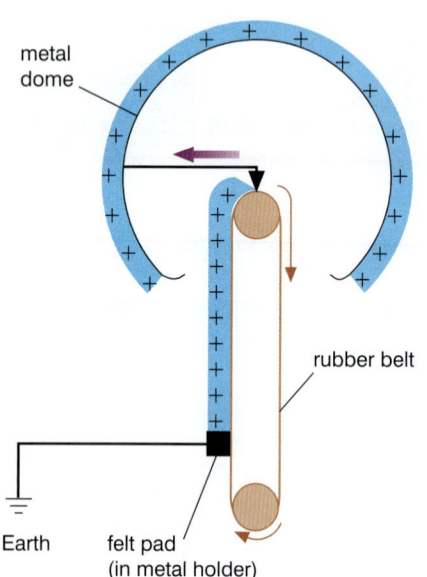

metal dome

rubber belt

Earth

felt pad (in metal holder)

Figure 1 *The Van de Graaff generator*

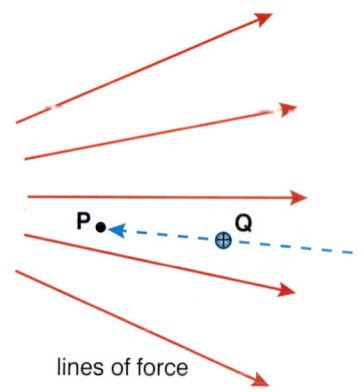

P

Q

lines of force

Figure 2 *Moving charge Q to position P*

The Van de Graaff generator

A Van de Graaff generator can easily produce sparks in air several centimetres in length. Figure 1 shows how a Van de Graaff generator works. Charge created when the rubber belt rubs against a pad is carried by the belt up to the metal dome of the generator. As charge gathers on the dome, the potential difference between the dome and Earth increases until sparking occurs.

A spark suddenly transfers energy from the dome. Work must be done to charge the dome because a force is needed to move the charge on the belt up to the dome. So the electric potential energy of the dome increases as it charges up. Some or all of this energy is transferred from the dome when a spark is created.

In general, work must be done to move a charged object A towards another object B that has the same type of charge. Their electric potential energy increases as A moves towards B.

The electric potential energy of A increases from zero if it is moved from infinity towards B. The electric field of B causes a force of repulsion to act on A and this force must be overcome to move A closer to B.

> The **electric potential** at a certain position in any electric field is defined as the work done per unit positive charge on a 'positive test charge' (i.e. a small positively charged object) when it is moved from infinity to that position.

By definition, the position of zero potential energy is infinity. Thus the electric potential at a certain position is the potential energy per unit positive charge of a 'positive test charge' at that position P.

The unit of electric potential is the volt (V), equal to $1\,\text{J}\,\text{C}^{-1}$.

Consider a positive test charge Q placed at a position in an electric field where its electric potential energy is E_P. The electric potential V at this position is given by

$$V = \frac{E_P}{Q}$$

Note that rearranging this equation gives $E_P = QV$

> **Note**
>
> If a test charge $+Q$ is moved in an electric field from one position where the electric potential is V_1 to another where the electric potential is V_2, the work done ΔW on it is given by
>
> $$\Delta W = Q(V_2 - V_1)$$

Potential gradients

Equipotentials are lines of constant potential. A test charge moving along an equipotential has constant potential energy. No work is done by the electric field on a test charge moving along an equipotential because the force due to the field is at right angles to the equipotential. In other words, the lines of force of the electric field cross the equipotential lines at right angles.

The equipotentials for an electric field are like equipotentials for a gravitational field; both are lines of constant potential energy for the appropriate test object, in one case a test charge and in the other case a test mass.

Figure 3 shows the equipotentials of the electric field due to two positively charged objects.

Suppose a +2.0 μC test charge is moved from X to Y.

The potential at X, V_X, is +1000 V so the test charge at X has potential energy equal to $+2.0 \times 10^{-3}$ J

$$(= QV_X = + 2.0 \times 10^{-6}\,C \times +1000\,V)$$

The potential at Y, V_Y, is +400 V so the test charge at Y has potential energy equal to $+8.0 \times 10^{-4}$ J

$$(= QV_Y = + 2.0 \times 10^{-6}\,C \times +400\,V)$$

Therefore, moving the test charge from X to Y lowers its potential energy by 1.2×10^{-3} J.

Note that if the test charge is moved from Y to Z along the +400 V equipotential, its potential energy remains constant at $+8.0 \times 10^{-4}$ J.

The **potential gradient** at any position in an electric field is the change of potential per unit change of distance in a given direction.

1 If the field is non-uniform as in Figure 3, the potential gradient varies according to position and direction. The closer the equipotentials are, the greater the potential gradient. The potential gradient is at right angles to the equipotentials.
2 If the field is uniform, such as the field between the two oppositely charged parallel plates shown in Figure 4, the equipotentials **between the plates** are equally spaced lines parallel to the plates. Figure 4 also shows how the potential relative to the negative plate changes with perpendicular distance x from the negative plate.

The graph shows that the potential relative to the negative plate is proportional to distance x. In other words, the potential gradient is constant (such that the potential increases in the opposite direction to the electric field) and equal to $\frac{\Delta V}{\Delta d}$.

The electric field strength E between the plates is equal to $\frac{\Delta V}{\Delta d}$ and is directed from the + to the − plate. In other words:

The electric field strength is equal to the negative of the potential gradient.

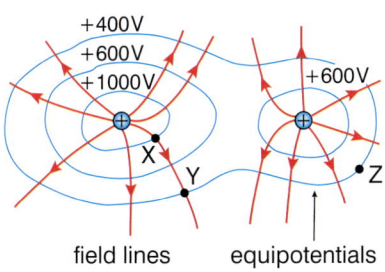

Figure 3 Equipotentials

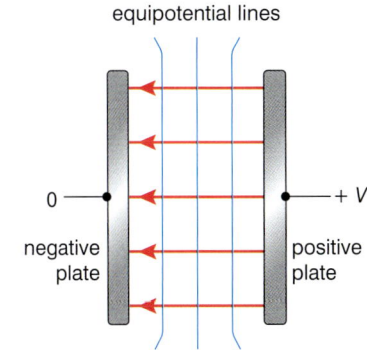

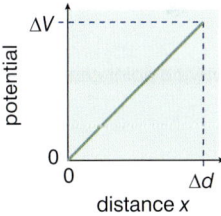

Figure 4 A uniform potential gradient

Summary test 14.3

$e = 1.6 \times 10^{-19}$ C

1 An electron in a beam is accelerated from a potential of −50 V to a potential of +450 V. Calculate:

    a the potential energy of the electron at:
      i −50 V,
      ii +450 V,

    b the change of potential energy of the electron.

2 In Figure 3 above, a test charge q is moved from X to Z. Calculate the change of potential energy of the test charge:

    a if $q = +3.0$ μC,

    b if $q = -2.0$ μC.

3 An oil droplet carrying a charge of $+2e$ is in air between two parallel metal plates separated by a distance of 20 mm. The p.d. between the plates is 5.0 V.

    a Calculate:
      i the potential gradient between the two plates,
      ii the force on the droplet.

    b Calculate the change of electrical potential energy of the oil droplet if it moves from the midpoint of the plates to the negative plate.

4 a Define electric potential and state its unit.

    b Two parallel horizontal metal plates are placed one above the other at a separation of 20 mm. A potential difference of +60 V is applied between the plates with the top plate positive.
      i Calculate the electric field strength between the plates.
      ii Sketch a graph to show how the electric potential V between the plates varies with height h above the lower plate.

Electric field strength and potential of point charges

Learning outcomes

On these pages you will learn to:

- recall the definition of electric field strength and use $F = QE$
- recall and use $E = Q/4\pi\varepsilon_0 r^2$ for the field strength of a point charge in free space or air
- use the equation $V = Q/4\pi\varepsilon_0 r$ for the potential in the field of a point charge

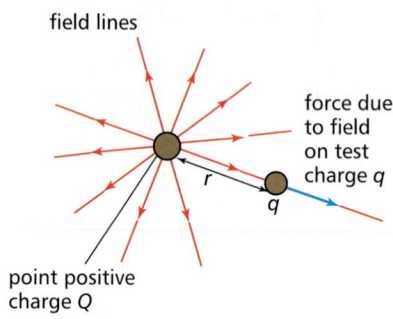

field lines

force due to field on test charge q

r

q

point positive charge Q

Figure 1 *Force near a point charge Q*

A point charge is a convenient expression for a charged object in a situation where distances under consideration are much greater than the size of the object. The same idea applies to a distant star which is considered as a point object because its diameter is much smaller than the distance to it from the Earth. A 'test' charge in an electric field is a point charge that does not alter the electric field in which it is placed. This would happen if an object with a sufficiently large charge is placed in an electric field and it causes a change in the distribution of charge that creates the field.

Consider the electric field due to a point charge $+Q$, as shown in Figure 1. The field lines radiate from the point charge because a test charge $+q$ in the field would experience a force directly away from Q wherever it was placed. Coulomb's law gives the force F on the test charge q as

$$F = \frac{1}{4\pi\varepsilon_0}\frac{Qq}{r^2}$$

Therefore, as electric field strength $E = \dfrac{F}{q}$ by definition, the electric field strength at distance r from Q is given by

$$E = \frac{Q}{4\pi\varepsilon_0 r^2}$$

Note that, if Q is negative, the above formula gives a negative value of E corresponding to the field lines pointing inwards towards Q.

Worked example

$\dfrac{1}{4\pi\varepsilon_0} = 9.0 \times 10^9\,\text{m F}^{-1}$, $e = 1.6 \times 10^{-19}\,\text{C}$

Calculate the electric field strength due to a nucleus of charge $+82\,e$ at a distance of $0.35\,\text{nm}$.

Solution

$E = \dfrac{Q}{4\pi\varepsilon_0 r^2} = \dfrac{9.0 \times 10^9 \times (+82 \times 1.6 \times 10^{-19})}{(0.35 \times 10^{-9})^2} = 9.6 \times 10^{11}\,\text{V m}^{-1}$

Electric field strength as a vector

If a test charge is in an electric field due to several point charges, each charge exerts a force on the test charge. The resultant force F on the test charge gives the resultant electric field strength at the position of the test charge. Consider the following situations:

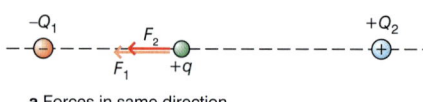

$-Q_1$ F_2 $+Q_2$

F_1 $+q$

a Forces in same direction

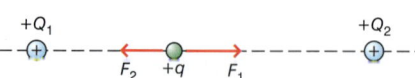

$+Q_1$ $+Q_2$

F_2 $+q$ F_1

b Forces in opposite direction

Figure 2 *Forces along the same line*

- **Forces in the same direction**: Figure **2a** shows a test charge $+q$ on the line between a negative point charge Q_1 and a positive point charge Q_2. The test charge experiences a force $F_1 = qE_1$ where E_1 is the electric field strength due to Q_1 and a force $F_2 = qE_2$ where E_2 is the electric field strength due to Q_2. The two forces act in the same direction because Q_1 attracts q and Q_2 repels q. So the resultant force $F = F_1 + F_2 = qE_1 + qE_2$. Therefore, the resultant electric field strength $E = \dfrac{F}{q} = \dfrac{qE_1 + qE_2}{q} = E_1 + E_2$

- **Forces in opposite directions**: Figure **2b** shows a test charge $+q$ on the line between two positive point charges Q_1 and Q_2. The test charge experiences a force $F_1 = qE_1$ where E_1 is the electric field strength due to Q_1 and a force $F_2 = qE_2$ where E_2 is the electric field strength due to Q_2. The two forces act in opposite directions because Q_1 repels q and Q_2 repels q. Assuming F_1 is greater than F_2, the resultant force $F = F_1 - F_2 = qE_1 - qE_2$.

  Therefore, the resultant electric field strength $E = \dfrac{F}{q} = \dfrac{qE_1 - qE_2}{q} = E_1 - E_2$

- **Forces at angle θ to each other**: Figure 3 shows a test charge $+q$ on perpendicular lines from two positive point charges Q_1 and Q_2. The test charge experiences a force $F_1 = qE_1$ where E_1 is the electric field strength due to Q_1 and a force $F_2 = qE_2$ where E_2 is the electric field strength due to Q_2. The two forces are at angle θ to each other, F_1 along the line between q and Q_1 and F_2 along the line between q and Q_2. The magnitude of the resultant force F is given by the formula $F^2 = F_1^2 + F_2^2 + 2F_1F_2 \cos \theta$ as explained in Topic 1.1. Therefore, the resultant electric field strength $E = \dfrac{F}{q}$.

In general, the resultant electric field strength is the vector sum of the individual electric field strengths.

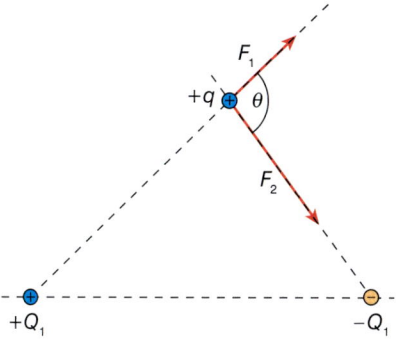

Figure 3 Forces at angle θ

Worked example

$$\frac{1}{4\pi\varepsilon_0} = 9.0 \times 10^9 \, \text{m F}^{-1}$$

A $+65\,\mu\text{C}$ point charge Q_1 is at a distance of 50 mm from a $+38\,\mu\text{C}$ charge Q_2.

A $+12\,\text{pC}$ charge q is placed at M, midway between Q_1 and Q_2.

Calculate:

a the resultant electric field strength at M,

b the magnitude and direction of the force on q.

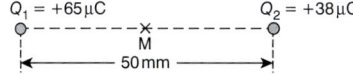

Figure 4

Solution

a The electric field strength due to Q_1 at M, $E_1 = \dfrac{Q_1}{4\pi\varepsilon_0 r_1^{\,2}}$, where $r_1 = 25$ mm.

Therefore, $E_1 = 9.0 \times 10^9 \times \dfrac{65 \times 10^{-6}}{(25 \times 10^{-3})^2} = 9.4 \times 10^8 \, \text{V m}^{-1}$ away from Q_1.

The electric field strength due to Q_2 at M, $E_2 = \dfrac{Q_2}{4\pi\varepsilon_0 r_2^{\,2}}$, where $r_2 = 25$ mm.

Therefore, $E_2 = 9.0 \times 10^9 \times \dfrac{38 \times 10^{-6}}{(25 \times 10^{-3})^2} = 5.5 \times 10^8 \, \text{V m}^{-1}$ away from Q_2.

As E_1 and E_2 are in opposite directions, the resultant electric field strength $E = E_1 - E_2 = 9.4 \times 10^8 - 5.5 \times 10^8 = 3.9 \times 10^8 \, \text{V m}^{-1}$ away from Q_1 towards Q_2.

b The resultant force on $q = qE = +12 \times 10^{-12} \times 3.9 \times 10^8 = 4.7 \times 10^{-3} \, \text{N}$.

Electric potential near a point charge

Because Coulomb's law $F = \dfrac{1}{4\pi\varepsilon_0} \dfrac{Q_1 Q_2}{r^2}$ and Newton's law $F = G\dfrac{m_1 m_2}{r^2}$ are both inverse-square relationships, the forces vary with distance in the same way. Therefore the formula for the electrical potential near a point charge Q is of the same form as the formula for the gravitational potential near a point mass (or spherical mass), i.e. $\phi = \dfrac{-GM}{r}$, which was derived in Topic 13.4.

Using this analogy and replacing M by Q and G by $\dfrac{1}{4\pi\varepsilon_0}$ in the gravitational potential formula above therefore gives the following formula for the electric potential V at distance r from a point charge Q:

$$V = \frac{Q}{4\pi\varepsilon_0 r}$$

The equation shows that the electric potential V is inversely proportional to the distance r. The curve is not an 'inverse-square law' curve as V is proportional to $1/r$. See Figure 6.

Note

A test charge inside the sphere would experience zero force due to the charge on the sphere, because the electric field due to the charge on the sphere is in all directions from any point inside the sphere.

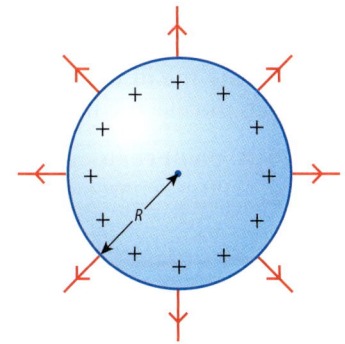

Figure 5 *The electric field near a charged hollow metal sphere*

More about radial fields

The electric field surrounding a charged metal sphere is radial, like the field due to a point charge. The charge on the metal surface is evenly distributed on its surface. The field has the same strength as if all the charge was concentrated at the centre of the sphere.

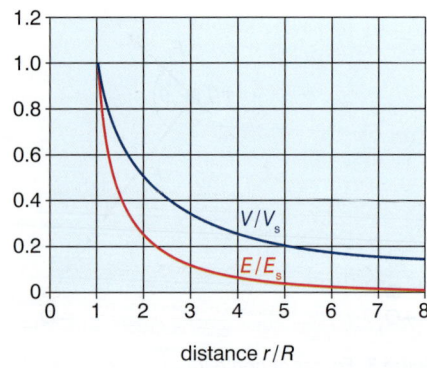

Figure 6 *The electric field strength and potential near a charged sphere*

Notes

1 Gravitational potential is always negative as the force is always attractive, whereas electric potential in the electric field near a point charge Q can be positive or negative depending on whether Q is a positive or a negative charge.

2 The potential energy E_P of two point charges Q_1 and Q_2 at distance d apart is given by

$$E_P = QV = \frac{Q_1 Q_2}{4\pi\varepsilon_0 d}$$

For a sphere of radius R with a charge Q,

- Outside the sphere at distance r from the centre of the sphere,

  the electric field strength $E = \dfrac{Q}{4\pi\varepsilon_0 r^2}$ and the electric potential $V = \dfrac{Q}{4\pi\varepsilon_0 r}$.

- At the surface of the sphere,

  the electric field strength $E_s = \dfrac{Q}{4\pi\varepsilon_0 R^2}$ and the electric potential $V_s = \dfrac{Q}{4\pi\varepsilon_0 R}$.

- If the sphere is hollow, there is no electric field inside it. Therefore the electric field strength is zero and the electric potential is V_s.

Figure 6 shows how the electric field strength and potential vary with distance from the centre of the sphere.

Outside the sphere:

- the field strength curve is an 'inverse-square law' curve as E is proportional to $\dfrac{1}{r^2}$.
- the potential curve is an 'inverse' curve as V is proportional to $\dfrac{1}{r}$.

Mathematical note

The electric field strength is equal to the negative of the potential gradient.

- The potential V at distance r from a point charge Q is $\dfrac{Q}{4\pi\varepsilon_0 r}$.
- The potential at distance $r + \Delta r$ is $\dfrac{Q}{4\pi\varepsilon_0 (r + \Delta r)}$.

Therefore from r to $r + \Delta r$, the change of potential $\Delta V = \dfrac{Q}{4\pi\varepsilon_0 (r + \Delta r)} - \dfrac{Q}{4\pi\varepsilon_0 r} = \dfrac{-Q\Delta r}{4\pi\varepsilon_0 r(r + \Delta r)}$.

For $\Delta r \ll r$, Δr can be neglected in the denominator so $\Delta V = \dfrac{-Q\Delta r}{4\pi\varepsilon_0 r^2} = -E\Delta r$.

Hence the electric field strength $E = -\dfrac{\Delta V}{\Delta r}$, which is the negative of the potential gradient.

The equation applies along a field line to small changes of distance in a non-uniform field and to any change of distance in a uniform field.

Summary test 14.4

$\dfrac{1}{4\pi\varepsilon_0} = 9.0 \times 10^9\,\mathrm{m\,F^{-1}}$

1 a Calculate the electric field strength at a distance of 3.2 mm from a +6.0 nC point charge.

  b Calculate the distance from the point charge in **a** at which the electric field strength is $5.4 \times 10^5\,\mathrm{V\,m^{-1}}$.

2 A $+25\,\mu\mathrm{C}$ point charge Q_1 is at a distance of 60 mm from a $+100\,\mu\mathrm{C}$ charge Q_2.

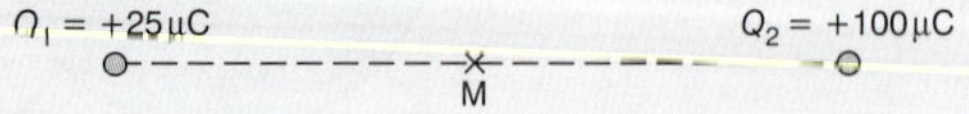

$Q_1 = +25\,\mu\mathrm{C}$ ⊶ - - - - - - - ✕ - - - - - - ⊶ $Q_2 = +100\,\mu\mathrm{C}$
M

Figure 7

  a A +15 pC charge q is placed at M, 25 mm from Q_1 and 35 mm from Q_2. Calculate:
  i the resultant electric field strength at M,
  ii the magnitude and direction of the force on q.

  b Show that the electric field strength due to Q_1 and Q_2 is zero at the point which is 20 mm from Q_1 and 40 mm from Q_2.

3 A $+15\,\mu\mathrm{C}$ point charge Q_1 is at a distance of 20 mm from a $+10\,\mu\mathrm{C}$ charge Q_2.

  a Calculate the resultant electric field strength:
  i at M, the midpoint between the two charges,
  ii at the point P along the line between Q_1 and Q_2 which is 25 mm from Q_1 and 45 mm from Q_2,
  iii at a point S which is 20 mm from Q_1 and 20 mm from Q_2.

  b i Explain why there is a point along the line between the two charges at which the electric field strength is zero.
  ii Calculate the distance from this point to Q_1 and to Q_2.

4 A $+15\,\mu\mathrm{C}$ point charge Q_1 is at a distance of 30 mm from a $-30\,\mu\mathrm{C}$ charge Q_2.

  a Calculate the electric potential at the midpoint of the two charges.

  b i Show that the electric potential is zero at a point between the two charges which is 10 mm from Q_1 and 20 mm from Q_2.
  ii Calculate the electric field strength at this position and state its direction.

Comparison between electric and gravitational fields

The similarities and differences between the two types of fields are listed in the table below. In the mid-nineteenth century, James Maxwell showed that electric and magnetic forces are different manifestations of the electromagnetic force. Towards the end of the twentieth century, physicists proved that the electromagnetic force and the nuclear force responsible for radioactive decay are different manifestations of a more fundamental force, the electroweak force. At the present time, the force of gravity remains outside this theoretical framework, despite repeated attempts to establish a unified framework. The fundamental nature of the force of gravity remains mysterious even though we use it in everyday situations more than any other force.

Table 1

	Gravitational force	Electric force
Similarities		
Line of force or a field line	Path of a free 'test' mass in the field	Path of a free 'test' charge in the field
Inverse-square law of force	Newton's law of gravitation $F = \dfrac{Gm_1 m_2}{r^2}$	Coulomb's law of force $F = \dfrac{1}{4\pi\varepsilon_0}\dfrac{Q_1 Q_2}{r^2}$
Field strength	Force per unit mass, $g = F/m$	Force per unit $+$ charge, $E = F/q$
Unit of field strength	$N\,kg^{-1}$ or $m\,s^{-2}$	$N\,C^{-1}$ or $V\,m^{-1}$
Potential	Gravitational potential energy per unit mass	Electric potential energy per unit $+$ charge
Unit of potential	$J\,kg^{-1}$	$V\ (= J\,C^{-1})$
Potential energy between two point masses or charges	$E_P = -\dfrac{Gm_1 m_2}{r}$	$E_P = -\dfrac{Q_1 Q_2}{4\pi\varepsilon_0 r}$
Uniform fields	g is the same everywhere, field lines parallel	E is the same everywhere, field lines are parallel
Radial fields	Due to a point mass or a uniform spherical mass M, $g = \dfrac{GM}{r^2}$ $\varphi = -\dfrac{Gm}{r}$	Due to a point charge or a charged metal sphere of charge Q, $E = \dfrac{Q}{4\pi\varepsilon_0 r^2}$ $V = \dfrac{Q}{4\pi\varepsilon_0 r}$
Differences		
Action at a distance	Between any two masses	Between any two charged objects
Force	Attracts only	Unlike charges attract, like charges repel
Constant of proportionality in force law	G	$\dfrac{1}{4\pi\varepsilon_0}$

Learning outcomes

On these pages you will learn to:

- recognise the analogy between certain qualitative and quantitative aspects of electric fields and gravitational fields

Chapter Summary

A **line of force** or a **field line** of an electric field is a line which a free positive 'test' charge would follow in the field.

The electric field strength, E, at a point in the field is defined as the force per unit charge on a positive test charge placed at that point.

Electric potential, V, at a point is defined as the work done per unit charge to move a positive test charge from infinity to that point.

A uniform electric field is one where the electric field strength has the same magnitude and direction everywhere between the plates.

Electric field strength $E = \dfrac{F}{Q}$

Coulomb's law of force
$$F = \frac{1}{4\pi\varepsilon_0}\frac{Q_1 Q_2}{r^2}$$

Uniform electric field $E = \dfrac{\Delta V}{\Delta d}$

where V is the p.d. between two oppositely charged parallel plates at separation d.

Radial electric field
$$E = \frac{Q}{4\pi\varepsilon_0 r^2} \qquad V = \frac{Q}{4\pi\varepsilon_0 r}$$

where r is the distance from a point charge Q or from the centre of a metal sphere carrying a charge Q.

Electric potential
$$V = \frac{E_p}{Q}$$

Potential energy between two point charges
$$E_P = -\frac{Q_1 Q_2}{4\pi\varepsilon_0 r}$$

14   Exam-style and Practice Questions

📖 Launch additional digital resources for the chapter

$\frac{1}{4\pi\varepsilon_0} = 9.0 \times 10^9 \, \text{m F}^{-1}$, $e = 1.6 \times 10^{-19} \, \text{C}$,

$g = 9.81 \, \text{N kg}^{-1}$, $G = 6.67 \times 10^{-11} \, \text{N m}^2 \, \text{kg}^{-2}$

1 **a** Explain what is meant by a *uniform electric field*.

   b A beam of electrons is directed horizontally into a uniform electric field which acts vertically downwards, as shown in Figure 1.1 below.

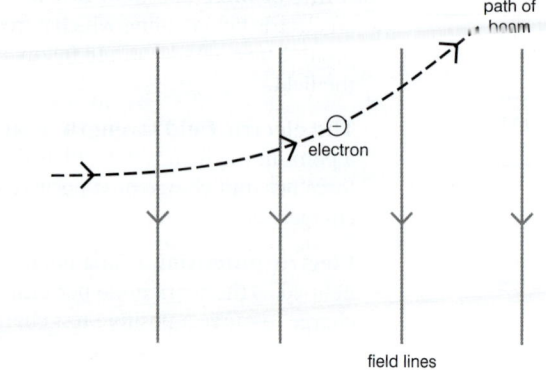

   path of beam

   electron

   field lines

   Figure 1.1

   i Explain why the beam follows a path which curves upwards.

   ii Each electron in the beam experiences a constant force of $1.8 \times 10^{-13} \, \text{N}$. Calculate the strength of the electric field.

2 A small metal sphere on an insulating rod is placed on a top pan balance, as shown in Figure 2.1 below. An identical metal sphere carrying a charge of $2Q$ on an insulating rod is brought into contact with the sphere on the balance so that each sphere acquires the same charge.

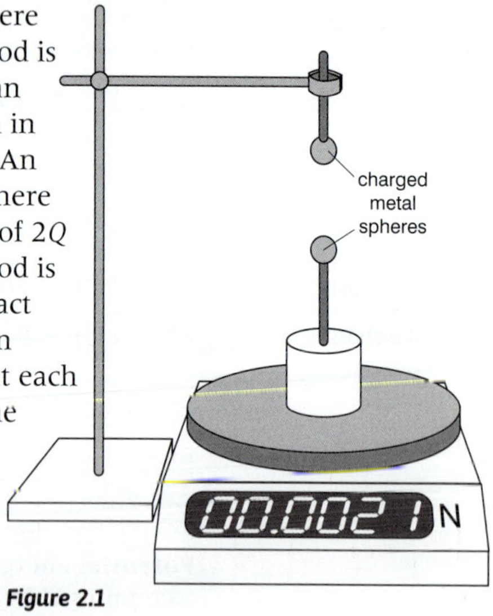

   a When the two spheres are at a separation of 68 mm, the top pan balance reading increases by $2.1 \times 10^{-3} \, \text{N}$. Calculate the charge, Q, on each sphere.

   charged metal spheres

   00.0021 N

   Figure 2.1

   b Explain why the electric field strength is zero at the midpoint between the two charges.

3 A +36 nC point charge X is at a distance of 75 mm from a −9 nC point charge Y.

a i Sketch the pattern of the electric field lines between the point charges.

   ii Calculate the electric field strength at the midpoint between X and Y.

   b At a certain position along the line XY, the resultant electric field strength is zero.

   i Explain why this position is more than 75 mm further from X than it is from Y.

   ii Calculate the distance from this position to X and to Y.

4 **a** Explain why an air molecule can be ionised by an electric field that is sufficiently strong.

   b The spherical dome of a Van de Graaff machine has a diameter of 0.30 m. Sparks are produced from the dome if the electric field strength at the surface reaches $40 \, \text{kV m}^{-1}$. Calculate:

   i the maximum charge that can be stored on the dome,

   ii the electric field strength at a distance of 1.0 m from the surface of the dome when the surface electric field strength is $40 \, \text{kV m}^{-1}$.

5 Two identical small conducting spheres are supported at either end of an insulating fibre which is attached at its midpoint to a fixed support. When the two spheres are in contact with each other, they are charged by contact with a charged rod. The two spheres repel each other as shown in Figure 5.1 opposite.

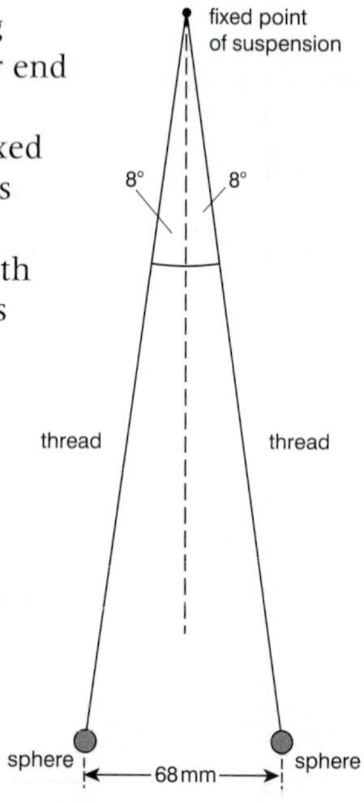

   a In equilibrium, each section of the thread makes an angle of 8° to the vertical. Show that the electrostatic force of repulsion on each sphere $F = mg \tan 8°$, where m is the mass of each sphere.

   b The mass of each sphere is $3.4 \times 10^{-3} \, \text{kg}$. Their separation at equilibrium is 68 mm. Calculate the charge on each sphere.

   fixed point of suspension

   8°   8°

   thread   **thread**

   sphere ⟵ 68 mm ⟶ **sphere**

   Figure 5.1

6 **a** State two similarities and two differences between an electric field and a gravitational field.

   b Show that the electrostatic force of repulsion between two protons is $1.2 \times 10^{36} \times$ the gravitational force between them at the same distance.

   Mass of a proton $= 1.67 \times 10^{-27} \, \text{kg}$

7 a Define *electric field strength*.

b Figure 7.1 shows how the electric potential near a positively charged metal sphere of radius 0.100 m varies with distance from the surface of the sphere when the sphere is at a potential of 32.0 kV.

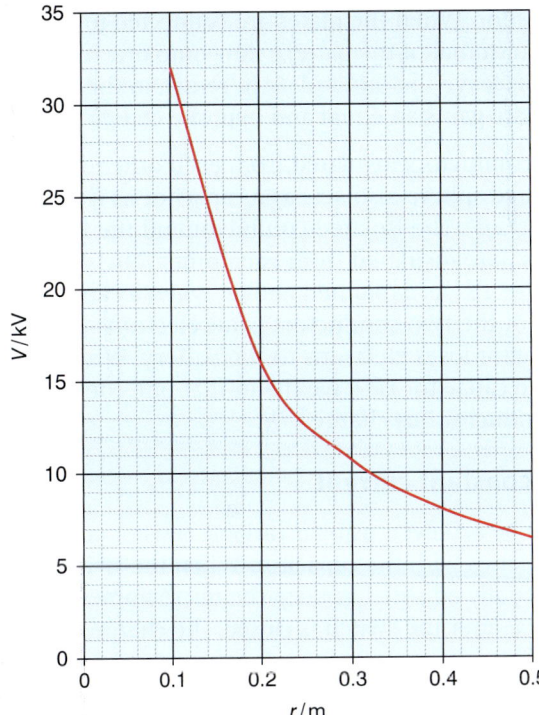

Figure 7.1

i Calculate the charge on the sphere and hence determine the charge per unit area on the sphere.

ii Calculate the electric field strength at a distance of 0.100 m from the surface of the sphere.

c An insulated uncharged metal rod XY of length 0.100 m is placed as shown in Figure 7.2, with X at a distance of 0.100 m from the surface of the sphere.

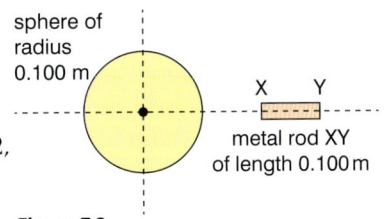

Figure 7.2

i Explain why there is a force of attraction between the rod and the sphere even though the rod is uncharged.

ii Sketch a graph on a copy of Figure 7.1 to show how the electric potential varies with position along the radial line from the sphere through XY to beyond Y.

iii Describe how the magnitude of the electric field strength along a radial line from the sphere through XY to beyond Y compares with the field strength when the rod was absent.

8 a Define the *electric field strength* of an electric field.

b Figure 8.1 shows electric field lines between a negatively charged metal plate A at a fixed distance from a larger positively charged metal plate B.

i Mark the direction of the field lines.

ii Mark a region where the electric field is uniform. Label this region U.

iii Describe how the electric field strength changes between A and B along the top curved field line.

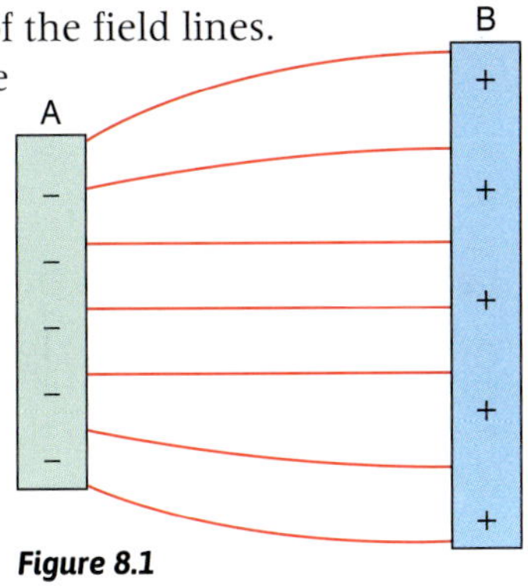

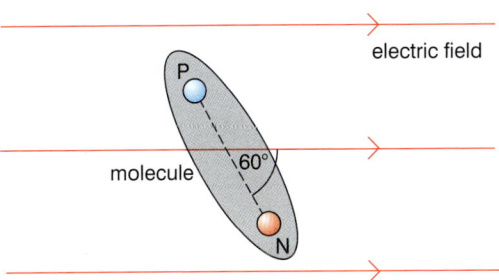

Figure 8.1

c A polar molecule is a molecule in which the centre of its negative charge (N) is not at the same position as the centre of its positive charge (P) and P and N are at a fixed distance apart, as in Figure 8.2.

Figure 8.2

In a particular polar molecule, the magnitude of the charge at P and N is 1.6×10^{-19} C and their distance apart, PN, is 3.5×10^{-9} m. The molecule is subjected to a uniform electric field of strength 50 kV m^{-1} such that the line PN is at an angle of 60° to the electric field, as shown in Figure 8.2.

i Draw an arrow at P and an arrow at N to show the direction of the electric force on P and on N.

ii Calculate the torque on the molecule due to these electric forces.

9 a Define *electric potential*.

b A hydrogen atom consists of a single proton and a single electron. In a simple model of the atom, the electron moves in a circular orbit around the proton. When the atom is in its lowest energy state, the electron is at a mean distance of 0.053 nm from the proton.

i Calculate the force of attraction between the proton and the electron in this state.

ii Calculate the potential energy of the atom in this state.

iii Assuming the proton is at rest and the electron moves at constant speed v on a circular orbit of radius r, write down an equation relating the electrostatic force on the electron to its centripetal acceleration. Hence show that the kinetic energy of the electron is equal to 0.5 times the magnitude of the potential energy of the atom.

iv Calculate the total energy E_0 of the atom in electron volts in its lowest energy state.

15.1 Capacitance

Learning outcomes

On these pages you will learn to:

- define capacitance and the farad, as applied to both isolated conductors and to parallel plate capacitors
- state and explain how the p.d. across a capacitor changes when the charging current is constant
- recall and use $C = Q/V$

A capacitor is a device designed to store charge. Two parallel metal plates placed near each other form a capacitor. When the plates are connected to a battery, electrons from the negative terminal of the battery flow onto one of the plates. An equal number of electrons leave the other plate to return to the battery via its positive terminal. So each plate gains an equal and opposite charge.

A capacitor consists of two conductors insulated from each other. An isolated conductor such as an insulated sphere also acts as a capacitor as it can store charge because it is not 'earthed'. The isolated conductor and the Earth (which acts as the second conductor) are insulated from each other. The symbol for a capacitor is shown in Figure 1b. As explained above, when a capacitor is connected to a battery, one of the two conductors gains electrons from the battery and the other conductor loses electrons to the battery. When we say that the charge stored by the capacitor is Q, we mean that one conductor has charge $+Q$ and the other conductor has charge $-Q$.

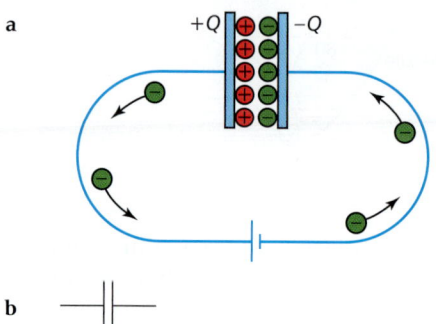

Figure 1 a Storing charge
b Capacitor symbol

Charging a capacitor at constant current

Figure 2 shows how this can be achieved using a variable resistor, a switch, a microammeter and a cell in series with the capacitor. When the switch is closed, the variable resistor is continually adjusted to keep the microammeter reading constant. At any given time, t, after the switch is closed, the charge Q on the capacitor can be calculated using the equation $Q = It$, where I is the current.

A high-resistance voltmeter connected in parallel with the capacitor enables the capacitor p.d. to be measured. To investigate how the capacitor p.d. changes with time for constant current, use the variable resistor to keep the current constant and either:

- use a stopwatch and measure the voltmeter reading at measured times, or
- use a datalogger as shown in Figure 2b.

Typical readings for a current of $15\,\mu A$ are shown in the following table. The charge Q has been calculated using $Q = It$.

Table 1 Current = 15 mA

Time, t/s	0	20	40	60	80	100
p.d., V/volts	0	0.29	0.62	0.90	1.22	1.50
charge, Q/μC	0	300	600	900	1200	1500

The graph of charge stored, Q, against p.d., V, for these measurements is shown in Figure 3. The measurements define a straight line passing through the origin. Therefore, the charge stored, Q, is proportional to the p.d., V. In other words, the charge stored per volt is constant.

> **The capacitance C of a capacitor is defined as the charge stored per unit p.d.**

The unit of capacitance is the farad (F), equal to 1 coulomb per volt. Note that $1\,\mu F = 10^{-6}\,F$.

For a capacitor which stores charge Q at p.d. V, its capacitance can be calculated using the equation

$$C = \frac{Q}{V}$$

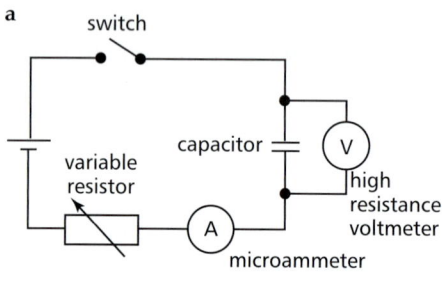

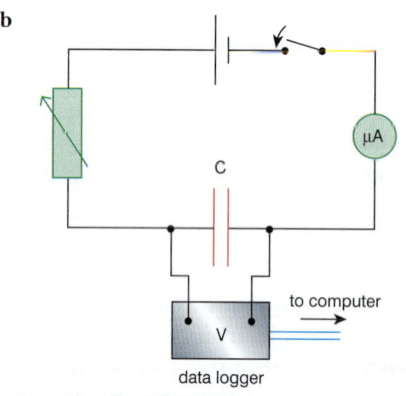

Figure 2 Investigating capacitors
a Circuit diagram   b Using a data logger

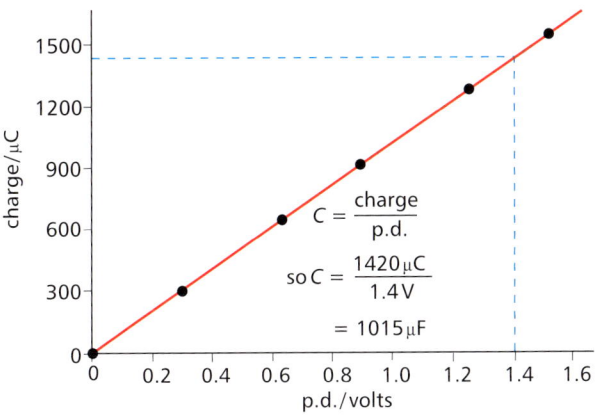

Figure 3 *Graph of results*

Graph annotations:
$$C = \frac{charge}{p.d.}$$
$$so\ C = \frac{1420\,\mu C}{1.4\,V}$$
$$= 1015\,\mu F$$

Worked example

A capacitor which is initially uncharged is charged at a constant current of $16\,\mu A$ for $40\,s$. At the end of this time, the capacitor p.d. increases to $5.8\,V$. Calculate:

a the charge stored by the capacitor,
b the capacitance of the capacitor.

Solution
a $Q = It = 16 \times 10^{-6} \times 40 = 6.4 \times 10^{-4}\,C$
b $C = \dfrac{Q}{V} = \dfrac{6.4 \times 10^{-4}}{5.8\,V} = 1.1 \times 10^{-4}\,F\ (-110\,\mu F)$

Summary test 15.1

1 Complete the following table

	a	b	c	d	e	f
charge/μC	60	330		6.30	52	
p.d./V	12		9.0	4.5		50
capacitance/μF		150	1100		4.7	68

2 A $22\,\mu F$ capacitor is charged by means of a constant current of $2.5\,\mu A$ to a p.d. of $12.0\,V$. Calculate:

a the charge stored on the capacitor at $12.0\,V$,
b the time taken.

3 A capacitor is charged by means of a constant current of $0.5\,\mu A$ to a p.d. of $5.0\,V$ in $55\,s$. Calculate:

a the charge stored,   **b** the capacitance of the capacitor.

4 A capacitor is charged by means of a constant current of $24\,\mu A$ to a p.d. of $4.2\,V$ in $38\,s$. The capacitor is then charged from $4.2\,V$ by means of a current of $14\,\mu A$ in $50\,s$. Calculate:

a the charge stored at a p.d. of $4.2\,V$,
b the capacitance of the capacitor,
c the extra charge stored at a current of $14\,\mu A$,
d the p.d. after the extra charge was stored.

Rearranging $C = \dfrac{Q}{V}$ gives $Q = CV$ or $V = \dfrac{Q}{C}$.

Extension

Practical capacitors
Most practical capacitors are designed using two strips of aluminium foil as plates separated by a thin layer of **dielectric** which is an insulating material that increases the charge that can be stored for a given p.d. Capacitances of less than about $10\,\mu F$ contain dielectrics such as mica or polyester.

Figure 4 shows the construction of such a capacitor in which the foil and the dielectric are rolled up into a convenient size. See Topic 15.2 for more about dielectrics.

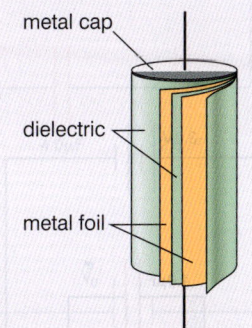

Figure 4 *A practical capacitor*

Electrolytic capacitors contain paper soaked with an electrolyte such as aluminium borate between the metal strips. When used in a d.c. circuit, a thin oxide layer which is a dielectric forms on the positive plate. As will be explained in Topic 15.3, because the dielectric layer is very thin, the capacitance can be much greater than that of non-electrolytic capacitors. Electrolytic capacitors must be connected with the correct polarity otherwise the dielectric is damaged.

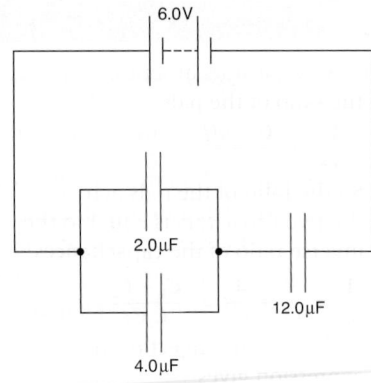

Figure 6

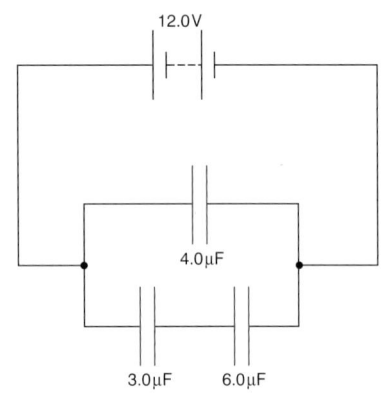

Figure 7

Worked example

A 2.0 μF capacitor and a 4.0 μF capacitor are in parallel with each other. The parallel combination is in series with a 12.0 μF capacitor and a 6.0 V battery, as shown in Figure 6. Calculate:

a the total capacitance of the combination,
b the charge stored by the combination,
c the charge and p.d. for each capacitor.

Solution

a $C' = 2.0 + 4.0 = 6.0\,\mu F$

∴ the total capacitance C is given by $\dfrac{1}{C} = \dfrac{1}{6.0} + \dfrac{1}{12.0} = 0.25\,\mu F^{-1}$

∴ $C = \dfrac{1}{0.25} = 4.0\,\mu F$

b The charge stored $Q = CV = 4.0 \times 6.0 = 24\,\mu C$

c C_3: charge stored $= Q = 24\,\mu C$. ∴ p.d. across $C_3 = \dfrac{Q}{C_3} = \dfrac{24\,\mu C}{12\,\mu F} = 2.0\,V$

C_1 and C_2: $\dfrac{C_1}{C_2} = \dfrac{2\,\mu F}{4\,\mu F} = 0.5$, then $\dfrac{Q_1}{Q_2} = \dfrac{C_1}{C_2} = 0.5$

As $Q = Q_1 + Q_2 = 24\,\mu C$, then $Q_1 = 8\,\mu C$ and $Q_2 = 16\,\mu C$

∴ $V_1 = \dfrac{Q_1}{C_1} = \dfrac{8\,\mu C}{2\,\mu F} = 4\,V$

$V_2 = \dfrac{Q_2}{C_2} = \dfrac{16\,\mu C}{4\,\mu F} = 4\,V$

Note $V_1 = V_2 = 6.0\,V - V_3$

Worked example

A 3.0 μF capacitor and a 6.0 μF capacitor are in series with each other. The series combination is in parallel with a 4.0 μF capacitor and a 12.0 V battery, as shown in Figure 7. Calculate:

a **i** the total capacitance of the combination,
   ii the charge stored by the combination,
b the charge and p.d. for each capacitor.

Solution

a **i** $\dfrac{1}{C'} = \dfrac{1}{3.0} + \dfrac{1}{6.0} = 0.5\,\mu F^{-1}$, ∴ $C' = \dfrac{1}{0.5} = 2.0\,\mu F$

∴ the total capacitance $C = C' + C_3 = 2.0 + 4.0 = 6.0\,\mu F$

   ii The charge stored $Q = CV = 6.0 \times 12.0 = 72\,\mu C$

b C_3: p.d. $V_3 = $ battery p.d. $= 12\,V$

charge stored $Q_3 = C_3 V_3 = 4.0\,\mu F \times 12.0\,V = 48\,\mu C$.

C_1 and C_2: charge stored by the combination $= Q - Q_3 = 72 - 48 = 24\,\mu C$

They store the same amount of charge as they are in series with each other.

∴ $Q_1 = Q_2 = 24\,\mu C$

∴ $V_1 = \dfrac{Q_1}{C_1} = \dfrac{24\,\mu C}{3.0\,\mu F} = 8.0\,V$

$V_2 = \dfrac{Q_2}{C_2} = \dfrac{24\,\mu C}{6.0\,\mu F} = 4.0\,V$

Note $V_1 + V_2 = V_3 = 12\,V$

Extension

More about dielectrics

When a dielectric is placed between oppositely charged metal plates, the electrons in each dielectric molecule are pulled a little towards the positive plate. Each molecule of the dielectric becomes **polarised**, as shown in Figure 8. The surface of the dielectric facing the positive plate gains a layer of negative charge due to polarisation and the surface facing the negative plate gains positive charge. The p.d. between the plates is equal to the battery p.d. and is constant. As a result:

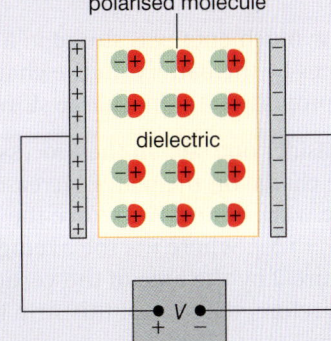

- the positive charge on the dielectric surface facing the negative plate attracts electrons from the battery onto the negative plate,
- the negative charge on the dielectric surface facing the positive plate pushes electrons back to the battery from the negative plate.

So the dielectric increases the charge stored on the plates without changing the plate p.d. In other words, the dielectric increases the capacitance of the parallel plates.

Figure 8 Dielectrics

The relative permittivity ε_r or dielectric constant, may be defined as

$$\varepsilon_r = \frac{C}{C_0}$$

where C is the capacitance with a dielectric completely filling the space between the plates and C_0 is the capacitance with empty space between the plates.

Typical values for ε_r are 7 for mica, 2.3 for polythene, 2.7 for paper and 81 for water.

The capacitance C of a pair of parallel metal plates depends on:

- the perpendicular distance Δd between the plates,
- the surface area A of the plates,
- the presence of a dielectric between the plates.

For a constant p.d. ΔV between the plates, it can be shown using Coulomb's law, as outlined in Topic 14.2, that in the absence of a dielectric between the plates, the electric field strength $E = \frac{Q}{A\varepsilon_0}$, where Q is the magnitude of the charge on each plate Q.

Since $E = \frac{\Delta V}{\Delta d}$, rearranging $\frac{\Delta V}{\Delta d} = \frac{Q}{A\varepsilon_0}$ gives the following formula for the capacitance C_0 of the 'empty' parallel plates:

$$C_0 = \frac{A\varepsilon_0}{\Delta d}$$

If the space between the plates is completed filled with a dielectric of relative permittivity ε_r, the capacitance $C = \varepsilon_r C_0 = \frac{A\varepsilon_r\varepsilon_0}{\Delta d} = \frac{A\varepsilon}{\Delta d}$, where the permittivity of the dielectric $\varepsilon = \varepsilon_r\varepsilon_0$.

Note ε_0 may be determined by measuring the capacitance C_0 of 'empty' parallel plates of known area A and known spacing Δd. then using the equation for C_0 above to calculate ε_0.

1 A 3.0 V battery is connected to a 2.0 μF capacitor in parallel with a 3.0 μF capacitor. Sketch the circuit diagram and calculate:

   a the combined capacitance of the two capacitors,

   b the charge stored and the p.d. across each capacitor.

2 a A 4.5 V battery is connected to a 6.0 μF capacitor in series with a 4.0 μF capacitor. Calculate:

   i the combined capacitance of the two capacitors,

   ii the charge stored and the p.d. across each capacitor.

   b A 10 μF capacitor is connected in parallel with the 6.0 μF capacitor.

   i Sketch the new circuit diagram and calculate the total capacitance of the three capacitors.

   ii Calculate the charge and p.d. across each capacitor.

3 A student is given three capacitors of values 10 μF, 22 μF and 47 μF. Sketch the eight different possible combinations using all three capacitors and calculate the capacitance of each combination.

4 A 4.0 μF capacitor in series with a 10.0 μF capacitor are connected to a 6.0 V battery. A 2.0 μF capacitor is then connected to the battery in parallel with the two capacitors in series.

   a Sketch the circuit diagram for this arrangement and calculate its total capacitance.

   b Calculate the charge and p.d. for each capacitor in the arrangement.

Learning outcomes

On these pages you will learn to:

- derive from the area under a potential–charge graph the equations $E = \frac{1}{2}QV = \frac{1}{2}CV^2$ for the energy stored in a capacitor
- solve problems using the above equations

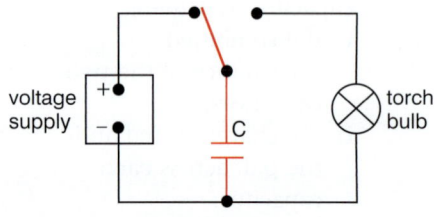

Figure 1 *Releasing stored energy*

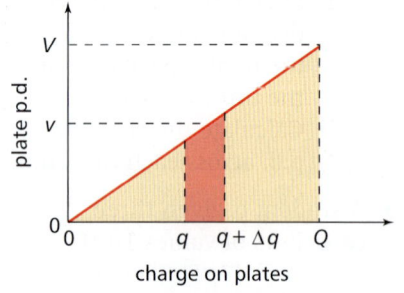

Figure 2 *Energy stored in a capacitor*

When a capacitor is charged, energy is stored in it. A charged capacitor discharged across a torch bulb will release its energy in a brief flash of light from the bulb, as long as the capacitor has been charged initially to the operating p.d. of the bulb. Charge flow is rapid enough to give a large enough current to light the bulb, but only for a brief time. The bulb could be replaced by a miniature electric motor which would spin briefly when the capacitor is discharged through it.

How much energy is stored in a charged capacitor? The charge is forced onto the plates by the battery. In the charging process, the p.d. across the plates increases in proportion to the charge stored, as shown in Figure 2.

Consider one step in the process of charging a capacitor of capacitance C when the charge on the plates increases by a small amount Δq from q to $q + \Delta q$. The work done ΔW to force the extra charge Δq on to the plates is given by $\Delta W = v \Delta q$, where v is the average p.d. during this step. $v \Delta q$ is represented in Figure 2 by the area of the vertical strip of width Δq and height v under the line. Therefore, the area of this strip represents the work done ΔW in this small step.

Now consider all the small steps from zero p.d. to the final p.d. V. The total work done W is obtained by adding up the work done for each small step. In other words, W is represented by the total area under the line from zero p.d. to p.d. V. As this area is a triangle of height V and base length Q (= CV), the total work done W = triangle area = $\frac{1}{2} \times$ height $\times$ base = $\frac{1}{2}VQ$

Because the energy stored in the capacitor, or **capacitor energy**, is equal to the work done on it to charge it, the energy stored = $\frac{1}{2}VQ$.

Energy stored by the capacitor, $W = \frac{1}{2}QV$

Note

Using $Q = CV$, the above equation may be written as $W = \frac{1}{2}CV^2 = \frac{1}{2}\frac{Q^2}{C}$

Measuring the energy stored in a charged capacitor

A joulemeter is used to measure the energy transfer from a charged capacitor to a light bulb when the capacitor discharges. The capacitor p.d., V, is measured and the joulemeter reading recorded before the discharge starts. When the capacitor has discharged, the joulemeter reading is recorded again. The difference of the two joulemeter readings is the energy transferred from the capacitor during the discharge process. This is the total energy stored in the capacitor before it discharged. This can be compared with the calculation of the energy stored using $W = \frac{1}{2}CV^2$.

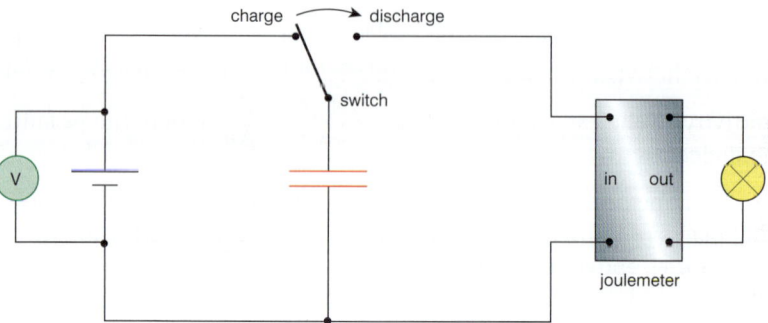

Figure 3 *Measuring energy stored*

The energy stored in a thundercloud

Imagine a thundercloud and the Earth below like a pair of charged parallel plates. Because the thundercloud is charged, a strong electric field exists between the thundercloud and the ground. The potential difference between the thundercloud and the ground, $V = Ed$, where E is the electric field strength and d is the height of the thundercloud above the ground.

- For a thundercloud carrying a constant charge Q, the energy stored $W = \frac{1}{2}QV = \frac{1}{2}QEd$.
- If the thundercloud is forced by winds to rise to a new height d', the energy stored $W' = \frac{1}{2}QEd'$.
- As the electric field strength is unchanged (since it depends on the charge per unit area), the increase in the energy stored
$= W' - W = \frac{1}{2}QEd' - \frac{1}{2}QEd = \frac{1}{2}QE\,\Delta d$, where $\Delta d = d' - d$.

This increase in the energy stored is because work is done by the force of the wind to overcome the electrical attraction between the thundercloud and the ground and make the charged thundercloud move away from the ground.

The insulating property of air breaks down if it is subjected to an electric field strength more than about $300\,\text{kV}\,\text{m}^{-1}$. Prove for yourself that, for every metre rise of the thundercloud carrying a maximum charge of $20\,\text{C}$, the energy stored would increase by $3\,\text{MJ}$. At a height of $500\,\text{m}$, the energy stored would be $1500\,\text{MJ}$.

Summary test 15.3

1 Calculate the charge and energy stored in a $10\,\mu\text{F}$ capacitor charged to a p.d. of:

 a 3.0 V,

 b 6.0 V.

2 A $50\,000\,\mu\text{F}$ capacitor is charged from a 9 V battery then discharged through a light bulb in a flash of light lasting 0.2 s. Calculate:

 a the charge and energy stored in the capacitor before discharge,

 b the average power supplied to the light bulb.

3 A $2.2\,\mu\text{F}$ capacitor is connected in series with a $10\,\mu\text{F}$ capacitor and a 3.0 V battery. Calculate the charge and energy stored in each capacitor.

4 In Figure 4, a $4.7\,\mu\text{F}$ capacitor is charged from a 12.0 V battery by connecting the switch to X. The switch is then reconnected to Y to charge a $2.2\,\mu\text{F}$ capacitor from the first capacitor.

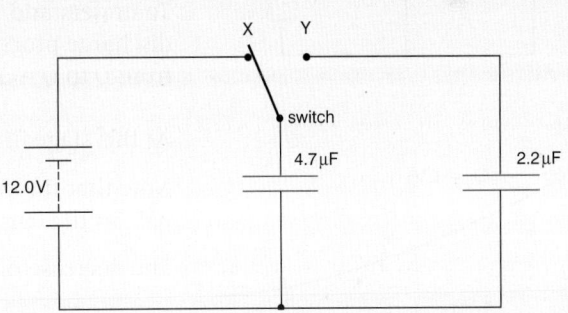

Figure 4

Calculate:

 a the initial charge and energy stored in the $4.7\,\mu\text{F}$ capacitor,

 b the combined capacitance of the two capacitors,

 c the final p.d. across the two capacitors,

 d the final energy stored in each capacitor. Account for the loss of energy stored.

Learning outcomes

On these pages you will learn to:

- draw a graph to show how the p.d. across a capacitor decreases as the capacitor discharges through a fixed resistor
- recall and use the equation $V = V_0 e^{-t/RC}$
- explain what is meant by the time constant of a circuit that has a capacitor in series with a resistor
- recall and use the time constant formula $\tau = RC$
- show an understanding of the function of capacitors in simple circuits

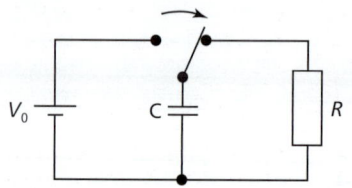

a

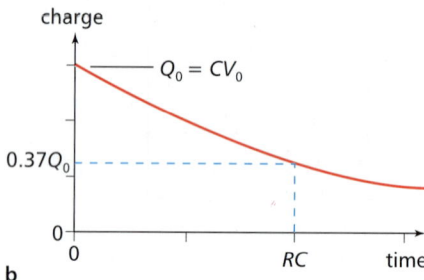

b

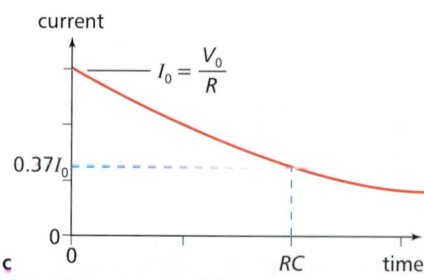

c

Figure 1 *Capacitor discharge*

Note

A graph of capacitor p.d. V against time t has the same shape as the graphs above. This is because V is proportional to the charge Q (because $Q = CV$) and it is also proportional to the current I (because $V = IR$)

When a capacitor discharges through a fixed resistor, the discharge current decreases gradually to zero. Figure 1 shows a circuit in which a capacitor is discharged through a resistor when the switch is changed over. The reason why the current decreases gradually is that the p.d. across the capacitor decreases as it loses charge. Because the resistor is connected directly to the capacitor, the resistor current (= p.d./resistance) decreases as the p.d. decreases.

The situation is not unlike water emptying through a pipe at the bottom of the container. When the container is full, the flow rate out of the pipe is high because the water pressure at the pipe is high. As the container empties, the water level falls so the water pressure at the pipe falls and the flow rate drops.

The graphs in Figure 1 show how the current and the charge decrease with time. Both curves have the same shape because both the current and charge (and p.d.) decrease **exponentially**. This means that any of these quantities decreases by the same factor in equal intervals of time. For example, for initial charge Q_0, if the charge is $0.9Q_0$ after a certain time t_1, the charge will be:

- $0.9 \times 0.9\, Q_0$ after time $2t_1$,
- $0.9 \times 0.9 \times 0.9\, Q_0$ after time $3t_1$ ...
- $0.9^n\, Q_0$ after time $n\, t_1$

About exponential decrease

To understand why the decrease is exponential, consider one small step in the discharge process of a capacitor C through a resistor R when the charge decreases from Q to $Q - \Delta Q$ in time Δt.

At this stage, the current $I = \dfrac{\text{p.d. across the plates, } V}{\text{resistance, } R} = \dfrac{Q}{CR}$ as $V = \dfrac{Q}{C}$

Note that the current is proportional to the charge which is proportional to the p.d. So the curves all have the same shape.

The decrease of charge $\Delta Q = -I\, \Delta t$ (− as Q decreases).

Therefore, $\Delta Q = -\dfrac{Q}{CR} \Delta t$, which gives

$$\frac{\Delta Q}{Q} = -\frac{\Delta t}{CR}$$

The equation tells us that the fractional drop of charge $\dfrac{\Delta Q}{Q}$ is the same in any short interval of time Δt during the discharge process. For example, suppose $\Delta t = 10\,\text{s}$ and $CR = 100$. Then $\dfrac{\Delta Q}{Q} = -0.1 \left(= -\dfrac{\Delta t}{CR}\right)$. So the charge decreases to 0.9 of its value at the start of the 10 s interval. So, if the initial charge is Q_0, the charge still on the plates will be:

- $0.9\, Q_0$ after 10 s,
- $0.9 \times 0.9\, Q_0$ after a further 10 s,
- $0.9 \times 0.9 \times 0.9\, Q_0$ after a further 10 s, etc.

In theory, the charge on the plates never becomes zero.

Exponential changes occur whenever the rate of change of a quantity is proportional to the quantity itself. Rearranging the equation $\dfrac{\Delta Q}{Q} = -\dfrac{\Delta t}{CR}$ gives

$$\frac{\Delta Q}{\Delta t} = -\frac{Q}{CR}$$

For very short intervals of time (i.e. $\Delta t \to 0$), $\dfrac{\Delta Q}{\Delta t}$ represents the rate of change of charge and is written $\dfrac{dQ}{dt}$.

Therefore,
$$\frac{dQ}{dt} = -\frac{Q}{CR}$$

The graphical solution to this equation is shown in Figure 1b.
The mathematical solution is

$Q = Q_0 e^{-t/RC}$, where Q_0 = initial charge, and
e = the exponential function (sometimes written 'exp')

The quantity RC is called the **time constant**, τ, for the circuit. At time τ after the start of the discharge, the charge falls to 0.37 (= e^{-1}) of its initial value.

Time constant = RC, where R = circuit resistance,
C = capacitance

The unit of RC is the second. This is because 1 ohm = $1 \frac{\text{volt}}{\text{ampere}}$ and 1 farad

= $1 \frac{\text{coulomb}}{\text{volt}}$ so the unit of $RC = \frac{\text{volt}}{\text{ampere}} \times \frac{\text{coulomb}}{\text{volt}} = \frac{\text{coulomb}}{\text{ampere}}$ = the second.

Notes

1 The current, the p.d. and the charge are all proportional to one another. All three quantities decrease exponentially in capacitor discharge in accordance with the equation $x = x_0 e^{\frac{-t}{CR}}$, where x represents either the current or the charge or the p.d.

2 The inverse function of e^x is $\ln x$, where ln is the natural logarithm. To calculate t, given x, x_0, R and C, use of the inverse function of e^x gives $\ln x = \ln x_0 - \left(\frac{t}{RC}\right)$.

3 The function $e^z = 1 + z + \frac{z^2}{2 \times 1} + \frac{z^3}{3 \times 2 \times 1} + \dots$, etc. It can be shown mathematically that the rate of change of this function with respect to z is the same function. This is why the function appears whenever the rate of change of a quantity is proportional to the quantity itself. Note that $z = 1$ gives $e = 1 + 1 + \frac{1}{2} + \frac{1}{6} + \frac{1}{24}$ etc. = 2.718. You can check this on your calculator by keying in 'e^x' then pressing 1 to give 2.718. Keying in −1 instead of 1 gives 0.37 for e^{-1}.

Worked example

A 47 µF capacitor charged to 12.0 V is then discharged through a 2.2 MΩ resistor.

a Calculate the time taken for the capacitor p.d. to decrease to below 3.0 V.
b i Calculate the percentage of the energy stored in the capacitor that remains when its p.d. is 3.0 V.
ii Discuss where the energy transferred from the capacitor is transferred to.

Solution

a $3.0 = 12.0 e^{-t/RC}$. Rearranging this gives $e^{-t/RC} = \frac{3.0}{12.0} = 0.25$. Taking

natural logarithms of both sides gives $\ln 0.25 = -\frac{t}{RC}$ so $\frac{t}{RC} = -\ln 0.25 = 1.386$.

Therefore $t = 1.386 \, RC = 1.386 \times 2.2 \, \text{M}\Omega \times 47 \, \mu\text{F} = 143 \, \text{s}$.

b i Energy stored initially = $\frac{1}{2}CV^2 = 0.5 \times 2.2 \, \mu\text{F} \times (12.0 \, \text{V})^2 = 158.4 \, \mu\text{J}$

Energy stored at 3.0 V = $0.5 \times 2.2 \, \mu\text{F} \times (3.0 \, \text{V})^2 = 9.9 \, \mu\text{J}$

% of initial energy stored = $\left(\frac{9.9 \, \mu\text{J}}{158.4 \, \mu\text{J}}\right) \times 100\% = 6.25\%$

ii Energy is transferred to the wires by the resistance heating effect of the current in the wires. The energy is then dissipated to the surroundings.

Worked example

A 2200 µF capacitor is charged to a p.d. of 9.0 V then discharged through a 100 kΩ resistor using a circuit as shown in Figure 1a.

a Calculate:
i the initial charge stored by the capacitor,
ii the time constant of the circuit.
b Calculate the p.d. after:
i a time equal to the time constant, ii 300 s.

Solution

a i $Q_0 = CV_0$
= $2200 \times 10^{-6} \, \text{F} \times 9.0 \, \text{V}$
= $2.0 \times 10^{-2} \, \text{C}$,
ii Time constant = RC
= $100 \times 10^3 \, \Omega \times 2.2 \times 10^{-3} \, \text{F}$
= 220 s

b i When $t = RC$, $V = V_0 e^{-1}$
= $0.37 \times 9.0 = 3.33 \, \text{V}$
ii When $t = 300 \, \text{s}$,
$\frac{t}{RC} = \frac{300}{220} = 1.36$
∴ $V = V_0 e^{-t/RC} = 9.0 e^{-1.36}$
= 2.3 V

See Chapter 24 Mathematical skills for more about exponential decrease.

Note

A quicker and easier method is to recognise that capacitor energy is proportional to the square of the capacitor p.d. As the capacitor p.d. decreases to $\frac{1}{4}$ of the initial p.d., the energy stored decreases to $\frac{1}{16}$ of the initial energy stored (i.e. 6.25%).

Investigating capacitor discharge

Figure 2 shows how to measure the p.d. across a capacitor as it discharges through a fixed resistor. An oscilloscope is used as it has a very high resistance so the discharge current from the capacitor passes only through the fixed resistor.

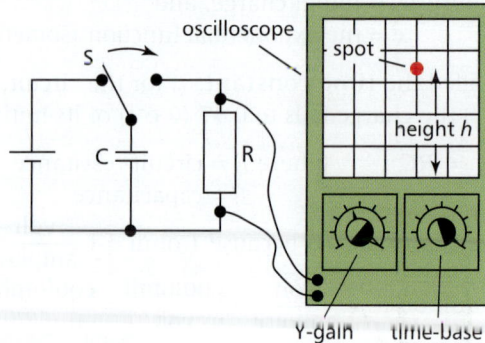

Figure 2 *Measuring capacitor discharge*

The oscilloscope is used to measure the capacitor p.d. at regular intervals. A data logger or a digital voltmeter could be used instead of the oscilloscope.

The measurements may be used to plot a graph of voltage against time. The time taken for the voltage to decrease to 37% (= 1/e) of the initial value can be measured from the graph and compared with the calculated value of RC.

The significance of the time constant

The time constant RC is the time taken, in seconds, for the capacitor to discharge to 37% of its initial charge. Given values of R and C, the time constant can be quickly calculated and used as an approximate measure of how quickly the capacitor discharges. In the worked example in the margin of the previous page, the time constant of 220 s gives a 'rule of thumb' estimate of the time taken to discharge significantly but not completely. Also, $5RC$ gives a 'rule of thumb' estimate for the time taken to discharge by over 99%. Prove for yourself that $t = 5RC$ gives a value which is less than 1% of the initial value.

Applications of capacitor discharge

1 **Any electronic timing circuit or time-delay circuit** makes use of capacitor discharge through a fixed resistor. Figure 3 shows an alarm circuit where the alarm rings if the input voltage to the electronic circuit drops below a certain value after the switch is reset. The time delay between resetting the switch and the alarm ringing can be increased by increasing the resistance R or the capacitance C. Such a change to the circuit would make the discharge of C through R slower so increasing the time for the capacitor voltage to decrease sufficiently to make the alarm ring.

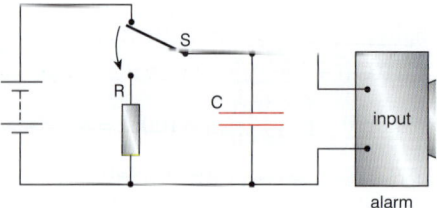

Figure 3 *A time-delayed alarm circuit*

2 **Capacitor smoothing** is used in applications where sudden voltage variations or 'glitches' can have undesirable effects. For example, mains appliances being switched on or off in a building could affect computers connected to the mains supply in the building. A large capacitor in a computer supplies current if the mains supply is interrupted so the computer circuits continue to function normally.

Summary test 15.4

1 A $50\,\mu F$ capacitor is charged by connecting it to a 6.0 V battery then discharged through a $100\,k\Omega$ resistor.

  a Calculate:

    i the charge stored in the capacitor immediately after it has been charged,

    ii the time constant of the circuit.

  b i Estimate how long the capacitor would take to discharge to about 2 V.

    ii Estimate the resistance of the resistor that you would use in place of the $100\,k\Omega$ resistor if the discharge is to be 99% completed within about 5 s.

2 A $68\,\mu F$ capacitor is charged to a p.d. of 9.0 V then discharged through a $20\,k\Omega$ resistor.

  a Calculate:

    i the charge stored by the capacitor at a p.d. of 9.0 V,

    ii the initial discharge current.

  b Calculate the p.d. and the discharge current 5.0 s after the discharge started.

3 A $2.2\,\mu F$ capacitor is charged to a p.d. of 6.0 V and then discharged through a $100\,k\Omega$ resistor. Calculate:

  a the charge and energy stored in this capacitor at 6.0 V,

  b the p.d. across the capacitor 0.5 s after the discharge started,

  c the energy stored at this time.

4 A $4.7\,\mu F$ capacitor is charged to a p.d. of 12.0 V and then discharged through a $220\,k\Omega$ resistor. Calculate:

  a the energy stored in this capacitor at 12.0 V,

  b the time taken for the p.d. to fall from 12.0 V to 3.0 V,

  c the energy lost by the capacitor in this time.

Chapter Summary

The **capacitance** of a capacitor is defined as the charge stored per unit p.d.

The **unit of capacitance** is the farad (F), equal to 1 coulomb per volt. Note that $1\,\mu F = 10^{-6}$ F.

Capacitor equation $C = \dfrac{Q}{V}$

Capacitor combination rules

1 Capacitors in parallel: combined capacitance
$C = C_1 + C_2 + C_3 + \dots$

2 Capacitors in series: combined capacitance
$\dfrac{1}{C} = \dfrac{1}{C_1} + \dfrac{1}{C_2} + \dfrac{1}{C_3} + \dots$

Energy stored by the capacitor,
$W = \tfrac{1}{2}QV = \tfrac{1}{2}CV^2$

Capacitor discharge

1 **Time constant**, $\tau = RC$

2 **Exponential decrease equation** for current or charge or p.d.: $x = x_0\,e^{-t/RC}$

$$\boxed{\text{📖 Launch additional digital resources for the chapter}}$$

1 a Define the capacitance of a capacitor.

  b i With the aid of a circuit diagram, describe how a capacitor may be charged using a battery and a variable resistor to keep the charging current constant.

    ii Figure 1.1 below shows how the p.d. changes with time when a capacitor is charged at a constant current of 16 μA. Calculate the charge stored after 20 s and calculate the capacitance of the capacitor.

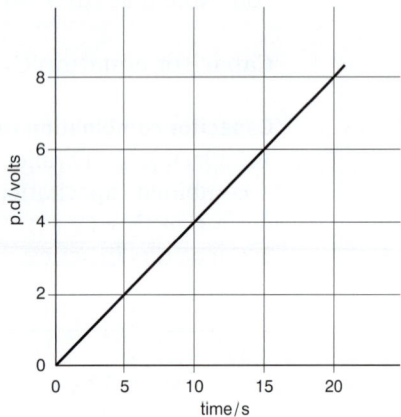

Figure 1.1

2 a A 2.0 μF capacitor is connected in series with a 4.0 μF capacitor and a 6.0 V battery.

    i Sketch the circuit diagram and calculate the combined capacitance of the two capacitors.

    ii Calculate the charge, p.d. and energy stored for the 2.0 μF capacitor.

  b A capacitor C is connected in parallel with the 4.0 μF capacitor in the circuit in **a**. The p.d. across the 2.0 μF capacitor changes to 5.0 V. Sketch the new circuit diagram and calculate the capacitance of C.

3 A 4.0 μF capacitor is charged by connecting it to a 9.0 V battery, as shown below. The capacitor is then disconnected from the battery and connected to an uncharged 6.0 μF capacitor.

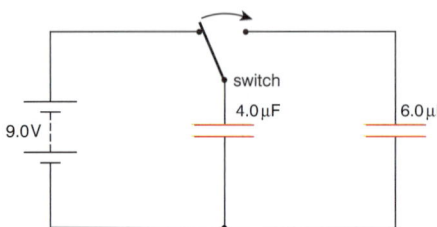

Figure 3.1

  a Calculate the charge and energy stored in the 4.0 μF capacitor when the p.d. across its terminals is 9.0 V.

b The p.d. across the two capacitors becomes the same when they are connected together.

  i Calculate the combined capacitance of the two capacitors.

  ii Show that the p.d. across the two capacitors is 3.6 V after they have been connected together.

  iii Calculate the total energy stored by the two capacitors after they have been connected together and explain why it differs from the initial energy stored.

4 The flashlight circuit of a camera includes two 47 000 μF capacitors in parallel with each other. The capacitors are charged using a 4.5 V battery. A flash of light is emitted from the flashbulb when the charged capacitors are connected across the flashbulb.

  a i Calculate the energy stored in the capacitor when the p.d. across its terminals is 4.5 V.

    ii The duration of the discharge is approximately 50 ms. Estimate the power of the flashlight, stating any assumptions made in your estimate.

  b Explain why less energy would be stored in the capacitors if they were in series with each other.

5 A capacitor C of capacitance 470 μF is connected to a 12 V battery then discharged through a 100 kΩ resistor R.

  a Calculate the charge and energy stored in C when the p.d. across its terminals is 12 V.

  b i Calculate the time constant of the discharge circuit.

    ii Show that the p.d. across the capacitor decreases to 3.3 V in 60 s.

    iii Calculate the energy transferred from the capacitor to the resistor when the p.d. decreases from 12 V to 3.3 V.

6 A capacitor circuit consists of 4 identical 10 mF capacitors connected in parallel to a 12 V voltage supply unit.

  a Calculate the total charge and energy stored.

  b The circuit is used as a 'back-up' supply in a data capture device in case the p.d. from the voltage supply unit is interrupted. The device switches off if the voltage supplied to it falls below 10 V. In normal operation, the voltage supply unit supplies a current of 80 mA at 12 V.

  i Show that the circuit resistance in normal operation is approximately 150 Ω.

  ii If the voltage supply unit is switched off, estimate how much longer the device would continue to operate for.

7 An insulated metal sphere X of radius R has a positive charge Q at potential V.

    a Show that the capacitance C of the sphere is equal to $4\pi\varepsilon_0 R$.

    b An insulated uncharged metal sphere Y of radius $0.20R$ at a fixed distance from X is charged from X using an insulated conductor, as shown in Figure 7.1. Y is then moved far away from X.

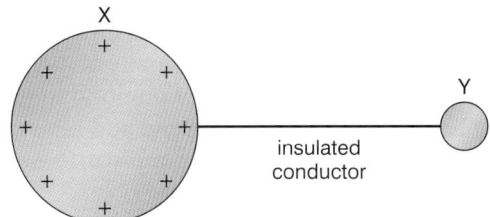

Figure 7.1

      i In terms of electrons, describe how Y becomes charged when the conductor is connected between X and Y.

      ii Explain why the potential of Y increases and that of X decreases until X and Y are at the same potential V_f.

      iii Show that $V_f = 0.83V$.

    c The energy stored by the charged sphere can be calculated using a capacitor energy equation.

      i Use this equation to show that the energy dissipated in the conductor when Y is charged from X is one-sixth of the initial energy stored in X.

      ii State the reason why energy is dissipated when Y becomes charged from X.

8 In an experiment to measure the capacitance C of a capacitor, the circuit in Figure 8.1 was used to charge the capacitor and then discharge it through a resistor of known resistance R.

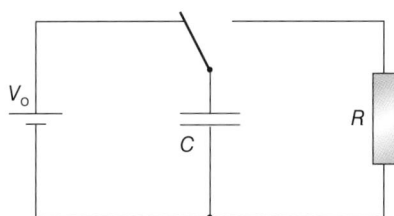

Figure 8.1

    a The capacitor p.d., V, at time t after the discharge commenced is given by $V = V_0 e^{-t/CR}$.

    Show that this equation can be rearranged into an equation of the form $\ln V = a - bt$, where a and b are constants, and determine expressions for a and b.

    b As the capacitor discharged, its p.d. was measured every 30 seconds using a digital voltmeter. The measurements were repeated twice as shown in Table 1.

Table 1

t / s	0	30	60	90	120	150	180	210	240	270	300
V / V	4.50	3.82	3.26	2.78	2.33	2.00	1.70	1.43	1.23	1.04	0.89
	4.51	3.81	3.25	2.77	2.35	2.10	1.72	1.43	1.25	1.02	0.90
	4.50	3.83	3.25	2.76	2.34	1.98	1.69	1.42	1.22	1.04	0.87
mean V / V	4.503	3.820	3.253	2.760	2.340	2.027	1.703				
ln V	1.505	1.340	1.180	1.017	0.850	0.707	0.532				

      i Write the missing entries for Table 1.

      ii Use the measurements to plot a graph of $\ln V$ on the y-axis against t on the x-axis.

      iii Use your graph to determine the time constant of the discharge circuit.

      iv The resistance R of the resistor was $68\,k\Omega$. Determine the capacitance C of the capacitor.

    c **i** Discuss the reliability of the measurements.

      ii Estimate the accuracy of your value of capacitance, given the resistor value is accurate to within 1%.

9 A $2.2\,\mu F$ capacitor is discharged from a $2.0\,V$ cell and the capacitor is then discharged through a $470\,k\Omega$ resistor.

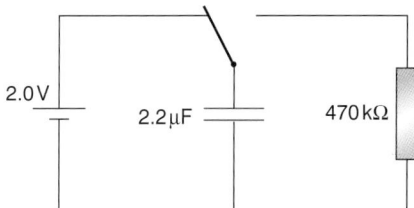

Figure 9.1

    a **i** Calculate the initial current in the resistor when the capacitor starts to discharge through the resistor.

      ii Explain why the current in the resistor decreases as the capacitor discharges.

    b **i** Calculate the time constant of the discharge circuit.

      ii Calculate the current in the resistor $5.0\,s$ after the discharge started.

    c **i** Sketch a graph to show how the current decreases with time in the first $5.0\,s$ of the discharge.

      ii Explain why the charge on the capacitor decreases at the same rate as the current decreases.

16.1 Magnetic field patterns

Learning outcomes

On these pages you will learn to:

- recognise that a magnetic field is an example of a field of force produced either by current-carrying conductors or by permanent magnets
- explain what is meant by a magnetic field line
- sketch magnetic field patterns due to a bar magnet, a long straight wire, a flat circular coil and a long solenoid

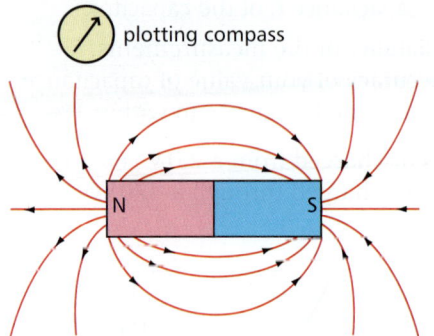

⊘ plotting compass

Figure 1 *The magnetic field near a bar magnet*

Lines of force

- A **magnetic field** is a force field produced either by a magnet or by moving charges such as when a current is in a wire. The force field acts on any other magnet or current-carrying wire placed in the field.
- The magnetic field of a bar magnet is strongest near its ends, which are referred to as '**poles**'. A bar magnet free to turn horizontally about its centre aligns itself with one end pointing north and the other pointing south. This occurs because the Earth's magnetic field attracts one end and repels the other end. The poles are referred to as **north-seeking** and **south-seeking**, according to the direction in which each pole points.
- A line of force of a magnetic field is a line along which a 'free' north pole would move in the field. Lines of force are often referred to as '**magnetic field lines**'. Note that the lines of force of a permanent magnet loop round from the north pole to the south pole of the magnet. A plotting compass points in the direction of a line of force.

The force between two magnets

Two bar magnets placed end-to-end attract or repel, depending on whether the nearest poles are:

- **like polarity**, in which case they repel, or
- **unlike polarity**, in which case they attract.

Electromagnetism

A magnetic field is created round a wire whenever **a current passes** along the wire. The pattern of the magnetic field lines for a long straight wire, a solenoid and a flat coil are shown opposite. Note that the lines of force are **complete loops**. Also, the **direction of the lines of force** depends on the direction of current. If the current is reversed, the direction of the lines of force is reversed.

- **For a long straight wire**, the lines of force are circles centred on the wire in a plane perpendicular to the wire. The direction of the lines of force depends on the current direction and can be worked out using the 'corkscrew rule' as shown in Figure 2.
- **For a solenoid**, the lines of force pass through the solenoid along its axis and loop round outside the solenoid. The direction of the lines of force depends on the current direction and can be worked out using the right hand grip rule or the solenoid rule, as shown in Figure 3.

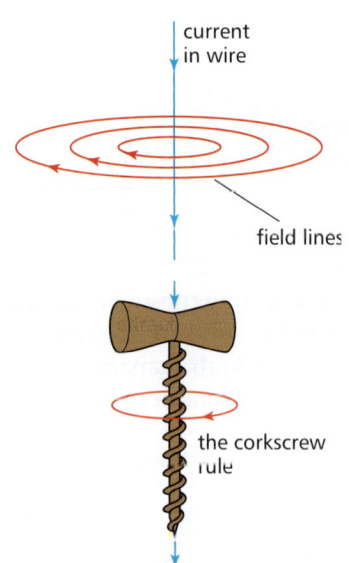

Figure 2 *The magnetic field near a wire*

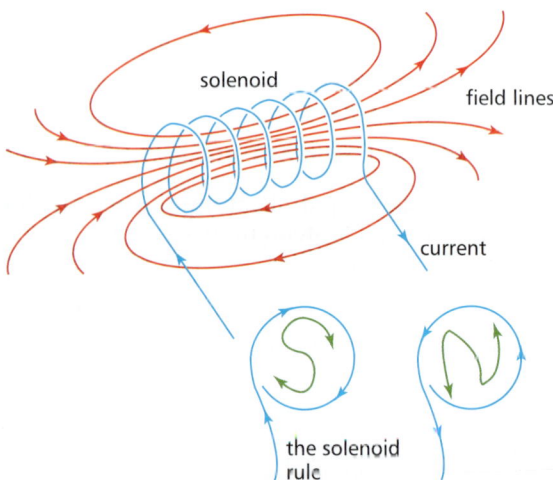

Figure 3 *The magnetic field of a solenoid*

- **For a flat coil**, the lines of force are lines that pass through the coil and loop round outside the coil. The direction of the lines of force can be worked out from the current direction using the solenoid rule, as explained on the previous page.

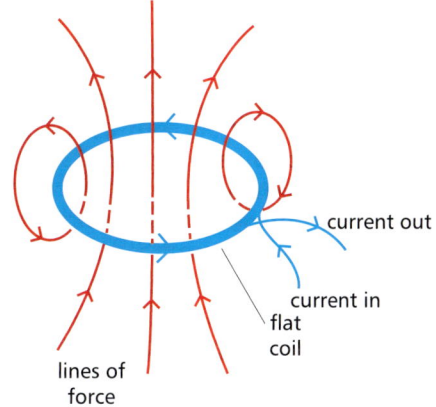

Figure 4 *The magnetic field of a flat coil*

Summary test 16.1

1 A plotting compass is placed at the intersection of two perpendicular lines drawn on a sheet of paper. The plotting compass points north along one of the lines. When a bar magnet is placed along the other line near the plotting compass, as shown in Figure 5, the plotting compass points north-east.

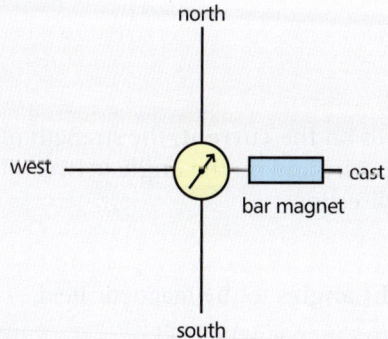

Figure 5

a What is the polarity of the pole of the magnet nearest the plotting compass?

b If the magnet is turned round, what direction will the plotting compass then point to?

2 An underground cable is aligned horizontally in an east-west direction. The cable carries a direct current from west to east, as shown in Figure 6.

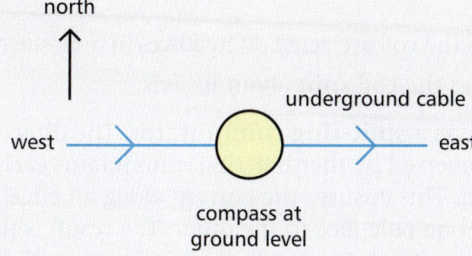

Figure 6

a What would be the direction of a magnetic compass directly above the wire, assuming the magnetic field due to the cable is much stronger than the Earth's magnetic field?

b How would the direction of the compass change if the current in the cable is gradually reduced to zero?

3 A plotting compass is placed at point P at the end of a solenoid. When a direct current is passed through the solenoid, the compass points into the solenoid, as shown in Figure 7.

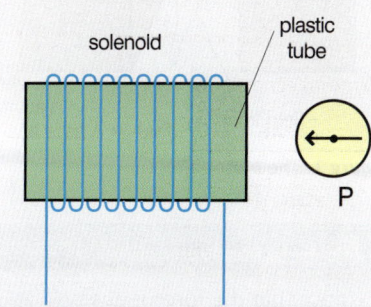

Figure 7

a State the direction of the current round the solenoid, as seen by an observer looking directly at the end of the solenoid at P.

b How would the direction of the plotting compass change if the current in the solenoid was reversed?

4 A student makes a model ammeter using a pair of flat coils in series with each other and a plotting compass, as shown in Figure 8.

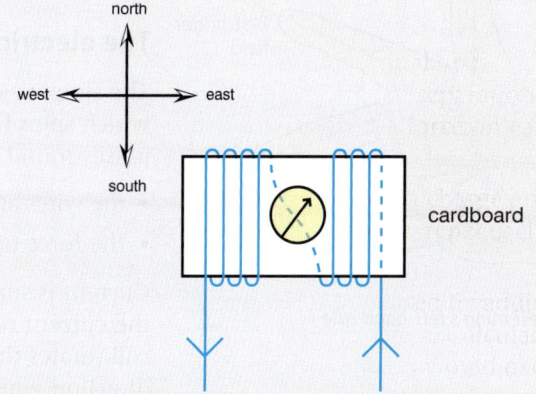

Figure 8

a Explain why the compass needle deflects when a direct current is passed through the coils.

b Explain why the compass needle cannot deflect more than 90° no matter how much current is passed through the coils.

Learning outcomes

On these pages you will learn to:

- recognise that a force acts on a current-carrying conductor placed in a magnetic field at a non-zero angle to the field lines
- describe the operation of a simple electric motor

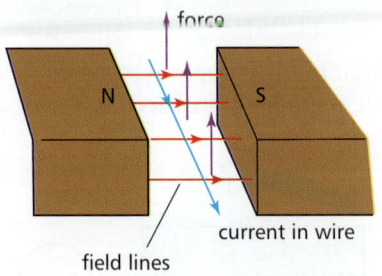

Figure 1 The motor effect

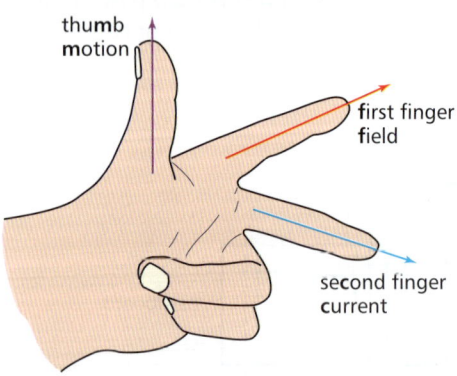

Figure 3 Fleming's left-hand rule

The force on a current-carrying wire in a magnetic field

A current-carrying wire placed at a non-zero angle to the lines of force of an external magnetic field experiences a force due to the field. This effect is known as the **motor effect.** The force is perpendicular to the wire and to the lines of force.

The motor effect can be tested using the simple arrangement shown in Figure 1. The wire is placed between opposite poles of a U-shaped magnet so it is at right angles to the lines of force of the magnetic field. When a current is passed through the wire, the section of the wire in the magnetic field experiences a force that pushes it out of the field. The combined magnetic field due to the wire and the magnet is stronger on one side of the wire than on the opposite side. The wire is pushed in the direction where the combined field is weakest, as shown in Figure 2.

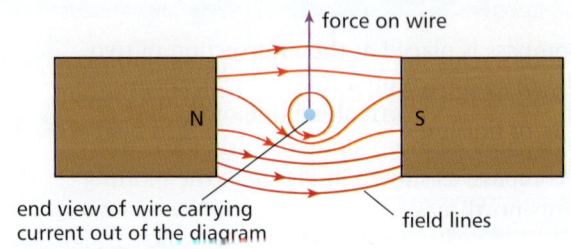

Figure 2 A field pattern

Force factors

The magnitude of the force depends on the **current**, the strength of the **magnetic field**, the **length** of the wire and on the **angle** between the lines of force of the field and the current direction.

The force is:

- greatest when the wire is at **right angles** to the magnetic field,
- zero when the wire is **parallel** to the magnetic field.

The direction of the force is **perpendicular** to the direction of the field and to the direction of the current, as indicated by Fleming's left-hand rule, shown in Figure 3. If the current is reversed, or if the magnetic field is reversed, the direction of the force is reversed.

The electric motor

The simple electric motor consists of a coil of insulated wire, the **armature**, which spins between the poles of a U-shaped magnet. When a direct current passes round the coil:

- the wires at opposite edges of the coil are acted on by forces in opposite directions,
- the force on each edge makes the coil **spin** about its axis.

Current is supplied to the coil via a **split-ring commutator**. The direction of the current round the coil is reversed by the split-ring commutator each time the coil rotates through half a turn. This ensures the current along an edge changes direction when it moves from one pole face to the other. The result is that the force on each edge continues to turn the coil in the same direction (Figure 4).

The **direction of rotation** of the motor is reversed by either reversing the current **or** the direction of the magnetic field. If both the current and the magnetic field are reversed, the direction of rotation is unchanged.

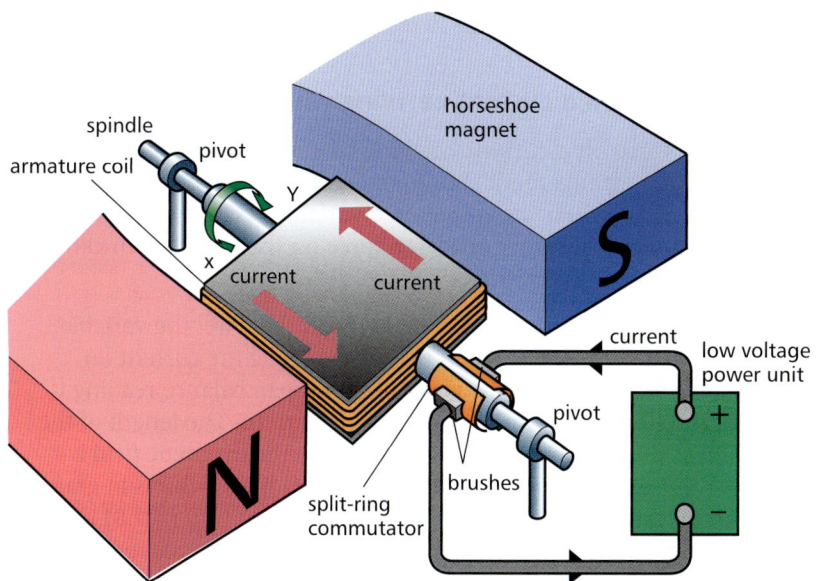

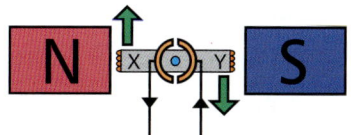

Initially, current is up side X and down side Y; therefore the coil turns clockwise.

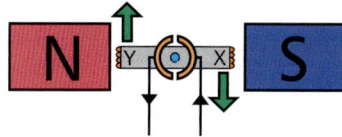

After half a turn, current is up side Y and down side X; therefore the coil continues to turn clockwise.

Figure 4 *The simple electric motor*

The **a.c. motor** has an electromagnet instead of a permanent magnet connected to the same a.c. supply as the armature. The direction of rotation does not change when the current reverses because the magnetic field of the electromagnet reverses as well.

The **speed of rotation** depends on the **current**, the **strength** of the magnet and the **number of turns** of the coil. The strength of the magnetic field is increased if an armature with an iron core is used.

A practical electric motor

A practical electric motor has an armature with several **evenly spaced coils** wound on it. Each coil is connected to its own section of the commutator. The result is that each coil in sequence experiences a turning effect when it is connected to the voltage supply, so the motor runs smoothly.

Figure 5 *A practical electric motor*

Summary test 16.2

1 A fixed vertical wire is in a horizontal magnetic field. State the direction of the force on the wire if:

 a the current is upwards and the magnetic field lines are from east to west,

 b the current is upwards and the magnetic field lines are from south to north,

 c the current is downwards and the magnetic field lines are from east to west.

2 A fixed horizontal wire lies along a line from east to west in a magnetic field. State the direction of the magnetic field lines if the current in the wire is from east to west and the force on it is:

 a vertically up,

 b horizontal and due north.

3 A rectangular coil carries a current in a magnetic field.
 a Explain why the coil experiences a turning effect when the plane of the coil is parallel to the magnetic field lines, as in Figure 6a.

 b Explain why the coil experiences no turning effect when the plane of the coil is perpendicular to the magnetic field lines, as in Figure 6b.

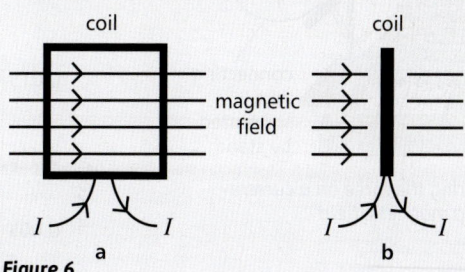

Figure 6

4 **a** What is the function of the split-ring commutator in a d.c. electric motor?

 b A simple electric motor containing a permanent magnet is connected to a battery and a variable resistor. What would be the effect on the motor of:
 i increasing the current,
 ii reversing the current,
 iii using an a.c. supply instead of the battery?

Learning outcomes

On these pages you will learn to:

- define magnetic flux density and the tesla
- describe how the force on a current-carrying conductor can be used to measure the flux density of a magnetic field using a current balance
- recall and solve problems using the equation $F = BIl\sin\theta$, with directions as interpreted by Fleming's left-hand rule
- recognise that the magnetic flux density in a solenoid is increased if a ferrous material is placed in the solenoid
- explain the forces between current-carrying conductors and predict the direction of the forces

Investigating the force on a current-carrying wire in a magnetic field

The magnitude of the force on a current-carrying wire in a magnetic field can be investigated using the arrangement shown in Figure 1. The stiff wire frame is connected in series with a switch, an ammeter, a variable resistor and a battery. When the switch is closed, the magnet exerts a force on the wire which can be measured from the change of the top-pan balance reading.

- To test the variation of force with the current through the wire, the variable resistor is adjusted to change the current. Before switching the current on, the top-pan balance reading should be noted. The top-pan balance reading is then measured for different measured values of the current. The length of the test wire in the field is kept the same. The force due to the magnetic field is worked out from the change of the top-pan balance reading. If this is in grams, the reading must be converted to kilograms then multiplied by g (= $9.81\,\mathrm{m\,s^{-2}}$) to give the force. For example, if the change of the top-pan balance reading is $20.5\,\mathrm{g}$, the force due to the magnetic field is $0.20\,\mathrm{N}$ (= $20.5 \times 10^{-3}\,\mathrm{kg} \times 9.81\,\mathrm{m\,s^{-2}}$).

A graph of a typical set of results is shown in Figure 2.

- To test the variation of force with the length of the wire, the current is kept the same throughout by using the variable resistor as necessary. The length of the wire in the field is changed by reconnecting the wires to the frame. For each length, the reading of the top-pan balance is noted and the force due to the magnetic field is calculated. A graph of a typical set of results is shown in Figure 3.

- To test the variation of the force with the angle between the wire and the magnetic field lines, the magnet can be turned gradually. This test shows that the force is a maximum when the wire is perpendicular to the magnetic field lines.

The tests above show that the force F on the wire is proportional to:

- the current I,
- the length l of the wire.

Magnetic flux density

The magnetic flux density, B, is defined as the force per unit current per unit length on a current-carrying wire placed perpendicular to the field lines.

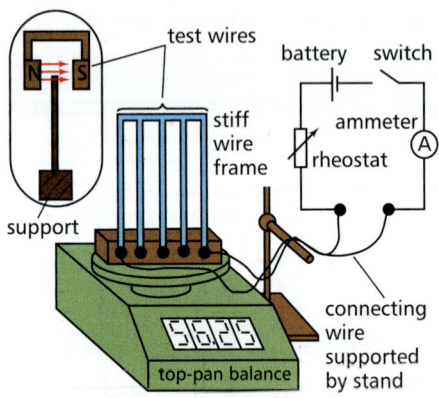

Figure 1 Measuring the force on a current-carrying wire in a magnetic field

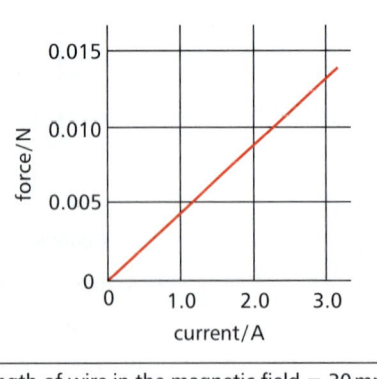

lenqth of wire in the magnetic field = 30 mm

Figure 2 Force v. current

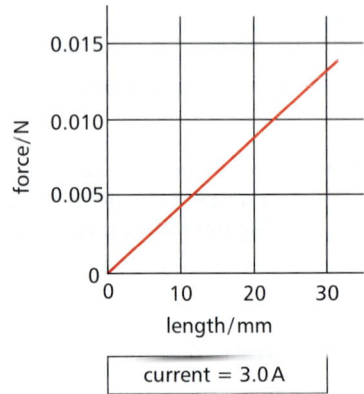

current = 3.0 A

Figure 3 Force v. length

The unit of B is the **tesla (T)**, equal to $1\,N\,A^{-1}\,m^{-1}$. The magnetic flux density is sometimes also referred to as the **magnetic field strength**.

For a wire of length l at right angles to a uniform magnetic field of flux density B, the force on the wire when current I passes through it is given by:

$$F = BIl$$

Worked example

A horizontal wire of length $0.050\,m$ is in a uniform magnetic field directed vertically upwards. The wire lies along a line from north to south. When a current of $4.0\,A$ is passed along the wire, a force of $5.6 \times 10^{-2}\,N$ is exerted on the wire, as shown in Figure 4.

a Calculate the magnetic flux density of the magnetic field.
b State the direction of the force on the wire if the current in the wire was from north to south.

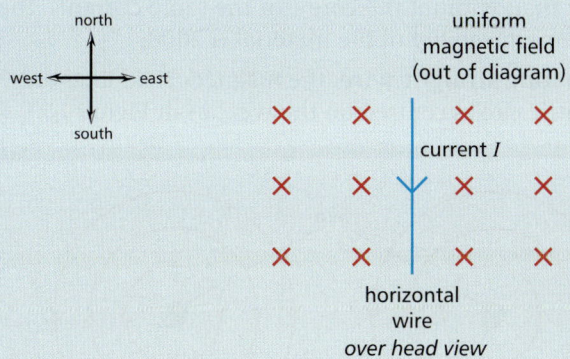

Figure 4

Solution

a $B = \dfrac{F}{Il} = \dfrac{5.6 \times 10^{-2}}{4.0 \times 0.050} = 0.28\,T$

b Using Fleming's left-hand rule (Topic 16.2, Figure 3), gives due west for the direction of the force.

For a straight wire that is not at right angles to the magnetic field lines, the force on the wire due to the field is determined using the component of the magnetic field at right angles to the wire.

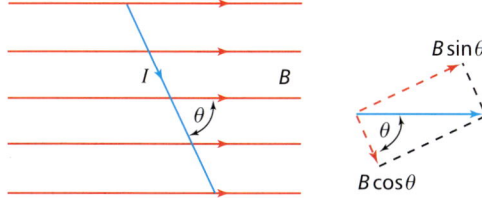

Figure 5 $F = BIl\sin\theta$

For angle θ between the wire and the field lines:

- the component of B perpendicular to the wire = $B\sin\theta$,
- the component of B parallel to the wire = $B\cos\theta$.

Because no force acts on the wire due to the parallel component,

1 the magnitude of the force on the wire, $F = (B\sin\theta)Il$
   $= BIl\sin\theta$
2 the direction of the force on the wire is given by Fleming's left hand rule, where the field direction is the direction of the perpendicular component of B at the wire.

For a wire of length l carrying a current I in a uniform magnetic field B at angle θ to the field lines,

the force on the wire, $F = BIl\sin\theta$

Worked example

A straight horizontal wire XY of length $5.0\,m$ is in a uniform horizontal magnetic field of magnetic flux density $120\,mT$. The wire is at an angle of $30°$ to the field lines, as shown in Figure 6. When the wire conducts a current of $14\,A$ from X to Y, calculate the magnitude of the force on the wire and state its direction.

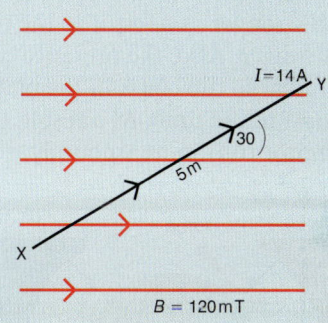

Figure 6

Solution
$F = BIl\sin\theta = 0.12\,T \times 14\,A \times 5.0\,m \times \sin 30 = 4.2\,N$

The force on the wire is vertically downwards.

The couple on a coil in a magnetic field

Consider a rectangular current-carrying coil in a uniform horizontal magnetic field, as shown in Figure 7. The coil has n turns of wire and can rotate about a vertical axis.

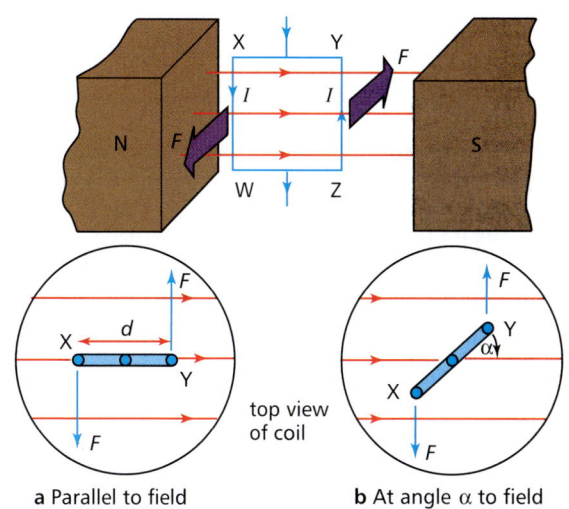

a Parallel to field   b At angle α to field

Figure 7 *Couple on a coil*

- The long sides of the coil are vertical and of length l. Each side therefore experiences a horizontal force $F = BIln$ in opposite directions at right angles to the field lines.
- The pair of forces acting on the long sides form a couple as the forces are not directed along the same line. The torque of the couple $= Fd$, where d is the perpendicular distance between the line of action of the forces on each side. If the plane of the coil is at angle α to the field lines, then $d = w\cos\alpha$ where w is the width of the coil.
- Therefore, the torque $= Fw\cos\alpha = BIlnw\cos\alpha$
  $= BIAn\cos\alpha$, where $A = lw =$ the coil area.
  If $\alpha = 0$ (i.e. the coil is parallel to the field) the torque $= BIAn$ since $\cos 0 = 1$.
  If $\alpha = 90°$ (i.e. the coil is perpendicular to the field) the torque $= 0$ since $\cos 90° = 0$.

In an electric motor, the coil is wound on an iron core. The field is much stronger as a result so the torque on the motor is much stronger. Also, the presence of the core makes the field radial so the coil is in the plane of the field (i.e. $\alpha = 0$) for most of the time. As a result, the torque is steady and the motor runs more smoothly.

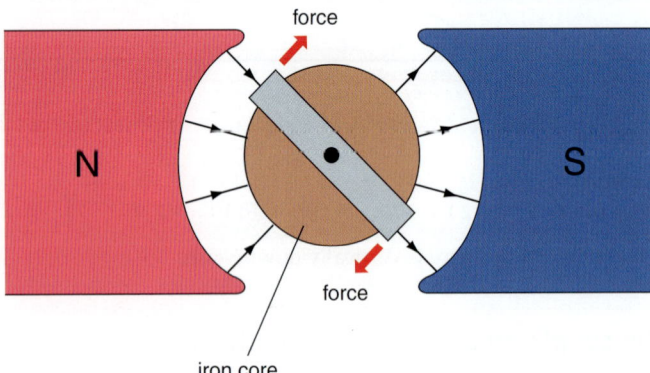

Figure 8 *In an electric motor*

Magnetic flux density due to current-carrying conductors

In a solenoid, experiments show that the magnetic flux density B inside a solenoid does not vary with position and is proportional to:

- the current I, and
- the number of turns per metre, n, on the solenoid.

Figure 9 shows how the magnetic flux density in a solenoid varies along its length. The magnetic flux density is constant at any position inside the solenoid away from its ends. At each end, the magnetic flux density is half the value given by the above formula.

The magnetic flux density of a solenoid is increased considerably if a ferrous material such as iron or steel is placed inside the solenoid. The **relative permeability** μ_r of such a material is defined as the ratio of magnetic flux

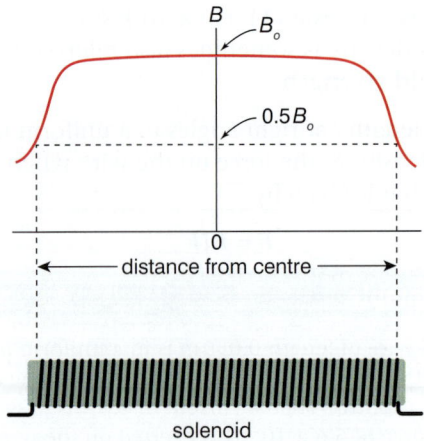

Figure 9 *B in a solenoid*

density with and without the solenoid core filled with the material. For example, if a solenoid with an iron core produces a magnetic flux density which is 2000 times greater than without the core (for the same current), the relative permeability of the material is 2000.

Near a long straight wire, the magnetic field lines are concentric circles centred on the wire, as in Figure 10.

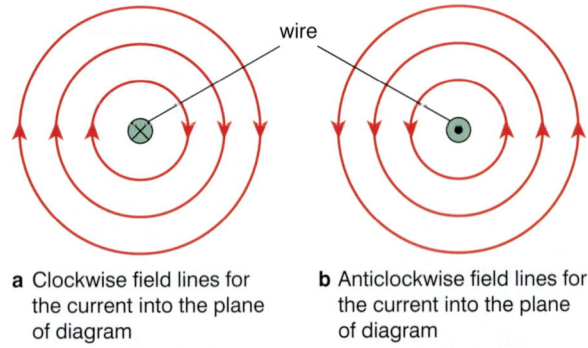

a Clockwise field lines for the current into the plane of diagram

b Anticlockwise field lines for the current into the plane of diagram

Figure 10 *Magnetic field lines near a current-carrying wire*

The magnetic field strength near the wire decreases with distance from the wire. The field lines are clockwise for a current into the plane of the diagram and anticlockwise for the opposite direction of current.

Two current-carrying wires parallel to each other exert a magnetic force on each other , as shown in Figure 11. The direction of the force on each wire can be worked out using Fleming's left-hand rule (see Topic 16.2). If their currents are:

- **in the same direction**, the wires attract each other. This is because each wire experiences a magnetic force towards the other one due to the magnetic field of the other wire. See Figure 12a.
- **in opposite directions**, the wires repel each other. This is because each wire experiences a magnetic force away from the other one due to the magnetic field of the other wire. See Figure 12b.

In Figure 12a, both currents are in the same direction and the wires attract each other. So why do the forces on both X and Y reverse in direction when the current in Y is reversed? This is because reversing the current reverses the magnetic field direction. As a result:

• The force on wire X reverses because the magnetic field acting on X has reversed and X's current is in the same direction.
• The force on wire Y reverses because its current has been reversed and the magnetic field acting on Y is unchanged in direction.

Note that the force F on each wire:

• decreases the greater the separation of the wires. This is because the magnetic flux density of each wire decreases with distance from the wire.
• is proportional to the product of their currents. In other words, F is proportional to $I_X \times I_Y$, where I_X is the current in X and I_Y is the current in Y.

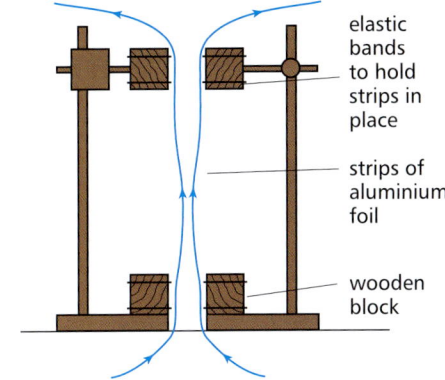

Figure 11 The force between two current-carrying conductors

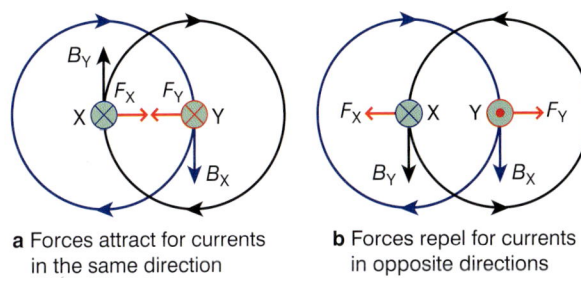

a Forces attract for currents in the same direction

b Forces repel for currents in opposite directions

Figure 12 The magnetic force between two parallel current-carrying wires

Summary test 16.3

1 a Use Figure 2 to work out the magnetic flux density of the magnet that was used in the test.

b Calculate the magnitude of the force on a straight horizontal wire of length 0.10 m carrying a current of 4.0 A in a uniform horizontal magnetic field of flux density 55 mT when the angle between the wire and the field lines is:
i 0, **ii** 30°, **iii** 90°.

2 Table 1 relates the force on a current-carrying wire, at right angles to the lines of force of a magnetic field, to the magnetic flux density and the current. Complete the table by working out the missing data in each column.

Table 1

	a	b	c	d
B/T	0.20 T vertically down	0.20 T vertically down	?	0.1 T horizontal due?
I/A	3.0 A horizontal due north	?	3.0 A horizontal due north	2.0 A vertically up
l/m	0.040 m	0.040 m	0.040 m	0.040 m
F/N	?	0.036 N horizontal due south	0.024 N horizontal due west	? horizontal due east

3 At a certain location on the Earth's surface, the magnetic flux density B of the Earth is 70 μT in a direction due north at 70° to the horizontal. A horizontal cable of length 52 m aligned from east to west carries a current of 28 A.

Calculate the magnitude of the magnetic force on the cable and state its direction.

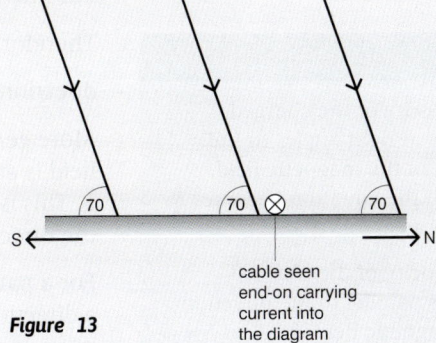

Figure 13

4 A rectangular coil of width 60 mm and of length 80 mm has 50 turns. The coil is placed horizontally in a uniform horizontal magnetic field of flux density 85 mT with its shorter side parallel to the field lines. A current of 8.0 A was passed through the coil. Sketch the arrangement and determine the force on each side of the coil.

243

16.4 Moving charges in a magnetic field

Learning outcomes

On these pages you will learn to:

- recognise that a charged particle moving at a non-zero angle to the lines of a magnetic field experiences a force
- predict the direction of the force on a charge moving in a magnetic field
- recall and solve problems using $F = BQv\sin\theta$
- derive and use the equation $V_H = \dfrac{BI}{ntq}$ for the Hall voltage

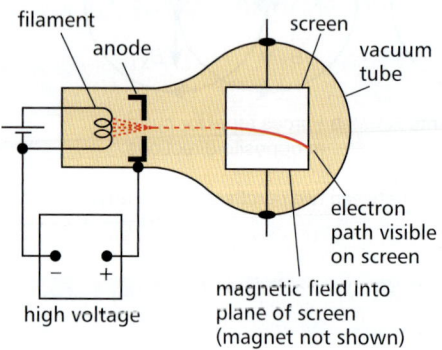

Figure 1 An electron deflection tube

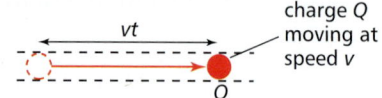

Figure 3 Force on a moving charge

Note

1 If the velocity of the charged particle is at right angles to the direction of the magnetic field, $\theta = 90°$ so the equation becomes $F = BQv$ because $\sin 90° = 1$.

2 If the velocity of the charged particle is parallel to the direction of the magnetic field, $\theta = 0$ so $F = 0$ because $\sin 0 = 0$.

Electron beams

Figure 1 shows a vacuum tube designed to show the effect of a magnetic field on an electron beam. The production of the electron beam is outlined in Topic 14.1. The path of the beam can be seen where it passes over the fluorescent screen in the tube. The beam is deflected downwards when a magnetic field is directed into the plane of the screen. Each electron in the beam experiences a force due to the magnetic field. The beam follows a circular path because the direction of the force on each electron is perpendicular to the direction of motion of the electron (and to the field direction). The direction of the force on an electron in the beam can also be worked out using Fleming's left hand rule, provided we remember the convention that the current direction is opposite to the direction in which the electrons move.

The reason why a current-carrying wire in a magnetic field experiences a force is that the electrons moving along the wire are pushed to one side by the field. If the electrons in Figure 2 had been confined to a wire, the whole wire would have been pushed downwards.

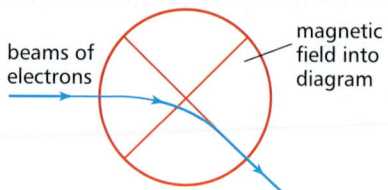

Figure 2 Electrons in a magnetic field

Force on a moving charge in a magnetic field

A beam of charged particles crossing a vacuum tube is an electric current across the tube. Suppose each charged particle has a charge Q and moves at speed v. In a time interval t, each particle travels a distance vt. Its passage is equivalent to current $I = \dfrac{Q}{t}$ along a wire of length $l = vt$. See Figure 3.

If the particles pass through a uniform magnetic field in a direction at right angles to the field lines, each particle experiences a force F due to the field. If the particles were confined to a wire, the force would be given by $F = BIl$.

For moving charges, the same equation applies where $I = \dfrac{Q}{t}$ and $l = vt$.

Therefore, for a charged particle moving across a uniform magnetic field in a direction at right angles to the field, $F = BIl = B \times \dfrac{Q}{t} \times vt = BQv$.

More generally, if the direction of motion of a charged particle in a magnetic field is at angle θ to the lines of the field, then $B\sin\theta$ is used in the equation for F. This is because $B\sin\theta$ is the component of the magnetic field perpendicular to the direction of motion of the charged particle.

For a particle of charge Q moving through a uniform magnetic field at speed v in a direction at angle θ to the field, the force on the particle is given by

$$F = BQv\sin\theta$$

The Hall probe

Hall probes are used to measure magnetic field strength and also as magnetic field sensors.

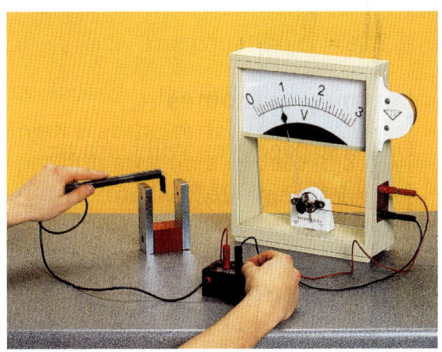

A Hall probe contains a slice of semiconducting material. Figure 4 shows the slice in a magnetic field with the field lines perpendicular to the flat side of the slice. A constant current passes through the slice, as shown. The **charge carriers** (which are electrons in an n-type semiconductor) are initially deflected by the field. As a result, a potential difference is created between the top and bottom edges of the slice. This effect is known as the **Hall Effect** after its discoverer.

The p.d., referred to as the Hall voltage, is proportional to the magnetic flux density, provided the current is constant. This is because each charge carrier passing through the slice is subjected to a magnetic force $F_{mag} = BQv$, where v is the speed of the charge carrier. Once the Hall voltage has been created, the magnetic deflection of a charge carrier entering the slice is opposed by the force on it due to the electric field created by the Hall voltage so each charge carrier passes through undeflected. The electric field force $F_{elec} = \dfrac{QV_H}{d}$, where V_H represents the Hall voltage and d is the distance between the top and bottom edges of the slice. Therefore, $\dfrac{QV_H}{d} = BQv$ gives $V_H = Bvd$.

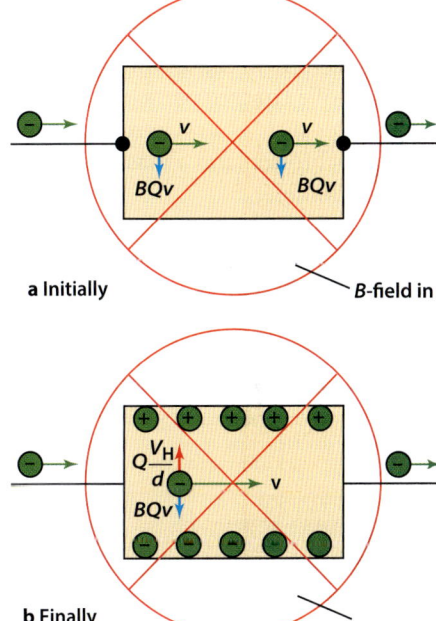

a Initially   B-field in

For constant current I passing through a slice of cross-sectional area A, $I = nAvq$, where n is the number of charge carriers per unit volume and v is their drift velocity. See Topic 6.2.

Combining this equation with the equation $V_H = Bvd$ gives:

$$V_H = Bvd = \frac{B(nAvq)d}{nAq} = \frac{BId}{nAq}$$

As $A = d \times t$, where t is the thickness of the slice:

$$V_H = \frac{BId}{nAq} = \frac{BId}{n(d \times t)q}$$

Therefore

$$\boldsymbol{V_H = \frac{BI}{ntq}}$$

b Finally   B-field in

Figure 4 The Hall Effect

Summary test 16.4

$e = 1.6 \times 10^{-19}\,C$

1 In Figure 2, how would the force on the electrons in the magnetic field differ if:

    i   the magnetic field was reversed in direction,

    ii the magnetic field was reduced in strength,

    iii the speed of the electrons was increased?

2 Calculate the force on an electron that enters a uniform magnetic field of flux density $150\,mT$ at a velocity of $8.0 \times 10^6\,m\,s^{-1}$ at an angle of:

    i $90°$,   **ii** $30°$ to the field.

3 A beam of protons moving at constant speed is directed into a uniform magnetic field in the same direction as the field.

    a Explain why the beam is not deflected by the field.

    b Describe and explain how the path of the beam in the field would have differed if the beam had been directed into the field at a slight angle to the field lines.

4 In a Hall probe, electrons passing through the semiconductor slice experience a force due to a magnetic field.

    a Explain why a potential difference is created across the slice as a result of the application of the magnetic field.

    b When the magnetic flux density is $90\,mT$, each electron moving through the slice experiences a force of $6.4 \times 10^{-20}\,N$ due to the magnetic field. Calculate:

        i the mean speed of the electrons passing through the slice,

        ii the force on each electron if the magnetic flux density is increased to $120\,mT$.

Learning outcomes

On these pages you will learn to:

- describe and analyse the deflection of beams of charged particles by uniform electric and uniform magnetic fields
- explain how electric and magnetic fields can be used in velocity selection
- recognise that a uniform magnetic field causes the circular orbit of a charged particle in devices such as a cyclotron, a synchrotron and a mass spectrometer

Magnetic fields are used to control beams of charged particles in many devices, from television tubes to high energy accelerators. The force of the magnetic field on a moving charged particle is at right angles to the direction of motion of the particle.

- No work is done by the magnetic field on the particle as the force acts at right angles to the velocity of the particle. Its direction of motion is changed by the force but not its speed. The kinetic energy of the particle is unchanged by the magnetic field.
- In accordance with Fleming's left-hand rule, the magnetic force is perpendicular to the velocity at any point along the path. The force therefore acts towards the centre of curvature of the circular path.
- The particle moves on a circular path. The force causes a centripetal acceleration because it is perpendicular to the velocity. Figure 1 shows the deflection of a beam of electrons in a uniform magnetic field. The path is a complete circle because the magnetic field is uniform and the particle remains in the field.

The radius, r, of the circular orbit in Figure 1 depends on the speed v of the particles and the magnetic flux density B.

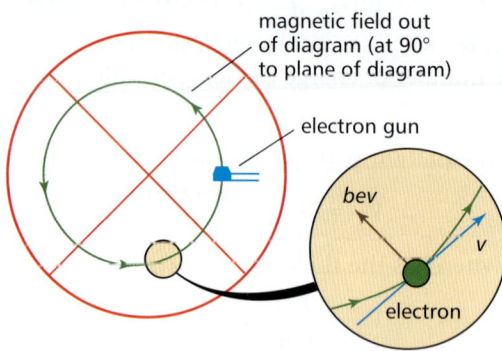

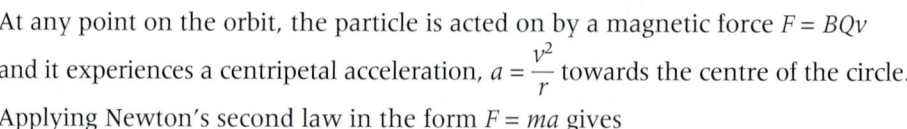

Figure 1 A circular orbit in a magnetic field

At any point on the orbit, the particle is acted on by a magnetic force $F = BQv$ and it experiences a centripetal acceleration, $a = \dfrac{v^2}{r}$ towards the centre of the circle.

Applying Newton's second law in the form $F = ma$ gives

$$BQv = \frac{mv^2}{r}$$

Rearranging this equation gives

$$r = \frac{mv}{BQ}$$

The equation for r shows that:

1. r decreases if B is increased,
2. r increases if v is increased,
3. particles in a beam with different values of specific charge, $\dfrac{Q}{m}$, are separated by a magnetic field.

Applications

The following applications make use of the essential principle that charged particles move on circular paths when in a magnetic field and moving at right angles to the field lines.

1 Electrons moving in a circle

The fine beam tube shown in Figure 3 contains hydrogen gas at low pressure. When a beam of electrons from the electron gun in the tube passes through

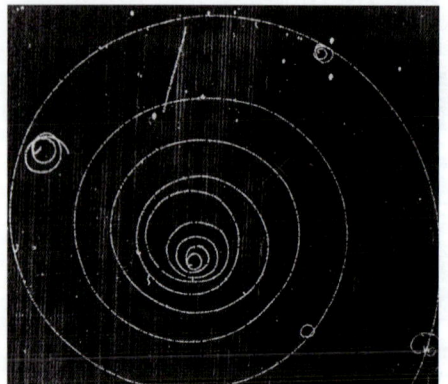

Figure 2 Charged particles in a magnetic field. The large spiral is due to a charged particle created by a collision (not shown) between a fast-moving incoming particle and the nucleus of an atom. The charged particle is forced onto a curved path by the magnetic field. It spirals inwards because its kinetic energy and momentum decreases gradually as it transfers energy to the atoms it passes through. Therefore it is deflected more and more by the field as its speed decreases.

the gas, the atoms along the beam path emit light due to collisions with electrons. So the path of the beam is seen as a fine trace of light in the tube. A pair of coils, placed either side of the tube, is supplied with a direct current to produce a uniform magnetic field through the tube.

Provided the initial direction of the beam is at right angles to the magnetic field lines, the beam curves round in a circle.

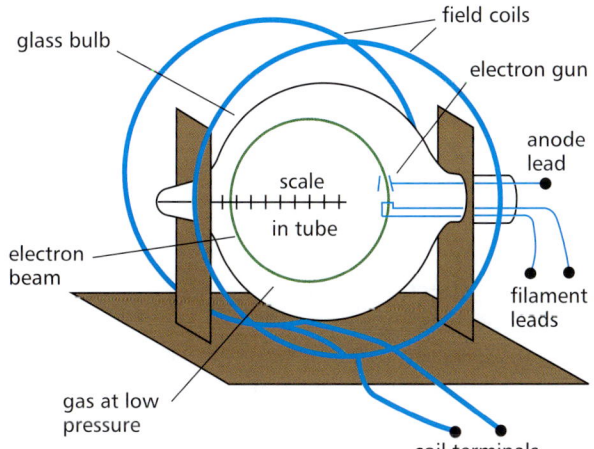

Figure 3 Using a fine beam tube

As explained opposite, the radius of the circle $r = \dfrac{mv}{Be}$, where B is the magnetic flux density and v is the speed of the electrons.

The equation shows that the radius of the circle can be reduced by:

- increasing the magnetic flux density, or
- reducing the anode voltage of the electron gun since this reduces the speed of the electrons.

The fine beam tube may be used to measure e/m, the specific charge of the electron. This was first determined by J. J. Thomson in 1895. Its value is $1.76 \times 10^{11}\,\text{C}\,\text{kg}^{-1}$.

Before Thomson made this measurement, the hydrogen ion was known to have the largest specific charge of any charged particle. Thomson showed that the electron's specific charge is 1860 times larger than that of the hydrogen ion. However, Thomson could not conclude that the electron has a much smaller mass than the hydrogen ion as the charge of the electron was not known at that time. The charge of the electron was measured by R. Millikan in 1915.

2 The cyclotron

The cyclotron was invented in 1930 by E.O. Lawrence. It consists of two hollow D-shaped electrodes (referred to as 'dees') in a vacuum chamber. A high-frequency alternating voltage is applied between the dees. A beam of charged particles is directed into one of the dees near the centre of the cyclotron. A uniform magnetic field is applied perpendicular to the plane of the dees.

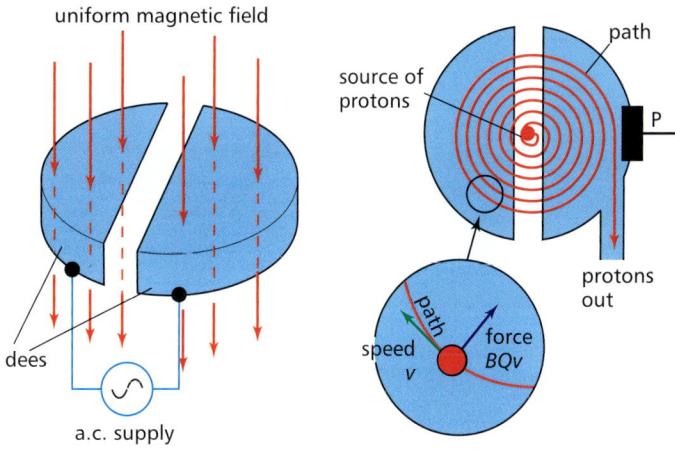

Figure 4 The cyclotron

The charged particles are forced on a circular path by the magnetic field, causing the particles to emerge from the dee they were directed into. The particles emerging from the dee when the alternating voltage reverses are accelerated into the other dee where they are forced on a circular path by the magnetic field. When they emerge from this dee, the alternating voltage reverses again and accelerates the particles into the first dee where the process is repeated. This occurs because the time taken, T, by a particle to move round its semi-circular path in the dee $= \dfrac{\pi r}{v}$, where r is the radius of the path and v is the particle speed.

As explained previously,

$$r = \frac{mv}{BQ} \text{ so } T = \frac{m\pi}{BQ}.$$

For an alternating voltage of frequency f, the time for one half cycle $= \dfrac{1}{2f}$.

Therefore,

$$\frac{1}{2f} = \frac{m\pi}{BQ} \text{ so } f = \frac{BQ}{2\pi m}.$$

The equation shows that the frequency is independent of the radius and the speed so the charged particles cross between the dees each time the voltage reverses.

The particles gain speed each time they are accelerated from one dee to the other. The radius of the circular path, therefore, is larger each time a particle travels into and out of a dee. The particles emerge from the cyclotron when the radius of orbit is equal to the dee radius R. As $v = \dfrac{BQr}{m}$ the speed of the particles on exit from the cyclotron $= \dfrac{BQR}{m}$.

The **synchrotron** accelerates charged particles to much higher energies than a cyclotron. The magnetic field is increased to keep the particles in an orbit of constant radius as they are boosted to higher and higher speeds.

247

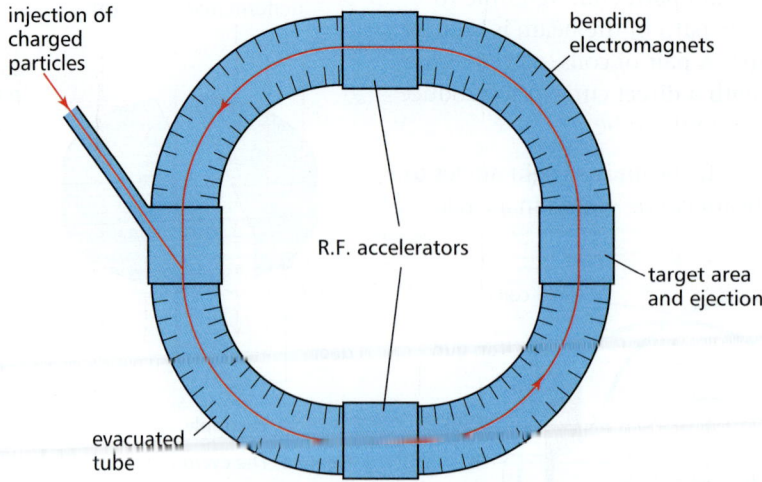

Figure 5 *The synchrotron*

As the radius of orbit is constant, semi-circular dees are not necessary and the particles move round a ring-shaped tube in a magnetic field created by a large number of electromagnets positioned round the ring. A high-frequency alternating voltage, applied between electrodes positioned in the ring, is used to accelerate the charged particles in the ring to high energies. In operation, the particles are injected into the ring and are boosted in 'bursts' to high energies each time the magnetic field is increased.

Extension

The Van Allen belts

Charged particles from space are trapped in belts above the Earth's atmosphere by the Earth's magnetic field. These belts were predicted by Van Allen and were discovered by Geiger counters aboard the US Explorer 1 satellite in 1958. The charged particles in the belts spiral around the field lines, bouncing back near the poles. The inner belt consists of protons and other charged particles with energies of 10 to 100 MeV that stream from the Sun in the 'solar wind' or as a result of cosmic rays (i.e. high energy particles from space) colliding with atoms in the upper atmosphere. The outer belt stretches out to about 20 000 km above the Earth and consists mostly of electrons with energies of 10 to 100 keV from the Sun.

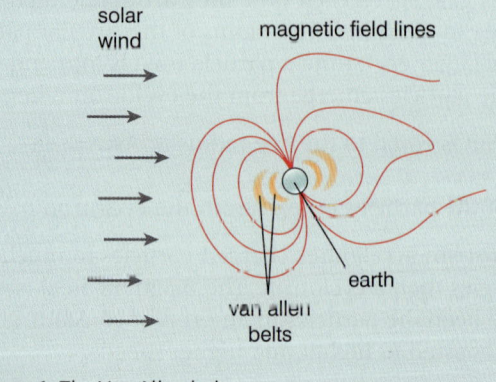

Figure 6 *The Van Allen belts*

3 The mass spectrometer

The mass spectrometer is used to analyse the type of atoms present in a sample. The atoms of the sample are ionised and directed in a narrow beam at constant velocity into a uniform magnetic field. Each ion is deflected in a semi-circle by the magnetic field onto a detector, as shown in Figure 7. The radius of curvature of the path of each ion depends on the specific charge $\frac{Q}{m}$ of the ion in accordance with the equation $r = \frac{mv}{BQ}$. Each type of ion is deflected by a different amount onto the detector. The detector is linked to a computer which is programmed to show the relative abundance of each type of ion in the sample.

The ions in the beam enter the magnetic field at the same speed. When they are produced from the sample, they have a continuous range of speeds. Before they enter the magnetic field, they are formed into a beam and directed through a **velocity selector**, as shown in Figure 7. The velocity selector consists of a magnet and a pair of parallel plates at spacing d and voltage V_p due to a high voltage supply. The magnet and the plates are aligned so each ion passing through the velocity selector is acted on by an electric field force, $F_{elec} = \frac{QV_P}{d}$, in the opposite direction to the magnetic field force $F_{mag} = B_S Q v$, where B_S is the magnetic flux density of the magnet in the velocity selector.

Ions moving at a certain speed such that $B_S Q v = \frac{QV_P}{d}$ experience equal and opposite forces and so pass through undeflected. All other ions are deflected and do not pass through the collimator slit. So the beam emerging from the collimator consists of different types of ions, all with the same speed $v = \frac{V_P}{B_S d}$.

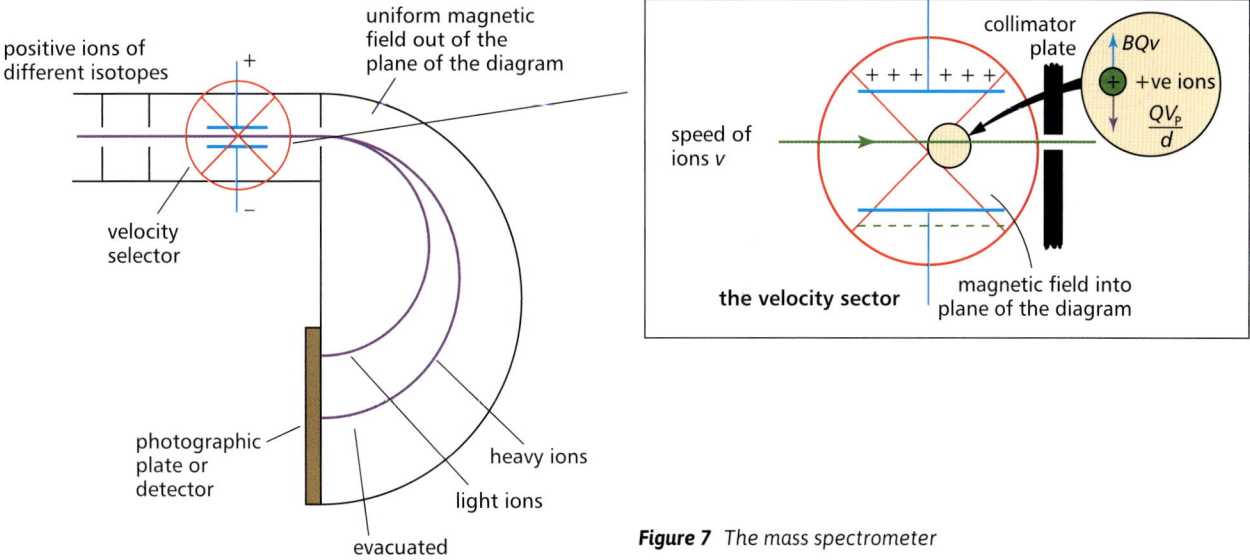

Figure 7 The mass spectrometer

Summary test 16.5

$e = 1.6 \times 10^{-19}\,\text{C}$, e/m for the electron $= 1.76 \times 10^{11}\,\text{C}\,\text{kg}^{-1}$

1 A beam of electrons at a speed of $3.2 \times 10^7\,\text{m}\,\text{s}^{-1}$ is directed into a uniform magnetic field of flux density 8.5 mT in a direction perpendicular to the field lines. The electrons move on a circular orbit in the field.

  a i Explain why the electrons move on a circular orbit.
    ii Calculate the radius of the orbit.

  b The flux density is adjusted until the radius of orbit is 65 mm. Calculate the flux density for this new radius.

2 In a fine beam tube, electrons are accelerated from rest through a certain p.d. before being directed at a speed of $2.9 \times 10^7\,\text{m}\,\text{s}^{-1}$ in a narrow beam into a uniform magnetic field.

  a The beam follows a circular path of radius 35 mm in the magnetic field. Calculate the flux density of the magnetic field.

  b The speed of the electrons in the beam was halved as a result of reducing the anode voltage. Calculate the new radius of curvature of the beam in the field.

3 The first cyclotron, used to accelerate protons, was 0.28 m in diameter and was in a magnetic field of flux density 1.1 T.

  a Show that protons emerged from this cyclotron at a maximum speed of $1.5 \times 10^7\,\text{m}\,\text{s}^{-1}$.

  b Calculate the maximum kinetic energy, in MeV, of a proton from this accelerator.
    The mass of a proton $= 1.67 \times 10^{-27}\,\text{kg}$.
    $1\,\text{MeV} = 1.6 \times 10^{-13}\,\text{J}$

4 In a mass spectrometer, a beam of different ions moving at a speed of $7.6 \times 10^4\,\text{m}\,\text{s}^{-1}$ was directed into a uniform magnetic field of flux density 680 mT, as shown in Figure 7.

  a An ion was deflected through 180° to a position on the detector which was 28 mm from where it entered the field. Calculate the specific charge of the ion.

  b A different type of ion was deflected onto the same position on the detector when the magnetic flux density was changed to 400 mT. Calculate the specific charge of this ion.

Chapter Summary

1 **a** $F = BIl\sin\theta$ gives the force F on a current-carrying wire of length l in a uniform magnetic field B at angle θ to the field lines, where I is the current.

  b The direction of the force is given by Fleming's left-hand rule where the field direction is the direction of the field component perpendicular to the wire.

2 **a** $F = BQv\sin\theta$ gives the force F on a particle of charge Q moving through a uniform magnetic field B at speed v in a direction at angle θ to the field.

  b If the velocity of the charged particle is perpendicular to the field, $F = BQv$.

  c The direction of the force is given by Fleming's left hand rule, provided the current is in the direction in which positive charge would flow.

3 $BQv = \dfrac{mv^2}{r}$ gives the radius of the orbit of a charge moving in a direction at right angles to the lines of a magnetic field.

(Launch additional digital resources for the chapter)

$e = 1.6 \times 10^{-19}$ C, e/m for the electron $= 1.76 \times 10^{11}$ C kg^{-1}
The mass of a proton is 1.67×10^{-27} kg

1 Figure 1.1 shows a plotting compass mid-way between a bar magnet and one end of a solenoid. The solenoid is connected in series with a battery, a variable resistor and a switch.

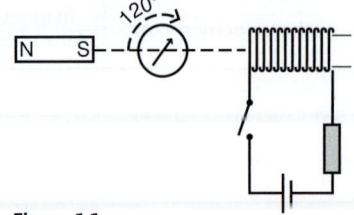

Figure 1.1

  a When the switch is open, the plotting compass points directly towards the bar magnet. When the switch is closed, the needle of the plotting compass turns through 120°. Explain why the needle turns when the switch is closed.

  b With the switch closed, the variable resistor is adjusted, making the needle turn back by 30°.

    i What must have been the effect of the adjustment of the variable resistor on the solenoid current?

    ii What would be the effect on the direction of the compass needle if the magnet is moved away from the plotting compass?

2 A U-shaped magnet is placed on the pan of a top-pan balance, as shown below. A straight wire is placed horizontally between the poles of the magnet, which are of length 32 mm. The wire is connected to a variable resistor, an ammeter, a switch and a battery.

  a When the switch is closed, the reading of the top-pan balance changes by 0.028 N when the ammeter reads 3.8 A. Calculate the magnetic flux density between the poles of the magnet.

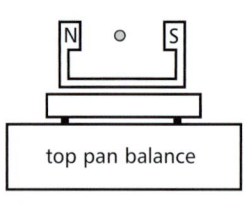

Figure 2.1

  b Calculate the force on the wire if the current is increased to 7.0 A.

3 The Earth's magnetic field at a certain location has a downward vertical component of 58 μT, and a horizontal component of 18 μT in a direction due north. A horizontal cable of length 50 m lies along a line from north to south. The cable carries a direct current of 26 A from south to north.

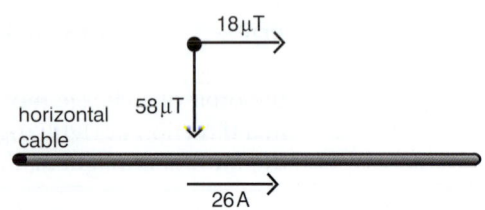

Figure 3.1

  a **i** Show that the magnitude of the force on the cable due to the Earth's magnetic field is 7.5×10^{-2} N.

    ii State the direction of the force on the cable.

  b Without further calculation, explain why the force on the cable for the same current would have been different, had the cable been aligned along a line from east to west instead of from north to south.

4 The armature coil of an electric motor has 100 turns and is of length 0.12 m, as shown below. The coil spins between the poles of a U-shaped magnet, where the magnetic flux density is 0.18 T.

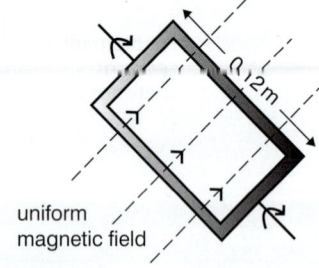

uniform magnetic field

Figure 4.1

  a **i** Calculate the force on each side of the coil when the current through it is 0.8 A.

    ii Discuss how the force on each side of the coil changes during one complete rotation of the coil.

  b Explain why the force acting on each side of the coil has its maximum turning effect when the plane of the coil is parallel to the lines of force of the magnetic field.

5 **a** The magnetic flux density between the poles of a U-shaped magnet was determined by measuring the force on a wire of length 32 mm positioned at right angles to the field lines. When a current of 4.5 A was passed along the wire, the force was 0.013 N. Calculate the magnetic flux density at the wire.

  b Calculate the angle the wire would need to be moved through to reduce the force on the wire by 5% when the current is 4.5 A.

6 The Earth's magnetic field at a certain location has a flux density of 0.070 mT in a direction due north at an angle of 70° to the surface, as shown in Figure 6.1. A straight wire of length 35 mm carrying a current of 8.2 A upwards is placed vertically in the field.

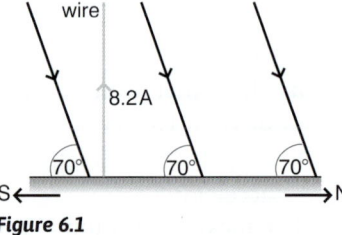

Figure 6.1

  a Show that the wire experiences a force of 6.9×10^{-6} N.

  b Determine the direction of the force.

7 A flat rectangular coil of length 65 mm and width 42 mm consists of 20 turns of insulated wire. The coil is placed in a horizontal uniform magnetic field of flux density 95 mT with its plane parallel to the field and with its shorter sides horizontal.

  a Sketch this arrangement and explain why the coil experiences a torque when it carries a direct current.

  b The current in the coil is adjusted to 4.2 A. Calculate the force on each of the longer sides due to the magnetic field and hence show that the torque on the coil is 0.022 N m.

8 In a d.c. electric motor, the coil consists of 60 turns of insulated wire wound on a rectangular frame of length 25 mm and width 20 mm. The coil spins in a magnetic field of magnetic flux density 110 mT.

  a Calculate the maximum torque on the coil due to the magnetic field when a current of 0.80 A passes through the coil.

  b A 1.0 N weight is attached to a thread wrapped round a pulley of diameter 5 mm fitted to the motor spindle. Discuss if this weight could be raised by the motor when the coil current is 0.80 A.

9 A beam of electrons in a uniform magnetic field of magnetic flux density 3.6 mT travel on a circular path of radius 55 mm.

  a Explain why the electrons travel at constant speed in the field.

  b Calculate the speed of the electrons.

10 A beam of electrons moving at a speed of $2.7 \times 10^7 \, \text{m s}^{-1}$ is directed horizontally into a uniform magnetic field of flux density 8.6 mT which is directed vertically upwards.

  a i Explain why the beam moves on a circular orbit in the field.

    ii Calculate the radius of curvature of the beam in the field.

  b The magnetic flux density is reduced steadily to zero then reversed and increased steadily to 8.6 mT in the opposite direction. Describe and explain the effect on the beam of these changes.

11 A cyclotron has a diameter 0.80 m. Protons are accelerated by the cyclotron to a maximum kinetic energy of 4.2 MeV.

  a Show that the maximum speed of the protons is $2.8 \times 10^7 \, \text{m s}^{-1}$.

  b i Calculate the magnetic flux density of the magnet used in this cyclotron.

    ii Show that the frequency of the alternating voltage applied to the cyclotron was 11 MHz.

12 In a mass spectrometer, a beam of different ions travelling at the same speed is directed into a uniform magnetic field at right angles to the field.

  a Explain why the magnetic field separates the ions according to their specific charge $\frac{Q}{m}$.

  b i In a test of a mass spectrometer, protons at a speed of $4.8 \times 10^6 \, \text{m s}^{-1}$ were directed into the field. The protons moved in the field in a semi-circular orbit of radius 60 mm. Calculate the magnetic flux density of the magnetic field.

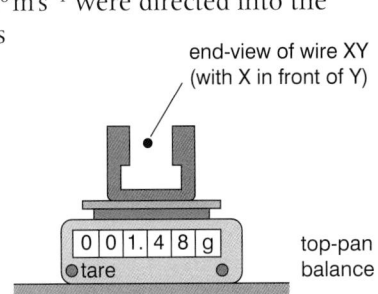

end-view of wire XY (with X in front of Y)

0 0 1.4 8 g

tare   top-pan balance

Figure 13.1

    ii Calculate the specific charge of an ion that moved on an orbit of radius 420 mm in the same flux density.

13 a Define the *magnetic flux density* of a magnetic field.

  b Figure 13.1 shows an arrangement used to measure the magnetic flux density B between the poles of a U-shaped magnet.

The wire is horizontal and perpendicular to the field which is also horizontal. The top-pan balance is used to measure the magnetic force on the length of wire between the poles when there is a current in the wire. The balance reading was set to zero (by pressing 'tare') before the current was switched on.

    i The following balance readings were recorded for different values of current in the wire.

Current/A	0	0.49	1.01	1.48	2.02	2.50
Balance reading/g	0	0.37	0.70	1.02	1.47	1.82
Force/10⁻³ N	0					

    i Copy and complete the table above by calculating the force on the wire.

    ii Plot a graph of the magnetic force on the wire against the current.

    iii Use the graph to calculate the force per unit current on the wire.

    iv The length of wire between the poles was 42 mm. Calculate the magnetic flux density between the poles.

  c i Assuming the readings did not fluctuate, the uncertainty in the current is ±0.01 A and the uncertainty in the balance reading is ±0.01 g. Estimate the uncertainty in your answer to **biii**.

    ii The uncertainty in the length measurement was ±2 mm. Estimate the uncertainty in the magnetic flux density.

17.1 Generating electricity

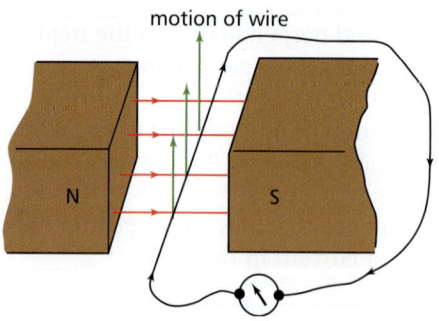

Figure 1 *Generating an electric current*

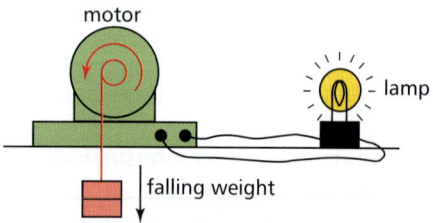

Figure 2 *A motor as a generator*

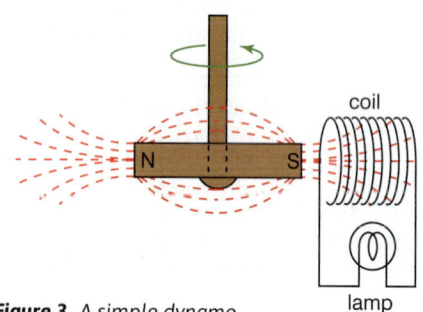

Figure 3 *A simple dynamo*

To generate electricity, all you need is a magnet and some wire which is connected to a sensitive meter, as shown in Figure 1. When the magnet is moved near the wire, a small current passes through the meter. This happens because an e.m.f. is **induced** in the wire. This effect, known as **electromagnetic induction**, occurs whenever a wire cuts across the lines of a magnetic field. If the wire is part of a complete circuit, the induced e.m.f. causes an induced current in the circuit as it forces electrons round the circuit. The induced e.m.f. can be increased by:

- moving the wire faster,
- using a stronger magnet,
- making the wire into a coil and pushing the magnet in or out of the coil, as in Figure 3 below. The more turns there are in the coil, the greater the induced e.m.f. is.

No e.m.f. is induced in the wire if the wire moves parallel to the magnetic field lines as it moves through the field. The wire must cut across the lines of the magnetic field for an e.m.f. to be induced in the wire.

Other methods of generating induced e.m.fs include:

1 **Using an electric motor in reverse**, as in Figure 2. The falling weight makes the motor coil turn between the poles of the magnet in the motor. The e.m.f. induced in the coil forces a current round the circuit and so causes the lamp to light. The faster the coil turns, the brighter the lamp is.
2 **Using a dynamo**, as in Figure 3. When the magnet in the dynamo spins, an e.m.f. is induced in the coil. If the coil is connected to a lamp, the lamp lights because the e.m.f. forces a current round the circuit.

In both examples above, an e.m.f. is induced because there is relative motion between the coil and the magnet. In the electric motor in reverse, the coil spins and the magnet is fixed. In the dynamo, the magnet spins and the coil is fixed.

Energy changes

When a magnet is moved relative to a conductor (e.g. a wire or a coil), an e.m.f. is induced in the circuit. If the conductor is part of a complete circuit which has no other sources of e.m.f., a current passes round the circuit just as if the circuit included a battery. However, unlike the e.m.f. of a battery which is constant, the induced e.m.f. becomes zero when the relative motion between the magnet and the wires ceases.

An electric current transfers energy from the source of the e.m.f. in a circuit to the other components in the circuit. For example, when a dynamo is used to light a lamp, energy is transferred from the dynamo to the lamp. The current through the dynamo coil causes a reaction force on the coil due to the magnet. Work must therefore be done to keep the magnet spinning. The energy transferred from the coil to the lamp is equal to the work done on the coil to keep it spinning.

The rate of transfer of energy from the source of e.m.f. to the other components of the circuit is equal to the product of the induced e.m.f. and the current. This is because:

- the induced e.m.f. is the energy transferred from the source per unit charge that passes through the source,
- the current is the charge flow per second.

So the induced e.m.f. × the current = energy transferred per unit charge from the source × the charge flow per second = energy transferred per second from the source.

Michael Faraday 1791–1861

Electromagnetic induction was discovered by Michael Faraday in 1831 at the Royal Institution, London. Faraday knew that a current passing along a wire produced a magnetic field near the wire and he wanted to know if a magnet could be used to produce a current. Using a magnetic compass near a loop of wire as a detector of current, he showed that the compass deflected whenever the magnet was moved in or out of the wire. He used the term **'electromotive force' (e.m.f.)** to describe the voltage induced in a wire. When he demonstrated his discoveries to an invited audience at the Royal Institution, he was asked the question 'What use is electricity, Mr Faraday?'. He replied with another question 'What use is a new baby?'. No one can tell what can grow from a new discovery.

Figure 4 Apparatus for electromagnetic induction

Understanding electromagnetic induction

When a beam of electrons is directed across a magnetic field, each electron experiences a force at right angles to its direction of motion and to the field direction. A metal rod is like a tube containing lots of free electrons. If the rod is moved across a magnetic field, as shown in Figure 5, the magnetic field forces the free electrons in the rod to one end away from the other end. So, one end of the rod becomes negative and the other end positive. In this way, an e.m.f. is induced in the rod. The same effect happens if the magnetic field is moved and the rod is stationary. As long as there is relative motion between the rod and the magnetic field, an e.m.f. is induced in the rod. If the relative motion ceases, the induced e.m.f. becomes zero because the magnetic field no longer exerts a force on the electrons in the rod. Note that when the rod is part of a complete circuit, the electrons are forced round the circuit. In other words, the induced e.m.f. drives a current round the circuit.

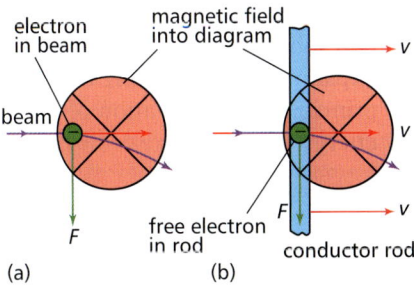

Figure 5 Deflection of electrons in a magnetic field

The dynamo rule

In Figure 5, the magnetic field is into the plane of the diagram and the motion of the conductor is towards the right. The electrons in the rod are forced downwards. The direction of the induced current can also be worked out using Fleming's right-hand rule, also referred to as the **dynamo rule**, as shown in Figure 6. The direction of the induced current is, in accordance with the current convention, opposite to the direction of the flow of electrons in the conductor.

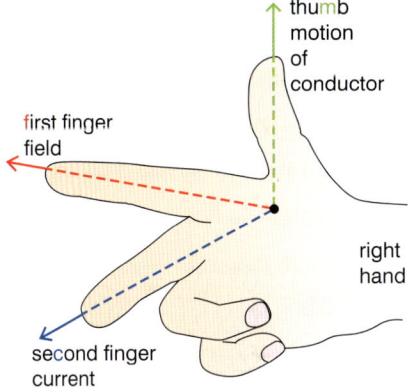

Figure 6 Fleming's right-hand rule

Summary test 17.1

1 A coil of wire is connected to a sensitive meter.

 a Explain why the meter shows a brief reading when a magnet is pushed into the coil.

 b State two ways in which the meter reading could be made larger.

2 An electric motor consists of a coil of wire between the poles of a magnet. The motor is connected to a lamp. A thread wrapped round the motor spindle is used to support a weight, as shown in Figure 2.

 a Explain why the lamp lights when the weight descends.

 b What difference would it have made if the magnet had been much stronger?

3 a Explain why a lamp connected to a dynamo lights when the dynamo turns.

 b Why is the dynamo easier to turn when the lamp is disconnected?

4 A horizontal rod aligned along a line from east to west is dropped through a horizontal magnetic field which is directed from south to north.

 a i What is the direction of the velocity of the rod?

   ii Determine which end of the rod is positive. Explain your answer.

 b Explain why no e.m.f. is induced in the rod if it is aligned from north to south then dropped in the field.

Learning outcomes

On these pages you will learn to:

- define magnetic flux and the weber
- recall and use $\Phi = BA$
- define magnetic flux linkage
- recall Faraday's law of electromagnetic induction and Lenz's law
- use the above laws to explain and solve problems
- explain Lenz's law using the principle of conservation of energy

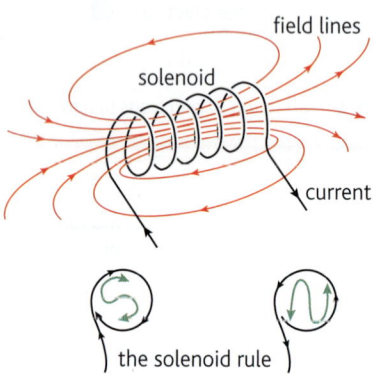

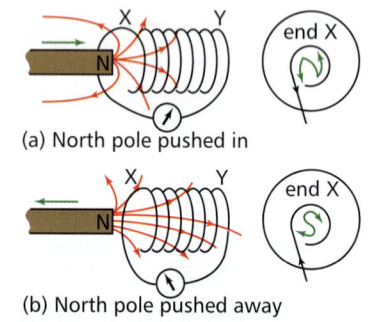

Figure 1 *The magnetic field near a solenoid*

Figure 2 *Lenz's law*

(a) North pole pushed in

(b) North pole pushed away

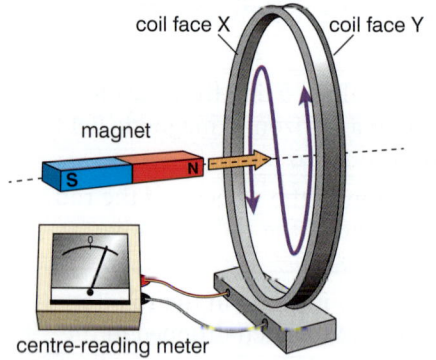

Figure 3 *Induced polarity of a coil*

Coils, currents and fields

A magnetic field is produced in and around a coil when an electromagnet is connected to a battery and a current is passed through it. A magnetic compass near the electromagnet is deflected when current passes through the coil. For a long coil or solenoid, the pattern of the magnetic field lines is like the pattern for a bar magnet – except the magnetic field lines near a bar magnet loop round from the north pole to the south pole of the magnet. Figure 1 shows the magnetic field pattern of a current-carrying solenoid. The field lines pass through the solenoid and loop round outside the solenoid from one end (the north pole) to the other end (the south pole). If each end in turn is viewed from outside the solenoid:

- current passes a**N**ticlockwise (or cou**N**terclockwise) round the '**N**orth pole' end,
- current passes clockwise round the 'south pole' end.

Lenz's law

When a bar magnet is pushed into a coil connected to a meter, the meter deflects. If the bar magnet is pulled out of the coil, the meter deflects in the opposite direction. What determines the direction of the induced current? Consider the north pole of a bar magnet approaching end X of a coil, as shown in Figure 2.

The induced current passing round the circuit creates a magnetic field due to the coil. The coil field must act against the incoming north pole, otherwise it would pull the north pole in faster, making the induced current bigger, pulling the north pole in even faster still, etc. Clearly, conservation of energy forbids this creation of kinetic and electrical energy from nowhere. So, the induced current creates a magnetic field in the coil which opposes the incoming north pole. The induced polarity of end X must therefore be a north pole so as to repel the incoming north pole. Therefore, the current must go round end X of the coil in an anticlockwise direction, as shown.

If the magnet is removed from inside the coil, the induced current passes round end X of the coil in a clockwise direction. This corresponds to an induced south pole at end X which, therefore, opposes the magnet moving away.

> **Lenz's law states that the direction of the induced current is always such as to oppose the change that causes the current.**

The explanation of Lenz's law is that energy is never created or destroyed. The induced current could never be in a direction to help the change that causes it; that would mean producing electrical energy from nowhere, which is forbidden!

An induced current is generated if the coil is moved instead of the magnet and the magnet is at rest. In general, whenever the magnet and the coil move relative to each other, an induced current is generated and Lenz's law can be used to work out the direction of the current.

Some further examples of the applications of Lenz's law are given below.

1 **A bar magnet moves towards and through a flat coil**, as shown in Figure 3.
   - When the bar magnet approaches face X of the coil with its north pole leading, the induced polarity of face X of the coil must be a north pole in order to oppose the movement of the bar magnet. So the induced current in the coil is anticlockwise, as seen in face X.
   - After passing through the coil, the bar magnet moves away from face Y of the coil with its south pole trailing. The induced polarity of face Y of the coil must be a north pole in order to oppose the south pole of the bar magnet moving away.

So the induced current in the coil is anticlockwise, as seen in face Y and therefore clockwise, as seen in face X, opposite in direction to when the magnet was approaching the coil.

2 **A flat coil or a solenoid (P) carrying a direct current moves towards another flat coil (Q)**, as shown in Figure 4. The direction of the current in P determines its magnetic polarity.

When P and Q are approaching each other, the induced polarity in Q is the same as the 'leading' polarity of P (and therefore opposite to the 'trailing' polarity of P). The same ideas apply if Q is replaced by a large solenoid.

Faraday's law of electromagnetic induction

Consider a conductor of length L, which is part of a complete circuit cutting through the lines of a magnetic field of flux density B.

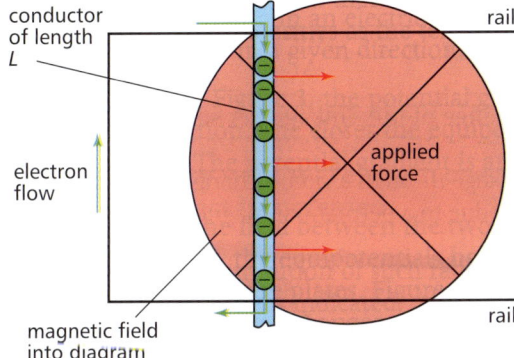

Figure 5 Induced e.m.f. in a conductor

An e.m.f., E is induced in the conductor and an induced current I passes round the circuit.

The conductor experiences a force $F = BIL$ due to carrying a current in a magnetic field. The force opposes the motion of the conductor and so an equal and opposite force must be applied to the conductor to keep it moving in the field. If the conductor moves a distance Δs in time Δt:

- the work done W by the applied force is given by $W = F\,\Delta s = BIL\,\Delta s$
- the charge transfer along the conductor in this time $Q = I\Delta t$.

Therefore, the induced e.m.f. $E = \dfrac{W}{Q} = \dfrac{BIL\,\Delta s}{I\Delta t} = \dfrac{BL\,\Delta s}{\Delta t}$

As $L\Delta s = \Delta A$ 'swept out' in time Δt,

$$E = \frac{B\Delta A}{\Delta t} = \frac{\Delta BA}{\Delta t} = \frac{\Phi}{\Delta t}$$

where $\Phi = BA$ = magnetic flux.

The product of the magnetic flux density, B, and the area, A, is called the **magnetic flux**. The concept of magnetic flux is very useful for calculating induced e.m.fs. The example of the conductor cutting across the field lines shows that the induced e.m.f. is equal to the magnetic flux swept out by the conductor each second. Michael Faraday was the first person to show how induced e.m.fs could be calculated from magnetic flux changes.

- **Magnetic flux**, $\Phi = BA$.
- **Flux linkage through a coil of N turns = $N\Phi = NBA$** where B is the magnetic flux density perpendicular to area A.
- The unit of magnetic flux is the **weber** (Wb), equal to $1\,\mathrm{T\,m^2}$.

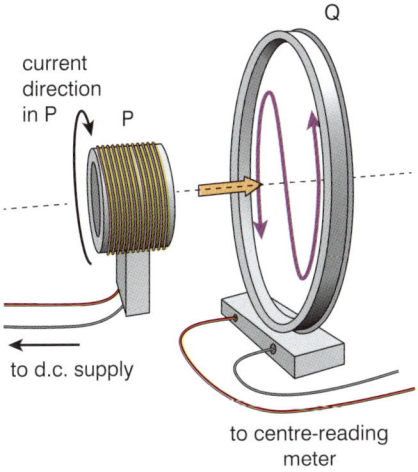

Figure 4 Induced polarity due to another coil

Figure 6 An electric car

Regenerative braking

A battery-powered electric vehicle contains an **alternator** that can be used as an electric motor or as a generator. When the alternator is used as an electric motor, it is driven by the batteries. When the brakes are applied, the alternator is used to generate electricity which is used to recharge the battery. Some of the kinetic energy is transferred to electrical energy in the battery. The induced current through the alternator coil creates a magnetic field that acts against the magnetic field of the alternator. So, the alternator experiences a braking force which helps to slow the vehicle down.

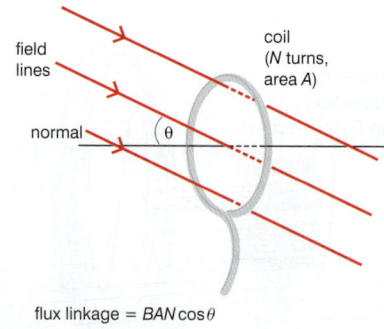

flux linkage = $BAN \cos \theta$

Figure 7 *Flux linkage*

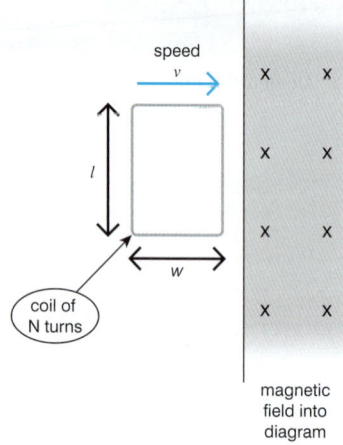

Figure 8 *Flux changes*

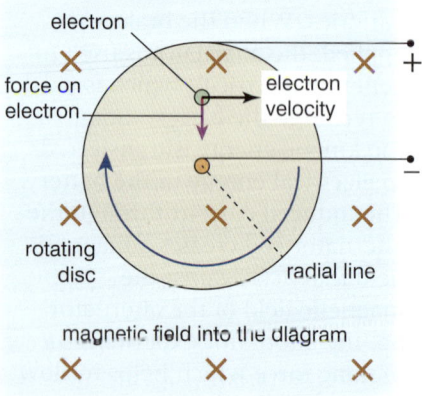

Figure 9 *The rotating disc*

Note that magnetic flux density B (in teslas) is the flux per unit area passing through an area at right angles to the area (i.e. normally). Therefore 1 tesla = 1 weber per square metre.

In general, when the magnetic field is at angle θ to the normal at the coil face, as shown in Figure 7, the flux linkage through the coil $N\Phi = BAN \cos \theta$

1 When the magnetic field is along the normal (i.e. perpendicular) to the coil face, the flux linkage = BAN.
2 When the coil is turned through 180°, the flux linkage = $-BAN$.
3 When the magnetic field is parallel to the coil area, the flux linkage = 0 as no field lines pass through the coil area.

Faraday's law of electromagnetic induction states that the induced e.m.f. in a circuit is proportional to the rate of change of flux linkage through the circuit.

$$\text{Induced e.m.f., } E = -N\frac{\Delta\Phi}{\Delta t}$$

where $N\dfrac{\Delta\Phi}{\Delta t}$ is the change of flux linkage per second.

Whenever the flux linkage through a circuit changes, an e.m.f. is induced in the circuit. The flux can be due to a permanent magnet or due to a current-carrying wire.

- If the flux is due to a permanent magnet, motion of the magnet relative to the circuit is necessary to cause an induced e.m.f. This is how an e.m.f. is generated in an a.c. generator or a dynamo.
- If the flux is due to a current-carrying wire, changing the current in the wire causes an induced e.m.f. in the circuit. This is how an e.m.f. is generated in a **transformer** or an induction coil.

Examples

1 A moving conductor in a magnetic field
An e.m.f. is induced in the conductor provided the conductor cuts across the lines of the magnetic field. The direction of motion of the conductor in Figure 5 is at right angles to the field lines.

As explained on the previous page, the magnitude of the induced e.m.f. $E = \dfrac{Bl\Delta s}{\Delta t}$,

where l is the length of the conductor and Δs is the distance it moves in time Δt. Note that the change of flux in this time, $\Delta\Phi = Bl\Delta s$ so the change of flux per second, $\dfrac{\Delta\Phi}{\Delta t}$, is equal to the magnitude of the induced e.m.f.

Because the speed of the conductor, $v = \dfrac{\Delta s}{\Delta t}$, the induced e.m.f. $E = \dfrac{Bl\Delta s}{\Delta t} = Blv$.

$$\text{Induced e.m.f., } E = Blv$$

2 A rectangular coil moving into a uniform magnetic field
Consider a rectangular coil of N turns, length l and width w moving into a uniform magnetic field of flux density B at constant speed v, as shown in Figure 8. Suppose the coil enters the field at time $t = 0$.

- The time taken by the coil to enter the field completely $= \dfrac{\text{coil width}}{\text{speed}} = \dfrac{w}{v}$.

  During this time, the flux linkage $N\Phi$ increases steadily from 0 to $BNlw$. Therefore, the induced e.m.f. = change of flux linkage per second,
  $$\frac{N\Delta\Phi}{\Delta t} = \frac{BNlw}{w/v} = BlvN$$

- When the coil is completely in the field, the flux linkage through it (= $BNlw$) does not change so the induced e.m.f. is zero.

3 A spinning disc in a uniform magnetic field

When the disc spins at constant frequency, a constant e.m.f. is induced between the centre of the disc and its edge. In Figure 9, the conduction electrons in the rotating disc are forced to the centre by the magnetic field so the induced e.m.f. is negative at the centre and positive at the edge. Reversing the field or rotation direction reverses the induced e.m.f. In one revolution, each radial line on the disc sweeps out an area $A = \pi r^2$ where r is the disc radius. Therefore, the magnetic flux swept out:

- in one revolution is $B \times \pi r^2$ ($= B \times A$),
- in one second is $B \times \pi r^2 \times f$, where f is the frequency of rotation of the disc.

Therefore, the induced e.m.f., E = change of magnetic flux per second = $B \times \pi r^2 \times f$

4 A fixed coil in a changing magnetic field

Figure 10 shows a small coil on the axis of a current-carrying solenoid. The magnetic field of the solenoid passes through the small coil. If the current in the solenoid changes, an e.m.f. is induced in the small coil. This is because the magnetic field through the coil changes so the flux linkage through it changes, causing an induced e.m.f. Because the induced e.m.f. is proportional to the rate of change of flux linkage through the coil and the flux linkage is proportional to the current I in the solenoid, the magnitude of the induced e.m.f is, therefore, proportional to the rate of change of current in the solenoid.

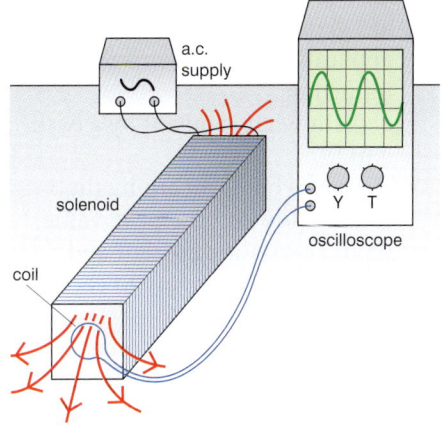

Figure 10 *A changing magnetic field*

Summary test 17.2

1 A uniform magnetic field of flux density 72 mT is confined to a region of width 60 mm, as shown in Figure 11. A rectangular coil of length 50 mm and width 20 mm has 15 turns. The coil is moved into the magnetic field at a speed of 10 mm s^{-1} with its longer edge parallel to the edge of the magnetic field.

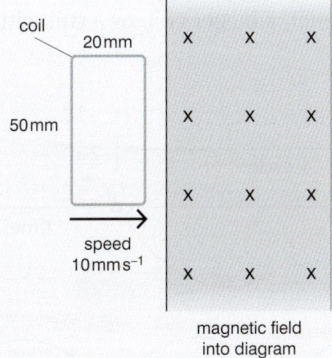

Figure 11

a Calculate:
   i the flux linkage through the coil when it is completely in the field,
   ii the time taken for the flux linkage to increase from zero to its maximum value,
   iii the induced e.m.f. in the coil as it enters the field.

b i Sketch a graph to show how the flux linkage through the coil changes with time from the instant the coil enters the field to when it leaves the field completely.
   ii Sketch a graph to show how the induced e.m.f. in the coil varies with time.

2 A rectangular coil of length 40 mm and width 25 mm has 20 turns. The coil is in a uniform magnetic field of flux density 68 mT.

a Calculate the flux linkage through the coil when the coil is at right angles to the field lines.

b The coil is removed from the field in 60 ms. Calculate the mean value of the induced e.m.f.

3 A circular coil of diameter 24 mm has 40 turns. The coil is placed in a uniform magnetic field of flux density 85 mT with its plane perpendicular to the field lines.

a Calculate:
   i the area of the coil in m^2,
   ii the flux linkage through the coil.

b The coil was reversed in a time of 95 ms. Calculate:
   i the change of flux linkage through the coil,
   ii the magnitude of the induced e.m.f.

4 A small circular coil of diameter 15 mm and 25 turns is placed in a fixed position on the axis of a solenoid, as shown in Figure 10. The magnetic flux density of the solenoid at this position varies with current according to the equation $B = kI$, where $k = 1.2 \times 10^{-3}\,\text{T A}^{-1}$.

a Calculate the flux linkage through the coil when the current in the solenoid is 1.5 A.

b The current in the solenoid was reduced from 1.5 A to zero in 0.20 s. Calculate the magnitude of the induced e.m.f. in the small coil.

Learning outcomes

On these pages you will learn to:

- describe and explain the a.c. generator as a simple application of electromagnetic induction
- explain what is meant by a back e.m.f.
- explain what is meant by the r.m.s. value and the peak value of an alternating current and know how they relate to each other
- deduce that the mean power in a resistive load is half the maximum power for a sinusoidal alternating current
- represent a sinusoidally alternating current or voltage by an equation of the form $x = x_0 \sin \omega t$
- describe rectifier circuits with and without a smoothing capacitor

Alternating current generators are used in power stations and in mobile and emergency generators. In this section, the simple a.c. generator is considered as an application of electromagnetic induction.

The simple a.c. generator

The simple a.c. generator consists of a rectangular coil that spins in a uniform magnetic field, as shown in Figure 1. When the coil spins at a steady rate, the flux linkage changes continuously. At an instant when the normal to the plane of the coil is at angle θ to the field lines, the flux linkage through the coil, $N\Phi = BAN\cos\theta$, where B is the magnetic flux density, A is the coil area and N is the number of turns on the coil.

For a coil spinning at a steady frequency, f, $\theta = 2\pi ft$ at time t after $\theta = 0$. So the flux linkage $N\Phi$ (= $BAN\cos 2\pi ft$) changes with time as shown in Figure 2.

- The gradient of the graph is the change of flux linkage per second, $N\dfrac{\Delta\Phi}{\Delta t}$, so it represents the induced e.m.f. It can be shown mathematically that the induced e.m.f. alternates in accordance with the equation $E = E_0 \sin 2\pi ft$, where f is the frequency of rotation of the coil and E_0 is the peak e.m.f. Substituting the angular frequency ω for $2\pi f$ therefore gives $E = E_0 \sin \omega t$. Figure 3 shows how the induced e.m.f., E, varies with time t.

- The induced e.m.f is zero when the sides of the coil move parallel to the field lines. At this position, the rate of change of flux is zero and the sides of the coil do not cut the field lines.

- The induced e.m.f. is a maximum when the sides of the coil cut at right angles across the field lines. At this position, the e.m.f. induced in each wire of each side = Blv, where v is the speed of each wire and l is its length. So, for N turns and two sides, the induced e.m.f. at this position $E_0 = 2NBlv$. The equation shows that the peak e.m.f. can be increased by increasing the speed (i.e. the frequency of rotation) or by using a stronger magnet, a bigger coil, or a coil with more turns.

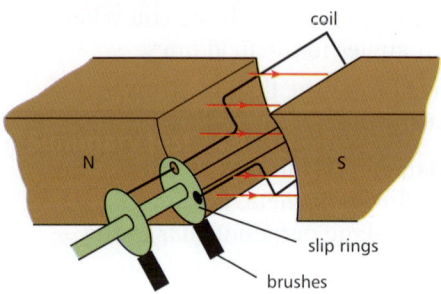

Figure 1 The a.c. generator

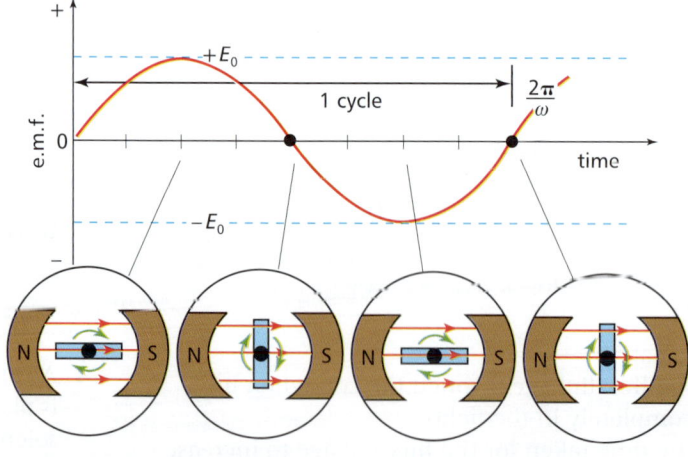

Figure 3 E.m.f. v. time for an a.c. generator

Note An a.c. generator in a power station has three coils at 120° to each other. Each coil produces an alternating voltage 120° out of phase with the voltage from the other coils.

a

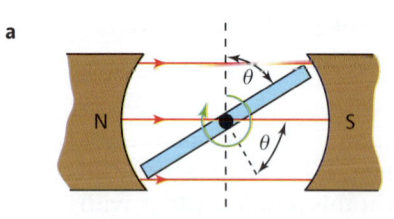

b

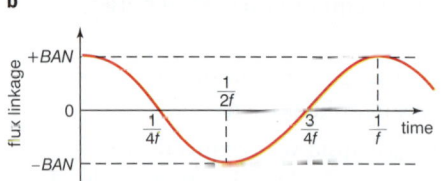

Figure 2 Flux linkage in a spinning coil

Back e.m.f.

An e.m.f. is induced in the spinning coil of an electric motor because the flux linkage through the coil changes. The induced e.m.f. E is referred to as a **back e.m.f.** because it acts against the p.d. V applied to the motor in accordance with Lenz's law. At any instant, $V - E = IR$, where I is the current through the motor coil and R is the circuit resistance.

Because the induced e.m.f. is proportional to the speed of rotation of the motor, the current changes as the motor speed changes.

- At low speed, the current is high because the induced e.m.f. is small.
- At high speed, the current is low because the induced e.m.f. is high.

Multiplying the equation $V - E = IR$ by I throughout gives $IV - IE = I^2R$. Rearranging this equation gives:

Electrical power supplied by the source		Electrical power transferred to mechanical power		Electrical power wasted due to the circuit resistance
IV	$=$	IE	$+$	I^2R

The above power equation shows that electrical power supplied to the motor that is not used as mechanical power is wasted due to the resistance heating effect of the current.

- When the motor spins without driving a **load**, it spins at high speed so the current is very small. Its speed is limited by friction in the bearings and air resistance. It uses little or no power.
- When the motor is used to drive a load, its speed is less than when it is 'off-load' so the current is much larger. The power it uses from the voltage source that is not transferred as mechanical power to the load is wasted due to the resistance heating effect of the current.

Rectifier circuits

Diodes are used to 'rectify' or convert alternating current to direct current.

Half-wave rectification is produced using a single diode, as shown in Figure 4. The resistor is needed to limit the current otherwise the diode would be destroyed by overheating. The diode conducts every other half-cycle when the a.c. supply across it is in the forward direction. The current is negligible when the supply is in the reverse direction.

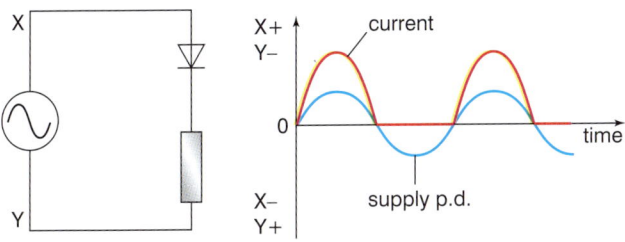

Figure 4 Half-wave rectification

Full-wave rectification is produced using four diodes in a **bridge rectifier** circuit as shown in Figure 5. Diodes D_1 and D_3 conduct when X is positive (and Y is negative); diodes D_4 and D_2 conduct when Y is positive (and X is negative). As a result, the direction of the current through R is the same in both half-cycles. The current waveform in Figure 5 shown in purple is the full-wave rectified wave form.

The waveform can be made smoother by connecting a capacitor across the output terminals of the bridge rectifier. The capacitor charges up from the bridge rectifier as the output p.d. increases then discharges through the resistor as the output p.d. decreases. In this way, the current through the resistor is smoothed out, as shown in red in Figure 5. Note that the capacitance C of the capacitor should be sufficiently large in relation to the load resistance R so that the time constant RC is much greater than the time period of the a.c. supply. The effect would then be to reduce the decrease in the current (from the peak value) so the ripple is much smaller.

The effect of increasing the time constant can be calculated using the capacitor discharge equation $I = I_0 e^{-t/RC}$. For example, in each half-cycle at 50 Hz, the discharge time would be 0.01 s ($= \frac{1}{2} \times 0.02$ s) so for $R = 100\,\Omega$ and

- $C = 1.0\,\text{mF}$, the time constant RC would be 0.1 s giving $I = I_0 e^{-0.01\,s/0.1\,s} = I_0 e^{-0.1} = 0.905\,I_0$, which is a decrease of $0.095\,I_0$ or a 9.5% drop.
- $C = 100\,\text{mF}$, the time constant RC would be 10 s giving $I = I_0 e^{-0.01\,s/10\,s} = I_0 e^{-0.001} = 0.999\,I_0$, which is a decrease of $0.001\,I_0$ or a 0.1% drop.

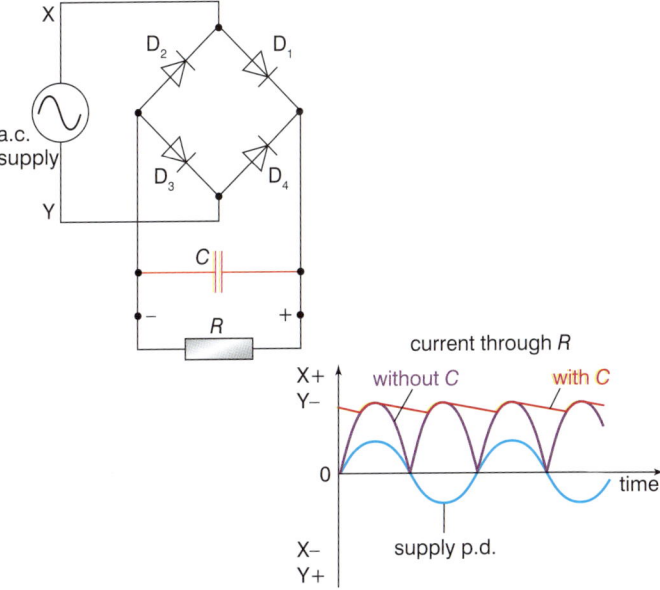

Figure 5 Full-wave rectification

The heating effect of an alternating current

Imagine an electric heater supplied with alternating current at a very low frequency. The heater would heat up and then cool down repeatedly as the current changed.

As explained in Topic 7.2, the heating effect of an electric current varies according to the square of the current. This is because the electrical power P supplied to the heater for a current I is given by

$P = IV = I^2R = I_0^2R \sin^2 \omega t$, where R is the resistance of the heater element and the current $I = I_0 \sin \omega t$.

Figure 6 shows how the power ($= I^2R$) varies with time.

- At the peak current I_0 in either direction, maximum power is supplied equal to I_0^2R.
- At zero current, zero power is supplied.

For a sinusoidal current, the **mean power** over a full cycle is half the peak (i.e. maximum) power. This can be seen from the symmetrical shape of the power curve in Figure 6 about the mean power. The mean power is therefore $\frac{1}{2}I_0^2R$.

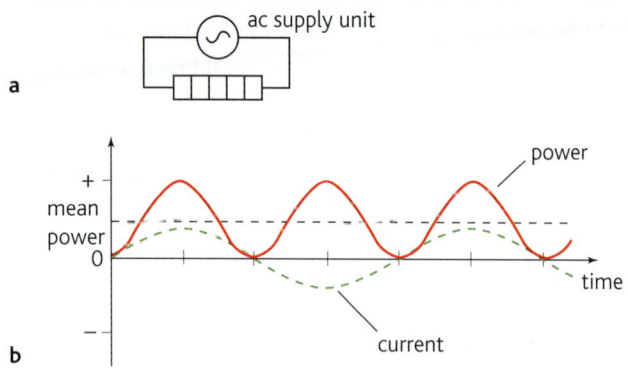

Figure 6 *Variation of power with time for an alternating current*

The direct current that would give the same power as the mean power is called the **root mean square value** of the alternating current, $I_{r.m.s.}$.

> **The root mean square value of an alternating current is the value of direct current that would give the same heating effect as the alternating current in the same resistor.**

Therefore, $I_{r.m.s.}^2R = 0.5I_0^2R$

Cancelling R from this equation and rearranging the equation gives $I_{r.m.s.}^2 = 0.5I_0^2$

Therefore $I_{r.m.s.} = \dfrac{1}{\sqrt{2}}I_0$

Also, the root mean square value of an alternating p.d. is given by $V_{r.m.s.} = \dfrac{1}{\sqrt{2}}V_0$

The root mean square value of an alternating current or p.d. = $\dfrac{1}{\sqrt{2}}$ × the peak value

For example, if the peak voltage of an a.c. supply is 50 V, the r.m.s. value will be 35 V.

> ### Note
>
> For any resistor of known resistance in an alternating circuit, if we know the r.m.s. p.d. or current for the resistor, we can calculate the mean power supplied to it using the r.m.s. values.
>
> $$P = I_{r.m.s.}^2R = V_{r.m.s.}^2/R = I_{r.m.s.}V_{r.m.s.}$$
>
> For example, if an alternating current of r.m.s. value 4 A is passed through a 5 Ω resistor, the mean power supplied to the resistor $= 4^2 \times 5 = 80$ W.

Summary test 17.3

1 a An a.c. generator produces an alternating e.m.f. with a peak value of 8.0 V and a frequency of 20 Hz. Sketch a graph to show how the e.m.f. varies with time.

   b The frequency of rotation of the a.c. generator in **a** is increased to 30 Hz. On the same axes, sketch a graph to show how the e.m.f. varies with time at 30 Hz.

2 A 230 V 1000 W electric heater has a single heating element.

   a Calculate: **i** the r.m.s. current, **ii** the peak current in the heating element when the heater is switched on.

   b Calculate the peak power supplied to the heater when it is on.

   The potential difference of the electricity supply decreases at times of high demand. Estimate the percentage drop in the power supplied to the heater when it is on and the p.d. drops by 5%.

3 The coil of an a.c. generator has 80 turns, a length of 65 mm and a width of 38 mm. It spins in a uniform magnetic field of flux density 130 mT at a constant frequency of 50 Hz.

   a Calculate the maximum flux linkage through the coil.

   b **i** Show that each side of the coil moves at a speed of 6.0 m s^{-1}.

      ii Show that the peak voltage is 8.1 V.

4 An electric motor is to be used to move a variable load. The motor is connected in series with a battery and an ammeter.

   a Explain why the motor current is very small when the load is zero.

   b Explain why the motor current increases when the load is increased.

The transformer

A transformer changes an alternating p.d. to a different peak value. Any transformer consists of two coils: the primary coil and the secondary coil. The two coils have the same iron core. When the primary coil is connected to a source of alternating p.d., an alternating magnetic field is produced in the core. The field passes through the secondary coil. So, an alternating e.m.f. is induced in the secondary coil by the changing magnetic field. The symbol for the transformer is shown in Figure 1.

The transformer rule

A transformer is designed so that all the magnetic flux produced by the primary coil passes through the secondary coil.

Let Φ = the flux in the core passing through each turn at an instant when an alternating p.d. V_P is applied to the primary coil.

- The flux linkage in the secondary coil = $N_S \Phi$, where N_S is the number of turns on the secondary coil. From Faraday's law, the induced e.m.f. in the secondary coil, $V_S = N_S \dfrac{\Delta \Phi}{\Delta t}$

- The flux linkage in the primary coil = $N_P \Phi$, where N_P is the number of turns on the primary coil. From Faraday's law, the induced e.m.f. in the primary coil = $N_P \dfrac{\Delta \Phi}{\Delta t}$. The induced e.m.f. in the primary coil opposes the p.d. applied to the primary coil, V_P.

Assuming the resistance of the primary coil is negligible, the applied p.d.

$$V_P = N_P \frac{\Delta \Phi}{\Delta t}$$

Dividing the equation for V_S by the equation for V_P gives $\dfrac{V_S}{V_P} = N_S \dfrac{\Delta \Phi}{\Delta t} \bigg/ N_P \dfrac{\Delta \Phi}{\Delta t}$

Cancelling $\dfrac{\Delta \Phi}{\Delta t}$ from this equation gives the **transformer rule**

$$\frac{V_S}{V_P} = \frac{N_S}{N_P}$$

- **A step-up transformer** has more turns on the secondary coil than on the primary coil. So the secondary voltage is stepped up compared with the primary voltage (i.e. $N_S > N_P$ so $V_S > V_P$)

- **A step-down transformer** has fewer turns on the secondary coil than on the primary coil. So the secondary voltage is stepped down compared with the primary voltage (i.e. $N_S < N_P$ so $V_S < V_P$).

Transformer efficiency

$$\frac{\text{The efficiency of}}{\text{a transformer}} = \frac{\text{power delivered by the secondary coil}}{\text{power supplied to the primary coil}} = \frac{I_S V_S}{I_P V_P} \ (\times 100\%)$$

When a device (e.g. a lamp) is connected to the secondary coil, because the efficiency of a transformer is almost equal to 100%,

$$\frac{\text{the electrical power supplied}}{\text{to the primary coil}} = \frac{\text{the electrical power supplied}}{\text{by the secondary}}$$

Therefore, the current ratio $\dfrac{I_S}{I_P} = \dfrac{V_P}{V_S} = \dfrac{N_P}{N_S}$

- In a step-up transformer, the voltage is stepped up and the current is stepped down.
- In a step-down transformer, the voltage is stepped down and the current is stepped up.

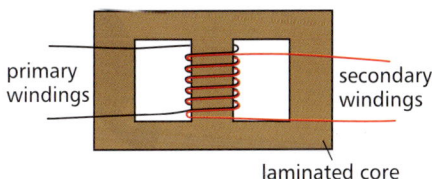

primary windings   secondary windings

laminated core

a Practical arrangement

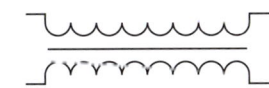

b Transformer symbol

Figure 1 The transformer

The grid system

Electricity from power stations in most countries is fed into a national grid system which supplies electricity to most parts of the country. The grid is a network of cables, either underground or on pylons. Each power station generates alternating current in three phases (see Topic 17.3) at a precise frequency of 50 Hz (or 60 Hz in some countries) at about 25 kV.

Step-up transformers at the power station increase the alternating voltage to 400 kV or more for long-distance transmission via the grid system. Step-down transformers operate in stages, as shown in Figure 2. Factories are supplied with all three phases at either 33 kV or 11 kV. Homes are supplied via a local transformer sub-station with single-phase a.c. at 220–240 V in most countries except in America (110–130 V), Japan (100 V) and a few other countries.

Transmission of electrical power over long distances is much more efficient at high voltage than at low voltage. The reason is that the current needed to deliver a certain amount of power is reduced if the voltage is increased. So power wasted due to the heating effect of the current through the cables is reduced. To deliver power P at voltage V, the current required $I = \dfrac{P}{V}$.

If the resistance of the cables is R, the power wasted through heating the cables is $I^2 R = \dfrac{P^2 R}{V^2}$. Therefore, the higher the voltage, the smaller the ratio of the wasted power to the power transmitted.

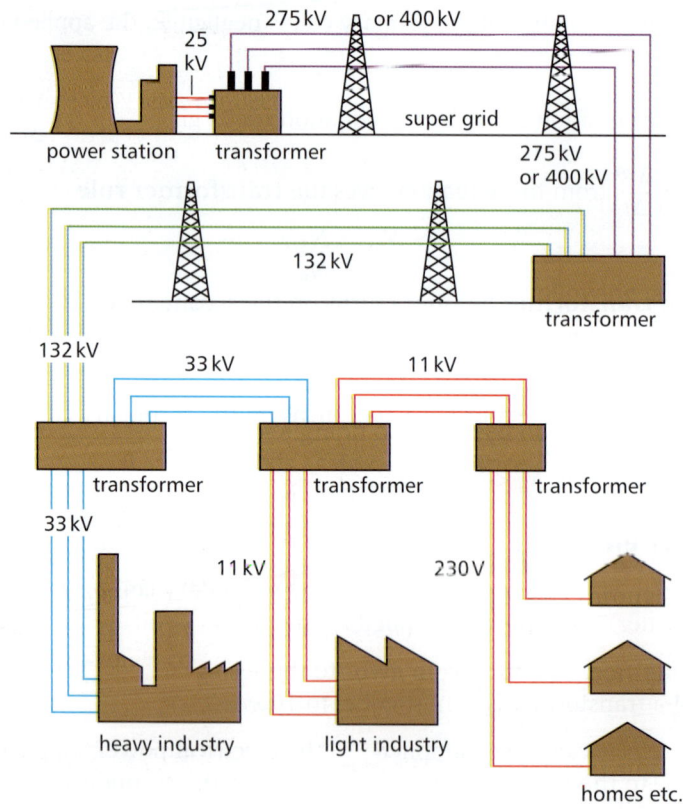

Figure 2 *The grid system*

For example, for transmission of 1 MW of power through cables of resistance 500 Ω at 25 kV, the current necessary would be 40 A (= 1 MW/25 kV) so the power wasted would be 0.8 MW (= $I^2 R = 40^2 \times 500$ W). Prove for yourself that at 400 kV, the power wasted would be about 3 kW.

Voltage adaptors

Portable electronic devices such as laptop computers contain batteries that need to be recharged using a voltage adaptor. Figure 3 shows the circuit of such an adaptor. It supplies a constant low voltage when it is plugged into a mains socket.

- The transformer steps down the alternating p.d. from the mains to an alternating p.d. with a much smaller peak p.d. The turns ratio of the transformer is chosen to give a suitable p.d. from the transformer.
- The bridge rectifier converts the alternating p.d. from the transformer to a full-wave direct p.d.
- The capacitor across the bridge rectifier smooths out the full-wave p.d. to give a constant output p.d, as explained in Topic 17.3. The greater the capacitance of C, the smaller the variation in the output p.d. as current is drawn from the voltage adaptor.

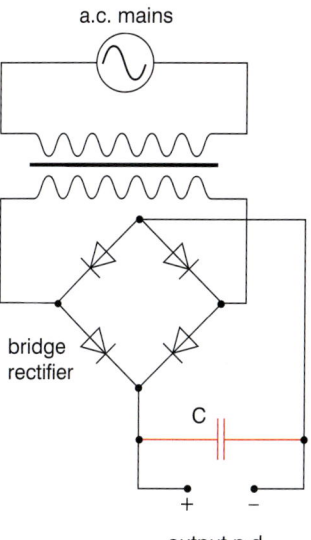

Figure 3 A voltage adaptor

1 a Explain why an alternating e.m.f. is induced in the secondary coil of a transformer when the primary coil is connected to an alternating voltage supply.

 b In terms of electrical power, explain why the current through the primary coil of a transformer increases when a device is connected to the secondary coil.

2 a Explain why a transformer is designed so that as much of the magnetic flux produced by the primary coil of a transformer as possible passes through the secondary coil.

 b Explain why a transformer works using alternating current but not using direct current.

3 A transformer has a primary coil with 120 turns and a secondary coil with 2400 turns.

 a Calculate the primary voltage needed for a secondary voltage of 230 V.

 b A 230 V, 60 W lamp is connected to its secondary coil. Calculate the current through:
 i the secondary coil,
 ii the primary coil. State any assumptions made in this calculation.

4 a Explain why transmission of electrical power over a long distance is more efficient at high voltage than at low voltage.

 b A power cable of resistance 200 Ω is to be used to deliver 2.0 MW of electrical power at 120 kV from a power station to an industrial estate. Calculate:
 i the current through the cable,
 ii the power wasted in the cable.

Magnetic flux, $\Phi = BA$

Flux linkage through a coil of N turns $= N\Phi = NBA$, where B is the magnetic flux density perpendicular to area A.

Lenz's law states that the direction of the induced current is always such as to oppose the change that causes the current.

Faraday's law of electromagnetic induction states that the induced e.m.f. in a circuit is proportional to the rate of change of flux linkage through the circuit.

Equation for Faraday's law; $E = -N\dfrac{\Delta \Phi}{\Delta t}$, where $N\dfrac{\Delta \Phi}{\Delta t}$ is the change in flux linkage per second.

For a moving conductor in a uniform magnetic field, the induced e.m.f. $= Blv$

For a changing magnetic field in a fixed coil,

induced e.m.f. $= NA\dfrac{\Delta B}{\Delta t}$

Units

The unit of magnetic flux density B is the **tesla** (T).

The unit of magnetic flux and of flux linkage is the **weber** (Wb), equal to $1\,T\,m^2$ or $1\,V\,s$.

The unit of rate of change of flux (or rate of change of flux linkage) is the weber per second ($Wb\,s^{-1}$), equal to $1\,V$.

17 Exam-style and Practice Questions

⌕ **Launch additional digital resources for the chapter**

1 In a ribbon microphone, a metal ribbon vibrates between the poles of a magnet when sound waves reach the microphone.

  a Explain why an e.m.f. is induced across the ends of the ribbon when it vibrates.

  b How would the e.m.f. be affected if:
    i a stronger magnet was used,
    ii a ribbon of greater mass was used?

2 **a** A bar magnet was positioned near a coil connected to a centre-reading meter. When the bar magnet was pushed into the coil, the meter pointer deflected briefly to the right.
    i Explain why the pointer deflected briefly.
    ii State and explain what is observed when the magnet is withdrawn from the coil.

  b In **a**, the flux density of the magnet was 25 mT and the area of the 30 turn coil was $4.0 \times 10^{-4}\,\text{m}^2$. The magnet was pushed into the coil in 0.20 s. Calculate:
    i the flux linkage through the coil,
    ii the mean induced e.m.f.

3 A U-shaped magnet was placed at the centre of a horizontal stretched steel wire such that the magnetic field was vertical. When the wire was plucked at its centre, an alternating e.m.f. was induced between the ends of the wire.

  a Explain why an alternating e.m.f. was induced between the ends of the wire.

  b The length of wire between the poles was 28 mm and the magnetic flux density of the magnet was 78 mT. The peak voltage produced was 3.4 mV. Calculate the maximum speed of the wire between the poles.

4 **a** A straight conducting rod PQ of length 0.10 m moves through a uniform magnetic field at a speed of 20 mm s⁻¹. The induced e.m.f. was 0.60 mV when the conductor was moving perpendicular to the field. Calculate:
    i the flux swept out in 5.0 s,
    ii the magnetic flux density.

  b In **a**, the field was vertically downwards and the rod was horizontal, as shown in Figure 4.1.
    i State the polarity of each end of the rod.
    ii If the rod was part of a complete circuit, state and explain the direction in which the induced current would pass through it.

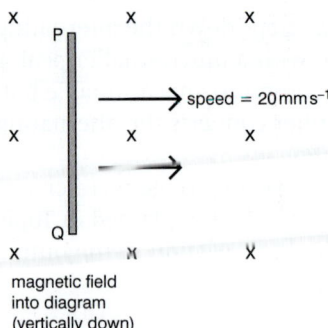

magnetic field into diagram (vertically down)

Figure 4.1

5 A rectangular coil of length 50 mm and width 20 mm has 25 turns. Figure 5.1 below shows the coil just before it was moved at a constant speed of 5 mm s⁻¹ into a uniform magnetic field of flux density 86 mT. The field lines are perpendicular to the plane of the coil.

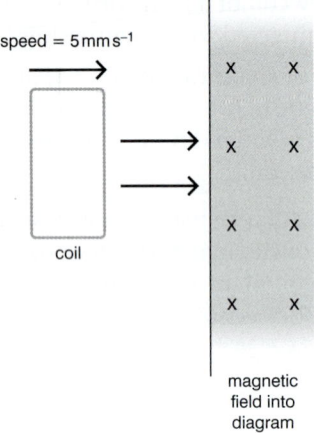

magnetic field into diagram

Figure 5.1

  a Show that the coil takes 4.0 s to enter the field and calculate the flux linkage through the coil when it is completely in the field.

  b Sketch a graph to show how:
    i the flux linkage through the coil changes with time,
    ii the induced e.m.f. varies with time t from $t = 0$ when the coil enters the field to $t = 5.0$ s.

6 An a.c. generator is used to provide electricity for a lighting circuit.

  a Explain why the generator is easier to turn when the lamps are switched off.

  b The generator coil has 120 turns on a rectangular coil of length 40 mm and width 30 mm. The coil spins in a magnetic field of flux density 220 mT.
    i When the generator spins at a frequency of 20 Hz, show that the peak voltage is 4.0 V.
    ii Calculate the peak voltage when the frequency is 25 Hz.

7 A bar magnet was held vertically in a horizontal coil connected to a data recorder, as shown in Figure 7.1 below. When the magnet was released, the data recorder was used to measure the voltage induced in the coil every 5 ms. The diagram shows how the voltage changed with time.

a *Arrangement*

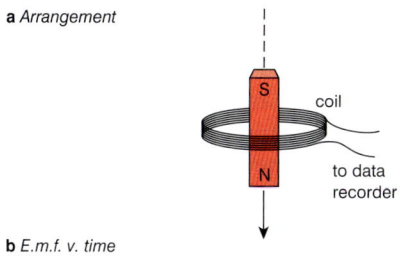

b *E.m.f. v. time*

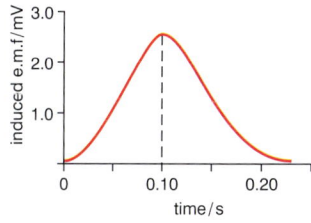

Figure 7.1

a i Without calculations, sketch a graph to show how the flux linkage through the coil changed with time.

ii Use your graph to explain the shape of the voltage v. time graph.

b Sketch the voltage v. time graph that would have been produced if the bar magnet had been released from a position above the coil so that it dropped through the coil.

8 A transformer is used to step down an alternating voltage of 230 V to 12 V. The transformer has a primary coil with 1000 turns.

a Calculate the number of turns on the secondary coil.

b The transformer is used to supply power to a 12 V, 60 W lamp. Calculate the current in:
i the secondary coil,
ii the primary coil when the lamp is on.

c The lamp is connected to the transformer by means of a cable of resistance 0.4 Ω.
i Estimate the power wasted due to the heating effect of the current in the cable.
ii Discuss whether or not it would be better to replace the lamp and transformer with a 230 V, 60 W lamp connected to the mains using the same cable.

9 a State Faraday's law of electromagnetic induction.

b The arrangement shown in Figure 9.1 can be used to test Faraday's law of electromagnetic induction. A cylindrical bar magnet is clamped horizontally on a wooden base which is designed to enable the wire to be rotated about the bar magnet when the handle is turned.

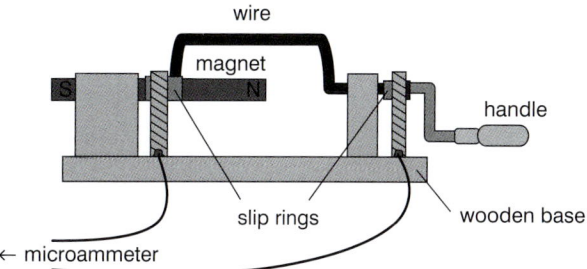

Figure 9.1

The wire is connected to a microammeter via two slip rings. When the handle is turned, the wire rotates about the magnet and cuts across the magnetic field lines, inducing a current in the circuit. The microammeter records the induced current in the circuit.

i Describe an experiment using the above apparatus and a stopwatch to investigate how the induced current varies with the frequency of rotation of the wire about the magnet.

ii Use your knowledge of Faraday's law to predict the relationship between the induced current and the frequency of rotation and how you would use your results to test your prediction.

10a i Describe the function of a transformer and explain how a step-up transformer works.

ii Explain why electrical power is transmitted more efficiently through cables at high voltage than at low voltage.

iii Cables of total resistance 5.0 Ω used to transfer electrical energy to a heating system are supplied with 5.0 kW of electrical power at a certain alternating voltage. Calculate and compare the output power and the efficiency of the transfer for root mean square input voltages of 100 V and 1000 V.

b i Describe, with the aid of a diagram, how a transformer and a bridge rectifier are used to convert alternating current to direct current.

ii Draw a graph to show how the output voltage and the input voltage vary over two cycles of the input voltage.

Key concepts: Fields

In studying Chapters 11 and 12, you will have gained a clear understanding of the force conditions necessary for an object to undergo uniform circular motion and simple harmonic motion. You should also appreciate that in uniform circular motion, because the velocity direction is always perpendicular to the resultant force, no work is done on the object and hence its kinetic energy and speed are constant. This consideration also is important where a satellite is in uniform circular motion about a spherical planet. When considering simple harmonic motion, you should appreciate that the energy of an oscillating system is constant provided dissipative forces such as air resistance are negligible. In addition, you will have developed your mathematical skills to a level where you can apply non-linear equations including sinusoidal equations and graphs to calculate the acceleration and the time period of the motion and the speed and position at any time. In addition, by investigating oscillating systems such as a loaded spring, you will have developed your practical and analytical skills by plotting straight-line graphs to test related non-linear theoretical equations.

In studying Chapters 13, 14 and 16, you should have gained awareness of important field concepts such as lines of force, uniform fields and field strength. Beyond these concepts, electric and gravitational fields share common features such as an inverse square law of force, similar equations for field strength and potential in radial fields and similar trajectories for objects moving across a uniform field. Magnetic fields differ from electric and gravitational fields in respect of these and other features. For example, charged objects moving across a uniform magnetic field follow a circular path because the magnetic force is perpendicular to the direction of the object's velocity (as well as being perpendicular to the field direction). The force on a moving charged object in a magnetic field is the reason why a current-carrying conductor in a magnetic field experiences a force when it is not aligned along the field lines, and it provides the basis of explanations of the motor effect and the Hall effect. Knowledge and understanding gained in Chapter 16 ought to have given you a sound basis for the study of electromagnetic induction in Chapter 17. Conservation of energy is also important in understanding and applying Lenz's law and Faraday's law to important applications such as the a.c. generator and the transformer. In all these chapters, you will have developed your mathematical skills by applying them to further non-linear equations and related graphs.

Your practical skills as well as your knowledge of electricity should have developed considerably in your studies of Chapter 15. Practical investigations into capacitors in Chapter 15 involve timing measurements as well as measurements of currents and potential differences. By making such measurements with and without a data logger, you should appreciate the benefits of using a data logger in these experiments. In addition, your mathematical knowledge and skills will have been enhanced by carrying out calculations and plotting graphs involving exponential decrease equations and graphs.

The following table provides an overview of the topics in Chapters 11 to 17, showing where key concepts have been developed.

Chapters 11–17 Key concepts

		Topic		Key concepts
11.1–11.4	uniform motion in a circle	centripetal force and centripetal acceleration	forces and fields	vectors
		calculations involving angular speed, centripetal acceleration and Newton's laws of motion	use of maths	problem-solving
12.1–12.4	principles and applications of simple harmonic motion	conditions for SHM	forces and fields	restoring force
		calculations involving SHM equations and sinusoidal functions	use of maths	problem-solving
		simple pendulum to measure g	testing	practical skills including analysis and evaluation
		oscillations of a loaded spring		
12.5–12.6	energy and simple harmonic motion	kinetic and potential energy of an object	energy and matter	conservation of energy
		damping, forced oscillations and resonance		work done by dissipative forces
		SHM calculations of kinetic and potential energies	use of maths	problem-solving
13.1–13.4	gravitational forces and fields	Newton's law of gravitation	forces and fields	inverse-square law
		gravitational field strength, uniform and radial fields		field strength
		gravitational potential	energy and matter	conservation of energy
		calculations involving formulae for gravitational field strength and potential	use of maths	problem-solving
13.5	satellite motion	Newton's law of gravitation applied to uniform circular motion leading to speed and time period equations	forces and fields	use of inverse-square law
		calculations involving above equations	use of maths	problem-solving
14.1–14.4	electric fields	electric charge, line of force, Coulomb's law, electric field strength, uniform and radial fields	forces and fields	inverse-square law for radial fields, uniform fields
		electric potential, equipotentials	energy and matter	conservation of energy
		calculations involving formulae for electric field strength and potential	use of maths	problem-solving
14.5	comparison of electric and gravitational fields	field strength and potential	forces and fields / energy and matter	inverse and inverse-square equations and graphs
15.1–15.4	capacitors	capacitance	forces and fields	electric charge
		capacitor combination rules		
		energy stored in a capacitor	energy and matter	conservation of energy
		capacitor charging and discharging	testing	practical skills including analysis
		calculations and graphs involving capacitor formulae, including capacitor discharge	use of maths	problem-solving

continued on p.268

Chapters 11–17 Key concepts (*continued*)

	Topic		Key concepts	
16.1–16.4	magnetic fields	magnetic field lines, magnetic flux density, the motor effect, magnetic field patterns	forces and fields	force perpendicular to conductor and to the field lines
		calculations and graphs involving magnetic field formula	use of maths	problem-solving
	moving charges in a magnetic field	force on a moving charge in a magnetic field, Hall effect		
		use of a Hall probe		
16.5	charged particles in circular orbits	circular motion of charged particles in a magnetic field	forces and fields	force perpendicular to velocity
		calculations involving charged particles in circular orbits and use of the Hall voltage equation	testing	practical skills; measurement of magnetic flux density
			use of maths	problem-solving
17.1–17.4	electromagnetic induction	origin of induced e.m.f., magnetic flux	forces and fields	
		Lenz's law and Faraday's law	energy and matter	conservation of energy
		calculations involving magnetic flux, Faraday's law, the a.c. generator and the transformer	use of maths	problem-solving
				transformer efficiency

Question

This question is about the oscillations of a loaded spring. It tests how we use models to make predictions about a physical system and how we then test these predictions with an experimental investigation. It also tests practical skills involving timing the oscillations and measuring the extension of a loaded spring. Systematic errors can arise when making extension measurements, for example as a result of not using a plane mirror correctly when reading a pointer against a millimetre scale and also when timing oscillations, for example by miscounting the number of oscillations. Random errors can be minimised when timing oscillations by making repeat measurements of at least 20 oscillations.

As in the previous key concepts questions on p.92 and 165, physics theory and mathematics are used to predict the relationship between the static extension e of the spring and the time period T of the oscillations. In this case, the predicted relationship is of the form $T^2 = ke$ (or $T = k^{1/2}e^{1/2}$). So by plotting a graph of T^2 against e (or T against $e^{1/2}$), we can tell if the relationship is valid from whether or not the graph is linear (i.e. a straight-line graph) and passes through the origin. If the graph is linear, we can also determine the constant k, as it is equal to the gradient of the graph. Also, the value of g can then be found from the result for k.

In some investigations, the relationship between two variables may not be easy to predict from the relevant theory or the theory might not give a straight-line graph. In these cases, assuming the relationship is of the form $y = kx^n$, a graph of log y against log x may be plotted; a straight-line graph would confirm that the relationship is of the form $y = kx^n$ and the power n could then be found, as it is equal to the gradient of the graph.

Before answering the question, students should be given the opportunity to do this or a similar investigation.

1 In an investigation, a small object was suspended from the lower end of a vertical steel spring, which was fixed at its upper end, as shown in Figure 1.

A horizontal marker pin P was attached to the lower end of the spring. The vertical position x of the pin was measured against the millimetre scale of a metre rule clamped vertically in a fixed position. The measurement was made three times without, then with, the small object suspended from the spring.

a The readings obtained are shown in Table 1.

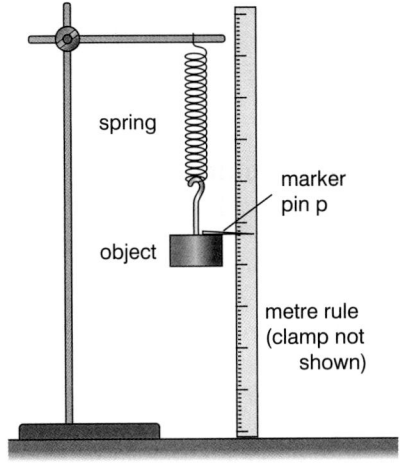

Figure 1

Table 1

	x/mm			mean x/mm	extension e/mm
without the object on the spring	2	2	2	2.0	0
with the object on the spring	71	72	73		

   i Copy and complete Table 1 by calculating the mean vertical position of P and the extension of the spring when the object was placed on it.
   ii The readings were taken to a precision of 0.5 mm using a millimetre ruler. Estimate the percentage 'uncertainty' in the extension. *(2 marks)*

b The time period, T, of small vertical oscillations of the object on the spring was also measured by timing 20 oscillations three times. The timing readings for 20 oscillations were 10.98 s, 11.11 s and 10.97 s.
   i Calculate the time period T.
   ii Use the readings to estimate the percentage 'uncertainty' in T. *(2 marks)*

c i Give an expression for the extension e of the spring in terms of the mass m of the object and the spring constant k of the spring.

   ii Hence show that $T = 2\pi\sqrt{\dfrac{e}{g}}$ *(3 marks)*

d The experiment was repeated with objects of different mass suspended from the spring. The measurements obtained are given in Table 2.

Table 2

object	e/mm	T/s
1	70	0.551
2	139	0.761
3	205	0.923
4	271	1.062
5	341	1.187
6	409	1.291

   Plot a suitable graph using the above measurements to confirm the equation and to determine g. *(9 marks)*
e Discuss the accuracy of your determination of g. *(4 marks)*

18 Thermal physics

18.1 Internal energy and temperature

Figure 1 *Infrared image showing heat transfer from a house*

When you are outdoors in a cool climate in winter, you need to wrap up well otherwise heat transfer from your body to the surroundings takes place and you lose energy. In hot weather, if you are in a very hot room, you gain energy from the room due to heat transfer.

Energy transfer between two objects takes place if:

- one object exerts a force on the other one and makes it move. In other words, one object does work on the other (see Topic 2.4 for more about work).
- one object is hotter than the other so heat transfer takes place by means of conduction, convection or radiation. In other words, heat transfer is energy transfer due to a temperature difference.

Internal energy

The brake pads of a moving vehicle become hot if the brakes are applied for long enough. The work done by the frictional force between the brake pads and the wheel heats the brake pads. The brake pads gain energy from the kinetic energy of the vehicle. The temperature of the brake pads increases as a result and the internal energy of each brake pad increases.

As explained below, the internal energy of an object is the energy of its molecules due to their individual movements. The internal energy of an object due to its temperature is sometimes referred to as **thermal energy**. However, some of the internal energy of an object might be due to other causes. For example, an iron bar that is magnetised has more internal energy than if it is unmagnetised because of the magnetic interaction between its atoms.

The internal energy of an object changes as a result of:

- energy transfer to or from the object by heating or by radiation, or
- work done on or by the object, including work done by electricity.

If q represents the energy transfer to an object by heating and W represents the work done on it (leaving out work done to make an object move faster or to raise it), then

the object's change of internal energy, $\Delta U = q + W$

The equation follows from the **principle of conservation of energy**. The above equation is known as the **first law of thermodynamics**.

If the internal energy of an object is constant, either:

- there is no heat transfer and no work is done, or
- heat transfer and work done 'balance' each other out.

For example, the internal energy of a lamp filament increases when the lamp is switched on because work is done by the electricity supply pushing electrons through the filament. The filament becomes hot as a result. When it reaches its operating temperature, heat transfer to the surroundings takes place and it radiates light. Work done by the electricity supply pushing electrons through the filament is balanced by heat transfer including light radiated from the filament.

About molecules

A molecule is the smallest particle of a pure substance that is characteristic of the substance. For example, a water molecule consists of two hydrogen atoms joined to an oxygen atom.

An atom is the smallest particle of an element that is characteristic of the element. For example, a hydrogen atom consists of a proton and an electron.

- In a solid, the atoms and molecules are held to each other by forces due to the electrical charges of the protons and electrons in the atoms. The molecules in a solid vibrate randomly about fixed positions. The higher the temperature of the solid, the more the molecules vibrate. The energy supplied to raise the temperature of a solid increases the kinetic energy of the molecules. If the temperature is raised sufficiently, the solid melts. This happens because its molecules vibrate so much that they break free from each other and the substance loses its shape. The energy supplied to melt a solid raises the potential energy of the molecules because they break free from each other.

- In a liquid, the molecules move about at random in contact with each other. The forces between the molecules are not strong enough to hold the molecules in fixed positions. The higher the temperature of a liquid, the faster its molecules move. The energy supplied to a liquid to raise its temperature increases the kinetic energy of the liquid molecules. Heating the liquid more and more causes it to vaporise. The molecules have sufficient kinetic energy to break free and move away from each other.

- In a gas or vapour, the molecules also move about randomly but much further apart on average than in a liquid. Heating a gas or a vapour makes the molecules speed up and so gain kinetic energy.

> **The internal energy of an object is the sum of the random distribution of the kinetic and potential energies of its molecules.**

Increasing the internal energy of a substance increases the kinetic and / or potential energy associated with the random motion and positions of its molecules. The kinetic and potential energy of each individual molecule differs between molecules and changes at random.

Summary test 18.1

1 Describe the energy transfers that occur when a low voltage heater connected to a battery is used to heat some water in a beaker.

2 a Explain why an electric motor becomes warm when it is used.

 b A battery is connected to an electric motor which is used to raise a weight at a steady speed. When in operation, the electric motor is at a constant temperature which is above the temperature of its surroundings. Describe the energy transfers that take place.

3 a State what is meant by internal energy.

 b Describe a situation in which the internal energy of an object is constant even though work is done on the object.

4 a State one difference between the motion of the molecules in a solid and the molecules in a liquid.

 b Describe how the motion of the molecules in a solid changes when the solid is heated.

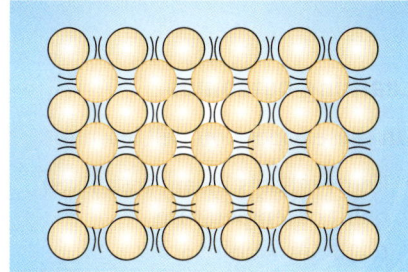

A solid is made up of particles arranged in a regular 3-dimensional structure. There are strong forces of attraction between the particles. Although the particles can vibrate, they cannot move out of their positions in the structure.

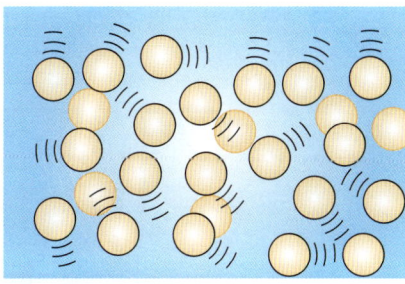

In a liquid the particles are free to move around. A liquid therefore flows easily and has no fixed shape. There are still forces of attraction between the particles.
 When a liquid is heated, some of the particles gain enough energy to break away from the other particles. The particles which escape from the body of the liquid become a gas.

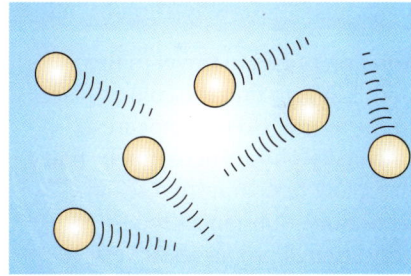

In a gas, the particles are far apart. There are almost no forces of attraction between them. The particles move about at high speed. Because the particles are so far apart, a gas occupies a very much larger volume than the same mass of liquid.
 The molecules collide with the container. These collisions are responsible for the pressure which a gas exerts on its container.

Figure 2 *Particles in a solid, a liquid and a gas*

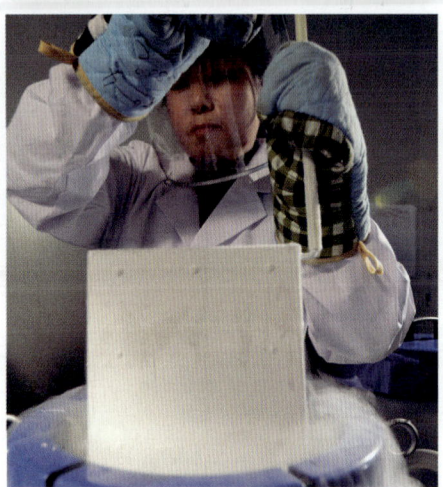

Figure 1 *A low temperature research laboratory*

📖 The coldest places in the world

You don't need to travel to the South Pole to find the coldest places in the world. Go to the nearest university physics department that has a low temperature research laboratory. Substances have very strange properties at very low temperatures. For example, metals cooled to a few degrees within absolute zero become superconductors which means they have zero electrical resistance.

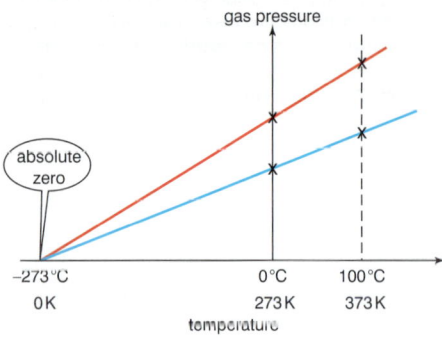

Figure 2 *Absolute zero*

The temperature of an object is a measure of the degree of hotness of the object. The hotter an object is, the more internal energy it has. Place your hand in cold water and it loses internal energy due to heat transfer. Place it in warm water and it gains internal energy due to heat transfer. If the water is at the same temperature as your hand, no overall heat transfer takes place. Your hand is in **thermal equilibrium** with the water. No overall heat transfer takes place between two objects at the same temperature.

The 'baby in the bath' rule

Before dipping the baby in the bath water, the parent tests the water by putting a hand (not the baby's hand!) in the water. If the baby is at the same temperature as the parent and the parent's hand is at the same temperature as the bath water, the baby will be at the same temperature as the bath water.

Practical temperature scales

A temperature scale is defined in terms of **fixed points** which are standard degrees of hotness which can be accurately reproduced.

- **The Celsius scale of temperature**, in °C, is defined in terms of:

    1 ice point, 0 °C, which is the temperature of pure melting ice,

    2 steam point, 100 °C, which is the temperature of steam at standard atmospheric pressure.

- **The thermodynamic (Kelvin) scale of temperature**, in kelvins (K) is defined in terms of:

    1 **absolute zero**, 0 K, which is the lowest possible temperature,

    2 the triple point of water, 273.16 K, which is the temperature at which ice, water and water vapour are in thermal equilibrium.

Because ice point on the thermodynamic scale is 273.15 K and steam point is 100 K higher,

> **temperature in °C = thermodynamic temperature in kelvins − 273.15**

About absolute zero

The thermodynamic scale of temperature, also referred to as the absolute or kelvin scale, is based on absolute zero, the lowest possible temperature. No object can have a temperature below absolute zero. **An object at absolute zero has minimum internal energy**, regardless of the substances the object consists of.

As explained in Topic 19.1, the pressure of a fixed mass of gas in a sealed container of fixed volume decreases as the gas temperature is reduced. If the pressure measured at ice point and at steam point is plotted on a graph, as shown in Figure 2, the line between the two points always cuts the temperature axis at −273 °C, regardless of which gas is used or how much gas is used.

The thermodynamic scale of temperature starts at absolute zero. Its unit, the kelvin, is defined so that a temperature change of 1 K is the same as a temperature change of 1 °C. The Kelvin scale depends on a fundamental feature of nature, namely, the lowest possible temperature. In comparison, the Celsius scale depends on the properties of a substance, water, chosen for convenience rather than for any fundamental reason.

Thermometer	Thermometric property	Advantages	Disadvantages
Liquid-in-glass	the liquid's density (which changes with temperature causing the length of the liquid thread to change)	easy to use, portable	fragile, limited range
Thermocouple	e.m.f., E, across the junction of two different metal wires in contact with each other	fast response, wide range, remote readings	millivoltmeter needed
Resistance	resistance, R of a thermistor (or a metal wire)	wide range, accurate, measures small changes accurately	slow response
Gas	pressure, p of an ideal gas at constant volume	wide range, accurate, standard thermometer	bulky, slow response

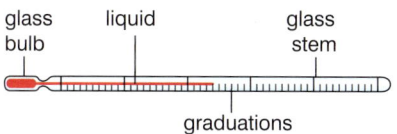

a A liquid-in-glass thermometer

Thermometers

The **thermometric property** of a thermometer is a physical property that varies smoothly with change of temperature and can therefore be used to measure temperature. Four different thermometers are compared in the table above and in Figure 3.

All thermometers are calibrated in terms of the temperature measured by a gas thermometer. This is a thermometer consisting of dry gas in a sealed container. The pressure of the gas is proportional to the thermodynamic temperature of the gas. In other words, equal increases of temperature cause equal increases of gas pressure.

By measuring the gas pressure, p_{Tr}, at the triple point of water (= 273.16 K by definition) and an unknown temperature T/K, the unknown temperature, in kelvins, can be calculated using

$$\frac{T}{273.16} = \frac{p}{p_{Tr}}$$

where p is the gas pressure at the unknown temperature.

To determine the temperature on the Celsius scale or the thermodynamic (i.e. Kelvin) scale using any other thermometer, it must be calibrated against a gas thermometer (or indirectly against a thermometer previously calibrated against a gas thermometer). This involves measuring the thermometric property at different temperatures, T, measured using the gas thermometer (or indirectly as above). A graph or 'calibration curve' can then be plotted of temperature T on the y-axis against the thermometric property on the x-axis.

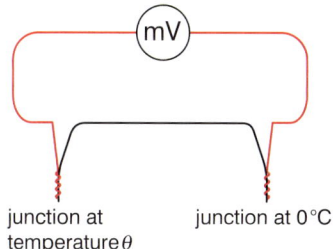

junction at temperature θ

junction at 0°C

b A thermocouple

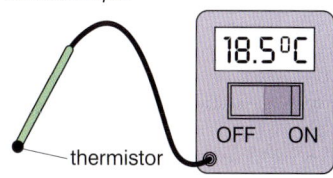

thermistor

c A resistance thermometer

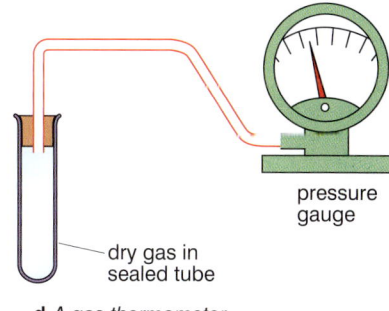

pressure gauge

dry gas in sealed tube

d A gas thermometer

Figure 3 Different types of thermometers

Note

If the thermometric property is X_0 at ice-point, X_{100} at steam point and X at an unknown temperature θ, the unknown temperature can be calculated on a centigrade scale (ie a scale of 100 equal intervals from ice point to steam point) using the equation $\theta = \dfrac{X - X_0}{X_{100} - X_0} \times 100$. This gives the values that would be obtained by dividing the interval between X_0 and X_{100} into 100 equal degrees. If a thermometer is graduated with such a scale, a calibration curve of the 'gas thermometer' temperature T against θ would need to be used to determine the temperature T from a thermometer reading, θ.

Summary test 18.2

1 a Define the *thermodynamic scale of temperature* and state its unit.

   b State each of the following temperatures to the nearest degree on the thermodynamic scale:
   i   the temperature of pure melting ice,
   ii 20°C,   iii −196°C.

2 The pressure of a constant-volume gas thermometer was 100 kPa at a temperature of 273 K.

   a Calculate the temperature of the gas when its pressure was 120 kPa.

   b Calculate the pressure of the gas at 100°C.

3 Explain why the 50°C mark on the stem of a liquid-in-glass thermometer is not exactly mid-way between the 0 and 100°C marks.

4 Explain why a gas thermometer is used to calibrate other types of thermometers.

18.3 Specific heat capacity

Learning outcomes

On these pages you will learn to:

- relate a rise in temperature of a body to an increase in its internal energy
- define and use the concept of specific heat capacity, and identify the main principles of its determination by electrical methods

Heating and cooling

Sunbathers on hot sandy beaches dive into the sea to cool off. Sand heats up much more readily than water does. Even when the sand is almost too hot to walk barefoot across, the sea water is refreshingly cool. The temperature rise of an object when it is heated depends on:

- the mass of the object,
- the amount of energy supplied to it,
- the substance or substances from which the object is made.

> **The specific heat capacity, c, of a substance is the energy needed to raise the temperature of unit mass of the substance by 1 K without change of state.**

The unit of c is $\text{J kg}^{-1}\text{K}^{-1}$.

Specific heat capacities of some common substances are shown in Table 1.

To raise the temperature of mass m of a substance from temperature T_1 to temperature T_2,

$$\text{the energy needed } \Delta Q = mc\,(T_2 - T_1)$$

For example, to calculate the heat that must be supplied to raise the temperature of 5.0 kg of water from 20 °C to 100 °C, using the above formula is $\Delta Q = 5.0 \times 4200 \times (100 - 20) = 1.7 \times 10^6$ J.

> **The heat capacity, C, of an object is the heat supplied to raise the temperature of the object by 1 K.**

Therefore, for an object of mass m made of a single substance of specific heat capacity c, its heat capacity $C = mc$. For example, the heat capacity of 5.0 kg of water is $21\,000\,\text{J K}^{-1} = 5.0\,\text{kg} \times 4200\,\text{J kg}^{-1}\text{K}^{-1}$.

Table 1 Some specific heat capacities

Substance	Specific heat capacity / J kg⁻¹ K⁻¹
Aluminium	900
Concrete	850
Copper	390
Iron	490
Lead	130
Oil	2100
Water	4200

The inversion tube experiment

In the inversion tube experiment, the gravitational potential energy of an object falling in a tube is converted into internal energy when it hits the bottom of a tube. Figure 1 shows the idea. The object is a collection of tiny lead spheres.

The tube is inverted each time the spheres hit the bottom of the tube. The temperature of the lead shot is measured initially and after a certain number of inversions.

Let m represent the mass of the lead shot.

For a tube of length L, the loss of gravitational potential energy for each inversion = mgL

Therefore, for n inversions, the loss of gravitational potential energy = $mgLn$

The gain of internal energy of the lead shot = $mc\Delta T$, where c is the specific heat capacity of lead and ΔT is the temperature rise of the lead shot.

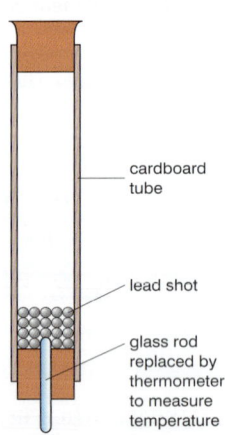

cardboard tube

lead shot

glass rod replaced by thermometer to measure temperature

Figure 1 The inversion tube experiment

Assuming all the gravitational potential energy lost is transferred to internal energy of the lead shot,

$$mc\Delta T = mgLn$$

$$\therefore \qquad c = \frac{gLn}{\Delta T}$$

The experiment can therefore be used to measure the specific heat capacity of lead with no other measurements than the length of the tube, the temperature rise of the lead and the number of inversions.

Specific heat capacity measurements using electrical methods

1 Measurement of the specific heat capacity of a metal

A block of the metal of known mass m in an insulated container is used. A 12 V electrical heater is inserted into a hole drilled in the metal and used to heat the metal by supplying a measured amount of electrical energy. A thermometer inserted into a second hole drilled in the metal is used to measure the temperature rise ΔT (= its final temperature – its initial temperature).

The electrical energy supplied = heater current, $I \times$ heater p.d., $V \times$ heating time, t

$\therefore$ assuming no heat loss to the surroundings, $\qquad mc\Delta T = IVt$

$$\therefore \quad c = \frac{IVt}{m\Delta T}$$

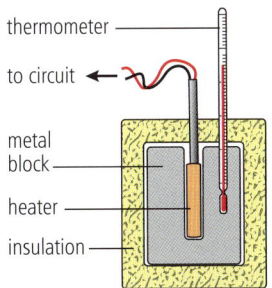

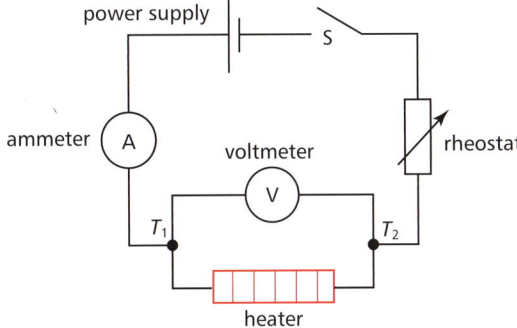

Figure 2 *Measuring c*

2 Measurement of the specific heat capacity of a liquid

A known mass of the liquid is used in an insulated calorimeter of known mass and known specific heat capacity. A 12 V electrical heater is placed in the liquid and used to heat it directly. A thermometer inserted into the liquid is used to measure the temperature rise, ΔT. Assuming no heat loss to the surroundings:

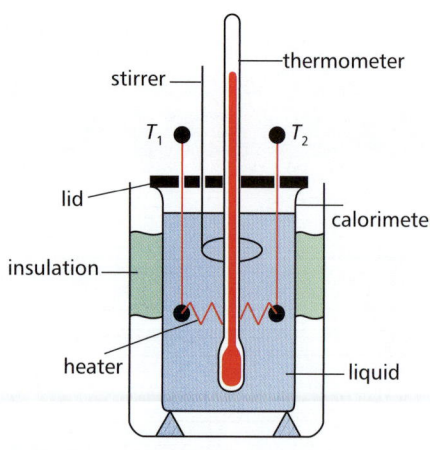

Figure 3 *Measurement of the specific heat capacity of a liquid*

- The electrical energy supplied = current I × voltage V × heating time t.
- The energy needed to heat the liquid =
  mass of liquid (m_l) × specific heat capacity of liquid (c_l) × temperature rise ΔT.

- The energy needed to heat the calorimeter =
  mass of calorimeter (m_{cal}) × specific heat capacity of calorimeter (c_{cal}) × temperature rise ΔT

$$\therefore \quad IVt = m_l c_l \Delta T + m_{cal} c_{cal} \Delta T$$

Hence, c_l can be calculated from this equation as all the other quantities are known.

Extension

Method of mixtures experiments can be carried out as follows to determine specific heat capacities. No electrical heating is required. An insulated calorimeter as shown in Figure 3 is used without the heater. The masses of the calorimeter (m_c) and of the liquid (m_l) in it must be measured.

To determine the specific heat capacity, c, of a small object of known mass m, the object is firstly suspended on a thread in a beaker of boiling water to raise its initial temperature T_1 to 100 °C. The initial temperature T_2 of the liquid (e.g. water or oil) in the calorimeter is measured then the object is quickly transferred from the boiling water into the liquid in the calorimeter. As a result, the liquid becomes warmer and it is then stirred until its temperature reaches a steady value, T_3.

Assuming no heat transfer to or from the surroundings:

- the heat energy transferred from the object (to the liquid and calorimeter) = $mc(T_1 - T_3)$, where $T = 100$ °C,
- the heat energy gained by the calorimeter = $m_c c_0 (T_3 - T_2)$, where c_0 is the specific heat capacity of the calorimeter,
- the heat energy gained by the liquid = $m_l c_l (T_3 - T_2)$, where c_l is the specific heat capacity of the liquid.

Therefore, because no heat energy is transferred to or from the surroundings, the heat energy transferred from the object = the heat energy gained by the object and the calorimeter.

$$mc(T_1 - T_3) = m_l c_l (T_3 - T_2) + m_c c_0 (T_3 - T_2)$$

Hence c can be calculated by substituting the measured values of all the other quantities in the above equation.

To determine the specific heat capacity of a liquid, the method above can be used with the liquid in the calorimeter and with an object of known specific heat capacity.

Note

To increase the accuracy of the experiments in the Extension:

- the initial temperature of the calorimeter and the liquid should be at room temperature so the temperature difference between T_1 (= 100 °C) and T_2 is as large as possible, and

- the masses should be chosen so the final temperature T_3 is approximately midway between T_1 and T_2.

Worked example

An insulated copper calorimeter of mass 0.127 kg contains 0.170 kg of water at an initial temperature of 20.0 °C. A stone of mass 0.050 kg suspended on a thread is transferred from boiling water at 100 °C to the water, causing the temperature of the water to increase to 31.5 °C. Calculate:

a the heat energy gained by the water and the calorimeter,
b the specific heat capacity of the stone.

specific heat capacities: copper 390 J kg^{-1} K^{-1}, water 4200 J kg^{-1} K^{-1}.

Solution

a Heat energy gained by calorimeter = 0.127 × 390 × (31.5 − 20.0) = 570 J
Heat energy gained by the water = 0.170 × 4200 × (31.5 − 20.0) = 8210 J
Heat energy gained by calorimeter and water = 8780 J

b Heat energy lost by the stone = 0.050 × c × (100 − 31.5) = 8780 J

$$\text{Therefore } c = \frac{8780}{0.050 \times (100 - 31.5)} = 2560 \text{ J kg}^{-1}\text{K}^{-1}$$

Continuous flow heating

In an electric shower, water passes steadily through copper coils heated by an electrical heater. The water is hotter at the outlet than at the inlet. This is an example of continuous flow heating. For mass m of fluid passing through the heater in time t at a steady flow rate, assuming no heat loss,

the electrical energy supplied per second $IV = \dfrac{mc\Delta T}{t}$

where ΔT is the temperature rise of the water.

Summary test 18.3

Use the data in Table 1, p.274, for the following calculations.

1 Calculate:

  a the energy needed to heat an aluminium pan of mass 0.30 kg from 15 °C to 100 °C,

  b the energy needed to heat 1.5 kg of water from 15 °C to 100 °C.

2 a Calculate the time taken to heat the water and pan in Question **1** from 15 °C to 100 °C using a 2.0 kW electric hot plate.

  b Calculate the energy needed to raise the temperature of 80 kg of water in an insulated copper tank of mass 20 kg from 20 °C to 50 °C.

3 In an inversion tube experiment, 0.50 kg of lead shot at an initial temperature of 18 °C was inverted fifty times in a tube of length 1.30 m. The final temperature of the lead shot was 23 °C. Calculate:

  a the total gravitational potential energy released by the lead,

  b the specific heat capacity of lead.

4 An electric shower is capable of heating water from 10 °C to 40 °C when the flow rate is 0.025 kg s^{-1}. Calculate the minimum power of the heater. The specific heat capacity of water = 4200 J kg^{-1} K^{-1}.

18.4 Change of state

Learning outcomes

On these pages you will learn to:

- explain in terms of molecules why melting and boiling take place without a change in temperature
- define specific latent heat capacity and carry out calculations involving specific latent heat values
- identify the main principles of electrical methods to determine specific latent heat values
- explain why the specific latent heat of vaporisation is higher than the specific latent heat of fusion for the same substance

When a solid is heated and heated, its temperature increases until it melts. If it is a pure substance, it melts at a well-defined temperature, its **melting point**. Once all the solid has melted, continued heating causes the temperature of the liquid to increase until the liquid boils. This occurs at a certain temperature, the **boiling point**. The substance turns to a vapour as it boils away.

The three physical states of a substance, solid, liquid and vapour, have different physical properties. For example:

- The density of a gas is much less than the density of the same substance in the liquid or the solid state. This is because the molecules of a liquid and of a solid are packed together in contact with each other. In contrast, the molecules of a gas are, on average, separated from each other by relatively large distances.
- Liquids and gases can flow but solids cannot. This is because the atoms in a solid are locked together by strong force bonds which the atoms are unable to break free from. In a liquid or a gas, the molecules are not locked together because they have too much kinetic energy and the force bonds are not strong enough to keep the molecules fixed to each other.

Latent heat

When a solid or a liquid is heated so its temperature increases, its molecules gain kinetic energy. In a solid, the atoms vibrate more about their mean positions. In a liquid, the molecules move about faster, still keeping in contact with each other but free to move about.

1 **When a solid is heated at its melting point**, its atoms vibrate so much that they break free from each other. The solid therefore becomes a liquid due to energy being supplied at the melting point. The energy needed to melt a solid at its melting point is referred to as **latent heat of fusion**. Fusion is used for the melting of a solid because the solid 'fuses' into a liquid as it melts. Latent heat is released when a liquid solidifies. This happens because the liquid molecules slow down as the liquid is cooled; at the melting point, the molecules move slowly enough for the force bonds to lock the molecules together.

2 **When a liquid is heated at its boiling point**, the molecules gain enough kinetic energy to overcome the bonds that hold them close together. The molecules therefore break away from each other to form bubbles of vapour in the liquid. The energy needed to vaporise a liquid is referred to as **latent heat of vaporisation**.

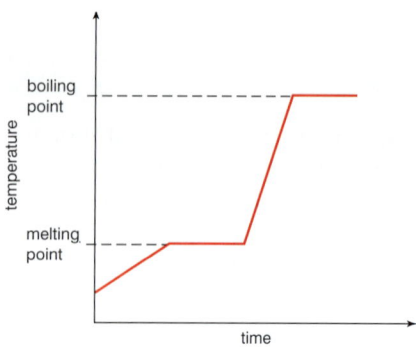

Figure 1 *Melting and boiling*

Latent heat is released when a vapour condenses. This happens because the vapour molecules slow down as the vapour is cooled; the molecules move slowly enough for the force bonds to pull the molecules together to form liquid.

Some solids vaporise directly when heated. This process is known as **sublimation**.

Specific latent heat

In general, much more energy is needed to vaporise a substance than to melt it. This is because the molecules need to move completely away from each other when vaporisation occurs (unlike when melting occurs). For example, 2.2 MJ is needed to vaporise 1 kg of water at its boiling point. In comparison, 0.34 MJ is needed to melt 1 kg of ice at its melting point.

The energy needed to change the state of 1 kg of a substance at its melting point (or its boiling point) is referred to as its specific latent heat of fusion (or vaporisation).

Note

'Latent' means 'hidden'; latent heat supplied to melt a solid or boil a liquid may be thought of as hidden because no temperature change takes place even though the substance is being heated. The energy supplied as latent heat increases the internal potential energy of the molecules.

- **The specific latent heat of fusion, L_f, of a substance** is the energy needed to change the state of unit mass of the substance from solid to liquid without change of temperature.
- **The specific latent heat of vaporisation, L_v, of a substance** is the energy needed to change the state of unit mass of the substance from liquid to vapour without change of temperature.

Therefore, the heat energy Q needed to change the state of mass m of a substance from solid to liquid or liquid to vapour without change of temperature is given by

$$Q = mL$$

where L is the specific latent heat of fusion or vaporisation as appropriate.

The unit of specific latent heat is $J\,kg^{-1}$.

Note Where a substance changes its state and changes its temperature, to calculate the heat energy transferred,

- use $Q = mL$ to calculate the heat energy transferred when its state changes
- use $Q = mc(T_2 - T_1)$ when its temperature changes.

Worked example

Calculate the energy needed to melt 5.0 kg of ice at 0 °C and heat the melted ice to 50 °C.

Specific latent heat of fusion of ice = $3.36 \times 10^5\,J\,kg^{-1}$

Specific heat capacity of water = $4200\,J\,kg^{-1}\,K^{-1}$

Solution
To melt 5.0 kg of ice, energy needed $Q_1 = mL = 5.0 \times 3.36 \times 10^5 = 1.68 \times 10^6\,J$

To heat 5.0 kg of melted ice (i.e. water) from 0 °C to 50 °C, energy needed

$$Q_2 = mc(T_2 - T_1) = 5.0 \times 4200 \times (50 - 0) = 1.05 \times 10^6\,J$$

Therefore, the total energy needed $= Q_1 + Q_2 = 2.73 \times 10^6\,J$

Measurement of the specific latent heat of fusion of ice
In this experiment, a low voltage heater is to be used to melt crushed ice in a funnel. The melted ice is collected using a beaker under the funnel, as shown in Figure 3.

To take account of heat transfer from the surroundings, the mass of ice melted in a certain time must be measured with the heater off then with it on. The difference in the two measurements gives the mass of ice melted due to the heater only.

1 With the heater off, water from the funnel is collected in the beaker for a measured time (e.g. 10 minutes). The mass of the beaker and water, m_1, is then measured. The beaker is then emptied for the next stage.
2 With the heater on, the procedure is repeated for exactly the same time. The ammeter and voltmeter readings are recorded with the heater switched on. After the heater is switched off, the mass of the beaker and the water, m_2, is measured once more.

Cooling by evaporation

If you have an injection, the doctor or nurse might 'numb' your skin by dabbing it with a liquid that easily evaporates. As the liquid evaporates, your skin becomes too cold to feel any pain.

Evaporation occurs at the open surface of a liquid. Relatively weak attractive forces exist between the particles in the liquid. Cooling by evaporation occurs because the faster particles in the liquid near the surface have enough kinetic energy to break away from the other particles and escape from the liquid. After they leave, the liquid is cooler because the average kinetic energy of the particles left in the liquid has decreased.

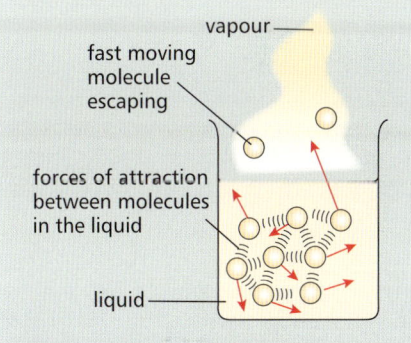

Figure 2 Cooling by evaporation

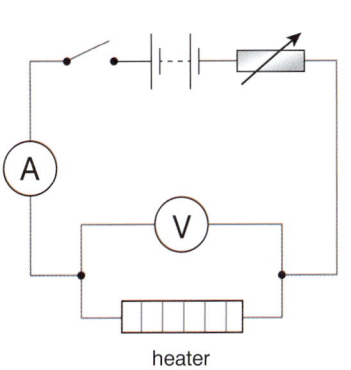

a Heater circuit

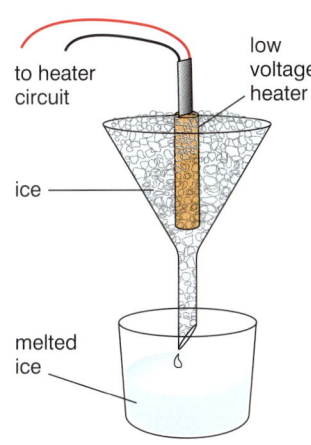

b Collecting melting ice

Figure 3 Measuring the specific latent heat of fusion of ice

279

To calculate the specific latent heat of fusion of ice, note that:

- the mass of ice melted due to the heater, $m = m_2 - m_1$
- the energy supplied E to the heater = current × p.d. × time
- the specific latent heat of fusion of ice, $l = \dfrac{E}{m} = \dfrac{E}{m_2 - m_1}$

Extension

Note The method of mixtures described in Topic 18.3 can also be used to measure the specific latent heat of fusion of ice. Pieces of ice dried with a tissue are melted one by one in warm water in the calorimeter. To compensate for heat loss to the surroundings ice is added until the temperature of the water is as much below room temperature as it was above room temperature before any ice was added.

The mass of the calorimeter, m_c, the initial mass of water, m_1, and the final mass of water, m_2, must be measured. The mass of ice, m, melted is therefore $(m_2 - m_1)$.

The initial temperature, T_1, and the final temperature, T_2, of the water must also be measured.
- The energy gained by the ice in melting and warming to temperature $T_2 = mL + mc_w(T_2 - 0)$, where L is the specific latent heat of fusion of ice and c_w is the specific heat capacity of water.
- The energy lost by the calorimeter and initial mass of water in it $= m_1 c_w(T_1 - T_2) + m_c c_c(T_1 - T_2)$
  where c_c is the specific heat capacity of the calorimeter.

Assuming no heat loss to the surroundings occurs, equating the heat loss to the heat gain enables L to be calculated:

$$mL + mc_w(T_2 - 0) = m_1 c_w(T_1 - T_2) + m_c c_c(T_1 - T_2)$$

Measurement of the specific latent heat of vaporisation of a liquid

In this experiment, a low voltage heater is used to heat the liquid in the inner flask to its boiling point. The heater circuit is the same as that shown in Figure 3 on the previous page. When the liquid is boiling, vapour from the liquid leaves the inner flask, condenses in the condenser and drips into the flask underneath the condenser. (See Figure 4.)

The ammeter and voltmeter readings are recorded at the normal current rating of the heater and the mass of liquid collected in a certain time is measured. The procedure is repeated with the liquid boiling at a reduced heater power.

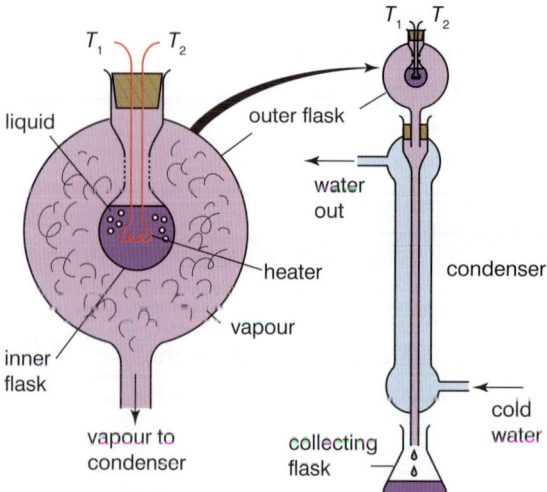

Figure 4 *Measuring the specific latent heat of steam*

In each collecting time, the energy supplied = current × p.d. × time.

Let E_1 and E_2 represent the energy supplied during each 'collecting' time, and let m_1 and m_2 represent the mass of liquid collected each time,

Therefore, $E_1 = m_1 L + Q$ and $E_2 = m_2 L + Q$, where L is the specific latent heat of the liquid and Q is the heat loss to the surroundings during the collecting time.

Hence $\quad\quad\quad E_1 - E_2 = m_1 L - m_2 L \quad$ so $\quad L = \dfrac{E_1 - E_2}{m_1 - m_2}$

Note The rate of heat loss to the surroundings is the same at normal power as at reduced power because the liquid is at the same temperature (i.e. its boiling point) in both situations. Since the collecting time is the same, the heat loss Q is therefore the same.

Temperature v. time graphs

If a pure solid is heated to its melting point and beyond, its temperature v. time graph will be as shown in Figure 5.

Assuming no heat loss occurs during heating:

- before the solid melts, the energy supplied each second $= mc_S \left(\dfrac{\Delta T}{\Delta t}\right)_S$

  where $\left(\dfrac{\Delta T}{\Delta t}\right)_S$ is the rise of temperature per second of the solid and c_S is the specific heat capacity of the solid,

- at the melting point, the energy supplied = energy supplied per second × time taken to melt,

- after the solid melts, the energy supplied each second $= mc_L \left(\dfrac{\Delta T}{\Delta t}\right)_L$

  where $\left(\dfrac{\Delta T}{\Delta t}\right)_L$ is the rise of temperature per second of the liquid and c_L is the specific heat capacity of the liquid.

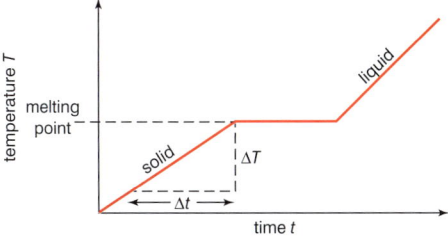

Figure 5 *Temperature v. time for heating a solid*

Chapter Summary

The internal energy of an object is the sum of the random distribution of the kinetic and potential energies of its molecules.

Absolute zero is the temperature at which an object has minimum internal energy.

Temperature in °C = thermodynamic temperature in kelvins − 273.15.

Specific heat capacity, c, of a substance is the energy needed to raise the temperature of 1 kg of the substance by 1 K without change of state.

Energy needed to raise the temperature of mass m of a substance from T_1 to $T_2 = mc (T_2 - T_1)$, where c is the specific heat capacity of the substance.

Energy to melt (or boil) mass m of a substance $= mL$, where L is the specific latent heat of fusion (or vaporisation) of the substance.

Summary test 18.4

1 a Explain why energy is needed to melt a solid.

  b Explain why the internal energy of the water in a beaker must be reduced to freeze the water.

2 Calculate the mass of water boiled away in a 3 kW electric kettle in 2 minutes, given that 2.2 MJ of energy must be supplied to boil away 1 kg of water at atmospheric pressure.

3 A plastic beaker containing 0.080 kg of water at 15 °C was placed in a refrigerator and cooled to 0 °C in 1200 s.

  a Calculate how much energy each second was removed from the water in this process. The specific heat capacity of water = 4200 J kg⁻¹ K⁻¹.

  b Calculate how long the refrigerator would take to freeze the water in **a**, given that 340 kJ of energy must be removed from 1.0 kg of water at 0 °C to freeze it.

4 The temperature v. time graph shown in Figure 6 was obtained by heating 0.12 kg of a substance in an insulated container. The specific heat capacity of the substance in the solid state is 1200 J kg⁻¹ K⁻¹. Calculate:

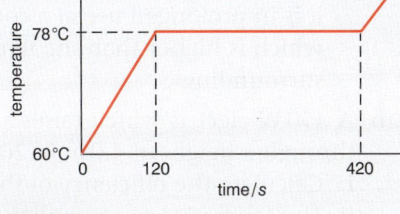

Figure 6

  a the energy per second supplied to the substance in the solid state if its temperature increased from 60 °C to its melting point at 78 °C in 120 s,

  b the energy needed to melt the solid if it took 30 s to melt with energy supplied at the same rate as in **a**.

Launch additional digital resources for the chapter

$g = 9.81\,\text{m s}^{-2}$

1 a In terms of molecules, explain why:
   i a solid has its own shape,
   ii a liquid and a gas can flow,
   iii a gas is much less dense than a solid or a liquid.
  b Describe the effect on the molecules of a solid of:
   1 supplying energy to it to raise its temperature,
   ii supplying energy to it to melt it.

2 When a certain object is heated, its *internal energy* increases. Explain, in molecular terms:
  a what is meant by internal energy,
  b why increasing the internal energy of the object differs from increasing its kinetic energy.

3 A vehicle of mass 1200 kg is fitted with brake pads of total mass 23 kg. When the car is travelling at a speed of $30\,\text{m s}^{-1}$, the brakes are applied and the car stops in a distance of 110 m. The temperature of the brake pads increases by 7.5 K during this time.
  a Calculate:
   i the initial kinetic energy of the car,
   ii the gain of internal energy of the brake pads, assuming no heat transfer from the brake pads occurs.
  b Show that the frictional force on the car due to the brakes was at least 3.9 kN.

  The specific heat capacity of the brake pad material $= 2500\,\text{J kg}^{-1}\text{K}^{-1}$.

4 a An electric motor in prolonged use becomes hotter than its surroundings.
   i State two reasons why some of the electrical energy transferred to the motor is wasted as heat.
   ii Describe the energy transfers in the motor when it is in prolonged use at a constant temperature which is higher than the temperature of its surroundings.
  b A 250 W electric winch raises a load of weight 110 N through a height of 4.0 m in 20 s.
   i Calculate the efficiency of the winch.
   ii The temperature of a pulley wheel in the winch increases by 2.0 K when the load in **i** is raised. The pulley wheel is made of steel and has a mass of 0.095 kg. Calculate the energy needed to raise the temperature of the wheel by 2.0 K and determine the increase in efficiency of the winch if friction at the pulley could be eliminated.

  Specific heat capacity of steel $= 460\,\text{J kg}^{-1}\text{K}^{-1}$

5 An insulated metal cylinder of mass 1.4 kg was heated by a 24 W electric heater placed in a slot in the cylinder, as shown in Figure 2 on p.275. The temperature of the metal was measured using a thermometer placed in a different slot in the metal. The measurements were used to plot a graph of the temperature v. time, as shown in Figure 5.1 below.

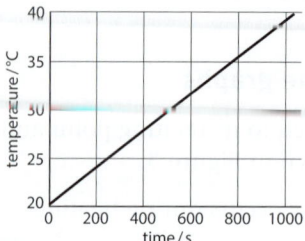

Figure 5.1

  a Determine the temperature rise per second of the cylinder.
  b Calculate the specific heat capacity of the metal.

6 A 3.0 kW electric kettle took 260 s to heat 1.8 kg of water from 18 °C to 100 °C.
  a Calculate:
   i the electrical energy supplied to the kettle in this time,
   ii the internal energy gained by the water in the kettle.
  b The mass of the kettle was 1.2 kg and it was made from aluminium.
   i Calculate the internal energy gained by the kettle.
   ii Account for the difference between the electrical energy supplied to the kettle and the internal energy gained by the water and the kettle.

  specific heat capacity of water $= 4200\,\text{J kg}^{-1}\text{K}^{-1}$,
  specific heat capacity of aluminium $= 900\,\text{J kg}^{-1}\text{K}^{-1}$

7 A shower fitted with an electric heater raised the temperature of water passing through the shower from 12 °C to 46 °C when the flow rate was $0.050\,\text{kg s}^{-1}$.
  a Calculate the power of the electric heater, assuming there is no heat loss to the surroundings.
  b The shower is fitted with a thermostat that restricts the temperature of the outflowing water to 50 °C. Calculate the flow rate that would give this outflow temperature when the temperature of the inflow is 12 °C.

  Specific heat capacity of water $= 4200\,\text{J kg}^{-1}\text{K}^{-1}$

8 **a** Define the *specific latent heat of vaporisation* of a liquid.

b In an experiment to measure the specific latent heat of vaporisation of a liquid, the liquid was heated electrically at its boiling point as shown in Figure 8.1. The vapour from the boiling liquid was condensed in a condenser and the condensate was collected in a flask below the condenser.

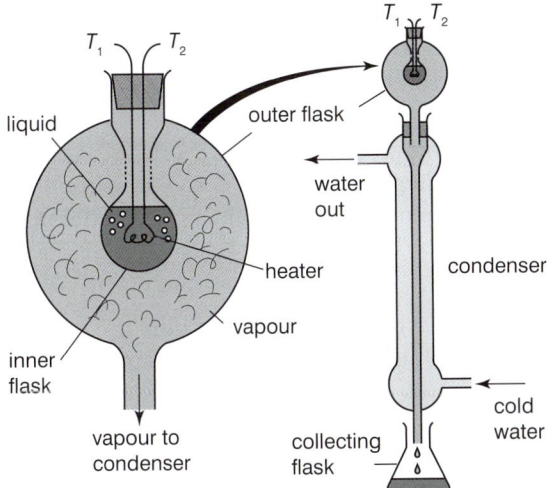

Figure 8.1

i Draw a circuit diagram of the circuit you would use to supply and measure the electrical power to the heating element from a 12 V power supply over a current range up to 2.0 A.

ii For a constant current, the mass of the liquid collected in the condenser in 180 s was measured for two different values of current. The results are listed below.

Current/A	1.20	2.00
Potential difference/V	5.2	11.5
Mass collected/g	0.80	4.35

Use this information to calculate the specific latent heat of vaporisation of the liquid.

9 **a** State what is meant by the *internal energy* of an object.

b In an experiment to measure the specific heat capacity of a rock sample, the sample was attached to a thread and immersed in boiling water. The sample was then transferred quickly to an insulated plastic beaker containing cold water at 15 °C. The water was then stirred, and its temperature was measured every 30 s for 5 minutes.

i Figure 9.1 shows how the temperature of the cold water changed after the sample was transferred to it.

ii The beaker contained 128 g of water. The mass of the rock sample was 30.5 g.

Calculate the specific heat capacity of the rock sample.

Assume that the specific heat capacity of water is 4200 J kg^{-1} °C^{-1}.

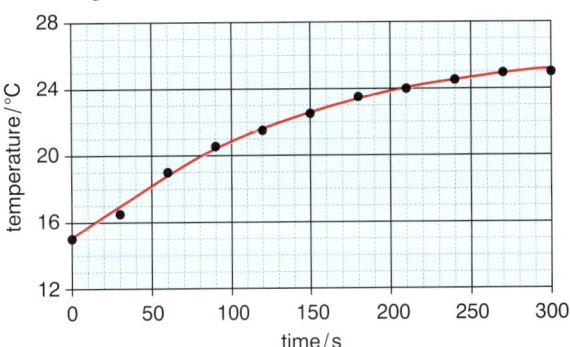

Figure 9.1

c The uncertainty in each mass measurement was ±0.2 g. The uncertainty in each thermometer reading was 0.25 °C. Calculate the percentage uncertainty for each measurement and hence calculate the uncertainty in the value of the specific heat capacity you obtained.

19.1 Experiments on gases

Learning outcomes

On these pages you will learn to:

- know what is meant by an isothermal change
- recall that temperature in kelvins = temperature in °C + 273(.15)
- recognise that for a fixed mass of gas at pressure p, volume V and temperature T in kelvins:
  - pV = constant at constant temperature T
  - V/T = constant at constant pressure
  - p/T = constant at constant volume

The experimental gas laws

When you use a cycle pump to inflate a tyre, you raise the air pressure in the tyre because the pump pushes air through a valve into the tyre. The valve lets the air in but does not allow it out. The tyre is a buffer between the wheel frame and the ground. If the tyre pressure is too low, the wheel frame will rub on the ground when cycling.

The pressure of a gas is the force per unit area that the gas exerts on a surface. Pressure is measured in pascals (Pa), where $1\,\text{Pa} = 1\,\text{N}\,\text{m}^{-2}$. The pressure of a gas depends on its temperature, the volume of the gas container and on the mass of gas in the container.

1 Boyle's law

The apparatus shown in Figure 1 may be used to investigate how the pressure of a fixed mass of gas depends on its volume when the temperature remains the same. Measurements using this apparatus show that the gas pressure times its volume is constant for a fixed mass of gas at constant temperature. This is known as Boyle's law, after Robert Boyle who first discovered the law in 1662. Note that a change at constant temperature is called an **isothermal change**.

Boyle's law states that for a fixed mass of gas at constant temperature,

$$pV = \textbf{constant},$$

where p = gas pressure and V = gas volume.

The measurements plotted as a graph of pressure versus $\dfrac{1}{\text{volume}}$ give a straight line through the origin.

This is because Boyle's law may be written as $p = \text{constant} \times \dfrac{1}{V}$

which represents the equation $y = mx$ for a straight-line graph through the origin if p is plotted on the y-axis and $\dfrac{1}{V}$ on the x-axis.

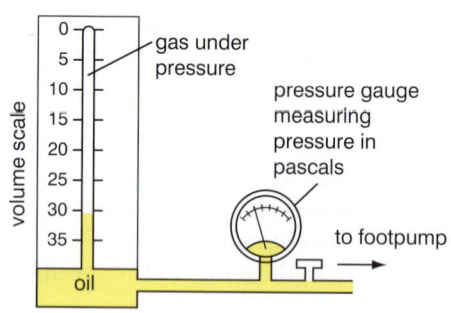

Figure 1 Testing Boyle's law

> ### Notes
>
> 1 A graph of pressure versus volume is a curve which tends towards each axis, as shown in Figure 2.
>
> 2 An ideal gas is a gas that obeys Boyle's law.

2 Charles' law

The volume of a fixed mass of gas at constant pressure increases with temperature. Plotting the measurements of the volume of the gas at 0 °C and 100 °C on a graph leads to the idea of absolute zero. No matter how much gas is used, provided the gas is an ideal gas, its volume would be zero at absolute zero which is   273.15 °C.

Figure 3 shows how the volume of a fixed mass of gas at constant pressure varies with thermodynamic temperature T in kelvins. The graph is a straight line. The relationship, known as **Charles' law**, between the gas volume V and the temperature T in kelvins can therefore be written as

$$\frac{V}{T} = \textbf{constant}$$

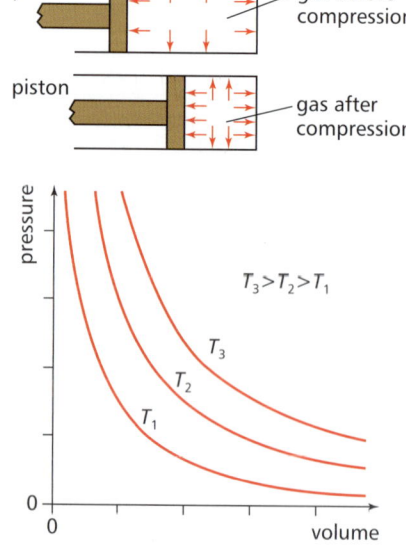

Figure 2 Boyle's law

3 Pressure law

Figure 4 shows how the pressure of a fixed mass of gas at constant volume can be measured at different temperatures. If the measurements are plotted on a graph of pressure against temperature in kelvins, they give a straight line through the origin. The relationship between pressure p and temperature T, in kelvins, can therefore be written as

$$\frac{p}{T} = \text{constant}$$

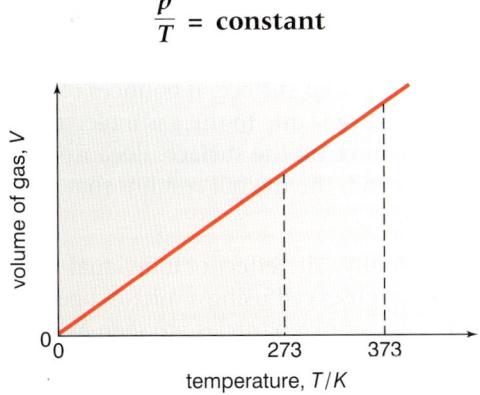

Figure 3 Charles' law

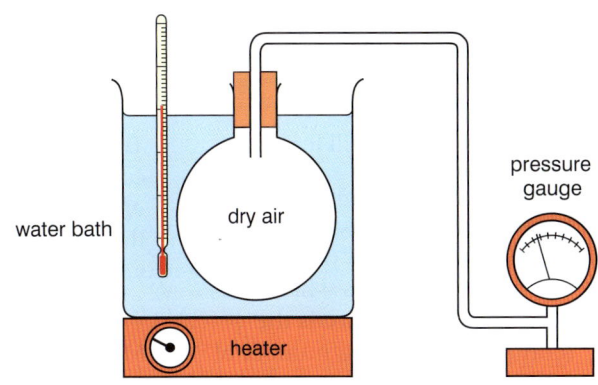

Figure 4 The pressure law

Extension

Measurement of the density of a powder using Boyle's law

A hand pump of volume V_P is used to compress the air in a sealed flask of known volume V_F. A pressure gauge is used to measure the pressure of the air in the flask before compression (p_0) and after compression. See Figure 5.

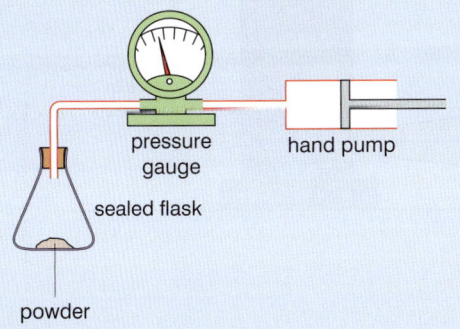

Figure 5 Measuring the volume of a powder

1 Without any powder in the flask; applying Boyle's law gives $p_1 V_F = p_0 (V_F + V_P)$, where p_1 is the measured pressure after compression.

$$\therefore \ p_0 V_F = p_1 V_F - p_0 V_P$$

2 With a measured mass of powder in the flask; applying Boyle's law gives $p_2 (V_F - v) = p_0 (V_F - v + V_P)$, where p_2 is the measured pressure after compression and v is the powder volume.

$$\therefore \ p_0 V_F = p_2 V_F - p_0 V_P - (p_2 - p_0)v$$

Subtracting the two equations gives

$$0 = p_1 V_F - p_2 V_F + (p_2 - p_0)v$$

$$\therefore \quad (p_2 - p_0) v = p_2 V_F - p_1 V_F$$

$$v = \frac{(p_2 - p_1)}{(p_2 - p_0)} V_F$$

Hence, the volume and the density of the powder can be determined.

Summary test 19.1

1 A hand pump of volume $2.0 \times 10^{-4} \text{m}^3$ is used to force air through a valve into a container of volume $8.0 \times 10^{-4} \text{m}^3$ which contains air at an initial pressure of 101 kPa. Calculate the pressure of the air in the container after one stroke of the pump, assuming the temperature is unchanged.

2 A sealed can contains air at a pressure of 101 kPa at 100 °C. The can is then cooled to a temperature of 20 °C. Calculate the pressure of the air in the can assuming the volume of the can is unchanged.

3 The volume of a fixed mass of gas at 15 °C was 0.085 m^3. The gas was then heated to 55 °C without change of pressure. Calculate the new volume of this gas.

4 In an experiment to measure the density of a powder, a hand pump was used to raise the pressure of the air in a flask of volume $1.20 \times 10^{-4} \text{m}^3$ first without, and then with the powder in the flask.

  1 Without the powder in the flask, the pressure increased from 110 kPa to 135 kPa.
  2 With 0.038 kg of powder in the flask, the pressure increased from 110 kPa to 141 kPa.

  a Show that the volume of air in the hand pump initially was $2.7 \times 10^{-5} \text{m}^3$.

  b Calculate the volume and the density of the powder.

Learning outcomes

On these pages you will learn to:

- know that the Avogadro constant N_A is the number of atoms in 0.012 kg of carbon-12
- use molar quantities where one mole of any substance is the amount containing a number of particles equal to the Avogadro constant N_A
- recall and solve problems using the ideal gas equation $pV = nRT$, where n = number of moles
- recognise the experimental basis of the ideal gas equation
- infer from a Brownian motion experiment the evidence for the movement of molecules

Molecules in a gas

The molecules of a gas move at random with different speeds. When a molecule collides with another molecule or with a solid surface, it bounces off without loss of speed. The pressure of a gas on a surface is due to the gas molecules hitting the surface. Each impact causes a tiny force on the surface. Because there are a very large number of impacts each second, the overall result is that the gas exerts a measurable pressure on the surface.

Molecules are too small to see individually. The effect of individual molecules in a gas can be seen if smoke particles are observed using a microscope. If a beam of light is directed through the smoke, the smoke particles are seen as tiny specks of light wriggling about unpredictably. This type of motion is called **Brownian motion** after Robert Brown who first observed it in 1827 with pollen grains in water. The motion of each particle is due to it being bombarded unevenly and at random by individual molecules. The particle is therefore subjected to a force due to the impacts which changes its magnitude and direction at random.

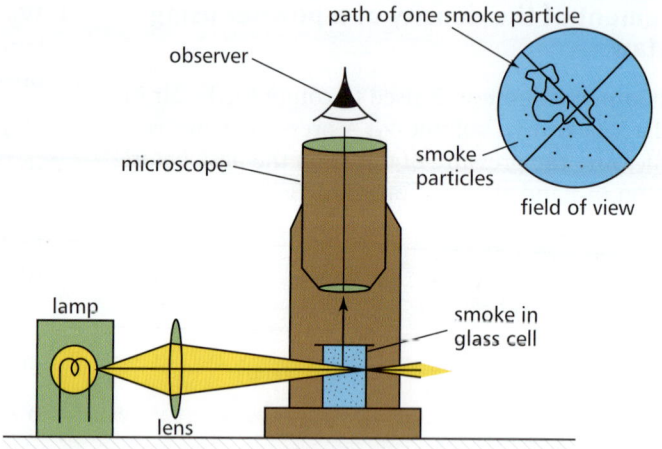

Figure 2 *Brownian motion*

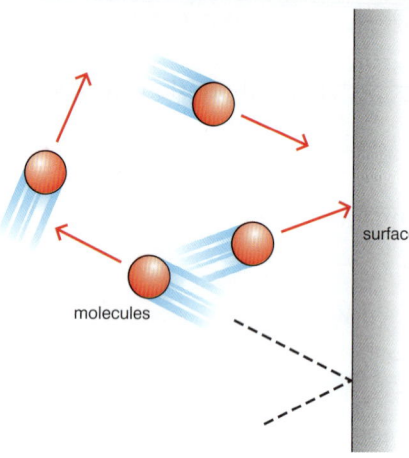

Figure 1 *Molecules in motion*

The Avogadro number

The density of oxygen gas is 16 times that of hydrogen gas at the same temperature. Therefore, the mass of a certain volume of oxygen is 16 times that of the mass of the same volume of hydrogen at the same temperature. When such measurements were first made in the nineteenth century, Amadeo Avogadro put forward the hypothesis that equal volumes of gases at the same temperature and pressure contain an equal number of molecules.

How many molecules are in a certain amount of gas? Avogadro thought of the idea of counting atoms and molecules in terms of the number of atoms in 1 gram of hydrogen. Now we use 12 grams of the carbon isotope $^{12}_{6}C$ as the standard amount as hydrogen contains a small proportion of the isotope of hydrogen $^{2}_{1}H$ which cannot easily be removed.

> **The Avogadro constant, N_A, is defined as the number of atoms in exactly 12 grams of the carbon isotope $^{12}_{6}C$.**

The value of N_A (to four significant figures) is 6.023×10^{23}. Therefore the mass of an atom of $^{12}_{6}C$ is 1.993×10^{-23} grams (= 12 grams/6.02×10^{23}).

> **One atomic mass unit (u) is $\frac{1}{12}$th of the mass of a $^{12}_{6}C$ atom.**

The mass of a carbon atom is 1.993×10^{-26} kg, so $1 u = 1.661 \times 10^{-27}$ kg.

Notes

1 The masses in atomic mass units of some different atoms (to 1 u) are:
hydrogen, H = 1 u, carbon, C = 12 u, nitrogen, N = 14 u, oxygen, O = 16 u, copper, Cu = 64 u

2 The masses in atomic mass units of some different molecules (to 1 u) are:
water H_2O = 18 u, carbon monoxide CO = 28 u, carbon dioxide CO_2 = 44 u, oxygen O_2 = 32 u

Molar mass

• **One mole** of a substance consisting of identical particles is defined as the quantity of substance that contains N_A particles. The number of moles in a certain quantity of a substance is its **molarity**. The unit of molarity is the mol.

• **The molar mass** of a substance is the mass of 1 mole of the substance. The unit of molar mass is kg mol^{-1}. For example, the molar mass of oxygen gas is 0.032 kg mol^{-1}. So 0.032 kg of oxygen gas contains N_A oxygen molecules.

Therefore:

1 the number of moles in mass m of a substance $= \dfrac{m}{M}$, where M is the molar mass of the substance,

2 the number of molecules in mass m of a substance $= N_A \dfrac{m}{M}$.

For example, because the molar mass of carbon dioxide is 44 grams:

2 moles of carbon dioxide has a mass of 88 grams and contains $2N_A$ molecules,

10 moles of carbon dioxide has a mass of 440 grams and contains $10N_A$ molecules,

n moles of carbon dioxide has a mass of $44n$ grams and contains nN_A molecules.

The ideal gas equation

An ideal gas is a gas that obeys the gas laws. The three experimental gas laws can be combined to give the equation

$$\frac{pV}{T} = \textbf{constant, for a fixed mass of ideal gas,}$$

where p is the pressure, V is the volume and T is the **thermodynamic** temperature. This equation takes in all situations where the pressure, volume and temperature of a fixed mass of gas change.

As explained above, equal volumes of ideal gases at the same temperature and pressure contain equal numbers of moles. Further measurements show that one mole of any ideal gas at 273 K and a pressure of 101 kPa has a volume of 0.0224 m^3.

Therefore, for 1 mole of any ideal gas, the value of $\dfrac{pV}{T}$ for 1 mole is equal to

$$8.31\,\text{J mol}^{-1}\text{K}^{-1} \left(= \frac{pV}{T} = \frac{101 \times 10^3\,\text{Pa} \times 0.0224\,\text{m}^3}{273\,\text{K}}\right).$$

This value is known as the **molar gas constant, R**.

Hence, the combined gas law may be written as

$$pV_m = RT$$

where V_m = volume of 1 mole of ideal gas at pressure p and temperature T.

Therefore, for n moles of ideal gas,

$$pV = nRT$$

where V = volume of the gas at pressure p and temperature T.

This equation is known as the **ideal gas equation**.

Note

The unit of R is the joule per mol per kelvin (J mol^{-1}K^{-1}) which is the same as the unit of $\dfrac{\text{pressure} \times \text{volume}}{\text{temperature}}$.

This is because the unit of pressure (the pascal = 1 N m^{-2}) × the unit of volume (m^3) is the joule (= 1 N m).

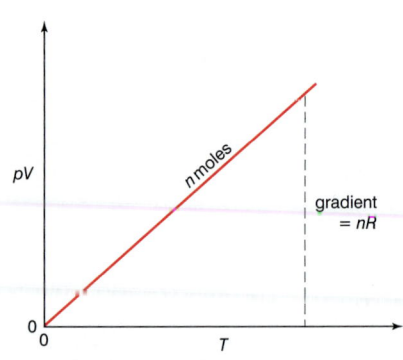

Figure 3 *A graph of pV against T for an ideal gas*

Notes

1 T is the temperature in kelvins.

2 The equation $pV = nRT$ can be written as $pV = NkT$ where N is the number of molecules in the gas, $n = \dfrac{N}{N_A}$ and $k = \dfrac{R}{N_A}$.

   k is referred to as the Boltzmann constant. See Topic 19.4.

3 A graph of pV against temperature T is a straight line through absolute zero and has a gradient equal to nR.

4 The density of an ideal gas of molar mass M, $\rho = \dfrac{\text{mass}}{\text{volume}} = \dfrac{nM}{V} = \dfrac{pM}{RT}$.

   Therefore, for an ideal gas at constant pressure, its density ρ is inversely proportional to its temperature T (as $\rho = \dfrac{pM}{RT} = \dfrac{\text{constant}}{T}$ for constant pressure).

Worked example

$R = 8.31\,\text{J}\,\text{mol}^{-1}\,\text{K}^{-1}$

Calculate the number of moles of air in a balloon when the air pressure in the balloon is 170 kPa, the volume of the balloon is $8.4 \times 10^{-4}\,\text{m}^3$ and the temperature of the air in the balloon is 17 °C.

Solution

$T = 273 + 17 = 290\,\text{K}$

Using $pV = nRT$ gives $n = \dfrac{pV}{RT} = \dfrac{170 \times 10^3 \times 8.4 \times 10^{-4}}{8.31 \times 290} = 5.9 \times 10^{-2}$ moles

Summary test 19.2

$N_A = 6.02 \times 10^{23}\,\text{mol}^{-1}$,
$R = 8.31\,\text{J}\,\text{mol}^{-1}\,\text{K}^{-1}$

1 A gas cylinder has a volume of 0.024 m³ and is fitted with a valve designed to release the gas if the pressure of the gas reaches 125 kPa. Calculate:

   a the maximum number of moles of gas that can be contained by this cylinder at 50 °C,

   b the pressure in the cylinder of this amount of gas at 10 °C.

2 In an electrolysis experiment, $2.2 \times 10^{-5}\,\text{m}^3$ of a gas is collected at a pressure of 103 kPa and a temperature of 20 °C. Calculate:

   a the number of moles of gas present,

   b the volume of this gas at 0 °C and 101 kPa.

3 a Sketch a graph to show how the pressure of 2 moles of gas varies with temperature when the gas is heated from 20 °C to 100 °C in a sealed container of volume 0.050 m³.

   b The molar mass of the gas in a is 0.032 kg. Calculate the density of the gas.

4 The molar mass of air is 0.029 kg.

   a Calculate the density of air at 20 °C and a pressure of 101 kPa.

   b Calculate the number of molecules in 0.001 m³ of air at 20 °C and a pressure of 101 kPa.

The kinetic theory of gases

The gas laws can be explained by assuming a gas consists of point molecules moving about at random, continually colliding with the container walls. Each impact causes a force on the container. The force of many impacts is the cause of the pressure of the gas on the container walls.

Explanation of Boyle's law: the pressure of a gas at constant temperature is increased by reducing its volume because the gas molecules travel a shorter distance between impacts at the walls due to the reduced volume. Hence there are more impacts per second so the pressure is greater.

Explanation of the pressure law: the pressure of a gas at constant volume is increased by raising its temperature. The average speed of the molecules is increased by raising the gas temperature so the impacts of the molecules on the container walls are harder and more frequent. Hence the pressure is raised as a result.

Molecular speeds

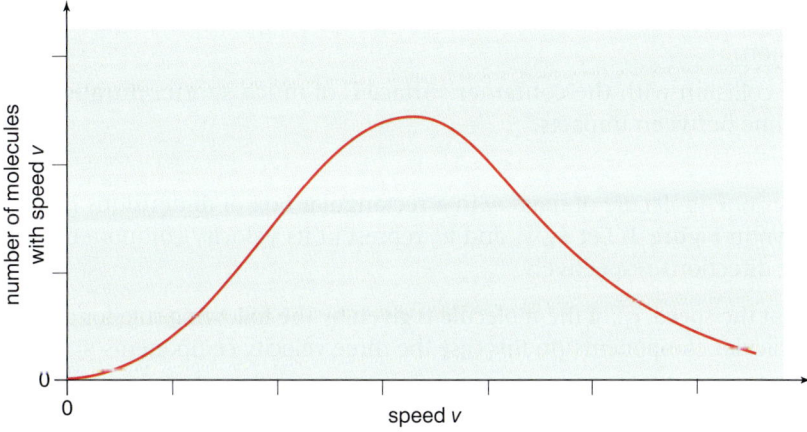

Figure 1 *Distribution of molecular speeds*

The molecules in an ideal gas have a continuous spread of speeds, as shown in Figure 1. The speed of an individual molecule changes when it collides with another gas molecule but the distribution stays the same, provided the temperature does not change.

The **root mean square speed** of the molecules, $c_{\text{r.m.s.}} = \dfrac{[(c_1^2 + c_2^2 + \ldots + c_N^2)]^{1/2}}{N}$

where $c_1, c_2, c_3, \ldots c_N$ represent the speeds of the individual molecules and N is the number of molecules in the gas.

Notes

1 The root mean square (r.m.s.) speed of the molecules of a gas is not the same as the mean speed which is the sum of the speeds divided by the number of molecules.

2 The symbols $<c^2>$ and $\bar{c^2}$ are sometimes used for the mean square speed, $c_{\text{r.m.s.}}^2$.

If the temperature of a gas is raised, its molecules move faster on average. The r.m.s. speed of the molecules increases. The distribution curve becomes flatter and broader as there are molecules at higher speeds (see Figure 2).

Learning outcomes

On these pages you will learn to:

* explain how molecular movement causes the pressure exerted by a gas
* state the basic assumptions of the kinetic theory of gases and derive the relationship $pV = \frac{1}{3}Nm<c^2>$ where N = number of molecules
* solve problems using the above equation and the equation $p = \frac{1}{3}\rho c_{\text{r.m.s.}}^2$, where ρ is the density of the gas

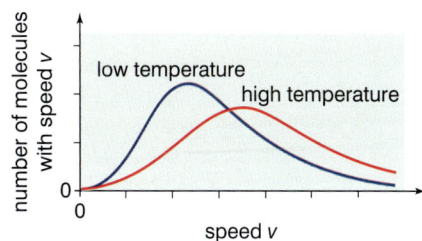

Figure 2 *The effect of temperature on the distribution of speeds*

The kinetic theory equation

For an ideal gas consisting of N identical molecules, each of mass m, in a container of volume V, the pressure p of the gas is given by

$$pV = \frac{1}{3}Nmc^2_{\text{r.m.s.}}$$

where $c_{\text{r.m.s.}}$ is the root mean square speed of the gas molecules.

We need to apply the laws of mechanics and statistics to the molecular model of a gas to derive the kinetic theory equation. In doing so, certain assumptions must be made about the molecules in a gas.

1 The molecules are point molecules. The volume of each molecule is negligible compared with the volume of the gas.

2 They do not attract each other. If they did, the effect would be to reduce the force of their impacts on the container surface.

3 They move about in continual random motion.

4 The collisions they undergo with each other and with the container surface are elastic collisions (i.e. there is no overall loss of kinetic energy in a collision).

5 Each collision with the container surface is of much shorter duration than the time between impacts.

Part 1

Consider one molecule of mass m in a rectangular box of dimensions l_x, l_y and l_z as shown in Figure 3. Let u_1, v_1 and w_1 represent its velocity components in the x, y, and z directions respectively.

Note that the speed, c_1, of the molecule is given by the following rule for adding perpendicular components (in this case the three velocity components u_1, v_1 and w_1),

$$c_1{}^2 = u_1{}^2 + v_1{}^2 + w_1{}^2,$$

We will need to use this rule in Part 2.

Each impact of the molecule with the shaded face in Figure 3 reverses the x-component of velocity thus changing the x-component of its momentum from $+mu_1$ to $-mu_1$.

Therefore, the change of its momentum due to the impact

$$= \text{final momentum} - \text{initial momentum}$$

$$= (-mu_1) - (mu_1) = -2mu_1$$

The time, t, between successive impacts on this face is given by

$$t = \frac{\text{the total distance to the opposite face and back}}{x\text{-component of velocity}} = \frac{2l_x}{u_1}$$

Using Newton's second law therefore gives:

$$\text{the force on the molecule} = \frac{\text{change of momentum}}{\text{time taken}} = \frac{-2mu_1}{(2l_x/u_1)} = \frac{-mu_1{}^2}{l_x}$$

Since the force F_1 of the impact on the surface is equal and opposite to the force on the molecule in accordance with Newton's third law, then $F_1 = \dfrac{+mu_1{}^2}{l_x}$

As pressure = force/area, the pressure p_1 of the molecule on the surface is given by

$$p_1 = \frac{\text{force}}{\text{area of the shaded face } (l_y \times l_z)} = \frac{mu_1{}^2}{l_x \times l_y \times l_z} = \frac{mu_1{}^2}{V}$$

where V = the volume of the box = $l_x \times l_y \times l_z$

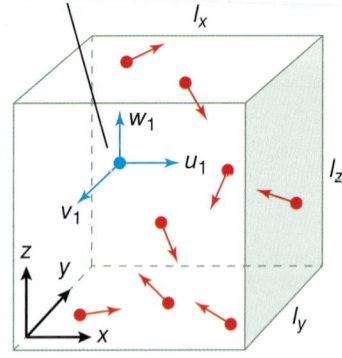

chosen molecule with its velocity components shown

l_x

w_1

u_1

v_1

l_z

l_y

z

y

x

Figure 3 Molecules in a box

Part 2

For N molecules in the box moving at different velocities, the total pressure p is the sum of the individual pressures $p_1, p_2, p_3, \ldots p_N$ where each subscript refers to each molecule.

$$\text{Hence} \quad p = \frac{mu_1^2}{V} + \frac{mu_2^2}{V} + \frac{mu_3^2}{V} + \ldots + \frac{mu_N^2}{V} = \frac{m}{V}(u_1^2 + u_2^2 + u_3^2 + \ldots + u_N^2)$$

$$= \frac{Nm\overline{u^2}}{V}$$

where $\overline{u^2} = (u_1^2 + u_2^2 + u_3^2 + \ldots + u_N^2)/N$

As the motion of the molecules is random, there is no preferred direction of motion. The equation above could equally well have been derived in terms of the y-components of velocity $v_1, v_2, v_3 \ldots v_N$ or the z-components of velocity $w_1, w_2, w_3 \ldots w_N$.

$$\text{i.e.} \qquad p = \frac{Nm\overline{v^2}}{V} \qquad \text{where } \overline{v^2} = (v_1^2 + v_2^2 + v_3^2 + \ldots + v_N^2)/N$$

$$p = \frac{Nm\overline{w^2}}{V} \qquad \text{where } \overline{w^2} = (w_1^2 + w_2^2 + w_3^2 + \ldots + w_N^2)/N$$

Therefore

$$p = \frac{Nm}{3V}(\overline{u^2} + \overline{v^2} + \overline{w^2})$$

The note below shows that, because the motion of the molecules is random, **the root mean square speed** of the gas molecules is given by the equation

$$c_{\text{r.m.s.}}^2 = \overline{u^2} + \overline{v^2} + \overline{w^2}$$

$$\text{Hence} \qquad p = \frac{Nmc_{\text{r.m.s.}}^2}{3V} \qquad \text{or} \qquad pV = \frac{1}{3}Nmc_{\text{r.m.s.}}^2$$

Note

As explained in Part 1, the speed c of each molecule is related to its velocity components according to equations of the form:

$$c_1^2 = u_1^2 + v_1^2 + w_1^2, \qquad c_2^2 = u_2^2 + v_2^2 + w_2^2, \qquad c_3^2 = u_3^2 + v_3^2 + w_3^2, \ldots$$
$$\ldots c_N^2 = u_N^2 + v_N^2 + w_N^2$$

The root mean square speed of the molecules, $c_{\text{r.m.s.}}$ is defined by:

$$c_{\text{r.m.s.}}^2 = (c_1^2 + c_2^2 + c_3^2 + \ldots + c_N^2)/N$$
$$= (u_1^2 + v_1^2 + w_1^2 + u_2^2 + v_2^2 + w_2^2 + u_3^2 + v_3^2 + w_3^2 + \ldots + u_N^2 + v_N^2 + w_N^2)/N$$
$$= \overline{u^2} + \overline{v^2} + \overline{w^2}$$

Pressure and density

In the kinetic theory equation $pV = \frac{1}{3}Nmc_{\text{r.m.s.}}^2$, Nm is the total mass of gas present because N is the number of molecules and m is the mass of each molecule.

Therefore the density of the gas, $\rho = \dfrac{\text{mass}}{\text{volume}} = \dfrac{Nm}{V}$

Rearranging the kinetic theory equation therefore gives $p = \dfrac{1}{3} \times \dfrac{Nm}{V} \times c_{\text{r.m.s.}}^2$

$$= \frac{1}{3}\rho c_{\text{r.m.s.}}^2$$

$$p = \frac{1}{3}\rho c_{\text{r.m.s.}}^2$$

Summary test 19.3

$N_A = 6.02 \times 10^{23}\,\text{mol}^{-1}$,
$R = 8.31\,\text{J}\,\text{mol}^{-1}\,\text{K}^{-1}$

1 Explain in molecular terms why the pressure of a gas in a sealed container increases when its temperature is raised.

2 Calculate the r.m.s. speed of the molecules of a gas when its pressure is $1.10 \times 10^5\,\text{Pa}$ and its density is $1.20\,\text{kg}\,\text{m}^{-3}$.

3 A sealed flask of volume $1.50 \times 10^{-4}\,\text{m}^3$ contains oxygen gas at a pressure of $1.2 \times 10^5\,\text{Pa}$ at a temperature of $15\,°\text{C}$. Calculate:

   a the number of moles of oxygen in the flask,

   b the r.m.s. speed of the oxygen molecules.

   Molar mass of oxygen $= 0.032\,\text{kg}$

4 An ideal gas of molar mass $0.028\,\text{kg}$ is in a container of volume $0.037\,\text{m}^3$ at a pressure of $100\,\text{kPa}$ and a temperature of $300\,\text{K}$. Calculate

   a the number of moles,

   b the mass of gas present,

   c the r.m.s. speed of the molecules of the gas.

Learning outcomes

On these pages you will learn to:

- recall that the Boltzmann constant $k = R/N_A$ and express the ideal gas equation as $pV = NkT$
- deduce that the average translational kinetic energy of a molecule is proportional to T using the equations $pV = \frac{1}{3}Nmc_{r.m.s.}^2$ and $pV = NkT$
- use the first law of thermodynamics, expressed in terms of the increase in internal energy, the heating of the system and the work done on the system ($\Delta U = q + W$), to solve problems

Molecules and kinetic energy

The mean **kinetic energy of a molecule of an ideal gas**

$$= \frac{\text{total kinetic energy of all the molecules}}{\text{total number of molecules}}$$

$$= (\tfrac{1}{2}mc_1{}^2 + \tfrac{1}{2}mc_2{}^2 + \tfrac{1}{2}mc_3{}^2 + \ldots + \tfrac{1}{2}mc_N{}^2)/N$$

$$= \tfrac{1}{2}m\,(c_1{}^2 + c_2{}^2 + c_3{}^2 + \ldots + c_N{}^2)/N = \tfrac{1}{2}mc_{r.m.s.}^2$$

The higher the temperature of a gas, the greater is the mean kinetic energy of a molecule of the gas.

For an ideal gas, by assuming the mean kinetic energy of a molecule $\frac{1}{2}mc_{r.m.s.}^2 = \frac{3}{2}kT$, where $k = R/N_A$, then $3kT = mc_{r.m.s.}^2$.

Substituting $3kT$ for $mc_{r.m.s.}^2$ in the kinetic theory equation $pV = \frac{1}{3}Nmc_{r.m.s.}^2$ therefore gives $pV = \frac{1}{3}N \times 3kT = NkT$

As $Nk = NR/N_A = nR$, we then obtain the ideal gas equation $pV = nRT$. So we have derived the ideal gas equation (which is an experimental law) from the kinetic theory equation by assuming that the mean kinetic energy of an ideal gas molecule $= \frac{3}{2}kT$. Therefore we can say for an ideal gas at thermodynamic temperature T, **the mean kinetic energy of a molecule of the gas $= \frac{3}{2}kT$, where $k = \dfrac{R}{N_A}$.**

The constant k is called the Boltzmann constant. Its value ($= \dfrac{R}{N_A}$) is $1.38 \times 10^{-23}\,\text{J K}^{-1}$

Note

- The total kinetic energy of 1 mole of an ideal gas $= N_A \times \frac{3}{2}kT = \frac{3}{2}RT$ (as $k = \dfrac{R}{N_A}$),
- The total kinetic energy of n moles of an ideal gas $= n \times \frac{3}{2}RT = \frac{3}{2}nRT$

> **The total kinetic energy of n moles of an ideal gas $= \frac{3}{2}nRT$**

Note

The idea that the mean kinetic energy of a particle is proportional to the thermodynamic temperature can be applied in many other situations, for example to explain why conduction electrons can't escape from a metal at room temperature. See Topic 20.2.

Worked example

$N_A = 6.02 \times 10^{23}\,\text{mol}^{-1}$, $k = 1.38 \times 10^{-23}\,\text{J K}^{-1}$.

Calculate the r.m.s. speed of oxygen molecules at $0\,°\text{C}$.

The molar mass of oxygen $= 0.032\,\text{kg mol}^{-1}$

Solution $T = 273\,\text{K}$

The mass of an oxygen molecule, $m = \dfrac{0.032}{6.02 \times 10^{23}}$

$$= 5.3 \times 10^{-26}\,\text{kg}$$

Rearranging $\frac{1}{2}mc_{r.m.s.}^2 = \frac{3}{2}kT$ gives

$$\therefore \quad c_{r.m.s.}^2 = \frac{3kT}{m} = \frac{3 \times 1.38 \times 10^{-23} \times 273}{5.3 \times 10^{-26}}$$

$$= 2.13 \times 10^5\,\text{m}^2\,\text{s}^{-2}$$

$\therefore$ root mean square speed, $c_{r.m.s.} = (2.13 \times 10^5)^{\frac{1}{2}} = 460\,\text{m s}^{-1}$

Using the first law of thermodynamics

As explained in Topic 18.1, the first law of thermodynamics tells us that the change of internal energy of a system is equal to the sum of the work done on or by it and the heat transfer to or from it.

The internal energy U of an ideal gas is due to the kinetic energy of its molecules. No internal forces of attraction exist between 'ideal gas' molecules so an ideal gas does not possess molecular potential energy. As explained above, for n moles of an ideal gas at temperature T, the total kinetic energy $= \frac{3}{2}nRT$. Therefore, its internal energy $= \frac{3}{2}nRT$.

To change the temperature of an ideal gas, its internal energy must be changed. To do this, work must be done on or by the gas or heat must be transferred to or from it.

Work W is done on a gas when its volume is reduced (i.e. it is compressed). At constant volume, no work is done because the forces due to pressure in the gas do not move the container walls.

- When a gas is **compressed**, work is done on it by the applied forces that push the pressure forces back.
- When a gas **expands**, the forces due to its pressure push the container walls back.

Consider when a gas in a piston expands at constant pressure as shown in Figure 1. When the piston moves a small distance Δs outwards:

- The work done W by the air in the cylinder is given by $W = F\Delta s$, where F is the force of the air on the piston.

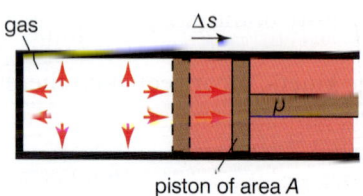

Figure 1

- The increase of volume of the air in the cylinder $V = A\Delta s$, where A is the cross-sectional area of the piston.
- Since pressure = force per unit area, then $F = pA$, where p is the air pressure (assumed constant as the movement of the piston is small). Hence the work done $W = F\Delta s = pA \, \Delta s = p\Delta V$.

More generally:

> **When a gas expands by a volume ΔV at constant pressure p, the work done by the gas, $W = p\Delta V$**

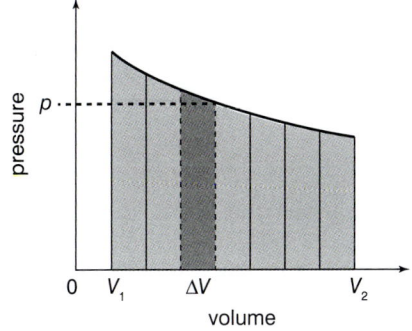

Figure 2 *Estimating work done*

Notes

1 The equation can be applied if the gas is compressed at constant pressure. In this case, energy is transferred to the gas by the force applied to it.

2 For an ideal gas, since $pV = nRT$, where n is the number of moles, then at constant pressure, $p\Delta V = nR\Delta T$. Therefore, the work done $= p\Delta V = nR\Delta T$.

3 **If the pressure changes,** the work done can be estimated from a graph of pressure against volume as shown in Figure 2. The work done is given by the area under the curve. This is because the work done in each small change of volume ΔV is represented by the area of the corresponding strip under the curve; so the total work done when the volume changes from V_1 to V_2 is represented by the total area under the curve from V_1 to V_2.

4 **Heat transfer** q to or from a gas may or may not cause a change of its temperature, depending on whether or not work is done on or by the gas. For example, if 1000 J of work is done on a gas by compressing it and 1200 J of heat energy is transferred from it by cooling it at the same time, its internal energy will decrease by 200 J and its temperature will decrease.

Summary test 19.4

$N_A = 6.02 \times 10^{23} \, \text{mol}^{-1}$, $R = 8.31 \, \text{J K}^{-1}$, $k = 1.38 \times 10^{-23} \, \text{J K}^{-1}$.

1 The molar mass of oxygen is 0.032 kg. A cylinder of volume 0.025 m³ contains oxygen gas at a pressure of 120 kPa and a temperature of 373 K. Calculate: **a** the number of moles of oxygen in the cylinder, **b** the mean kinetic energy of an oxygen molecule in the cylinder, **c** the total kinetic energy of all the gas molecules in the container.

2 Calculate: **a** the mean kinetic energy of a hydrogen molecule at 0 °C, **b** the root mean square speed of a hydrogen molecule at 0 °C. The molar mass of hydrogen gas = 0.002 kg.

3 Air consists mostly of nitrogen and oxygen in proportions 1 : 4 by mass.

  a Explain why the mean kinetic energy of a nitrogen molecule in air is the same as that of an oxygen molecule in the same sample of air.

  b Show that the r.m.s. speed of a nitrogen molecule in air is 1.07 × that of an oxygen molecule in the same sample of air.

  Molar mass nitrogen = 0.028 kg; oxygen = 0.032 kg

4 An ideal gas in a cylinder at an initial temperature of 290 K is compressed and cooled from a volume of 0.025 m³ to a volume of 0.019 m³ at a constant pressure of 1.45×10^5 Pa. Calculate:

  a i the number of moles of the gas,

    ii the temperature of the gas after the change,

  b the work done on the gas,

  c the decrease of internal energy of the gas,

  d the heat transferred from the gas.

Chapter Summary

The ideal gas law $pV = nRT$, where n is the number of moles of gas, T is the absolute temperature and R is the molar gas constant.

The Avogadro constant, N_A is defined as the number of atoms in 12 grams of the carbon isotope $^{12}_{6}\text{C}$

One mole of a substance consisting of identical particles is the quantity of substance that contains N_A particles of the substance.

The molar mass of a substance is the mass of one mole of that substance.

The kinetic theory of gases equation $pV = \frac{1}{3}Nmc^2_{\text{r.m.s.}}$, where $c_{\text{r.m.s.}}$ is the root mean square speed of the gas molecules

The mean kinetic energy of a molecule in a gas at absolute temperature $T = \frac{3}{2}kT$, where k is the Boltzmann constant $(= R/N_A)$

Launch additional digital resources for the chapter

$N_A = 6.02 \times 10^{23}\,\text{mol}^{-1}$, $R = 8.3\,\text{J}\,\text{mol}^{-1}\,\text{K}^{-1}$

1 a In molecular terms, explain why the pressure of a gas increases:
  i if the temperature of the gas is raised at constant volume,
  ii if the amount of gas is increased at constant volume.

  b Sketch a graph to show how the pressure of a gas in a sealed cylinder increases with the absolute temperature of the gas if the cylinder contains:
  i 1 mole of gas,
  ii 2 moles of gas.

2 a Assuming that air at atmospheric pressure consists of 80% nitrogen and 20% oxygen, show that the molar mass of air is 0.029 kg.

  The molar mass of nitrogen = 0.028 kg,
  The molar mass of oxygen = 0.032 kg.

  b A rectangular room has a length of 5.0 m, a width of 4.0 m and a height of 2.5 m. The air in the room has a pressure of 100 kPa and a temperature of 290 K. Calculate:
  i the number of moles of air in the room,
  ii the number of gas molecules in the room.

3 A vehicle air bag inflates rapidly when an impact causes the production and release of a large quantity of nitrogen in a chemical reaction. In a test of an air bag, the bag inflates to a volume of 1.2 m³ and a pressure of 103 kPa at a final temperature of 280 K. Calculate:

  a the number of moles of gas in the bag,

  b the initial pressure of the gas if it was released from a container of volume $5.6 \times 10^{-4}\,\text{m}^3$ at the same temperature.

4 A hot air balloon rises from the ground because it is partly filled with helium which is less dense than air.

  a A certain hot air balloon contains 370 m³ of helium gas at a temperature of 300 K. Calculate:
  i the number of moles of helium gas in the balloon when the pressure of the gas is 110 kPa,
  ii the density of the gas.

  b The helium gas in the balloon fills most of the balloon from the top down. Heating the gas causes it to expand and displace some of the air in the lower part of the balloon.
  i Explain why heating the gas causes its density to decrease.
  ii Explain why the balloon ascends when the helium gas is heated.

  The molar mass of helium = $4.0 \times 10^{-3}\,\text{kg}$

5 A hollow steel cylinder fitted with a pressure gauge is used to store nitrogen gas. The cylinder has an internal volume of $5.0 \times 10^{-3}\,\text{m}^3$.

  The molar mass of nitrogen = 0.028 kg

  a i Calculate the mass of nitrogen gas that must be stored in the cylinder at 300 K to give a gas pressure of 150 kPa.
  ii Describe how the mass of gas remaining in the cylinder could be estimated from the pressure reading of the pressure gauge.

  b The cylinder is fitted with a valve that releases gas from the cylinder if the gas pressure exceeds 160 kPa.

  Calculate the maximum mass of nitrogen that could be stored in the cylinder at 350 K.

6 Argon is an inert gas. The gas is composed of single atoms that do not combine with each other. A certain light bulb contains $8.0 \times 10^{-5}\,\text{m}^3$ of argon gas. When the light bulb was at a temperature of 300 K, the pressure of the gas in the light bulb was 15 kPa.

  a i Calculate the number of moles of argon gas in the light bulb.
  ii Calculate the number of argon atoms in the light bulb.
  iii The mean kinetic energy of an argon atom at 300 K is $6.2 \times 10^{-21}\,\text{J}$. Calculate the speed of an argon atom which has $6.2 \times 10^{-21}\,\text{J}$ of kinetic energy.

  b When the light bulb was switched on, the temperature of the gas in the bulb increased to 350 K. Calculate:
  i the pressure in the light bulb at this temperature,
  ii the increase of kinetic energy of the gas molecules in the light bulb.

  The molar mass of argon = 0.040 kg.

7 Helium gas released with oil from an oil well was collected and stored in a sealed underground cavern of volume 26 000 m³ at a temperature of 280 K. As a result of storing the helium gas, the pressure of the gas in the cavern increased from 100 kPa to 125 kPa.

  a Calculate:
  i the mass of helium gas stored in the cavern,
  ii the mass of a helium molecule.

  b The mean kinetic energy of a helium gas molecule at 280 K is $5.8 \times 10^{-21}\,\text{J}$. Calculate:
  i the speed of a helium gas molecule which has $5.8 \times 10^{-21}\,\text{J}$ of kinetic energy,
  ii the kinetic energy of all the helium gas molecules in the cavern at 280 K.

  The molar mass of helium = 0.004 kg

8 In a vehicle braking system, pressure applied by the driver to the footbrake is transmitted from the master cylinder to the brake pads via sealed pipes filled with brake fluid.

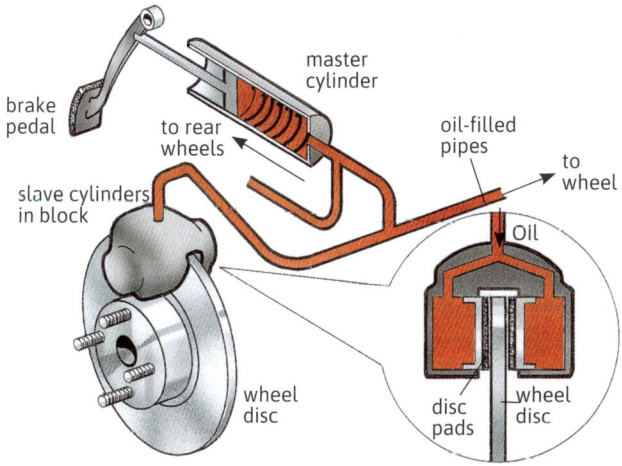

Figure 8.1

a Explain why a bubble of air in the brake fluid makes the brakes less effective than they should be.

b A vehicle brake system is filled with a volume of $1.2 \times 10^{-4} \, m^3$ of brake fluid. When a force of 20 N is applied to the piston in the master cylinder, the pressure in the brake system increases from 105 kPa to 155 kPa.

   i Calculate the area of cross-section of the master cylinder.

   ii Air leaks into the system when the brake fluid is replaced. When a force of 20 N is applied to the piston in the master cylinder, it travels 5 mm more than in **a** and the pressure in the system increases from 105 kPa to 145 kPa. Show that the volume of the air bubble before the brakes were applied was $7.3 \times 10^{-6} \, m^3$.

   iii The temperature of the brake fluid was 290 K. Calculate the mass of air in the bubble.

   The molar mass of air = 0.029 kg

9 An ideal gas in a sealed container of volume $2.71 \times 10^{-4} \, m^3$ is at a pressure of 125 kPa when the temperature is 20.0 °C.

a Calculate the number of moles of the gas.

b The temperature of the gas is reduced to 11.5 °C.

   i Calculate the change of internal energy of the gas as a result of this change of temperature.

   ii Calculate the pressure of the gas at 11.5 °C.

c **i** Draw a graph to show how the pressure of the gas in the sealed container changes between 0 °C and 20 °C.

   ii On the same graph, show how the pressure of the gas would change if the number of moles in the container is half of the value you obtained in **a**.

10a **i** Define the *root mean square (r.m.s.) speed* of the molecules of a gas.

   ii Calculate the r.m.s. speed of the molecules of an ideal gas which has a molar mass of 0.028 kg and is at a temperature of 290 K.

   iii Show that the mean square speed of the molecules of an ideal gas is proportional to the absolute temperature of the gas.

b The graph in Figure 10.1 shows how the pressure and volume of the fuel in a cylinder of a petrol engine changes during part of an engine cycle.

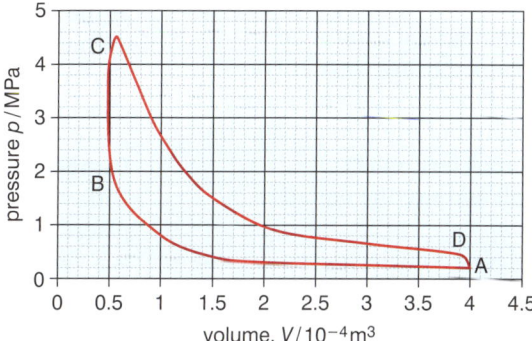

Figure 10.1

In simplified terms:

- From A to B, compression occurs as the fuel (petrol and air) in the cylinder is compressed by the piston in the cylinder.
- For B to C, combustion of the fuel occurs rapidly after the fuel is ignited at B by a spark from a spark plug. The pressure in the cylinder increases until no fuel remains in the cylinder.
- From C to D, expansion of the gas occurs as the piston is forced to reverse by the high pressure in the cylinder enabling the engine to do work until the exhaust valve opens at D.
- From D to A, the gaseous combustion products are expelled from the cylinder and the pressure drops back to its initial value.

   i Explain why no work is done by the engine between B and C.

   ii Explain why the internal energy of the gas in the cylinder decreases between C and D.

   iii Show that the net work done by the engine in one cycle is approximately 330 J.

   iv Combustion between B and C releases 790 J as thermal energy. Calculate the efficiency of the cycle.

   v State two reasons why the efficiency is not higher than the value you calculated in **iv**.

Photoelectricity

Learning outcomes

On these pages you will learn to:

- describe the photoelectric effect and its principal features
- recall the significance of threshold frequency and explain why it cannot be explained using wave theory
- state Einstein's photon theory of light and use it to explain the photoelectric effect
- recall and use the equation $E = hf$
- explain photoelectric phenomena in terms of photon energy and work function energy

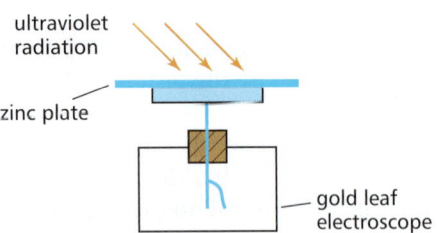

ultraviolet radiation

zinc plate

gold leaf electroscope

Figure 1 *Demonstrating photoelectricity*

> **Note**
>
> The **wavelength** of the incident light must be less than a **maximum** value equal to the speed of light/the threshold frequency. This value of the wavelength is called the **threshold wavelength** of the light.

The discovery of photoelectricity

A metal contains conduction electrons which move about freely inside the metal. These electrons collide with each other and with the positive ions of the metal. Heinrich Hertz discovered how to produce and detect radio waves. He found that the sparks produced in his spark gap detector, when radio waves were being transmitted, were stronger when ultraviolet radiation was directed at the spark gap. Further investigations of the effect of electromagnetic radiation on metals showed that electrons are emitted from the surface of a metal when electromagnetic radiation above a certain frequency is directed at the metal. This effect is known as the **photoelectric effect**.

Demonstration of the photoelectric effect

Ultraviolet radiation from a UV lamp is directed at the surface of a zinc plate placed on the cap of a gold leaf electroscope, as shown in Figure 1. This device is a very sensitive detector of charge. When it is charged, the thin gold leaf of the electroscope rises: it is repelled from the metal 'stem', because they are both charged with a like charge.

- If the electroscope is **charged negatively**, the leaf rises and stays in position. However, if ultraviolet light is directed at the zinc plate, the leaf gradually falls. The leaf falls because conduction electrons at the zinc surface leave the zinc surface when ultraviolet light is directed at it. The emitted electrons are referred to as **photoelectrons**.
- If the electroscope is **charged positively**, the leaf rises and stays in position, regardless of whether or not ultraviolet light is directed at the zinc plate. This is because the conduction electrons in the zinc plate are held on the zinc plate, as the plate is charged positive and electrons carry negative charge.

Puzzling problems

The following observations were made about **photoelectricity** after Hertz's discovery. These observations were a major problem, because they could not be explained using the wave theory of electromagnetic radiation.

- Photoelectric emission of electrons from a metal surface does not take place if the frequency of the incident electromagnetic radiation is below a certain value, known as the **threshold frequency**. This minimum frequency depends on the type of metal.
- The number of electrons emitted per second is proportional to the **intensity** of the incident radiation, provided the frequency is greater than the threshold frequency. If the frequency of the incident radiation is less than the threshold frequency, no photoelectric emission from that metal surface can take place, no matter how intense the incident radiation is.
- Photoelectric emission occurs without delay, as soon as the incident radiation is directed at the surface, provided the frequency of the radiation exceeds the threshold frequency. No matter how weak the intensity of the incident radiation is, electrons are emitted as soon as the source of radiation is switched on.

The wave theory of light cannot explain either the existence of a threshold frequency, or why photoelectric emission occurs without delay. According to wave theory, each conduction electron at the surface of a metal should gain some energy from the incoming waves, regardless of how many waves arrive each second. Wave theory therefore predicted that:

- Emission should take place with waves of any frequency.
- Emission would take longer using low intensity waves than using intensity waves.

The discovery of radio waves and X-rays confirmed the prediction by Maxwell that electromagnetic radiation exists beyond the known spectrum from ultraviolet to infrared radiation. Using Maxwell's theory of electromagnetic waves, physicists were very successful in predicting the properties of electromagnetic waves until the discovery of the photoelectric effect.

Einstein's explanation of photoelectricity

The photon theory of light was put forward by Einstein in 1905 to explain photoelectricity. Einstein assumed that light is composed of wave packets, or **photons**, each of energy equal to hf, where f is the frequency of the light and h is the Planck constant. The accepted value for h is $6.63 \times 10^{-34}\,\text{J s}$.

$$\text{Energy of a photon} = hf$$

For electromagnetic waves of wavelength λ, the energy of each photon $E = hf = \dfrac{hc}{\lambda}$ where c is the speed of the electromagnetic waves.

To explain photoelectricity, Einstein said that:

- When light is incident on a metal surface, an electron at the surface absorbs a **single** photon from the incident light and therefore gains energy equal to hf, where hf is the energy of a light photon.
- An electron can leave the metal surface, if the energy gained from a single photon exceeds **the work function, ϕ,** of the metal. This is the minimum energy needed by an electron to escape from the metal surface.

Hence the maximum kinetic energy of an emitted electron:

$$E_{\text{K(max)}} = hf - \phi$$

Emission can take place from a surface at zero potential, provided $E_{\text{Kmax}} > 0$, i.e. $hf > \phi$. Thus the threshold frequency of the metal:

$$f_{\text{min}} = \frac{\phi}{h}$$

Figure 2 Albert Einstein 1879–1955

In 1905 Einstein published three papers, each of which changed existing ideas. His paper on Brownian motion showed that atoms must exist, and he established the photon theory of light in the second paper. Perhaps he is best remembered for his theories of relativity which he stated in the third paper.

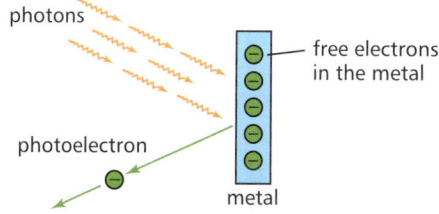

photons

free electrons in the metal

photoelectron

metal

Figure 3 Explaining photoelectricity

Summary test 20.1

$h = 6.63 \times 10^{-34}\,\text{J s}$, $c = 3.00 \times 10^{8}\,\text{m s}^{-1}$

1 a What is meant by 'photoelectric emission' from a metal surface?

b Explain why photoelectric emission from a metal surface only takes place if the frequency of the incident radiation is greater than a certain value.

2 a Calculate the frequency and energy of a photon of wavelength:

   i 450 nm,   **ii** 1500 nm.

b A metal surface at zero potential emits electrons from its surface if light of wavelength 450 nm is directed at it, but not if light of wavelength 650 nm is used. Explain why photoelectric emission happens with light of wavelength 450 nm but not with light of wavelength 650 nm.

3 The work function of a certain metal plate is $1.1 \times 10^{-19}\,\text{J}$. Calculate:

a the threshold frequency of incident radiation,

b the maximum kinetic energy of photoelectrons emitted from this plate when light of wavelength 520 nm is directed at the metal surface.

4 Light of wavelength 635 nm is directed at a metal plate at zero potential. Electrons are emitted from the plate with a maximum kinetic energy of $1.5 \times 10^{-19}\,\text{J}$. Calculate:

a the energy of a photon of this wavelength,

b the work function of the metal,

c the threshold frequency of electromagnetic radiation incident on this metal.

Learning outcomes

On these pages you will learn to:

- explain why the maximum photoelectric energy is independent of intensity, whereas the photoelectric current is proportional to intensity
- recall, use and explain the significance of $hf = \Phi + \frac{1}{2}mv_{max}^2$

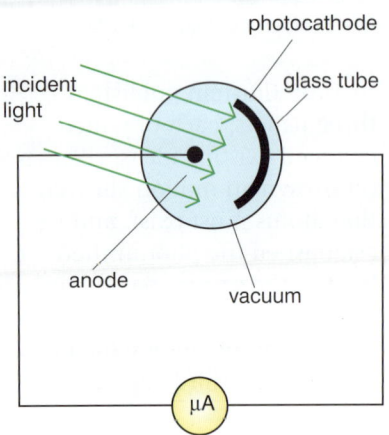

Figure 1 *Using a vacuum photocell*

 The electron volt

The electron volt (eV) is a unit of energy equal to the work done when an electron is moved through a p.d. of 1 V.

For a charge q moved through a p.d., V:

Work done $= qV$

Therefore, the work done when an electron moves through a potential difference of 1 V is equal to **1.6×10^{-19} J** $(= 1.6 \times 10^{-19}$ C $\times 1$ V$)$. This amount of energy is defined as **1 electron volt.**

Examples
The work done on:

- an electron when it moves through a potential difference of 1000 V = 1000 eV,
- an ion of charge +2e when it moves through a potential difference of 10 V = 20 eV.

More about conduction electrons

The **average kinetic energy** of a conduction electron in a metal depends on the **temperature** of the metal. As the conduction electrons move about at random in the metal, they can be likened to the molecules of a gas. The average kinetic energy of a gas molecule is proportional to the absolute temperature of the gas. It can be shown that the average kinetic energy of an conduction electron in a metal at 300 K is therefore about 6×10^{-21} J.

- The **work function of a metal** is the **minimum** energy needed by a conduction electron to escape from the metal surface when the metal is at zero potential. The work function of a metal is of the order of 10^{-19} J, which is about 20 times greater than the average kinetic energy of a conduction electron in a metal at 300 K. In other words, a conduction electron in a metal at about 20 °C does not have sufficient kinetic energy to leave the metal.

- When a conduction electron **absorbs a photon**, its kinetic energy increases by an amount equal to the energy of the photon. Provided the energy of the photon exceeds the work function of the metal, the conduction electron can leave the metal. If the electron does not leave the metal, it collides repeatedly with other electrons and positive ions, and it quickly loses its extra kinetic energy.

Photoelectricity investigations

The vacuum photocell
A **vacuum photocell** is a glass tube that contains a metal plate, referred to as the **photocathode**, and a smaller metal electrode referred to as the **anode**. Figure 1 shows a vacuum photocell in a circuit. When light of frequency greater than the threshold frequency for the metal is directed at the photocathode, electrons emitted from the cathode transfer to the anode. The microammeter in the circuit can be used to measure the photoelectric current, which is proportional to the number of electrons per second that transfer from the cathode to the anode.

- For a photoelectric current I, the **number of photoelectrons per second** that transfer from the cathode to the anode is I/e, where e is the charge of the electron.

- The photoelectric current is proportional to the **intensity** of the light incident on the cathode. This is because the intensity of the incident light is a measure of the **energy per second** carried by the incident light, which is proportional to the number of photons per second incident on the cathode. Because each photoelectron must have absorbed one photon to escape from the metal surface, the number of photoelectrons emitted per second (i.e. the photoelectric current) is therefore proportional to the intensity of the incident light.

- The intensity of the incident light does **not** affect the maximum kinetic energy of a photoelectron. No matter how intense the incident light is, the energy gained by a photoelectron is due to the absorption of one photon only. Therefore, the maximum kinetic energy of a photoelectron is still given by $E_{Kmax} = hf - \phi$, as explained in Topic 20.1.

Measurement of the work function of a metal
Photoelectric emission can be stopped by making the photocathode sufficiently positive. Figure 2 shows how a potential divider can be used to make the photocathode increasingly positive, relative to the anode. As the potential difference is increased from zero, the microammeter reading decreases to zero. This happens because each photoelectron leaving the metal surface needs to do extra work, as the plate is at a positive potential. The kinetic energy of

a photoelectron is therefore reduced, because each electron has to do work to overcome the attraction of the plate. When the microammeter reading is zero, photoelectric emission stops: this is because the kinetic energy of a photoelectron is reduced to zero before the electron can escape from the attraction of the plate.

- As explained earlier, the work done by a charged particle when it moves through a potential difference V is qV, where q is the charge of the particle. Therefore, to escape from a metal plate at positive potential V, the **extra work** needed to be done by a photoelectron is eV, where e is the charge of the electron.

- At zero potential, the **maximum kinetic energy** of an emitted photoelectron is $hf - \phi$, where f is the frequency of the incident radiation and ϕ is the work function of the metal.

Therefore, photoelectric emission is stopped when the potential V is such that

$$eV_S = hf - \phi$$

where V_S is the potential needed to stop emission.

In other words, the maximum kinetic energy of a photoelectron from a surface at zero potential:

$$E_{K(max)} = eV_S$$

By measuring V_S for different frequencies f, $E_{K(max)}$ can be calculated for each frequency; then a graph of $E_{K(max)}$ against frequency f can be plotted. Because $E_{K(max)} = hf - \phi$, the graph is a straight line with:

- a gradient h,
- a y-intercept equal to $-\phi$, and
- an x-intercept equal to the threshold frequency, $f_{min} = \dfrac{\phi}{h}$

The above equation can be written as $hf = \phi + \frac{1}{2}mv^2_{max}$ where v_{max} is the maximum speed of the emitted photoelectrons and $\frac{1}{2}mv^2_{max}$ is the maximum kinetic energy $E_{K(max)}$.

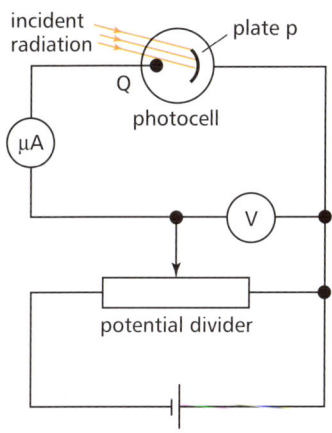

Figure 2 Investigating photoelectricity

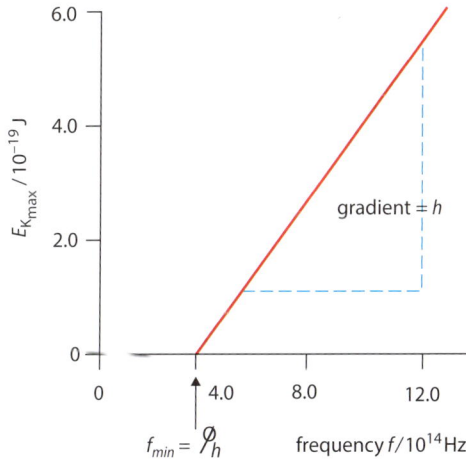

Figure 3 A graph of $E_{K(max)}$ against frequency

Summary test 20.2

$h = 6.63 \times 10^{-34}\,J\,s$, $c = 3.00 \times 10^8\,m\,s^{-1}$, $e = 1.6 \times 10^{-19}\,C$

1 A vacuum photocell is connected to a microammeter. Explain the following observations:

  a When the cathode was illuminated with blue light of low intensity, the microammeter showed a non-zero reading.

  b When the cathode was illuminated with an intense red light, the microammeter reading was zero.

2 A vacuum photocell is connected to a microammeter. When light is directed at the photocell, the micro-ammeter reads $0.25\,\mu A$.

  a Calculate the number of photoelectrons emitted per second by the photocathode of the photocell.

  b Explain why the microammeter reading is doubled if the intensity of the incident light is doubled.

3 A narrow beam of light of wavelength 590 nm and of power 0.5 mW is directed at the photocathode of a vacuum photocell, which is connected to a micro-ammeter that reads $0.4\,\mu A$. Calculate:

  a the energy of a single light photon of this wavelength,

  b the number of photons per second incident on the photocathode,

  c the number of electrons emitted per second from the photocathode.

4 a Use Figure 3 to estimate:
    i the threshold frequency,
    ii the work function of the photocathode that gave the results used to plot the graph.

  b A metal surface has a work function of $1.9 \times 10^{-19}\,J$. Light of wavelength 435 nm is directed at the metal surface. Calculate the maximum kinetic energy of the photoelectrons emitted from this metal surface.

Learning outcomes

On these pages you will learn to:

- recognise the existence of discrete electron energy levels in isolated atoms (e.g. atomic hydrogen) and deduce how this leads to spectral lines
- distinguish between emission and absorption line spectra
- recall and solve problems using the relation $hf = E_1 - E_2$

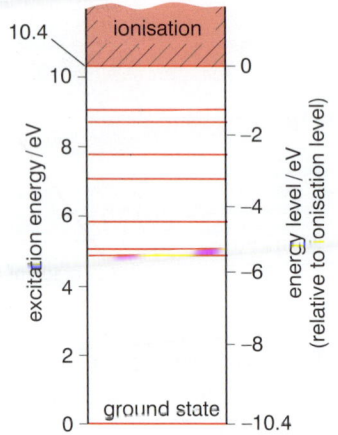

Figure 1 *The energy levels of the mercury atom*

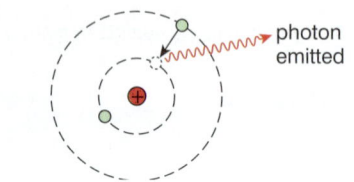

Figure 2 *De-excitation by photon emission*

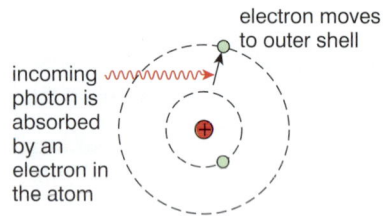

Figure 3 *Excitation by photon absorption*

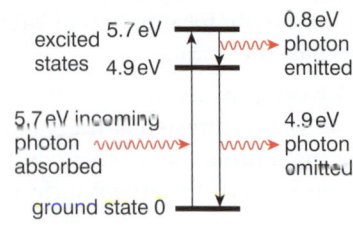

Figure 4 *Fluorescence*

Electrons in atoms

The electrons in an atom are trapped by the electrostatic force of attraction of the nucleus. They move about the nucleus in allowed orbits or 'shells' surrounding the nucleus. The energy of an electron in a shell is constant. An electron in a shell near the nucleus has less energy than an electron in a shell further away from the nucleus. Each shell can only hold a certain number of electrons. For example, the innermost shell (that is, the shell nearest to the nucleus) can only hold two electrons and the next nearest shell can only hold eight electrons.

Each type of atom has a certain number of electrons. For example, a helium atom has two electrons and a lithium atom has three electrons. Thus, in its lowest energy state:

- a helium atom has both electrons in the innermost shell,
- a lithium atom has two electrons in the innermost shell and one in the next shell.

The lowest energy state of an atom is called its **ground state**. When an atom in the ground state absorbs energy, one of its electrons moves to a shell at higher energy, so the atom is now in an **excited state**. We can use the excitation energy measurements to construct an energy level diagram for the atom, as shown in Figure 1. This shows the allowed energy values of the atom. Each allowed energy corresponds to a certain electron configuration in the atom. Note that the ionisation level may be considered as the 'zero' reference level for energy, instead of the ground state level.

De-excitation

Did you know that gases at low pressure emit light when they are made to conduct electricity? For example, a neon tube emits red-orange light when it conducts. The gas-filled tube used to measure excitation energies emits light when excitation occurs. This happens because the atoms absorb energy as a result of excitation by collision but they do not retain the absorbed energy permanently.

The electron configuration in an excited atom is unstable because an electron that moves to an outer shell leaves a vacancy in the shell it moves from. Sooner or later, the vacancy is filled by an electron from an outer shell transferring to it. When this happens, the electron emits a photon of energy equal to the energy lost by the electron. The atom therefore moves to a lower energy level. See Figure 2.

The energy of the photon is equal to the energy lost by the electron and therefore by the atom. For example, when a mercury atom at an excitation energy level of 4.9 eV de-excites to the ground state, it emits a photon of energy 4.9 eV.

An atom in an excited state may de-excite to the ground state indirectly. For example, a mercury atom at an excitation energy of 5.7 eV may de-excite:

- to the 4.9 eV level, emitting a 0.8 eV photon in the process, then
- to the ground state from the 4.9 eV level, emitting a 4.9 eV photon in this process.

In general, when an electron moves from energy level E_1 to a lower energy level E_2, **the energy of the emitted photon $hf = E_1 - E_2$**

Excitation using photons

An electron in an atom can absorb a photon and move to an outer shell where a vacancy exists – but only if the energy of the photon is exactly equal to the gain in the electron's energy (see Figure 3). In other words, the photon energy must be exactly equal to the difference between the final and initial energy levels of the atom. If the photon's energy is smaller or larger than the difference between the two energy levels, it will not be absorbed by the electron.

Fluorescence

An atom in an excited state can de-excite directly or indirectly to the ground state, regardless of how the excitation took place. Therefore, an atom can absorb photons of certain energies and then emit photons of the same or lesser energies. For example, a mercury atom in the ground state could be excited to its 5.7 eV energy level by absorbing a photon of energy 5.7 eV; the mercury could then de-excite to its 4.9 eV energy level by emitting a photon of energy 0.8 eV; then de-excite to the ground state by emitting a photon of energy 4.9 eV.

Figure 4 represents these changes on an energy level diagram.

This overall process explains why certain substances **fluoresce** or glow with visible light when they absorb ultraviolet radiation. Atoms in the substance absorb ultraviolet photons and become excited. When the atoms de-excite, they emit visible photons. When the source of ultraviolet radiation is removed, the substance stops glowing.

Line spectra

A rainbow is a natural display of the colours of the spectrum of sunlight. Raindrops split sunlight into a continuous spectrum of colours. Figure 5 shows the continuous spectrum produced by passing a beam of white light from a filament lamp through a glass prism. The wavelength of the light photons that produce the spectrum increases across the spectrum from deep violet at less than 400 nm to deep red at about 650 nm.

If we use a tube of glowing gas as the light source instead of a filament lamp, we see a spectrum of discrete lines of different colours, as shown in Figure 6. This type of spectrum is referred to as a **line emission spectrum**.

If the spectrum of light from a filament lamp is observed after passing it through a glowing gas, we see dark vertical lines in the continuous spectrum, as shown in Figure 7. The lines are due to absorption of photons of certain energies by gas atoms. This type of spectrum is referred to as a **line absorption spectrum**.

The wavelengths of the lines of a line spectrum of an element are characteristic of the atoms of that element. By measuring the wavelengths of a line spectrum, we can therefore identify the element that produced the light. No other element produces the same pattern of light wavelengths. This is because the energy levels of each type of atom are unique to that atom. So the photons emitted are characteristic of the atom.

- Each line in a line spectrum is due to light of a certain colour and therefore a certain wavelength.
- The photons that produce each line all have the same energy, which is different from the photons that produce any other line.
- Each photon is emitted when an atom de-excites due to one of its electrons moving from an inner shell.
- As explained above, if the electron moves from energy level E_1 to a lower energy level E_2, the energy of the emitted photon $hf = E_1 - E_2$.

For each wavelength λ, we can calculate the energy of a photon of that wavelength as its frequency $f = c/\lambda$, where c is the speed of light. Given the energy level diagram for the atom, we can therefore identify on the diagram the transition that causes a photon of that wavelength to be emitted.

Worked example

$c = 3.0 \times 10^8 \, m s^{-1}$, $e = 1.6 \times 10^{-19} \, C$, $h = 6.63 \times 10^{-34} \, Js$
A mercury atom de-excites from its 4.9 eV energy level to the ground state. Calculate the wavelength of the photon released.

Solution

$E_1 - E_2 = 4.9 - 0 = 4.9 \, eV = 4.9 \times 1.6 \times 10^{-19} \, J = 7.84 \times 10^{-19} \, J$

Therefore, $f = \dfrac{E_1 - E_2}{h} = \dfrac{7.84 \times 10^{-19}}{6.63 \times 10^{-34}} = 1.12 \times 10^{15} \, Hz$

$\lambda = \dfrac{c}{f} = \dfrac{3.0 \times 10^8}{1.12 \times 10^{15}} = 2.52 \times 10^{-7} \, m$

Figure 5 *Observing a continuous spectrum*

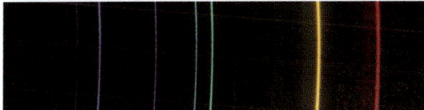
Figure 6 *A line emission spectrum*

Figure 7 *A line absorption spectrum*

Summary test 20.3

$e = 1.6 \times 10^{-19} \, C$

1 Figure 1 shows some of the energy levels of the mercury atom.

 a Estimate the energy needed to excite the atom from the ground state to the highest excitation level shown in the diagram.

 b Mercury atoms in an excited state at 5.7 eV can de-excite directly or indirectly to the ground state. Show that the photons released could have six different energies.

2 a In terms of electrons, state two differences between excitation and de-excitation.

 b A certain type of atom has excitation energies of 1.8 eV and 4.6 eV.
  i Sketch an energy level diagram for the atom using these energy values.
  ii Calculate the possible photon energies from the atom when it de-excites from the 4.6 eV level, indicating on your diagram, by downward arrows, the energy change responsible for each photon energy.

3 Explain why the line spectrum of an element is unique to that element and can be used to identify it.

301

Wave–particle duality

Learning outcomes

On these pages you will learn to:

- show an understanding that the photoelectric effect provides evidence for a particulate nature of electromagnetic radiation while phenomena such as interference and diffraction provide evidence for a wave nature
- describe and interpret qualitatively the evidence provided by electron diffraction for the wave nature of particles
- recall and use the relation for the de Broglie wavelength $\lambda = h/p$

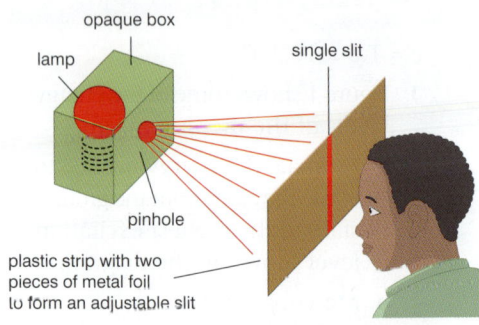

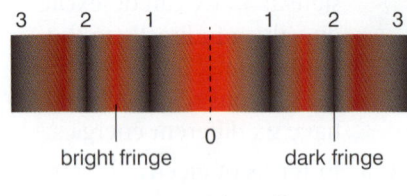

bright fringe   dark fringe

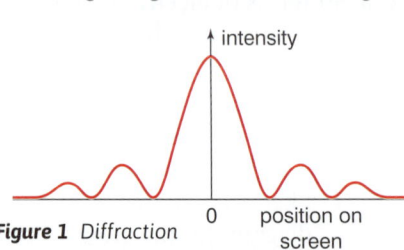

Figure 1 *Diffraction*

Worked example

Calculate the momentum of an X-ray photon of wavelength 2.0×10^{-11} m.

Solution

For a photon of wavelength λ, its energy $E = hf$, where its frequency $f = \dfrac{c}{\lambda}$.

Therefore its momentum,

$p = \dfrac{E}{c} = \dfrac{hf}{c} = \dfrac{hc}{\lambda c} = \dfrac{h}{\lambda}$

$= 6.63 \times 10^{-34}\,\text{J s}/2.0 \times 10^{-11}\,\text{m}$

$= 3.3 \times 10^{-23}\,\text{kg m s}^{-1}$

The dual nature of light

Light is part of the electromagnetic spectrum of waves. The theory of electromagnetic waves predicted the existence of electromagnetic waves beyond the visible spectrum. The subsequent discovery of X-rays and radio waves confirmed these predictions and seemed to show that the nature of light had been settled. Many scientists in the late nineteenth century reckoned that all aspects of physics could be explained using Newton's laws of motion, and the theory of electromagnetic waves. They thought that the few minor problem areas, such as photoelectricity, would be explained sooner or later using Newton's laws of motion and Maxwell's theory of electromagnetic waves. However, photoelectricity was not explained until Einstein put forward the radical theory that light consists of photons, which are 'particle-like' packets of electromagnetic waves. Light has a dual nature, in that it can behave as a wave or as a particle, according to circumstances.

- The **wave-like nature** is observed when **diffraction** of light takes place. This happens, for example, when light passes through a narrow slit. The light emerging from the slit spreads out in the same way as water waves spread out after passing through a gap. Although you were introduced to diffraction in Topic 8.3, you can see in Figure 1 how to observe diffraction of light. The light spreads out from the slit because the slit is very narrow. Notice that the pattern shows bright and dark fringes. This is because the wavelets from each wavefront that passes through the slit reinforce each other in certain directions only and cancel each other out in other directions.

- The **particle-like nature** is observed, for example, in the photoelectric effect. When light is directed at a metal surface and an electron at the surface absorbs a photon of frequency f, the kinetic energy of the electron is increased from a negligible value by hf. The electron can escape if the energy it gains from a photon exceeds the work function of the metal.

The particle-like nature of electromagnetic radiation was reinforced when it was discovered that X-rays can undergo collisions with electrons, causing the electrons to scatter by transferring momentum to them. Thus photons must possess momentum even though they are massless! From these experiments, it was deduced that a photon of energy E possesses momentum p equal to $\dfrac{E}{c}$, where c is the speed of light in a vacuum.

$$\textbf{Photon momentum } p = \frac{E}{c}$$

where E is the photon energy.

Matter waves

If light has a dual wave–particle nature, perhaps particles of matter also have a dual wave–particle nature. Electrons in a beam can be deflected by a magnetic field. This is evidence that electrons have a particle-like nature. The idea that matter particles also have a wave-like nature was first considered by de Broglie in 1923. By extending the ideas of duality from photons to matter particles, de Broglie put forward the hypothesis that:

- matter particles have a dual wave–particle nature,
- the wave-like behaviour of a matter particle is characterised by a wavelength, its **de Broglie wavelength**, λ, which is related to the momentum, p, of the particle by means of the equation

$$\lambda = \frac{h}{p}$$

Since the momentum of a particle is defined as its mass × its velocity, according to de Broglie's hypothesis, a particle of mass m moving at velocity v has a de Broglie wavelength given by

$$\lambda = \frac{h}{mv}$$

Note that the de Broglie wavelength of a particle can be altered by changing the velocity of the particle.

Evidence for de Broglie's hypothesis

The wave-like nature of electrons was discovered when, three years after de Broglie put forward his hypothesis, it was demonstrated that a beam of electrons can be diffracted. Figure 2 shows in outline how this is done. After this discovery, further experimental evidence, using other types of particles, confirmed the correctness of de Broglie's theory.

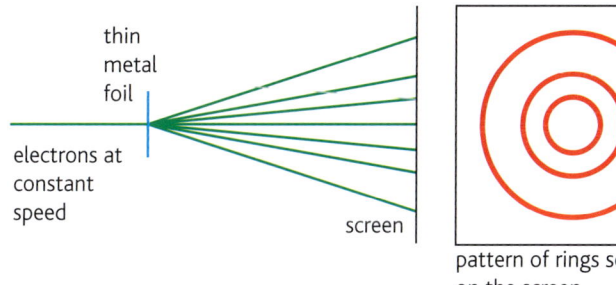

Figure 2 Diffraction of electrons

- A narrow beam of electrons in a vacuum tube is directed at a thin metal foil. A metal is composed of many tiny crystalline regions. Each region or 'grain' consists of positive ions arranged in fixed positions in rows in a regular pattern. The rows of atoms cause the electrons in the beam to be diffracted, just as a beam of light is diffracted by the slits.

- The electrons in the beam pass though the metal foil and are diffracted in certain directions only, as shown in Figure 2. They form a pattern of rings on a fluorescent screen at the end of the tube. Each ring is due to electrons diffracted by the same amount from grains of different orientations, at the same angle to the incident beam.

- The beam of electrons is produced by attracting electrons from a heated filament wire to a positively charged metal plate, which has a small hole at its centre. Electrons that pass through the hole form the beam. The speed of these electrons can be increased by increasing the potential difference between the filament and the metal plate. This makes the diffraction rings smaller, because the increase of speed makes the de Broglie wavelength smaller. So less diffraction occurs and the rings become smaller.

Energy levels explained

An electron in an atom has a fixed amount of energy that depends on the shell it occupies. Its de Broglie wavelength has to fit the shape and size of the shell. This is why its energy depends on the shell it occupies.

For example, an electron in a spherical shell moves round the nucleus in a circular orbit. The circumference of its orbit must be equal to a whole number of de Broglie wavelengths (circumference = $n\lambda$, where $n = 1$ or 2 or 3, etc.). You don't need to know this for your exam but this condition can be used to derive the energy level formula for the hydrogen atom – and it gives you a deeper insight into quantum physics.

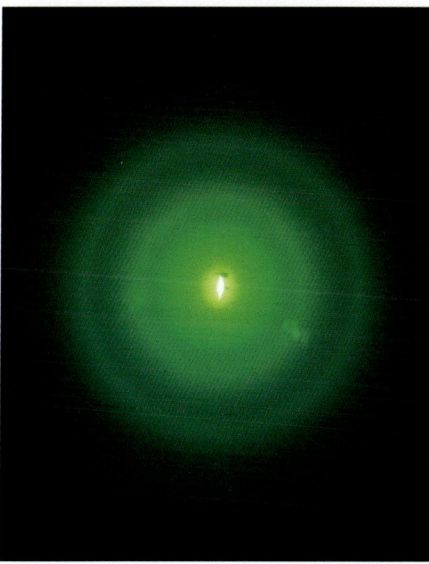

Figure 3 Electron diffraction

Summary test 20.4

$h = 6.63 \times 10^{-34}\,\text{J s}$, the mass of an electron $= 9.11 \times 10^{-31}\,\text{kg}$, the mass of a proton $= 1.67 \times 10^{-27}\,\text{kg}$

1 With the aid of an example in each case, explain what is meant by the dual wave–particle nature of:

   a light, **b** matter particles.

2 State whether each of the following experiments demonstrates the wave nature or the particle nature of matter or of light: (a) photoelectricity, (b) electron diffraction.

3 Calculate the de Broglie wavelength of:

   a an electron moving at a speed of $2.0 \times 10^7\,\text{m s}^{-1}$

   b a proton moving at the same speed.

4 Calculate the momentum and speed of:

   a an electron that has a de Broglie wavelength of 500 nm,

   b a proton that has the same de Broglie wavelength.

303

Launch additional digital resources for the chapter

1 When light at sufficiently high frequency, f, is incident on a metal surface, the maximum kinetic energy, E_{Kmax}, of a photoelectron emitted from the surface is given by:

$E_{Kmax} = hf - \phi$, where ϕ is the work function of the metal

  a State what is meant by the *work function* ϕ.

  b The following results were obtained using a certain metal X in an experiment to measure E_{Kmax} for different frequencies, f.

$f/10^{14}$ Hz	5.6	6.2	6.8	7.3	8.3	8.9
$E_{Kmax}/10^{-19}$ J	0.8	1.2	1.6	2.0	2.6	3.0

    i Use these results to plot a graph of E_{Kmax} against f.

    ii Determine the gradient of your graph.

    iii Explain why your graph confirms the equation above.

    iv Use your graph to determine the work function of the metal and hence calculate the threshold frequency of light for this metal.

  b The metal was replaced by a different metal Y with a known work function of 3.3×10^{-19} J. Calculate the threshold frequency of this metal and draw and label a line on your graph to show the results you would expect to obtain for this metal.

2 a For a metal, state what is meant by:

    i the *work function* of the metal,

    ii the *threshold frequency* for photoelectric emission from the metal.

  b State and explain the relationship between the work function, ϕ, and the threshold frequency, f, of a metal.

  c i Explain why the existence of the threshold frequency for photoelectric emission from a metal supports the photon theory of light in favour of the wave theory of light.

    ii State and explain one other experimental observation about photoelectric emission that supports the photon theory of light over the wave theory.

3 Electrons can escape from a certain metal surface when it is illuminated by blue light, but not when it is illuminated by red light.

  a Explain why blue light causes emission of electrons from this metal, whereas red light does not

  b What difference would be made to the emission of electrons from this metal surface if:

    i the intensity of the blue light were increased,

    ii the surface were at a positive potential?

4 Light of wavelength 600 nm was used to illuminate a metal surface. The maximum kinetic energy of an electron emitted as a result was 1.3×10^{-19} J. Calculate:

  a the energy of a photon of this wavelength,

  b the work function of the metal surface.

5 a Explain the meaning of the term 'work function' of a metal surface.

  b i Calculate the wavelength of a photon of energy 2.0×10^{-19} J.

    ii Explain why photons of wavelength longer than the value calculated in part i could not cause photoelectric emission from a metal surface which has a work function of 2.0×10^{-19} J.

6 A vacuum photocell connected to a microammeter is shown below.

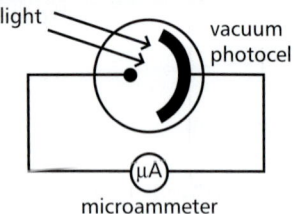

Figure 6.1

  a Explain why:

    i the microammeter registers a current when light of a certain wavelength is directed at the photocell,

    ii the current increases when the light intensity is increased.

  b When light of a longer wavelength is directed at the photocell, no current is registered.

    i Explain why there is no current in this situation.

    ii Explain why making the light more intense would not cause a current with light of this longer wavelength.

7 The spectrum of light from a sodium lamp has two prominent closely spaced lines at wavelengths of 589.0 nm and 589.6 nm. These lines are due to electron transitions from two closely spaced energy levels X and Y to the same lower energy level Z, as shown below (not to scale).

  a Copy the diagram and show the electron transition that causes the emission of photons of wavelength:

    i 589.0 nm,   ii 589.6 nm.

——————— X
——————— Y

  b Calculate the energy difference between the two closely spaced energy levels.

——————— Z

8 The energy level diagram for a particular type of atom is shown in Figure 8.1.

a i Calculate the potential difference which electrons, initially at rest, must be accelerated through in order to ionise such an atom.

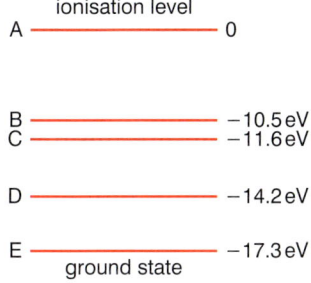

ionisation level
A ———————— 0

B ———————— −10.5 eV
C ———————— −11.6 eV

D ———————— −14.2 eV

E ———————— −17.3 eV
ground state

Figure 8.1

ii Calculate the wavelength of the electromagnetic radiation emitted by this type of atom when it de-excites from energy level D.

b i Determine the energy, in eV, of each photon that could be emitted from the atom as a result of de-exciting from level B, and state the electron transition that releases each photon.

ii Determine the maximum wavelength of a photon that could cause excitation from the ground state as a result of being absorbed by the atom in its ground state.

9 Protons from a cyclotron emerge at a maximum speed of $1.7 \times 10^7\,\text{m s}^{-1}$. Ignore relativistic effects in this question.

a For a proton moving at this speed, calculate:
i its kinetic energy,
ii its momentum and its de Broglie wavelength.

b Calculate the momentum of a photon with the same energy as the kinetic energy of a proton moving at a speed of $1.7 \times 10^7\,\text{m s}^{-1}$.

10 A narrow beam of electrons moving at a speed of $9.8 \times 10^7\,\text{m s}^{-1}$ is directed at a thin crystal. A fluorescent screen on the other side of the crystal shows a pattern of rings due to diffraction of electrons by the crystal, as shown in Figure 10.1.

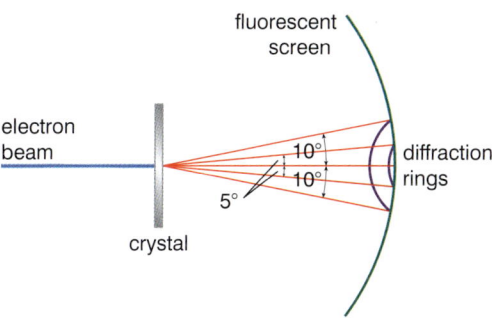

fluorescent screen

electron beam

crystal

diffraction rings

10°
10°
5°

Figure 10.1

a Calculate the de Broglie wavelength, λ, of the electrons.

b Measurements show that the rings are produced by electrons diffracted through 5° and 10°. The angle of diffraction θ of each ring is given by the equation $2d\sin\theta = n\lambda$, where n is the order number of the ring and d is the spacing between the layers of atoms in the crystal. Use this information to calculate a value for d.

c State and explain how the diameter of the rings would change if the speed of the electrons was increased.

11 State whether each of the following experiments or observations using light demonstrates the wave-like or the particle-like nature of light.
i When light passes through a narrow gap, it diffracts.
ii When light is directed at a metal surface, electrons are emitted from the surface only if the light wavelength is less than a certain value.

12 a An electron in a beam has a speed of $1.5 \times 10^7\,\text{m s}^{-1}$. Calculate:
i the kinetic energy,
ii the momentum and de Broglie wavelength of this electron.

b Calculate the wavelength of a photon with the same energy as the electron in part **a**.

21.1 Radioactive decay

Half-life

When a nucleus of a radioactive **isotope** emits an α- or a β-particle, it becomes a nucleus of a different isotope because its proton number changes. The number of nuclei of the initial radioactive isotope therefore decreases. The mass of the initial isotope decreases gradually as the number of nuclei of the isotope decreases. Figure 1 shows how the mass decreases with time. The curve is referred to as a decay curve. The mass of the isotope decreases at a slower and slower rate. Measurements show that the mass decreases exponentially which means that the mass drops by a constant factor (e.g. ×0.8) in equal intervals of time. For example, if the initial mass of the radioactive isotope is 100 g and the mass decreases by a factor of ×0.8 every 1000 seconds, then:

• after 1000 s, the mass remaining = 80 g (= 0.8 × 100 g),

• after 2000 s, the mass remaining = 64 g (= 0.8 × 0.8 × 100 g),

• after 3000 s, the mass remaining = 51 g (= 0.8 × 0.8 × 0.8 × 100 g).

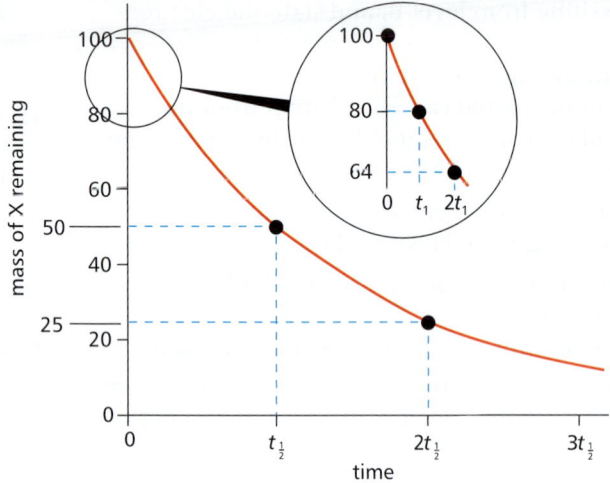

Figure 1 *A radioactive decay curve*

A convenient measure for the rate of decrease is the time taken for a decrease by half. This is the half-life of the process.

> The half-life, $t_{\frac{1}{2}}$, of a radioactive isotope is the time taken for the mass of the isotope to decrease to half the initial mass.

This is the same as the time taken for the number of nuclei of the isotope to decrease to half the initial number.

Consider a sample of a radioactive isotope X which initially contains 100 g of the isotope.

• After 1 half-life, the mass of X remaining = 0.5 × 100 = 50 grams

• After 2 half-lives from the start, the mass of X remaining = $0.5^2 \times 100$ = 25 grams

• After 3 half-lives from the start, the mass of X remaining = $0.5^3 \times 100$ = 12.5 grams

• After n half-lives from the start, the mass of X remaining = $0.5^n m_0$, where m_0 = the initial mass

The mass of X decreases exponentially. This is because radioactive decay is a **random** process. We know this because the fluctuations in the count rate of a Geiger counter are random variations. So the number of nuclei that decay in a certain time is in proportion to the number of nuclei of X remaining.

Randomness

To understand the idea of randomness, consider a game of dice starting with 1000 dice, each representing a nucleus of X. The throw of a dice is a random process in which each face has a 1 in 6 chance of being uppermost.

After each throw, suppose all the dice showing '1' uppermost are removed. For example, after the first throw 167 dice (= 1000/6) would be removed and 833 (= 1000 − 167) would remain. The table below shows how many dice remain after each throw.

Throw	Number of dice		
	initially	removed	remaining
1st	1000	167	833
2nd	833	139	694
3rd	694	116	578
4th	578	96	482
5th	482	80	402

The analysis shows that 4 throws are needed to reduce the number of dice remaining to less than half the initial number. Prove for yourself that a further 4 throws would reduce the number of dice remaining to 25% of the initial number. Figure 2 shows how the number of dice remaining decreases with time. The curve has the same shape as Figure 1. The half-life of the process is 3.8 'throws'.

Activity

The activity A of a radioactive isotope is the number of nuclei of the isotope that disintegrate per second. In other words, it is the rate of change of the number of nuclei of the isotope. The unit of activity is the **becquerel (Bq)**, where $1\,\text{Bq} = 1$ disintegration per second.

The activity of a radioactive isotope is proportional to the mass of the isotope. Because the mass of a radioactive isotope decreases with time due to radioactive decay, the activity decreases with time. Figure 3 shows an experiment in which the activity of a radioactive isotope of protoactinium $^{234}_{91}\text{Pa}$ is measured and recorded using a Geiger tube and a counter. This isotope is a β-emitter produced by the decay of the radioactive isotope of thorium $^{234}_{90}\text{Th}$. In this experiment, an organic solvent in a sealed bottle is used to separate protoactinium from thorium to enable the activity of the protoactinium to be monitored.

Before the experiment is carried out, the background count rate is measured without the bottle present. The bottle is then shaken to mix the aqueous and solvent layers and then placed near the end of the Geiger tube. The layers are allowed to separate as shown in Figure 3. The protoactinium is collected by the solvent and the thorium by the aqueous layer. The Geiger tube detects β-particles emitted by the decay of the protoactinium nuclei in the solvent layer.

The counter is used to measure the number of counts every 10 seconds. The count rate is the number of counts in each ten-second interval divided by 10 s. The background count rate is subtracted to give the corrected count rate. Since the activity is proportional to the corrected count rate, a graph of the corrected count rate against time, as in Figure 4, shows how the activity of the protoactinium decreases with time.

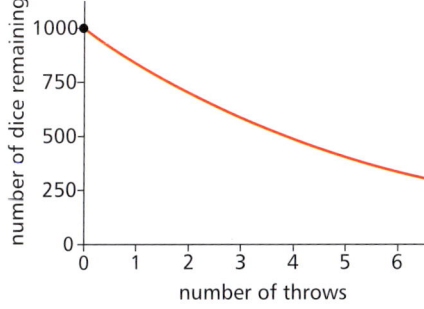

Figure 2 *Exponential decrease*

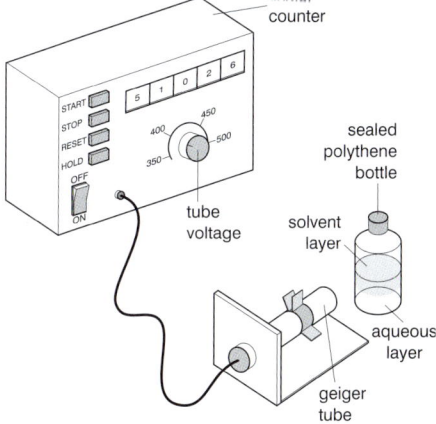

Figure 3 *Measuring the activity of protoactinium*

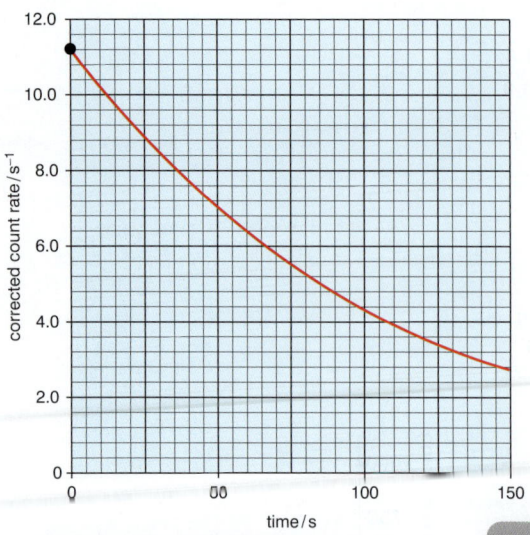

Figure 4 *A radioactive decay curve*

Note

Reminder about molar mass and the Avogadro constant

As explained in Topic 19.2 for an element with a mass number A:

- its molar mass is the mass of 1 mole,
- one mole of the element contains N_A atoms where N_A is the Avogadro constant,
- mass m of the element contains $\left(\dfrac{m}{M}\right) N_A$ atoms.

Activity and power

For a radioactive source of activity A that emits particles (or photons) of the same energy E, the energy per second released by radioactive decay in the source by the radiation is the product of its activity and the energy of each particle. In other words, the power of the source $= AE$.

> **The energy transfer per second from a radioactive source $= AE$**

If the source is in a sealed container and emits only α-particles which are all absorbed by the container, the container gains thermal energy from the absorbed radiation equal to the energy transferred from the source. For example, for a source that has an activity of 30 MBq and emits particles of energy 2.5 MeV, the energy transfer per second from the source is given by

$$30 \times 10^6\,\text{Bq} \times 2.5\,\text{MeV} = 7.5 \times 10^7\,\text{MeV s}^{-1} = 1.2 \times 10^{-5}\,\text{J s}^{-1}.$$

Summary test 21.1

$N_A = 6.02 \times 10^{23}\,\text{mol}^{-1}$, $1\,\text{MeV} = 1.6 \times 10^{-13}\,\text{J}$

1 Figure 4 shows how the activity of protoactinium $^{234}_{91}\text{Pa}$ decreases with time.

   a Use the graph to work out the half-life of this isotope,

   b If the initial mass of the isotope was 48 g, calculate the mass of the isotope remaining after three half-lives.

2 A freshly prepared sample of a radioactive isotope X contains 1.8×10^{15} atoms of the isotope. The half-life of the isotope is 8.0 hours. Calculate:

   a the number of atoms of this isotope remaining after:
      i 8 hours,
      ii 24 hours.

   b the number of atoms of X that would have decayed after:
      i 8 hours,
      ii 24 hours.

   c the energy transfer from the sample in 24 hours if the isotope emits α-particles of energy 5 MeV.

3 $^{131}_{53}\text{I}$ is a radioactive isotope of iodine which has a half-life of 8.0 days. A sample of this isotope has an initial activity of 38 kBq. Calculate the activity of this sample:

   a 8.0 days later,

   b 32 days later.

4 $^{137}_{55}\text{Cs}$ is a radioactive isotope of caesium which has a half-life of 35 years. A sample of this isotope has a mass of $1.0 \times 10^{-3}\,\text{kg}$.

   a Calculate the number of atoms in $1.0 \times 10^{-3}\,\text{kg}$ of this isotope.

   b Calculate the number of atoms of the isotope remaining in the sample after 70 years.

The theory of radioactive decay

The random nature of radioactive decay

An unstable nucleus becomes stable by emitting an α- or a β-particle or a γ-photon. This is an unpredictable event. It happens at random and is spontaneous in the sense that it is a change without any external cause. Every nucleus of a radioactive isotope has an equal probability of becoming stable in any given time interval. Therefore, for a large number of nuclei of a radioactive isotope, the number of nuclei that disintegrate in a certain time interval depends only on the total number of nuclei present. The same idea was considered in the dice experiment. The greater the number of dice used, the more likely it is that 1 in every 6 dice show a particular number on the upper face.

Consider a sample of a radioactive isotope X that initially contains N_0 nuclei of the isotope.

Let N represent the number of nuclei of X remaining at time t after the start.

Suppose in time Δt, the number of nuclei that disintegrate is ΔN.

Because radioactive disintegration is a random process, ΔN is proportional to:

1 N, the number of nuclei of X remaining at time t,
2 the duration of the time interval Δt.

Therefore, $\Delta N = -\lambda N \Delta t$, where λ is a constant referred to as the **decay constant**. The minus sign is necessary because ΔN is a decrease.

So, the rate of disintegration, $\dfrac{\Delta N}{\Delta t} = -\lambda N$

For a given radioactive isotope, its activity is the rate of disintegration $\dfrac{\Delta N}{\Delta t}$

Therefore, the activity A of N atoms of a radioactive isotope is given by:

$$A = \lambda N$$

The solution of the equation $\dfrac{\Delta N}{\Delta t} = -\lambda N$ is $N = N_0 e^{-\lambda t}$

where e^t is the exponential function.

Figure 1 shows that a graph of N against t gives a decay curve. The number of nuclei N decreases exponentially with time. In other words:

- in one half-life, the remaining number of nuclei $N_1 = 0.5\,N_0$
- in two half-lives, the remaining number of nuclei $N_2 = 0.25\,N_0$
- in n half-lives, the remaining number of nuclei $N = 0.5^n N_0$

The graph of the number of nuclei N against time t as represented by the equation $N = N_0 e^{-\lambda t}$ is shown in Figure 1 above. It is a curve with exactly the same shape as Figure 1 of Topic 21.1.

The mass, m, of a radioactive isotope decreases from initial mass, m_0, in accordance with the equation $m = m_0 e^{-\lambda t}$ because the mass, m, is proportional to the number of nuclei, N, of the isotope.

The activity, A, of a sample of N nuclei of an isotope decays in accordance with the equation

$$A = A_0 e^{-\lambda t}$$

where A_0 is the initial activity (or the activity at $t = 0$).

This is because the activity, A = the magnitude of the number of disintegrations per second = λN. Hence, $A = \lambda N_0 e^{-\lambda t} = A_0 e^{-\lambda t}$, where $A_0 = \lambda N_0$.

Learning outcomes

On these pages you will learn to:

- solve problems using the relationship $x = x_0 e^{-\lambda t}$, where x could represent activity, number of undecayed nuclei or count rate
- sketch a graph to show the exponential nature of radioactive decay
- define half-life and solve related problems using the equations above and the relationship $t_{1/2} = 0.693/\lambda$

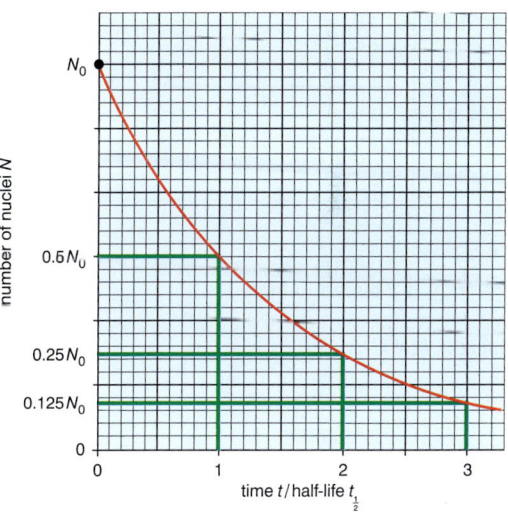

Figure 1 $N = N_0 e^{-\lambda t}$

The count rate, **C**, due to a sample of a radioactive isotope at a fixed distance from a Geiger tube is proportional to the activity of the source. Therefore, the count rate decreases with time in accordance with the equation $C = C_0 e^{-\lambda t}$, where C_0 is the count rate at time $t = 0$.

The above equations for the number of nuclei, N, the activity, A, and the count rate, C, received from the source (i.e. the corrected count rate) are all of the same general form, namely $x = x_0 e^{-\lambda t}$, where x represents N or A or C and x_0 represents the initial value.

1 The exponential function appears in any situation where the rate of change of a quantity is in proportion to the quantity itself. This is because the rate of change of each term in the function sequence is equal to the previous term in the sequence.

2 The exponential function, $e^x = 1 + x + \dfrac{x^2}{2!} + \dfrac{x^3}{3!} + \dots$ (See Chapter 24, Mathematical skills)

Differentiating e^x with respect to x gives e^x $\left(\text{i.e. } \dfrac{d(e^x)}{dx} = e^x\right)$ because differentiating each term in the expression for e^x gives the previous term. The exponential function is indicated on a calculator as 'exp' or 'e^x' or 'inv ln'. (See Chapter 24, Mathematical skills)

3 The natural logarithm function, $\ln x$, is the inverse exponential function. In other words, if $y = e^x$, then $\ln y = x$. Therefore, $N = N_0 e^{-\lambda t}$ may be written $\ln N = \ln N_0 - \lambda t$.

The graph of $\ln N$ against t is therefore a straight line with:
- a gradient $= -\lambda$, and
- a y-intercept $= \ln N_0$.

4 The exponential decrease formula is also used in the theory of capacitor discharge. (See Topic 15.4)

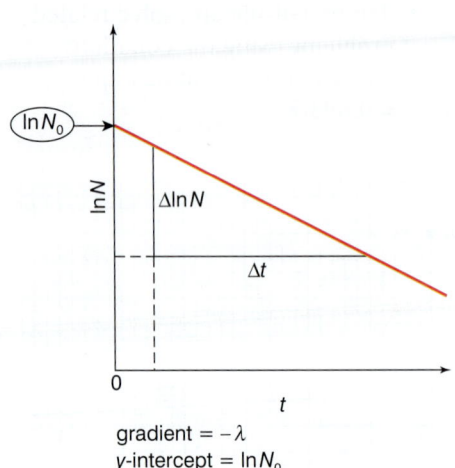

gradient $= -\lambda$
y-intercept $= \ln N_0$

Figure 2 $\ln N$ v. t

Worked example

A sample of a radioactive isotope initially contains 1.2×10^{20} atoms of the isotope. The decay constant for the isotope is $3.6 \times 10^{-3}\,\text{s}^{-1}$. Calculate:

a the number of atoms of the isotope remaining after 1000 s,
b the activity of the sample after 1000 s.

Solution
a $N_0 = 1.2 \times 10^{20}$, $\lambda = 3.6 \times 10^{-3}\,\text{s}^{-1}$, $t = 1000\,\text{s}$,
$\lambda t = 3.6 \times 10^{-3}\,\text{s}^{-1} \times 1000\,\text{s} = 3.6$
$\therefore N = N_0 e^{-\lambda t} = 1.2 \times 10^{20}\,e^{-3.6} = 1.2 \times 10^{20} \times 2.7 \times 10^{-2} = 3.2 \times 10^{18}$
b Activity, $A = \lambda N = 3.6 \times 10^{-3} \times 3.2 \times 10^{18} = 1.2 \times 10^{16}\,\text{Bq}$

The decay constant

The decay constant λ is the probability of an individual nucleus decaying per second. If there are 10 000 nuclei present and 300 decay in 20 seconds, the decay constant is $0.0015\,\text{s}^{-1}$ $\left(= \dfrac{\left(\frac{300}{10\,000}\right)}{20\,\text{s}}\right)$.

In general, if the change of the number of nuclei ΔN in time Δt is given by $\Delta N = -\lambda N \Delta t$, then, the probability of decay, $\dfrac{\Delta N}{N} = \lambda \Delta t$ (the minus sign is not needed here as reference is made to decay).

So, the probability per unit time $-\dfrac{\frac{\Delta N}{N}}{\Delta t} = \lambda$

As explained on p.306, the **half-life**, $t_{1/2}$, of a radioactive isotope is the time taken for half the initial number of nuclei to decay. The longer the half-life, the smaller the decay constant because the probability of decay per second is smaller.

The half-life $t_{1/2}$ is related to the decay constant λ according to the equation

$$t_{1/2} = \frac{\ln 2}{\lambda}$$

As $\ln 2 = 0.693$, this equation may be written as $t_{1/2} = \frac{0.693}{\lambda}$

Proof of $t_{1/2} = \frac{\ln 2}{\lambda}$

The proof of this equation is not part of this specification. It is provided below to help you develop a better understanding of the topic.

Let the number of nuclei $N = N_0$ at time $t = 0$, so at time $t = t_{1/2}$, $N = 0.5 N_0$

Inserting $t = t_{1/2}$, $N = 0.5 N_0$ into $N = N_0 e^{-\lambda t}$ gives $0.5 N_0 = N_0 e^{-\lambda t_{1/2}}$

Cancelling N_0 and taking the natural logarithm (ln) of each side gives $\ln 0.5 = -\lambda t_{1/2}$

Because $\ln 0.5 = -\ln 2$, then $\ln 2 = \lambda t_{1/2}$.

Rearranging this equation gives $t_{1/2} = \frac{\ln 2}{\lambda}$

Note

To calculate N at time t, given values of N_0 and $t_{1/2}$,

- **either** calculate λ using $\lambda = \frac{\ln 2}{t_{1/2}}$ then use the equation $N = N_0 e^{-\lambda t}$,
- **or** calculate the number of half-lives, n, using $n = \frac{t}{t_{1/2}}$ then use $N = 0.5^n N_0$

Summary test 21.2

$N_A = 6.02 \times 10^{23}\,\text{mol}^{-1}$, $1\,\text{MeV} = 1.6 \times 10^{-13}\,\text{J}$

1 $^{131}_{53}\text{I}$ is a radioactive isotope of iodine which has a half-life of 8.0 days. A fresh sample of this isotope contains 4.2×10^{16} atoms of isotope. Calculate:

 a the decay constant of this isotope,

 b the number of atoms of this isotope remaining after 24 hours.

2 A radioactive isotope has a half-life of 35 years. A fresh sample of this isotope has an activity of 25 kBq. Calculate:

 a the decay constant in s^{-1},

 b the activity of the sample after 10 years.

3 a Calculate the number of atoms present in 1.0 kg of $^{226}_{88}\text{Ra}$.

 b The isotope $^{226}_{88}\text{Ra}$ has a half-life of 1620 years. For an initial mass of 1.0 kg of this isotope, calculate:

  i the mass of this isotope remaining after 1000 years,

  ii how many atoms of the isotope will remain after 1000 years.

4 A fresh sample of a radioactive isotope has an initial activity of 40 kBq. After 48 hours, its activity has decreased to 32 kBq. Calculate:

 a the decay constant of this isotope,

 b its half-life.

Radioactive isotopes in use

Learning outcomes

On these pages you will learn to:

- show an understanding of the uses of radioactive isotopes in terms of the properties of the isotopes used and the radiation emitted
- use the equations in Topic 21.2 to solve problems in connection with the use of radioactive isotopes

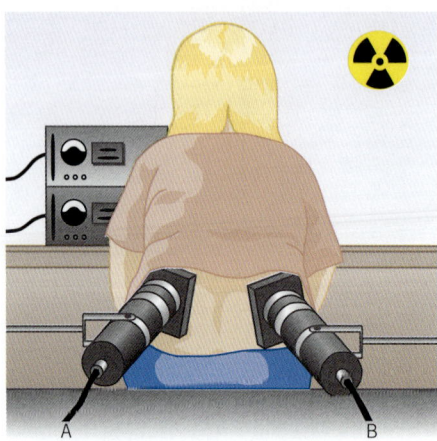

a Chart recorder a

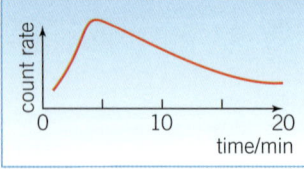

b Chart recorder B

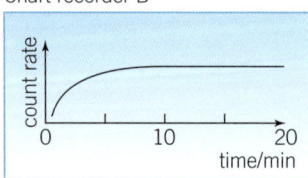

Figure 1 *Thyroid monitoring*

Radioactive isotopes are used for many purposes. The choice of an isotope for a particular purpose depends on its half-life and on the type of radiation it emits. For some uses, the choice also depends on how the isotope is obtained and on whether or not it produces a stable decay product. The following examples are intended to provide a wider awareness of important uses of radioactive substances and to set contexts in which knowledge and understanding of radioactivity is developed further. Read again Topics 10.3 and 10.4 about α, β and γ radiation if necessary.

Radioactive tracers

A radioactive tracer is used to follow the path of a substance through a system, as illustrated below.

- **Environmental uses**: for example, the detection of leaks in underground pipes that carry water or oil. Such a leak can be detected by injecting a radioactive tracer into the flow. Geiger tubes on the surface above the pipeline can then be used to detect leakage. The radioactive isotope used should have a half-life short enough so that it decays quickly after use and long enough so that the test can be completed before its activity becomes too low. In addition, it should be a β-emitter as α-radiation would be absorbed by the pipes and γ radiation would pass through the pipes without absorption.
- **Geological uses**: for example, to improve the recovery of oil from an underground reservoir. Water containing a radioactive tracer is injected into an oil reservoir at high pressure, forcing some of the oil out. Detectors at the production wells monitor breakthrough of the radioactive isotope. The results are used to build up a model of the reservoir to improve and control recovery. Because the time from injection to breakthrough can be many months, the tracer must have a suitably long half-life. A suitable tracer is 'tritiated' water ^3H_2O, a β-emitter with a half-life of 12 years.
- **Medical uses**: for example, to monitor the uptake of iodine by the thyroid gland (see Figure 1). The thyroid gland absorbs iodine to maintain its function of producing a hormone. The rate of uptake is measured by giving the patient a solution containing sodium iodide which includes a small quantity of the radioactive isotope of iodine, $^{131}_{53}I$, which emits beta and gamma radiation with a half-life of 8 days. The activity of the patient's thyroid and the activity of an identical sample prepared at the same time are measured 24 hours later. The percentage uptake by the patient is then calculated from (the corrected count rate of the thyroid / the corrected count rate of the identical solution) × 100%. A normal thyroid has a percentage uptake of 20–50% after 24 hours.
- **Agricultural research**: for example, to investigate the uptake of fertilisers by plants. This can be done by using a fertiliser which contains the radioactive isotope of phosphorus, $^{32}_{15}P$, which is a β-emitter with a half-life of 14 days. By measuring the radioactivity of the leaves, the amount of fertiliser reaching them can be determined.

Radioactive dating

- **Carbon dating**: living plants and trees contain a small percentage of the radioactive isotope of carbon, $^{14}_6C$, which is formed in the atmosphere as a result of cosmic rays knocking out neutrons from nuclei. These neutrons then collide with nitrogen nuclei to form carbon-14 nuclei.

$$^1_0n + ^{14}_7N \rightarrow ^{14}_6C + ^1_1p$$

Carbon dioxide from the atmosphere is taken up by living plants as a result of photosynthesis. So a small percentage of the carbon content of any plant is carbon-14. This isotope has a half-life of 5570 years so there is negligible decay during the life-time of a plant. Once a tree has died, no further carbon is taken in so the proportion of carbon-14 in the dead tree decreases as the carbon-14 nuclei decay. Because activity is proportional to the number of atoms still to decay, measuring the activity of the dead sample enables its age to be calculated, provided the activity of the same mass of living wood is known.

Worked example

A certain sample of dead wood is found to have an activity of 0.28 Bq. An equal mass of living wood is found to have an activity of 1.3 Bq. Calculate the age of the sample.

The half-life of carbon-14 is 5570 years.

Solution
The half-life, $t_{1/2}$, in seconds = 5570 × 365 × 24 × 3600 s = 1.76×10^{11} s

∴ the decay constant of carbon-14, $\lambda = \dfrac{0.693}{t_{1/2}} = \dfrac{0.693}{1.76 \times 10^{11}} = 3.95 \times 10^{-12}\,\text{s}^{-1}$

Using activity $A = A_0 e^{-\lambda t}$, where $A = 0.28$ Bq and $A_0 = 1.30$ Bq gives

$0.28 = 1.3 e^{-\lambda t}$ so $e^{-\lambda t} = \left(\dfrac{0.28}{1.30}\right) = 0.215$

∴ $\lambda t = 1.535$

$t = \dfrac{1.535}{\lambda} = \dfrac{1.535}{3.95 \times 10^{-12}\,\text{s}} = 3.88 \times 10^{11}\,\text{s} = 12\,300$ years

Note

A useful check is to estimate the number of half-lives needed for the activity to decrease from 1.30 Bq to 0.28 Bq. You should find that just over 2 half-lives are needed, corresponding to about 11 000 years.

- **Argon dating**; ancient rocks contain trapped argon gas as a result of the decay of the radioactive isotope of potassium, $_{19}^{40}\text{K}$ into the argon isotope $_{18}^{40}\text{Ar}$. This happens when its nucleus captures an inner shell electron. As a result, a proton in the nucleus changes into a neutron and a neutrino is emitted. The equation for the change is

$$_{19}^{40}\text{K} + _{-1}^{0}\text{e} \rightarrow {}_{18}^{40}\text{Ar} + \nu$$

The potassium isotope $_{19}^{40}\text{K}$ also decays by β-emission to form the calcium isotope $_{20}^{40}\text{Ca}$. This process is 8 times more probable than electron capture.

$$_{19}^{40}\text{K} \rightarrow {}_{-1}^{0}\beta + {}_{20}^{40}\text{Ca} + \bar{\nu}$$

The effective half-life of the decay of $_{19}^{40}\text{K}$ is 1250 million years. The age of the rock (i.e. the time from when it solidified) can be calculated by measuring the proportion of argon-40 to potassium-40. For every n potassium-40 atoms now present, if there is 1 argon-40 atom present, there must have originally been $n + 9$ potassium atoms. (i.e. 1 that decayed into argon-40 + 8 that decayed into calcium-40 + n remaining). The radioactive decay equation $N = N_0 e^{-\lambda t}$ can then be used to find the age of the sample. For example, suppose for every 4 potassium-40 atoms now present, a certain rock now has 1 argon-40 atom. Therefore, $N = 4$ and $N_0 = 13$. Substituting these values into the equation $N = N_0 e^{-\lambda t}$ gives $4 = 13 e^{-\lambda t}$.

Therefore, $e^{-\lambda t} = \dfrac{4}{13} = 0.308$, which gives $t = \dfrac{-\ln 0.308}{\lambda}$. Substituting $\dfrac{0.693}{t_{1/2}}$

for λ into this equation gives $t = \left(\dfrac{-\ln 0.308}{0.693}\right) t_{1/2} = 1.70\, t_{1/2}$. The age of the sample is therefore 2120 million years.

Note

A useful check is to estimate the number of half-lives needed for N to decrease from 13 to 4. You should find that between 1 and 2 half-lives are needed, corresponding to an age of between 1250 and 2500 million years.

Figure 2 *Pistons (and piston rings) fit into this engine block*

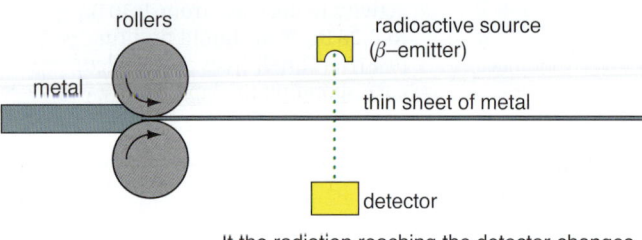

It the radiation reaching the detector changes the detector makes the rollers move further apart or closer together

Figure 3 *The manufacture of metal foil*

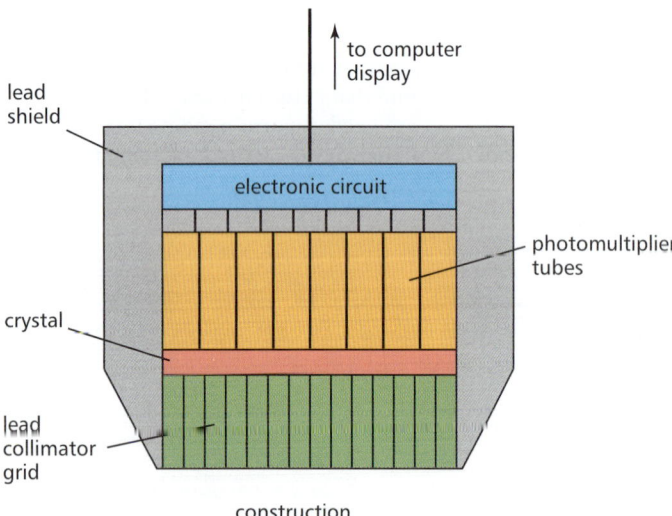

Figure 4 *The gamma camera*

Industrial uses

The examples below are just two of a wide range of applications of radioactivity in industry.

- **Engine wear**
  The rate of wear of a piston ring in an engine can be measured by fitting a ring that is radioactive. As the ring slides along the piston compartment, radioactive atoms transfer from the ring to the engine oil. By measuring the radioactivity of the oil, the mass of radioactive metal transferred from the ring can be determined and the rate of wear calculated. A metal ring can be made radioactive by exposing it to neutron radiation in a nuclear reactor. Each nucleus that absorbs a neutron becomes unstable and disintegrates by β emission.

- **Thickness monitoring**
  Metal foil is manufactured by using rollers to squeeze plate metal on a continuous production line, as shown in Figure 3. A detector measures the amount of radiation passing through the foil. If the foil is too thick, the detector reading drops. A signal from the detector is fed back to the control system to make the rollers move closer together and so make the foil thinner. The source used is a β-emitter with a long half-life. α radiation would be absorbed completely by the foil and γ radiation would pass straight through without absorption.

- **Power for remote devices** such as satellites and weather sensors can be obtained using a radioactive isotope in a thermally insulated sealed container which absorbs all the radiation emitted by the isotope. A thermocouple attached to the container produces electricity as a result of the container becoming warm through absorbing radiation. For mass, m, of the isotope, its activity, $A = \lambda N$, where N is the number of radioactive atoms present in mass m. If each disintegration of a nucleus releases energy E, the energy transfer per second from the source $= \lambda NE$. The source needs to have a reasonably long half-life so it does not need to be replaced frequently but a very long half-life would require too much mass to generate the necessary power.

Medical and health uses

As explained earlier, radioactive isotopes are used as medical tracers. Further uses of radioactive isotopes in medicine include the gamma camera used to form images of joints and organs, the PET scanner and gamma therapy to destroy cancerous tissues.

- **The gamma camera** is designed to detect γ radiation from sites inside the body where a γ-emitting isotope is located. For example, bone deposits can be located using a phosphate tracer containing the radioactive isotope of technetium, $^{99}_{52}$Te which is a γ-emitter that has a half-life of 6 hours.
  The γ-photons from inside the body are absorbed by a lead collimator grid unless they travel parallel to narrow

channels through the collimator. Each γ-photon that passes through the grid strikes a large sodium iodide crystal, causing a flash of light which is detected by a photomultiplier tube in an array of tubes. The tubes are connected to a computer which displays an image of the γ-emitting sources in the body.

- **Gamma therapy** is used to destroy tumours inside the body. A narrow beam of γ radiation from the radioactive isotope of cobalt, $^{60}_{27}$Co, is directed at the tumour from different directions by moving the source or by moving the patient. This movement is necessary to ensure healthy tissue in the path of the beam is exposed much less than the target tissue. The cobalt-60 source has a half-life of 5.3 years and emits γ-photons of energies 1.17 MeV and 1.33 MeV. The source is enclosed in a thick lead container. When the source is to be used, it

is rotated to the inner end of an exit channel so that a beam of γ radiation emerges after passing along the exit channel. When the source is not in use, it is rotated away from the inner end of the exit channel so that no γ radiation can emerge from the container.

- **Food preservation** can be achieved by irradiating food with γ radiation. About 30% of the world's food is lost through spoilage. The major cause is bacteria, moulds and yeast which grow on food. Some bacteria produce toxic waste products that cause food poisoning. Irradiation of food with γ radiation kills 99% of the disease-carrying organisms in the food, such as *Salmonella* which infects poultry and *Clostridium*, the cause of botulism. The treatment is not suitable for all foods. Red meat turns brown and develops an unpleasant taste, eggs develop a smell and tomatoes go soft.

Summary test 21.3

$N_A = 6.02 \times 10^{23}\,\text{mol}^{-1}$, $1\,\text{MeV} = 1.6 \times 10^{-13}\,\text{J}$

1 a Explain why living wood is slightly radioactive.

b A sample of ancient wood of mass 0.5 g is found to have an activity of 0.11 Bq. A sample of living wood of the same mass has an activity of 0.13 Bq. Calculate the age of the sample of wood. The half-life of radioactive carbon $^{14}_{6}$C is 5570 years.

2 The radioactive isotope of iodine, $^{131}_{53}$I, is used for medical diagnosis of the kidneys. The isotope has a half-life of 8 days. A sample of the isotope is given to a patient in a glass of water. The passage of the isotope through each kidney is then monitored using two detectors outside the body. The isotope is required to have an activity of 800 kBq at the time it is given to the patient.

a Calculate:
  i the activity of the sample 24 hours after it was given to the patient,
  ii the activity of the sample when it was prepared 24 hours earlier,
  iii the mass of $^{131}_{53}$I in the sample when it was prepared.

b The reading from the detector near one of the patient's kidneys rises then falls. The reading from the other detector which is near the other kidney rises and does not fall. Discuss the conclusions that can be drawn from these observations.

3 a i In the manufacture of metal foil, describe how the thickness of the foil is monitored using a radioactive source and a detector.
  ii Explain why the source needs to be a β-emitter, not an α-emitter or a γ-emitter.

b i Explain why a cobalt-60 source used for γ-therapy is enclosed in a thick lead-lined container.
  ii Explain why a beam of γ radiation used to destroy a tumour inside a patient is directed at the tumour from different directions during treatment.

4 a A cardiac pacemaker is a device used to ensure that a faulty heart beats at a suitable rate. The required electrical energy in one type of pacemaker is obtained from the energy released by a radioactive isotope. The radiation is absorbed inside the pacemaker. As a result, the absorbing material gains thermal energy and heats a thermocouple attached to the absorbing material. The voltage from the thermocouple provides the source of electrical energy for the pacemaker.
  i Discuss whether the radioactive source should be an α-emitter, a β-emitter or a γ-emitter.
  ii The radioactive source needs to have a reasonably long half-life, otherwise it would need to be replaced frequently. Discuss the disadvantages of using a radioactive source with a very long half-life.

b The energy source for a remote weather station is the radioactive isotope of strontium $^{90}_{38}$Sr, which has a half-life of 28 years. It emits β-particles of energy 0.40 MeV. For a mass of 10 g of this isotope, calculate:
  i its activity,
  ii the energy released per second.

Learning outcomes

On these pages you will learn to:

- show an appreciation of the association between energy and mass as represented by $E = mc^2$ and recall and use this relationship
- show an understanding of the energy released in radioactive decay
- solve problems relating the energy released in radioactive decay to the change of total mass

Energy and mass

In 1905, Einstein published the theory of special relativity in which he showed that moving clocks run more slowly than stationary clocks, fast-moving objects appear shorter than when stationary, the mass of a moving object changes with its speed and no material object can travel as fast as light. He also showed that the mass of an object increases (or decreases) when it gains (or loses) energy, E, in accordance with the equation

$$E = mc^2$$

where m is the change of its mass and c is the speed of light in free space which is $3.0 \times 10^8 \, \text{m s}^{-1}$.

For example:

- A sealed torch that radiates $10 \, \text{W}$ of light for $10 \, \text{h}$ ($= 36\,000 \, \text{s}$) would lose $0.36 \, \text{MJ}$ of energy ($= 10 \, \text{W} \times 36\,000 \, \text{s}$). Its mass would therefore decrease by $4.0 \times 10^{-12} \, \text{kg}$ ($= 0.36 \, \text{MJ}/c^2$), an insignificant amount compared with the mass of the torch.
- A mass of a $1000 \, \text{kg}$ car that speeds up from a standstill to $30 \, \text{m s}^{-1}$ would gain $450 \, \text{kJ}$ of kinetic energy so its mass when moving at $30 \, \text{m s}^{-1}$ would be $5.0 \times 10^{-12} \, \text{kg}$ ($= 450 \, \text{kJ}/c^2$) more than when it is rest.
- An unstable nucleus that releases a $5 \, \text{MeV}$ γ photon would lose $8.0 \times 10^{-13} \, \text{J}$ of energy. Its mass would therefore decrease by $8.9 \times 10^{-30} \, \text{kg}$ ($= 8.0 \times 10^{-13} \, \text{J}/c^2$), which is not an insignificant amount compared with the mass of a nucleus.

The equation applies to all energy changes of any object. These two examples show that such changes are important in nuclear reactions but are not usually significant otherwise. A century after Einstein published his theory, the reason why the mass of an object changes when energy is transferred to or from it is still not clearly understood. However, it is known for every type of particle, there is a corresponding antiparticle with the same mass and opposite charge (if charged). We also know that:

- When a particle and its corresponding antiparticle meet, they **annihilate** each other and two gamma (γ) photons are produced, each of energy mc^2 where m is the mass of the particle or antiparticle.
- A single γ photon of energy in excess of $2mc^2$ can produce a particle and an antiparticle, each of mass m, in a process known as pair production.

Energy changes in reactions

Reactions on a nuclear or sub-nuclear scale do involve significant changes of mass. For example, in radioactive decay, if we know the exact rest mass of each particle involved, we can calculate the energy released (Q) from the difference Δm in the total mass before and after the reaction. In general, for a spontaneous reaction in which no energy is supplied,

the energy released, $Q = \Delta mc^2$

In any change where energy is released such as radioactive decay, the total mass after the change is always less than the total mass before the change. This is because, in the change, some of the mass is converted to energy which is released.

1. In α decay, the nucleus recoils when the α-particle is emitted so the energy released is shared between the α-particle and the nucleus. Applying conservation of momentum to the recoil, you should be able to show that the energy released is shared between the α-particle and the nucleus in inverse proportion to their masses.
2. In β decay, the energy released is shared in variable proportions between the β-particle and the neutrino or antineutrino released in the decay. When the β-particle has maximum kinetic energy, the neutrino or antineutrino has negligible kinetic energy in comparison. The maximum kinetic energy of the β-particle is very slightly less than the energy released in the decay because of recoil of the nucleus.

Note

1. To calculate the energy corresponding to a mass difference of 1 atomic mass unit ($1 \, \text{u} = 1.6605 \times 10^{-27} \, \text{kg}$), using $E = mc^2$ gives
$$E = 1.6605 \times 10^{-27} \, \text{kg} \times (2.9979 \times 10^8 \, \text{m s}^{-1})^2$$
$$= 1.4923 \times 10^{-10} \, \text{J} = 931.5 \, \text{MeV using exact values.}$$
2. When calculating Q in beta decay, assume that the mass of the neutrino is negligible.
3. If the mass of each atom is given instead of the mass of its nucleus, calculate the mass of each nucleus by subtracting the mass of the electrons ($= Z_m$) in the atom from the mass of each atom.

Worked example

The polonium isotope $^{210}_{84}$Po emits α-particles and decays to form the stable isotope of lead $^{206}_{82}$Pb. Write down an equation to represent this process and calculate the energy released when a $^{210}_{84}$Po nucleus emits an α-particle.

mass of $^{210}_{84}$Po nucleus = 209.93667 u

mass of $^{206}_{82}$Pb nucleus = 205.92936 u

mass of α-particle = 4.00150 u

1 u is equivalent to 931.5 MeV

Solution

$$^{210}_{84}\text{Po} \longrightarrow {}^{4}_{2}\alpha + {}^{206}_{82}\text{Pb} \ (+ \text{ energy released } Q)$$

mass difference = total initial mass − total final mass

$$= 209.93667 - (205.92936 + 4.00150)$$

$$= 5.81 \times 10^{-3} \text{ u}$$

energy released Q = mass difference in u × 931.5

$$= 5.41 \text{ MeV}$$

More about the strong nuclear force

The fact that most nuclei are stable tells us there must be an attractive force, the **strong nuclear force**, between any two protons or neutrons in the nucleus.

- The strength of the strong nuclear force can be estimated by working out the force of repulsion between two protons at a separation of 1 fm (= 10^{-15} m), the approximate size of the nucleus. The strong nuclear force must be greater in magnitude that this force of repulsion.

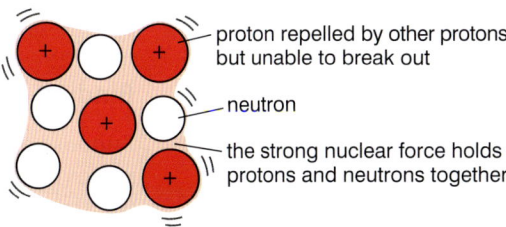

Figure 1 The strong nuclear force

Prove for yourself, using Coulomb's law of force (see Topic 14.1), that the force of repulsion between two protons at a separation of 10^{-15} m is of the order of 200 N. So the strong nuclear force is at least 200 N.

- The range of the strong nuclear force is no more than about 2 to 3×10^{-15} m. The diameter of a nucleus can be measured from high-energy electron scattering experiments. The results show that nucleons are evenly spaced at about 10^{-15} m in the nucleus and therefore the strong nuclear force acts only between nearest neighbour nucleons.

- The energy needed to pull a nucleon out of the nucleus is of the order of millions of electron volts (MeV). This can be deduced because the strong nuclear force is at least about 200 N and it acts over a distance of about 2 to 3×10^{-15} m. The work done by the strong nuclear force over this distance is therefore about 5×10^{-13} J (= 200 N × 2.5×10^{-15} m) which is about 3 MeV as 1 MeV = 1.6×10^{-13} J.

- The strong nuclear force between two nucleons must become repulsive at separations of about 0.5 fm or less, otherwise nucleons would pull each other closer and closer together and be much smaller than it is.

Summary test 21.4

Magnitude of the charge of the electron = 1.60×10^{-19} C, rest mass of an electron = 9.11×10^{-31} kg, 1 u = 931.5 MeV, $g = 9.81 \text{ m s}^{-2}$

1 Calculate the increase of mass of:

   a a 10 kg object when it is raised through a height of 2.0 m,

   b an electron when it is accelerated from rest through a p.d. of: **i** 5000 V,   **ii** 5 MV.

2 The bismuth isotope $^{212}_{83}$Bi emits α-particles and decays to form the stable isotope of thallium $^{208}_{81}$Tl.

   a Write down an equation to represent this process and calculate the energy released.
   Mass of $^{212}_{83}$Bi nucleus = 211.94562 u,
   Mass of $^{208}_{81}$Tl nucleus = 207.93746 u,
   Mass of α-particle = 4.00150 u

   b Explain without calculation why the thallium nucleus in the above decay gains a small proportion of the energy released.

3 The strontium isotope $^{90}_{38}$Sr emits β⁻ particles and decays to form the stable isotope of yttrium $^{90}_{39}$Y.

   a Write down an equation to represent this process and calculate the energy released.
   Mass of $^{90}_{38}$Sr nucleus = 89.88640 u,
   Mass of $^{90}_{39}$Y nucleus = 89.88525 u,
   Mass of β⁻ particle = 0.00055 u

   b Explain without calculation why the kinetic energy of the β-particle released when the strontium nucleus decays varies from zero up to a maximum.

4 The sodium isotope $^{25}_{11}$Na emits β⁻ particles and decays to form the stable isotope of magnesium $^{25}_{12}$Mg.

   Calculate the Q-value of this decay.

   Mass of $^{25}_{11}$Na nucleus = 24.98931 u,

   Mass of $^{25}_{12}$Mg nucleus = 24.98528 u

21.5 Binding energy

Learning outcomes

On these pages you will learn to:

- define and understand the terms mass defect and binding energy
- calculate the binding energy per nucleon of a nuclide
- sketch the variation of binding energy per nucleon with nucleon number
- explain the relevance of binding energy per nucleon to nuclear fusion and to nuclear fission

Binding energy and mass defect

Suppose all the nucleons in a nucleus were separated from one another, removing each one from the nucleus in turn. Work must be done to overcome the strong nuclear force and separate each nucleon from the others. The potential energy of each nucleon is therefore increased when it is removed from the nucleus.

The binding energy of the nucleus is the work that must be done to separate a nucleus into its constituent neutrons and protons.

When a nucleus forms from separate neutrons and protons, energy is released as the strong nuclear force does work pulling the nucleons together. The energy released is equal to the binding energy of the nucleus. Because energy is released when a nucleus forms from separate neutrons and protons, the mass of a nucleus is less than the mass of the separated nucleons.

The mass defect Δm of a nucleus is defined as the difference between the mass of the separated nucleons and the combined mass of the nucleus.

- Calculation of the mass defect of a nucleus of known mass: a nucleus of an isotope $_Z^A X$ is composed of Z protons and $(A - Z)$ neutrons. Therefore, for a nucleus $_Z^A X$ of mass M_N, **its mass defect $\Delta m = Zm_p + (A - Z)m_n - M_N$.**

- Calculation of the binding energy of a nucleus: the mass defect Δm is due to energy released when the nucleus formed from separate neutrons and protons. The energy released in this process is equal to the binding energy of the nucleus. Therefore, **the binding energy of a nucleus $= c^2(\Delta m)$.**

Worked example

The mass of a nucleus of the bismuth isotope $_{83}^{212} Bi$ is $211.80012\,u$. Calculate the binding energy of this nucleus in MeV.

The mass of a proton, $m_p = 1.00728\,u$; the mass of a neutron, $m_n = 1.00866\,u$

$1\,u$ is equivalent to $= 931.5\,MeV$

Solution
Mass defect $\Delta m = 83m_p + (212 - 83)m_n - M_N = 1.92126\,u$

$\therefore$ binding energy $= 1.92126\,u \times 931.5\,MeV/u = 1790\,MeV$

Note

1. The mass of an atom of an isotope $_Z^A X$ is measured using a mass spectrometer. The mass of a nucleus can then be calculated by subtracting the mass of Z electrons from the atomic mass.
2. The atomic mass unit, $1\,u = 1.661 \times 10^{-27}\,kg$. This is defined as $\frac{1}{12}$th of the mass of an atom of the carbon isotope $_6^{12} C$.
3. The energy corresponding to a mass of $1\,u = 931.5\,MeV$. See Topic 23.4 if necessary.

Nuclear stability

The binding energy of each nuclide is different. The **binding energy per nucleon** of a nucleus is the work done to remove a **nucleon** from a nucleus; it is therefore a measure of the stability of a nucleus. For example, the binding energy per nucleon of the $_{83}^{212} Bi$ nucleus is $8.4\,MeV$ per nucleon ($= 1790\,MeV/212$ nucleons).

If the binding energy per nucleon of two different nuclides are compared, the nucleus with more binding energy per nucleon is the more stable of the two nuclei. Figure 1 shows a graph of the binding energy per nucleon v. mass number A for all the known nuclides. This graph is a curve which has a maximum value of 8.7 MeV per nucleon between $A = 50$ and $A = 60$. Nuclei with mass numbers in this range are the most stable nuclei. As explained below, energy is released in:

- **nuclear fission**, the process in which a large unstable nucleus splits into two fragments which are more stable than the original nucleus. The binding energy per nucleon increases in this process, as shown in Figure 1.

- **nuclear fusion**, the process of making small nuclei fuse together to form a larger nucleus. The product nucleus has more binding energy per nucleon than the smaller nuclei. So the binding energy per nucleon also increases in this process, provided the nucleon number of the product nucleus is no greater than about 50.

Note The change of binding energy per nucleon is about 0.5 MeV in a fission reaction and can be more than 20 times as much in a fusion reaction.

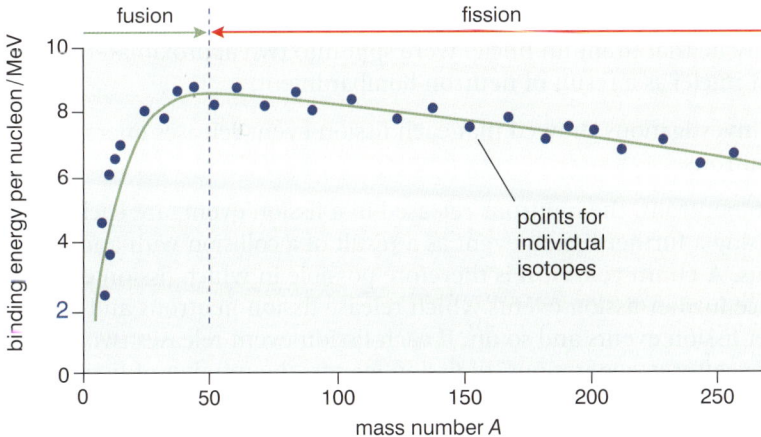

Figure 1 *Binding energy per nucleon for all known nuclides*

Summary test 21.5

mass of a proton, m_p = 1.00728 u; mass of a neutron, m_n = 1.00866 u; 1 u is equivalent to = 931.5 MeV

1 **a** Explain what is meant by the *binding energy* of a nucleus.

  b Sketch a curve to show how the binding energy per nucleon of a nucleus varies with its mass number A, showing the approximate scale on each axis.

2 Calculate the binding energy per nucleon, in MeV per nucleon, of:

  a a $^{12}_{6}$C nucleus (mass = 12 u by definition),

  b a $^{56}_{26}$Fe nucleus (mass = 55.92067 u).

3 **a** Calculate the binding energy per nucleon in MeV per nucleon, of:
    i an α-particle,   **ii** a $^{3}_{2}$He nucleus.
    Mass of an α-particle = 4.00150 u; mass of a $^{3}_{2}$He nucleus = 3.01493 u

  b Use the results of your calculations in **a** to explain why an α-particle rather than a $^{3}_{2}$He nucleus is emitted by a large unstable nucleus.

4 Calculate the binding energy per nucleon, in MeV, of the $^{2}_{1}$H nucleus. Mass of $^{2}_{1}$H nucleus = 2.01355 u

On these pages you will learn to:

- explain what is meant by nuclear fission and by a chain reaction
- calculate the energy released in a fission or fusion event from the masses of the nuclei and other particles involved
- explain what is meant by nuclear fusion and explain in terms of forces the necessary condition to fuse two nuclei

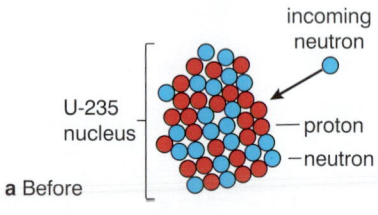

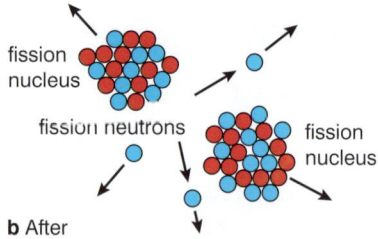

Figure 1 Induced fission

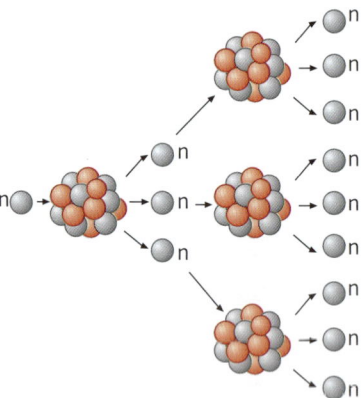

Figure 2 A chain reaction in a nuclear reactor

Note

The fragment nuclei have excess neutrons compared with stable nuclei of the same proton number. They become stable by emitting β^- particles. See Topic 10.4.

Induced fission

Fission of a nucleus occurs when a nucleus splits into two approximately equal fragments. This happens when the uranium isotope $^{235}_{92}U$ is bombarded with neutrons, a discovery made by Hahn and Strassmann in 1938. The process is known as induced fission. The plutonium isotope, $^{239}_{94}Pu$, is the only other isotope that is fissionable. This isotope is an artificial isotope formed by bombarding nuclei of the uranium isotope $^{238}_{92}U$ with neutrons.

Hahn and Strassmann knew that bombarding different elements with neutrons produces radioactive isotopes. Uranium is the heaviest of all the naturally occurring elements; scientists thought that neutron bombardment could turn uranium nuclei into even heavier nuclei. Hahn and Strassmann undertook the difficult work of analysing chemically the products of uranium after neutron bombardment to try to discover any new elements heavier than uranium. Instead, they discovered that many lighter elements such as barium were present after bombardment, even though the uranium was pure before. The conclusion could only be that uranium nuclei were split into two approximately equal fragment nuclei as a result of neutron bombardment.

Further investigations showed that each fission event releases energy and two or three neutrons.

- Fission neutrons, the neutrons released in a fission event, are each capable of causing a further fission event as a result of a collision with another $^{235}_{92}U$ nucleus. A **chain reaction** is therefore possible in which fission neutrons produce further fission events which release fission neutrons and cause further fission events and so on. If each fission event releases two neutrons on average, after n 'generations' of fission events, the number of fission neutrons would be 2^n. Prove for yourself that fission of 6×10^{23} $^{235}_{92}U$ nuclei (i.e. 235 g of the isotope) would happen in 79 generations. As explained below, each fission event releases about 200 MeV of energy. Because each event takes no more than a fraction of a second, a huge amount of energy is released in a very short time. Using the above figures, complete fission of 235 g of $^{235}_{92}U$ would release about 10^{13} J (= $6 \times 10^{23} \times 200$ MeV). This is about a million times more than the energy released as a result of burning a similar mass of fossil fuel.

- Energy is released when a fission event occurs because the fragments repel each other (as they are both positively charged) with sufficient force to overcome the strong nuclear force trying to hold them together. The fragment nuclei and the fission neutrons therefore gain kinetic energy. The two fragment nuclei are smaller and therefore more tightly bound than the original $^{235}_{92}U$ nucleus. In other words, they have more binding energy so they are more stable than the original nucleus. The energy released is equal to the change of binding energy. The binding energy of each nucleon increases from about 7.5 MeV to about 8.5 MeV as a result of the fission event. As there are about 240 nucleons in the original nucleus, the energy released in a fission event is of the order of 200 MeV (= 240 × about 1 MeV).

- Many fission products are possible when a fission event occurs. For example, the equation below shows a fission event in which a $^{235}_{92}U$ nucleus is split into a barium $^{144}_{56}Ba$ nucleus and a krypton $^{90}_{36}Kr$ nucleus and two neutrons are released.

$$^{235}_{92}U + ^{1}_{0}n \longrightarrow ^{144}_{56}Ba + ^{90}_{36}Kr + 2\,^{1}_{0}n + \text{energy released, } Q$$

- The energy released, Q, can be calculated using $E = mc^2$ in the form $Q = c^2(\Delta m)$, where Δm is the difference between the total mass before and after the event.
- In the above equation, the mass difference

$$\Delta m = M_{\text{U-235}} - M_{\text{Ba-144}} - M_{\text{Kr-90}} - m_{\text{n}}$$

where M represents the appropriate nuclear mass and m_{n} is the mass of the neutron.

Nuclear fusion

Fusion takes place when two nuclei combine to form a bigger nucleus. The binding energy curve Topic 21.5 shows that if two light nuclei are combined, the individual nucleons become more tightly bound together. The binding energy per nucleon of the product nucleus is greater than of the initial nuclei. In other words, the nucleons become even more trapped in the nucleus when fusion occurs. As a result, energy is released equal to the increase of binding energy.

Nuclear fusion can only take place if the two nuclei that are to be combined collide at high speed. This is necessary to overcome the electrostatic repulsion between the two nuclei so that they can become close enough to interact through the strong nuclear force. Some examples of nuclear fusion reactions are shown in Figure 4 and described below.

1 The fusion of two protons produces a nucleus of deuterium (the hydrogen isotope $_1^2\text{H}$), a β^+ particle and a neutrino.

$$_1^1\text{p} + _1^1\text{p} \longrightarrow _1^2\text{H} + _{+1}^{\;\;0}\beta + \nu$$

2 The fusion of a proton and a deuterium nucleus $_1^2\text{H}$ produces a nucleus of the helium isotope $_2^3\text{He}$ and 5.5 MeV of energy.

$$_1^2\text{H} + _1^1\text{p} \longrightarrow _2^3\text{He}$$

3 The fusion of two nuclei of helium isotope, $_2^3\text{He}$, produces a nucleus of the helium isotope $_2^4\text{He}$, two protons and 12.9 MeV of energy.

$$_2^3\text{He} + _2^3\text{He} \longrightarrow _2^4\text{He} + 2_1^1\text{p}$$

In each case, the energy released in the reaction may calculated using $E = mc^2$ in the form $Q = c^2(\Delta m)$, where Δm is the difference between the total mass before and after the event.

Solar energy is produced as a result of fusion reactions inside the Sun. The temperature at the centre of the Sun is thought to be 10^8 K or more. At such temperatures, atoms are stripped of their electrons. Matter in this state is referred to as 'plasma'. The nuclei of the plasma move at very high speeds because of the enormous temperature. When two nuclei collide, they fuse together because they overcome the electrostatic repulsion due to their charge and approach each other closely enough to interact through the strong nuclear force. Protons (i.e. hydrogen nuclei) inside the Sun's core fuse together in stages (corresponding to equations 1, 2 and 3 above) to form helium $_2^4\text{He}$ nuclei. For each helium nucleus formed, 25 MeV of energy is released. This corresponds to 7 MeV per proton, considerably more than the energy released per nucleon in a fission event.

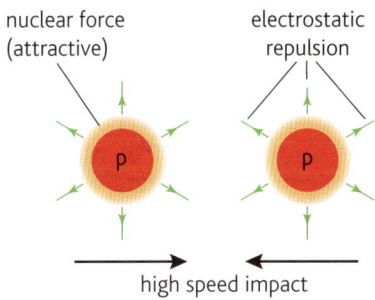

Figure 3 *Fusion of two protons*

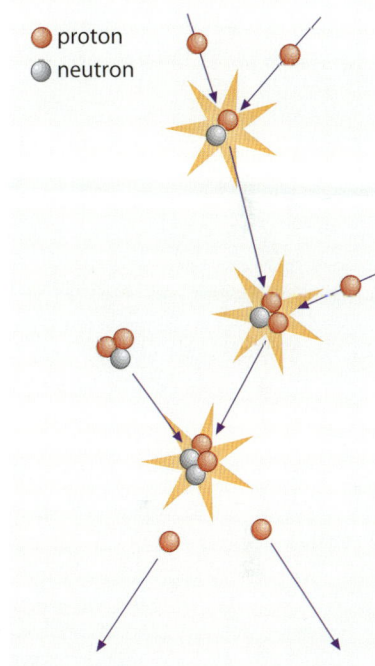

Figure 4 *Fusion reactions inside the Sun*

321

Fusion power

Fusion reactors are still at the prototype stage even though scientific teams in several countries have been working on fusion research for more than 50 years. Prototype fusion reactors such as JET, the Joint European Torus, in the United Kingdom have produced large amounts of power but only for short periods of time. JET produces less power than it uses but the less powerful International Thermonuclear Experimental Reactor (ITER) due to start up in 2025 is designed to produce several times more power than it uses.

Energy is released in JET by fusing nuclei of deuterium ^2_1H and tritium ^3_1H to produce nuclei of the helium isotope ^4_2He and neutrons, as below.

$$^2_1H + {}^3_1H \longrightarrow {}^4_2He + {}^1_0n + 17.6\,MeV$$

The neutrons are absorbed by a 'blanket' of lithium surrounding the reactor vessel. The reaction between the neutrons and the lithium nuclei, as shown below, produces tritium which is then used in the main reaction. Deuterium occurs naturally in water as it forms 0.01% of naturally occurring hydrogen.

$$^6_3Li + {}^1_0n \longrightarrow {}^4_2He + {}^3_1H + 4.8\,MeV$$

The plasma is contained in a doughnut-shaped steel container and is heated by passing a very large current through it. A magnetic field is used to confine the plasma so it does not touch the sides of its steel container, otherwise it would lose its energy. In theory, the energy released per second should be more than is needed to heat the plasma so the reactor ought to give a continuous output of power. However, at the present time, power can only be produced for a short time as the plasma becomes unstable at such high temperatures.

Figure 5 *The JET fusion reactor*

Summary test 21.6

1 u is equivalent to 931.5 MeV

1 a Explain why the protons in a nucleus do not leave the nucleus even though they repel each other.

b Explain why the mass of a nucleus is less than the mass of the separated protons and neutrons from which the nucleus is composed.

2 a What is meant by *nuclear fission*?

b i The incomplete equation below represents a reaction that takes place when a neutron collides with a nucleus of the uranium isotope $^{235}_{92}\text{U}$. Determine the values of a and b in this equation.

$$^{235}_{92}\text{U} + ^{1}_{0}\text{n} \longrightarrow ^{136}_{a}\text{Xe} + ^{b}_{36}\text{Kr} + 2^{1}_{0}\text{n} + \text{energy released}, Q$$

ii Calculate the energy, in MeV, released in this fission reaction.

Masses:
$^{235}_{92}\text{U}$ nucleus 234.993 u, $^{136}_{a}\text{Xe}$ nucleus 135.877 u,
$^{b}_{36}\text{Kr}$ nucleus 97.886 u, neutron 1.00866 u

3 a What is meant by *nuclear fusion*?

b Hydrogen nuclei fuse together to form helium nuclei in the Sun. Two stages in this process are represented by the following equations:

$$^{1}_{1}\text{p} + ^{1}_{1}\text{p} \longrightarrow ^{2}_{1}\text{H} + ^{0}_{+1}\beta$$

$$^{2}_{1}\text{H} + ^{1}_{1}\text{p} \longrightarrow ^{3}_{2}\text{He}$$

i Describe the reactions that these equations represent.

ii Calculate the energy released in each reaction.

Masses: β-particle 0.00055 u, proton 1.00728 u, $^{2}_{1}\text{H}$ nucleus 2.01355 u, $^{3}_{2}\text{He}$ nucleus 3.01493 u

4 a Explain why light nuclei do not fuse when they collide unless they are moving at a sufficiently high speed.

b Calculate the energy released in the following fusion reaction:

$$^{3}_{2}\text{He} + ^{3}_{2}\text{He} \longrightarrow ^{4}_{2}\text{He} + 2^{1}_{1}\text{p}$$

Masses: proton 1.00728 u, $^{3}_{2}\text{He}$ nucleus 3.01493 u, α-particle 4.00150

c Show that about 25 MeV of energy is released when a $^{4}_{2}\text{He}$ nucleus is formed from 4 protons.

Chapter Summary

The **half-life**, $t_{\frac{1}{2}}$, of a radioactive isotope is the time taken for the mass of the isotope to decrease to half the initial mass. This is the same as the time taken for the number of nuclei of the isotope to decrease to half the initial number.

The decay constant, λ, is the probability of an individual nucleus decaying per second.

The **activity**, A, of a radioactive isotope is the number of nuclei of the isotope that disintegrate per second. The unit of activity is the becquerel (Bq), equal to one disintegration per second.

Binding energy of a nucleus is the work that must be done to separate a nucleus into its constituent neutrons and protons.

Binding energy = mass defect $\times c^2$

Binding energy/nucleon is greatest for nuclei of mass number 57.

Fission is the splitting of a $^{235}_{92}\text{U}$ nucleus or a $^{239}_{94}\text{Pu}$ nucleus into two approximately equal fragments. Induced fission is fission caused by an incoming neutron colliding with a $^{235}_{92}\text{U}$ nucleus or a $^{239}_{94}\text{Pu}$ nucleus.

Fusion is the fusing together of light nuclei to form a heavier nucleus.

Equations

1 $A = \lambda N$

2 $N = N_0 e^{-\lambda t}$, $A = A_0 e^{-\lambda t}$, $C = C_0 e^{-\lambda t}$

3 Binding energy = mass defect $\times c^2$

Launch additional digital resources for the chapter

1 u is equivalent to 931.5 MeV

1 a A radioactive isotope of polonium, $^{210}_{84}$Po, has a half-life of 140 days. Calculate:
  i the decay constant in s^{-1},
  ii the mass of a sample of this isotope which has an activity of 1.0×10^{12} Bq.
b The isotope emits α-particles and forms a stable isotope of lead. Each disintegration in the isotope releases 5.3 MeV of energy. The isotope is to be used to supply power for a remote weather station for at least 1 year.
  i Show that 0.72 g of this isotope will release 100 W of power.
  ii Calculate the power supplied after 1 year by an initial mass of 0.72 g of this isotope.

2 a Calculate the number of atoms present in 1.0 kg of $^{226}_{88}$Ra.
b The radioactive isotope $^{226}_{88}$Ra has a half-life of 1620 years. Calculate the activity of 1 milligram of this isotope.

3 a Radioactive disintegration is a random process yet it is possible to calculate reasonably accurately the number of atoms in a radioactive source of known activity and half-life. Explain why.
b A radioactive isotope X with a half-life of 44 hours disintegrates to form a stable product. A pure sample of X is prepared with an activity of 60 kBq. Calculate:
  i the activity of the sample after 24 hours,
  ii the time taken for the activity of the sample to decrease to 10% of its initial activity.

4 The radioactive isotope of cobalt, $^{60}_{27}$Co, has a half-life of 5.3 years. It emits γ-photons of energy 1.3 MeV.
a Calculate:
  i the activity of a sample of mass of 10 g of this isotope,
  ii the energy transfer per second from this sample.
b The sample is in a thick lead container of mass 920 kg. Calculate the temperature rise of the container in 24 hours, assuming no energy transfer occurs from the container.
  Specific heat capacity of lead = 130 J kg^{-1}K^{-1}.
c $^{60}_{27}$Co is used for gamma therapy to destroy tumours.
  i What property of γ radiation is made use of in this application?
  ii What precautions are taken to ensure the patient is not exposed to the beam unnecessarily?

5 a Explain what is meant by the binding energy of a nucleus.
b Calculate the binding energy per nucleon, in MeV, of:
  i an α-particle,
  ii a deuterium $^{2}_{1}$H nucleus.
  Masses: neutron = 1.00866 u, proton = 1.00728 u, $^{2}_{1}$H nucleus = 2.01355 u, α-particle = 4.00150 u
c Discuss why an α-particle does not break up into two $^{2}_{1}$H nuclei.

6 a In a nuclear reaction, a neutron collided with a nucleus of the lithium isotope $^{6}_{3}$Li. A tritium nucleus $^{3}_{1}$H and another nucleus X was formed as a result.
  Write down an equation that represents this reaction and identify the nucleus X.
b The mass loss in the above reaction was 0.00514 u.
  i Calculate the energy released, in J, in this reaction.
  ii Calculate the mass of X, given the masses of the other nuclei are as follows: $^{3}_{1}$H 3.0155 u, $^{6}_{3}$Li 6.01348 u, neutron 1.00866 u

7 a i State two properties of the strong nuclear force.
  ii Explain why energy is released when a $^{235}_{92}$U nucleus undergoes induced fission.
b i Copy and complete the induced fission equation below:
  $$^{235}_{92}U + ^{1}_{0}n \rightarrow ^{140}_{?}Xe + ^{93}_{38}Sr + ? ^{1}_{0}n$$
  ii Calculate the energy released, in J, in the above reaction.
  Masses: $^{235}_{92}$U 235.0439 u, $^{1}_{0}$n 1.00866 u, $^{140}_{?}$Xe 139.9216 u, $^{93}_{38}$Sr 92.9140 u

8 a State one difference and one similarity between a fusion reaction and a fission reaction.
b i Explain why fusion in a plasma containing light nuclei only takes place if the temperature of the plasma is of the order of 10^8 K.
  ii Two deuterium ($^{2}_{1}$H) nuclei fuse together to form a tritium ($^{3}_{1}$H) nucleus and a proton. Write down the equation which represents this reaction and use the information below to calculate the energy released, in MeV, in this reaction.
  Masses $^{2}_{1}$H 2.0136 u, $^{3}_{1}$H 3.0155 u, proton 1.0073 u

9 a Explain why the spent fuel rods from a nuclear reactor are more radioactive after removal from the reactor than they were before they were used in the reactor.

b The radioactive isotope, $^{90}_{38}$Sr, is a β^--emitter which has a half-life of 28 years. It is produced as a fission product in a fuel rod in a nuclear reactor. If it escapes into the environment, it can be absorbed by the body in place of calcium. Calculate:

i the activity of 1 mg of this isotope,

ii the activity in 100 years of a sample of this isotope that has a mass of 1 mg at the present time.

10 a State what is meant by the binding energy of a nucleus.

b The mass of the hydrogen $^2_1$H nucleus is 2.01355 u.

i State how many protons and how many neutrons are in this nucleus.

ii The mass of a proton is 1.00728 u and the mass of a neutron is 1.00867 u. Calculate the binding energy per nucleon of the hydrogen $^2_1$H nucleus.

c The binding energy per nucleon of the hydrogen $^3_1$H nucleus is 2.832 MeV.

i Calculate the ratio of the binding energy of the hydrogen $^3_1$H nucleus to that of the hydrogen $^2_1$H nucleus.

ii Student A suggests the ratio should be 1.5 on the grounds that the ratio of the number of nucleons in the two nuclei is 3 : 2. Student B suggests the ratio should be 3 on the grounds that there are three force bonds between the nucleons in the $^3_1$H nucleus and only one in the $^2_1$H nucleus. State which suggestion is closer to the binding energy ratio calculated in **ci** and discuss how this suggestion could be applied to other known light nuclei such as the helium $^4_2$He nucleus which has a binding energy of 28 MeV.

d A helium $^4_2$He nucleus is formed when a hydrogen $^3_1$H nucleus and a hydrogen $^2_1$H nucleus are fused. In this process, a particle Y is released.

i Identify particle Y and write an equation to represent this reaction.

ii The binding energy of a hydrogen $^3_1$H nucleus is about 9 MeV. Estimate the energy released in the above reaction.

11 a A thermal nuclear reactor is designed to release energy at a steady rate as a result of induced nuclear fission.

i Explain what is meant by induced *nuclear fission*.

ii State an isotope that undergoes fission in a thermal nuclear reactor.

iii Explain what is meant by a *chain reaction* in nuclear fission.

iv Explain why the mass of fissile material in a nuclear reactor must be greater than a minimum mass in order for a chain reaction to occur.

b The fuel rods in the core of a nuclear reactor are surrounded by a moderator. Control rods in the core are used to control the rate of release of energy from the fuel rods. A coolant is pumped through the core to transfer energy from the reactor core to a heat exchanger.

i State the purpose of the moderator.

ii Name a substance which control rods are made from and describe how the control rods are used to keep the rate of release of energy constant.

iii State two physical properties of the coolant.

22.1 Ultrasonic imaging

Learning outcomes

On these pages you will learn to:

- explain the principles of the generation and detection of ultrasonic waves using piezo-electric transducers
- explain the main principles behind the use of ultrasound to obtain diagnostic information about internal structures
- define specific acoustic impedance and explain the importance in relation to the intensity reflection coefficient at a boundary
- recall and solve problems by using the equation $I = I_0 e^{-\mu x}$ for the attenuation of ultrasound in matter

Producing ultrasonic waves

Ultrasonics are sound waves at frequencies of more than about 18 kHz above the range of the human ear. Unlike X-rays, ultrasonic radiation is non-ionising radiation and therefore does not damage living tissue. For medical imaging, ultrasonics at frequencies between about 1 and 10 MHz are used as diffraction would be significant at lower frequencies and reduced intensity due to absorption would be significant at higher frequencies.

An **ultrasonic probe** used in ultrasonic scanning is a hand-held device placed in contact with the body surface to direct pulses of ultrasound into the body. Each emitted pulse is partially reflected by internal boundaries in the body. The reflected pulses are then detected by the probe before the next pulse is emitted.

Figure 1 shows the construction of an ultrasonic probe.

- The probe contains a piezo-electric **transducer** in the shape of a disc. Piezo-electricity is a property of certain solids whereby a p.d. applied between opposite faces causes a change of distance between the two faces. When an alternating p.d. is applied between the faces of the disc, the disc vibrates due to its changing thickness.

- By applying an alternating p.d. of frequency equal to the resonant frequency of vibration of the disc, the disc vibrates at resonance and creates ultrasonic waves in the surrounding medium. The thickness of the disc determines its resonant frequency.

- An absorber pad of 'backing material' behind the disc prevents ultrasonic waves created at the two surfaces of the disc from cancelling each other out. The pad also damps the vibrations of the disc rapidly after each pulse is emitted.

More about piezo-electricity

A **transducer** is any device that is designed to convert energy from one form to another. A piezo-electric transducer generates a p.d. when it is squeezed. The piezo-electric material contains positive and negative ions, which are held together by the electrostatic forces they exert on each other. The centre of the negative charge of each molecule is in the same position as the centre of positive charge. When pressure is applied to opposite surfaces of the material, the centres of charge of the positive and negative ions are displaced slightly in opposite directions, causing a potential difference between the two surfaces. The effect is known as the **piezo-electric effect** and is displayed by crystals such as quartz.

To apply a p.d. across a quartz crystal, opposite surfaces to which the pressure is applied must be coated with metal so that an electrical connection can be made to each surface.

The piezo-electric effect is reversible in that the application of a p.d. across a piezo-electric material causes the distance across the material to increase or decrease according to the polarity of the p.d.

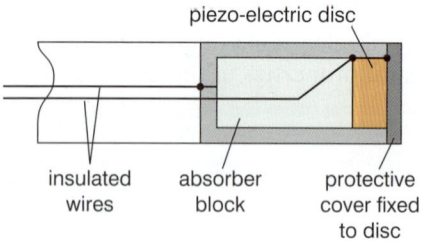

piezo-electric disc

insulated wires   absorber block   protective cover fixed to disc

a Probe construction

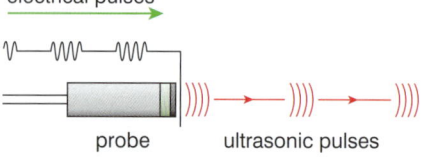

electrical pulses

probe   ultrasonic pulses

b Ultrasonic pulses

Figure 1 An ultrasonic probe

Absorption and reflection of ultrasonic waves

Ultrasonic waves can be reflected and refracted just like sound waves. When ultrasonic waves reach a boundary between two substances, some of the wave energy is reflected and some is transmitted as shown in Figure 3.

Considering the energy reaching the boundary in 1 second, as energy cannot be created or destroyed, the sum of the reflected energy and the transmitted energy is equal to the incident energy. Hence,
incident intensity I = the reflected intensity I_R + the transmitted intensity I_T.

The fraction of the incident energy that is reflected or transmitted depends on:

- the angle of incidence θ of the incident waves,
- the densities ρ_1 and ρ_2 of the two substances,
- the wave speeds c_1 and c_2 of ultrasonic waves in the two substances.

The **acoustic impedance** of a substance, Z is defined by the equation $Z = \rho c$

The unit of Z is given by the product of the unit of density (i.e. $kg\,m^{-3}$) and the unit of speed (i.e. $m\,s^{-1}$). Hence the unit of Z is $kg\,m^{-2}\,s^{-1}$.

Specific values for different substances are given in Table 1. Notice the very small value for air and the large values for quartz and bone compared with the other substances listed in the table.

When ultrasonic waves are incident on a boundary between two substances with acoustic impedances Z_1 and Z_2, the ratio of the reflected intensity to the incident intensity $\dfrac{I_R}{I}$, the reflection coefficient, is given by the equation

$$\frac{I_R}{I} = \frac{(Z_2 - Z_1)^2}{(Z_2 + Z_1)^2}$$

Table 1

Substance	Acoustic impedance $Z/kg\,m^{-2}\,s^{-1}$
air	430
blood	1.59×10^6
bone	6.80×10^6
fat	1.38×10^6
muscle	1.70×10^6
quartz	1.52×10^7
soft tissue	1.63×10^6
water	1.50×10^6

- If Z_1 and Z_2 are almost equal, the ratio is close to zero and the reflected intensity is very small compared with the incident intensity. In other words, most of the wave energy is transmitted.
- If Z_1 and Z_2 are very different, the ratio is close to 1 so most of the wave energy is reflected. The transmitted intensity is very small compared with the incident intensity. In other words, most of the wave energy is reflected.

The intensity of the reflected pulses from the ultrasonic probe when they return to it depends on:

1 the ratio of the reflected intensity to the incident intensity at each boundary, and
2 the absorption of the ultrasonic waves by each substance they pass through.

When ultrasound is directed from a probe into the body, any air trapped between the probe and the body will cause most of the ultrasound energy to be reflected from the body. This is because the acoustic impedances of air and soft tissue are very different so the ratio I_R/I is close to 1. To eliminate such trapped air, a **coupling medium** such as a gel is applied between the probe and the body surface. Such substances have similar acoustic impedances to soft tissue so the ratio I_R/I is close to zero. In other words, most of the wave energy is transmitted into the body.

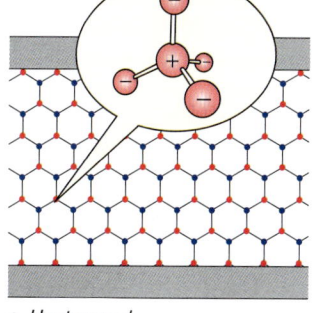
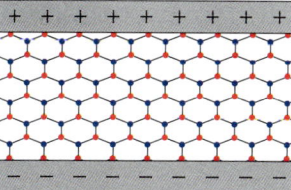
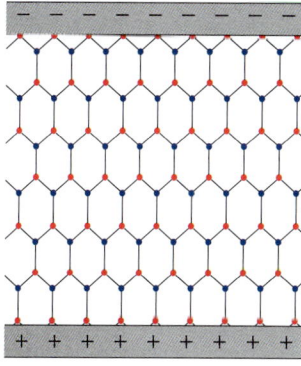

a Unstressed

b Compressed

c Extended

Figure 2 *Piezo-electricity*

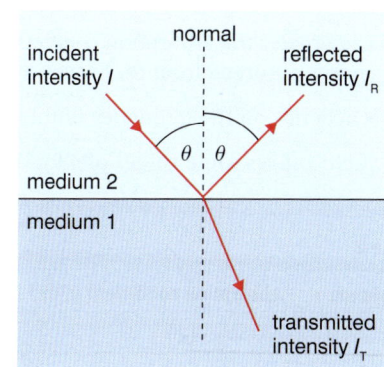

Figure 3 *Reflection and transmission at a boundary*

Reflection at tissue boundaries is significant and can't be avoided. For example, the ratio $\frac{I_R}{I}$ for a boundary between soft tissue and fat is 6.9×10^{-3} (using values from Table 1).

- The reflected pulses are therefore much weaker than the pulses leaving the probe.
- Also, the further a boundary is from the probe, the further the pulses reflected from that boundary travel and the weaker the reflected pulses from it will be due to absorption (see below).

Absorption of ultrasonic waves depends on the substances the waves pass through and the distance travelled through each substance. The energy absorbed by the substance causes the temperature of the substance to increase.

When a parallel beam of ultrasonic waves travels through a substance, the intensity of the waves decreases exponentially with distance.

The intensity I of the waves after travelling through distance x of a substance is given by the equation

$$I = I_0 e^{-\mu x}$$

where I_0 is the incident intensity and μ is the absorption coefficient of the substance. The unit of μ is m^{-1}.

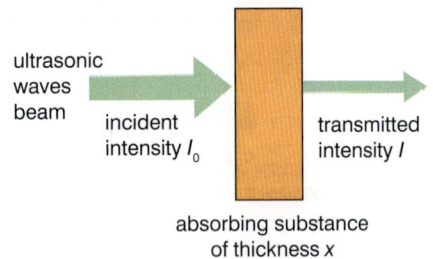

Figure 4 Absorption

Note

1 The greater the absorption coefficient μ, the greater the energy absorbed from the waves over any given distance.

2 μ depends on the frequency of the ultrasonic waves.

3 Table 2 gives some values of μ for different substances.

Table 2

Substance	Absorption coefficient μ / m^{-1}
air	120
bone	130
muscle	23
water	0.02

Ultrasonic scans

An **ultrasonic scanner** consists of an ultrasonic probe connected to a control unit and a visual display unit. In the simplest scan system, referred to as the **A-scan system**, a pulse generator is used to supply electrical pulses to the probe and to trigger the oscilloscope time base each time a pulse is generated. Figure 5 shows an A-scan of an eye.

- In each scan, a pulse is generated by the probe and, before the next pulse is generated, the probe detects reflected pulses from the boundaries in the path of the pulse. The reflected pulses or 'echoes' detected by the probe are amplified and displayed on the oscilloscope.
- Each time the time base is triggered, the oscilloscope beam sweeps across the screen from left to right. The time base of the oscilloscope is adjusted to display on the screen all the reflected pulses for each transmitted pulse. As a result, each pulse on the screen will be very narrow as the duration of each pulse is much shorter than the time for each 'sweep' of the oscilloscope screen.

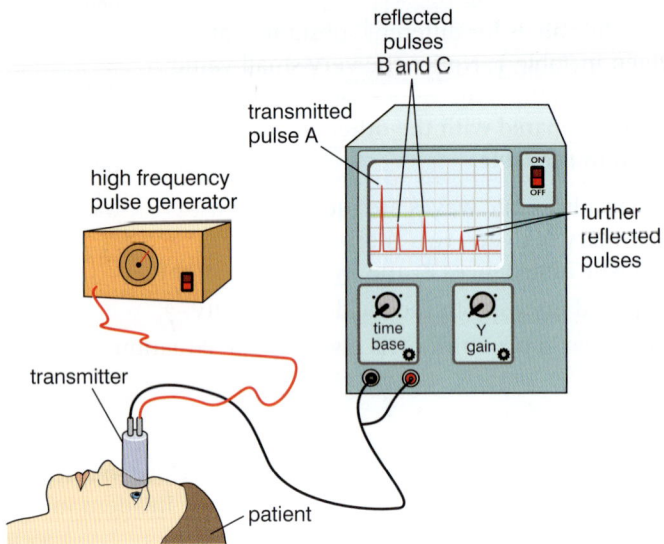

Figure 5 The A-scan system

1 The position of each reflected pulse on the screen depends on the transit time of the ultrasonic pulse (i.e. the time taken by the ultrasonic pulse to travel from the probe to the internal boundary that reflected the pulse and back).

2 The further a reflected pulse on the screen appears from the transmitted pulse:
- the longer the transit time of the pulse in the body and the further away the boundary is from the probe,
- the smaller the pulse height will be because the ultrasonic pulse is partially reflected by the boundaries it passes through and the substances it passes through absorb some of its energy.

3 The transit time is proportional to the distance from the probe to the boundary. Therefore, the greater the distance from the probe to the boundary, the further the pulse appears across the screen. The oscilloscope can be used to measure the transit time, t, of a pulse. The distance travelled by the pulse $s = vt$, where v is the speed of ultrasonic waves in the body. Therefore, the distance from the probe to the internal boundary causing the pulse $= \frac{1}{2}vt$. Using this equation, the screen could be calibrated in terms of distance from the probe so the distance between a boundary and any other boundary or the probe can be measured directly.

In the **B-scan system**, the probe has a number of ultrasonic transducers side by side, each one sending out ultrasonic pulses in a slightly different direction to the others. The signals from the transducers due to the reflected pulses are processed by a computer such that each reflected pulse is displayed as a bright spot on the screen in the correct direction and at the correct distance from the probe. As the probe is moved over the body surface, the bright spots on the screen build up a two-dimensional image of the reflecting boundaries scanned. The image may be enhanced and stored electronically.

Comparison with X-rays

• B-scans are used for pre-natal scans (i.e. to observe unborn babies in the womb) rather than X-ray CT scans. This is because ultrasonic waves are non-ionising and, at the intensities used in scanning, do not damage human tissue.

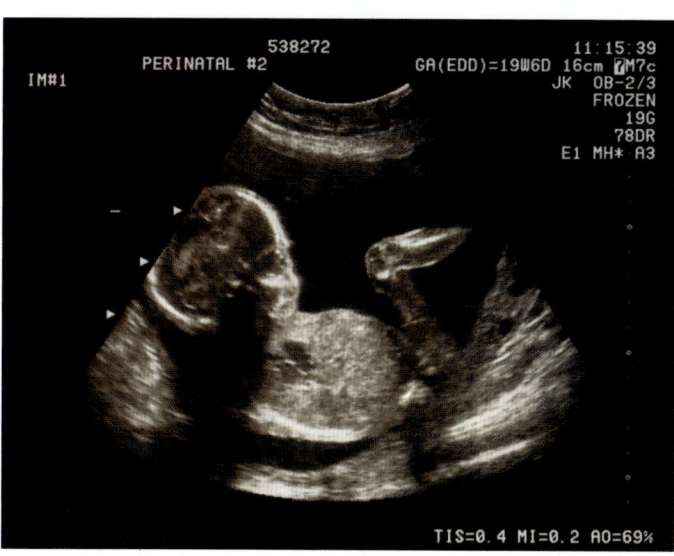

Figure 6 A B-scan image of an unborn baby

• Ultrasound reflects at bone/tissue boundaries as well as at internal boundaries between soft substances such as fat, muscle and tissue because the acoustic impedance of such substances differs. Therefore, ultrasonic images show such 'soft' boundaries whereas X-ray images do not (because X-rays are not reflected at such boundaries).

Summary test 22.1

Use the data in Table 1, p.327, where necessary.

1 An ultrasonic probe generates ultrasonic waves at a frequency of 2.5 MHz. The speed of ultrasound in air $= 350\,\mathrm{m\,s^{-1}}$ and in soft tissue $= 1550\,\mathrm{m\,s^{-1}}$.

  a Calculate the wavelength of the ultrasonic waves from this probe: **i** in air, **ii** in soft tissue.

  b Explain why ultrasonic waves of much lower frequency are unsuitable for medical imaging.

2 a **i** With the aid of a diagram, describe the construction of an ultrasonic probe and how it produces ultrasonic waves.

    ii Explain the function of the backing block in an ultrasonic probe.

  b In the A-scan arrangement shown in Figure 5, on the previous page, the furthest boundary from the probe is the retina.

    i Explain the presence of each pulse on the screen in terms of the cross-section of the patient's eye.

    ii Calculate the distance between the boundary responsible for pulse B and the retina in Figure 5, if the distance from the probe to the furthest boundary is 24 mm.

3 a Use the data in Table 1 to calculate the reflection coefficient of the boundary between: **i** air and skin, **ii** water and skin. Assume skin is soft tissue.

  b Use the results of your calculation to explain why a gel must be applied between an ultrasonic probe and the skin when the probe is used.

  c A body organ has a density of $1040\,\mathrm{kg\,m^{-3}}$ and the speed of sound through it is $1580\,\mathrm{m\,s^{-1}}$.

    i Calculate the acoustic impedance of the organ tissue.

    ii Use the data in Table 1 to calculate the reflection coefficient of the boundary between the organ and the surrounding soft tissue.

4 a State the main differences between an A-scan and a B-scan.

  b Ultrasonic waves and X-rays are both used for medical imaging. Explain why an ultrasonic scan rather than an X-ray scan is used for scanning a baby in the womb.

Learning outcomes

On these pages you will learn to:

- explain the principles of the production of X-rays by electron bombardment of a metal target
- describe the main features of a modern X-ray tube, including control of the intensity and hardness of the X-ray beam
- show an understanding of the use of X-rays in imaging internal body structures, including a simple analysis of the causes of sharpness and contrast in X-ray imaging
- recall and solve problems by using the equation $I = I_0 e^{-\mu x}$ for the attenuation of X-rays in matter
- show an understanding of the purpose and principles of computed tomography (CT scanning)
- show an understanding of how the image of an 8-voxel cube can be developed using CT scanning

The production and properties of X-rays

X-ray imaging in medicine is an example of a diagnostic technique that is non-invasive. X-rays are electromagnetic waves of wavelength of the order of 0.1 nm or less. Figure 1 shows how a diagnostic X-ray tube works. The current through the filament wire heats the wire, which causes electrons to be emitted from the wire. These electrons are attracted from the filament or 'cathode' to the **anode** when the anode is positive relative to the filament, typically 20–100 kV for X-ray imaging. The electrons are stopped when they collide with the anode and they emit X-rays in the process.

For an anode potential V, the maximum energy of an X-ray photon $= eV$ so the maximum frequency $f_{max} = eV/h$
Hence the minimum wavelength λ_{min} is given by:

$$\lambda_{min} = \frac{hc}{eV}$$

Note

X-ray tubes used for therapy to destroy tumours are designed differently because they need to produce photons at higher energies. Such tubes operate at voltages from 250 kV to an upper limit (due to insulation breakdown) of 300 kV. X-ray photons from such tubes can destroy tumours no deeper than 5 cm beneath the skin. For deeper tumours, gamma photons with energies of the order of 1 MeV from radioactive isotopes are used.

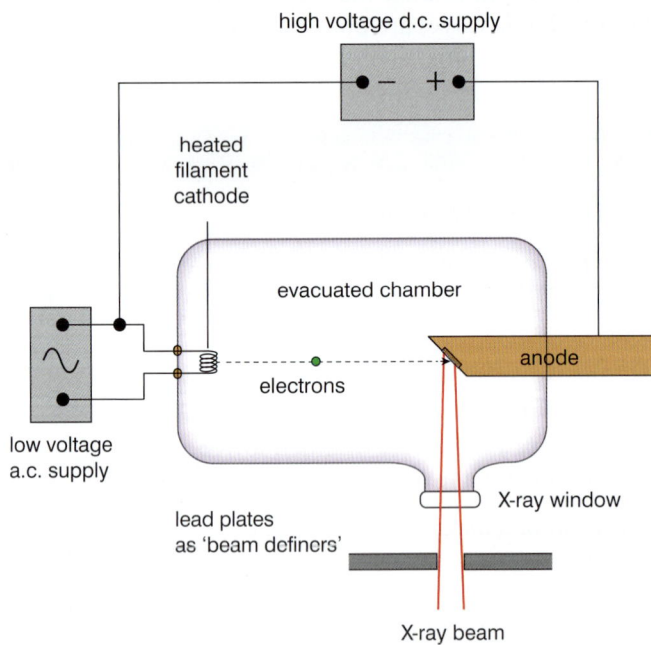

Figure 1 *The X-ray tube*

The spectrum of photon energies from an X-ray tube

An X-ray tube produces a continuous spectrum of photon energies up to the maximum value of eV, where V is the maximum tube voltage, as shown in Figure 2. Raising the tube voltage increases the intensity at all photon energies up to the maximum photon energy as well as increasing the maximum photon energy.

In addition, intensity 'spikes' are produced, which are characteristic of the atoms of the anode and do not change position when the tube voltage is altered. However, if the tube voltage is reduced sufficiently, each spike will disappear when the maximum photon energy is less than the energy of the photons at the spike.

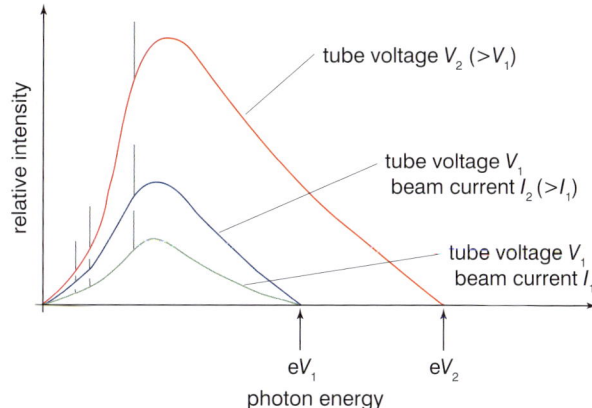

Figure 2 *The energy spectrum of an X-ray tube*

The spikes are caused by the excitation of atoms in the anode when electrons from the filament collide with them. As a result, electrons in the atoms move temporarily from the innermost shells of the atom to higher energy levels. When these electrons return to their original levels, they emit X-ray photons at energies which are characteristic of the anode atoms. These emitted X-rays form patterns of line spectra, each pattern corresponding to electrons returning from the outer energy levels to a particular electron shell. The pattern for electrons returning to:

- the innermost energy shell ($n = 1$) is referred to as the K-series,
- the second energy shell ($n = 2$) is referred to as the L-series,
- the third shell ($n = 3$) is referred to as the M-series.

Notes

1 The energy of an X-ray photon is often expressed in electronvolts (eV) where $1\,\text{eV} = 1.6 \times 10^{-19}\,\text{J}$.

2 The power supplied to an X-ray tube $= IV$, where I is the beam current. The % efficiency of an X-ray tube is the percentage of the power supplied emitted as X-radiation. A typical X-ray tube has an efficiency of about 1%. The wasted energy is dissipated as heat at the anode.

X-ray imaging

When an X-ray picture is made, X-rays from the X-ray tube are directed for a specified time at the relevant area of the body with a film cassette on the other side of the body. Bones, teeth and other dense matter in the path of the X-rays absorb X-rays much more than muscle and body tissue does. When the film is developed, the areas of the film exposed to X-rays are darker than the unexposed areas so a negative image of the bones, teeth, etc. is formed on the developed film.

For any given application, the following factors must be taken into account before the X-ray tube is used.

1 **The penetrating power or 'hardness' of the X-ray beam** is increased by increasing the tube voltage. The higher the energy of the X-ray photon, the further it can travel through matter. The X-rays used to give an image of the bones of a broken arm do not need to penetrate as far as the X-rays used to give an image of an organ in the body. Increasing the tube voltage increases the maximum energy of the photons emitted by the tube so the beam is more penetrating. Low energy photons are easily absorbed. An **aluminium filter** placed between the X-ray tube and the patient is used to absorb

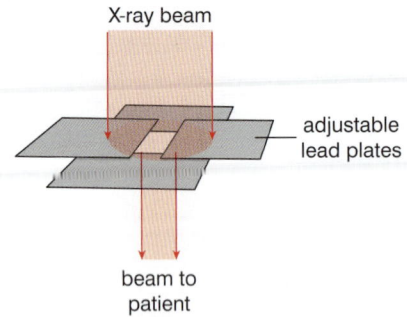

Figure 3 *Beam definers*

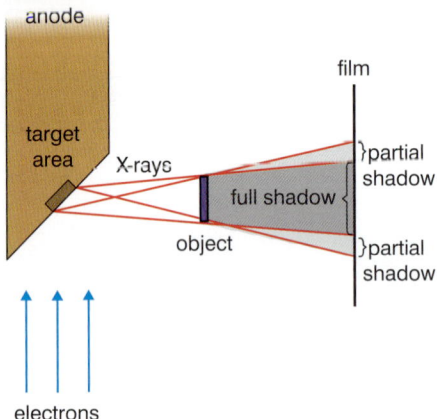

Figure 4 *Sharpness*

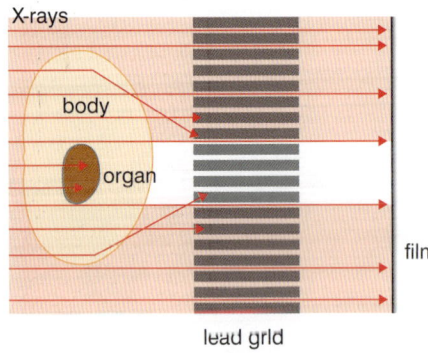

Figure 5 *Using a lead grid*

such photons which would otherwise be absorbed by the body and cause unnecessary exposure of the body to X-rays.

2 **The intensity of the X-ray beam** is the radiation energy per second passing through unit area at right angles to the area. The darkening of an X-ray film depends on the intensity of the X-radiation as well as on the duration of exposure. The greater the intensity or the longer the duration of exposure, the darker the exposed parts of the film will be. An organ that moves would need a shorter duration of exposure and therefore greater intensity than a bone in the arm which can be held still. The intensity depends on the number of electrons per second reaching the anode and therefore on the tube current (since the tube current is a measure of the number of electrons per second reaching the anode). The tube current is controlled by the current through the filament wire. If the filament current is increased, the intensity of the X-ray beam is increased because:

• the filament becomes hotter and emits more electrons per second,

• more electrons per second hit the anode so more X-ray photons are released each second.

3 **The width of the beam** is set using 'beam definer' lead plates to ensure that only the part of the patient to be X-rayed is exposed to X-rays. Lead plates surrounding the X-ray tube are used to prevent people other than the patient being exposed to X-rays (see Figure 3).

Image quality

Image sharpness

The sharpness of an X-ray image is determined by how clearly the edges of structures in the image are defined. A sharp image is one in which such edges can clearly be seen.

To form a sharp image on the film, the X-rays need to originate from a small area of the anode and X-rays scattered by body organs and tissues need to be stopped from reaching the film.

If the area of the anode is too large, the images will be blurred at the edges by large partial shadows as shown in Figure 4. However if the area is too small, the intense concentration of electrons in this area of the target area will damage the anode. To prevent overheating of the anode, it usually consists of a tungsten metal block set in a copper cylinder which is kept cool by pumping water or oil through it. Tungsten is chosen as it has a high melting point and copper is chosen as it is an excellent conductor of heat.

• X-ray photons may be scattered by atoms in the body tissues which they pass through. Some scattered X-ray photons may be scattered into the shadow areas on the film of bones or body organs. This would lessen the contrast between the images on the film of bones and body organs and the surrounding tissues. To eliminate scattered X-rays, a lead grid is placed between the patient and the film, as shown in Figure 5. Lead is used because it is a very effective absorber of X-rays.

The grid holes are aligned with the direction in which the unscattered X-rays are travelling so unscattered X-rays that enter the holes of the grid pass straight through it. However, scattered X-rays are absorbed by the grid because they travel mostly through lead after reaching the grid.

Contrast

An X-ray image with good contrast has areas where the film is very dark due to exposure to X-rays and other areas that are hardly darkened by X-rays. Bones and teeth are good absorbers of X-rays so they give images that stand out in good contrast with the surrounding tissue.

Body organs such as the stomach are not as effective at absorbing X-rays as bones and teeth. In order to obtain good X-ray images of an organ, a **contrast medium** is used. For example, a patient about to undergo a stomach X-ray is given a drink containing barium sulphate. Because barium is a good absorber of X-rays, X-rays that would otherwise pass through the stomach are absorbed so the contrast between the image of the stomach and its surroundings is vastly improved. A contrast medium is also used to obtain X-ray images of blood vessels where the contrast medium is injected into the bloodstream.

Contrast is lessened if:

- **The duration of exposure is too long**. The light areas and the dark areas of the film both become darker but the increase of darkness is greater in the light areas of the film. So the difference in darkness between the light and dark areas is reduced.
- **The X-rays are too penetrating**. Increasing the energy of the photons by increasing the tube voltage would increase their penetrating power. So more X-rays would pass through the organ and reach the shadow area of the film.
- **Too much scattering of X-ray photons** occurs when they pass through the tissue surrounding the organ.

Contrast can be improved if the film in its cassette is covered with a sheet of fluorescent substance. X-rays directed at a fluorescent substance cause the atoms of the substance to emit light photons. Each X-ray photon might cause many light photons to be emitted, darkening the film more in the areas which are exposed to X-rays. In addition to improving the contrast, the exposure of the patient to X-rays can be reduced as the film is more sensitive to darkening.

Absorption of X-rays by matter

When a beam of X-rays spreads out in a vacuum from a source, the intensity decreases as the distance from the source increases. If the beam spreads out in all directions, the intensity at distance r from the source is proportional to $1/r^2$. This is the same rule as for gamma photons from a point source. See Topic 10.4 for an explanation.

When a beam of X-rays passes through matter, some X-ray photons are absorbed and some are scattered. The transmitted beam is said to be **attenuated** because it is less intense than the incident beam.

For a parallel beam of intensity I_0 directed at normal incidence at an absorber of thickness x, the transmitted intensity I is given by:
$$I = I_0 e^{-\mu x}$$

where μ is the **attenuation** (or absorption) **coefficient** of the absorber.

The variation of intensity with thickness is shown in Figure 7. The curve decreases exponentially with increase of thickness.

The half thickness, $X_{1/2}$, of an absorber is the thickness required to reduce the intensity of the beam to half its initial value.

For thickness $x = X_{1/2}$, $I = \frac{1}{2}I_0$.

Substituting these values into the above equation gives
$$\tfrac{1}{2}I_0 = I_0 e^{-\mu X_{1/2}}$$

Cancelling I_0, rearranging and taking natural logs of both sides of this equation gives:
$$\mu X_{1/2} = \ln 2$$

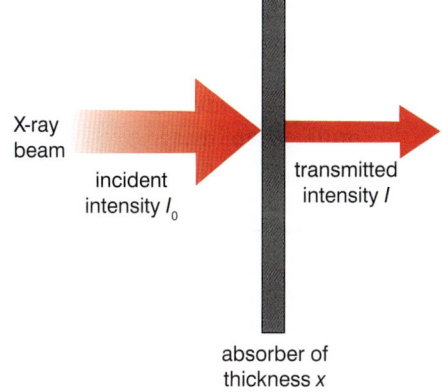

X-ray beam

incident intensity I_0

transmitted intensity I

absorber of thickness x

Figure 6 *Attenuation*

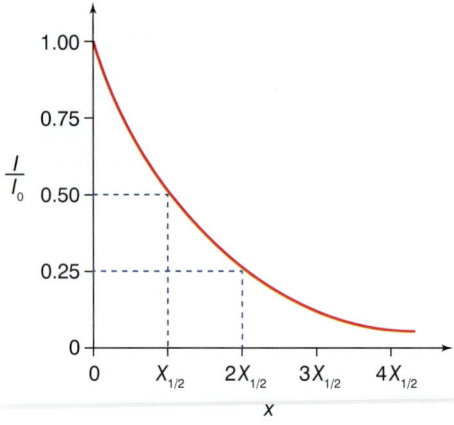

Figure 7 *Transmitted intensity against thickness*

Computed tomography (CT)

In a computed tomography (CT) scan, a narrow collimated beam of X-rays is directed through the patient at an array of detectors. The X-ray tube and the detectors are rotated in small steps around an axis along the length of the patient. The detectors are connected to a computer and the signals from the **detectors** are processed to form an image of the cross-section of the body which is exposed to the beam.

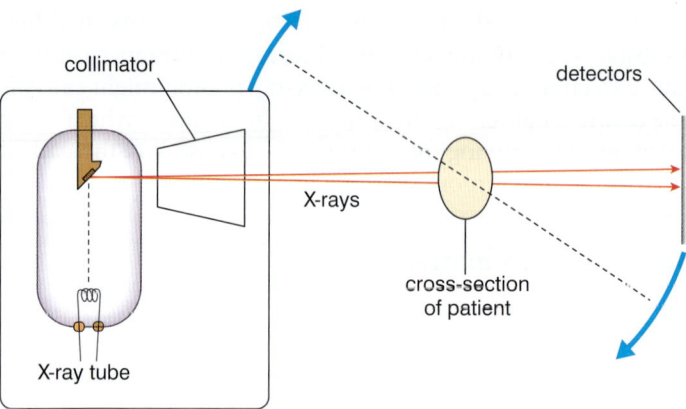

Figure 8 *CT scanning*

- The tube and the detectors are mounted in a gantry which is rotated through 360° about the patient lying in a stationary position on a flat couch between the tube and the detector.
- Each detector receives X-rays that have travelled along a straight path between the tube and the detector.

The intensity detected by each detector depends on the absorption of X-rays along the path from the tube to the detector. This absorption depends on the different densities and thicknesses of the tissue along the path.

By considering the body divided into small volume elements, referred to as 'voxels', each voxel would contribute a certain 'pixel' value to the intensity reduction along any given path.

- Each detector records the sum of the pixel values along each path,
- Each voxel contributes its pixel value to the detector reading (i.e. the detected intensity) many times as the tube and detectors are rotated round the patient.

The signals from the detectors can be processed by a computer to work out each pixel value for each voxel and hence display a 2-D 'mosaic' image of the cross-section. The contrast and sharpness of the pixels can be adjusted to give the best possible image. In addition, the image can be stored and transmitted electronically.

Note

To obtain a 2D image, a complete scan of a single cross-section takes about 10 s. By taking further scans in parallel planes along an axis, the computer can produce a 3D image of an internal structure by combining the 2D images of multiple sections. In addition, the computer can produce an image of a cross-section through the patient in any plane, not just in the scanning planes.

A model scan

Figure 9 shows a simplified version of the process where four voxels are scanned from four different directions which are horizontal, diagonal, vertical and the opposite diagonal.

The pixel value of each voxel is worked out by reconstructing an image consisting of four image voxels and assigning the detector readings to the image voxels according to the path the X-rays pass through in each direction. In this 2×2 example:

- Each image voxel receives a contribution four times (once for each direction) from its corresponding object voxel and once from each of the other three object voxels. The total value of all the image voxels after the complete scan is therefore seven times the sum of the object voxel values.
- The sum of the object voxel values is therefore the total value of all the image voxels ÷ 7. This sum is effectively a background value for each of the four image voxels.

Subtracting the background value from the final value of each image voxel leaves a value equal to three times the contribution from the corresponding object voxel. The value of each object voxel can then be calculated.

Suppose the image voxel values shown in Figure 9 after a complete scan are 22, 16, 25 and 28.

- Their total value = 22 + 16 + 25 + 28 = 91.
- The sum of the object voxel values = 91 ÷ 7 = 13.
- Subtracting this background value from the values of the image voxels gives 9, 3, 12 and 15.
- Dividing these 'corrected' values by 3 gives 3, 1, 4 and 5 for the object voxel values.

object   image voxels

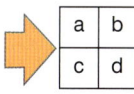
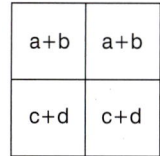
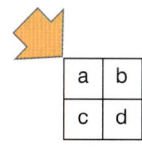
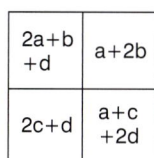
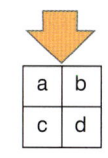
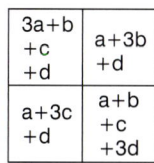
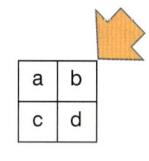

Figure 9 *A simplified scan*

Summary test 22.2

1 An X-ray tube operates at an anode potential of 50 kV.

  a i Show that the minimum wavelength of the X-rays from this tube is 0.025 nm.

    ii When the beam current at 50 kV is 0.2 mA, the tube operates at an efficiency of 1.5%. Calculate the radiation energy per second produced by the tube.

  b State and explain what change in the X-ray beam occurs if:

    i the tube current is increased,

    ii the anode voltage is increased.

2 a What is meant by: **i** the sharpness, **ii** the contrast of an X-ray image.

  b State the function of a scattering grid and explain why it is necessary when an X-ray image is obtained.

3 a A metal plate of thickness 1.3 mm placed in the path of a collimated X-ray beam reduces the beam intensity by 70%. Show that the absorption coefficient of the metal plate is 9.3×10^{-2} m^{-1}.

  b Explain why a contrast medium is used when an X-ray picture of the stomach is made.

4 a With the aid of a diagram, outline the principle of operation of a CT scanner.

  b In the simplified model in Figure 9, the values of the image voxels after a complete scan were 21, 15, 21 and 27. Calculate: **i** the background value of the image voxels, **ii** the pixel value of each object voxel.

22.3 PET scanning

Learning outcomes

On these pages you will learn to:

- explain what is meant by a positron emission from a suitable radioactive nucleus and that a positron is the antimatter particle of the electron
- recall that a positron and an electron annihilate each other when they collide in a process that causes two gamma-ray photons to be released from a single point
- understand why two gamma rays from a positron emission travel in opposite directions and that if the radioactive nucleus is inside the body, the gamma-ray photons can be detected outside the body
- understand that a medical tracer is a substance that contains radioactive nuclei that is introduced into the body and absorbed by the tissue under investigation
- recall that the point of origin of each pair of gamma-ray photons can be determined by using signals from a ring of gamma-ray detectors around the body and the information from such signals can be used to trace an image of the radioactive nuclei in the body

About annihilation

The positron is the antimatter counterpart or 'antiparticle' of the electron. It has the same rest mass as an electron, but it has an equal and opposite charge to that of the electron. If an antiparticle meets its particle counterpart, they **annihilate** each other and radiation is released in the form of two gamma-ray photons. Because momentum and energy are conserved in the process, the two photons must travel in opposite directions each with the same energy. The creation of a single photon only would violate the principle of conservation of momentum because it would possess far more momentum than the initial momentum of the electron and positron would have possessed. The two photons possess equal and opposite momentum and, since their energy is proportional to their momentum, they move in opposite directions with equal energy.

The rest mass of a positron or an electron is 9.11×10^{-31} kg. When an electron and a positron with negligible kinetic energy annihilate each other, radiation energy is created due to the complete loss of mass of $2 \times 9.11 \times 10^{-31}$ kg in accordance with Einstein's equation $E = mc^2$. Using this equation gives energy equal to 1.64×10^{-13} J or 1.02 MeV. Each of the two gamma-ray photons is therefore released with 0.51 MeV of energy, assuming that the electron and positron lose most their kinetic energy before they annihilate each other.

A positron emitted from an unstable 'proton-rich' nucleus in body tissue travels about 1–2 mm before it collides with an electron and both the electron and the positron are annihilated in the interaction, producing two gamma photons in the process. Positrons referred to as β^+ particles are emitted by unstable nuclei that have a greater proton to neutron ratio than stable nuclei of the same element. In comparison, β^- particles are electrons emitted by unstable nuclei that have a smaller proton to neutron ratio than stable nuclei of the same element. See Topic 10.4.

Note

Positrons are emitted with an initial kinetic energy characteristic of the positron-emitting isotope. For example, the positrons from the fluorine isotope $^{18}_{9}$F are emitted with initial kinetic energies of 0.6 MeV. If the isotope is inside the body, the positrons lose most of their kinetic energy within a distance of a few millimetres before they are annihilated. So each annihilation event from this isotope causes two 0.51 MeV gamma photons to be released in opposite directions.

Positron-emitting isotopes as tracers in medicine

Certain radioactive isotopes are suitable for use as tracers in medicine because they can be attached to substances that are absorbed by specific organs or types of tissue such as cancer cells in the body. Positron-emitting tracers differ from other gamma-emitting tracers in that each positron releases two gamma photons in opposite directions from the decay of each unstable nucleus, whereas other gamma-emitting tracers release a single photon from the decay of each unstable nucleus. As explained later, this key difference enables the tracer to be scanned and mapped more precisely than, for example, with a CT or a gamma camera scan.

The positron-emitting fluorine isotope $^{18}_{9}$F is used to detect cancer cells because such cells have a higher uptake of glucose than normal cells. Glucose molecules containing an atom of the positron-emitting fluorine isotope $^{18}_{9}$F instead of an oxygen atom attach themselves preferentially to cancer cells. So when a nucleus of this isotope decays, two gamma photons are released close to the cancer cell. The gamma photons released in this way can be detected outside the body and the location of the tracer nuclei can be traced as described on the next page.

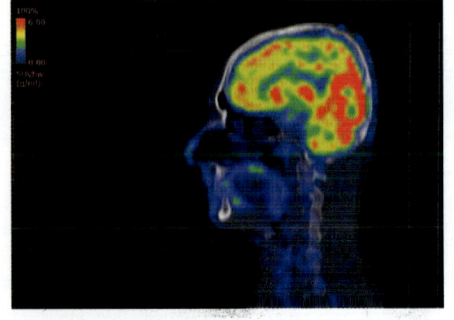

Figure 1 *A PET image*

Production of positron-emitting isotopes

Positron-emitting isotopes suitable for medical use are produced by exposing stable isotopes to high-energy protons from, for example, a cyclotron. For instance, if nuclei of the stable oxygen isotope $^{18}_{8}O$ are exposed to high energy protons, some of them may have a neutron knocked out and replaced by a proton to become nuclei of the fluorine isotope $^{18}_{9}F$ which is a positron emitter with a half-life of 110 minutes. Using chemical techniques including ion exchange, atoms of this particular isotope can be attached to glucose molecules by replacing oxygen atoms to form 18-FDG (fluorodeoxyglucose) molecules. In the body, these molecules preferentially attach themselves to cancer cells due to the higher uptake of glucose by cancer cells. With a half-life of just under 2 hours, 18-FDG needs to be prepared just a few hours before being administered to a patient in a PET scanner.

The PET scanner

A small quantity of the tracer such as the fluorine isotope $^{18}_{9}F$ is injected into the patient via a saline drip. To allow the tracer to be distributed in the body, the PET scan starts about 40 minutes later.

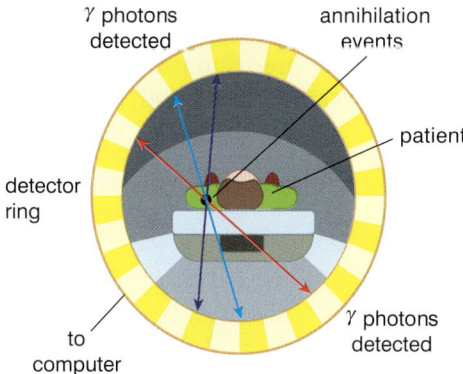

Figure 2 *A PET scanner*

Each positron travels less than a millimetre in the patient before it meets an electron and they annihilate each other to produce two γ photons travelling in opposite directions.

- A ring of detectors connected to a computer registers a positron emission when two of its detectors each detect a γ photon within a very short time interval (called the 'coincidence time' window) of about 5–10 ns.
- Such an emission must be from a point very close to, or along a straight line between, the two detectors as the γ photons travel at the same speed in opposite directions and are created at the same time. The position along the line, called a 'line of response' or LOR, can be found by mapping this and other LORs (from other nearby $^{18}_{9}F$ decays) that cross each other at the same point.
- Although each detector is narrow, there is a small degree of uncertainty about the LOR between two detectors and therefore of the two γ photons. In addition, there is some uncertainty about the point of origin in relation to the position of the positron-emitting nucleus.

• Using a computer to process the arrival times of the γ photons (see Figure 2) enables an image of the tracer in the tissue to be created. PET scans can take 20 minutes or more to gather sufficient data to map out the location of the positron-emitting isotope in the body.

Summary test 22.3

1 The fluorine isotope $^{18}_{9}F$ emits positrons and becomes stable. The isotope is prepared by irradiating the oxygen isotope $^{18}_{8}O$ with protons.

    a Write down an equation to represent the production of a $^{18}_{9}F$ nucleus from an $^{18}_{8}O$ nucleus.

    b Write down an equation to represent the decay of a nucleus of $^{18}_{9}F$.

2 A sample of the fluorine isotope $^{18}_{9}F$ in a solution has an activity of 220 MBq. This isotope has a half-life of 110 minutes.

    a Calculate the mass of the isotope in the sample when its activity is 220 MBq.

    b Estimate the activity of this sample after 24 hours.

3 When a positron-emitting nucleus in a substance decays, two gamma-ray photons are emitted from the substance.

    a Explain why gamma photons are emitted each time a positron-emitting nucleus in a substance decays.

    b Explain why two gamma photons are always emitted in opposite directions from each decay.

4 **a** State what is meant by a *radioactive tracer*.

    b Explain why a PET scanner needs to use positron-emitting isotopes to trace radioactive substances in the body rather than other gamma-emitting isotopes.

$$\boxed{\text{📖 Launch additional digital resources for the chapter}}$$

1 An ultrasound probe generates ultrasound waves at a frequency of 2.5 MHz.

    a Calculate the wavelength of the ultrasound waves from this probe:
        i in air,   **ii** in soft tissue.

    b **i** Explain why ultrasound waves of much lower frequency are unsuitable for medical imaging.
        ii With the aid of a diagram, describe the construction of an ultrasound probe and how it produces ultrasound waves.

    c **i** Define the *specific acoustic impedance* of a substance.
        ii Use the data in Table 2, p.328, to calculate the reflection coefficient of the boundary between: (1) air and skin and (2) water and skin. Assume that skin is soft tissue.

    d Use the results of your calculation in **cii** to explain why a gel is applied between an ultrasound probe and the skin when the probe is used.

2 **a** An X-ray tube operates at a potential difference of 40 kV.
        i Calculate the minimum wavelength of the X-rays from this X-ray tube.
        ii Sketch a graph to show how the intensity of the X-rays from this tube varies with the energy of the photons.

    b **i** What is meant by a *contrast medium* as used in X-ray imaging?
        ii Give an example of a contrast medium, stating an organ it is used for.

    c Describe a CT scanner and explain how it works.

3 **a** **i** State the type of radiation used to form an image in a PET scanner.
        ii State the type of substance used to produce the above radiation and describe how the above radiation is produced.

    b In a PET scanner, the detectors are arranged radially and equally spaced along a ring. The detectors are linked to a computer which registers any two detectors that are each triggered by radiation in the same short time interval. Figure 1 represents the ring of detectors when two detectors X and Y have been triggered in the same time interval.

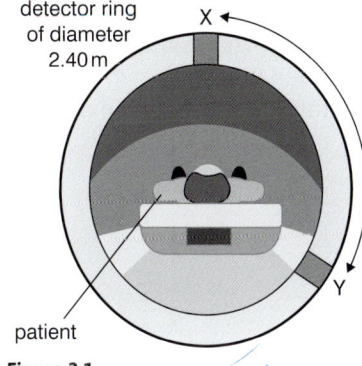

detector ring of diameter 2.40 m

X

Y

patient

Figure 3.1

        i Describe how the radiation from the substance in a patient undergoing a PET scan could have triggered the two detectors within the same short time.

        ii The ring has a diameter of 2.40 m. The arc XY subtends an angle of 135° at the centre of the ring. Use this information to determine the distance along a straight line from X to Y.

        iii Explain why the source of the radiation must be at a location on or close to the straight line from X to Y.

        iv Discuss what further information could be used to locate the source between X and Y.

23.1 Astronomical distances

Stars and galaxies

On a clear night, the stars we see are pinpoints of light that may or may not differ in brightness. Each one is a massive glowing ball of gas, mostly hydrogen, with a core at its centre where nuclear fusion takes place. The radiation from the core heats the star's outer layers and causes them to emit electromagnetic radiation into space in all directions. The Sun is our nearest star. Light from the Sun takes about 500 seconds to reach the Earth. In comparison, light takes 4.3 years to reach us from Proxima Centauri, the next nearest star. All the stars except the Sun appear as pinpoints of light in the night sky because they are so far away even though each one is a glowing ball of gas emitting radiation into space in all directions.

Figure 1 *The Sun*

Galaxies are vast collections of stars held together by their own gravity. A galaxy is like an 'island' of stars. The Sun is one of about 100 000 million stars in the Milky Way galaxy, which is a spiral galaxy about 100 000 light years across. The Milky Way galaxy is the second largest of a cluster of 'local' nearby galaxies, the largest being the Andromeda galaxy twice as large as the Milky Way galaxy and about 2.5 million light years away. There are thought to be over 100 000 million galaxies in the Universe, the furthest being over 13 000 million light years away. In this chapter, you will discover how astronomers used the laws of physics to measure these enormous distances.

Luminosity and intensity

The luminosity L of a star is the total power of the radiation emitted by a star.

Luminosity is measured in watts. The luminosity of the Sun is about 4×10^{26} W. In other words, the Sun emits radiant energy at a rate of about 4×10^{26} J/s. The luminosity of Proxima Centauri is about 10 000 times greater than that of the Sun. The Earth is over a quarter of a million times further from Proxima Centauri than it is from the Sun. So even though Proxima Centauri emits 10 000 times more radiant energy than the Sun, because it is much further away than the Sun, the intensity of the radiation from it at the Earth is much less than the intensity of sunlight at the Earth.

The intensity of the radiation from a star in the night sky depends on the **radiant flux intensity, F,** of its radiation at the Earth. This is the radiation energy per second per unit surface area received from the star at normal incidence on a surface.

Consider a star of luminosity L at the centre of an imaginary sphere of radius d. The radiation from the star spreads out equally in all directions. The radiant flux intensity F of the radiation at the sphere is given by the equation

$$F = \frac{\text{the energy per second emitted by the star}}{\text{the area of the sphere}}$$

Note the unit of F is the watt per square metre (W m^{-2}) or joules per second per square metre (J s^{-1} m^{-2}). Since the energy per second emitted by the star is its luminosity L and the surface area of the sphere is $4\pi d^2$:

$$F = \frac{L}{4\pi d^2}$$

The equation shows that:

the radiant flux intensity of the radiation from a star at a distance d from the centre of a star is proportional to:

- **its luminosity L**

- $\frac{1}{d^2}$, **the inverse of the square of the distance d from the sphere to the centre of the star.**

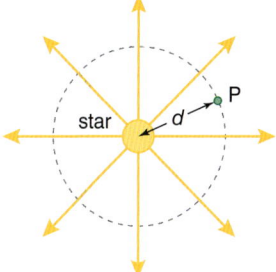

Figure 2 Radiation from a star

Worked example

The radiant flux intensity of solar radiation at the Earth's surface is about 1400 W m^{-2}. The mean distance from the Earth to the Sun is 1 AU which is 1.50×10^{11} m. Estimate the luminosity of the Sun.

Solution

Rearranging the equation $F = \frac{L}{4\pi d^2}$ gives $L = F \times 4\pi d^2$

Hence: $L = 1400 \text{ W m}^{-2} \times 4\pi \times (1.5 \times 10^{11} \text{ m})^2 = 4.0 \times 10^{26}$ W

To calculate the distance d to a nearby star, consider Figure 3 which shows the 'six month' angular shift of a nearby star's position relative to stars much further away.

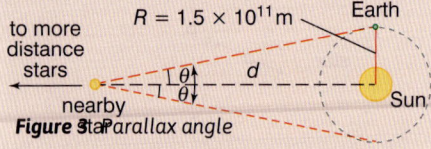

Figure 3 Parallax angle

The parallax angle θ is defined as the angle subtended by the star to the line between the Sun and the Earth, as shown in Figure 3. This angle is half the angular shift of the star's line of sight over six months. From the triangle consisting of the three lines between the Sun, the star and the Earth as shown in Figure 1.3,

$\tan\theta = R / d$.

and $d = R / \tan\theta$ where R is the mean distance from the centre of the Sun to the centre of the Earth.

Since the parallax angle θ is always less than 10°, using the small angle approximation:

$\tan\theta \approx \theta$ gives

$\theta = \dfrac{R}{d}$,

where θ is in radians and $d = \dfrac{R}{\theta}$

Standard candles

The stars in the night sky might seem to be in fixed positions relative to each other, but accurate measurements of their positions show that some of them change their relative position slightly against the background of the other stars. This 'parallax' effect is due to the Earth's orbital motion round the Sun, and is because the direction of line of sight from the Earth to a nearby star changes slightly as the Earth orbits the Sun, as shown in Figure 3. The effect can be measured and used to calculate the distance to a nearby star. By measuring the radiant flux intensity of a star, its luminosity can then be calculated if its distance from the Earth is known.

An object of known luminosity is known as a **standard candle**, because it can be used to determine the distance to more distant stars and to galaxies. For example, a supernova event is a star that explodes and becomes much brighter before fading away. Type 1a supernovae are known to reach a peak luminosity of about 10^{36} W, about ten thousand million times more than the Sun's luminosity. By measuring the maximum radiant flux intensity of such a supernova and knowing its peak luminosity, its distance from the Earth can be calculate.

Summary test 23.1

$c = 3.00 \times 10^8\,\mathrm{m\,s^{-1}}$

1 The luminosity of Proxima Centauri is about $10\,000$ times greater than the luminosity of the Sun. Proxima Centauri is about $250\,000$ times further from the Sun than the Earth is. Use this information to estimate the radiant flux intensity at the Earth of the radiation from Proxima Centauri. Assume that the radiant flux intensity of solar radiation at the Earth is $1400\,\mathrm{W\,m^{-2}}$.

2 Sirius is the brightest star in the sky. Its luminosity is about 25 times that of the Sun. It is 8.1×10^{16} m from the Earth.

   The luminosity of the Sun = 4.0×10^{26} W

   a Calculate how long light takes to travel from Sirius to the Earth.

   b Estimate the radiant flux intensity at the Earth of the radiation from Sirius.

3 A supernova is a star that explodes and outshines its host galaxy before it fades away. In 2014, a supernova event occurred in a distant galaxy. Its luminosity peaked at about 4.4×10^{36} W. The maximum radiant flux intensity at the Earth from this supernova was $0.21\,\mathrm{pW\,m^{-2}}$. Estimate the distance from the Earth to this supernova.

4 The Earth's orbit around the Sun is slightly elliptical. The mean distance from the Earth to the Sun is 1.5×10^{11} m. The percentage change of the radiant flux intensity of solar radiation incident on the Earth is about 1.7% when it moves from its least distance to its greatest distance from the Sun. Estimate the change of the distance from the Earth to the Sun when the Earth moves from its greatest distance to its least distance from the Sun.

Stellar radii

Starlight

Stars differ in colour as well as brightness. Viewed through a telescope, stars that appear to be white to the unaided eye appear in their true colours. This is because a telescope collects much more light than the unaided eye, thus activating the colour-sensitive cells in the retina. **Charge coupled devices** (CCDs) with filters and colour-sensitive photographic film show that stars vary in colour from red to orange and yellow to white, to bluish-white.

Like any glowing object, a star emits thermal radiation which includes visible light and infrared radiation. For example, if the current through a torch bulb is increased from zero to its working value, the filament glows dull red then red then orange-yellow, as the current increases and the filament becomes hotter. The spectrum of the light emitted shows that there is a continuous spread of colours which change their relative intensities as the temperature is increased. This example shows that:

- **The thermal radiation from a hot object at constant temperature consists of a continuous range of wavelengths.**
- **The distribution of intensity with wavelength changes as the temperature of the hot object is increased.**

Figure 1 shows how the intensity distribution of such radiation varies with wavelength for different temperatures, as indicated.

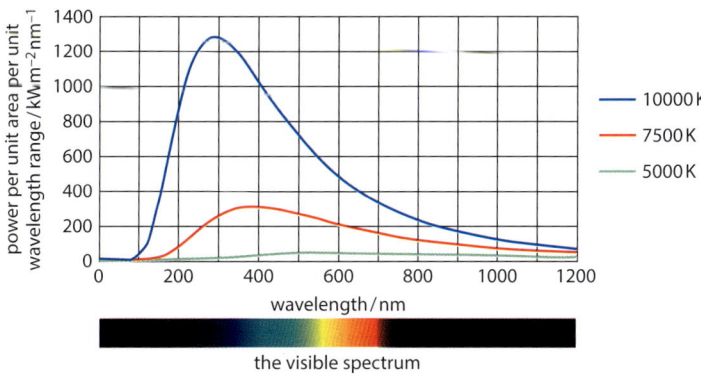

Figure 1 *Black body radiation curves*

Note

The curves are referred to as **black body** radiation curves, a black body being defined as a body that is a perfect absorber of radiation (i.e. absorbs 100% of radiation incident on it at all wavelengths) and therefore emits a continuous spectrum of wavelengths. A small hole in the door of a furnace is an example of a black body, because any thermal radiation that enters the hole from outside would be completely absorbed by the inside walls. We can assume that a star is a black body because any radiation incident on it would be absorbed, and none would be reflected or transmitted by the star. In addition, the spectrum of thermal radiation from a star is a continuous spectrum with an intensity distribution that matches the shape of a black body radiation curve.

Learning outcomes

On these pages you will learn to:

- recall and use Wien's law of radiation to estimate the peak surface temperature of a star
- use the Stefan–Boltzmann's law of radiation to relate the luminosity of a star to its radius and its surface temperature
- use the above laws of radiation to estimate the radius of a star

The laws of thermal radiation

The radiation curves in Figure 1 are obtained by measuring the intensity of the thermal radiation from a black body at different constant temperatures. Each curve has a peak which is higher and at shorter wavelength than the curves at lower temperatures. The following two laws of thermal radiation were obtained by analysing the radiation curves:

1 Wien's displacement law: The wavelength at peak intensity, λ_P, is inversely proportional to the thermodynamic temperature T of the object, in accordance with the following equation known as **Wien's law**:

$$\lambda_{\text{max}}\, T = 0.0029\,\text{m K}$$

Therefore, if λ_{max} for a given star is measured from its spectrum, the above equation can be used to calculate the temperature T of the light-emitting outer layer, the **photosphere**, of the star. The photosphere is sometimes referred to as the **surface** of a star.

Notice that the unit symbol 'm K' stands for 'metre kelvin' not milli kelvin!

Worked example

The peak intensity of thermal radiation from the Sun is at a wavelength of 500 nm. Calculate the surface temperature of the Sun.

Solution

Rearranging $\lambda_{\text{max}}\, T = 0.0029\,\text{m K}$ gives $T = \dfrac{0.0029\,\text{m K}}{500 \times 10^{-9}\,\text{m}} = 5800\,\text{K}$

2 Stefan–Boltzmann's law: The luminosity L (i.e. total energy per second emitted) of a black body at temperature T is proportional to its surface area A and to T^4, in accordance with the following equation known as **Stefan–Boltzmann's law**:

$$L = \sigma A T^4$$

where σ is the Stefan constant which has a value of $5.67 \times 10^{-8}\,\text{W}\,\text{m}^{-2}\,\text{K}^{-4}$. In effect, L is the power output of the star.

Therefore, if the temperature T of a star and its luminosity L are known, the surface area A and hence the radius R of the star can be calculated.

Worked example

$\sigma = 5.67 \times 10^{-8}\,\text{W}\,\text{m}^{-2}\,\text{K}^{-4}$

A star has a luminosity of $6.0 \times 10^{28}\,\text{W}$ and a surface temperature of 3400 K.

a Show that its surface area is $7.8 \times 10^{21}\,\text{m}^2$.

b Calculate: **i** its radius, **ii** the ratio of its radius to the radius of the Sun. (Radius of Sun $= 7.0 \times 10^8\,\text{m}$)

Solution

a Rearranging $L = \sigma A T^4$ gives: $A = \dfrac{L}{\sigma T^4}$

Hence $A = \dfrac{6.0 \times 10^{28}}{5.67 \times 10^{-8} \times (3400)^4} = 7.9 \times 10^{21}\,\text{m}^2$

b i For a sphere of radius R, its surface area $A = 4\pi R^2$

Rearranging this equation gives: $R^2 = \dfrac{A}{4\pi} = \dfrac{7.9 \times 10^{21}}{4\pi} = 6.3 \times 10^{20}\,\text{m}^2$

Hence $R = 2.5 \times 10^{10}\,\text{m}$

ii Ratio of radius to Sun's radius $= \dfrac{2.5 \times 10^{10}\,\text{m}}{7.0 \times 10^8\,\text{m}} = 36$

Notes

1 For two stars X and Y that have the same luminosity:

- Luminosity of X, $L_X = \sigma A_X T_X^4$, where A_X = surface area of X and T_X = surface temperature of X.

- Luminosity of Y, $L_Y = \sigma A_Y T_Y^4$, where A_Y = surface area of Y and T_Y = surface temperature of Y.

Therefore $\dfrac{L_X}{L_Y} = \dfrac{\sigma A_X T_X^4}{\sigma A_Y T_Y^4} = \dfrac{A_X T_X^4}{A_Y T_Y^4}$

2 For equal luminosity: $\sigma A_X T_X^4 = \sigma A_Y T_Y^4$ hence $\dfrac{A_X}{A_Y} = \dfrac{T_X^4}{T_Y^4}$

Therefore, if X and Y have equal surface temperatures, they must have the same radius. If their surface temperatures are unequal, the cooler star must have a bigger radius than the hotter star.

Summary test 23.2

1 The spectrum of light from a star has its peak intensity at a wavelength of 620 nm. Calculate the temperature of the star's light-emitting surface.

2 A star has a power output of 6.0×10^{28} W and a surface temperature of 3400 K.

  a Show that it surface area is 7.9×10^{21} m^2.

  b Calculate:
    i its radius,
    ii the ratio of its radius to the radius of the Sun.
    (Radius of Sun = 7.0×10^8 m)

3 A star has a surface temperature which is twice that of the Sun and a diameter that is four times as large as the Sun's diameter. Show that it emits approximately 250 times as much energy per second as the Sun.

4 Two stars, X and Y, have the same surface temperature of 5400 K. Star X emits 100 times more power that star Y.

  a State and explain which star, X or Y, has the bigger diameter.

  b X has a diameter of 2.0×10^9 m. Calculate the luminosity of star X.

  c Calculate the diameter of star Y.

Learning outcomes

On these pages you will learn to:

- explain why the emission spectra from distant galaxies show an increase in wavelength compared with the spectra from laboratory and other sources
- understand what is meant by a red shift and calculate its value using an appropriate equation from suitable data
- explain why red shift leads to the idea that the Universe is expanding
- recall and use Hubble's law and explain how Hubble's law leads to the Big Bang theory

Red shift

The wavelengths of the electromagnetic radiation from a star or galaxy moving towards Earth are shorter than they would be if the star or galaxy was stationary. If the star or galaxy had been moving away from the Earth, the wavelengths would be longer than if the star or galaxy was stationary. As explained in Topic 8.9, this effect applies to all waves and is known as the **Doppler effect**. For electromagnetic waves, the effect can be seen by comparing the line emission spectrum of the light from a laboratory sources with that from a galaxy moving away from us, as shown in Figure 1.

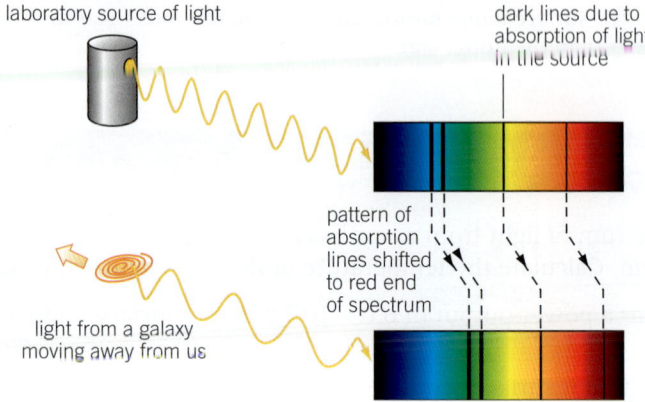

laboratory source of light

dark lines due to absorption of light in the source

pattern of absorption lines shifted to red end of spectrum

light from a galaxy moving away from us

Figure 1 *Red shift*

The lines of the line emission spectrum of the light from the galaxy are shifted towards the red part of the visible spectrum because the light is lengthened in wavelength. This shift in the wavelength due to the light source moving away (i.e. receding) from the Earth is called a **red shift**. As explained in Topic 8.9, for light of wavelength, λ, from a star or galaxy receding at speed v, the change of the wavelength $\Delta\lambda = (v/c)\,\lambda$. Hence:

$$\frac{\Delta\lambda}{\lambda} = \frac{v}{c} \qquad \text{where } c = \text{the speed of light in a vacuum}$$

Notes

1 The equation can also be written in terms of frequency f as: $\dfrac{\Delta f}{f} = \dfrac{v}{c}$

2 The formulae above can only be applied to electromagnetic waves emitted by stars travelling at speeds much less than the speed of light c.

3 By measuring the shift in wavelength of a line of a star or galaxy's line spectrum, the speed v of the star or galaxy relative to Earth can be found using the rearranged equation: $v = c\left(\dfrac{\Delta\lambda}{\lambda}\right)$

Worked example

$c = 3.0 \times 10^8\,\text{m s}^{-1}$

A spectral line of a star is found to be displaced from its laboratory value of 434 nm by + 0.087 nm. State whether the star is moving towards or away from the Earth and calculate its speed relative to the Earth.

Solution

The star is moving away from the Earth because the wavelength of its light is increased.

Rearranging $\Delta\lambda = \dfrac{v\lambda}{c}$ gives $v = \dfrac{c\Delta\lambda}{\lambda} = \dfrac{3.0 \times 10^8 \times 0.087 \times 10^{-9}}{434 \times 10^{-9}} = 6.0 \times 10^4\,\text{m s}^{-1}$

Hubble's law

The Universe consists of galaxies, each containing millions of stars, separated by vast empty spaces. Edwin Hubble and other astronomers studied the light spectra of many galaxies and were able to identify prominent spectral lines as in the spectra of individual stars but 'red-shifted' to longer wavelengths. Hubble studied galaxies which were close enough to be resolved into individual stars; for each galaxy, he measured:

- **its red shift and then calculated its** *speed of recession* **(i.e. the speed at which it was moving away),**
- **its distance from Earth by observing individual stars of known luminosity as standard candles.**

Hubble's results showed that galaxies are receding from us, each moving at speed v, that is directly proportional to the distance, d, to it. This discovery, referred to as **Hubble's law**, is usually expressed as the following equation:

$$v = Hd$$

where H, the constant of proportionality, is referred to as the **Hubble constant**. H represents the speed of recession per unit distance from Earth. The value of H is $2.2 \times 10^{-18}\,\text{s}^{-1}$ or $21\,\text{km s}^{-1}$ per million light years.

Figure 2 shows the pattern of typical measurements of the speed of recession v and distance d plotted on a graph is a straight line through the origin. Note that the gradient of the graph is equal to the Hubble constant H.

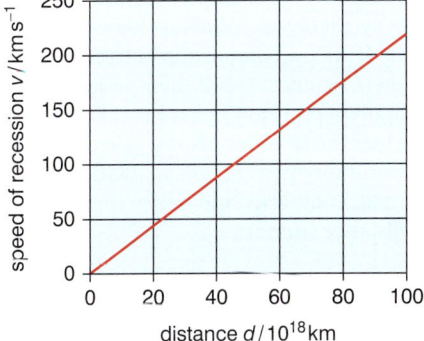

Figure 2 *Speed of recession v. distance*

Worked example

$c = 3.0 \times 10^8\,\text{m s}^{-1}$, $H = 2.2 \times 10^{-18}\,\text{s}^{-1}$

The wavelength of a spectral line in the spectrum of light from a distant galaxy was measured at 398.6 nm. The same line measured in the laboratory has a wavelength of 393.3 nm. Calculate:

a the speed of recession of the galaxy,

b the distance to the galaxy.

Solution

a $\Delta\lambda = 398.6 - 393.3 = 5.3\,\text{nm}$

Rearranging, $\Delta\lambda = \dfrac{v\lambda}{c}$ gives $v = \dfrac{c\Delta\lambda}{\lambda} = \dfrac{3.0 \times 10^8\,\text{m s}^{-1} \times 5.3 \times 10^{-9}\,\text{m}}{393.3 \times 10^{-9}\,\text{m}}$

$$= 4.0 \times 10^6\,\text{m s}^{-1}$$

b Rearranging, $v = Hd$ gives $d = \dfrac{v}{H} = \dfrac{4.0 \times 10^6\,\text{m s}^{-1}}{2.2 \times 10^{-18}\,\text{s}^{-1}} = 1.8 \times 10^{24}\,\text{m}$

The Big Bang theory

Hubble's law tells us that the distant galaxies are receding from us. The conclusion we must draw from this discovery is that the galaxies are all moving away from each other and **the Universe must therefore be expanding**. At first, some astronomers thought this expansion is because the Universe was created in a massive 'primordial' explosion and has been expanding ever since. This theory was referred to by its opponents as the **Big Bang theory**.

With no evidence for a primordial explosion other than an explanation of Hubble's law, many astronomers supported an alternative theory that the Universe is unchanging, the same now as it ever was. This theory, known as the **Steady State theory**, explained the expansion of the Universe by supposing matter entering the Universe at 'white holes' pushes the galaxies apart as it

Note

Since Hubble's discovery, astronomers have used other methods to measure the distances to distant stars and galaxies. One such method used 'Cepheid variable' stars which are stars that vary periodically in luminosity with a period of variability that increases with luminosity. So, by timing the variability of such stars at known distances in the Milky Way galaxy, astronomers were able to deduce a relationship between their luminosity and their period. By applying this to Cepheid variables observed in other nearby galaxies, astronomers were able to deduce their luminosity and hence their distance from us.

enters. The Big Bang theory was accepted in 1965 when radio astronomers discovered microwave radiation from all directions in space. Steady state theory could not explain the existence of this **cosmic microwave background radiation (CMBR)**, but the Big Bang theory could.

Evidence for the Big Bang theory

The spectrum of microwave radiation from space matched the theoretical spectrum of thermal radiation from an object at a temperature of 2.7 K. Because the radiation was detected from all directions in space with little variation in intensity, it was realised it must be universal or 'cosmic' in origin.

This background cosmic microwave radiation is explained readily by the Big Bang theory as radiation that was created in the Big Bang and has been travelling through the Universe ever since the Universe became transparent. As the Universe expanded after the Big Bang, its mean temperature has decreased and is now about 2.7 K. The expansion of the Universe has gradually increased the background cosmic microwave radiation to its present range of wavelengths.

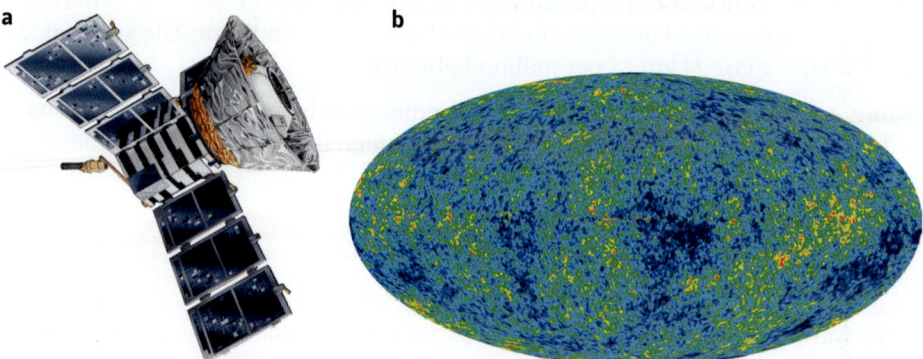

Figure 3 *Cosmic background microwave radiation*
a *The Cosmic Background Explorer (COBE)* **b** *Microwave map of the Universe*

Estimate of the age of the Universe

The speed of light in free space, c, is $300\,000\,\text{km s}^{-1}$. No material object can travel as fast as light. Therefore, even though the speed, v, of a galaxy increases with its distance d, no galaxy can travel as fast as light. Therefore the furthermost galaxy could not be any further than a distance equal to cT where T is the age of the Universe.

Applying this distance to Hubble's law, $v = Hd$, therefore gives $c = H(cT)$.

Hence, $T = 1/H = (2.2 \times 10^{-18}\,\text{s}^{-1})^{-1} = 4.5 \times 10^{17}\,\text{s}$. Prove for yourself that this gives about 14 000 million years (i.e. 14 billion years) as an estimate of the age of the Universe.

Note

Measurements of cosmic microwave background radiation using different satellites and other methods give values of H between 2.1 and $2.3 \times 10^{-18}\,\text{s}^{-1}$.

Dark energy

Astronomers in 1998 studying type Ia supernova were astounded when they discovered very distant supernovas much further away than expected. To reach such distances, they must have been accelerating. The astronomers concluded that the expansion of the Universe is accelerating and has been for about the past 5000 million years. Before this discovery, most astronomers expected that the Universe is decelerating, as very distant objects would be slowed down by the force of gravity from other galaxies. Many more observations since then have confirmed the Universe is accelerating. Scientists think that no known force could cause an acceleration of the expansion of the Universe and that a hitherto unknown type of force must be releasing hidden energy referred to as dark energy.

Evidence for accelerated expansion of the Universe

Evidence for accelerated expansion of the Universe is based on differing distance measurements to type Ia supernova by two different methods:

1 The red shift method: Measurement of the red shift of each of these distant type Ia supernova and use of Hubble's law gives the distance to each one.

2 The luminosity method: Type Ia supernova at peak intensity are known to be 10^9 times more luminous that the Sun. The distance to such a supernova can be calculated from its luminosity L and its radiant flux intensity F. See Topic 23.1.

The two methods give results that are different and indicate that the distant type Ia supernova are dimmer and therefore further away than their red shift indicates.

The nature of dark energy

The nature of dark energy is unclear. It is thought to be a form of background energy present throughout space and time. It is more prominent than gravity at very large distances, because gravity becomes weaker and weaker with increased distance whereas the force associated with dark energy is thought to be constant. Current theories suggest it makes up about 70% of the total energy of the Universe. The search for further evidence of dark energy will continue with observations using larger telescopes and more sensitive microwave detectors on satellites.

Summary test 23.3

$c = 3.0 \times 10^8 \, \text{m s}^{-1}$, $H = 2.2 \times 10^{-18} \, \text{s}^{-1}$

1 a State Hubble's law.

b What conclusion was drawn by astronomers when Hubble's law was discovered?

2 The wavelength of a spectral line in the spectrum of light from a distant galaxy was measured at 597.2 nm. The same line measured in the laboratory has a wavelength of 589.6 nm. Calculate:

a the speed of recession of the galaxy,

b the distance to the galaxy.

3 A supernova in a distant galaxy was observed to have a wavelength of 525 nm for a certain spectral line in the spectrum of its light. The same line measured in the laboratory has a wavelength of 486 nm.

a i Calculate the speed of recession of this galaxy.
 ii Estimate the distance to this galaxy.

b The peak radiant flux intensity of the supernova at the Earth was measured at $2.7 \times 10^{-15} \, \text{W m}^{-2}$. Estimate the peak luminosity of the supernova.

4 a State what is meant by *cosmic microwave background radiation* (CMBR).

b What is thought to be the origin of CMBR?

Launch additional digital resources for the chapter

$\sigma = 5.67 \times 10^{-8}\,\mathrm{W\,m^{-2}\,K^{-4}}$

1 a i State Wien's displacement law.
  ii The black body radiation curve for a certain star has its peak intensity at a wavelength of 740 nm. Calculate the surface temperature of this star.

b The star is 6.0×10^{17} m from the Earth. The radiation flux intensity of the radiation from this star at the Earth is $3.5 \times 10^{-8}\,\mathrm{W\,m^2}$. Calculate the luminosity of the star.

c i Show that the radiant energy per second per unit surface area emitted from the star is equal to σT^4, where T is the surface temperature of the star.
  ii Hence calculate the surface area of the star and determine its diameter.

2 a Define the *radiant flux intensity* at the Earth of the electromagnetic radiation from a star.

b The table below gives the luminosity of three stars A, B and C in terms of the Sun's luminosity, their distances from the Earth and the wavelength of their peak intensity.

Star	Luminosity of star / Luminosity of the Sun	Distance from Earth / 10^{17} m	Peak intensity wavelength / nm
A	10.6	1.58	380
B	130	7.39	230
C	0.0035	0.57	940

  i State and explain which star is hottest and which star is coolest, giving a reason for your answer.
  ii List the stars in order of increasing radiant flux intensity at the Earth.

c The luminosity of the Sun is 4.0×10^{26} W. For star A, calculate:
  i its luminosity,
  ii its surface temperature,
  iii the radius of its surface.

d i The surface temperature of the Sun is 5800 K. Show that the radius R of a star is proportional to $L\,\lambda_{max}^4$, where L is its luminosity and λ_{max} is its peak intensity wavelength.
  ii List the three stars in order of increasing radius, giving reasons for their order.

3 A supernova is an event that occurs when a large star explodes. In 1987, a supernova was observed in a galaxy 1.5×10^{23} m from the Earth.

a When the supernova was at its brightest, the radiant flux intensity at the Earth from it was $1.4 \times 10^{-13}\,\mathrm{W\,m^{-2}}$. Calculate the luminosity of the supernova when it was at its brightest.

b The star that exploded was identified from past records as a blue star.
  i Assuming the peak intensity wavelength of the radiation from the blue star was about 300 nm, estimate the surface temperature of this star.
  ii The luminosity of the blue star was about 4×10^{29} W. Show that its diameter was about ten times larger than that of the Sun. The diameter of the Sun is 1.4×10^9 m.

4 a i Explain what is meant by a *red shift* in the light from a star or a galaxy.
  ii Describe how a red shift is used to determine the speed of recession of a star or galaxy moving away from the Earth.

b The table below shows some speeds of recession of six distant galaxies at different distances from the Earth.

Galaxy	Distance, d / 10^{19} km	Speed, v / km s^{-1}
A	4.6	63
B	10.5	328
C	20.9	437
D	25.3	719
E	32.0	609
F	38.5	1030

  i Plot a graph of the speed of recession v on the y-axis against the distance d on the x-axis.
  ii Use your graph to determine a value of the Hubble constant, H.
  iii Estimate the uncertainty in the value of H you obtained from your graph.

c Hubble discovered that the speed of recession of galaxies at known distances is proportional to their distance from Earth. The measurements he used gave a value of H which was seven times greater than the currently accepted value. Later distance measurements using different methods since then have led to the present accepted value. Discuss the continuing *validity* of Hubble's law, even though the value of the Hubble constant is very different to the value estimated by Hubble.

Key concepts: Thermal and nuclear physics, medical imaging and astrophysics and cosmology

In studying Chapter 18, you will have gained a sound understanding of internal energy and absolute temperature and a good grasp of the practical skills involved in measuring specific heat capacities and specific latent heats. In addition, by considering conservation of energy, you will have learned how to analyse problems involving specific heat and latent heat and to carry out the necessary calculations in order to solve such problems.

After studying Chapter 19 on gases, you should appreciate the experimental basis of the ideal gas equation and how the kinetic theory of gases provides a full explanation of the properties of an ideal gas. The random motion of the molecules of an ideal gas is a key concept here and you should recognise that the observation of Brownian motion provides direct evidence of this random motion. In studying gases, you will have developed your mathematical skills in carrying out calculations using the ideal gas equation and the kinetic theory of gases equation. In addition, you should be able to combine these two equations to show that the mean kinetic energy of the molecules of an ideal gas is proportional to the absolute temperature of the gas.

In Chapter 20, you should be aware that the results of investigations in photoelectricity could not be explained using the wave theory of light and consequently led to the establishment by Einstein of the photon theory of light. The success of the photon theory should be evident to you from your studies of energy levels and spectra as well as from the successful explanation of photoelectricity and from the discovery that photons scattered by electrons lose momentum to the electrons and therefore possess momentum even though they have no rest mass. In contrast to the properties described above, where experiments and investigations preceded important theoretical work, the wave nature of matter was predicted by de Broglie and only accepted when experimental evidence of electron diffraction was discovered. Throughout Chapter 20, in addition to furthering your knowledge and understanding of the structure and properties of matter, you should have further developed your problem-solving skills as well as your analytical and mathematical skills.

Your studies of nuclear physics in Chapter 21 build on previous studies on radioactivity in Chapter 10. After studying Chapter 21, you should be aware that radioactive decay is a random process that gives a full explanation of the exponential decrease in the activity of a radioactive isotope. In addition, you should be able to appreciate how the concept of binding energy and mass defect enables us to explain why energy is released in nuclear changes including fission and fusion. In studying radioactive decay, you will have further developed your mathematical skills in solving problems and carrying out calculations involving exponential decreases and nuclear reactions.

Chapter 22, on medical imaging, provides awareness of the important role of physics in medicine. The three topics, ultrasonic imaging, X-ray imaging and positron emission tomography (PET scanning) are not the only imaging techniques used, but they do involve techniques applicable to different situations.

Chapter 23 Astrophysics and cosmology builds on previous studies on waves in order to develop understanding and awareness of how astronomers can use data from their observations to work out the size of a star and the distances to stars and galaxies.

The following table provides an overview of the topics in Chapters 18 to 23, showing where key concepts have been developed.

Chapters 18–23: Key concepts

	Topic		Key concepts	
18.1–18.2	internal energy and temperature	first law of thermodynamics	energy and matter	conservation of energy
		absolute temperature		internal energy
18.3–18.4	specific heat capacity and specific latent heat	measuring specific heat capacity and specific latent heat	testing	practical skills
		heat capacity, latent heat	energy and matter	conservation of energy
		calculations involving specific heat capacity and specific latent heat	use of maths	problem-solving
19.1–19.2	ideal gases	testing the gas laws	testing	practical skills
		Brownian motion molar mass	energy and matter	random motion molecular theory
		calculations involving the ideal gas equation	use of maths	problem-solving
19.3–19.4	kinetic theory of gases	derivation of the kinetic theory of gases equation	models	kinetic theory of gases model
		random motion, root mean square speed, internal energy of an ideal gas	energy and matter	random motion
		calculations involving the kinetic theory of gases equation and the ideal gas equation	use of maths	problem-solving
20.1–20.3	electrons and photons	electromagnetic waves, photoelectricity, threshold frequency, work function energy levels and spectra	energy, matter and waves	nature of light
		the photon model used to explain photoelectricity, Einstein's photoelectric equation	models	photon model
		calculations on photoelectricity and energy levels	use of maths	problem-solving
20.4	wave particle duality	wave properties of electrons, electron diffraction	energy, matter and waves	dual nature of matter
		de Broglie's hypothesis and equation, photon momentum		
		calculations involving de Broglie's equation photon momentum	use of maths	problem-solving

continued on p.353

Chapters 18–23: Key concepts (*continued*)

	Topic		Key concepts	
21.1–21.3	radioactive decay	activity, half-life	energy, matter and waves	
		random nature of radioactive decay	models	randomness
		theory of radioactive decay		exponential decrease
		calculations using the radioactive decay equation	use of maths	problem-solving
21.4–21.6	energy from the nucleus	binding energy, mass defect	energy and matter	conservation of energy
		binding energy curve		
		fission and fusion	models	chain reaction
		induced fission		
		calculations of binding energy per nucleon and of energy released in nuclear reactions	use of maths	problem-solving
22.1–22.3	medical imaging	ultrasonic imaging	energy, matter and waves	use of waves to probe matter
		X-ray imaging in medicine		
		PET scanning		ionising and non-ionising radiation
		CT scans	models	3D images
		calculations involving acoustic impedance, reflection coefficients, absorption using $I = I_0e^{-\mu x}$ and minimum photon wavelength from an X-ray tube	use of maths	problem-solving
23.1–23.3	astrophysics and cosmology	luminosity and radiation, intensity,	energy and waves	Intensity of radiation from a point source
		astronomical distances	energy and waves	
		stellar radii	energy and waves	radiation laws applied to observations
		Hubble's law and expansion of the Universe	energy and waves	proportionality (of red shift and distance) and prediction of expansion of Universe
		the Big Bang	models	explanation of expansion of the Universe
		Calculations using the inverse square law, Wien's displacement law and the Stefan–Boltzmann's law	use of maths	problem-solving

353

Question

The first part of the question is about the nuclear model of the atom and why scientists predicted that the nucleus is composed of neutrons as well as protons before there was any direct evidence of the existence of neutrons. It also gives some insight into how such direct evidence was obtained using alpha radiation. The question also tests the application of the principles of conservation of energy and of momentum to the emission of α-particles from an α-emitting isotope to explain why the α-particles are emitted with well-defined energy.

This well-defined energy is the reason why α-particles from an α-emitting isotope have a well-defined range in air, and the final part of the question also tests the relationship between the range R in air of α-particles and their initial kinetic energy E using experimental data from several α-emitting isotopes. A graph of $\log R$ against $\log E$ is plotted on the assumption that the relationship is of the form $R = kE^n$ where n is an unknown constant. A straight-line graph would confirm that the relationship is of this form, enabling the power n to be found as it is equal to the gradient of the graph. Knowledge of this value of n provided physicists with a further challenge beyond the question to develop a theoretical model that 'confirms' the value of n found using the graphical method. The question thus shows how we can use data from experimental investigations to find relationships that can then be investigated by devising theoretical models that explain the relationships. In addition, knowing the value of n enables E to be calculated for other α-emitting isotopes by measuring the range R and using the now-known relationship between E and R.

Before answering the question, students should be given the opportunity to do this or a similar investigation.

1 The neutron–proton model of the nucleus was first put forward by Rutherford to explain the general composition of the nucleus. The existence of the neutron was not proved experimentally until some years later.

   a Rutherford knew from his alpha-scattering experiments that the charge of a nucleus is equal to $+Ze$, where Z is the atomic number of the nucleus and e is the magnitude of the charge of the electron. He also knew that the mass of a nucleus is approximately equal to $A\,m_u$, where A is the mass number of the nucleus and m_u is the mass of a hydrogen nucleus.

     i By reference to the isotopes ${}^1_1\text{H}$ and ${}^4_2\text{He}$, give **two** reasons why Rutherford's neutron–proton model was considered more than an untested hypothesis when it was first put forward. *(2 marks)*

     ii Explain why Rutherford concluded that the hydrogen ${}^1_1\text{H}$ nucleus is a single proton. *(1 mark)*

b Evidence of the existence of neutrons was found when it was discovered that α-particles knocked uncharged particles out of beryllium foil. Investigations into the nature of the uncharged radiation showed it consisted of particles each uncharged and of about the same mass as a proton.

   i Complete the nuclear reaction below to show the result of an inelastic collision between an α-particle and a beryllium $^{9}_{4}$Be nucleus.

$$^{4}_{2}\alpha + ^{9}_{-}\text{Be} \longrightarrow ^{}_{6}\text{C} + ^{1}_{0}\text{n} \qquad \textit{(1 mark)}$$

   ii Explain why the above change only takes place if the α-particle has sufficient kinetic energy. *(3 marks)*

c The α-particles from any α-emitting isotope have the same initial kinetic energy and a well-defined range in air at atmospheric pressure.

   i The isotope $^{210}_{84}$Po emits α-particles with initial kinetic energy of 5.3 MeV. Show that the nucleus recoils with an initial kinetic energy of 0.1 MeV and hence calculate the total energy released by the nucleus when it emits an α-particle. *(4 marks)*

   ii The table below shows the range R in air and the initial kinetic energy E of α-particles from several α-emitting isotopes.

R/mm	39	48	53	57	66	78
E/MeV	5.3	6.0	6.5	6.8	7.4	8.3

Plot a suitable graph to find out if the relationship between R and E is of the form $R = kE^n$, where k and n are constants, and determine a value for n. Explain your choice of graph. *(9 marks)*

24.1 Data handling

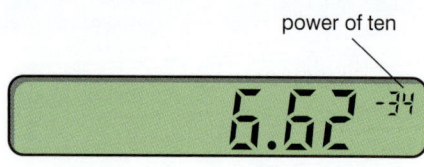

number displayed = 6.62 × 10⁻³⁴

Figure 1 *Displaying powers of ten*

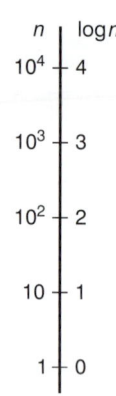

Figure 2 *A logarithmic scale*

> **Note**
>
> The label for each function on your calculator may differ from the label shown here. If you are unsure which button to press, check with your teacher.

Using a calculator

(EXP) (or (EE) on some calculators)

This is the calculator button you press to key in a **power of ten**. To key in a number in standard form (e.g. 3.0×10^8), the steps are as follows:

Step 1 Key in the number between 1 and 10 (e.g. ③ ⊙ ⓪).

Step 2 Press the calculator button (EXP) (or (EE) on some calculators).

Step 3 Key in the power of ten (e.g. ⑧).

The display should now read 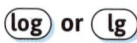 which should be read as 3.0×10^8 (not 3.0^8 which means 3.0 multiplied by itself 8 times). If the power of ten is a negative number (e.g. 10^{-8} not 10^8), press the calculator button (⁺/₋) after Step 3 to change the sign of the power of ten.

(Inv)

This is the button you press if you want the calculator to give the value of the **inverse of a function**. For example, if you want to find out which angle has a sine of 0.5, you key in ⓪⊙⑤, then (Inv), then (sin), to obtain the answer of 30°. Some calculators have a 'second function' button that you press instead of the 'inv' button.

(log) or (lg)

This is the button you press to find out what a number is as a power of ten. For example, press (log), then key in ① ⓪ ⓪ and the display will show ②, because $100 = 10^2$. **Logarithmic scales** have equal intervals for each power of ten.

(deg/rad)

Angles can be expressed in degrees or **radians** where 2π radians = 360 degrees. A scientific calculator has a button you can press to use either degrees or radians. Make sure you know how to switch your calculator from one of these two modes to the other. Many marks have been lost in examinations as a result of forgetting to use the correct mode. For example:

- sin 30° = 0.50 whereas sin 30 rad = −0.99,
- inv sin 10° = 0.17 whereas inv sin 10 rad = −0.54

Also, take care when you calculate the sine, cosine or tangent of the product of a number and an angle. For example, if you forget about the brackets in a calculation of sin (2 × 30°), the answer 1.046 is obtained instead of the correct answer of 0.866. The reason for the error is that, unless you insert the brackets, the calculator is programmed to work out sin 2° then multiply the answer by 30.

(yˣ)

This is used to raise any number to any power, for example, if you want to work out the value of 2^8, key in ② onto the display, then press the (yˣ) button, then ⑧ and press ⊜. The display should then show ② ⑤ ⑥ as the decimal value of 2^8.

The (yˣ) button can be used to find roots. For example, given the equation $T^4 = 5200$, you can find T by keying in ⑤ ② ⓪ ⓪ onto the display, then pressing the y^x button, followed by (1 ÷ 4) which will give the answer 8.49.

Worked example

Calculate the cube root of 2.9×10^6.

Solution

Step 1 Key in 2.9×10^6 as explained earlier.

Step 2 Press the (y^x) button.

Step 3 Key in $( )\ (1)\ (\div)\ (3)\ ( )$.

Step 4 Press $(=)$.

The display should show $\boxed{1.426 \quad 02}$, so the answer is 142.6.

Significant figures

In general, always write a numerical answer to the same number of significant figures as the data used for the calculation. Because a calculator display shows a large number of digits, you should always round up, or round down, the answer displayed on a calculator to achieve the appropriate number of significant figures. For example, a calculator will show $\boxed{9.0630777870 \times 10^{-1}}$ for the sine of 65°. Because the data used (65°) is given to two significant figures, the sine of 65° should then be rounded off to 9.1×10^{-1}, and written as 0.91 because 9.06 to two significant figures is rounded up to 9.1.

In calculations where there is more than one stage, carry forward the results of intermediate calculations without rounding off the data (or rounding it off to at least one more significant figure than the data has) and round off the final calculation only. Rounding off the data at the intermediate stage may result in an incorrect 'rounded-off' final answer.

For example, consider the calculation of the density ρ of the material in a solid cylinder of radius r, length l and mass m using the equations volume $V = \pi r^2 l$ and density $\rho = m \div V$, where $r = 2.30 \times 10^{-3}\,m$, $l = 0.152\,m$ and $m = 0.223\,kg$.

- Rounding off to three significant figures at the final stage only gives $\rho = 8830\,kg\,m^{-3}$. The same answer is obtained if the value of the volume is rounded off to four significant figures before the density is calculated and rounded off to three significant figures.
- Rounding off the calculation of V to three significant figures before the density is calculated and rounded off to three significant figures gives $8810\,kg\,m^{-3}$.

Summary test 24.1

Write your answers to each of the following questions in standard form where appropriate, and to the same number of significant figures as the data.

1 Copy and complete the following conversions:

   a i 500 mm = ... m
     ii 3.2 m = ... cm
     iii 9560 cm = ... m

   b i 0.45 kg = ... g
     ii 1997 g = ... kg
     iii 54 000 kg = ...g

   c i 20 cm² = ...m²
     ii 55 mm² = ...m²
     iii 0.05 cm² = ...m²

2 **a** Write the following values in standard form:

    i 150 million km in metres
    ii 365 days in seconds
    iii 630 nm in metres
    iv 25.7 mg in kilograms
    v 150 m in millimetres
    vi 1.245 μm in metres

   b Write the following values with a prefix instead of in standard form:

    i $3.5 \times 10^4\,m = ...$ km
    ii $6.5 \times 10^{-7}\,m = ...$ nm
    iii $3.4 \times 10^6\,g = ...$kg
    iv $8.7 \times 10^8\,W = ...$ MW $= ...$ GW

3 **a** Use the equation 'average speed $= \dfrac{\text{distance}}{\text{time}}$', to calculate the average speed in $m\,s^{-1}$ of:

    i a vehicle that travels a distance of 9000 m in 450 s,
    ii a vehicle that travels a distance of 144 km in 2 h,
    iii a particle that travels a distance of 0.30 nm in a time of 2.0×10^{-18} s,
    iv the Earth on its orbit of radius 1.5×10^{11} m, given the time taken per orbit is 365.25 days.

   b Use the equation resistance $= \dfrac{\text{potential difference}}{\text{current}}$, to calculate the resistance of a component for the following values of current I and p.d. V:

    i $V = 15\,V$, $I = 2.5\,mA$
    ii $V = 80\,mV$, $I = 16\,mA$
    iii $V = 5.2\,kV$, $I = 3.0\,mA$
    iv $V = 250\,V$, $I = 0.51\,\mu A$
    v $V = 160\,mV$, $I = 53\,mA$

4 **a** Calculate each of the following:

    i 6.7^3 **ii** $(5.3 \times 10^4)^2$
    iii $(2.1 \times 10^{-6})^4$ **iv** $(0.035)^2$
    v $(4.2 \times 10^8)^{1/2}$ **vi** $(3.8 \times 10^{-5})^{1/4}$

   b Calculate each of the following:

    i $\dfrac{2.4^2}{3.5 \times 10^3}$ **ii** $\dfrac{3.6 \times 10^{-3}}{6.2 \times 10^2}$

    iii $\dfrac{8.1 \times 10^4 + 6.5 \times 10^3}{5.3 \times 10^4}$

    iv $7.2 \times 10^{-3} + \dfrac{6.2 \times 10^4}{2.6 \times 10^6}$

More about the rules of trigonometry

Angles and arcs

- Angles are measured in **degrees or radians**. The scale for conversion is $360° = 2\pi$ radians. The symbol for the radian is rad so $1\,\text{rad} = \dfrac{360}{2\pi} = 57.3°$ (to three significant figures).
- The circumference of a circle of radius $r = 2\pi r$. So the circumference can be written as the angle in radians (2π) round the circle $\times\, r$.
- For a segment of a circle, the length of the arc of the segment is in proportion to the angle θ which the arc makes at the centre of the circle. This is shown in Figure 1. Because the arc length is $2\pi r$ (i.e. the circumference) for an angle of $360°$ ($= 2\pi$ radians):

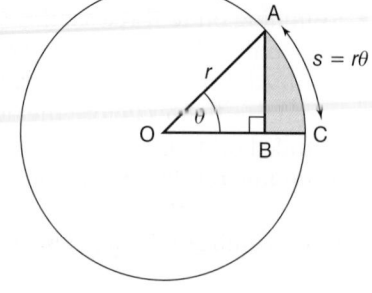

Figure 1 Arcs and segments

$$\frac{\text{arc length, } s}{2\pi r} = \frac{\theta \text{ (in degrees)}}{360} = \frac{\theta \text{ (in radians)}}{2\pi}$$

Equating $\dfrac{\text{arc length, } s}{2\pi r}$ to $\dfrac{\theta \text{ in radians}}{2\pi}$ gives

$$\textbf{arc length } s = r\theta$$

where θ is the angle subtended in radians.
Note that for $s = r$, $\theta = 1\,\text{rad}$ ($= 360/2\pi = 57.3°$)

The small angle approximation

For angle θ less than about $10°$,

$$\sin \theta \approx \tan \theta \approx \theta \text{ in radians,} \quad \text{and} \quad \cos \theta \approx 1$$

To explain these approximations, consider Figure 1 again. If angle θ is sufficiently small, then the segment OAC will be almost the same as triangle OAB, as shown in Figure 2.

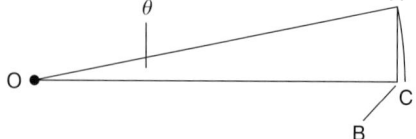

Figure 2 The small angle approximation

- AB $\approx$ arc length s so $\sin\theta = \dfrac{\text{AB}}{\text{OA}} \approx \dfrac{s}{r} = \theta$ in radians.
  $\therefore \sin\theta \approx \theta$ in radians.

- OB $\approx$ radius r, so $\tan\theta = \dfrac{\text{AB}}{\text{OB}} \approx \dfrac{s}{r} = \theta$ in radians.
  $\therefore \tan\theta \approx \theta$ in radians.
  and $\cos\theta = \dfrac{\text{OB}}{\text{OA}} \approx \dfrac{r}{r} = 1$
  $\therefore \cos\theta \approx 1$

Use a calculator to prove for yourself that for $\sin 10° = 0.1736$, $\tan 10° = 0.1763$ and $10° = 0.1745$ rad. Also, $\cos 10° = 0.9848$. So the small angle approximation is almost 99% accurate up to $10°$. Figure 3 shows how $\sin\theta$ and $\cos\theta$ change as θ increases.

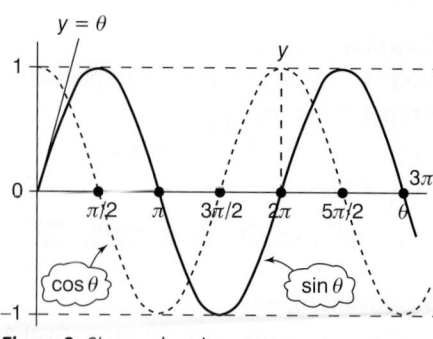

Figure 3 Sine and cosine curves

More triangle rules

In addition to the 'right-angle triangle' rules described on page x, the following rules apply to any triangle:

- Area of a triangle $= \frac{1}{2} \times$ its height (h) $\times$ its base (b). See Figure 4a.
- For a triangle with sides of lengths a, b and c and angles A, B and C opposite sides a, b and c, as shown in Figure 4b.

$$\frac{a}{\sin A} = \frac{b}{\sin B} = \frac{c}{\sin C} \qquad \text{(the sine rule)}$$

$$a^2 = b^2 + c^2 - 2bc\cos A \qquad \text{(the cosine rule)}$$

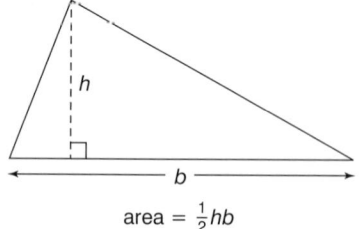

area $= \frac{1}{2}hb$

a The area of a triangle

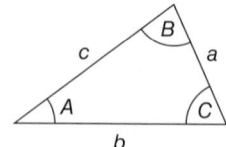

b The sine rule and the cosine rule

Figure 4

Pythagoras' theorem and trigonometry

Pythagoras' theorem states that for any right-angled triangle:

The square of the hypotenuse = the sum of the squares of the other two sides

Applying Pythagoras' theorem to the right-angled triangle in Figure 5 gives:

$$h^2 = o^2 + a^2$$

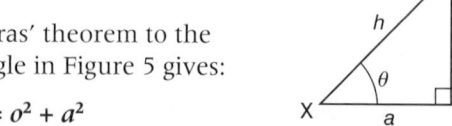

Figure 5 A right-angled triangle

Since $o = h\sin\theta$ and $a = h\cos\theta$, the above equation may be written:
$$h^2 = h^2\sin^2\theta + h^2\cos^2\theta$$

Cancelling h^2 therefore gives the following useful link between $\sin\theta$ and $\cos\theta$:

$$1 = \sin^2\theta + \cos^2\theta$$

Vector rules

Resolving a vector

As explained in Topic 1.1, any vector can be resolved into two perpendicular components in the same plane as the vector, as shown by Figure 6. The force vector F is resolved into a horizontal component $F \cos \theta$ and a vertical component $F \sin \theta$, where θ is the angle between the line of action of the force and the horizontal line.

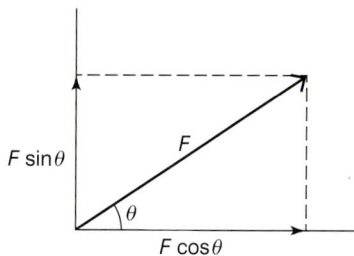

Figure 6 *Resolving a vector*

Note The component at angle θ to the direction of the vector is always $\cos \theta \times$ the magnitude of the vector.

Adding two vectors at angle θ to each other

Consider an object, O, acted on by forces F_1 and F_2 at angle θ to each other, as shown in Figure 7a. The magnitude and direction of the resultant force F_R can be found by resolving one of the forces into components that are parallel and perpendicular to the other force, as explained in Topic 1.1.

The method described in Topic 1.1 gives

$$F_R{}^2 = [(F_1 \cos \theta + F_2)^2 + (F_1 \sin \theta)^2].$$

Squaring the first term on the right-hand side gives

$$F_R{}^2 = F_1{}^2 \cos^2 \theta + 2F_1F_2 \cos \theta + F_2{}^2 + F_1{}^2 \sin^2 \theta$$

Because $\cos^2 \theta + \sin^2 \theta = 1$, then $F_1{}^2 \cos^2 \theta + F_1{}^2 \sin^2 \theta = F_1{}^2$

Hence

$$\boxed{F_R{}^2 = F_1{}^2 + 2F_1F_2 \cos \theta + F_2{}^2}$$

This equation above can also be obtained by applying the cosine rule to the force triangle shown in Figure 7b with $a = F_R$, $b = F_1$ and $c = F_2$ and $A = 180 - \theta$.

This gives $F_R{}^2 = F_1{}^2 + F_2{}^2 - 2F_1F_2 \cos (180 - \theta)$. The equation above then follows since $\cos (180 - \theta) = -\cos \theta$.

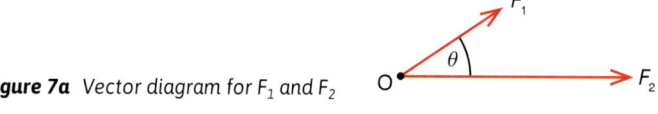

Figure 7a *Vector diagram for F_1 and F_2*

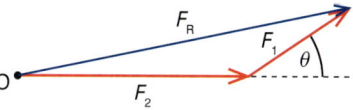

Figure 7b *Using the components to find the resultant force F_R*

Summary test 24.2

1 a Calculate the circumference of a circle of radius 0.25 m.

b Calculate the length of the arc of a circle of radius 0.25 m for the following angles, between the arc and the centre of the circle:
  i 360°, **ii** 240°, **iii** $\pi/3$ rad, **iv** 0.6π rad.

c An aircraft travels a distance of 30 km due north from an airport P to an airport Q. It then travels due east for a distance of 18 km, to an airport R.

Calculate:
  i the distance from P to R, **ii** the angle QPR.

2 a Measure the diameter of a coin to the nearest mm. Calculate the angle subtended at your eye, in degrees, by the coin held at a distance of 50 cm from your eye.

b i Estimate the angular width of the Moon, in degrees, at your eye by holding a millimetre scale at 50 cm from your eye and measuring the distance on the scale covered by the lunar disc.
  ii The diameter of the Moon is 3500 km. The average distance to the Moon from the Earth is 380 000 km. Calculate the angular width of the Moon as seen from the Earth and compare the calculated value with your estimate in **b i**.

3 a Use the small angle approximation to calculate $\sin \theta$ for: **i** $\theta = 2.0°$, **ii** $\theta = 8.0°$.

b A triangle has sides of lengths a, b, and c and internal angles A, B and C opposite sides a, b and c respectively. Sketch each of the following triangles and use the cosine rule to calculate a for:
  i $b = 80$ mm, $c = 60$ mm and $A = 60°$,
  ii $b = 75$ mm, $c = 40$ mm and $A = 70°$,
  iii $b = 120$ mm, $c = 45$ mm and $A = 120°$.

c For each triangle in **b**, use the sine rule as appropriate to calculate B and C.

4 a Calculate the perpendicular components A and B of each of the vectors in Figure 8.

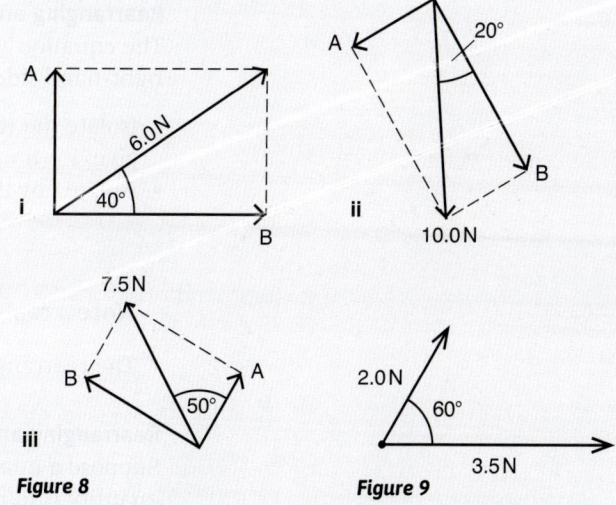

Figure 8

Figure 9

b Calculate the magnitude and direction of the resultant of the two vectors shown in Figure 9.

Signs and symbols

Signs you need to recognise

- **Inequality signs** are often used in physics. You need to be able to recognise the meaning of the signs in Table 1. For example, the inequality $I \geqslant 3$ A means that the current is greater than or equal to 3 A. This is the same as saying that the current is not less than 3 A.
- The **approximation sign** is used where an estimate or an order-of-magnitude calculation is made, rather than an accurate calculation. For an order-of-magnitude calculation, the final value is written with one significant figure only, or even rounded up or down to the nearest power of ten. Order-of-magnitude calculations are useful as a quick check after using a calculator. For example, if you are asked to calculate the density of a 1.0 kg metal cylinder of height 0.100 m and diameter 0.071 m, you ought to obtain a value of 2530 kg m^{-3} using a calculator. Now let's check the value:

$$\text{Volume} = \pi \, (\text{radius})^2 \times \text{height} \approx 3 \times (0.04)^2 \times 0.1 \approx 48 \times 10^{-5} \, \text{m}^3$$

$$\text{Density} = \frac{\text{mass}}{\text{volume}} \approx \frac{1.0}{50 \times 10^{-5}} \approx 2000 \, \text{kg m}^{-3}$$

This confirms our 'accurate' calculation.

- **Proportionality** is represented by the $\propto$ sign. A simple example of its use in physics is for Hooke's law: the tension in a spring is proportional to its extension.

$$\text{Tension, } T \propto \text{extension, } e$$

By introducing a constant of proportionality k, the link above can be made into an equation:

$$T = ke$$

where k is defined as the spring constant. See Topic 5.4. With any proportionality relationship, if one of the variables is increased by a given factor (e.g. $\times 3$), the other variable is increased by the same factor. So in the above example, if T is trebled, then extension e is also trebled. A graph of tension T on the y-axis against extension e on the x-axis would give a straight line through the origin.

More about equations and formulas

Rearranging an equation with several terms

The equation $v = u + at$ is an example of an equation with two terms on the right-hand side. These terms are u and at. To make t the subject of the equation:

- Isolate the term containing t on one side, by subtracting u from both sides, to give: $v - u = at$
- Isolate t by dividing both sides of the equation $v - u = at$ by a to give:

$$\frac{v - u}{a} = \frac{at}{a} = t$$

Note a cancels out in the expression $\dfrac{at}{a}$.

- The rearranged equation may now be written: $t = \dfrac{v - u}{a}$

Rearranging an equation containing powers

Suppose a quantity is raised to a power in a term in an equation and that quantity is to be made the subject of the equation. For example, consider the equation $V = \frac{4}{3}\pi r^3$, where r is to be made the subject of the equation:

Table 1 Signs

Sign	Meaning
>	greater than
<	less than
$\geqslant$	greater than or equal to
$\leqslant$	less than or equal to
$\gg$	much greater than
$\ll$	much less than
$\approx$	approximately equals
$\langle x \rangle$	mean value
$\langle x^2 \rangle$	mean square value
$\propto$	is proportional to
Δ	change of
$\sqrt{}$	square root

- Isolate r^3 from the other factors in the equation by dividing both sides by 4π then multiplying both sides by 3 to give: $\dfrac{3V}{4\pi} = r^3$

- Take the cube root of both sides to give: $\left(\dfrac{3V}{4\pi}\right)^{1/3} = r$

- Rewrite the equation with r on the left-hand side if necessary.

More about powers

- Powers add for identical quantities when two terms are multiplied together. For example, if $y = ax^m$ and $z = bx^n$, then: $yz = ax^m bx^n = abx^{m+n}$
- An equation of the form $y = \dfrac{k}{z^x}$ may be written in the form $y = kz^{-n}$
- The nth root of an expression is written as the power $\dfrac{1}{n}$.

  For example, the square root of x is $x^{1/2}$. Therefore, rearranging $y = x^n$ to make x the subject gives $x = y^{1/n}$

Quadratic equations

Quadratic equations occur in physics where a formula contains the square of a variable. The equation $s = ut + \frac{1}{2}at^2$ for displacement at constant acceleration is an example. See Topic 1.4.

Any quadratic equation can be written in the form $ax^2 + bx + c = 0$, where a, b and c are constants. The general solution of this equation is

$$x = \frac{-b \pm \sqrt{b^2 - 4ac}}{2a}$$

Note that every quadratic equation has two solutions, one given by the + sign before the square root sign in the above expression, and the other given by the − sign.

For example, consider the solution of the equation $2x^2 + 5x - 3 = 0$.

As $a = 2$, $b = 5$ and $c = -3$, the solution is

$$x = \frac{-5 \pm \sqrt{5^2 - (4 \times 2 \times -3)}}{2 \times 2} = \frac{-5 \pm \sqrt{49}}{4}$$

$$= \frac{-5 \pm 7}{4} = +0.50 \text{ or } -3$$

A graph of $y = 2x^2 + 5x - 3$ is shown in Figure 1. Note that the two solutions above are the values of the x-intercepts, which is where $y = 0$.

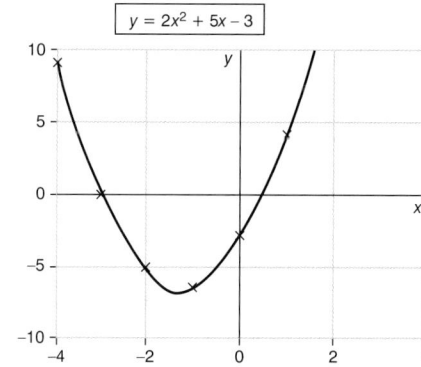

Figure 1 $y = 2x^2 + 5x - 3$

How to check a physics equation

Derived units written in terms of their **base units** can be used to check equations. The physical quantities on each side of an equation must match in terms of base units. If they don't match, the equation cannot be correct. For example, consider:

- the equation $v = \sqrt{(2gR)}$, which is used to calculate the escape speed v of an object from the surface of a planet of radius R and surface gravitational field strength g:

  left-hand side base units = m s^{-1}
  right-hand side base units = $\sqrt{\text{m s}^{-2} \times \text{m}}$ = m s^{-1}

The equation has the same combination of base units on each side, so it is correct; we say it is **homogeneous** in terms of the base units.

- the equation $W = QV$ is used to calculate the work done W to move a charge Q through a potential difference V:

  left-hand side base units (see Table 2) = kg m^2 s^{-2}
  right-hand side base units = (A s) × (kg m^2 s^{-3} A^{-1}) = kg m^2 s^{-2}

The equation has the same combination of base units on each side, so it is homogeneous.

Note

For simple equations such as this, homogeneity can sometimes be checked faster by recalling basic relationships between physical quantities. In the second example, one volt is one joule per coulomb, so the unit of QV is the joule per coulomb × the coulomb, which is the joule.

Table 2 *Links between units*

Quantity	Symbol	Unit	Unit symbol	Other forms of unit	Base unit	Chapter reference
acceleration	a	metre per second^2	m s^{-2}		m s^{-2}	1
activity of radioactive source	A	becquerel	Bq	s^{-1}	s^{-1}	23
angle	θ	radian or degree	rad or °			1,11
angular displacement	θ	radian or degree	rad or °			11
angular frequency	ω	radian per second	rad s^{-1}			11
angular speed or velocity	ω	radian per second	rad s^{-1}			11
capacitance	C	farad	F	C V^{-1}	kg^{-1} m^{-2} s^4 A^2	15
charge	q, Q	coulomb	C		A s	6
decay constant	λ	second $^{-1}$	s^{-1}		s^{-1}	23
density	ρ	kilogram per cubic metre	kg m^{-3}		kg m^{-3}	5
electric field strength	E	newton per coulomb	N C^{-1}	V m^{-1}	kg m s^{-3} A^{-1}	6
electric potential	V	volt	V	J C^{-1}	kg m^2 s^{-3} A^{-1}	14
energy, work	E, U	joule	J		kg m^2 s^{-2}	2
force	F	newton	N		kg m s^{-2}	2
frequency	f	hertz	Hz	s^{-1}	s^{-1}	8
gravitational field strength	g	newton per kilogram	N kg^{-1}		m s^{-2}	1,2
gravitational potential	ϕ	joule per kilogram	J kg^{-1}		m^2 s^{-2}	13
magnetic flux density	B	tesla	T	N A^{-1} m^{-1} Wb m^{-2}	kg s^{-2} A^{-1}	18
magnetic flux	Φ	weber	Wb	T m^2, V s	kg m^2 s^{-2} A^{-1}	18
momentum	p	kilogram metre per second	kg m s^{-1}	N s	kg m s^{-1}	4
permeability of free space	μ_0	henry per metre	H m^{-1}		kg m s^{-2} A^{-2}	18
permittivity of free space	ε_0	farad per metre	F m^{-1}		kg^{-1} m^{-3} s^4 A^2	14
phase difference	ϕ	radian or degree	rad or °			8
potential difference, e.m.f.	V, E	volt	V	J C^{-1}	kg m^2 s^{-3} A^{-1}	6, 7
power	P	watt	W	J s^{-1}	kg m^2 s^{-3}	2
pressure	p	pascal	Pa	N m^{-2}	kg m^{-1} s^{-2}	5
resistance	R	ohm	Ω	V A^{-1}	kg m^2 s^{-3} A^{-2}	6
resistivity	ρ	ohm metre	Ω m		kg m^3 s^{-3} A^{-2}	6
spring constant	k	newton per metre	N m^{-1}		kg s^{-2}	5
stress, Young modulus	σ, E	pascal	Pa	N m^{-2}	kg m^{-1} s^{-2}	5
velocity, speed	v	metre per second	m s^{-1}		m s^{-1}	1
wavelength	λ	metre	m		m	8

Summary test 24.3

1 a Complete each of the following statements:

    i If $x > 5$, then $\dfrac{1}{x} < \ldots$

    ii If $4 < x < 10$, then $\ldots < \dfrac{1}{x} < \ldots$

    iii If $x^2 > 100$ then $\dfrac{1}{x} \ldots$

b Make t the subject of each of the following equations:

    i $v = u + at$ **ii** $s = \frac{1}{2}at^2$

    iii $y = k\,(t - t_0)$ **iv** $F = \dfrac{mv}{t}$

c Solve each of the following equations:

    i $2z + 6 = 10$ **ii** $2\,(z + 6) = 10$

    iii $\dfrac{2}{z - 4} = 8$ **iv** $\dfrac{4}{z^2} = 36$

2 a Make x the subject of each of the following equations:

    i $y = 2x^{1/2}$ **ii** $2y = x^{-1/2}$

    iii $yx^{1/3} = 1$ **iv** $y = \dfrac{k}{x^2}$

b Solve each of the following equations:

    i $x^{-1/2} = 2$ **ii** $3x^2 = 24$

    iii $\dfrac{8}{x^2} = 32$ **iv** $2(x^{1/2} + 4) = 12$

3 a Use the data given with each equation below to calculate:

    i The surface area of cross-section A, of a wire of radius $r = 0.34\,\text{mm}$ and length $L = 0.840\,\text{m}$, using the equation $A = \pi r^2 L$.

    ii The radius r of a sphere, of volume $V = 1.00 \times 10^{-6}\,\text{m}^3$, using the formula $V = \frac{4}{3}\pi r^3$.

    iii The time period T of a simple pendulum of length $L = 1.50\,\text{m}$, using the formula $T = 2\pi(L/g)^{0.5}$, where $g = 9.81\,\text{m s}^{-2}$.

    iv The speed v of an object, of mass $m = 0.20\,\text{kg}$ and kinetic energy $E_k = 28\,\text{J}$, using the formula:
$$E_k = \tfrac{1}{2}mv^2$$

b Express each of the following combinations of physical quantities in SI base units:

    i pressure × volume,

    ii momentum^2 ÷ energy,

    iii resistance × capacitance.

4 a Solve each of the following quadratic equations.

    i $2x^2 + 5x - 3 = 0$

    ii $x^2 - 7x + 8 = 0$

    iii $3x^2 + 2x - 5 = 0$

b Use the data and the given equation to write down a quadratic equation and so determine the unknown quantity in each case:

    i $s = ut + \frac{1}{2}at^2$, where $s = 20\,\text{m}$, $u = 4\,\text{m s}^{-1}$ and $a = 6\,\text{m s}^{-2}$; find t.

    ii $P = \dfrac{V^2 R}{(R + r)^2}$ where $P = 16\,\text{W}$, $V = 12\,\text{V}$, $r = 2.0\,\Omega$; find R.

Straight-line graphs

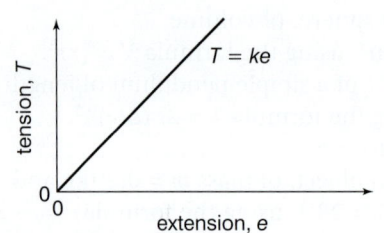

Figure 1

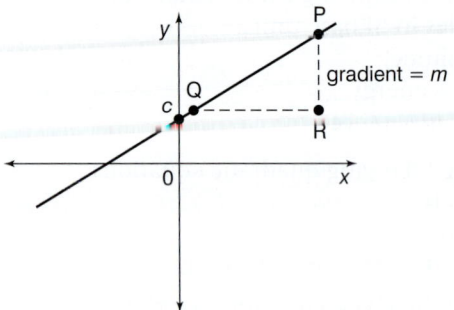

Figure 2 $y = mx + c$

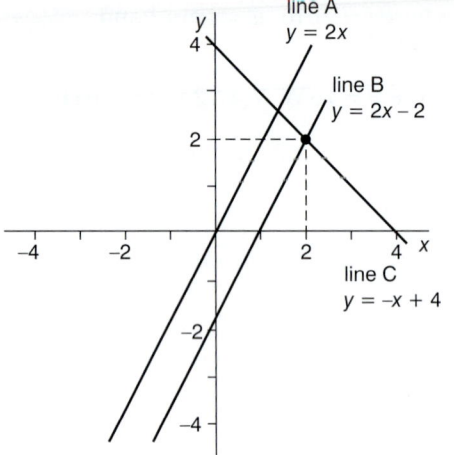

Figure 3 *Straight-line graphs*

The general equation for a straight-line graph

Links between two physical quantities can be established most easily by plotting a graph. One of the physical quantities is represented by the vertical scale (the 'ordinate', often called the y-axis) and the other quantity by the horizontal scale (the 'abscissa', often called the x-axis). The **coordinates** of a point on a graph are the x- and y-values, usually written (x, y) of the point.

The simplest link between two physical variables is where the plotted points define a straight line. For example, Figure 1 shows the link between the tension in a spring and the extension of the spring; the gradient of the line is constant and the line passes through the origin. Any situation where the y-variable is proportional to the x-variable gives a straight line through the origin. For Figure 1, the gradient of the line is the spring contant k. The relationship between the tension T and the extension e may therefore be written as $T = ke$.

The general equation for a straight-line graph is usually written in the form:

$$y = mx + c$$

where m = the gradient of the line, and c = the y-intercept.

- The **gradient** m can be measured by marking two points, P and Q, as far apart as possible on the line. The triangle PQR as shown in Figure 2 is then used to find the gradient. If (x_P, y_P) and (x_Q, y_Q) represent the x- and y-coordinates of points P and Q respectively, then

$$\text{gradient } m = \frac{y_P - y_Q}{x_P - x_Q}$$

- The **y-intercept**, c, is the point at $x = 0$, where the line crosses the y-axis. To find the y-intercept of a line on a graph that does not show $x = 0$, measure the gradient as above then use the coordinates of any point on the line with the equation $y = mx + c$ to calculate c. For example, rearranging $y = mx + c$ gives $c = y - mx$. Therefore, using the coordinates of point Q in Figure 2, the y-intercept $c = y_Q - mx_Q$.

Examples of straight-line graphs
In Figure 3

- **Line A**: $c = 0$, so the line passes through the origin; its equation is $y = 2x$.
- **Line B**: $m > 0$, so the line has a positive gradient; its equation is $y = 2x - 2$.
- **Line C**: $m < 0$, so the line has a negative gradient; its equation is $y = -x + 4$.

Straight-line graphs and physics equations

You need to be able to work out gradients and intercepts for equations you meet in physics that generate straight-line graphs. Some further examples are described below:

Motion at constant acceleration
The velocity v of an object moving at constant acceleration a at time t is given by the equation $v = u + at$, where u is its speed at time t. Figure 4 shows the corresponding graph of velocity v on the y-axis against time t on the x-axis.

Rearranging the equation as $v = at + u$ and comparing this with $y = mx + c$ shows that:

- the gradient, m = acceleration a, and
- the y-intercept, c = the initial velocity u.

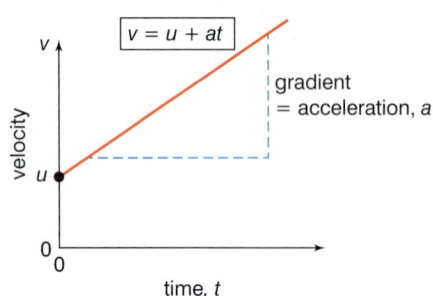

Figure 4 *Motion at constant acceleration*

P.d. and current for a battery

The p.d. V across the terminals of a battery, of e.m.f. E and internal resistance r, varies with current in accordance with the equation $V = E - Ir$ (see Topic 7.3). Figure 5 shows the corresponding graph of p.d. V on the y-axis against current I on the x-axis.

Rearranging the equation as $V = -rI + E$ and comparing this with $y = mx + c$ shows that:

- the gradient, $m = -r$, and
- the y-intercept, $c = E$ so the intercept on the y-axis gives the e.m.f. E of the battery.

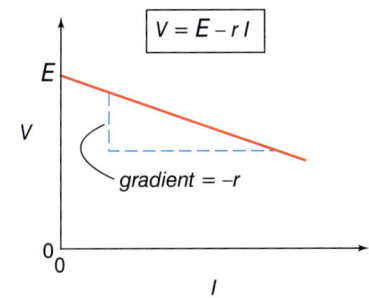

Figure 5 *P.d. v. current for a battery*

Photoelectric emission

The maximum kinetic energy $E_{K(max)}$ of a photoelectron, emitted from a metal surface of work function ϕ, varies with frequency f of the incident radiation in accordance with the equation $E_{K(max)} = hf - \phi$ (see Topic 20.1). Figure 6 shows the corresponding graph of $E_{K(max)}$ on the y-axis against f on the x-axis.

Comparing the equation $E_{K(max)} = hf - \phi$ with $y = mx + c$ shows that:

- the gradient, $m = h$, and
- the y-intercept, $c = -\phi$.

> ### Note
>
> The x-intercept is where $y = 0$ on the line. Let the coordinates of the x-intercept be $(x_0, 0)$. Therefore $mx_0 + c = 0$ so $x_0 = \dfrac{-c}{m}$. In Figure 6, the x-intercept is therefore $\dfrac{\phi}{h}$. This is the threshold frequency $f_0, = \dfrac{\phi}{h}$.

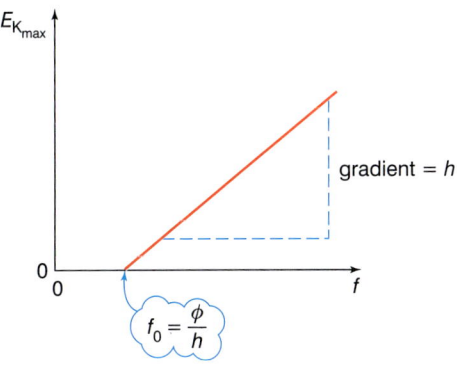

Figure 6 *Photoelectric emission*

Simultaneous equations

Two equations with two variable quantities, x and y, in each can be solved to find the values of x and y. Such a pair of equations is referred to as **simultaneous equations** because they have the same solution. They are described as **linear** because they contain terms in x and y and do not contain any higher order terms such as x^2 or y^2.

The general equation for a straight-line graph is $y = mx + c$, as explained in the previous section. Two straight lines on a graph can be represented by two such equations. Provided the two lines are not parallel to one another, they cross each other at a single point. The coordinates of this point are the values of x and y that fit both equations. In other words, these coordinates are the solution of a pair of simultaneous equations representing the two straight lines. For example, the two straight lines in Figure 7 $y = 2x - 2$ and $y = -x + 4$ meet at the point $(2, 2)$ so $x = 2$, $y = 2$ are the only values of x and y that fit both equations.

The graphical approach to finding the solution of a pair of simultaneous equations takes time and is not as accurate as a systematic algebraic method. This method can best be explained by considering an example, as follows

$$
\begin{aligned}
2x - y &= 2 &&\text{(equation 1)} \\
x + y &= 4 &&\text{(equation 2)}
\end{aligned}
$$

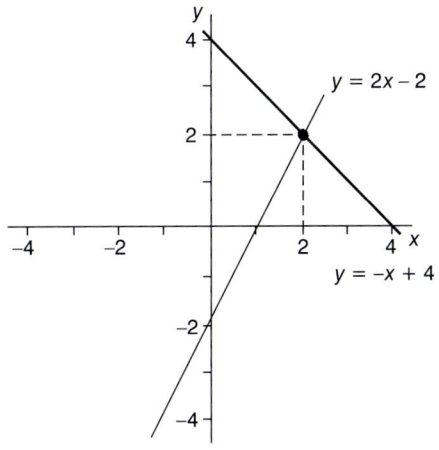

Figure 7 *A graphical solution*

Make the coefficient of x the same in both equations by multiplying one or both equations by a suitable number. In the above equation, this is most easily achieved by multiplying equation 2 throughout by 2 to give $2x + 2y = 8$.

The two equations to be solved are now

$$
\begin{aligned}
2x - y &= 2 &&\text{(equation 1)} \\
2x + 2y &= 8 &&\text{(modified equation 2)}
\end{aligned}
$$

Subtracting modified equation 2 from equation 1 gives

$$(2x - y) - (2x + 2y) = 2 - 8$$
$$\therefore -y - 2y = -6$$
$$-3y = -6$$
$$y = \frac{-6}{-3} = 2$$

Substituting this value into equation 1 or equation 2 enables the value of x to be determined. Using equation 2 for this purpose gives $x + 2 = 4$ and hence $x = 4 - 2 = 2$.

The solution of the two equations is, therefore, $x = 2$, $y = 2$.

Simultaneous equations with two unknown quantities can occur in several parts of the A level physics course, for example:

- $v = u + at$ in kinematics (see Topic 1.4)
- $V = \varepsilon - Ir$ in electricity (see Topic 7.3)
- $E_{K(max)} = hf - \phi$ (see Topic 20.1)

Summary test 24.4

1 For each of the following equations that represent straight-line graphs, write down:
  i the gradient,
  ii the y-intercept,
  iii the x-intercept:

  a $y = 3x - 3$

  b $y = -4x + 8$

  c $y + x = 5$

  d $2y + 3x = 6$

2 a A straight line on a graph has a gradient $m = 2$ and passes through the point (2, −4). Work out:
  i the equation for this line,
  ii its y-intercept.

  b The velocity v (in m s^{-1}) of an object varies with time t (in s) in accordance with the equation
  $v = 5 + 3t$.

  Determine:
  i the acceleration of the object,
  ii the initial velocity of the object.

3 a Plot the equations $y = x + 3$ and $y = -2x + 6$ over the range from $x = -3$ to $x = +3$. Write down the coordinates of the point P where the two lines cross.

  b Write down the equation for the line OP, where O is the origin of the graph.

4 Solve the following pairs of simultaneous equations, after making y the subject of each equation if necessary:

  a $y = 2x - 4$, $y = -x + 2$

  b $y = 3x - 4$, $x + y = 8$

  c $2x + 3y = 4$, $x + 2y = 2$

Gradients

The gradient of a straight line

The gradient of a straight line = $\frac{\Delta y}{\Delta x}$, where Δy is the change of the quantity plotted on the y-axis and Δx is the change of the quantity plotted on the x-axis. The gradient of a straight line is obtained by drawing as large a gradient triangle as possible and measuring the height Δy and the base Δx of this triangle, using the scale on each axis (Figure 1).

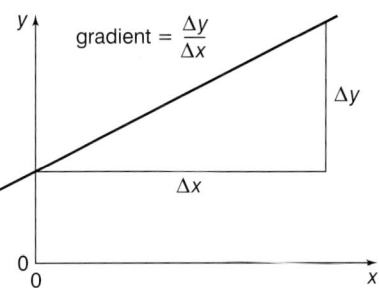

Figure 1 Constant gradient

> **Note**
>
> As a rule, when you plot a straight-line graph, always choose a scale for each axis that covers at least half the length of each axis. This will enable you to draw the line of best fit as accurately as possible, as explained on Practical skills p.xxiv. The measurement of the gradient of the line will therefore be more accurate. If the y-intercept is required and it cannot be read directly from the graph, it can be calculated by substituting the value of the gradient and the coordinates of a point on the line into the equation $y = mx + c$.

The gradient at a point on a curve

The gradient of a point on a curve is equal to the gradient of the tangent to the curve at that point.

- The **tangent to the curve at a point** is a straight line that touches the curve at that point without cutting across it. To see why, mark any two points on a curve and join them by a straight line. The gradient of the line is $\frac{\Delta y}{\Delta x}$, where Δy is the vertical separation of the two points and Δx is the horizontal separation. Now repeat with one of the points closer to the other; the straight line is now closer in direction to the curve. If the points are very close, the straight line between them is almost along the curve. The gradient of the line is then virtually the same as the gradient of the curve at that position (Figure 2). In other words, the gradient of the straight line $\frac{\Delta y}{\Delta x}$ becomes equal to the gradient of the curve as $\Delta x \rightarrow 0$. The curve gradient is written as:

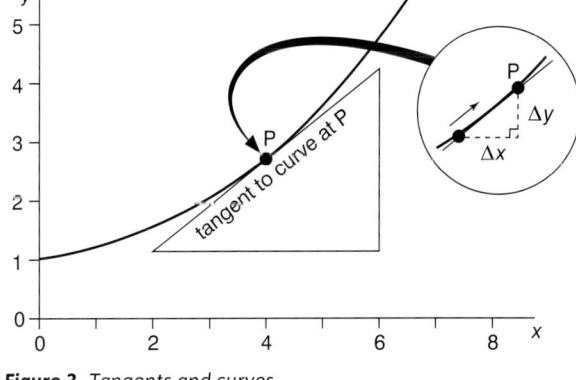

Figure 2 Tangents and curves

$$\frac{dy}{dx}$$

where $\frac{d}{dx}$ means 'rate of change'.

- The **gradient of the tangent** is a straight line and is obtained as explained above. Drawing the tangent to a curve requires practice. This skill is often needed in practical work. The **normal** at the point where the tangent touches the curve is the straight line perpendicular to the tangent at that point. An accurate technique for drawing the normal to a curve using a plane mirror is shown in Figure 3. At the point where the normal intersects the curve, the curve and its mirror image should join smoothly without an abrupt change of gradient where they join. After positioning the mirror surface correctly, the normal can be drawn and then used to draw the tangent to the curve.

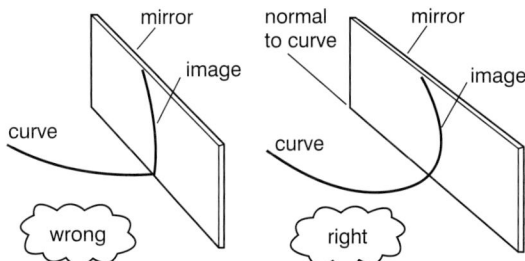

Figure 3 Drawing the normal to a curve

Turning points

A **turning point on a curve** is where the gradient of the curve is zero. This happens where a curve reaches a **peak** with a fall either side (i.e. a **maximum**) or where it reaches a **trough** with a rise either side (i.e. a **minimum**).

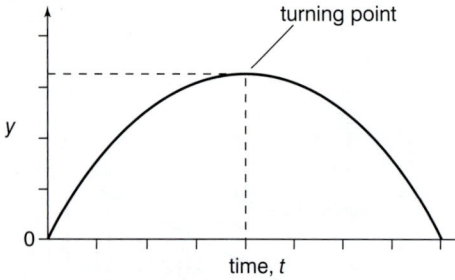

Figure 4 *Turning points*

Where the gradient represents a physical quantity, a turning point is where that physical quantity is zero. Figure 4 shows an example of a curve with a turning point. This is a graph of the vertical height against time for a projectile that reaches a maximum height, then descends as it travels horizontally. The gradient represents the vertical component of velocity. At maximum height, the gradient of the curve is zero so the vertical component of velocity is zero at that point.

Areas under graphs

The **area** under a line on a graph can give useful information. For example, consider Figure 5a which is a graph of the tension in a spring against its extension. Since 'tension × extension' is 'force × distance', which equals work done, then the area under the line represents the work done to stretch the spring.

Figure 5b shows a tension v. extension graph for a rubber band. Unlike Figure 5a, the area under the curve is not a triangle, but it still represents work done, in this case the work done to stretch the rubber band. (See Topic 5.6.)

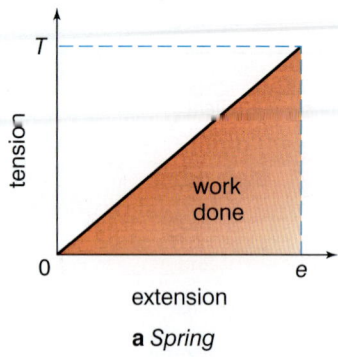

a *Spring*

> **Note**
>
> The **product** of the y-variable and the x-variable must represent a physical variable with a physical meaning if the area is to be of use.

Even where the area does represent a physical variable, it may not have any physical meaning. For example, for a graph of p.d. against current, the product of p.d. and current represents power – but this physical quantity has no meaning in this situation.

Examples
More examples of curves where the area is useful include:

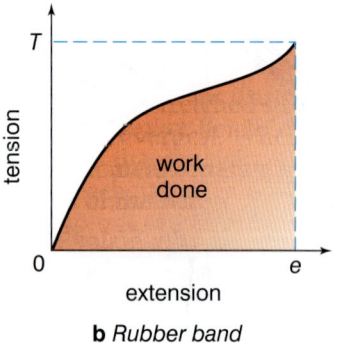

b *Rubber band*

Figure 5 *Tension v. extension*

- Velocity against time, where the area under the line and the time axis represents **displacement**.
- Acceleration against time, where the area under the line and the time axis represents **change of velocity**.
- Power against time, where the area between the curve and the time axis represents **energy**.
- Charge and potential difference, where the area between the curve and the p.d. axis represents **energy**.

Summary test 24.5

1 a Sketch a velocity against time graph (with time on the x-scale) to represent the equation $v = u + at$, where v is the velocity at time t.

b What feature of the graph represents:
  i the acceleration,   **ii** the displacement?

2 a Sketch a graph of current (on the y-axis) against p.d. (on the x-axis) to show how the current through an ohmic conductor varies with p.d.

b How can the resistance of the conductor be determined from the graph?

3 An electric motor is supplied with energy at a constant rate:

a Sketch a graph to show how the energy supplied to the motor increases with time.

b Explain how the power supplied to the motor can be determined from the graph.

4 A steel ball bearing was released in a tube of oil and it fell to the bottom of the tube.

a Sketch graphs to show how the following changed with time for the ball bearing, from the instant of release to the point of impact at the bottom of the tube:
  i the velocity,   **ii** the acceleration.

b i What is represented on graph **ai** by 1 the gradient, 2 the area under the line?
  ii What is represented on graph **aii** by the area under the line?

Logarithms

Any number can be expressed as any other number raised to a particular power. You can use the $x^\square$ key on a calculator to show, for example that $8 = 2^3$ and $9 = 2^{3.17}$ In these examples, 2 is referred to as the base number and is raised to a different power in each case to generate 8 or 9. The power is defined as the **logarithm** of the number generated.

In general, for a number $n = b^p$ where b is the base number, then $p = log_b\, n$ where $\log_b$ means a logarithm using b as the base number.

Note that $\log_b(b^p) = p$ as $b^p = n$ and $\log^b n = p$.

Applying the general definition above gives the following rules to remember when working with logs:

1 For any two numbers m and n,
$\log(nm) = \log n + \log m$
Let $p = \log n$ and let $q = \log m$ so $n = b^p$ and $m = b^q$.
$\therefore nm = b^p\, b^q = b^{p+q}$ so $\log(nm) = p + q = \log m + \log n$

2 For any two numbers m and n,
$$\log\left(\frac{n}{m}\right) = \log n - \log m$$
Let $p = \log n$ and let $q = \log m$ so $n = b^p$ and $m = b^q$. Therefore $\dfrac{1}{m} = \dfrac{1}{b^q} = b^{-q}$
$\therefore \dfrac{n}{m} = b^p\, b^{-q} = b^{p-q}$ so $\log\left(\dfrac{n}{m}\right) = p - q = \log n - \log m$.

3 For any number m raised to a power p,
$\log(m^p) = p\, \log m$

This is because $m^p = m$ multiplied by itself p times.

$$\overleftarrow{\hspace{1cm}}\; p \text{ terms} \;\overrightarrow{\hspace{1cm}}$$
Therefore, $\log m^p = \{\log m + \log m + \ldots + \log m\} = p\, \log m$

The following particular bases are used extensively in physics.

1 Base 10 logs, written as $\log_{10}$ or lg
For example,

- $100 = 10^2$ so $\log_{10} 100 = 2$,
- $50 = 10^{1.699}$ so $\log_{10} 50 = 1.699$,
- $10 = 10^1$ so $\log_{10} 10 = 1$,
- $5 = 10^{0.699}$ so $\log_{10} 5 = 0.699$

This illustrates the product rule for logs (i.e. $\log(nm) = \log n + \log m$) since $\log_{10} 50 = \log_{10} 5 + \log_{10} 10 = 0.699 + 1 = 1.699$.

Uses of base 10 logs
In graphs where a logarithmic scale is necessary to show the full range of a variable that covers a very wide range, as shown in Figure 1. Notice in Figure 1 that the frequency increases by ×10 in equal intervals along the horizontal axis.

In data analysis where a relationship between two variables is of the form $y = kx^n$ and k and n are unknown constants. Applying the above rules to an equation of the form $y = kx^n$,

$$\log_{10} y = \log_{10} k + \log_{10} x^n = \log_{10} k + n\, \log_{10} x$$

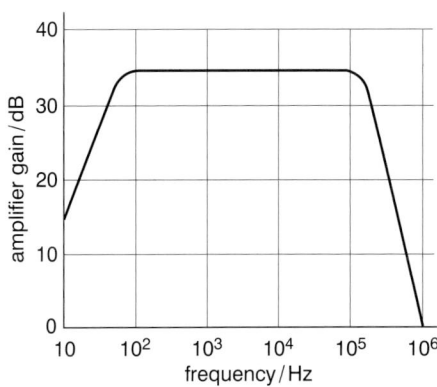

Figure 1 *Logarithmic scale*

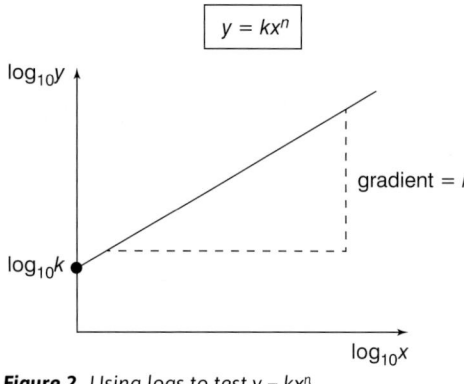

$$y = kx^n$$

Figure 2 *Using logs to test $y = kx^n$*

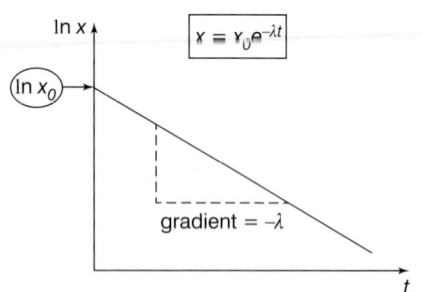

$$x = x_0 e^{-\lambda t}$$

Figure 3 *Using logs to test $x = x_0 e^{-lt}$*

The graph of $\log_{10} y$ (on the vertical axis) against $\log_{10} x$ is, therefore, a straight line of gradient n with an intercept equal to $\log k$ (see Figure 2).

In certain formulae where a ×10 scale is used. For example, the gain of an amplifier in decibels (dB) is a ×10 scale defined by the formula:

$$\text{voltage gain/dB} = 10 \log_{10} \left(\frac{V_{out}}{V_{in}} \right),$$

where V_{out} and V_{in} are the output and input voltages respectively.

If $V_{out} = 50 V_{in}$, the gain of the amplifier is 17 dB ($= 10 \log_{10} 50$).

2 Natural logs, written as $\log_e$ or $\ln$

Here, e is the exponential number used as the base of natural logarithms and is equal to 2.718. For example,

- $2.718 = e^1$ so $\ln 2.718 = 1$
- $7.389 = e^2$ so $\ln 7.389 = 2$
- $20.009 = e^3$ so $\ln 20.009 = 3$
- In general, for any number n, if p is such that $n = e^p$, then $\ln n = p$.

Uses of natural logarithms

Natural logs are used in the equations for radioactive decay (Topic 21.2) and capacitor discharge (Topic 15.4) or any other process where the rate of change of a quantity is proportional to the quantity itself. For example, the rate of decrease of p.d. across a capacitor discharging through a resistor is proportional to the p.d. across the capacitor. This type of change is described as an exponential decrease because the quantity decreases by the same factor in equal intervals of time.

Applying the general rule (that if p is such that $n = e^p$, then $p = \ln n$) to the equation $x = x_0 e^{-\lambda t}$ gives $\ln x = \ln x_0 - \lambda t$.

Therefore, a graph of $\ln x$ (on the vertical axis) against t is a straight line with a gradient equal to $-\lambda$ and a y-intercept equal to $\ln x_0$. See Figure 3.

Comparing the equation for capacitor discharge $Q = Q_0 e^{-t/RC}$ with the equation for radioactive decay $N = N_0 e^{-\lambda t}$:

- for capacitor discharge, $\ln Q = \ln Q_0 - \dfrac{t}{RC}$ so a graph of $\ln Q$ (on the vertical axis) against t is a straight line which has a gradient $\dfrac{-1}{RC}$ and $\ln Q_0$ as its y-intercept,

- for radioactive decay, $\ln N = \ln N_0 - \lambda t$ so a graph of $\ln N$ (on the vertical axis) against t is a straight line which has a gradient $-\lambda$ and $\ln N_0$ as its y-intercept.

Summary test 24.6

1 a Use your calculator to work out:
   i $\log_{10} 3$,   **ii** $\log_{10} 15$.

  b Use your answers in **a** to work out:
   i $\log_{10} 45$,   **ii** $\log_{10} 5$.

2 The gain of an amplifier, in decibels, is given by the formula $10 \log_{10} \left(\dfrac{V_{out}}{V_{in}} \right)$.

  a Calculate the gain, in decibels (dB), for:
   i $V_{out} = 12 V_{in}$,   **ii** $V_{out} = 5 V_{in}$.

  b Show that the gain, in decibels, of an amplifier for which $V_{out} = 60 V_{in}$ is equal to the sum of the gain in **i**, and the gain in **ii** above.

3 Write down the gradient and the y-intercept of a line on a graph representing the equation $\log_{10} y = n \log_{10} x + \log_{10} k$ for:
  a $y = 3x^5$,
  b $y = \frac{1}{2} x^3$,
  c $y = x^2$.

4 a Use your calculator to work out:
   i $\ln 3$,
   ii $\ln 15$.

  b Use your answers in **a** to work out:
   i $\ln 45$,
   ii $\ln 5$.

Exponential decrease

Rates of change

Consider a variable quantity y that changes with respect to a second quantity x as shown in Figure 1. The gradient of the curve at any point is the rate of change of y with respect to x at that point. This can be worked out from the graph by drawing a tangent to the curve at that point and measuring the gradient of the tangent. Figure 1 shows the idea. The rate of change of y with respect to x at point P is equal to the gradient of the tangent to the curve at P which is $\frac{\Delta y}{\Delta x}$.

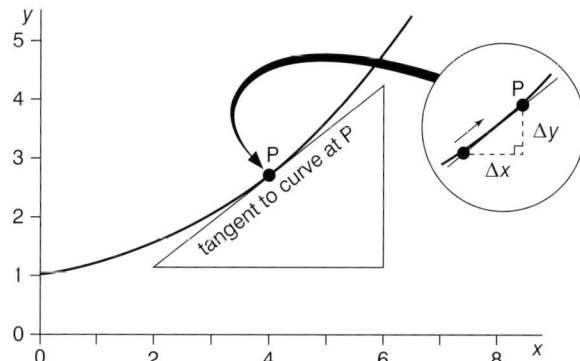

Figure 1 *Tangents and curves*

The rate of change of y with respect to x can be worked out algebraically if the equation relating y and x is known. This process is known as differentiation. For example:

- For $y = x^2$, then increasing x to $x + \Delta x$ increases y to $y + \Delta y$ where $y + \Delta y = (x + \Delta x)^2$.

  Multiplying out $(x + \Delta x)^2$ gives $y + \Delta y = x^2 + 2x\Delta x + \Delta x^2$

  Subtracting $y = x^2$ from this equation gives $\Delta y = 2x\Delta x + \Delta x^2$

  Dividing by Δx therefore gives $\frac{\Delta y}{\Delta x} = \frac{2x\Delta x + \Delta x^2}{\Delta x} = 2x + \Delta x$

  Therefore, as $\Delta x \to 0$, $\frac{\Delta y}{\Delta x} \to 2x$ which is therefore the formula for the gradient at x.

  This is written $\frac{dy}{dx} = 2x$, where $\frac{dy}{dx}$ is the mathematical expression for the rate of change of y with respect to x.

- For the general expression $y = x^n$, it can be shown that $\frac{dy}{dx} = nx^{n-1}$

  For example, if $y = 3x^5$, then $\frac{dy}{dx} = 15x^4$.

Note that differentiation of functions is not required in the A level specification. The information on differentiation is provided to help you develop your understanding of exponential change in the next section.

Exponential change happens when the rate of change of a quantity is proportional to the quantity itself. Such a change can be an increase (i.e. exponential growth) or a decrease (i.e. exponential decay). In both cases, the quantity changes by a fixed proportion in equal intervals of time. The A level specification requires knowledge and understanding of exponential decrease but not of exponential growth. The notes below will, therefore, concentrate on exponential decrease.

Summary test 24.7

1 a For each exponential decrease equation, write down the initial value at $t = 0$ and the decay constant:

   i $x = 2e^{-3t}$,
   ii $x = 12e^{-t/5}$,
   iii $x = 4e^{-0.02t}$.

b For each exponential decrease equation above, work out the half-life.

2 A radioactive isotope has a half-life of 720 s and it decays to form a stable product. A sample of the isotope is prepared with an initial activity of 12.0 kBq. Calculate the activity of the sample after:

   a 1 minute,
   b 5 minutes,
   c 1 hour.

3 A capacitor of capacitance 22 μF discharged from a p.d. of 12.0 V through a 100 kΩ resistor.

   a Calculate:
     i the time constant of the discharge circuit,
     ii the half-life of the exponential decrease.

   b Calculate the capacitor p.d.:
     i 2.0 s,
     ii 5.0 s after the discharge started.

4 A certain exponential decrease process is represented by the equation $x = 1000e^{-5t}$.

   a **i** Calculate the half-life of the process.
     ii Calculate N when $t = 0.5$ s.

   b Show that the above equation can be rearranged as an equation of the form $\ln x = a + bt$ and determine the values of a and b.

In your studies of capacitor discharge (Topic 15.4) and of radioactive decay (Topic 21.2), you will have met and used the equation $\dfrac{dx}{dt} = -\lambda x$ and the solution of this equation $x = x_0 e^{-\lambda t}$.

- Let's consider why the equation $\dfrac{dx}{dt} = -\lambda x$ represents an exponential decrease which is a change where the variable quantity x decreases with time at a rate in proportion to the quantity.

  If x decreases by Δx in time Δt, the rate of change is $\dfrac{\Delta x}{\Delta t}$. This is written as $\dfrac{dx}{dt}$ in the limit $\Delta t \to 0$.

  For an exponential decrease, the rate of change is negative and is proportional to x, therefore, $\dfrac{dx}{dt} = -\lambda x$, where λ is referred to as the decay constant.

- Now consider why the solution of this equation is $x = x_0 e^{-\lambda t}$, where x_0 is a constant.

  Applying the rules of differentiation to the function
  $$x = x_0\left(1 + t + \frac{t^2}{2 \times 1} + \frac{t^3}{3 \times 2 \times 1} + \frac{t^4}{4 \times 3 \times 2 \times 1} + \text{similar higher order terms}\right) \text{ gives}$$
  $$\frac{dx}{dt} = x_0\left(0 + 1 + t + \frac{t^2}{2 \times 1} + \frac{t^3}{3 \times 2 \times 1} + \text{similar higher order terms}\right)$$
  which is the same as x.

  So, $\dfrac{dx}{dt} = x$ if x is the above function.

  It can be shown that the function in brackets may be written as e^t, where e is referred to as the exponential number.

  Therefore,
  $$e^t = 1 + t + \frac{t^2}{2 \times 1} + \frac{t^3}{3 \times 2 \times 1} + \frac{t^4}{4 \times 3 \times 2 \times 1} + \text{similar higher order terms} \ldots$$
  The value of e, the exponential number, can be worked out by substituting $t = 1$ in the above expression for e^t, giving $e = 1 + 1 + \dfrac{1}{2} + \dfrac{1}{6} + \text{etc.} = 2.718$ to four significant figures.

- To show that the solution of the equation $\dfrac{dx}{dt} = -\lambda x$ is $x = x_0 e^{-\lambda t}$, divide both sides of the equation by $-\lambda$ to give $\dfrac{1}{-\lambda}\dfrac{dx}{dt} = x$.

  Substituting z for $-\lambda t$ therefore gives $\dfrac{dx}{dz} = x$ which has the solution $x = x_0 e^z = x_0 e^{-\lambda t}$

The half-life, $t_{1/2}$ of an exponential decrease is the time taken for x to decrease from x_0 to $\dfrac{x_0}{2}$.

Substituting $x = \dfrac{x_0}{2}$ and $t = t_{1/2}$ into $x = x_0 e^{-\lambda t}$ gives $\dfrac{x_0}{2} = x_0 e^{-\lambda t_{1/2}}$.

Applying logs to both sides gives $\ln x_0 - \ln 2 = \ln x_0 - \lambda t_{1/2}$ which simplifies to $\lambda t_{1/2} = \ln 2$

$\therefore t_{1/2} = \dfrac{\ln 2}{\lambda} = \dfrac{0.693}{\lambda}$

The time constant, τ of an exponential decrease is the time taken for x to decrease from x_0 to $\dfrac{x_0}{e}$ ($= 0.368\,x_0$ as $\dfrac{1}{e} = 0.368$).

Substituting $x = \dfrac{x_0}{e}$ and $t = \tau$ into $x = x_0 e^{-\lambda t}$ gives $\dfrac{x_0}{e} = x_0 e^{-\lambda \tau}$.

Applying logs to both sides gives $\ln x_0 - \ln e = \ln x_0 - \lambda \tau$ which simplifies to $\tau = \dfrac{1}{\lambda}$ (as $\ln e = 1$)

For capacitor discharge, $\lambda = \dfrac{1}{CR}$; therefore $\tau = \dfrac{1}{\lambda} = CR$.

As explained on the previous page, $\ln(e^{-\lambda t}) = -\lambda t$.

Therefore, $\ln x = \ln(x_0 e^{-\lambda t}) = \ln x_0 + \ln(e^{-\lambda t}) = \ln x_0 - \lambda t$.

Chapter Summary

Trigonometry
- 1 radian $= \dfrac{360}{2\pi}$ degrees.
- Arc length $s = r\theta$ where θ is the angle in radians.
- For small angles (i.e. $\theta <$ about $10°$) $\sin\theta \approx \tan\theta \approx \theta$ in radians, and $\cos\theta \approx 1$.
- The cosine rule; $a^2 = b^2 + c^2 - 2bc\cos A$
- The sine rule: $\dfrac{a}{\sin A} = \dfrac{b}{\sin B} = \dfrac{c}{\sin C}$.

Algebra
- Linear simultaneous equations: to solve a pair of simultaneous equations with two unknown variables, x and y,
  1. make the coefficient of x the same in both equations by multiplying one or both equations by a suitable number, then
  2. combine the two equations to eliminate x and so find y, then
  3. substitute the value of y into either equation to find x.

- The general solution of the quadratic equation $ax^2 + bx + c = 0$

  is $x = \dfrac{-b \pm \sqrt{(b^2 - 4ac)}}{2a}$

Logarithms
- For a number $n = b^p$, where b is the base number, $p = \log_b n$.
- $\log(nm) = \log n + \log m$.
- $\log(n/m) = \log n - \log m$.
- $\log(m^p) = p\log m$.
- For $y = kx^n$, $\log_{10} y = \log_{10} k + n\log_{10} x$:
  the graph of $\log_{10} y$ (on the vertical axis) against $\log_{10} x$ is a straight line of gradient n with an intercept equal to $\log k$.
- For $n = e^p$, $\ln n = p$.
- For $x = x_0 e^{-\lambda t}$, $\ln x = \ln x_0 - \lambda t$:
  the graph of $\ln x$ (on the vertical axis) against t is a straight line with a gradient equal to $-\lambda$ and a y-intercept equal to $\ln x_0$.

Exponential decrease
- Exponential change happens when the rate of change of a quantity is proportional to the quantity itself.
- For an exponential decrease, the rate of change is negative and is proportional to x,

  therefore, $\dfrac{dx}{dt} = -\lambda x$, where λ is referred to as the decay constant.
- The solution of this equation is $x = x_0 e^{-\lambda t}$, where x_0 is a constant.
- Half-life, $t_{1/2} = \dfrac{0.693}{\lambda}$ ($=$ time for x to decrease from x_0 to $\dfrac{x_0}{2}$).
- Time constant, $\tau = \dfrac{1}{\lambda}$ ($=$ time for x to decrease from x_0 to $\dfrac{x_0}{e}$).

Glossary

A-scan system ultrasound scan in which pulses from an ultrasonic transducer are detected by the probe after being reflected by internal boundaries and then displayed on an oscilloscope

absolute temperature, T, in kelvins = temperature in °C + 273(.15)

absolute zero is the temperature at which an object has minimum internal energy

acceleration change of velocity per unit time

acceleration of free fall acceleration of an object acted on only by the force of gravity

acoustic impedance the product of the density of a substance and the speed of the ultrasonic or sound waves through it

activity, A, of a radioactive isotope is the number of nuclei of the isotope that disintegrate per second. The unit of activity is the becquerel (Bq), equal to 1 disintegration per second

air resistance the force of the air opposing the motion of an object moving through the air (see drag force)

alpha radiation consists of particles that are each composed of two protons and two neutrons. An alpha (α) particle is emitted by a heavy unstable nucleus which is then less unstable as a result. Alpha radiation is easily absorbed by paper, has a range in air of no more than a few centimetres and is more ionising than beta (β) or gamma (γ) radiation

aluminium filter aluminium plate used in X-radiography to absorb 'soft' (i.e. low energy) X-rays

amplifier output p.d. that is proportional to the input p.d. supplied to it

amplitude maximum displacement of a vibrating particle; for a transverse wave, it is the distance from the middle to the peak of the wave

angular displacement, in radians, in time $t = 2\pi ft = \dfrac{2\pi t}{T}$ for an object in uniform circular motion

angular frequency $2\pi \times$ frequency of oscillating motion. The unit of ω is the radian per second (rad s^{-1})

annihilation process whereby a particle and its corresponding antiparticle collide, annihilate each other and produce photons

anode positive terminal of an electrical device

antineutrino an uncharged antiparticle with a very low rest mass compared with an electron and which is emitted from a nucleus when it emits a β$^-$ particle

antinode fixed point in a stationary wave pattern where the amplitude is a maximum

antiquark the antiparticle of a quark

Archimedes' principle The upthrust on a body wholly or partially immersed in a fluid is equal to the weight of liquid displaced by the cylinder

atomic mass unit, u, (correctly referred to as the unified atomic mass constant) is equal to 1.66×10^{-27} kg. It is defined as $\frac{1}{12}$ of the mass of an atom of the carbon isotope $^{12}_{6}C$

atomic number number of protons in the nucleus of the atom. It is also the order number of the element in the Periodic Table

attenuation reduction of intensity of a beam due to absorption or scattering

attenuation coefficient measure of the reduction of intensity per unit distance of a beam travelling through a substance and is equal to ln 2/half thickness of the substance

audio wave electrical waves produced from sound waves for example by a microphone

Avogadro constant, N_A, is defined as the number of atoms in 12 grams of the carbon isotope $^{12}_{6}C$

background radioactivity is radioactivity due to radioactive substances which may be in the ground or in building materials or elsewhere in the environment. Background radioactivity is also caused by cosmic radiation

baryons matter particles and antiparticles that each consist of three quarks or three antiquarks

base units the five units that define the SI system (the metre, the kilogram, the second, the ampere and the kelvin)

beta minus (β$^-$) particle an electron created and emitted by an unstable neutron-rich nucleus when a neutron in the nucleus changes into a proton – an antineutrino is created and emitted at the same time

beta plus (β$^+$) particle a positron created and emitted by an unstable proton-rich nucleus when a proton in the nucleus changes into a neutron – a neutrino is created and emitted at the same time

beta radiation consists of beta-particles (β) which are electrons emitted by unstable nuclei with too many neutrons compared to protons. Beta radiation has a range in air of about a metre and is less ionising than alpha (α) radiation and more ionising than gamma (γ) radiation

Big Bang theory The Universe was created in a massive 'primordial' explosion and has been expanding ever since

binding energy of a nucleus:
- the work that must be done to separate a nucleus into its constituent neutrons and protons,
- binding energy = mass defect $\times c^2$,
- binding energy/nucleon is greatest for nuclei of mass number 57

biofuel fuel obtained from biomass; see biomass

biomass biological material from living or recently living organisms used as fuel (eg animal waste or woodchip)

black body an object that absorbs all the radiation incident on it; the radiation emitted by a black body is called *black body radiation*

bonds forces that hold atoms or molecules together

brittle snaps without stretching or bending when subject to stress

Brownian motion is the random and unpredictable motion of a particle such as a smoke particle and is caused by molecules of the surrounding substance (which are all much smaller than smoke particles) colliding at random with the particle

capacitance of a capacitor is defined as the charge stored per unit p.d. The unit of capacitance is the farad (F), equal to 1 coulomb per volt

For a capacitor of capacitance C at p.d. V, the charge stored, $Q = CV$

capacitor combination rules
1 **capacitors in parallel**; combined capacitance
$C = C_1 + C_2 + C_3 + \ldots$
2 **capacitors in series**; combined capacitance is given by
w $\frac{1}{C} = \frac{1}{C_1} + \frac{1}{C_2} + \frac{1}{C_3} + \ldots$

capacitor energy; energy stored by the capacitor,
$W = \frac{1}{2}QV = \frac{1}{2}CV^2$

capacitor discharge through a fixed resistor R;
1 time constant $= RC$
2 exponential decrease equation for current or charge or p.d.; $x = x_0 e^{-t/RC}$

carrier wave electromagnetic waves used to carry a signal

centre of gravity point where the weight of a body may be considered to act

centripetal acceleration:
1 for an object moving at speed v in uniform circular motion, its centripetal acceleration $a = \frac{v^2}{r}$ towards the centre of the circle
2 for a satellite in a circular orbit, its centripetal acceleration $\frac{v^2}{r} = g$

centripetal force is the resultant force on an object that moves along a circular path. For an object of mass m moving at speed v along a circular path of radius r, the centripetal force $= \frac{mv^2}{r}$

chain reaction a series of reactions in which each reaction causes a further reaction. In a nuclear reactor, each fission event is due to a neutron colliding with a $^{235}_{92}U$ nucleus which splits and releases two or three further neutrons which can go on to produce further fission. A steady chain reaction occurs when one fission neutron on average from each fission event produces a further fission event

charge carriers charged particles that move through a substance when a p.d. is applied across it

coherence two sources of waves are coherent if they emit waves with a constant phase difference

collisions an **elastic** collision is one in which the total kinetic energy after the collision is the same as before the collision. A **totally inelastic** collision is where the colliding objects stick together

contrast medium substance passed through a body organ or blood vessel to enhance the X-ray image of the body part by increasing the absorption of X-rays

cooling by evaporation the decrease in the temperature of a liquid due to evaporation from its open surface

cosmic microwave background radiation (CMBR) radiation that was created by the Big Bang and has been travelling through the Universe ever since the Universe became transparent

Coulomb's law of force between two point charges states that the force is proportional to the product of the charges and inversely proportional to the square of the distance between the charges. For two point charges Q_1 and Q_2 at distance apart r, the force F between the two charges is given by the equation $F = \frac{Q_1 Q_2}{4\pi\varepsilon_0 r^2}$, where ε_0 is the absolute permittivity of free space

couple pair of equal and opposite forces acting on a body but not along the same line

coupling medium gel applied between the body surface and the surface of an ultrasonic probe to exclude air at the interface to ensure the ultrasonic waves are not reflected at the body surface

critical mass is the minimum mass of the fissile isotope (e.g. the uranium isotope $^{235}_{92}U$) in a nuclear reactor necessary to produce a chain reaction. If the mass of the fissile isotope in the reactor is less than the critical mass, a chain reaction does not occur because too many fission neutrons escape from the reactor or are absorbed without fission

cycle interval for a vibrating particle (or a wave) from a certain displacement and velocity to the next time the particle (or the next particle) has the same displacement and velocity

damped oscillations of an oscillating system are due to the presence of resistive forces due to friction and drag. For a lightly damped system, the amplitude of oscillations decreases gradually. For a heavily damped system displaced from equilibrium then released, the system slowly returns to equilibrium without oscillating. For a critically damped system, the system returns to equilibrium in the least possible time without oscillating

Dark energy unknown force releasing hidden energy thought to be causing the expansion of the Universe to accelerate

de Broglie wavelength A particle of matter has a wave-like nature because it can behave as a wave. For example, electrons directed at a thin crystal are diffracted by the crystal. The de Broglie wavelength, λ, of a matter particle depends on its momentum, p, in accordance with de Broglie's equation

$$\lambda = \frac{h}{p} = \frac{h}{mv}$$

decay constant, λ is the probability of an individual nucleus decaying per second

density of a substance mass per unit volume of the substance

diffraction is the spreading of waves when they pass through a gap or round an obstacle. X-ray diffraction is used to determine the structure of crystals, metals and long molecules. Electron diffraction and neutron diffraction are also used to probe the structure of materials. High-energy electron scattering is used to determine the diameter of the nucleus

diffraction grating plate with many close equally-spaced parallel slits that diffracts light at normal incidence into a direction that depends on its wavelength

diode an electrical device that conducts in one direction only

dispersion splitting of a beam of white light by a glass prism into the colours of a spectrum

displacement distance in a given direction

Doppler effect the effect of relative motion between a source of waves and an observer on the observed frequency, causing it to differ from the emitted frequency

Doppler shift the difference between the observed frequency and the emitted frequency of the waves from a source due to relative motion between the source and the observer

drag force the force of fluid resistance on an object moving through the fluid

ductile stretches easily without breaking

efficiency = $\dfrac{\text{useful energy transferred}}{\text{total energy supplied}}$

effort the force applied to a machine to make it move

elastic deformation regain of shape of an object after it has been deformed

elastic limit point beyond which a wire is permanently stretched

electric potential (at a point) work done in bringing unit positive charge from infinity to the point

electrolysis process of electrical conduction in a solution or molten compound due to ions moving to the oppositely charged electrode

electrolyte a solution or molten compound that conducts electricity

electromotive force (e.m.f.) the amount of electrical energy per unit charge produced inside a source of electrical energy

electron fundamental particle with a fixed negative charge equal and opposite to that of a proton and a mass approximately $\dfrac{1}{1840}$th of the mass of a proton

electron volt amount of energy equal to 1.6×10^{-19} J, defined as the work done when an electron is moved through a p.d. of 1 V

engine force the force that drives a vehicle

equilibrium state of an object when at rest or in uniform motion

equipotential a line or a surface of constant gravitational potential

error bar representation of an uncertainty on a graph

error of measurement uncertainty of a measurement

evaporation the process by which a liquid turns to vapour below its boiling point

excited state an energy state of an atom with more energy than the lowest energy state

explosion In an explosion where two objects fly apart, the two objects carry away equal and opposite momentum

exponential decrease Exponential change happens when the rate of change of a quantity is proportional to the quantity itself. For an exponential decrease of a quantity x, $\dfrac{dx}{dt} = -\lambda x$, where λ is referred to as the decay constant
The solution of this equation is $x = x_0 e^{-\lambda t}$, where x_0 is a constant

Faraday's law of electromagnetic induction states that the induced e.m.f. in a circuit is proportional to the rate of change of magnetic flux linkage through the circuit

For a changing magnetic field in a fixed coil, induced e.m.f. $= -NA\dfrac{\Delta B}{\Delta t}$

fission is the splitting of a $^{235}_{92}$U nucleus or a $^{239}_{94}$Pu nucleus into two approximately equal fragments. Induced fission is fission caused by an incoming neutron colliding with a $^{235}_{92}$U nucleus or a $^{239}_{94}$Pu nucleus

fluoresce light emitted from a substance as a result of high energy radiation or particles being directed at the substance

force resultant force on an object
= rate of change of its momentum
= $\dfrac{\text{change of momentum}}{\text{time taken}}$
(= mass × acceleration for fixed mass)

forward biased the direction of a diode in a circuit in order for the diode to conduct

forward voltage the potential difference necessary across a diode to make it conduct in its forward direction

free and forced oscillations
- **free oscillations** are oscillations where there is no damping and no periodic force acting on the system so the amplitude of the oscillations is constant
- **forced oscillations** are oscillations of a system that is subjected to an external periodic force

frequency the number of cycles of a wave that pass a point per second

fusion is the fusing together of light nuclei to form a heavier nucleus

galvanometer a centre-reading electrical meter used to detect and measure an electric current.

geothermal energy energy transferred towards the Earth's surface from underground rocks heated by radioative substances deep within the Earth

gravitational field strength, g, is the force per unit mass on a small test mass placed in the field

- $g = \dfrac{F}{m}$, where F is the gravitational force on a small mass m
- at distance r from a point mass M, $g = \dfrac{GM}{r^2}$
- at or beyond the surface of a sphere of mass M, $g = \dfrac{GM}{r^2}$ where r is the distance to the centre
- at the surface of a sphere of mass M and radius R, $g_s = \dfrac{GM}{r^2}$

gravitational potential, ϕ, (at a point) the work done per unit mass to move a small object from infinity to the point

gravitational potential energy energy due to position in a gravitational field; and is equal to the work done to move a small object from infinity to a point in the field

ground heat heat flow from underground

ground state lowest energy state of an atom

hadrons matter particles and antiparticles that can interact through the strong interaction. Hadrons are subdivided into baryons (which consist of three quarks or three antiquarks) and mesons (which consist of a quark and an antiquark). Protons and neutrons are baryon hadrons

half-life, $t_{1/2}$, of a radioactive isotope is the time taken for the mass of the isotope to decrease to half the initial mass. This is the same as the time taken for the number of nuclei of the isotope to decrease to half the initial number

Hall effect the effect of causing a potential difference between the opposite sides of a conductor or semiconductor when a magnetic field is used to deflect charge carriers passing though the material

Hall voltage the potential difference between the opposite sides of a conductor or semiconductor when a magnetic field is used to deflect charge carriers passing though the material

Hooke's law the extension of a spring is proportional to the force needed to extend it

Hubble's law the galaxies are receding from us, each moving at speed v, that is directly proportional to the distance, d, to it

Hubble constant, H, the speed of recession of a galaxy per unit distance from Earth

ideal gas law, $pV = nRT$, where n is the number of moles of gas, T is the absolute temperature and R is the molar gas constant

intensity of radiation at a surface is the radiation energy per second per unit area at normal incidence to the surface. The unit of intensity is $\mathrm{J\,s^{-1}\,m^{-2}}$ or $\mathrm{W\,m^{-2}}$

intensity of waves the power per unit area that waves transfer through an area perpendicular to the direction of the waves

interference formation of points of cancellation and reinforcement where two coherent waves pass through each other

internal energy of an object is the sum of the random distribution of the kinetic and potential energies of its molecules

internal resistance resistance inside a source of electrical energy

inverse-square laws
- **force** Newton's law of gravitation and Coulomb's law of force between electric charges are described as inverse-square laws because the force between two point objects (masses in the case of gravitation and charge in the case of charges) is inversely proportional to the square of the distance between the two objects. Because the two laws above are inverse-square laws, the field strength due to a point mass or a point charge varies with distance according to the inverse of the square of the distance to the point object
- **intensity** the intensity of γ radiation from a point source varies with the inverse of the square of the distance from the source. The same rule applies to radiation from any point source that spreads out equally in all directions and is not absorbed

ion a charged atom

ionising radiation produces ions in substances it passes through. It destroys cell membranes and damages vital molecules such as DNA directly or indirectly by creating 'free radical' ions which react with vital molecules

isotopes forms of the same element with different numbers of neutrons and the same number of protons in their nuclei

kaon a meson that contains a strange quark and an antiquark or a strange antiquark and a quark

kinetic energy is the energy of a moving object due to its motion. For an object of mass m moving at speed v, its kinetic energy $E_K = \frac{1}{2}mv^2$, provided $v \ll c$ (the speed of light in free space)

kinetic energy of a molecule of an ideal gas The mean kinetic energy of a molecule of an ideal gas $= \frac{3}{2}kT$, where the Boltzmann constant $k = \dfrac{R}{N_A}$

kinetic theory of a gas
- Assumptions: a gas consists of identical point molecules which do not attract one another. The molecules are in continual random motion colliding elastically with each other and with the container
- It can be shown that the pressure p of N molecules of such a gas in a container of volume V is given by the equation $pV = \frac{1}{3}Nm<c^2>$, where m is the mass of each molecule and $<c^2>$ is the mean square speed of the gas molecules
- Assuming that the mean kinetic energy of a gas molecule $\frac{1}{2}m<c^2> = \frac{3}{2}kT$, where $k = \frac{R}{N_A}$, it can be shown from $pV = \frac{1}{3}Nm<c^2>$ that $pV = nRT$ which is the ideal gas law

Kirchhoff's first law at a junction, the total current in = the total current out

Kirchhoff's second law the sum of the e.m.fs round a complete loop in a circuit = the sum of the p.ds round the loop

latent heat of fusion (energy to melt a solid) is used to break the bonds that lock the molecules of the solid into fixed positions

latent heat of vaporisation (energy to boil a liquid) is used to break the bonds that prevent molecules moving away from each other

Lenz's law states that the direction of the induced current is always such as to oppose the change that causes the current

lepton matter particles and antiparticles that can interact through the weak interaction. They cannot interact through the strong interaction. Leptons are thought to be elementary. Electrons, positrons, neutrinos and antineutrinos are examples of leptons

light-dependent resistor (LDR) resistor which is designed to have a resistance that changes with change of intensity

light-emitting diode (LED) a diode that emits light when it conducts

line of force or a field line of a gravitational field (or electrical field) A line followed by a small mass (or small charge) acted on by no other forces than the force of the field.
Line of force or a field line of magnetic field is the line along which a free north pole would move

linear two quantities are said to have a linear relationship if the change of one quantity is proportional to the change of the other

load the force to be overcome by a machine when it shifts or raises an object

logarithmic scale This is a scale such that equal intervals correspond to a change by a constant factor or multiple (e.g. ×10)

logarithms For a number $n = b^p$, where b is the base number, $p = \log_b n$
- $\log(nm) = \log n + \log m$ and $\log(n/m) = \log n - \log m$
- $\log(m^p) = p\log m$
- natural logs; for $n = e^p$, then $\ln n = p$
- base 10 logs; for $n = 10^p$, then $\log_{10} n = p$

log graphs
1 For $y = kx^n$, $\log_{10}y = \log_{10}k + n\log_{10}x$; the graph of $\log_{10}y$ (on the vertical axis) against $\log_{10}x$ is therefore a straight line of gradient n with an intercept equal to $\log k$
2 For $x = x_0e^{-\lambda t}$, $\ln x = \ln x_0 - \lambda t$; the graph of $\ln x$ (on the vertical axis) against t is a straight line with a gradient equal to $-\lambda$ and a y-intercept equal to $\ln x_0$

longitudinal waves waves with a direction of vibration of the particles parallel to the direction of energy transfer by the waves

luminosity the total power of the radiation emitted by a star

magnetic field region near a magnet or a current-carrying wire in which another magnet or current-carrying wire experiences a force

magnetic flux, Φ = BA for a uniform magnetic field that is perpendicular to an area A

magnetic flux density is defined as the force per unit length per unit current on a current-carrying conductor at right angles to the field lines. The unit of magnetic flux density B is the tesla (T). B is sometimes referred to as the magnetic field strength

magnetic flux linkage through a coil of N turns = $N\Phi$ = NBA, where B is the magnetic flux density perpendicular to area A. The unit of magnetic flux and of flux linkage is the **weber** (Wb), equal to $1\,T\,m^2$ or $1\,V\,s$

magnetic force
- $F = BIl\sin\theta$ gives the force F on a current-carrying wire of length l in a uniform magnetic field B at angle θ to the field lines, where I is the current. The direction of the force is given by Fleming's left-hand rule where the field direction is the direction of the field component perpendicular to the wire
- $F = BQv\sin\theta$ gives the force F on a particle of charge Q moving through a uniform magnetic field B at speed v in a direction at angle θ to the field. If the velocity of the charged particle is perpendicular to the field, $F = BQv$. The direction of the force is given by Fleming's left-hand rule, provided the current is in the direction in which positive charge would flow
- $BQv = \frac{mv^2}{r}$ gives the radius of the orbit of a charge moving in a direction at right angles to the lines of a magnetic field

Malus's law For polarised light passed through a polariser, its transmitted intensity $I = I_0\cos^2\theta$ where θ is the angle the polariser is rotated through from its orientation at the maximum transmitted intensity I_0

mass defect the difference between the mass of the separated nucleons (ie protons and neutrons from which the nucleus is composed) and the nucleus

mass measure of the inertia or resistance to change of motion of an object

matter waves the wave-like behaviour of particles of matter

mean kinetic energy of a molecule in a gas at absolute temperature $T = \frac{3}{2}kT$, where k is the Boltzmann constant $\left(= \frac{R}{N_A}\right)$

meson matter particles and antiparticles that each consist of a quark and an antiquark

microphone an electrical device that converts sound waves into electrical waves

mole One mole of a substance consisting of identical particles is the quantity of substance that contains N_A particles of the substance. The **molar mass** of a substance is the mass of one mole

molecule the smallest particle in a substance that can be identified as belonging to the substance

moment of a force about a point force × perpendicular distance from line of action of force to the point

momentum for an object, its momentum is defined as its mass × its velocity. For a photon, its momentum is equal to its energy $/c$ (the speed of light in a vacuum). The unit of momentum is kg m s^{-1}

muon (symbol μ) a negatively charged particle with a rest mass over 200 times the rest mass of the electron. Muons and antimuons decay into electrons and antineutrinos or positrons and neutrinos, respectively

neutrino an uncharged particle with a very low rest mass compared with an electron and which is emitted from a nucleus when it emits a β^+ particle

neutron a particle with no charge and a mass approximately the same as a proton. One or more neutrons are in every atomic nucleus except that of the smallest hydrogen nucleus

Newton's law of gravitation the gravitational force F between two point masses m_1 and m_2 at distance r apart is given by $F = G m_1 m_2/r^2$

Newton's laws of motion

- **first law**: an object continues at rest or in uniform motion unless it is acted on by a resultant force
- **second law**: the rate of change of momentum of an object is proportional to the resultant force on it
- **third law**: when two objects interact, they exert equal and opposite forces on one another

Newton's second law may be written as $F = \frac{\Delta p}{\Delta t}$, where p is the momentum ($= mv$) of the object and F is the force in newtons. For constant mass, $\Delta p = m\Delta v$ so $F = \frac{m\Delta v}{\Delta t} = ma$

node fixed point of no displacement

nuclear fission splitting of certain large nuclei such as a $^{235}_{92}\text{U}$ nucleus or a $^{239}_{94}\text{Pu}$ nucleus into two approximately equal fragments. Induced fission is fission caused by an incoming neutron colliding with a $^{235}_{92}\text{U}$ nucleus or a $^{239}_{94}\text{Pu}$ nucleus

nuclear fusion fusing together of light nuclei to form a heavier nucleus

nucleon a neutron or a proton in the nucleus

nucleon number the number of neutrons and protons in a nucleus

nuclide a nucleus of a certain isotope

nuclide of an isotope ^A_ZX is a nucleus composed of Z protons and $(A - Z)$ neutrons, where Z is the proton number (and also the atomic number of element X) and A is the number of protons and neutrons in a nucleus

Ohm's law the p.d. across a metallic conductor is proportional to the current through it provided the physical conditions do not change

pair production process whereby a photon of sufficient energy produces a particle and its corresponding antiparticle

particle-like nature properties that are characteristic of particles such as momentum or deflection by electric or magnetic fields

path difference the difference in distances from two coherent sources to an interference fringe

period of a wave time for one complete cycle of a wave to pass a point

phase difference the fraction of a cycle between the vibrations of two vibrating particles, measured either in radians or degrees

phase difference, in radians, $= \frac{2\pi\Delta t}{T_P}$, for two objects oscillating with the same time period, T_P, where Δt is the time between successive instants when the two objects are at maximum displacement in the same direction

photoconduction electrical conduction due to light

photoelectricity emission of electrons from a metal surface when the surface is illuminated by light of frequency greater than a minimum value, known as the threshold frequency

photon electromagnetic radiation consists of photons. Each photon is a wave packet of electromagnetic radiation. The energy of a photon, $E_{\text{ph}} = hf$, where f is the frequency of the radiation and h is the Planck constant. The momentum of a photon $= \frac{E_{\text{ph}}}{c}$

piezo-electric effect property of certain solids whereby a p.d. applied between opposite faces causes a change of distance between the two faces and where applied stress causes a potential difference

piezo-electric transducer the component in an ultrasonic probe that produces ultrasonic waves when an alternating p.d. is applied to it

pion a meson that contains an up or down quark and an up or down antiquark

plastic deformation deformation of a solid beyond its elastic limit

polarised waves transverse waves that vibrate in one plane only

positron the antimatter counterpart of the electron. It is positively charged and its rest mass is the same as that of the electron

potential difference the energy transferred per unit charge to a component when electric charge passes through it

potential divider two or more resistors in series connected to a source of fixed potential difference. The source p.d. is divided between the resistors as they are in series with each other

potential energy the energy of an object due to its position

potential gradient change of potential per unit change of distance in a given direction

power = rate of transfer of energy = $\dfrac{\text{energy transferred}}{\text{time taken}}$

pressure force per unit area acting on a surface perpendicular to the surface

principle of conservation of energy states that in any change, the total amount of energy after the change is always equal to the total amount of energy before the change

principle of conservation of momentum states that when two or more bodies interact, the total momentum is unchanged, provided no external forces act on the bodies

probable error estimate of the uncertainty of a measurement

projectile a projected object in motion acted on only by the force of gravity

proton a particle with a fixed positive charge of $+1.60 \times 10^{-19}$ C and a mass of 1.663×10^{-27} kg . One or more protons are in every atomic nucleus

quark protons and neutrons and all other hadrons consist of quarks and/or antiquarks. There are six types of quarks (up, down, strange, charm, top, bottom) referred to as quark flavours.

quark model (or standard model) a quark can join with an antiquark to form a meson or with two other quarks to form a baryon. An antiquark can join with two other antiquarks to form an antibaryon

radian measure of an angle defined such that 2π radians = $360°$

radiant flux intensity, F radiation energy per second per unit surface area received from the star at normal incidence on a surface. The unit of radiant flux intensity is W m^{-2}

random error error of measurement with no obvious cause

red shift increase of wavelength of electromagnetic radiation from a receding star or galaxy due to its receding motion

relative speed the difference between the speeds of two objects moving along the same straight line in the same direction (or the sum of their speeds if they are moving in opposite directions).

renewable energy This is energy from a source that is continually renewed. Examples include hydroelectricity, tidal power, geothermal power, solar power, wave power and wind power

resistance $\dfrac{\text{p.d.}}{\text{current}}$

resistivity resistance per unit length × area of cross-section

resonance large-amplitude oscillations that occur in a lightly-damped system when the frequency of the applied force is equal to the natural frequency of oscillation of the system

reverse biased the direction of a diode in a circuit in order for the diode not to conduct

Rutherford's α-particle scattering experiment demonstrated that every atom contains a positively charged nucleus which is much smaller than the atom and where all the positive charge and most of the mass of the atom is located

satellite motion a satellite is a small object in orbit round a larger object. For a satellite moving at speed v in a circular orbit of radius r round a planet, its centripetal acceleration, $\dfrac{v^2}{r} = g$. Substituting $v = \dfrac{2\pi r}{T}$, where T is its time period, and $g = \dfrac{GM}{r^2}$, where M is the mass of the planet, it can be shown that $T^2 = \left(\dfrac{4\pi^2}{GM}\right)r^3$. A satellite in a geostationary orbit is always directly above the same point on the Equator. This is because it is in a circular orbit in the same plane as the Equator and it has a time period of exactly 24 hours

scalar a physical quantity with magnitude only

semiconductor a substance in which the number of charge carriers increases when its temperature is raised

SI system the scientific system of units

simple harmonic motion an object oscillates in simple harmonic motion if its acceleration is proportional to the displacement of the object from equilibrium and is always directed towards equilibrium:
- the acceleration, a, of an object oscillating in simple harmonic motion is given by the equation $a = -(2\pi f)^2 x$, where x = displacement from equilibrium, and f = frequency of oscillations
- the solution of this equation depends on the initial conditions. If $x = 0$ and the object is moving in the + direction at time $t = 0$, then $x = A \sin(2\pi f t)$. If the object is at maximum displacement, $+A$, at time $t = 0$, then $x = A \cos(2\pi f t)$

sinusoidal wave a wave that has the same shape as a sine wave

specific heat capacity, c, of a substance is the energy needed to raise the temperature of 1 kg of the substance by 1 K without change of state. The energy needed to raise the temperature of mass m of a substance from T_1 to $T_2 = mc(T_2 - T_1)$, where c is the specific heat capacity of the substance

spectrometer optical instrument used to measure the wavelengths of the lines in a spectrum

speed change of distance per unit time

standard candle an object of known luminosity

stationary waves wave pattern with nodes and antinodes formed when two or more progressive waves of the same frequency and amplitude pass through each other

strain extension per unit length of a solid when deformed

stress force per unit area of cross-section in a solid perpendicular to the cross-section

strong interaction/strong nuclear force the force in a nucleus responsible for holding the protons and neutrons in the nucleus together and for interactions between hadrons

superposition the effect of two waves adding together when they meet

systematic error error of measurement with a known cause

terminal velocity the maximum speed reached by an object when the drag force on it is equal and opposite to the force causing the motion of the object

thermistor resistor which is designed to have a resistance that changes with temperature

thermodynamic temperature, T, in kelvins = temperature in °C + 273(.15)

threshold frequency minimum frequency of light that can cause photoelectric emission

threshold wavelength maximum wavelength of light that can cause photoelectric emission

time period (or period) is the time taken for one complete cycle of oscillations

torque of a couple force × perpendicular distance between the lines of action of the forces

total internal reflection a light ray travelling in a substance is totally internally reflected at a boundary with a substance of lower refractive index, if the angle of incidence is greater than a certain value known as the critical angle

transducer any device designed to convert energy from one form to another. A piezo-electric transducer generates a p.d. when it is squeezed

transformer a transformer converts the amplitude of an alternating p.d. to a different value. It consists of two insulated coils, the primary coil and the secondary coil, wound round a soft iron laminated core

- The transformer rule states that the ratio of the secondary voltage to the primary voltage is equal to the ratio of the number of secondary turns to the number of primary turns
- For a transformer that is 100% efficient, the output power (= secondary voltage × secondary current) = the input power (= primary voltage × primary current)

transverse waves waves with a direction of vibration of the particles perpendicular to the direction of energy transfer by the waves

ultrasonic probe hand-held device used in medicine to direct ultrasonic pulses into the body

ultrasonics sound waves at frequencies above the range of the human ear, which is about 18 kHz

uniform field a region where the field strength is the same in magnitude and direction at every point in the field

- the electric field between two oppositely charged parallel plates is uniform. The electric field strength $E = \dfrac{V}{d}$, where V is the p.d. between the plates and d is the perpendicular distance between the plates
- the gravitational field of the Earth is uniform over a region which is small compared to the scale of the Earth
- the magnetic field inside a solenoid carrying a constant current is uniform

upthrust the upward force on a body in a fluid due to the pressure in the fluid

vector a physical quantity with magnitude and direction

velocity change of displacement per unit time

velocity selector arrangement with perpendicular electric and magnetic fields that only allow charged particles at a particular speed through

voltage gain the ratio of the output p.d. from an amplifier to its input p.d.

W-boson a charged particle that is created in a nucleus when it undergoes beta decay. The W⁻ boson then decays into an electron and an antineutrino. The W⁺ boson then decays into a positron and a neutrino

wave-like nature properties that are characteristic of waves such as interference or diffraction

wave particle duality

- matter particles have a wave-like nature as well as a particle-like nature. For example, electrons directed at a thin crystal are diffracted by the crystal. This is wave-like behaviour in contrast with the particle-like behaviour of electrons in a beam which is deflected by a magnetic field. See de Broglie wavelength
- photons have a particle-like nature, as shown in the photoelectric effect, as well as a wave-like nature as shown in diffraction experiments

wavelength distance between two adjacent wave peaks

weak nuclear force/weak interaction the force in a nucleus responsible for beta decay and for interactions between leptons

weight the force of gravity acting on an object

work force × distance moved in the direction of the force

work done work is energy transferred when a force moves its point of application in the direction of the force. The work done W by a force F when its point of application moves through displacement s at angle θ to the direction of the force is given by $W = Fs\cos\theta$

work function of a metal minimum amount of energy needed by an electron to escape from a metal surface

X-rays electromagnetic radiation of wavelength less than about 1 nm. X-rays are emitted from an X-ray tube as a result of fast-moving electrons from a heated filament (as the cathode) being stopped on impact with the metal anode. X-rays are ionising and they penetrate matter. Thick lead plates are needed to absorb a beam of X-rays

Answers to summary tests

Chapter 1

1.1

1 **a** 63 km   **b** 18°
2 **a** 40 m s^{-1} East, 69 m s^{-1} North   **b** 21 km
3 6.1 kN vertically up, 2.2 kN horizontal
4 **a i** 10.4 km   **ii** 6.0 km
   b i 20.4 km   **ii** 10.0 km
   c 22.7 km
5 **a** 3.7 N at 33° to 3.1 N
   b 17.1 N at 21° to 16 N
   c 1.4 N at 45° to 3 N and to 1 N
6 **a** 14.0 N   **b** 6.0 N   **c** 10.8 N at 22° to the 10 N force
   d 13.5 N at 10° to the 10 N force

1.2

1 **a** 80 km h^{-1} **b** 22 m s^{-1}
2 **a** 2.5×10^4 km h^{-1} **b** 7.0×10^3 m s^{-1}
3 **a** 45 000 m   **b** 28.3 m s^{-1}
4 **b i** 4.0 km
   ii 30 m s^{-1} then, 25 m s^{-1} in the opposite direction

1.3

1 **a** 1.5 m s^{-2} **b** −2.5 m s^{-2}
2 **a** 0.45 m s^{-2} **b** 7.9 m s^{-1}
3 **a** 0–20 s: straight line from the origin to 12 m s^{-1};
   20–60 s: flat line; 60–90 s straight line from 12 m s^{-1}
   to zero speed at 90 s.
   b 0.60 m s^{-2}, 0, −0.40 m s^{-2}
4 **a** The velocity increases with time at a decreasing rate
   and reaches a constant value.
   b The acceleration decreases with time (at a
   decreasing rate) to zero.

1.4

1 **a** 2.0 m s^{-2} **b** 221 m
2 **a** 43 s   **b** −0.93 m s^{-2}
3 **a i** 0.2 m s^{-2} **ii** 90 m
   b i −0.75 m s^{-2} **ii** 8.0 s   **d** 3.0 m s^{-1}
4 **a** 5.0 m s^{-2} **b** 7.5 m   **c** 18 m   **d** 6.4 m s^{-1}

1.5

1 **a** 4.0 m   **b** 8.8 m s^{-1}
2 **a** 3.2 s   **b** 31 m s^{-1}
3 **a i** 3.9 s   **ii** 38 m s^{-1}
4 **a** 1.6 m s^{-2} **b** 3.6 m s^{-1} **c** 0.64 m

1.6

1 **a i** 83 s   **ii** 127 s
   b i The graph should show a straight line from
   0 to 100 m at 83 s then a straight line down to
   zero displacement at 210 s.

ii The graph should show a straight line from
0 to 100 m at 83 s then a straight line for 127 m to
a distance of 200 m at 210 s.
c The graph should be a flat line at 1.20 m s^{-1} from
0 to 83 s then a flat line at −0.79 m s^{-1} (= 100 m / 127 s)
from 83 s to 210 s.
2 **a** 600 s
   b i The line should be flat at 8.8 m s^{-1} from 0 to 200 s
   the at zero velocity from 200 to 800 s.
   ii The line should be flat at 2.2 m s^{-1} from 0 to 800 s
3 **a i** The displacement increases from O to A at an
   increasing rate for 1200 s then at a decreasing
   rate over 300 s to a constant displacement at B
   then at this value for 600 s to C before decreasing
   at an increasing rate for a further 900 s when it
   suddenly stops at D.
   ii See graph below

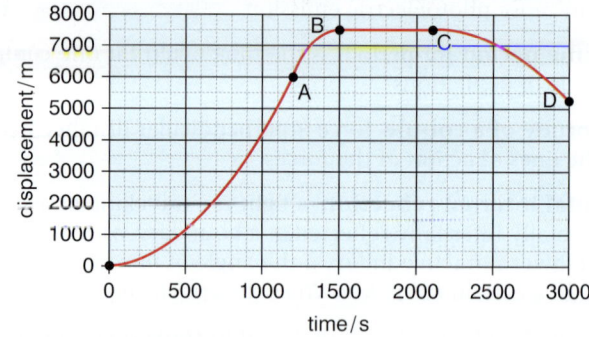

   b i 5250 m   **ii** 9750 m
4 **a i** 0.61 s   **ii** 5.9 m s^{-1} **iii** 0.43 s   **iv** 4.2 m s^{-1}

1.7

1 **a i** 52 s   **ii** 0.49 m s^{-2} **b i** 406 m   **ii** −1.04 m s^{-2}
2 **a** 15 m   **b** −0.13 m s^{-2}
   c 0.67 m s^{-1} downwards, 13.4 m from the start
3 **a i** 80 m   **ii** 8.0 m s^{-1} **b i** 65 s   **ii** −0.12 m s^{-2}
4 **a i** 180 m s^{-1} **ii** 2.7 km
   b 4.4 km
   c 290 m s^{-1}

1.8

1 **a** 32 m s^{-1} **b** 2.8 s   **c** 39 m
2 **a** 3.0 s   **b** 49 m   **c** 34 m s^{-1} (at 62° to the horizontal)
3 **a** 0.20 s   **b** 11.7 m s^{-1}
4 **a** 354 m   **b i** 1020 m   **ii** 1020 m   **c** 146 m s^{-1}

1.9

1 **a** 470 mm   **b** 3.0 m s^{-2}
   c 2.7 m s^{-1} (at 79° to the vertical)
2 **a** 25.8 m   **b** 2.3 s   **c** 4.6 s   **d** 179 m
3 **a** 3.5 m s^{-1}, 3.0 m s^{-1} **b i** 150 m   **ii** 20 m
4 **a** 21 m s^{-1}, 10 m s^{-1} **b** 8.7 m   **c** 27 m s^{-1}

d The range would have been less because it would not reach the same maximum height so would not be in the air for as long and the horizontal component of velocity would progressively decrease from its initial value.

Chapter 2

2.1
1 **a** $0.24\,\text{m}\,\text{s}^{-2}$ **b** $190\,\text{N}$ **c** 0.024
2 **a** $-2.4\,\text{m}\,\text{s}^{-2}$ **b** $12\,000\,\text{N}$
3 **a** $360\,\text{N}$ **b** $23\,\text{s}$
4 **a** $-1.3 \times 10^5\,\text{m}\,\text{s}^{-2}$ **b** $260\,\text{N}$

2.2
1 **a** $5400\,\text{N}$ **b** $7700\,\text{N}$
2 **a** $60\,\text{N}$ **b** $270\,\text{N}$
3 **a** $11.8\,\text{kN}$ **b** $11.8\,\text{kN}$ **c** $12.2\,\text{kN}$ **d** $12.2\,\text{kN}$
4 **a** $1.0\,\text{m}\,\text{s}^{-2}$ **b** $12.5\,\text{N}$

2.3
1 **a i** $0.04\,\text{m}\,\text{s}^{-1}$ **ii** $1.5\,\text{N}$
  b The drag force would be smaller but its weight would be even smaller so its rate of descent would be less.
2 Crouching reduces the drag force so the cyclist with the same 'driving force' could reach a higher speed before the drag force equals the driving force.
3 **a** $0.14\,\text{m}\,\text{s}^{-2}$ **b** $520\,\text{m}$
4 The component of weight down the slope increases the resultant force on the vehicle but the drag force at any given speed is unchanged. So the speed at which the resultant force is zero is higher.

2.4
1 **a** $200\,\text{J}$ **b** $4.5\,\text{J}$
2 **a** $48\,\text{J}$ **b** $24\,\text{J}$ **c** 0
3 **a** $1000\,\text{J}$ **b** $600\,\text{J}$ **c** $400\,\text{J}$
4 **a** $2.4\,\text{N}$ **b** $0.12\,\text{J}$

2.5
1 **a** $9.0\,\text{J}$ **b** $9.0\,\text{J}$ **c** $1.8\,\text{m}$
2 **a i** $2.9\,\text{J}$ **ii** $2.4\,\text{J}$ **b** $0.5\,\text{J}$
3 **a i** $16\,\text{kJ}$ **ii** $5.8\,\text{kJ}$ **b i** $10.2\,\text{kJ}$ **ii** $20\,\text{N}$
4 **a** $590\,\text{kJ}$ **b** $2.4\,\text{kJ}$ **c** $470\,\text{kJ}$ **d** $122\,\text{kJ}$ **e** $1.6\,\text{kN}$

2.6
1 **a** $1.13\,\text{kJ}$ **b** $62.5\,\text{J}\,\text{s}^{-1}$
2 $500\,\text{MW}$
3 **a** $156\,\text{MJ}$ **b** $140\,\text{MJ}$ **c** $12\,\text{MW}$
4 $122\,\text{m}$

2.7
1 **a** $450\,\text{J}$ **b** $1800\,\text{J}\,\text{s}^{-1}$
2 **a** $480\,\text{J}$ **b** $50\,\text{J}$ **c** 10%
3 **a** $570\,\text{MJ}\,\text{s}^{-1}$ **b** $6.2 \times 10^5\,\text{kg}$
4 **a** $600\,\text{s}$ **b** $3.7\,\text{MJ}$ **c** 8%

2.8
1 $1.0 \times 10^7\,\text{m}^2$
2 $4\,\text{MW}$

3 $125\,\text{m}^3\,\text{s}^{-1}$
4 **a** $6.3 \times 10^{11}\,\text{kg}$ **b** $430\,\text{MW}$

Chapter 3

3.1
1 **a** $7.3\,\text{N}$ **b** $7.3\,\text{N}$ at $31.5°$ to vertical
2 **a** See 3.1 Fig 3 with $\theta = 30°$
  b i $2.7\,\text{N}$ **ii** $4.7\,\text{N}$
3 **a** $139\,\text{N}$ **b** $95\,\text{N}$
4 **a** $73°$ **b** $7.0\,\text{N}$

3.2
1 $300\,\text{N}$
2 **a** See 3.2 Fig 2 with an extra weight shown and all the force arrows for the weights labelled and at the correct distances from the pivot.
  b $6.2\,\text{N}$
3 $27\,\text{cm}$
4 $6.75\,\text{N}$

3.3
1 $0.51\,\text{N}$ at $100\,\text{mm}$ mark, $0.69\,\text{N}$ at $800\,\text{mm}$ mark
2 **a** $122\,\text{N}$ at $1.0\,\text{m}$ end and $108\,\text{N}$ from the other end, both vertically upwards
  b $122\,\text{N}$ at $1.0\,\text{m}$ end and $108\,\text{N}$ at the other end, both vertically downwards
3 $620\,\text{kN}$, $640\,\text{kN}$
4 **a** $100\,\text{N}$, $50\,\text{N}$ **b** $150\,\text{N}$

3.4
1 The centre of gravity of the bookcase and books on it is higher than if the books were on the bottom shelf. It would be more unstable because the line of action of its weight would reach the outside its base more readily if it was tilted too far at the top.
2 $89\,\text{N}$
3 **a** $48°$
  b Yes, they will raise the overall centre of mass so it will topple on a less steep slope
4 The centre of gravity of a fully loaded lorry is higher than when it is unloaded. It would be more unstable because the line of action of its weight would reach the outside its base more readily when a side wind acts on it.

3.5
1 **a** $50\,\text{N}$ **b** $250\,\text{N}$
2 **a** $1800\,\text{N}\,\text{m}$ **b** $1800\,\text{N}$
3 $6\,\text{kN}$
4 $10.8\,\text{kN}$

3.6
1 **a** $15\,\text{N}$ **b** $3.0\,\text{N}$ **c** $10.8\,\text{N}$
2 $7\,\text{N}$
3 **a** $6.8\,\text{N}$ **b** $52°$
4 $18.0\,\text{N}$
5 **a** $17\,\text{kN}$ **b** $17\,\text{kN}$
6 **a** $6.2\,\text{N}$ **b** $11.2\,\text{N}$
7 $50\,\text{mm}$ away from pivot

8 a 6.8 N **b** 9.8 N

9 a 2200 N **b** 3100 N

10 a The diagram should show a uniform horizontal beam XY acted on by a labelled vertical force at each end and two labelled vertical downward forces at the correct positions to represent the weights of the beam and the person.

b 950 N at X, 750 N at Y

11 a 2820 kN **b** 1660 kN and 1540 kN

12 a 8.0 N and 16 N **b** 38 N and 76 N

13 a The angle of the beam to the horizontal should be labelled 8.1°. Three labelled vertical force arrows should be shown acting on the beam representing its weight, the tension in the cable and the support force on it from the ground.

b 11 kN, 11 kN

14 a The diagram should show a labelled tension force acting on each top corner of the picture and a force vector for its weight acting vertically downwards at the centre of the picture.

b 28.4 N

Chapter 4

4.1

1 a i $1.2 \times 10^{-18}\,\mathrm{kg\,m\,s^{-1}}$ **ii** $0.050\,\mathrm{kg\,m\,s^{-1}}$
iii $14\,\mathrm{kg\,m\,s^{-1}}$

b i 6.0 kg **ii** $20\,\mathrm{m\,s^{-1}}$

2 a $3.6 \times 10^{5}\,\mathrm{kg\,m\,s^{-1}}$ **b** 60 s

3 a $5.4 \times 10^{6}\,\mathrm{kg\,m\,s^{-1}}$ **b** 45 s

4 a 4.1 Fig 3 with F and t replaced by 400 N and 20 s and an extra step of 20 N for a further 20 s.

b i $9.0 \times 10^{3}\,\mathrm{kg\,m\,s^{-1}}$ **ii** $-8.4 \times 10^{3}\,\mathrm{kg\,m\,s^{-1}}$
iii $1.0\,\mathrm{m\,s^{-1}}$

4.2

1 a $1600\,\mathrm{kg\,m\,s^{-1}}$ **b** 3200 N

2 a $3000\,\mathrm{kg\,m\,s^{-1}}$ **b** 7.5 kN

3 a $4.2 \times 10^{-23}\,\mathrm{kg\,m\,s^{-1}}$ **b** $1.9 \times 10^{-13}\,\mathrm{N}$

4 a $2.1 \times 10^{-23}\,\mathrm{kg\,m\,s^{-1}}$ **b** $9.5 \times 10^{-14}\,\mathrm{N}$

4.3

1 $0.72\,\mathrm{m\,s^{-1}}$,

2 $0.70\,\mathrm{m\,s^{-1}}$ in the same direction

3 $0.050\,\mathrm{m\,s^{-1}}$ in the direction the 1.0 kg trolley was moving in

4 $0.63\,\mathrm{m\,s^{-1}}$ in the opposite direction to its initial direction.

4.4

1 a i Its total energy is the same at the end as at the start as its energy changes from potential to kinetic to elastic energy in the descent and from elastic to kinetic to potential during the ascent.

ii Its total energy is less at the end than at the start as some of its elastic energy is transferred to thermal energy of the ball in the impact and then dissipated to the surroundings. So it has less energy in its ascent than in its descent.

b Assuming air resistance is negligible, its KE just before impact = its loss of PE on the descent from height = $mg \times 1.2\,\mathrm{m}$ and its KE just after impact = its gain of PE on the ascent = $mg \times 0.9\,\mathrm{m}$. So its loss of KE due to the impact = $mg \times 0.3\,\mathrm{m}$. Therefore its % loss of KE = $(mg \times 0.3\,\mathrm{m}\,/\,mg \times 1.2\,\mathrm{m}) \times 100\% = 25\%$

2 a $9.0\,\mathrm{m\,s^{-1}}$ in the same direction **b** 24 kJ

3 a $1.1\,\mathrm{m\,s^{-1}}$ in the reverse direction **b** 20 J

4 a i $1.0\,\mathrm{m\,s^{-1}}$
b i $0.9\,\mathrm{m\,s^{-1}}$ **ii** $0.9\,\mathrm{m\,s^{-1}}$

4.5

1 $0.35\,\mathrm{m\,s^{-1}}$

2 a $0.25\,\mathrm{m\,s^{-1}}$; the mass of A and X was greater than the mass of B, so B moved away faster.

b After they separated, they have equal and opposite momentum therefore $(0.50\,\mathrm{kg} + m_X) \times 0.25\,\mathrm{m\,s^{-1}}$ $= 0.50\,\mathrm{kg} \times 0.30\,\mathrm{m\,s^{-1}}$ so $0.25 m_X = 0.150 - 0.125$ which gives $m_X = (0.150 - 0.125) \times 4 = 0.10(0)\,\mathrm{kg}$.

3 a i $0.10\,\mathrm{m\,s^{-1}}$ **ii** 15 mJ **b** $0.19\,\mathrm{m\,s^{-1}}$

4 a $8.9\,\mathrm{m\,s^{-1}}$ **b i** 1.1 J **ii** 79 J

Chapter 5

5.1

1 a $8.0 \times 10^{-4}\,\mathrm{m^3}$ **b** $3.1 \times 10^{3}\,\mathrm{kg\,m^{-3}}$

2 a 6.3 kg **b** $2.0 \times 10^{-3}\,\mathrm{m^3}$ **c** $3.1 \times 10^{3}\,\mathrm{kg\,m^{-3}}$

3 a $9.6 \times 10^{-6}\,\mathrm{m^3}$ **b** $7.5 \times 10^{-2}\,\mathrm{kg}$

4 a i 0.29 kg **ii** 0.12 kg **b** $2.3 \times 10^{3}\,\mathrm{kg\,m^{-3}}$

5.2

1 a The pressure of the water on the lower half of the ball is greater than the water pressure on the top half. So the ball experiences an upthrust. The volume of liquid displaced by the ball is equal to the volume of the ball. Because water is much denser than air, the weight of the liquid displaced is much greater than the weight of the ball. So the upthrust on the ball is much greater than its weight.

b Cork is used because it doesn't absorb water and it has a much smaller density than water has. When a cork object is in water, the upthrust on it doesn't need to be very large to support its weight so it doesn't need to be fully immersed.

2 a When it is in water, there is an upthrust acting on it. The reading on the newton-meter is therefore less than when it is in air as the upthrust helps to support it in water.

b i The upthrust = 5.2 N – 3.3 N = 1.9 N. The weight of water displaced = upthrust = 1.9 N. The mass of water displaced = weight of water displaced/g = $1.9\,\mathrm{N}/9.81\,\mathrm{N\,kg^{-1}} = 0.194\,\mathrm{kg}$. The volume of the object = the volume of water displaced = mass of water/density of water = $0.194\,\mathrm{kg}/1000\,\mathrm{kg\,m^{-3}}$ $= 1.94 \times 10^{-4}\,\mathrm{m^3}$

ii The mass of the object = its weight/g = 5.2 N/9.81 N kg^{-1} = 0.530 kg. Therefore the density of the object = mass/volume = 0.531 kg/1.94 × 10^{-4} m^{-3} = 2740 kg m^{-3} = 2700 kg m^{-3} to 2 significant figures.

3 a i A has the greatest density because it sinks and B and C both float. A must be more dense than water whereas B and C must be less dense than water.

ii Both B and C have a density less than that of water. Because B floats higher in the water than C, the difference between its density and the density of water must be greater than the corresponding difference for C. So the density of B must be less than the density of C. Therefore B has the lowest density.

4 a The extra weight causes the tube to float lower in the water. Therefore the length of the tube above the water decreases (*linearly*) as the total weight is increased.

b Remove the tube from the water. (Place a soft pad at the bottom of the tube.) Add a suitable metal object (e.g. a steel nail) of known weight to the tube. Use a millimetre ruler to measure the length L of the tube (and cork) above the water. Repeat the procedure five more times, each time adding another object of known weight. Record the measurements in a table including a column for the total weight added, W. Plot a graph of $y = L$ against $x = W$. If the (*linear*) prediction is correct, the graph should have a negative gradient (*and be a straight line with a constant gradient*).

5.3
1 1.2 kPa
2 120 kN
3 13 kN
4 **a** 10.3 m
   b i 1.47 kPa   **ii** This gas pressure is 1.5% ((= 1.47 kPa/101 kPa) × 100%) above atmospheric pressure so it is not normal.

5.4
1 **a** 0.40 m   **b** 12.5 N
2 **a** 20 N   **b** 100 mm   **c** 200 N m^{-1}
3 **a** 40 N   **b** 200 mm
4 **a** 12.3 N m^{-1} **b** 8.8 × 10^{-2} J   **c** 2.2 N

5.5
1 1.0 × 10^{9} Pa
2 1.3 × 10^{11} Pa
3 **a** 9.4 × 10^{8} Pa   **b** 1.2 × 10^{-2} m
4 **a** No, the UTS for glass is greater than for copper.
   b No, the initial gradient for steel is greater than for glass.
   c Yes, the copper curve extends more than the glass curve.

5.6
1 **a** 3.3 × 10^{6} Pa   **b** 2.8 × 10^{-4} m   **c** 0.21 J
2 **a** 2.3 mm   **b** 1.7 × 10^{-2} J

3 **a** 470 kN   **b** 47 J
4 **a** 10.5 J   **b** 3.5 J

Chapter 6

6.1
1 **a** Electrons are negatively charged so free electrons in the can are attracted towards the positively charged rod. Some of the free electrons transfer through the point of contact from the can onto the rod so the can is left with a positive charge.
   b Electrons transfer through the thread from the sphere to the ground.
2 **a** Free electrons on the conductor transfer to the ground through the wire when the conductor is earthed.
   b The conductor therefore loses negative change and is left with a positive charge when the wire is removed.
3 **a** The positive charge of the object attracts free electrons in the metal plate to the surface of the plate.
   b There is a force of attraction between the object and the electrons at the surface because the object is positively charged and the electrons are negatively charged. There is also a force of repulsion between the object and the positive ions in the plate but this force is less because the ions are further from the object than the electrons at the surface are.
4 **a and b** 6.1 Fig 4 with full lines not dashed and with the field direction arrows from + to −.

6.2
1 **a i** 3.5 C   **ii** 210 C   **b i** 3.0 A   **ii** 0.15 A
2 **a** 3.8 × 10^{14} **b** 1.9 × 10^{21}
3 **a** 800 C   **b i** 1600 s   **ii** 8000 s
4 1.0 mm s^{-1}

6.3
1 **a** 29 kJ   **b** 720 J
2 **a** 2 A   **b** 22 kJ
3 **a i** 48 kJ   **ii** 3.5 A   **b** 5 A
4 **a** 12 kJ   **b** 4.5 W   **c** 2700 s

6.4
1 **a** 6.0 Ω, 9.9 V, 0.125 mA, 160 Ω, 2.5 mA   **b** 7.5 Ω
2 31 Ω
3 0.11 mΩ
4 **a** 1.8 × 10^{-6} Ω m   **b** 33 mm

6.5
1 **a** 0.25 A, 12 Ω
   b The filament would become brighter and hotter until it melts and breaks as a result.
2 **a** 0.03 mA   **b** 0.38 mA
4 **a** 30.4 Ω   **b** 46 °C

Chapter 7

7.1

1 a 1.0 A, 4.0 A   b 5.0 A   c 30 W
2 a 7.1 Fig 5 with the battery pd changed and without the other pd values shown.
   b i 2.0 V   ii 0.20 A
3 a 7.1 Fig 3 with a resistor changed to an ammeter and the values / labels changed.
   b i 4.0 V   ii 2.0 V   iii 10 Ω
4 a 3.6 V   b 30 Ω

7.2

1 a 16 Ω   b 3.0 Ω   c 4 Ω
2 a 2 Ω   b 6 Ω   c 1.0 A   d 4.0 W
3 a 3.6 Ω   b 0.83 A
   c 2 Ω: 0.5 W; 4 Ω: 1.0 W; 9 Ω: 1.0 W,   d 2.5 W
4 a 14.4 W   b 2.4 Ω

7.3

1 a 6.0 Ω   b 2.0 A   c 3.0 V   d 9.0 V
2 a 0.5 A   b 1.25 V   c 0.63 W   d 0.13 W
3 a 2.0 Ω   b 1.5 V
4 12 V, 2 Ω

7.4

1 a 12.0 Ω   b 0.25 A
   c 4 Ω: 0.25 A, 1.0 V; 24 Ω: 0.08 A, 2.0 V; 12 Ω: 0.17 A, 2.0 V
2 a i 20.0 Ω   ii 1.05 A   iii 1.05 A, 15.8 V
   b i 2.49 A   ii 3.17 A
3 a i 2.0 W   ii 2.0 W   b i 2.0 W   ii 8.0 W
4 a Q: 0.6 V, 0.06 mA; P: 2.4 V, 0.48 mA
   b P: 0.6 V, 0.12 mA; Q: 2.4 V, 0.24 mA

7.5

1 a 7.5 Fig 1 with a battery instead of a single cell and the values/labels changed; 1 kΩ: 0.75 V; 5.0 kΩ: 3.75 V
   b 1 kΩ: 1.3 V; 5.0 kΩ: 3.2 V
2 Sketch: 7.5 Fig 3c with a light bulb connected between C and B and the labels / values changed; as the contact is moved up from B, the light bulb filament begins to glow and becomes increasingly bright until it reaches its normal brightness when the contact is at A.
3 a i 0.5 A   ii 8.0 Ω: 4.0 V; 4.0 Ω: 2.0 V
   b i 3.0 V   ii 4.0 V
4 a i 2.8 V   ii 6.4 kΩ
   b The voltmeter reading increases because the LDR resistance decreases so the pd across the LDR drops as its share of the 5.0 V cell pd decreases.

Chapter 8

8.1

1 Sound waves are longitudinal; the other 3 are transverse.
2 See 8.1 Fig 2.

3 As 8.1 Fig 4a, with the unnecessary labels removed and arrows to show the reversal of the direction of motion at the peaks and troughs.
4 Q moves from zero displacement to the right then back to zero after half a cycle, as in Fig 7.5b, then it moves to the left of zero displacement then back to zero in the next half-cycle.

8.2

1 a 0.10 m   b 1.9 × 10^{-2} m
2 a 10 GHz   b 5.0 × 10^{14} Hz
3 1.0 V, 1.0 kHz
4 a i amplitude = 9 mm   ii 180°   iii 270°
   b +9 mm

8.3

1 The reflected wavefront is at 30° to the reflector with its top end near the right hand side of the reflector and its direction arrows normal to the wavefront and pointing away from the reflector.
2 Label the wavefronts below the boundary A, B and C from left to right. The 2 missing sections of B and C above the boundary should be both parallel to the section of A above the boundary and the lower end of each missing section should join the corresponding wavefront below the boundary. The direction arrows of the wavefront sections above the boundary should be perpendicular to the wavefronts and pointing away from the boundary.
3 a decreased diffraction
   b increased diffraction
   c increased diffraction compared with b
   d little change
4 a Waves from the transmitter are diffracted at the gap and they spread out beyond the gap so they are detected by the detector.
   b If the detector was made a little wider, there would be less diffraction so the detector signal would be reduced. Making the gap too wide would enable waves to reach the detector directly so the signal would increase.

8.4

1 a An electromagnetic wave is a transverse wave that consists of an oscillating electric field perpendicular to and in phase with an oscillating magnetic field of the same frequency.
   b i In a polarised electromagnetic wave, the oscillations of each field are always in the same perpendicular plane. The plane of polarisation is defined by the plane of the electric field oscillations.
      ii In an unpolarised electromagnetic wave, the plane of polarisation is not constant and it changes at random.
2 The intensity of the observed light decreases and is a minimum when the angle of rotation is 90°. The intensity then increases as the angle of rotation is increased and reaches maximum intensity at 180°.

3 The radio waves are polarised. The signal becomes weaker as the aerial is rotated away from the plane of polarisation of the radio waves and is weakest when it is perpendicular to the plane of polarisation. Further rotation of the aerial causes the signal to become stronger as it is rotated nearer to the plane of polarisation.

4 a Malus's law states that when polarised light is passed through a polariser, the intensity of the transmitted light is given by the equation $I = I_0 \cos^2 \theta$ where θ is the angle the polariser is rotated through from its orientation at the maximum transmitted intensity I_0.

b For $\theta = 10°$, $I = I_0 \cos^2 10° = 0.970\, I_0$ therefore $I = 0.970 I_0 = 97\% \times I_0$. Therefore the % reduction in the transmitted intensity is 3%.

8.5

1 a When the peak is opposite the trough, the rope would be momentarily flat.

b The peak would be near the right end of the rope and the trough would be at the other end, both travelling away from the centre.

2 a The lines of the gaps would be closer together.

b The lines of the gaps would be further apart (because the wavelength would be greater).

3 a The signal would decrease gradually (as the intensity of the waves would decrease).

b The signal may increase due to increased diffraction but a considerable reduction in the gap width would cause the signal to decrease due to the reduced intensity of the diffracted waves.

4 a The signal decreases midway between A and B because the detector is then at a point of cancellation where the intensity is a minimum. The signal then increases as the detector reaches a point of reinforcement at B.

b The intensity is a minimum midway between A and B and a maximum at B. So the signal increases as the detector moves towards B from the midway position.

8.6

1 a 8.6 Fig 1 a and c superimposed with an appropriate arrows to indicate the vibration direction of the node over 1 cycle.

b 8.0 m

2 a 2.0 m **b i** 180° **ii** 225° **iii** 0

3 Progressive wave; all particles vibrate with the same amplitude and a phase difference (over each wavelength) that increases with distance apart. Stationary wave; all particles between adjacent nodes vibrate in phase with amplitudes that increase from zero at the nodes to a maximum midway between the nodes.

4 a The reflected waves and the incident waves form a stationary wave pattern with nodes as the zero signal positions which are 15 mm apart.

b 30 mm

8.7

1 a 1.6 m **b** 410 m s^{-1}

2 a 0.4 m **b** 0.53 m

3 a The frequency of each harmonic would be greater (because the wavelength of each harmonic on the wire would be shorter) so the pitch of each note (or frequencies of the sound waves in each notes created) would be higher.

b The frequency of each harmonic would be greater due to the increase of tension so the pitch of each note (or frequencies of the sound waves in each notes created) would be higher.

4 The notes are different in pitch because the mass per unit length of the two wires differs so the frequency of corresponding harmonics differs. The steel wire has a greater mass per unit length so the speed and therefore each of its harmonic frequencies is lower than that of the corresponding harmonic of the nylon wire.

8.8

1 a 2.40 m, 140 Hz **b** 425 Hz

2 a 2.0 m **b** 57 Hz

3 a 71 Hz **b** 142 Hz

4 68–680 Hz

8.9

1 a 1200 Hz

b When the car is directly under the bridge, its direction of motion is perpendicular to the straight line between an overhead observer and the car. So at that instant, the car is neither approaching or moving away from the bridge. So the observed frequency is equal to the emitted frequency, which is 1100 Hz.

2 a Towards Y, because Y observes a higher frequency than the frequency emitted by the horn

b 1060 Hz

3 a 2040 m s^{-1} **b** 3.4 GHz

4 a The reflected pulses would be too weak to be detected, as they would diffract too much on reflection. Also, the variable speed of the air would affect the speed of the waves.

b The small particles in the air would be very poor reflectors of microwaves because they are so small compared with the wavelength of the microwaves.

Chapter 9

9.1

1 a The fringes would be closer together because the fringe spacing would be smaller.

b The interference fringes would disappear although a diffraction pattern from the remaining slit would be seen.

2 550 nm

3 0.9 mm

4 0.75 m

9.2

1 **a** See 9.1 Fig 2

    b The fringe pattern would consist of alternate bright and dark fringes. The bright fringes would be yellow-orange, the same colour as the sodium lamp.

2 **a** See 9.1 Q1b answer

    b The fringe spacing would be unchanged but the bright fringes would be wider and the dark fringes narrower.

3 1.1 mm

4 The central fringe would be white and the other inner fringes would be blue at the edges nearest the central fringe, and red at the edges furthest from the central fringe. The edges of the outer fringes would overlap so these fringes would be less distinct.

9.3

1 **a** 10.9°, 22.2° **b** 5

2 **a** 2 **b** 0.58° (= 35′)

3 **a** 1090 **b** 69.9°

4 **a** 599 mm^{-1} **b i** 3 **ii** 50° 40′

Chapter 10

10.1

1 **a** Most of an atom is empty space.

    b There is a positively charged nucleus at the centre of the atom. The nucleus is much smaller than the atom and is where most of the atom's mass is located.

2 **a** Alpha particles from the source would collide with air molecules and be stopped. So the chamber needs to be evacuated.

    b If the foil is not thin enough, each alpha particle would be scattered by the nuclei of the atoms in the foil more than once.

    c Alpha particles of different speeds moving along the same initial path would be scattered differently so the scattered alpha particles at each angle of deflection would not have followed the same path and the measurements would not confirm Rutherford's theory.

    d If the beam was too wide, the scattering at nearby angles of deflection would overlap and the measurements may not confirm Rutherford's theory.

3 **a** See 10.1 Fig 5; the force on the particle at X should be shown as an arrow acting on the particle pointing away from the centre of the nucleus.

    b i The KE decreases as the particle approaches X where its KE is a minimum and then increases as it moves away from X.

      ii Its PE increases from zero as the particle approaches X where its PE is a maximum and then decreases to zero at infinity as it moves away from X.

4 **a** The activity of the source decreases because more and more nuclei become stable and the number of radioactive nuclei in the source decreases. If the source did not have a long half-life, later readings would be significantly lower than earlier readings due to the decrease of the source activity.

    b Alpha particles with different kinetic energies approaching a nucleus along the same path would be deflected by different amounts so the reading at each angle of deflection would not be due to alpha particles of the same kinetic energy.

10.2

1 **a** 6p, 6n **b** 8p, 8n **c** 92p, 143n

    d 11p, 13n **e** 29p, 34n

2 **a** neutron **b** electron **c** neutron

3 **a i** $+3.2 \times 10^{-19}$ C **ii** 63

    b 8 neutrons and 10 electrons

4 Most of the hydrogen atoms are $^{1}_{1}$H atoms and therefore have a single proton as a nucleus. Less energy is needed for a neutron to knock a $^{1}_{1}$H nucleus (i.e. a proton) out compared with knocking a carbon nucleus out. This is because a neutron hitting a carbon nucleus is likely to recoil and retain some kinetic energy, whereas a neutron hitting a $^{1}_{1}$H nucleus is likely to be stopped and use most of its kinetic energy to eject the $^{1}_{1}$H nucleus.

10.3

1 **a** Beta particles

    b Use a magnet to deflect a narrow beam of them; they should deflect as in 10.3 Fig 1 if they are beta particles.

2 **a** Alpha radiation

    b The radiation affects the film and causes it to blacken when the film is developed. The key prevented the radiation reaching the film underneath the key so the image of the key was seen on the film when it was developed.

3 **a i** gamma **ii** alpha

    b Alpha radiation in air has a certain range. When the alpha source was moved beyond the range of the source, the current dropped to zero because the alpha particles could not reach the ionisation chamber.

4 **a** 42% **b** 58%

10.4

1 **a** $^{238}_{92}\text{U} \rightarrow {}^{234}_{90}\text{Th} + {}^{4}_{2}\alpha$ **b** $^{228}_{90}\text{Th} \rightarrow {}^{224}_{88}\text{Ra} + {}^{4}_{2}\alpha$

2 **a** $^{64}_{29}\text{Cu} \rightarrow {}^{64}_{30}\text{Zn} + {}^{0}_{-1}\beta \ (+ \bar{\nu})$

    b $^{32}_{15}\text{P} \rightarrow {}^{32}_{16}\text{S} + {}^{0}_{-1}\beta \ (+ \bar{\nu})$

3 **a** $^{213}_{84}\text{Po}$, $^{209}_{82}\text{Pb}$, $^{209}_{83}\text{Bi}$

    b i 83p + 130n **ii** 83p + 126n

4 **a i** $^{205}_{84}\text{Th} \rightarrow {}^{201}_{82}\text{Pb} + {}^{4}_{2}\alpha$

      ii $^{201}_{82}\text{Pb} \rightarrow {}^{201}_{81}\text{Th} + {}^{0}_{+1}\beta + \nu$

    b 1 They have equal and opposite charge.

      2 The positron is an antimatter particle whereas the β⁻ particle is a matter particle (or the positron is the antiparticle of the β⁻ particle (or electron)).

10.5

1 a The creation of charged atoms (ions) by adding or removing electrons to or from uncharged atoms.

b Alpha radiation from outside the body is absorbed by the layer of dead skin at the body suface whereas beta radiation can penetrate the skin.

2 a Ionising radiation affects living cells by damaging or destroying cell membranes or damaging DNA molecules in the cells, causing the cells to divide uncontrollably.

b i A film badge provides a record of how much ionising radiation of each type its wearer has been exposed to.

ii See 10.5 Fig 3 and related text.

3 a Lead is very dense and a thick lead plate absorbs α, β and γ radiation. A lead-lined storage box absorbs the radiation from the radioactive sources inside it and so prevents the radiation from leaving the box.

b Long handles ensure the user is as far away as possible from thc radioactive source while it is being moved.

4 Solid sources should only be moved by robots or using long-handles tongs and should be out of its storage box for as little time as possible. Liquid or gas or powdered sources should be in sealed containers and only moved as above. No sources should be allowed to make contact with the skin. The eyes should never be exposed to ionising radiation from a radioactive source.

10.6

1 a A W boson is charged; a photon is uncharged. A W boson has a very short range; a photon has an unlimited range.

b lifetime $\cong$ range/speed of light = 3×10^{-15}m$/3 \times 10^{8}$s = 10^{-23}s

2 a Leptons interact through the weak interaction; hadrons interact through the strong interaction.

b i a proton or a neutron **ii** a π meson **iii** a K meson

3 a i uud **ii** udd

b One of the up quarks in the proton changes into a down quark and emits a W^+ boson, which then decays into a β^+ particle and a neutrino.

4 a −1

b $\pi^- = \overline{u}d$, X = ssu. X must contain a u quark rather than a d quark to ensure conservation of charge.

Chapter 11

11.1

1 a 1.75×10^{-3}rad **b** 0.105rad **c** 6.28rad

2 a 20ms **b i** 0.31rad **ii** 310rad

3 a 470ms^{-1} **b i** 0.0042° **ii** 7.3×10^{-5}rad

4 a 7.0kms^{-1} **b i** 0.050° **ii** 8.7×10^{-4}rad

11.2

1 a 0.23ms^{-1} **b i** 7.9×10^{-4}ms^{-2} **ii** 5.1×10^{-2}N

2 a 0.53ms^{-1}, 0.66ms^{-2} **b** 9.9×10^{-2}N

3 a i 3.0×10^{4}ms^{-1} **ii** 5.9×10^{-3}ms^{-2}

b i 7.9×10^{3}ms^{-1} **ii** 5.1×10^{3}s

4 a 8.4ms^{-1} **b** 88ms^{-2} **c** 175N

11.3

1 a 6.7ms^{-2} **b** 3.8kN

2 a 4.1ms^{-2} **b** 3.0kN

3 Sprinters run much faster than marathon runners so on a banked rather than a flat circular track, sprinters would be less likely to slip as there would be a component of their weight acting parallel to the track to maintain their circular motion.

4 a 40ms^{-1}

11.4

1 a 30ms^{-1} **b i** 11.3ms^{-2} **ii** 690N

2 a 25ms^{-1} **b** 20ms^{-2} **c** 2000N

3 a 13ms^{-1} **b** 13ms^{-2} **c** 240N

4 −0.04N

Chapter 12

12.1

1 Her/his velocity would increase until the bungee rope is vertical and starts to stretch, causing the velocity to decrease gradually to zero at the lowest point of the descent. The stretched rope would then pull the jumper upwards, increasing their velocity until the rope becomes slack after which the jumper's velocity would decrease to zero at or below the platform.

2 a No frictional forces are present and the oscillations are at constant amplitude.

b Depress the free end of the ruler and record its initial position against a vertical millimetre scale then release it and record its lowest position after every five cycles. The readings should be unchanged if it oscillates freely.

3 a 0.48s **b** 2.1Hz

4 a $\frac{\pi}{2}$ radians **b** π radians

12.2

1 a +25mm, changing direction from up to down

b 0, moving down

c −25mm, changing direction from down to up

d 0, moving up

2 a 0.5Hz **b i** −0.25ms^{-2} **ii** 0 **iii** 0.25ms^{-2}

3 a 0.5Hz **b** −0.32ms^{-2}

4 a −32mm, 0.32ms^{-2} **b** 0, 0

12.3

1 a 0.33Hz **b** 0.25ms^{-2}

2 a i 12mm **ii** 0.63s **b** 10.1mm

3 a 2.1Hz **b** 0.057m

4 a 3.7Hz

b i 8.7mm, −191mms^{-1}

ii −12mm, −16mms^{-1}

12.4

1 a i 0.33 s   ii 3.1 Hz
   b i 0   ii −3.7 m s^{-2} iii −7.5 m s^{-2}
2 a i 3.0 Hz   ii 0.33 s   b $f_2 < f_1$ ∴ $m_2 > m_1$
3 a i 70 mm   ii 21 N m^{-1}
   b i   Calculate ω as $\omega^2 = k/m$ then calculate the
        frequency f using $f = \omega/2\pi$
    ii 0.53 s
4 a i 1.25 N   ii 2.5 m s^{-2}
   b i $\omega^2 = k/m = 25$ N m^{-1}/0.50 kg = 50 rad^2 s^{-2}
        therefore $a = -\omega^2 x = -50x$
    ii 1.1 Hz, +46 mm
5 a i 2.0 s   ii 1.0 s   b 5.0 s
6 The mass–spring system would have the same time
   period on the Moon as it has on the Earth. The simple
   pendulum would have larger time period on the Moon
   than it has on the Earth.

12.5

1 a   In the first quarter cycle, its KE decreases to zero
     and its PE increases to maximum. In the next
     quarter cycle, its KE increases to a maximum and its
     PE decreases to a minimum. The above sequence is
     repeated in the next 2 quarter cycles.
   b See 12.5 Fig 2
2 a 60 N m^{-1} b i 7.5 × 10^{-2} J   ii 7.5 × 10^{-2} J
3 a i   In each half-cycle starting at maximum
       displacement, its PE decreases and its KE
       increases from zero then its PE increases and its
       KE decreases to zero. Air resistance causes the
       total energy to decrease gradually.
    ii   The water in the tube oscillates with a decreasing
       amplitude between the two sides of the U
       tube. In each half-cycle starting at maximum
       displacement, its PE decreases and its KE increases
       from zero then its PE increases and its KE
       decreases to zero. Fluid friction (viscosity) causes
       the kinetic energy of the water and the amplitude
       of the oscillations to decrease rapidly to zero.
   b   The suspension would be slow to respond and
     therefore less effective giving an less comfortable
     ride.
4 a 82 mm   b 44 mm

12.6

1 a   Resonance is when the amplitude of an oscillating
     system becomes very large as a result of being
     subjected to a periodic force of the same frequency
     as the natural frequency of the system.
   b   The periodic force is then in phase with the
     displacement of the oscillating system so the
     amplitude becomes very large.
2 a   It would be lower because the increased mass
     makes the time period longer so the frequency is
     lower.
   b   It would be higher because the spring constant of
     stiffer springs is greater and so the time period is
     shorter and the frequency is higher.

3   The rotation of the drum causes a periodic force to act on
   the panel. At a certain frequency of rotation of the drum,
   the panel resonates because the frequency of rotation is
   equal to the panel's natural frequency of vibration.
4 a   When the vehicle passes over the speed bumps, it
     experiences an upward force at each speed bump.
     If the frequency of this periodic force is equal to
     the natural frequency of the suspension system,
     resonance occurs and the chassis moves up and
     down violently because amplitude of the oscillations
     becomes very large.
   b i   Resonance would not occur because the periodic
       force frequency would no longer be equal to the
       natural frequency.
    ii   The mass of the system would be greater so the
       natural frequency of the system would be lower
       and resonance would occur at a lower speed than
       before.

Chapter 13

13.1

1 a   A line along which a small mass would move if no
     other forces acted on it.
   b See 13.1 Fig 2 and related text.
2 a i 33 N   ii 160 N   b i 16 N kg^{-1} ii 4.0 N kg^{-1}
3   See 13.1 page 1.
4 a   The field should be radial as in 13.5 Fig 2.
   b   The field should be mostly radial except near the
     surface above the mass where the line would be
     closer together.

13.2

1 a 235 J
   b $\Delta V = 235$ J/12 kg = 19.6 J kg^{-1}
2 a 2.0 MJ kg^{-1} b i −61 MJ kg^{-1} ii 2.1 × 10^9 J
3 a i −250 J   ii −200 J   iii −200 J
   b i 50 J   ii 0
4 a $\Delta\phi$ between X and the equipotential 1 km away
     = 5 kJ kg^{-1} so $\Delta\phi$ between X and an equipotential
     10 m away = 50 J kg^{-1} (= (5000 J kg^{-1} × 10 m/1000 m)
     so $W = m\Delta\phi$ = 1 kg × 50 J kg^{-1} = 50 J
   b 5 N kg^{-1} c 25 MJ

13.3

1 a 1.3 × 10^{-6} N   b 5.4 mm
2 a 780 N   b 6.0 × 10^{24} kg
3 a 54 N   b 0.24 N
4 a i 16.6 N   ii 0.2 N
   b 16.4 N towards the centre of the Earth

13.4

1 a 7.35 × 10^{22} kg
   b   At the Earth, $g_{Moon} = GM_{Moon}/d^2$ (where d = Earth–
     Moon distance) = 6.67 × 10^{-11} N m^2 kg^{-2} ×
     7.35 × 10^{22} kg/(3.8 × 10^8 m)^2 = 3.4 × 10^{-5} N kg^{-1}.
     Therefore, g_{Moon}/g_s = 3.4 × 10^{-5} N kg^{-1}/9.81 N kg^{-1}
     = 3 × 10^{-6}.

2 a i $68\,\mathrm{N\,kg^{-1}}$ **ii** $5.9 \times 10^{-3}\,\mathrm{N\,kg^{-1}}$
  b Use $g = GM/r^2$ to calculate the Earth's gravitational field strength at $260\,000\,\mathrm{km}$ from the Earth's centre and to calculate the Sun's gravitational field strength at $1.5 \times 10^{11}\,\mathrm{m}$ ($-2.6 \times 10^8\,\mathrm{m}$) from the Sun.

3 a $0.028\,\mathrm{N\,kg^{-1}}$
  b At height h above the Earth, $g = g_s R^2 / (R + h)^2$. For $h = 10\,\mathrm{km}$, $h \ll R$ (the Earth's radius) so at $10\,\mathrm{km}$ above the surface, $g = g_s$.
  c $7.1 \times 10^6\,\mathrm{J}$

4 $-2.8\,\mathrm{MJ\,kg^{-1}}$, $1410\,\mathrm{MJ}$

13.5

1 a X moves faster across the sky so its time period of less and therefore its radius of orbit is less.
  b Satellite TV dishes need to point to a geostationary satellite which is a satellite that stays in the same position directly above a point on the surface at the equator. If the dish is not aligned correctly, the signal it receiver from the satellite will be too weak to detect.

2 a $3.4 \times 10^6\,\mathrm{m}$ **b** $3.0\,\mathrm{N\,kg^{-1}}$ **c** $5.2 \times 10^{23}\,\mathrm{kg}$

3 a i v^2/r
    ii centripetal force mv^2/r = force of gravity on the satellite mg where m is the mass of the satellite therefore $v^2 = gr$.
  b i $9.5\,\mathrm{N\,kg^{-1}}$ **ii** $7.9\,\mathrm{km\,s^{-1}}$ **iii** $5200\,\mathrm{s}$

4 a See 13.5 page 1
  b i Insert appropriate values into the equation in part a and calculate v.
    ii $7100\,\mathrm{s}$

Chapter 14

14.1

1 a $1.4 \times 10^{-3}\,\mathrm{N}$ **b** $4.0 \times 10^4\,\mathrm{V\,m^{-1}}$

2 a i negative   **ii** $1.3 \times 10^{-7}\,\mathrm{C}$
  b i $7.3 \times 10^{-3}\,\mathrm{N}$ **ii** towards the metal surface

3 a i $9.0 \times 10^4\,\mathrm{V\,m^{-1}}$ **ii** $7.2 \times 10^{-14}\,\mathrm{N}$ **b** $80\,\mathrm{mm}$

4 The acceleration of each electron towards the positive plate, $a = F/m = eE/m$. The time taken, t, by each electron to cross the field $= x/v$. Since $y = \frac{1}{2}at^2$, $y = \frac{1}{2}(eE/m) \times (x/v)^2 = \frac{1}{2}kx^2$, where $k = eE/mv^2$

14.2

1 a $3.7 \times 10^{-11}\,\mathrm{N}$ **b** $2.6 \times 10^{-10}\,\mathrm{N}$

2 a i $69\,\mathrm{mm}$ **ii** $3.6 \times 10^{-6}\,\mathrm{N}$ **b** $2.5 \times 10^{-5}\,\mathrm{N}$ repulsion

3 a $6.1\,\mathrm{nC}$, negative   **b** $2.2 \times 10^{-2}\,\mathrm{N}$

4 a $2.7\,\mathrm{nC}$, attract   **b** $6.2 \times 10^{-2}\,\mathrm{m}$, repel

14.3

1 a i $-8.0 \times 10^{-18}\,\mathrm{J}$ **ii** $+7.2 \times 10^{-17}\,\mathrm{J}$
  b $+8.0 \times 10^{-17}\,\mathrm{J}$

2 a $-1.8 \times 10^{-3}\,\mathrm{J}$ **b** $+1.2 \times 10^{-3}\,\mathrm{J}$

3 a i $250\,\mathrm{V\,m^{-1}}$
    ii $8.0 \times 10^{-17}\,\mathrm{N}$ (towards the negative plate)
  b $-8.0 \times 10^{-19}\,\mathrm{J}$

4 a See Topic 14.3
  b i $3000\,\mathrm{V\,m^{-1}}$
    ii As 14.3 Figure 4 with the x-axis labelled 'distance h/mm' and the y-axis labelled 'potential $/\mathrm{V}$', and with the values 60 and 20 in place of ΔV and Δd respectively.

14.4

1 a $5.3 \times 10^6\,\mathrm{V\,m^{-1}}$ **b** $10\,\mathrm{mm}$

2 a i $3.7 \times 10^8\,\mathrm{V\,m^{-1}}$ **ii** $5.6 \times 10^{-3}\,\mathrm{N}$ towards Q_1
  b Insert appropriate values into the equation for the electric field strength near a point charge to calculate the electric field strength of each charge at the midpoint. The resultant value of E should be zero because the two calculated field strength values should be of equal strength and they are in opposite directions.

3 a i $4.5 \times 10^8\,\mathrm{V\,m^{-1}}$ towards Q_2
    ii $2.6 \times 10^8\,\mathrm{V\,m^{-1}}$ away from Q_1
    iii $4.90 \times 10^8\,\mathrm{V\,m^{-1}}$ at $83.4°$ to the line between the two charges
  b i Q_1 and Q_2 are both positive charge so a test charge at any point on the line between them would experience two forces in opposite directions. At some point along the line nearer Q_2 than Q_1 the two forces would be equal in magnitude and opposite in direction so the resultant electric field strength at that point would be zero.
    ii $11\,\mathrm{mm}$ from Q_1, $9\,\mathrm{mm}$ from Q_2

4 a $-9.0 \times 10^6\,\mathrm{V}$
  b i The electric potential near a point charge Q is proportional to Q/r where r is the distance from Q. For zero electric potential at a point P due to two point charges, $Q_1/r_1 + Q_2/r_2 = 0$ therefore $r_2/r_1 = -Q_2/Q_1 = -(-30\,\mathrm{mC})/(+15\,\mathrm{mC}) = 2.0$ which is in agreement with the given distances of $20\,\mathrm{mm}$ for r_2 and $10\,\mathrm{mm}$ for r_1.
    ii $2.0 \times 10^9\,\mathrm{V\,m^{-1}}$ directly towards Q_2

Chapter 15

15.1

1 a $5.0\,\mu\mathrm{F}$ **b** $2.2\,\mathrm{V}$ **c** $9.9\,\mathrm{mC}$
  d $1.4\,\mu\mathrm{F}$ **e** $11\,\mathrm{V}$ **f** $3.4\,\mathrm{mC}$

2 a $264\,\mu\mathrm{C}$ **b** $106\,\mathrm{s}$

3 a $27.5\,\mu\mathrm{C}$ **b** $5.5\,\mu\mathrm{F}$

4 a $910\,\mu\mathrm{C}$ **b** $220\,\mu\mathrm{F}$ **c** $700\,\mu\mathrm{C}$ **d** $7.4\,\mathrm{V}$

15.2

1 a $5.0\,\mu\mathrm{F}$
  b $2.0\,\mu\mathrm{F}$; $6.0\,\mu\mathrm{C}$, $3.0\,\mathrm{V}$; $3.0\,\mu\mathrm{F}$; $9.0\,\mu\mathrm{C}$, $3.0\,\mathrm{V}$

2 a i $2.4\,\mu\mathrm{F}$
    ii $6.0\,\mu\mathrm{F}$; $11.0\,\mu\mathrm{C}$, $1.8\,\mathrm{V}$, $4.0\,\mu\mathrm{F}$; $11.0\,\mu\mathrm{C}$, $2.7\,\mathrm{V}$
  b i $3.2\,\mu\mathrm{F}$
    ii $4\,\mu\mathrm{F}$; $14.4\,\mu\mathrm{C}$, $3.6\,\mathrm{V}$, $10\,\mu\mathrm{F}$; $9.0\,\mu\mathrm{C}$, $0.90\,\mathrm{V}$, $6.0\,\mu\mathrm{F}$; $5.4\,\mu\mathrm{C}$, $0.90\,\mathrm{V}$

3 1. All in series $6.0\,\mu$F   2. All in parallel $79\,\mu$F
   3. Two in series with the third in parallel, $25\,\mu$F, $30\,\mu$F, $54\,\mu$F
   4. Two in parallel in series with the third, $9\,\mu$F, $16\,\mu$F, $19\,\mu$F
4 **a** $4.9\,\mu$F
   b $2\,\mu$F; $12\,\mu$C, $6.0\,$V, $4\,\mu$F; $17\,\mu$C, $4.3\,$V, $10\,\mu$F; $17\,\mu$C, $1.7\,$V

15.3
1 **a** $30\,\mu$C, $45\,\mu$J   **b** $60\,\mu$C, $180\,\mu$J
2 **a** $0.45\,$C, $2.0\,$J   **b** $10\,$W
3 $2.2\,\mu$F; $5.4\,\mu$C, $6.6\,\mu$J, $10\,\mu$F; $5.4\,\mu$C, $1.5\,\mu$J
4 **a** $56\,\mu$C, $340\,\mu$J   **b** $6.9\,\mu$F
   c $8.2\,$V   **d** $4.7\,\mu$F; $160\,\mu$J, $2.2\,\mu$F; $73\,\mu$J

15.4
1 **a i** $300\,\mu$C   **ii** $5.0\,$s   **b i** $5\,$s approx   **ii** $20\,$kΩ
2 **a i** $0.61\,$mC   **ii** $0.45\,$mA   **b** $0.23\,$V, $11\,\mu$A
3 **a** $13\,\mu$C, $40\,\mu$J   **b** $0.62\,$V   **c** $0.42\,\mu$J
4 **a** $0.34\,$mJ   **b** $1.4\,$s   **c** $0.32\,$mJ

Chapter 16

16.1
1 **a** S   **b** NW
2 **a** S   **b** sudden switch to N
3 **a** clockwise   **b** reversed
4 **a** field due to coil along axis
   b $90°$ if coil field $>>$ Earth's field

16.2
1 **a** S   **b** W   **c** N
2 **a** N $\rightarrow$ S   **b** vertical up
3 **a** The current in the wires down the two long sides of the coil is in opposite directions. The two long sides of the coil experience a force due to the magnetic field which is equal in magnitude and opposite in direction. These two forces act as a couple because they are in opposite directions and they do not act along the same line (except when the plane of the coil is perpendicular to the field lines).
   b When the coil is perpendicular to the field lines, the force on each long side acts along the same line as the force on the other long side so they have no turning effect.
4 **a** It reverses the direction of the current in the coil when the plane of the spinning coil moves through the position where it is perpendicular to the field lines. As a result, the coil continues to spin in the same direction.
   b i It would spin faster.
   ii It would spin in the opposite direction.
   iii It would oscillate about the position where its plane is perpendicular to the field lines.

16.3
1 **a** $0.14\,$T   **b i** 0 **ii** $11\,$mN   **iii** $22\,$mN

2 **a** $2.4 \times 10^{-2}\,$N west
   b $4.5\,$A west
   c $0.20\,$T vertically down
   d South, $8.0 \times 10^{-3}\,$N
3 $0.10\,$N due south at $20°$ below the horizontal
4 Long sides: $2.7\,$N on each side perpendicular to the plane of the coil and in opposite directions; Short sides: 0

16.4
1 **i** The force would reverse in direction.
   ii The force would be reduced in magnitude.
   iii The force would increase.
2 **i** $1.9 \times 10^{-13}\,$N   **ii** $9.6 \times 10^{-14}\,$N
3 **a** The velocity of each electron is parallel to the field so the field does not exert a force on the electrons.
   b The initial velocity has a component parallel to the field and a component perpendicular to the field. Therefore the parallel component of velocity is constant and the perpendicular component continually changes direction without changing its magnitude. So the electrons move along a helical path (ie spiral) around the field lines.
4 **a** The magnetic field exerts a force on the conduction electrons that pushes them towards one edge of the slice, making that edge negative which causes the opposite edge to become positive and thus creating a potential difference between the negative edge and the opposite edge.
   b i $4.4\,$ms^{-1} **ii** $8.5 \times 10^{-20}\,$N

16.5
1 **a i** Each electron in the beam is acted on by a magnetic force perpendicular to their direction of motion. The force causes each electron to move on a circular path because it is always perpendicular to the direction of motion of the electron, so it acts as a centripetal force.
   ii $21\,$mm   **b** $2.8\,$mT
2 **a** $4.7\,$mT   **b** $17.5\,$mm
3 **a** Use the equation on 16.5 page 1 with the given data including the values of e and m.
   b $1.1\,$MeV
4 **a** $8.0 \times 10^{6}\,$Ckg^{-1} **b** $1.4 \times 10^{7}\,$Ckg^{-1}

Chapter 17

17.1
1 **a** The motion of the magnet into the coil causes an induced e.m.f in the coil that creates a current in the circuit, which is detected by the meter.
   b Any two of the following: Move the magnet faster or use a stronger magnet or wind more turns of wire on the coil.
2 **a** The coil in the motor spins between the poles of the magnets in the motor and so an e.m.f is induced in the coil which creates a current in the circuit that passes through the lamp and lights it.

b The lamp would be brighter because the induced e.m.f would be greater so the current in the lamp would be greater.

3 a A dynamo contains a magnet and a coil. In 14.5 Fig 14.5.3, when the dynamo turns, the magnet spins and the coil is fixed. As a result an e.m.f. is induced in the coil. Because a lamp is connected to the coil, the induced e.m.f. creates a current in the circuit which passes through the lamp and lights it.

b When the lamp lights, the forces needed to turn the dynamo transfer energy to the dynamo to generate the electric current and to overcome friction in the dynamo. Less force is needed when the lamp is disconnected because less energy is transferred to the dynamo as no current is generated.

4 a i Vertically downwards

ii The eastern end: each conduction electron in the rod is pushed by the magnetic field towards the westward end so the western end becomes negative and the eastern end positive.

b The rod does not cut across the field lines because it is parallel to them so no e.m.f. is induced in it.

17.2

1 a i 1.1 mWb **ii** 2.0 s **iii** 0.54 mV

b i The graph should have its y-axis labelled 'flux linkage / mWb' and its horizontal axis labelled 'time / s'. The line should be a sloped straight line for the 2 s from the origin to 1.08 mWb then flat for the next 4 s at 1.08 mWb then for the next 2 s, it is a sloped straight line down to the x-axis.

ii The graph should show the negative as well as the positive parts of the y-axis with its y-axis labelled 'emf / V' and its horizontal axis labelled 'time / s'. The line is flat at 0.54 mV for the first 2 s then zero for the next 4 s then −0.54 mV for the next 2 s then zero from 8 s after the start.

2 a 1.4 mWb **b** 23 mV

3 a i $4.5 \times 10^{-4}\,\text{m}^2$ **ii** 1.5(4) mWb

b i 3.1 mWb **ii** 32 mV

4 a 8.0 μWb **b** 40 μV

17.3

1 a As 17.3 Fig 3 with appropriate values and labels shown on both axes and the other labels removed. Note the time period is 50 ms.

b As above with a peak emf of 12 V (because the induced emf is proportional to the frequency) and with a time period of 33 ms.

2 a i 4.3 A (=1000 W/230 V = 4.347 A),

ii 6.1 A (= 4.347 A × $\sqrt{2}$ = 6.148 A)

b Peak power = 2 × mean power = 2000 W

c Power P is proportional to the square of the pd (assuming the resistance is unchanged) so a 5% drop in the pd V causes an approximate 10% drop in the power supplied.

3 a 26 mWb

b i The speed of rotation of each side = the angular speed of the coil × the radius of rotation r (which is half its width) = $2\pi f / r = 2\pi \times 50\,\text{Hz} / 0.019\,\text{m}$ = 6.0 m s^{-1}.

ii $E_0 = 2NBlv = 2 \times 80 \times 0.13\,\text{T} \times 0.065\,\text{m} \times 6.0\,\text{m s}^{-1}$ = 8.1 V

4 a A back emf is induced in the motor when it spins. The faster the motor spins, the greater the back e.m.f so the motor current is small because the back emf acts against the battery e.m.f.

b When the load is increased, the motor spins slower so the back emf is smaller. The current increases because the back emf acting against the battery e.m.f is less.

17.4

1 a The alternating current in the primary coil creates an alternating magnetic field (via the core) in the secondary coil which induces an alternating e.m.f. in the secondary coil.

b The current in the secondary coil increases when a device is connected across it and this reduces the magnetic flux in the core and therefore reduces the back emf in the primary coil so the primary current increases.

2 a The induced e.m.f. in the secondary coil is proportional to the rate of change of magnetic flux through it. If the magnetic flux in it is less than in the primary coil due to the transformer design, the peak secondary e.m.f is less than it could be.

b Direct current in the primary coil of a transformer would not create changing magnetic flux so there would not be an induced emf in the secondary coil. Alternating current does create changing magnetic flux in a transformer so does cause an induced emf in the secondary coil.

3 a 11.5 V

b i 0.26 A **ii** 5.2 A

4 a Electrical power = current × voltage so the current needed to deliver a certain amount of power is much reduced if the voltage applied to the transmission cables is increased much more. The power dissipated in the cables due to the resistance heating effect of the current is therefore much reduced

b i 17 A (16.7 A to 3 sig. fig.) **ii** 56 kW

Chapter 18

18.1

1 The electric current in the circuit transfers energy from the battery to the heater which increases its store of thermal energy so it becomes hot. The heater transfers energy by heating the water which increases its store of thermal energy. The current also heats the circuit wires due to their resistance, causing energy transfer to the wires and the surroundings.

2 a Friction in the motor bearings and resistance heating due to the current in the wires cause the motor to become warm.

b The current in the circuit transfers energy to the motor coil from the battery. The current in the coil does work on the coil to make it turn. In the process, energy is transferred from the coil to the weight which gains potential energy.

3 a Internal energy is the energy of its molecules due to their individual movements.

b An electric lamp at constant brightness has energy transferred to it by the electric current in it and transfers energy at the same rate to the surroundings as it heats the surroundings and radiates light. So its internal energy is constant.

4 a The molecules in a solid are held to each other in fixed positions and they vibrate about these fixed positions whereas the molecules in a liquid move about at random but still in contact with each other.

b The molecules vibrate more and more about their fixed positions as the solid's temperature increases to its melting point. At this temperature, substance changes from a solid to a liquid as the molecules break free from each other and move about at random state. Above this temperature, the molecules move about more and more as the temperature is increased.

18.2

1 a See 18.2 page 1
b i 273 K **ii** 293 K **iii** 77 K

2 a 328 K **b** 137 kPa

3 The liquid thread does not expand by equal lengths for equal increases of temperature because the volume of the thermometer bulb changes with temperature.

4 Any ideal gas in a gas thermometer always gives the same temperature reading at any temperature not just at the fixed points.

18.3

1 a 23 kJ **b** 536 kJ
2 a 270 s **b** 10.3 MJ
3 a 320 J **b** 130 J kg^{-1} K^{-1}
4 3.2 kW

18.4

1 a The vibrating molecules in the solid need energy in order to increase their kinetic energies so they can break free from each other.

b The internal energy of the water molecules needs to be reduced, so their kinetic energy is reduced so that they move slower and form strong bonds in fixed positions with each other.

2 0.16 kg
3 a 4.2 J s^{-1} **b** 6500 s
4 a 22 J s^{-1} **b** 6.5 kJ

Chapter 19

19.1

1 126 kPa
2 79 kPa
3 0.097 m^3
4 a Apply Boyle's law to the data without the powder present and with the final volume as $(1.20 \times 10^{-4}\,\text{m}^3 - V_\text{P})$ where V_P is the volume of the pump.
b $2.33 \times 10^{-5}\,\text{m}^3$, 1600 kg m^{-3}

19.2

1 a 1.1 moles **b** 109 kPa,
2 a 9.3×10^{-4} moles **b** $2.1 \times 10^{-5}\,\text{m}^3$
3 a Use the ideal gas equation to calculate the pressure at each temperature in kelvins. The temperature axis should be labelled in °C. A straight line should be drawn between the two plotted points.
b 1.3 kg m^{-3},
4 a 1.2 kg m^{-3} **b** 2.5×10^{22}

19.3

1 Increasing the temperature of the gas causes the root mean square speed of the molecules to increase. Therefore the molecules have more momentum on average so their change of momentum on impact is greater. Also, because the container volume is constant, the molecules move faster on average so they collide with the container walls more frequently. Therefore, the mean force exerted by each molecule is greater.

2 524 m s^{-1}
3 a 7.52×10^{-3} **b** 474 m s^{-1}
4 a 1.48 moles **b** 4.2×10^{-2} kg **c** 5.2×10^2 m s^{-1}

19.4

1 a 0.97 moles **b** 7.7×10^{-21} J **c** 4.5 kJ
2 a 5.7×10^{-21} J **b** 1.8×10^3 m s^{-1}
3 a The mean kinetic energy of a molecule of an ideal gas depends only on the temperature of the gas. The gas temperature is the same throughout the gas so the mean kinetic energy of any gas molecule is the same as that of any other molecule.

b In any sample of air, the mean kinetic energy of a nitrogen molecule is equal to that of an oxygen molecule in the same sample. The kinetic energy of a gas molecule is equal to $\frac{1}{2} m\, c_{\text{r.m.s.}}^2$, where m is its mass and $c_{\text{r.m.s.}}^2$ is its mean square speed. Therefore $\frac{1}{2} m_n (c_{\text{r.m.s.}})_n^2 = \frac{1}{2} m_\text{o} (c_{\text{r.m.s.}})_\text{o}^2$ where the subscripts n and o are used respectively to denote the masses and mean square speeds of a nitrogen molecule and an oxygen molecule.

Rearranging this equation gives
$$\frac{\text{the mean square speed of a nitrogen molecule, } (c_{\text{r.m.s.}})_n^2}{\text{the mean square speed of an oxygen molecule, } (c_{\text{r.m.s.}})_\text{o}^2}$$
$$= \frac{\text{mass of an oxygen molecule, } m_\text{o}}{\text{mass of a nitrogen molecule, } m_n}$$

The ratio of molecular masses m_o/m_n = the ratio of the molar masses = $\frac{0.032}{0.028}$ = 1.14

Therefore the ratio of mean square speeds = 1.14

Hence the r.m.s. speed of a nitrogen molecule = $1.14^{0.5}$ = 1.07 × the r.m.s. speed of an oxygen molecule.

4   a   i   1.50   ii   220K   b   870J   c   1310J   d   2180J

Chapter 20

20.1

1   a   The emission of electrons from the surface of a metal when light above a certain frequency is directed at the surface.

    b   The work function of a metal is the minimum amount of energy an electron at the surface of a metal needs to leave the surface. Light consists of photons each of energy equal to hf where f is the frequency of the light. When light is directed at the surface of a metal, an electron at the surface could absorb a photon and leave the metal surface if the energy it gains from the photon is greater than the work function of the metal.

2   a   i 6.7×10^{14} Hz, 4.4×10^{-19} J
       ii 2.0×10^{14} Hz, 1.3×10^{-19} J

    b   The energy of a 450 nm photon is greater than the work function of the metal, so if an electron at the metal surface absorbs a 450 nm photon, the electron can leave the surface. An electron at the metal surface that absorbs a 650 nm photon could not leave the surface as a 650 nm photon has less energy than the work function.

3   a 1.7×10^{14} Hz   b 2.7×10^{-19} J

4   a 3.1×10^{-19} J   b 1.6×10^{-19} J   c 2.5×10^{14} Hz

20.2

1   a   The blue light had a frequency greater than the threshold frequency of the metal so the energy of each of its photons was greater than the work function of the metal. Hence an electron at the metal surface could leave the surface if it absorbed a blue photon.

    b   The red light had a frequency below the threshold frequency of the metal so the energy of each of its photons was less than the work function of the metal . Hence an electron at the metal surface that absorbed a red photon would not have had sufficient energy to leave the surface and it would have quickly lost the energy it gained in collisions with other electrons no matter how many photons were incident on the surface.

2   a 1.6×10^{12}

    b   The number of electrons per second leaving the surface (ie the photoelectric current) is proportional to the number of photons per second incident on the surface. So if the intensity is doubled, the number of photons per second incident on the surface is doubled and hence the photoelectric current is doubled.

3   a 3.4×10^{-19} J
    b 1.5×10^{15}
    c 2.5×10^{12}

4   a   i 4.0×10^{14} Hz   ii 2.7×10^{-19} J,
    b 2.7×10^{-19} J

20.3

1   a   9.0 eV

    b   There are three energy levels below the 5.7 eV level. There are three possible transitions to the ground state, two to the first excited state and one to the second excited state making six transitions in total.

2   a   1. The energy of an electron in an atom increases in an excitation and decreases in a de-excitation.

       2. Excitation can occur through photon absorption or electron collision. De-excitation only occurs through photon emission.

    b   i   The diagram should show the ground state and two excited levels at 1.8 eV and 4.6 eV

       ii   photon energies 1.8 eV, 2.8 eV, 4.6 eV

3   The energy levels of each type of atom are unique to that atom. So the photons emitted are characteristic of that atom.

20.4

1   a   Light has wave-like properties, such as diffraction. It consists of wave packets called photons, which have particle-like properties such as in the photoelectric effect, in which a single photon is absorbed by a single electron.

    b   Matter particles transfer momentum when they collide hence have particle-like properties and they also have wave-like properties, for example they can be diffracted by thin crystals.

2   a   The particle-like nature of light.
    b   The wave-like nature of particles.

3   a 3.6×10^{-11} m   b 1.9×10^{-14} m

4   a 1.3×10^{-27} kg m s^{-1}, 1.5×10^{3} m s^{-1}
    b 1.3×10^{-27} kg m s^{-1}, 0.78 m s^{-1}

Chapter 21

21.1

1   a   73 s   b   6 g

2   a   i 9.0×10^{14} ii $2.2(5) \times 10^{14}$
    b   i 9.0×10^{14} ii 15.8×10^{14} c 1.3×10^{3} J

3   a   19 kBq   b   2.4 kBq

4   a 4.4×10^{21} b 1.1×10^{21} atoms

21.2

1   a 1.0×10^{-6} s^{-1} b 3.9×10^{16}

2   a 6.3×10^{-10} s^{-1} b   20.5 kBq

3   a 2.7×10^{24} b   i   0.65 kg   ii 1.7×10^{24}

4   a 1.3×10^{-6} s^{-1} b   149 hours

21.3

1 a The carbon dioxide content of atmosphere contains a small percentage of radioactive carbon isotope $^{14}_{6}C$. Living wood contains this isotope because trees and plants absorb carbon dioxide from the atmosphere.

 b 1340 years

2 a i 730 kBq ii 870 kBq iii 1.9×10^{-13} kg

 b The flow through the first kidney is normal because the radioactive isotope flows through it. The second kidney is blocked because the radioactive isotope enters it but does not leave it.

3 a i The foil is made by using rollers either side of a metal plate to squeeze the plate as it passes between the rollers. The foil then passes between a source of beta radiation and a detector which supplies a feedback signal to the rollers. If the foil becomes too thin, the amplitude of the signal increases and the pressure of the rollers on the foil is reduced. The opposite happens if the foil becomes too thick.

 ii The intensity of the beta radiation passing through the foil decreases with increasing thickness of the foil. Alpha radiation would be absorbed by the foil. Gamma radiation would almost all pass through the foil.

 b i Gamma radiation is absorbed by thick lead. The thick lead lining in the container prevents gamma radiation from the source harming people near the container.

 ii A narrow hole in the lead container allows a narrow beam of gamma radiation to be directed at the tumour in the patient in order to destroy it. Directing the beam at the tumour from different directions reduces the damage caused by gamma radiation to the normal tissues outside the tumour.

4 a i Human tissues are damaged by alpha, beta, and gamma radiation. An alpha source would be most suitable as alpha radiation from the source inside the pacemaker would all be absorbed inside the pacemaker and would not reach the tissues outside the pacemaker. Radiation from a beta or gamma source would reach the tissues outside the pacemaker and damage them.

 ii The longer the half-life of a radioactive isotope is, the less its activity per mole is. A greater amount of the radioactive isotope in the pacemaker would therefore be needed to provide sufficient power if a source with a much longer half-life was used.

 b i 5.2×10^{13} Bq ii $3.3 \, \mathrm{J\,s^{-1}}$

21.4

1 a 2.18×10^{-15} kg

 b i 8.89×10^{-33} kg ii 8.89×10^{-30} kg

2 a $^{212}_{83}Bi \rightarrow \,^{208}_{81}Tl + \,^{4}_{2}\alpha$; 6.2 MeV

 b When the alpha particle is emitted, the nucleus recoils with equal and opposite momentum. The energy released in the process is shared between the nucleus and the alpha particle in inverse proportion to their masses. So the nucleus gains a small proportion of the energy released.

3 a $^{90}_{38}Sr \rightarrow \,^{90}_{39}Y + \,^{0}_{-1}\beta + \bar{v}$; 0.56 MeV

 b When the beta particle is emitted, a neutrino or an antineutrino is emitted so the energy released is shared between the nucleus and the two emitted particles. The nucleus takes a small proportion of the energy released and the rest is shared between the two emitted particles so the beta particle's kinetic energy can be any value from zero to a maximum.

4 3.2 MeV

21.5

1 a The binding energy is the work that must be done to separate a nucleus into its constituent neutrons and protons.

 b See 21.5 Figure 1, without the individual points and unnecessary labels.

2 a 7.4 MeV b 8.8 MeV

3 a i 7.1 MeV ii 2.6 MeV

 b The binding energy per nucleon of an alpha particle is greater than that of the other particle. Both particles need kinetic energy to escape from the nucleus. An alpha particle has more kinetic energy after it forms than the other particle would have because more binding energy is released when the alpha particle forms

4 1.1 MeV

21.6

1 a They do not have sufficient kinetic energy to overcome the strong nuclear force holding them in the nucleus.

 b When a nucleus is formed from separate proton and neutrons, the strong nuclear force between them binds them together and their potential energy decreases so their total mass decreases in accordance with $E = mc^2$.

2 a Nuclear fission is the process that occurs when a large unstable nucleus splits into two approximately equal fragments and energy is released. This can happen spontaneously or can be induced in certain nuclei when a free neutron collides with such a nucleus.

 b i $a = 56$, $b = 98$ ii 206 MeV

3 a Nuclear fusion is the process that occurs when two nuclei combine to form a larger nucleus and energy is released. This can only happen if the nucleon number of the nucleus formed is no greater than about 50.

 b i In the first interaction, one of the two protons changes into a neutron and a positron is released. The neutron and the other proton form a hydrogen $^{2}_{1}H$ nucleus. In the second interaction, a third proton collides with the hydrogen $^{2}_{1}H$ nucleus to form a $^{3}_{2}He$ nucleus.

 ii 0.43 MeV, 5.5 MeV

4 a The nuclei need to have sufficient kinetic energy to overcome the mutual coulomb repulsion force between them so that they can approach each other closely enough for the strong nuclear force between them to attract them together.

b 12.9 MeV

c The mass difference between 4 separate protons and an alpha particle and 2 released positrons
= $(4 \times 1.00728\,u) - (4.00150\,u + 0.00110\,u)$
= $2.65 \times 10^{-2}\,u$. Therefore the energy released
= $2.65 \times 10^{-2}\,u \times 931.5\,MeV/u \approx 25\,MeV$

Chapter 22

22.1

1 a i 0.14 mm **ii** 0.62 mm

b At this frequency, their wavelength in the body is significant compared with the transducer and with body structures so diffraction is significant. As a result, they spread out and weaken too much so they do not form distinct images.

2 a i See Topic 22.1, Figure 1 and related notes

ii The backing block prevents ultrasonic waves created at the two surfaces of the disc from cancelling each other out. The pad also damps the vibrations of the disc rapidly after each pulse is emitted.

b i A is due to reflection at the cornea; B is due to reflection at the front surface of the eye lens; C is due to reflection at the back surface of the eye lens; the furthest pulse (at the right side of the screen) is due to reflection at the retina.

ii about 13 mm

3 a i 0.999 **ii** 1.73×10^{-3}

b Gel and skin have similar acoustic impedances so a gel–skin interface hardly reflects any ultrasonic waves from the probe. If the gel was not used, trapped air between the probe and the body would reflect most of the ultrasonic waves because air and skin have very different acoustic impedances.

c i $1.64 \times 10^{6}\,kg\,m^{-2}\,s^{-1}$ **ii** 1.6×10^{-5}

4 a A B-scan gives a two-dimensional image whereas an A-scan only gives information about the distances from the probe to reflecting boundaries in one direction. A B-scan requires a multi-transducer probe whereas an A-scan uses a probe with a single transducer.

b Ultrasonic waves are non-ionising unlike X-rays so they would not harm the baby whereas X-rays might.

22.2

1 a ii 0.15 W

b i The intensity increases because more electrons strike the anode each second so more X-ray photons are produced.

ii The beam becomes more penetrating because the maximum energy of the X-ray photons in the beam is increased. This happens because the electrons strike the anode with more kinetic energy when the tube voltage is increased.

2 a i Sharpness is to do with how clearly the edges of structures in an X-ray image can be seen.

ii Contrast is to do with the relative darkening of an X-ray film in different areas of the film. Poor contrast occurs if the lightest areas are not much lighter than the darkest areas.

b A scattering grid placed between the patient and the film prevents X-rays scattered in the body of the patient from reaching the film. Without the grid, such X-rays would darken the shadow areas of the film which would reduce the contrast and sharpness of the image.

3 b A contrast medium is necessary when an X-ray image of a soft body organ such as the stomach is being created. If the organ is filled with the contrast medium, it absorbs X-rays more effectively. This improves the contrast of the image so the edges and structures of the image can be seen more clearly.

4 a See Topic 22.2, Figure 8 for a suitable diagram. The X-ray tube produces a narrow beam which is detected by electronic detectors on the other side of the patient. The tube and the detectors are on a gantry which is rotated in steps about the patient so that the beam always passes through the same cross-section of the patient. The detector signals are recorded at each step and used to construct a cross-sectional image of the patient.

b i 12 **ii** 3, 1, 3, 5

22.3

1 a $^{18}_{8}O + ^{1}_{1}p \rightarrow ^{18}_{9}F + ^{1}_{0}n$

b $^{18}_{9}F \rightarrow ^{18}_{8}O + ^{0}_{1}\beta + \nu$

2 a No. of nuclei N = activity A/decay constant λ
= $A \times$ half-life/ln 2 = $220 \times 10^6 \times 110 \times 60\,s$/ln 2
= 2.09×10^{12} nuclei. Mass of each nucleus
= $0.018\,kg/6.02 \times 10^{23} = 3.00 \times 10^{-26}\,kg$. Mass of N nuclei = $2.09 \times 10^{12} \times 3.00 \times 10^{-26}\,kg = 6.2 \times 10^{-14}\,kg$

b $A = 220 \times 10^6 \times e^{-(\ln 2/110 \times 60) \times (24 \times 60)} = 25\,kBq$
(Alternative method: No. of half-lives = 24×60 mins/110 mins = 13.1 half-lives. Therefore activity = $220\,MBq/2^{13.1} = 25\,kBq$

3 a The positron is the antiparticle of the electron. When a positron-emitting nucleus decays, it emits a positron which is annihilated when it collides with an electron. The loss of mass and energy is converted into radiation energy in the form of gamma photons.

b Two gamma photons are produced in each annihilation event. Because momentum is conserved and the total momentum before the event is negligible, the two photons are created with equal and opposite momentum. Because energy is conserved and they have equal and opposite momentum, they each have the same energy after the event.

4 a A radioactive tracer is a radioactive emitter of gamma (or beta) radiation attached to the molecules of a fluid introduced into a system to obtain information about the system by using external detectors to detect the emissions.

b A PET scanner is designed to locate positron emissions from positron-emitting substances by using a ring of detectors to detect two gamma photons emitted in opposite directions within a very short time of each other. Other gamma-emitting isotopes emit gamma photons in random directions so are unlikely to trigger the PET detectors.

Chapter 23

23.1

1 For Proxima Centauri, $F = L/4\pi d^2$
$= 10\,000\,L_{SUN}/4\pi\,(250\,000\,d_{SUN})^2$
$= \{10\,000/(250\,000)^2\} \times (L_{SUN}/4\pi d_{SUN}^2)$
$= 1.6 \times 10^{-7}\,F_{SUN}$. Therefore, for Proxima Centauri,
$F = 1.6 \times 10^{-7} \times 1400\,W\,m^{-2} = 0.22\,mW\,m^{-2}$.

2 a $2.7 \times 10^8\,s$

b For Sirius, $F = L/4\pi d^2$
$= 4.0 \times 10^{26}\,W/(4\,\pi \times (8.1 \times 10^{16}\,m)^2$
$= 4.9 \times 10^{-9}\,W\,m^{-2}$

3 Rearranging $F = L/4\,\pi\,d^2$ hence $d^2 = L/4\pi F$
$= 4.4 \times 10^{36}\,W/(4\,\pi \times (2.1 \times 10^{-13}\,W\,m^{-2}))$
$= 1.7 \times 10^{48}\,m^2$. Therefore $d = 1.3 \times 10^{24}\,m$.

4 Since F is inversely proportional to the square of the distance from the Earth to the Sun, a change of F by 1.7% is due to a change of distance of about 3.4% ($= 2 \times 1.7\%$). Therefore the distance changes by about $5.1 \times 10^9\,m$ ($= 3.4\% \times 1.5 \times 10^{11}\,m$)

23.2

1 Rearranging $\lambda_{max}\,T = 0.0029\,m\,K$ gives
$T = 0.0029\,m\,K/620 \times 10^{-9}\,m = 4700\,K$

2 a Rearranging $L = \sigma AT^4$ gives $A = L/\sigma T^4$
Hence $A = \dfrac{6.0 \times 10^{28}}{5.67 \times 10^{-8} \times (3400)^4} = 7.9 \times 10^{21}\,m^2$

b i For a sphere of radius R, its surface area $A = 4\pi R^2$
Rearranging this equation gives
$R^2 = \dfrac{A}{4\pi} = \dfrac{7.9 \times 10^{21}}{4\pi} = 6.3 \times 10^{20}\,m^2$.
Hence $R = 2.5 \times 10^{10}\,m$

ii Ratio of radius to Sun's radius $= \dfrac{2.5 \times 10^{10}\,m}{7.0 \times 10^8\,m} = 36$

3 Since the surface area is proportional to the diameter squared, the surface area of the star is 16 × the surface area of the Sun. Therefore
$\dfrac{L_{star}}{L_{Sun}} = \dfrac{A_{star}\,T_{star}^4}{A_{Sun}\,T_{Sun}^4} = \dfrac{16A_{Sun}\,(2T_{Sun})^4}{A_{Sun}\,T_{Sun}^4} = 16 \times 2^4 = 256 \approx 250$.

4 a Since they have the same surface temperature and X has greater luminosity than Y, the surface area of X and therefore its diameter must be greater than that of Y.

b $L_X = \sigma A_X T_X^4 = 5.67 \times 10^{-8}\,W\,m^{-2}\,K^{-4} \times 4\pi \times (0.5 \times 2.0 \times 10^9\,m)^2 \times (5400\,K)^4 = 6.0 \times 10^{26}\,W$

c Since $T_X = T_Y$, $\dfrac{L_X}{L_Y} = \dfrac{\sigma\,A_X\,T_X^4}{\sigma\,A_Y\,T_Y^4} = \dfrac{A_X}{A_Y}$, therefore
$A_X/A_Y = L_X/L_Y = 100$.
Hence the ratio of the square of their diameters is 100 so the diameter of X is 10 × the diameter of Y. Hence the diameter of Y is $2.0 \times 10^8\,m$.

23.3

1 a Hubble's law states that the distant galaxies are receding and that their speed of recession is proportional to their distance from Earth.

b The Universe is expanding.

2 a $v = c\,(\Delta\lambda/\lambda) = 300\,000\,km\,s^{-1} \times 7.6\,nm/589.6\,nm$
$= 3900\,km\,s^{-1}$

b Rearranging $v = Hd$ gives $d = \dfrac{v}{H} = \dfrac{3.9 \times 10^6\,ms^{-1}}{2.2 \times 10^{-18}\,s^{-1}}$
$= 1.8 \times 10^{24}\,m$

3 a i $v = c(\Delta\lambda/\lambda) = 300\,000\,km\,s^{-1} \times 39\,nm/486\,nm$
$= 24\,000\,km\,s^{-1}$

ii Rearranging $v = Hd$ gives $d = \dfrac{v}{H} = \dfrac{2.4 \times 10^7\,ms^{-1}}{2.2 \times 10^{-18}\,s^{-1}}$
$= 1.1 \times 10^{25}\,m$

b $L = F \times 4\pi d^2$
$= 2.7 \times 10^{-15}\,W\,m^{-2} \times 4\pi \times (1.1 \times 10^{25}\,m)^2$
$= 4.1 \times 10^{36}\,W$

4 a CMBR is background microwave radiation that is detected from all directions in space.

b CMBR is thought to be electromagnetic radiation that has been travelling through space after being created in the Big Bang when the Universe was created in a massive explosion.

Chapter 24

24.1

1 a i 0.500 m   **ii** 320 cm   **iii** 95.6 m
b i 450 g   **ii** 1.997 kg   **iii** 5.4×10^7 g
c i $2.0 \times 10^{-3}\,m^2$ **ii** $5.5 \times 10^{-5}\,m^2$ **iii** $5 \times 10^{-6}\,m^2$

2 a i $1.5 \times 10^{11}\,m$ **ii** $3.15 \times 10^7\,s$
iii $6.3 \times 10^{-7}\,m$ **iv** $2.57 \times 10^{-8}\,kg$
v $1.50 \times 10^5\,mm$ **vi** $1.245 \times 10^{-6}\,m$
b i 35 km   **ii** 650 nm   **iii** $3.4 \times 10^3\,kg$
iv 870 MW (= 0.87 GW)

3 a i $20\,ms^{-1}$ **ii** $20\,ms^{-1}$ **iii** $1.5 \times 10^8\,ms^{-1}$
iv $3.0 \times 10^4\,ms^{-1}$
b i $6.0 \times 10^3\,\Omega$ **ii** $5.0\,\Omega$ **iii** $1.7 \times 10^6\,\Omega$
iv $4.9 \times 10^8\,\Omega$ **v** $3.0\,\Omega$

4 a i 301   **ii** 2.8×10^9 **iii** 1.9×10^{-23}
iv 1.2×10^{-3} **v** 2.0×10^4 **vi** 7.9×10^{-2}
b i 1.6×10^{-3} **ii** 5.8×10^{-6} **iii** 1.7   **iv** 3.1×10^{-2}

24.2

1 a 1.57 m
b i 1.57 m   **ii** 1.05 m   **iii** 0.26 m   **iv** 0.47 m
c i 35 km   **ii** 31°

2 a The angle subtended in degrees should be equal to $0.115 \times$ the diameter in mm.   **b ii** 0.5°

3 a i 0.035   **ii** 0.140
  b i 72 mm   **ii** 72 mm   **iii** 148 mm
  c i $B = 74°$, $C = 46°$ **ii** $B = 78°$, $C = 32°$
    iii $B = 45°$, $C = 15°$
4 a i 3.9 N, 4.6 N,   **ii** 3.4 N, 9.4 N   **iii** 4.8 N, 5.7 N
  b 4.8 N at 21° to the 3.5 N

24.3

1 a i 0.2   **ii** 0.1, 0.25   **iii** <0.1
  b i $t = \dfrac{v - u}{a}$ **ii** $t = \sqrt{\dfrac{2s}{a}}$
    iii $t = \dfrac{y}{k} + t_0$ **iv** $t = \dfrac{mv}{F}$
  c i 2   **ii** −1   **iii** 4.25   **iv** $\dfrac{1}{3}$
2 a i $x = \dfrac{y^2}{4}$ **ii** $x = \dfrac{1}{4y^2}$
    iii $x = \dfrac{1}{y^3}$ **iv** $x = \sqrt{\dfrac{k}{y}}$
  b i 0.25   **ii** ±2.8   **iii** ±0.5   **iv** 4
3 a i $3.1 \times 10^{-7}\,\mathrm{m}^2$ **ii** $6.2 \times 10^{-3}\,\mathrm{m}$
    iii 2.5 s   **iv** 17 m s^{-1}
  b i kg m^2 s^{-2} **ii** kg   **iii** s
4 a i ½ or −3   **ii** 1.44 or 5.56   **iii** 1 or −1.67
  b i $t = -\dfrac{20}{6}$ s or 2 s   **ii** $R = 1.0\,\Omega$ or $4.0\,\Omega$

24.4

1 a i 3   **ii** −3   **iii** 1
  b i −4   **ii** 8   **iii** 2
  c i −1   **ii** 5   **iii** 5
  d i −1.5   **ii** 3   **iii** 2
2 a i $y = 2x - 8$ **ii** −8
  b i 3 m s^{-2} **ii** 5 m s^{-1}
3 a (1, 4)   **b** $y = 4x$

4 a $x = 2$, $y = 0$
  b $x = 3$, $y = 5$
  c $x = 2$, $y = 0$

24.5

1 a See 1.3 Fig 2
  b i gradient   **ii** area under line
2 a See 6.5 Fig 3a
  b Resistance $= \dfrac{1}{\text{gradient}}$
3 a y = energy, x = time; the graph should be a straight line with a positive constant gradient from the origin.
  b The power is constant and is represented by the gradient of the line.
4 a i As end of chapter 2, Q11 Fig 11.1, with the line becoming flat and the y-axis labelled 'velocity'.
    ii The graph is an exponential decrease curve similar to 22.2 Figure 7 with the y-axis labelled acceleration and the x-axis labelled time.
  b i **1** acceleration   **2** distance fallen
    ii velocity

24.6

1 a i 0.477   **ii** 1.176   **b i** 1.653   **ii** 0.699
2 a i 10.8 dB   **ii** 7.0 dB   **b** 17.8 dB
3 a $n = 5$, $k = 3$ **b** $n = 3$, $k = \frac{1}{2}$ **c** $n = 2$, $k = 1$
4 a i 1.10   **ii** 2.71   **b i** 3.81   **ii** 1.61

24.7

1 a i 2, 3   **ii** 12, 0.2   **iii** 4, 0.02
  b i 0.23 s   **ii** 3.5 s   **iii** 35 s
2 a 11.3 kBq   **b** 9.0 kBq   **c** 0.38 kBq
3 a i 2.20 s   **ii** 1.52 s   **b i** 4.83 V   **ii** 1.24 V
4 a i 0.14 s   **ii** 82   **b** $a = 6.9$, $b = -5$

Chapter 1

1 a i A vector has magnitude and direction. A scalar has magnitude only.
  ii velocity (or acceleration) **iii** mass or volume
  b 7.8 N
2 a 1.48×10^3 s **c** $15.8\,\mathrm{m\,s^{-1}}$
3 a $40\,\mathrm{m\,s^{-1}}$ **b** $80\,\mathrm{m\,s^{-1}}$ **c** $2.0\,\mathrm{m\,s^{-2}}$
4 b $0.375\,\mathrm{m\,s^{-2}}$, 0, $0.75\,\mathrm{m\,s^{-2}}$; 300 m, 900 m, 150 m
  c $11.25\,\mathrm{m\,s^{-1}}$
5 a 24 m **b** $22\,\mathrm{m\,s^{-1}}$
6 a i 10 m **ii** $-0.20\,\mathrm{m\,s^{-2}}$ **b iii** $0.80\,\mathrm{m\,s^{-1}}$ downhill
7 a i $14.7\,\mathrm{m\,s^{-1}}$ **ii** 12.6 m **iii** $15.7\,\mathrm{m\,s^{-1}}$ downwards
8 a $20\,\mathrm{m\,s^{-1}}$ **c** 5.2 m
9 a i $150\,\mathrm{m\,s^{-1}}$, 1.88 km **b i** 25 s **ii** $243\,\mathrm{m\,s^{-1}}$
10 a 1.31 m **b i** 3.62 m **ii** $23.9\,\mathrm{m\,s^{-1}}$, 16° below the horizontal
  c i The speed on impact would have been the same because its loss of potential energy (from start to finish) would have been unchanged so its gain of kinetic energy would be the same.
  ii The angle (A) of its direction to the horizontal would be greater because the ratio of its vertical component of velocity to its horizontal component (which is equal to tan A) would be greater as the sum of the squares of these components is unchanged (equal to its speed squared) and its horizontal component is slightly less so its vertical component is greater than before. (OR Because it would be in the air for a longer time and its vertical displacement from maximum height would be greater. Therefore, its vertical component of velocity would be greater at impact and its horizontal component of velocity would be slightly less.)

Chapter 2

1 a Friction acts on the object as it slides on the floor. The force pushing the object must be equal and opposite to the frictional force to maintain steady speed in a constant direction.
  b Friction is absent on ice so no applied force is needed to maintain constant velocity.
2 a $0.25\,\mathrm{m\,s^{-2}}$ **b** 110 N **c** 0.026
3 a 1.5 ms **b** $-7.8 \times 10^4\,\mathrm{m\,s^{-2}}$ **c** 190 N
4 a i 3.3×10^4 kg **ii** $0.18\,\mathrm{m\,s^{-2}}$ **iii** 110 s **iv** 1.1 km
  b i $0.15\,\mathrm{m\,s^{-2}}$, **ii** 129 s
5 a i the gradient of the line,
  ii The speed increased from zero and gradually became constant.

iii

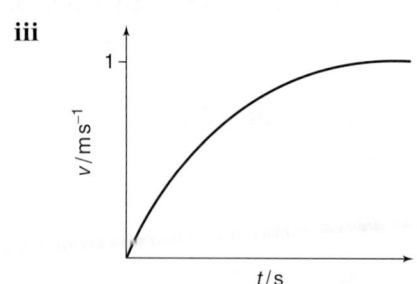

Figure 1

b As its speed increased, the drag force opposing its motion gradually increased. Therefore the resultant force which is the difference between its weight and the drag force gradually decreased until it became zero and the velocity is then constant.
6 a 1.7 kJ **b** $38\,\mathrm{m\,s^{-1}}$ **c** 73 m
7 a 61 J **b** 61 J **c** $7.0\,\mathrm{m\,s^{-1}}$
8 a i 4.7 J **ii** 4.7 J **iii** $4.8\,\mathrm{m\,s^{-1}}$
  b i 3.1 J **ii** 3.1 J **iii** $4.0\,\mathrm{m\,s^{-1}}$
9 a i 6.9 MJ **ii** 7.9 MJ **b i** 740 kW **ii** 0.99 kg
10 a 1.8 kN **b** 2.3 kN **c i** 16 kW **ii** 3.5 kW
  d KE gain each second
11 a i Resultant force = weight (or mg) – resistive force but resistive force – 0 initially (as initial speed = 0). So resultant force initially = mg hence initial acceleration = resultant force/mass = $mg/m = g$ = 9.81 m/s².
  ii Graph of acceleration (a) on the y-axis and time (t) on the x-axis; curve with a negative decreasing gradient starting at $a = 9.8\,\mathrm{m\,s^{-2}}$ (or g) at $t = 0$; non-zero gradient at $t = 5.6$ s (or object does not reach terminal (or constant speed).
  iii Resistive force of air on object increases as object's speed increases so the resultant force on the object (= weight – resistive force) decreases with increasing speed. Since acceleration = resultant force / mass, acceleration decreases as the object's speed increases so the object's rate of increase of speed decreases.
  b i Weight = 8.2(4) N
  ii Acceleration, a, at midpoint = gradient of speed–time graph at $t = 2.8$ s gives $a = 4.0$ to $4.5\,\mathrm{m\,s^{-2}}$, so resultant force = 3.4 to 3.8 N.
  c i Kinetic energy gained = 306 J
  ii Distance fallen estimated from area under the curve = 100 to 105 m; Loss of potential energy = 845 J [830 to 860 J acceptable]
  iii Work done against air resistance = loss of P.E. – gain of K.E. = average resistive force × distance fallen so average resistive force = (loss of P.E. – gain of K.E.)/distance fallen = 5.0 N to 5.5 N (using values above)

Chapter 3

1 a 2.5 kN **b** 4.5 kN
2 a **i** 270 N **ii** 225 N
   b The moment of the wind force about the direction of motion tends to overturn the boat. By leaning as shown, the weight of the crew provides an equal and opposite moment to prevent overturning.
3 a **i** 570 N **ii** 100 N **b i** 540 N **ii** 6180 N
   c 30 N down theslope
4 7.5 kN, 6.5 kN
5 a **i** The point where the weight of the body may be considered to act.
   ii force × perpendicular distance from the line of action of the force to the point about which moments are being considered.
   b i 60 N
6 **b** 12.7 N
7 a 152 N, 106 N
   b The paraglider is acted on by three forces which are the tension T in the cable, the weight W of the paraglider and the force F of the parachute. F cannot be in the opposite direction to T because the three force vectors form a triangle with a vertical side representing W. See Topic 3.5 Figure 1.
8 a 130 kN **b** 147 kN, 113 kN
9 a **i** A *couple* is a pair of equal and opposite forces acting on a body but not along the same line.
   ii The *torque* of a couple is the magnitude of one force multiplied by the perpendicular distance between the lines of action of the forces.
   b i Torque = force × distance therefore force = torque/perpendicular distance = 84 N m/0.44 m = 190 N
   ii Perpendicular distance (from the axis of rotation of the nut) to the line of action of the force would be less than 0.44 m so the force would need to be greater to give the same torque. OR Force has a component parallel to the handle and a component perpendicular to the handle. Perpendicular component would need to be 190 N to achieve the same torque so magnitude of the force would need to be greater than 190 N.

Chapter 4

1 a 140 N s **b** 20 m s^{-1}
2 a 9000 kg m s^{-1} **b** 60 s
3 **b** 19.2 m s^{-1} **c** 4.8 N
4 a 7.5×10^{-23} kg m s^{-1} **b** 1.5×10^{-19} N
5 a 2.6×10^{-23} kg m s^{-1} **b** 1.3×10^{-20} N
6 a 1.5 m s^{-1} **b i** 2.7 kJ
   b ii kinetic energy of the wagons is changed to sound energy and heat energy of the colliding parts.
7 a 24 m s^{-1} **b i** 163 kJ **ii** 91 kJ **iii** 72 kJ
8 a 2.5×10^5 m s^{-1} **b i** 1.3×10^{-14} J **ii** 7.5×10^{-13} J

9 a Duration of impact = 0.68 s (±0.2 s)
   Impact force = $\dfrac{\text{change of momentum}}{\text{time taken}}$
   $= \dfrac{25\,000 \text{ kg} \times 1.2 \text{ m s}^{-1}}{0.68 \text{ s}} = 37 \text{ kN}$
   b i For a system of interacting objects, the total momentum remains constant provided no external resultant force acts on the system.
   ii The total momentum is conserved, the total final momentum = the total initial momentum
   Hence $(25\,000 \text{ kg} \times 1.2 \text{ m s}^{-1}) + (15\,000 \text{ kg} \times v)$
   $= 27\,000 \text{ kg m s}^{-1}$, where v = velocity of P after the impact
   $15\,000 v = 27\,000 - (25\,000 \times 1.2)$
   $= 27\,000 - 30\,000 = -3000$
   so $v = \dfrac{-3000 \text{ kg m s}^{-1}}{15\,000} = -0.20 \text{ m s}^{-1}$
   c i A *perfectly elastic collision* is one in which there is no loss of kinetic energy.
   ii For P, its loss of K.E. = initial K.E. − final K.E. = $0.5 \times 15\,000 \text{ kg} \times (1.8 \text{ m s}^{-1})^2 - 0.5 \times 15\,000 \text{ kg} \times (-0.2 \text{ m s}^{-1})^2 = 24.3 \text{ kJ} - 0.3 \text{ kJ} = 24 \text{ kJ}$
   For Q, its gain of K.E. = $0.5 \times 25\,000 \text{ kg} \times (1.2 \text{ m s}^{-1})^2 = 18 \text{ kJ}$
   iii The collision is not elastic because P has lost more kinetic energy than Q has gained, so the total kinetic energy has decreased.
10 a **i** Before emission, the nucleus was stationary so it had no momentum. No external forces acted on the nucleus so, in accordance with the principle of conservation of momentum, the total momentum was also zero immediately after the emission. Therefore the momentum of the nucleus after emission must be equal in magnitude and opposite in direction to the momentum of the α-particle.
   ii The nucleus and the α-particle have different masses and, since momentum is mass × velocity, their velocities must differ in magnitude and hence their speeds must differ.
   b i 1.6×10^7 m s^{-1} **ii** speed = 3.1×10^5 m s^{-1}; K.E. = 1.6×10^{-14} J
   c i 2.4 ns **ii** 3.3×10^{15} m^2 s^{-2}
11 a **i** The force of the rubber band on each trolley changes but at any instant the rubber band pulls on each trolley with the same amount of force (but in opposite directions) even though the force changes. The duration of the force on each trolley is the same. Since the force on each trolley is equal to its rate of change of momentum, the two trolleys gain an equal magnitude of momentum but in opposite directions. (OR The total initial momentum is zero so the total final momentum at any instant is zero. Therefore the two trolleys gain equal and opposite momentum.)

ii The two trolleys gain equal and opposite momentum and, since momentum is mass × velocity, the velocity of the trolley with lower mass must be greater in magnitude than that of the other trolley. So, at any instant, the trolley with the lower mass must have a different (or greater) speed than that of the other trolley.

b $KE = \frac{1}{2}mv^2 = \frac{1}{2}(mv)^2/m = \frac{1}{2}$ (momentum)^2/mass. At any instant, their momentum is equal in magnitude so the trolley with the lower mass must have more kinetic energy than the other trolley.

Chapter 5

1 **a** $2.0 \times 10^{-7}\,m^3$ **b** $1.8 \times 10^{-3}\,kg$ **c** $7.2 \times 10^{-4}\,kg\,m^{-1}$
2 **a** 28kN **b i** 52kJ **ii** 88kJ **iii** P.E. of the platform
3 **a i** $33\,Nm^{-1}$ **ii** 480mm **b** 1.5N
4 **a** See Topic 5.6. **b** 58N
5 **a i** The limit to which it can be stretched and regain its original length when the applied force is removed,
  ii Plastic behaviour occurs when the strip is stretched beyond its elastic limit so it does not regain its original length when the stretching force is removed.
  b i See Topic 5.6 Figure 2 **ii** As each part of the tyre rolls over the road surface, some of the work done when that part of the tyre is squashed and stretched increases the internal energy of the rubber molecules so the tyre becomes warm.
6 **a** 5200N **b i** $1.0 \times 10^7\,Pa$
  ii $e = \dfrac{Fl}{AE}$, where $F = 4500\,N$ and the values of A, E and l are given
  iii 15J
7 **a** $9.4 \times 10^{10}\,Pa$ **b** $1.0 \times 10^{-2}\,J$
8 **a i** The extension beyond which the spring would not return to its original length when it is unloaded.
  ii The tension (or force) per unit extension needed to stretch the spring, provided its limit of proportionality is not exceeded.
  iii $\frac{1}{2}W^2/k$
  b P **i** $\dfrac{3e}{2}$ **ii** $\dfrac{2k}{3}$ Q **i** $\dfrac{e}{3}$ **ii** $3k$
9 **a** 10.0J ($\pm 0.1\,J$) **b** stiff at first then much less stiff, then stiffer but not as stiff as at the start
10 **a i** The water exerted an upthrust on the cylinder so the support force on the cylinder from the newton-meter was reduced from 1.95N. There was an equal and opposite force to the upthrust on the water from the cylinder. The weight of water in the beaker was unchanged. So the force on the top-pan balance increased.
  ii 0.15N

b i

Upthrust, U/N	0	0.03	0.07	0.10	0.13	0.15

ii See Figure 2.

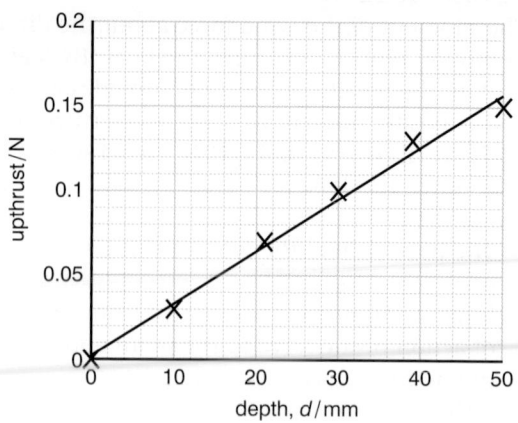

Figure 2

iii Gradient = $0.157\,N/50\,mm = 3.14 \times 10^{-3}\,N\,mm^{-1}$
c i At a depth d of 80mm, the upthrust would be equal to the graph gradient × 0.080m. Hence the upthrust at this depth = $80\,mm \times 3.14 \times 10^{-3}\,N\,mm^{-1} = 0.25\,N$ (to 2 significant figures)
  ii The weight of the cylinder is 1.95N and its length is 80mm. The weight of water displaced at $d = 80\,mm$ is equal to the upthrust which is 0.25N. The ratio of the cylinder's weight to the weight of water displaced at $d = 80\,mm$ is 7.80 (= 1.95N/0.25N) and this is equal to their density ratio as they have the same volume. So the density of the cylinder is $7.8 \times 1000\,kg\,m^{-3} = 7800\,kg\,m^{-3}$.

Chapter 6

1 **a** 450C **b** 675J
2 **a** 0.33A **b** 150J
3 **a** See Topic 6.5 Figure 2a or b **b i** See Topic 6.5 Figure 3b **ii** The filament resistance increases as the bulb becomes brighter **iii** As the current in the bulb increases, the bulb becomes brighter and hotter. The filament is a metal wire and the resistance of a metal wire increases as its temperature increases. This is because the metal atoms in the filament vibrate more and they make it harder for conduction electrons to pass through the filament.
4 **b i** 2.4V **ii** 0.60V **c** 0.048A
5 **a** Set up the circuit shown (as Topic 6.4 Figure 1) and measure the resistance R of different lengths L of the wire (as explained in Topic 6.4). Use a metre ruler to measure the length of wire connected in the circuit. Plot a graph of resistance R on the y-axis against length L on the x-axis. Determine the gradient of the line which is the resistance per unit length of the wire. The resistivity is calculated by

multiplying the resistance per unit length by the area of cross-section of the wire.

b i $5.5 \times 10^{-7}\,\Omega\,\text{m}$ **ii** $4.4\,\Omega$

6 a

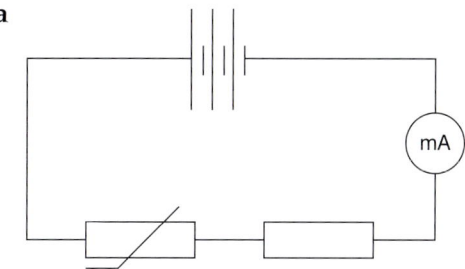

Figure 4

b i The ammeter reading would increase as the resistance of the thermistor would decrease so the current in the circuit would increase.

ii Connect an extra $270\,\Omega$ resistor in series

7 a i $0.21\,\Omega$ **ii** $2.8\,\text{V}$ **iii** $36\,\text{W}$ **b** 28A

8 a $530\,\Omega$ **b** $18.5\,\text{m}$

9 a $3.25\,\text{V}$ **ii** $1.25\,\text{V}$ **iii** $250\,\Omega$

b The milliammeter reading would increase as the resistance of the LDR would decrease so the current in the circuit would increase.

10 b $5.1 \times 10^{-7}\,\Omega\,\text{m}$

11 a As Topic 7.1, Figure 5, without the voltage values under the variable resistor and X and with the battery labelled $18\,\text{V}$

b $I = P/V = 24\,\text{W}/12\,\text{V} = 2.0\,\text{A}$ **c** Power supplied by the battery $= IV = 2.0\,\text{A} \times 18\,\text{V} = 36\,\text{W}$, power dissipated = power supplied by battery – power supplied to X $= 36\,\text{W} - 24\,\text{W} = 12\,\text{W}$

12 a i Resistance per unit length for 1 aluminium wire

$$= \frac{R}{l} = \frac{\rho}{A} = \frac{2.5 \times 10^{-8}\,\Omega\,\text{m}}{\pi \times (0.5 \times 4.0 \times 10^{-3}\,\text{m})^2}$$
$$= 1.99 \times 10^{-3}\,\Omega\,\text{m}^{-1}$$

Since the 50 wires are identical and in parallel, their effective cross-sectional are is $50A$ where A is the area of cross-section of one wire. The resistance per metre of all 50 wires = one-fiftieth of the resistance per unit length of one wire = $1.99 \times 10^{-3}\,\Omega\,\text{m}^{-1}/50 = 3.98 \times 10^{-5}\,\Omega\,\text{m}^{-1}$

ii Resistance per unit length for 1 steel strand

$$= \frac{R}{l} = \frac{\rho}{A} = \frac{1.6 \times 10^{-7}\,\Omega\,\text{m}}{\pi \times (0.5 \times 3.0 \times 10^{-3}\,\text{m})^2}$$
$$= 2.26 \times 10^{-2}\,\Omega\,\text{m}^{-1}$$

Since the 7 strands are identical and in parallel, their effective cross-sectional is $7A$ where A is the area of cross-section of one strand. The resistance per metre of all 7 strands = one-seventh of the resistance per unit length of one strand = $2.26 \times 10^{-2}\,\Omega\,\text{m}^{-1}/7 = 3.23 \times 10^{-3}\,\Omega\,\text{m}^{-1}$

b i Current in each component = p.d./resistance of 1000 m. The p.d. across each component is the same so the ratio

$$\frac{\text{current in the aluminium wires}}{\text{the current in the steel strands}}$$
$$= \frac{\text{resistance of the steel strands}}{\text{resistance of the aluminium wires}}$$
$$= \frac{3.23 \times 10^{-3}\,\Omega\,\text{m}^{-1} \times 1000\,\text{m}}{3.98 \times 10^{-5}\,\Omega\,\text{m}^{-1} \times 1000\,\text{m}} = 81 \text{ so the steel}$$

strands conduct $\dfrac{1}{82}$ of the total current.

Therefore the current in the steel strands $= 100\,\text{A}/82 = 1.22\,\text{A}$ and the current in the aluminium strands $= 100\,\text{A} - 1.22\,\text{A} = 98.78\,\text{A}$

ii p.d. across the cable = current × resistance of either component $= 98.78\,\text{A} \times 3.98 \times 10^{-5}\,\Omega\,\text{m}^{-1} \times 1000\,\text{m}$ (for the aluminium wires) $= 3.93\,\text{V}$

For the aluminium wires, the power dissipated = current × p.d. $= 98.78\,\text{A} \times 3.93\,\text{V} = 388.2\,\text{W}$

For the steel strands, the power dissipated = current × pd $= 1.22\,\text{A} \times 3.93\,\text{V} = 4.8\,\text{W}$

13 a See Figure 3.

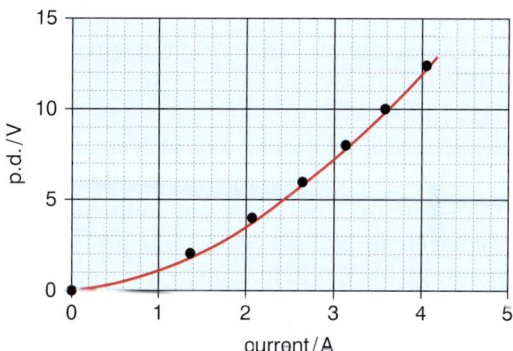

Figure 3

b i $1.50\,\Omega$ **ii** $1.70\,\Omega$ **iii** $3.06\,\Omega$

c % uncertainty for $2.00\,\text{A}$ measurement $= (\pm 0.02/2.00) \times 100\% = 1.0\%$
% uncertainty in the corresponding p.d. of $3.4\,\text{V} = (\pm 0.02/3.4) \times 100\% = 0.6\%$
% uncertainty in the resistance $= 1.6\%$, so the uncertainty in the resistance $= 1.6\%$ of $1.70\,\Omega$ $= 0.03\,\Omega$ rounded up from $0.027\,\Omega$.

d i As the current increases, the wire becomes hotter so the resistance of the filament increases. Because the length and area of cross-section of the wire do not change (significantly), the resistivity therefore increases as the filament becomes hotter.

ii The drift velocity $v = I/Anq$, where A is the area of cross-section of the filament, n is the number of charge carriers per unit volume in the filament and q is the charge of each charge carrier. Assuming A, n and q do not change significantly when the wire becomes hotter, v increases when the current increases.

Chapter 7

1 **a** $6.0\,\Omega$ **b** $2.0\,\Omega$: **i** $1.0\,A$ **ii** $2.0\,V$ **iii** $2.0\,W$;
$6.0\,\Omega$: **i** $0.67\,A$ **ii** $4.0\,V$ **iii** $2.7\,W$;
$12.0\,\Omega$: **i** $0.33\,A$ **ii** $4.0\,V$ **iii** $1.3\,W$

2 **a** The brightness increases from zero as the resistance is increased from zero.
 b When the variable resistor has zero resistance, no current passes through the light bulb as the variable resistor 'short-circuits' the bulb and there is no pd across the bulb.

3 **a i** $20\,\Omega$ **ii** $15\,\Omega$, $2.25\,V$, $3\,\Omega$, $0.45\,V$ **iii** $2.7\,V$
 b i $15\,\Omega$, $0.34\,W$; $3\,\Omega$, $0.068\,W$ **ii** $0.045\,W$

4 **a i** $0.75\,A$ **ii** $4.5\,V$ **b i** $3.4\,W$ **ii** $4.5\,W$
 c The power supplied to each heating element and to the internal resistance would be greater as the current in each would be greater. The battery would 'run down' faster as the energy stored in the battery would be used at a faster rate.

5 **a** $2.0\,\Omega$ **b** $1.5\,A$

6 **a i** $0.40\,A$ **ii** $200\,W$ **iii** $0.86\,A$ **b i** $0.40\,A$
 ii $105\,W$

7 **a** $5.0\,k\Omega$ **b** $3.0\,V$
 c The thermistor resistance would decrease so the proportion of the battery pd across the thermistor would decrease. Therefore, the voltmeter reading would decrease.

8 **a** Opposite D_1 $0.6\,V$, $0.12\,mA$; opposite D_2 $4.4\,V$, $0.88\,mA$
 b Opposite D_1 $0.12\,mA$, opposite D_2 $1.7\,mA$

9 **a** $1.08\,V$ **b i** $\pm0.67\%$ **ii** $\pm0.41\%$, $\pm0.57\%$ **c** $\pm0.02\,V$

10 **a i** With S across Y, the emf of Y, $E_Y = I(R + r)$ and the p.d. across cell Y, $V_Y = IR$ where I is the current through S. So $\frac{E_Y}{V_Y} = \frac{(R + r)}{R}$. Since $V_Y = kl_2$ and $E_Y = kl_1$ where k is the p.d. per unit length along the potentiometer slide wire, substituting these into the equation above gives $\frac{l_1}{l_2} = \frac{(R + r)}{R}$.
 ii $\frac{l_1}{l_2} = \frac{(R + r)}{R}$
 Substituting in the values gives: $\frac{741}{625} = \frac{5.00 + r}{5.00}$
 Rearranging in terms of r gives: $r = \frac{741 \times 5.00}{625} - 5 = 0.928\,\Omega$
 b (i) The driver cell's e.m.f. decreased during the measurements **(ii)** The measurements all need to be repeated.

11 **a i** $2.75R$ **ii** $E_X/11R$
 b i The p.d. across resistor A and across resistor B are both equal to $I_X R$. The current through resistor C is $(I_X + I_Y)$ in accordance with Kirchhoff's 1st law so the p.d. across C is $(I_X + I_Y)R$. For the loop consisting of E_X and resistors A, C and B, applying Kirchhoff's 2nd law gives $E_X =$ p.d. across A + p.d. across C + p.d. across B $= I_X R + (I_X + I_Y)R + I_X R = (3I_X + I_Y)R$.

 ii For the loop consisting of E_Y and resistors D, C and E, applying Kirchhoff's 2nd law gives $E_Y =$ p.d. across D + pd across C + pd across E $= I_Y R + (I_X + I_Y)R + I_Y R = (I_X + 3 I_Y)R$.
 iii Substituting $E_X = E_Y = 6.0\,V$ and $R = 2.0\,\Omega$ into the two equations above gives $6.0\,V = (3 I_X + I_Y) \times 2.0\,\Omega$ and $6.0\,V = (I_X + 3 I_Y) \times 2.0\,\Omega$. Subtracting the two equations gives $0 = (2 I_X - 2 I_Y) \times 2.0\,\Omega$ which means $I_X = I_Y$. Therefore $4 I_X = 6.0\,V/2.0\,\Omega = 3.0\,A$. So $I_X = I_Y = 0.75\,A$. The p.d. across C $= (I_X + I_Y)R = 1.5\,A \times 2.0\,\Omega = 3.0\,V$.

Chapter 8

1 **a i** The vibrations of a transverse wave are perpendicular to the direction of propagation of the waves whereas in a longitudinal wave, the vibrations are parallel to the direction of propagation.
 ii sound **iii** electromagnetic waves
 b i Diffraction is the spreading of a wave after it passes through a gap or round an obstacle.
 ii Refraction is the change of direction of a wave when it crosses a boundary at which its speed changes.

2 **a** Polarised transverse waves vibrate in one plane only. As with any transverse wave, the vibrations are perpendicular to the direction of propagation.
 b Sound waves are longitudinal and longitudinal waves vibrate parallel to the direction of propagation and therefore can never be polarised.

3 **a** $250\,Hz$
 b Two complete wave cycles of the same amplitude should be drawn from one side to the other side of the screen

4 **a** See Topic 8.5.
 b i At maximum loudness, the observer is at a position where the sound waves from each loudspeaker reinforce each other. At minimum loudness, the observer has a position where the waves cancel each other. The maxima and minima are equally spaced along XY.
 ii Reducing the frequency increases the wavelength of the waves. As a result the path difference to each maximum and minimum increases except for the maximum at the centre. Successive minima are therefore further apart.

5 **a i** Reinforcement occurs at both P and Q because the path difference in both cases is a whole number of wavelengths, different in each case.
 ii Cancellation occurs at the midpoint of P and Q because the path difference at this position is a whole number + half a wavelength so the microwaves from the two gaps are out of phase by 180°.

b The detector signal would increase as the waves from one gap would not reach the transmitter to cancel the waves from the other gap.

6 a i The amplitude is the same as at the midpoint. The frequency is the same. There is a phase difference of 180°.

ii The amplitude is the same. The frequency is the same. The vibrations are in phase.

b i 0.40 m

ii The pattern should show 4 equally spaced loops.

7 a At each resonant length, sound waves directly from the speaker reinforce sound waves from the speaker that have travelled down the tube and reflected at the water surface then partially reflected again at the open end. This happens when the length of the air column in the tube is an odd multiple of a quarter of the sound wavelength.

b i See Topic 8.7 Figure 3 **ii** 330 ms⁻¹

8 a i 1.65 m **ii** See Topic 8.8 Figure 4a.

b i 400 Hz

ii At each resonant length, sound waves directly from the speaker reinforce sound waves from the speaker that have travelled along the tube and reflected at the far end then partially reflected again at the open end where the speaker is. This happens when the length of the air column in the tube is a multiple of a half of the sound wavelength.

9 a i Every point along a progressive wave vibrates with the same amplitude, whereas the amplitude of a stationary wave differs along the wave and is zero at equally spaced points called *nodes*.

ii For a progressive wave, the phase difference between any two points increases with their separation. For a stationary wave, any two points between adjacent nodes (or separated by an even number of nodes) vibrate in phase (or any two points separated by an odd number of nodes vibrate out of phase by half a cycle).

b i Both points vibrate with the same amplitude, but their displacement at any given time is different. When one of the particles is at zero displacement, the other particle is at maximum displacement (either positive or negative). They vibrate with a phase difference of 270° (or three-quarters of a cycle or 3π/2).

ii The two progressive waves must pass through each other.

1. in opposite directions with the same amplitude.
2. and at the same frequency and speed.

Chapter 9

1 a i Light from the two slits reinforce at a bright fringe

ii Light from the two slits cancels at a dark fringe.

b i The two slits emit waves with a constant phase difference,

ii Two separate light sources emit waves at random so they are not coherent sources. Therefore, the points of cancellation and reinforcement continually move about so no dark fringes can be observed.

c 530 nm

2 a i See Figure 6; the diffracted wavefronts should both spread out and be at the same spacing as the incident wavefronts.

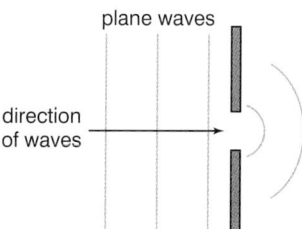

Figure 6

ii The two wavefronts should spread out less and the central part of each should be flatter.

b i Number of orders each side of the zero order = 2

ii 49.1°

3 a i fringe spacing, x = 4.8 mm/4 = 1.2 mm
Therefore $\lambda = ax/D$ = 0.40 mm × 1.2 mm/0.810 m = 5.9 × 10⁻⁹ m

ii % uncertainty in each measurement:
ΔD = 5 mm/810 mm × 100% = 0.6%;
Δa = 0.02 mm/0.40 mm × 100% = 5.0%:
Δx = 0.5 mm/ 4.8 mm × 100% = 10.4%. The measurement of the slit spacing is the least precise because it has the greatest uncertainty.

b If white light had been used, the centre of each fringe at and near the middle, M, of fringe pattern would be white but their edges would be tinged with red at the edge furthest from M and with blue at the edge nearest M. Further from the centre, the edges of the fringes would overlap and be less distinct.

4 a White light consists of light of all wavelengths in the visible spectrum. Each colour of the visible spectrum is due to a band of wavelengths.

b Grating line spacing d = 1/600 000 m⁻¹
= 1.667 × 10⁻⁶ m; order number n = 2

i For red light: $n\lambda = d \sin\theta$ gives
$\lambda = (d \sin\theta)/2$ = (1.667 × 10⁻⁶ m × sin 48°)/2
= 619 nm

ii For blue light: $n\lambda = d \sin\theta$ gives
$\lambda = (d \sin\theta)/2$ = (1.667 × 10⁻⁶ m × sin 31°)/2
= 429 nm

c Applying the condition $\theta \le 90°$ and the equation $n\lambda = d \sin\theta$ to third order diffraction gives $3\lambda \le d$. Therefore wavelengths greater than $3d$ ($= 500\,nm$ as $d = 1/600\,000\,m$) cannot undergo 3rd order diffraction with this grating. Therefore only wavelengths below $500\,nm$ are seen in the third order spectrum.

d i For red light in glass: $n\lambda' = d \sin\theta$ gives
$\lambda' = (d \sin\theta)/2 = (1.667 \times 10^{-6}\,m \times \sin 30°)/2$
$= 417\,nm$
For blue light in glass: $n\lambda' = d \sin\theta$ gives
$\lambda' = (d \sin\theta)/2 = (1.667 \times 10^{-6}\,m \times \sin 20°)/2$
$= 285\,nm$

ii Ratio for red light $= 619\,nm/417\,nm = 1.48$; Ratio for blue light $= 429\,nm/285\,nm = 1.51$

iii A $1°$ difference in the above calculations would mean for blue light, for example, using values of sin 32 not sin 31 for the wavelength in air and sin 19 not sin 20 for the wavelength in glass to calculate an upper limit for the ratio. In this case, the ratio would therefore be 1.63 not 1.51. The uncertainty in the ratio therefore is significantly greater than the difference between the calculated ratio values in **di** for blue light, so at this level of uncertainty it cannot be concluded that the ratios for red and blue light are different.

5 a Slit spacing, $a = 0.12\,mm$.
Therefore fringe spacing, $x = \lambda D/a$
$= 560 \times 10^{-9}\,m \times 1.800\,m/1.2 \times 10^{-4}\,m$
$= 8.4 \times 10^{-3}\,m$

b The slit spacing with the 2 outer slits only would be $3 \times 0.12\,mm$ so the fringe spacing would be one-third of the fringe spacing in **a** (i.e. $8.4 \times 10^{-3}\,m/3$ or $2.8 \times 10^{-3}\,m$). Therefore instead of a wave with a central peak and two peaks either side, the sketch should show 15 peaks (i.e. a central one with 7 equally spaced peaks either side). The peaks would be lower, as more fringes would be created per unit distance so the light would be less intense at each bright fringe.

c Further fringes patterns due to different pairs of slits would be superimposed on the pattern described in **a** or **b**. Suppose the slits are labelled A, B ,C and D from left to right, the two slits on each side (AB and CD) would give fringes with the same spacing as the two inner slits BC but their fringe patterns would be displaced slightly in opposite directions from the inner slits fringe pattern (OR A further fringe pattern with twice the fringe spacing would be caused due to interference of light from AC and BD and these would also be displaced from the fringes due to BC.)

6 a Grating line spacing, $d = 1/500\,000\,m^{-1}$
$= 2.00 \times 10^{-6}\,m$; order number $n = 4$
i For the blue line: $n\lambda = d \sin\theta$ gives $\lambda = (d \sin 59°25')/4 = (2.000 \times 10^{-6}\,m \times \sin 59.42°)/4$
$= 430\,nm$

ii For the orange line: Assume $n = 3$. Therefore $3\lambda = (d \sin\theta)$ gives $\lambda = (2.000 \times 10^{-6}\,m \times \sin 62.25°)/3 = 590\,nm$. Assuming $n = 2$ would give a wavelength of $885\,nm$ which is beyond the visible spectrum. Assuming $n = 4$ would give a wavelength of $442\,nm$ which would not be orange. Therefore, the wavelength of the orange line is $590\,nm$ and it is a third order line.

b A white light spectrum consists of a continuous distribution of colours of the spectrum from deep red through to deep blue, corresponding to a continuous range of wavelengths which increase across the spectrum from blue to red. A line emission spectrum consists of narrow discrete lines of different colours against a black background. Each line is due to light of a particular wavelength corresponding to the colour of the line.

Chapter 10

1 a i The nucleus is very small in diameter compared with the diameter of an atom. Most α particles do not pass close enough to the nucleus to be affected by the electrostatic force of repulsion of the nucleus.

ii α particles that pass close to the nucleus experience strong repulsion from the nucleus which causes a significant deflection. If an a particle approaches a nucleus very closely, the force of repulsion from the nucleus deflects the α particle by more than $90°$.

b See Topic 10.1 Figure 3.

2 a As the particle approaches, the kinetic energy decreases to zero at the closest approach then increases as the particle moves away becoming equal to the initial kinetic energy at infinity.

b i Factors include its initial speed or kinetic energy, its direction and the size of the charge on the nucleus. See Topic 10.1

ii It could penetrate the nucleus and cause a rearrangement of the nucleons with a different particle (e.g. a neutron being emitted)

3 $^{6}_{3}\text{Li} + ^{1}_{0}\text{n} \rightarrow ^{3}_{1}\text{H} + ^{4}_{2}\alpha$

4 a γ

b i The reading would be unchanged as the radiation would pass through the second plate with negligible absorption.

ii The reading would decrease significantly as lead absorbs γ radiation much more effectively than aluminium does.

5 a i $7.7(3)\,s^{-1}$ **ii** 710

b See Topic 10.5.

6 a i 90 p, 144 n **ii** 91 p, 143 n

b $X = ^{64}_{29}\text{Cu}, ^{64}_{30}\text{Zn}$

7 a Proton number increased by 1, nucleon number unchanged.

b γ radiation is much more penetrating than β radiation

c By the inverse-square law $\frac{I_4}{I_1} = \frac{1.0^2}{4.0^2} = \frac{1}{16}$ (or 0.0625)

8 a X experiences a greater force of repulsion by the nucleus than Y does because its initial direction takes it closer to the nucleus. So X's rate of change of momentum is greater and so it is deflected more than Y.

b Most of the incident α-particles pass through atoms with little or no deflection so most of the atom is empty space. The nucleus is very small in size and most of the mass of the atom is located there. The nucleus must be positively charged because it repels α-particles which are positively charged.

c i If the foil was too thick, individual α-particles would be scattered several times so the deflections would be random. The foil needs to be thin enough so that individual α-particles are scattered only once.

ii If their kinetic energies differed, α-particles with differing kinetic energies moving along the same initial path would be deflected by different amounts. The measured count rate at any particular angle would be due to α-particles with different kinetic energies moving along different paths.

9 a i 83 **ii** 126

b $2.3 \times 10^{17}\,\text{kg}\,\text{m}^{-3}$

c In bismuth metal, the atoms are in contact with each other with little space between them. Almost all the mass of a bismuth atom is in its nucleus which is about 10^5 times smaller in diameter than that of an atom. So the volume of the nucleus is about 10^{15} times smaller than the volume of the atom. Since density = mass/volume, the density of the nucleus is about 10^{15} times larger than that of the atom.

10 a $a = 2$, $b = 228$, $c = 88$, $d = -1$, $e = 89$

b 3

Chapter 11

1 a Its direction of motion keeps changing so its velocity keeps changing as velocity is speed in a certain direction.

b It has an acceleration towards the centre of its circular path because its rate of change of velocity is towards the centre.

2 a $1.0\,\text{km}\,\text{s}^{-1}$ **b** $2.7 \times 10^{-3}\,\text{m}\,\text{s}^{-2}$

3 a i $2.3\,\text{m}\,\text{s}^{-1}$ **ii** $430\,\text{m}\,\text{s}^{-2}$ **b i** $12\,\text{Hz}$ **ii** $170\,\text{m}\,\text{s}^{-2}$

4 a i $5.6\,\text{Hz}$ **ii** $530\,\text{m}\,\text{s}^{-2}$

b i $34\,\text{m}\,\text{s}^{-2}$

ii The chain will come off the gear wheel if it moves too fast round the gear wheel and the tension in the chain is too small to provide the necessary centripetal force on the chain at this speed.

5 a Friction between the tyres and the road is not sufficient to provide the necessary centripetal force to the car at this speed.

b i $3.2\,\text{m}\,\text{s}^{-2}$ **ii** $11\,\text{m}\,\text{s}^{-1}$

6 a i $29\,\text{m}\,\text{s}^{-1}$

b ii The force of gravity provides the centripetal force because $v^2/r = 9.81\,\text{m}\,\text{s}^{-2}$. Therefore $r = 23^2/9.81 = 54\,\text{m}$

7 a The resultant of the train weight and the normal reaction force of the track on the train provides the centripetal force needed to maintain the circular motion along the horizontal curve of the track.

b The resultant force would not be large enough to maintain the circular motion. The train would tilt outwards as its centre of mass would move away from the centre of curvature of the track.

8 a The tension in the springs pulling the brake pads onto the shaft would not be large enough to keep the pads on the shaft. The pads would move away from the shaft until they press on the collar.

b $19\,\text{Hz}$

9 a Angular velocity of an object about a fixed point is its angular displacement per second.

b i $15.1\,\text{rad}\,\text{s}^{-1}$

ii $10.3\,\text{m}\,\text{s}^{-1}$ (Note the radius of rotation = length of the string + the radius of ball)

iii The centripetal force for constant frequency is constant. At the lowest point, the tension T acts upwards on the ball and the force of gravity of the ball acts downwards. So, at this point, the tension in the string supports the weight of the ball as well as providing the centripetal force so the tension is greatest at this point. For tension T in the string, when the ball is at the lowest point $T - mg = mv^2/r$ therefore $T = mv^2/r + mg = (0.16\,\text{kg} \times (10.3\,\text{m}\,\text{s}^{-1})^2/0.682 + (0.25\,\text{kg} \times 9.81\,\text{m}\,\text{s}^{-2}) = 26.5\,\text{N}$

iv $2.86\,\text{Hz}$

Chapter 12

1 a $0.71\,\text{Hz}$ **ii** $-1.6\,\text{m}\,\text{s}^{-2}$

b The rate of decrease of the amplitude would increase.

2 a See Topic 12.5.

b i $0.57\,\text{Hz}$ **ii** $-0.64\,\text{m}\,\text{s}^{-2}$

3 a π/2 radians

b The graphs should show 2.25 cycles of two sinusoidal waves with the wave for X starting at maximum displacement of +60 mm and the wave for Y starting at zero displacement and reaching +60 mm displacement in the first quarter cycle.

4 a i The inertia of the ring causes it to stay momentarily at rest so it is displaced relative to its initial equilibrium position. This causes the trailing spring to be compressed and the other one stretched. The springs exert a resultant force on the ring and accelerate it.

ii The greater the acceleration, the greater the displacement of the ring.

b The inertia of the ring causes it to continue moving forward when the supports stop moving. As a result the ring oscillates about the position where the springs exert zero resultant force on it.

c i For a given acceleration, the displacement would be greater.

ii For a given acceleration, the displacement would be greater.

5 a 24 N m^{-1} **b i** 2.0 Hz **ii** 0.50 s

6 a amplitude, angular frequency

b 36 mm, 0.29 Hz **c** −15 mm **d** 0.12 m s^{-2}

7 a The engine vibrations subject the panel to a periodic force at the resonant frequency of the panel.

b The panel would be stiffer so its resonant frequency would be higher. This would solve the problem only if the resonant frequency is above the highest engine vibration frequency.

8 a i The vibrations cause the spring to oscillate and the oscillations are damped by the frictional force of the damper pads.

ii The frictional force on the dampers transfers energy from the vibrations to the dampers which gain internal energy so become warm.

b The vibrations caused large oscillations of the spring at or near the resonant frequency. The increase of the damping force reduces the amplitude of the oscillations.

9 a i 2.21 s, 0.453 Hz **ii** 0.325 mJ **iii** Measure the maximum horizontal or vertical displacement every 10 cycles for 50 or more cycles. If the displacement decreases and the average decrease per cycle is more than the uncertainty in the initial displacement, the decrease is significant.

b If P stops, then Q must gain all the momentum that P had. Since they are identical, Q must therefore have the same initial speed as P had immediately before the impact so Q's initial kinetic energy must the same as the kinetic energy P had immediately before the impact. Therefore the collision must be elastic.

10 a i Suitable scales and axes labelled correctly; points plotted correctly; correct graph shape, see Figure 7.

ii 18 mm, 1.8 Hz **iii** 0.20 m s^{-1}, 4.1 mJ

b i Curve should start and end at same points as first curve; curve should be below 1st curve by an decreasing amount from the peak to each end; peak should be slightly to the left of peak of first curve.

ii Energy transfers to the oscillating system from the oscillator and energy transfers from the system to the surroundings (at the same rate). (Note: The energy of the oscillating system is constant as long as the oscillator is on.)

c 1.3 Hz (= 1.8 Hz/2$^{0.5}$ as f is inversely proportional to $m^{0.5}$.)

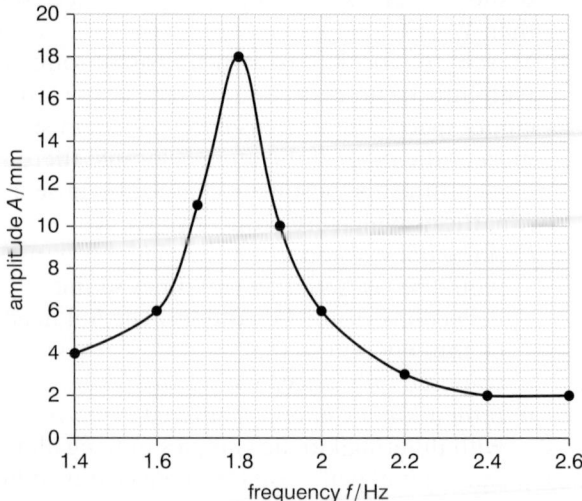

Figure 7

Chapter 13

1 a See Topic 13.1 Figure 2.

b i X and Y should be diametrically opposite at the same distance from the centre at or above the surface.

ii Z should be twice as far from the centre as X.

2 a 1.9 × 10^{18} N **b** 3.2 × 10^{-7} N kg^{-1}

3 b 3.1 N kg^{-1}

4 a 2.8 × 10^{-3} N kg^{-1}

b i 2.8 × 10^{-3} m s^{-2} **ii** 1.0 km s^{-1} **iii** 2.4 × 10^6 s

5 a 2.3 × 10^{10} m

b An object would experience equal and opposite gravitational forces from Jupiter and the Sun at a certain position between Jupiter and the Sun. These two forces would tend to stretch the object and may eventually pull it apart.

6 a 670 N

b The astronaut is acted on by the force of gravity due to the Earth so is not weightless. However, as this force provides the centripetal force needed to maintain circular motion, no other force acts on the astronaut so the astronaut is unsupported.

7 a 8.9 N kg^{-1} **b ii** 5400 s

8 a i See Topic 13.5.

ii Such a satellite remains at the same relative position above the equator as the Earth rotates. Satellite dishes on the ground pointed at the satellite do not need to be moved to continue to point at the satellite.

b i $8.4\,\mathrm{N\,kg^{-1}}$

iii Adjacent satellites are separated by 30° along their orbit. Each satellite takes about 8 minutes to move through 30°. As a satellite moves away from a mobile phone with which it is contact, the signal is switched to the nearest satellite moving towards it to maintain the signal. This happens about once every 8 minutes.

9 a Gravitational field strength at a point in a gravitational field is the force per unit mass on a small test mass placed at that point in the field.

b i $17.3\,\mathrm{km\,s^{-1}}$ **ii** Let m represent the mass of Io. The gravitational force F on Io $= GMm/r^2$ where M is the mass of Jupiter and r is the (mean) radius of the orbit. The centripetal acceleration of Io, $a = v^2/r$. Using $F = ma$ gives $mv^2/r = GMm/r^2$ which gives $v^2 = GM/r$ and hence $v = \sqrt{\dfrac{GM}{r}}$. Rearranging this equation gives

$$M = \frac{rv^2}{G} = \frac{4.22 \times 10^8 \times (17.3 \times 10^3)^2}{6.67 \times 10^{-11}}$$
$$= 1.89 \times 10^{27}\,\mathrm{kg}$$

iii $1320\,\mathrm{kg\,m^{-3}}$

c The least distance between Io and Europa is $2.49 \times 10^5\,\mathrm{km}$. The force on Io due to Jupiter (or Europa) $\propto m/d^2$ where $m =$ Jupiter (or Europa) and d is the relevant distance from Io to Jupiter (or Europa). The ratio of (m/d^2) for Europa to that of Jupiter gives a measure of the required significance. For Europa $m/d^2 = (4.92 \times 10^{22}\,\mathrm{kg}/(2.49 \times 10^8\,\mathrm{m})^2 = 7.94 \times 10^5\,\mathrm{kg\,m^{-2}}$. For Jupiter $m/d^2 = (1.89 \times 10^{27}\,\mathrm{kg}/(4.22 \times 10^8\,\mathrm{m})^2 = 1.06 \times 10^{10}\,\mathrm{kg\,m^{-2}}$. The ratio of the m/d^2 values is 7.2×10^{-5}. Although this may seem insignificant, the fact that the three bodies are in line every few days stretches Io periodically creating resonant disturbances in Io which heat Io internally and cause volcanic activity.

10 a i Newton's law of gravitation states that there is a gravitational force of attraction between any two point objects and that this force is proportional to the mass of each object and inversely proportional to the square of their distance apart.

ii At height h above the surface of the Earth, the force on a small object of mass $m = \dfrac{GMm}{(R + h)^2}$, where the mass of the Earth $= M$ and its radius $= R$.

Therefore $g = \dfrac{F}{m} = \dfrac{GM}{(R + h)^2} = \dfrac{GM}{R^2\left(1 + \dfrac{h}{R}\right)^2}$
$$= \frac{g_S}{\left(1 + \dfrac{h}{R}\right)^2} \text{ since } g_S = GM/R^2.$$

iii The graph should be a smooth curve from the point $(h/R = 0, g = g_S)$ through the following points $(h/R = 1, g = g_S/4)$, $(h/R = 2, g = g_S/9)$, $(h/R = 3, g/g_S = 1/16)$, $(h/R = 4, g = g_S/25)$,

$(h/R = 5, g = g_S/36)$ in accordance with the inverse square law. No line should be shown between the origin and the first point.

b i Centripetal acceleration $= \omega^2 r = (2\pi\,\mathrm{rad}/(24 \times 3600\,\mathrm{s}))^2 \times (0.5 \times 12756 \times 10^3\,\mathrm{m})$
$= 0.034\,\mathrm{m\,s^{-2}}$.
Since the difference in g at the Equator and the poles is $0.05\,\mathrm{m\,s^{-2}}$, centripetal acceleration is significant.

ii Using $g = GM/R^2$ with appropriate values for the radii and including $G = 6.67 \times 10^{-11}\,\mathrm{N\,m^2\,kg^{-2}}$ gives $9.80\,\mathrm{m\,s^{-2}}$ for g at the poles and $9.86\,\mathrm{m\,s^{-2}}$ for g at the Equator. This is also a significant factor although, since the two effects taken together exceed the observed difference of $0.05\,\mathrm{m\,s^{-2}}$, other factors (e.g. geological, solar, lunar) need to be taken into account.

11 a Gravitational potential at a point is defined as the work done per unit mass to move a small object from infinity to that point.

b i $\phi = -\dfrac{GM}{r}$, $\phi_S = -\dfrac{GM}{R}$

ii $\phi/\phi_S = R/r$ therefore ϕ/ϕ_S is inversely proportional to r. The graph should be a curve starting at $\phi/\phi_S = 1$, $r/R = 1$ then decreasing smoothly, passing through the following points:

r/R	1.5	2.0	2.5	3.0	3.5	4.0	4.5	5.0
ϕ/ϕ_S	0.667	0.500	0.400	0.333	0.286	0.250	0.220	0.200

iii $\phi = -\dfrac{GM}{r}$ and $g = -\dfrac{GM}{r^2}$ therefore multiplying both sides of $g = -\dfrac{GM}{r^2}$ by r gives $gr = -\dfrac{GM}{r^2} \times r = -\dfrac{GM}{r}$; hence $\phi = gr$.

c i $-2.82\,\mathrm{J\,kg^{-1}}$ **ii** $-2.88\,\mathrm{J\,kg^{-1}}$

iii The difference is due in part to the gravitational potential of the Earth at the Moon's surface.

Chapter 14

1 a The electric field strength has the same magnitude and direction at all positions; the lines of force are parallel lines.

b i The electron carries a negative charge and therefore the direction of the electric force on it is in the opposite direction to the field direction.

ii $1.1 \times 10^6\,\mathrm{V\,m^{-1}}$

2 a $33\,\mathrm{nC}$

b The electric field of the spheres at the midpoint is equal in magnitude and opposite in direction. A positively charged particle at the midpoint therefore experiences equal and opposite electric forces of attraction to the spheres.

3 a i The sketch should show lines from the + to the – charge, straight across between the centres and curved above and below.

  ii $2.9 \times 10^5\,\text{V m}^{-1}$ towards Y

 b i The position needs to be 75 mm further from X than from Y in order that the electric field strength of X is equal and opposite to that of Y.

  ii 75 mm to Y, 150 mm to X

4 a A sufficiently strong electric field pulls an electron attached to an atom away from the atom and so ionises the atom.

 b i 100 nC **ii** 680 V m^{-1}

5 b 49 nC

6 a See Topic 14.5.

7 a Electric field strength at a point in an electric field is the force per unit charge on a positive test charge at that point.

 b i $3.56 \times 10^{-7}\,\text{C}$, $2.83 \times 10^{-6}\,\text{C m}^{-2}$ **ii** 80 kV m^{-1}

 c i The electric field of the sphere attracts conduction electrons in the rod towards X, so X becomes positive and Y becomes negative. The electrical force of attraction on end X due to the sphere is greater than the electrical force of repulsion on end Y due to the sphere. So there is an overall force of attraction on the rod towards the sphere.

  ii See Figure 8. (Note: The rod is at constant potential with the potential at its midpoint approximately unchanged. The potentials at X and Y are at the same potential as the midpoint of XY. The surface potential of the sphere is unchanged. So the potential should decrease smoothly from the sphere to X then it is constant from X to Y then it decreases smoothly from Y.)

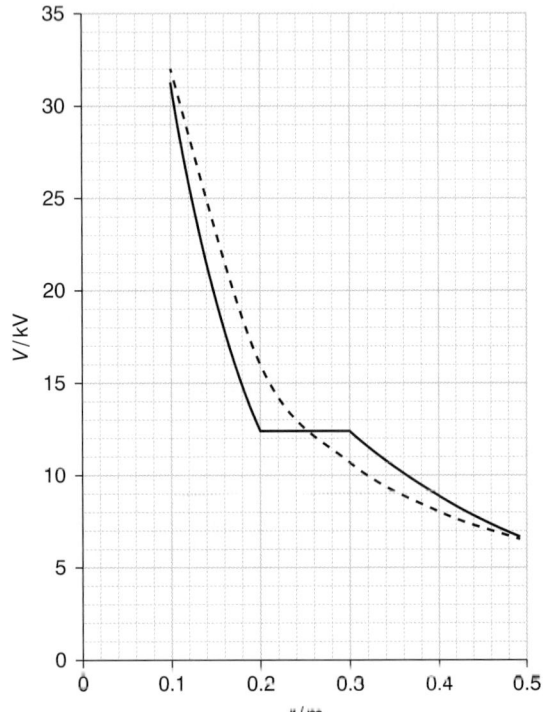

Figure 8

iii The electric field strength is stronger than before and it decreases between the sphere and X. From X to Y, the field strength is zero, whereas it decreased over the same distance before the rod was present. Beyond Y, the field is stronger and decreases gradually to zero at infinity.

8 a Electric field strength at a point in an electric field is the force per unit charge on a positive test charge placed at that point.

 b i The arrow on each line should be from plate B to plate A.

  ii U should be anywhere between the parallel lines between A and B.

  iii The magnitude of the field strength decreases from A to B. The direction of the field changes along the line as it is always tangential to the line.

 b i The arrow at P should be horizontal pointing in the same direction as the field lines. The arrow at Q should be horizontal and pointing in the opposite direction to the field lines.

  ii The perpendicular distance d between the two force arrows = PN sin 60 = 3.5×10^{-9} sin 60 = $3.0(3) \times 10^{-9}\,\text{m}$
  Torque = $Fd = qEd = 1.6 \times 10^{-19}\,\text{C} \times 50\,000\,\text{V m}^{-1} \times 3.03 \times 10^{-9}\,\text{m} = 2.4 \times 10^{-24}\,\text{N m}$

9 a The electric potential at a point in an electric field is the work done per unit charge on a positive test charge when it is moved from infinity to that point.

 b i $8.20 \times 10^{-8}\,\text{N}$

  ii $-4.35 \times 10^{-18}\,\text{J}$

  iii $\dfrac{mv^2}{r} = \dfrac{e^2}{4\pi\varepsilon_0 r^2}$ where m is the mass of the electron. Therefore K.E. $= \frac{1}{2}mv^2 = \frac{1}{2} \times \dfrac{e^2}{4\pi\varepsilon_0 r}$ which is 0.5 × the magnitude of the potential energy of the atom.

  iv P.E. = $-27.2\,\text{eV}$ so the total energy = K.E. + P.E. = $(0.5 \times 27.2\,\text{eV}) - 27.2\,\text{eV} = -13.6\,\text{eV}$

Chapter 15

1 a Charge stored per unit p.d.

 b i See Topic 15.1.

  ii 320 μC, 40 μF

2 a i 1.33 μF **ii** 8.0 μC, 4.0 V, 16 μJ **b** 6.0 μF

3 a 36 μC, 160 μJ **b i** 10 μF **iii** 65 μJ

4 a i 0.95 J **ii** 19 W

 b The combined capacitance is $0.5 \times 47\,000\,\mu\text{F}$ compared with $2 \times 47\,000\,\mu\text{F}$ in parallel. The battery p.d. is the same as before. Since charge stored = capacitance × p.d., the charge stored is less than when they are in parallel.
 The energy stored is $\frac{1}{2}CV^2$, so the energy stored when the capacitors are connected in series is $\frac{1}{4}$ of the energy stored when they are connected in parallel.

5 **a** 5.6 mC, 34 mJ   **b i** 47 s   **iii** 31 mJ
6 **a** 480 mC, 2.88 J   **b ii** 1.1 s
7 **a** The potential V of X is given by the equation
$V = \dfrac{Q}{4\pi\varepsilon_0 R}$. Since its capacitance $C = Q/V$, rearranging

$V = \dfrac{Q}{4\pi\varepsilon_0 R}$ gives $C = \dfrac{Q}{V} = 4\pi\varepsilon_0 R$

  b i Conduction electrons in Y are attracted to X because X is positively charged so they transfer to X via the conductor.

  ii X gains the negative charge of the electrons transferred to it so its overall charge becomes less positive and its potential decreases. Y loses negative charge so it becomes positively charged (as it was initially uncharged). Therefore the potential of Y increases. The flow of electrons from Y to X continues until X and Y are at the same potential.

  iii Let V_f be the potential of X and Y when the flow of electrons ceases and let q represent the positive charge on Y. Therefore the potential of Y = $\dfrac{q}{4\pi\varepsilon_0 \times 0.2R}$. The charge now on X = $Q - q$ so the potential of X = $\dfrac{Q - q}{4\pi\varepsilon_0 R}$. Since the two potentials are equal, $Q - q = \dfrac{q}{0.2} = 5q$ so $q = Q/6$. Therefore the potential of X = $\dfrac{Q - \frac{1}{6}Q}{4\pi\varepsilon_0 R} = \dfrac{\frac{5}{6}Q}{4\pi\varepsilon_0 R} = \dfrac{5}{6}$ or $0.83 \times$ the original potential of X.

  c i The energy stored on a capacitor = $\frac{1}{2}QV$. All the charge on both spheres is at a potential of $0.83 \times$ the original potential. The total charge of both spheres is unchanged at Q. So the total energy stored is 0.83 x the original energy stored on X.

  ii Energy is dissipated due to the resistance heating of the current in the conductor.
8 **a** Taking natural logs on both sides of $V = V_0 e^{-t/CR}$ gives $\ln V = \ln V_0 + \ln(e^{-t/CR})$

As $\ln(e^{-t/CR}) = -\dfrac{t}{CR}$, then $\ln V = \ln V_0 - \dfrac{t}{CR} = a - bt$

Hence $a = \ln V_0$ and $b = \dfrac{1}{CR}$
  b i

t/s	210	240	270	300
mean V/V	1.427	1.233	1.033	0.887
ln V	0.356	0.209	0.032	−0.120

  ii For correct labels on each axis, for suitable scales, for correctly plotted points, see Figure 9.

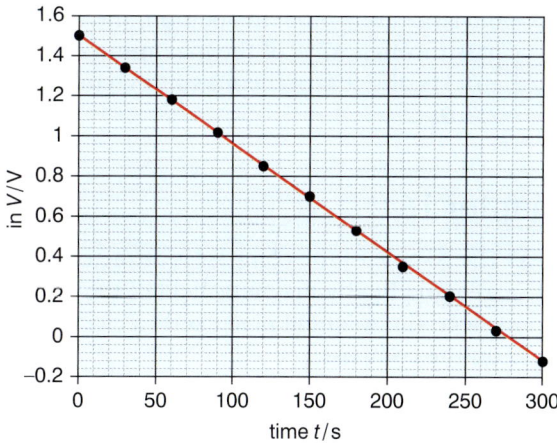

Figure 9

  iii time constant $(= RC) = \dfrac{1}{\text{gradient of graph}}$
gradient of graph = $5.40 \times 10^{-3}\,\text{s}^{-1}$
time constant = $\dfrac{1}{5.40 \times 10^{-3}} = 185\,\text{s}$

  iv $C = \dfrac{\text{time constant}}{R} = \dfrac{185}{6.8 \times 10^4}$
= $2.72 \times 10^{-3}\,\text{F} = 2720\,\mu\text{F}$

  c i The range of each set of readings is no more than 0.03 V, except for the reading at $t = 150$ s which is 0.12 V. This exception is due to an anomalous reading.
The readings are therefore reliable because the range of each set of readings is very small compared with the mean value.

  ii Apart from the exception at $t = 150$ s, the precision of the readings is therefore no more than ±0.02 V.
Using the smallest mean value (i.e. 0.887 V), for 0.887 ± 0.02 V, $\ln V = -0.143$ for $V = 0.867$ V and -0.098 for $V = 0.907$ V. The range of this value of $\ln V$ is therefore 0.045.
Thus the % uncertainty in this value of $\ln V$ is about 3% $(= \frac{1}{2} \times 0.045/0.887 \times 100\%)$.
Given R is accurate to 1%, the value of C is accurate to within 4% (= 3% + 1%)
Note: If time permits, you could estimate the random error in V and hence in $\ln V$ for every point and represent them as error bars on the graph. This would allow you to draw lines of maximum and minimum gradient and so determine maximum and minimum values for the time constant to give an uncertainty value for C.

9 **a** 4.7 μA **ii** As the charge on the capacitor decreases, the capacitor p.d. decreases. Since this p.d. acts directly across the resistor, the current through the resistor decreases.

  b i 1.0(3) s   **ii** $I = I_0 e^{-(t/RC)}$ and $t/RC = 5.0\,\text{s}/1.03\,\text{s} = 4.85$
Therefore $I = I_0 e^{-(t/RC)} = 4.7\,\mu\text{A} \times e^{-4.85} = 0.037\,\mu\text{A}$

  c ii The graph should be an exponential decrease curve from 4.7 μA at $t = 0$ to just above zero (approx

411

0.04 μA) at $t = 5.0$ s. In addition, the curve should pass through 37% of the maximum current at 1.0(3) s and 9% of the maximum current at 2.0(6) s.

Chapter 16

1   a   The N-pole of the compass is attracted to the solenoid by the magnetic field of the solenoid. In effect, the solenoid field exerts a clockwise couple on the compass.
   b   i   The solenoid field becomes weaker so the couple it exerts on the compass becomes smaller and the compass turns back towards the bar magnet.
      ii   The compass would turn clockwise.
2   a   0.23 T   b   0.052 N
3   a   ii   West
   b   A horizontal force due to the 18 μT component of the Earth's field acts on the cable as well as the force due to the vertical component.
4   a   i   1.7 N
      ii   The force reverses direction every half-turn. The magnitude of the force is the same before and after reversal but the turning effect of the force decreases to zero as the coil turns perpendicular to the field.
   b   The perpendicular distance between the lines of action of the force on each side is greatest when the coil is parallel to the field so the couple on the coil due to the forces is a maximum.
5   a   90 mT   b   18°
6   b   Horizontal due west
7   a   See Topic 16.3 Figure 7.   b   0.52 N
8   a 2.6×10^{-3} N m
   b   The weight exerts a moment of 2.5×10^{-3} N m (= weight × spindle radius). The magnetic couple is 2.6×10^{-3} N m, which is greater than the couple exerted by the weight (2.5×10^{-3} N m), so the motor should just be able to lift the weight.
9   a   See Topic 16.5.
   b 3.5×10^7 m s^{-1}
10 a   i   See Topic 16.5.   ii   18 mm
   b   The beam radius increases until the beam is straight when the field is zero. As the reversed field increases from zero, the beam curves in the opposite direction on a circular path with a decreasing radius of curvature.
11 b   i   0.73 T
12 a   The radius of curvature r of the path of an ion is given by $r = mv/BQ$. Since the speed of the ions is the same and they are in the same uniform field, radius r is proportional to m/Q. Hence ions with different specific charges move along different paths.
   b   i   0.84 T   ii 1.4×10^7 C kg^{-1}

13 a   Magnetic flux density is the force per unit length per unit current on a current-carrying wire perpendicular to the field lines.
   b   i   Multiplying each balance reading in grams by 9.81×10^{-3} N/g gives 0, 3.63, 6.87, 10.00, 14.42, 17.85 mN to two decimal places.
      ii   See Figure 10.

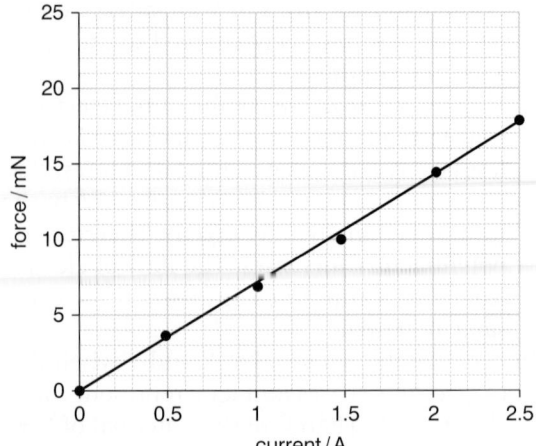

Figure 10

      iii   Force per unit current = the gradient of the line: answer should be between 7.0 and 7.1×10^{-3} N m^{-1}
      iv B = force per unit length/wire length = 0.17 T (= 7.05×10^{-3} N m^{-1}/0.042 m to 2 s.f.)
   c   i   An estimate of the maximum and minimum gradient values gives about ±0.2 mN m^{-1} which gives a % uncertainty in the force per unit length of 2.8%
      ii   The % uncertainty in the wire length = 4.8% (= 2 mm/42 m × 100%)
      The % uncertainty in B = 7.6% therefore the uncertainty in B = ±0.01 T

Chapter 17

1   a   The ribbon is a conductor that cuts across the field lines when it moves backwards and forwards. As a result an alternating e.m.f. is induced in the ribbon.
   b   i   The amplitude of the e.m.f. would be greater,
      ii   The ribbon would vibrate with a smaller amplitude.
2   a   i   The movement of the magnet into the coil induced an emf in the coil (when the coil was in motion) due to the changing magnetic flux in the coil. The induced emf caused a current briefly in the coil and the galvanometer.
      ii   The pointer deflects briefly to the left
   b   i   0.30 mWb   ii   1.5 mV
3   a   The wire is a conductor that cuts across the field lines when it vibrates. As a result an alternating emf is induced in the wire.
   b   1.6 m s^{-1} (1.56 m s^{-1} to three significant figures)

4 a i 3.0 mWb   ii 0.30 T
   b i P +, Q −   ii from Q to P in the rod (and from P to Q round the rest of the circuit)
5 a 2.2 mWb (2.15 mWb to three significant figures)
   b i The graph should show a steady increase in the flux linkage from zero to 2.2 mWb in 4.0 s then constant flux linkage.
      ii The graph should show the induced e.m.f. is 0.54 mV in the first 4 seconds then it is zero.
6 a When the lamps are on, the current through the circuit causes reaction forces on the generator coil which create a couple opposing the applied torque. When the lamps are off, there is no current in the circuit so there is no opposing couple.
   b ii 5.0 V
7 a i The graph should show the flux linkage falling from an initial value and decreasing to zero.
      ii The induced e.m.f. = − the gradient of the graph in ai. The gradient changes from zero and becomes increasingly negative before becoming less and less negative until it is zero.
   b The graph should have a positive peak as in the graph shown then becoming negative to give a negative peak then becoming zero.
8 a 52   b i 5.0 A   ii 0.26 A
   c i 10 W,
      ii The current through the cable would be much smaller so less power would be dissipated in the cable.
         However, the cable might not be intended for use at 230 V so would be unsafe.
9 a Faraday's law of electromagnetic induction states that the induced e.m.f. in a circuit is proportional to the rate of change of magnetic flux linkage through the circuit.
   b i The induced current is measured when the handle is turned at different constant frequencies. For each frequency, a steady rate of rotation is maintained and (with the aid of another student) the time taken for twenty turns is measured at that frequency. The measurements should be recorded in a table and each frequency calculated (= 20/t where t is the time for 20 turns). A graph of the induced current (on the y-axis) against the frequency (on the x-axis) should be plotted.
      ii The induced current is proportional to the induced e.m.f. and the rate of change of the flux linkage is proportional to the frequency of rotation. The induced current should be proportional to the frequency. A graph of the induced current (on the y-axis) against the frequency (on the x-axis) should give a straight line through the origin.

10 a i A transformer increases or decreases the peak voltage of an alternating voltage. The magnetic flux produced by the primary coil passes through the secondary coil. The ratio of the e.m.f. induced in the secondary coil to the alternating voltage applied to the primary coil is equal to the ratio of number of turns of the secondary coil to the number of turns on the primary coil. In a step-up transformer, there are more turns on the secondary coil than on the primary coil so the peak e.m.f. induced in the secondary coil is greater than the peak voltage applied to the primary coil.
       ii For transmission of a certain amount of power, the greater the voltage of the cables, the smaller the current through the cables is. Therefore the greater the voltage, the smaller the power wasted due to the heating effect of the resistance of the cables.
       iii For 5.0 kW of power at 100 V, the current in the cables would be 50 A (= 5000 W/100 V). Therefore the power wasted in the cables due to resistance heating would be 12 500 W = 12.5 kW (= 50^2 × 5.0 Ω). For the same amount of power at 1000 V, the current would be 5.0 A and the power wasted would be 125 W (= 5.0^2 × 5.0 Ω) which is 1/100 th of the power wasted at 100 V.

Chapter 18

1 a See Topic 18.1 Figure 2.   b See Topic 18.1.
2 a The sum of the random distribution of kinetic and potential energies of the molecules,
   b Increasing the internal energy at the melting or boiling point enables the molecules to break free from each other and therefore increases their potential energies. As the temperature is constant, the sum of the kinetic energies of the molecules is unchanged.
3 a i 540 kJ   ii 430 kJ
4 a i Some of the electrical energy is wasted as heat due to friction at the bearings and due to resistance heating by the electric current passing through the motor coil.
      ii Electric energy from the power supply is transferred to the load as work is done on the load and also transferred to the surroundings as heat as explained above and as sound.
   b i 8.8%   ii 87 J, 1.7%
5 a 1.9 × 10^{-2} K s^{-1} b 900 J kg^{-1} K^{-1}
6 a i 780 kJ   ii 620 kJ
   b i 89 kJ   ii Heat loss to surroundings, energy used to vaporise water below boiling point
7 a 7.1 kW   b 0.045 kg s^{-1}

8 a *Specific latent heat of vaporisation* is the energy needed to change the state of unit mass of the substance from liquid to vapour without change of temperature.

   b i See Figure 11.

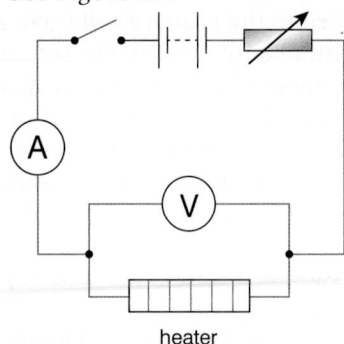

a *Heater circuit*
Figure 11

   ii $850\,\mathrm{J\,kg^{-1}}$

9 a The *internal energy* of an object is the sum of the random distribution of the kinetic and potential energies of its molecules.

   b i $25.0\,^{\circ}\mathrm{C}$ **ii** $2350\,\mathrm{J\,kg^{-1}\,^{\circ}C^{-1}}$

   c % uncertainties: mass of water 0.3%; mass of rock 1.3%; temperature change of water 0.7%; temperature change of rock 5.0%; total % uncertainty = 7.3%

   Uncertainty in specific heat capacity of the rock $= \pm 170\,\mathrm{J\,kg^{-1}\,^{\circ}C^{-1}}$

Chapter 19

1 a i and ii See Topic 19.3.

   b See Figure 12 below.

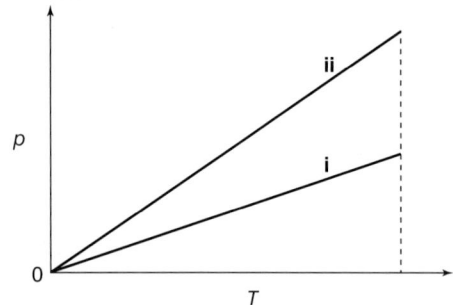

Figure 12

2 b i $2.1 \times 10^{3}\,\mathrm{moles}$

   a ii 1.3×10^{27} (1.25×10^{27} to three significant figures)

3 a 53 moles   **b** $220\,\mathrm{MPa}$

4 a i $1.6 \times 10^{4}\,\mathrm{moles}$ **ii** $0.18\,\mathrm{kg\,m^{-3}}$

   b i The mass of helium is constant and the volume increases so the density decreases as it is equal to the mass per unit volume.

   ii The helium gas in the balloon displaces air. As air is more dense than helium, an upthrust acts on the balloon which is greater than the weight of the balloon.

5 a i $8.4 \times 10^{-3}\,\mathrm{kg}$

   ii The pressure is proportional to the mass of gas provided the temperature is constant. Therefore at pressure p, the mass remaining $= (p/\mathrm{kPa})/150 \times$ the mass at $150\,\mathrm{Pa}$ (assuming the temperature is $300\,\mathrm{K}$ when the reading is taken).

   b $7.7 \times 10^{-3}\,\mathrm{kg}$

6 a i $4.8 \times 10^{-4}\,\mathrm{moles}$ **ii** 2.9×10^{20} **iii** $430\,\mathrm{m\,s^{-1}}$

   b i $17.5\,\mathrm{kPa}$ **ii** $0.30\,\mathrm{J}$

7 a i $1.1 \times 10^{3}\,\mathrm{kg}$ **ii** $6.7 \times 10^{-27}\,\mathrm{kg}$

   b i $1.3 \times 10^{3}\,\mathrm{m\,s^{-1}}$ **ii** $9.8 \times 10^{8}\,\mathrm{J}$

8 a When the brakes are applied, the pressure is not transmitted through the oil as effectively as it compresses the bubble. So the braking force is reduced.

   b i $4.0 \times 10^{-4}\,\mathrm{m^{2}}$ **iii** $9.2\,\mathrm{mg}$

9 a $0.0139\,\mathrm{mole}$ **b i** $1.47\,\mathrm{J}$ **ii** $121\,\mathrm{kPa}$

   c i The graph should be a straight line passing through $(20\,^{\circ}\mathrm{C}, 125\,\mathrm{kPa})$, $(11.5\,^{\circ}\mathrm{C}, 121\,\mathrm{kPa})$ and $(0\,^{\circ}\mathrm{C}, 116\,\mathrm{kPa})$.

   ii The graph should be a straight line passing through $(20\,^{\circ}\mathrm{C}, 62.5\,\mathrm{kPa})$, $(11.5\,^{\circ}\mathrm{C}, 60.5\,\mathrm{kPa})$ and $(0\,^{\circ}\mathrm{C}, 58\,\mathrm{kPa})$.

10 a The *r.m.s. speed* is the square root of the mean value of the square of the speeds of the molecules.

   b $508\,\mathrm{m\,s^{-1}}$ **ii** Combining the ideal gas equation and the kinetic theory equation for 1 mole of ideal gas gives $RT = \frac{1}{3}N_{A}mc_{\mathrm{r.m.s.}}^{2}$ $(= pV)$ and, since the molar mass $M = N_{A}m$, then $c_{\mathrm{r.m.s.}}^{2} = \left(\frac{3R}{M}\right)T$.

   Therefore the mean square speed $c_{\mathrm{r.m.s.}}^{2}$ is proportional to the absolute temperature T because $\left(\frac{3R}{M}\right)$ is constant.

   c i The volume of the gas does not change, so no work is done.

   ii No energy is transferred by heating to or from the gas between C and D and the gas does work since it expands from C to D so its internal energy decreases.

   iii The area of the loop represents the net work done in a cycle. There are about 165 small squares in the loop and each small square represents energy transfer of $2\,\mathrm{J}$ $(= 0.2 \times 10^{6}\,\mathrm{Pa} \times 0.1 \times 10^{-4}\,\mathrm{m^{3}})$ so the work done is about $330\,\mathrm{J}$.

   iv 42%

   v Energy is transferred to the surroundings by the hot exhaust gases. The engine block becomes hot and transfers energy by heating to the cooling system. Sound waves created by the vibrations of the engine transfer energy to the surroundings.

Chapter 20

1 a The *work function* of a metal is the minimum energy needed by an electron to escape from the metal's surface.

b i See Figure 13.

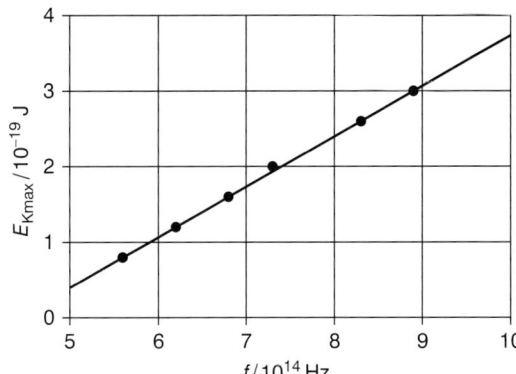

Figure 13

ii Gradient = (2.40 − 0.40/8.0 − 5.0) = 6.67 × 10^{-34} J Hz^{-1}

iii The graph is a straight line so it fits the straight line equation $y = mx + c$ equation where $y = E_{Kmax}$, $x = f$, m = the Planck constant h and $c = -\phi$, where ϕ is the work function of the metal. The graph gradient is very close to the accepted value of h.

iv The work function = c as it is the y-coordinate at $x = 0$. Since $E_{Kmax} = 0.40 \times 10^{-19}$ J at $f = 5.0 \times 10^{14}$ Hz, then substituting these values and the gradient value above into $y = mx + c$ gives 0.40×10^{-19} J = (6.67 × 10^{-34} J Hz^{-1} × 5.0 × 10^{14} Hz) + c. Hence $c = 0.40 \times 10^{-19}$ J − (6.67 × 10^{-34} J Hz^{-1} × 5.0 × 10^{14} Hz) = −2.94 × 10^{-19} J. Therefore the work function is equal to 2.94 × 10^{-19} J.
(Note: a non-graphical value can be obtained by recognising the line rises from the y-intercept by 3.4 squares over 5 squares along the x-axis. Therefore extrapolating backwards to zero frequency would mean a descent of 3.4 squares to reach $x = 0$ which would be 3.0 below the x-axis, corresponding to a work function of 3.0 × 10^{-19} J.

b The threshold frequency = ϕ/h
= 3.3 × 10^{-19} J/6.63 × 10^{-34} J s = 5.0 × 10^{14} Hz
The line should be parallel to the existing line and pass through the point on the x-axis, where $f = 5.0 \times 10^{14}$ Hz

2 a i The work function of a metal is the minimum amount of energy needed by a conduction electron to escape from the surface of a metal when the metal is at zero potential.

ii The threshold frequency f_0 of a metal is the minimum frequency of incident light that will cause photoelectric emission from the surface of the metal.

b $\phi = hf_0$

c i According to wave theory, photoelectric emission from a metal surface should take place at *any* frequency of light as a conduction electron at the surface of the metal will repeatedly absorb energy from the incident waves until it has sufficient energy to leave the surface. The fact that there is a threshold frequency of light below which photoelectric emission does not take place contradicts the wave theory. The photon theory of light assumes light consists of photons, each with energy in proportion to the frequency of light according to Einstein's photon equation $E = hf$. The photon theory therefore explains the existence of the threshold frequency because photoelectric emission can only take place if a conduction electron at the surface of the metal absorbs of photon of energy hf greater than or equal to the work function of the metal (i.e. $hf \geq \phi$).

ii Photoelectric emission takes place instantly provided the frequency of light is greater than or equal to the threshold frequency of the metal. This supports the photon theory because conduction electrons at the surface are emitted as soon as the light is incident on the surface. In comparison, according to wave theory, conduction electrons at the surface would need to wait to accumulate sufficient energy from the incident light waves so emission would not be instant.

3 a Light consists of photons of energy proportional to the light frequency. A photon of blue light therefore has more energy than a photon of red light. Each photon directed at the metal surface that is absorbed by a single electron at or near the metal surface gives all its energy to the electron. The energy gained by an electron from a photon of blue light is sufficient to enable it to leave the metal surface. The energy gained by an electron from a photon of red light is not sufficient to enable it to leave the metal surface.

b i Increasing the intensity of blue light causes more photons of blue light to be directed at the surface each second so more electrons are released from the surface each second.

ii If the metal surface is at a positive potential, the maximum kinetic energy of the electrons emitted from the surface would be less as each electron leaving the surface would need to do work to move away from the positively charged surface. If the positive potential is too large, no electrons would have sufficient energy to leave the surface so photoelectric emission would be stopped.

4 a 3.3×10^{-19} J **b** 2.0×10^{-19} J

5 a The work function of a metal surface is the minimum energy needed by an electron to escape from the surface.

  b i 9.9×10^{-7} m

    ii A photon of longer wavelength would have less energy than 2.0×10^{-19} J and would not therefore be able to give an electron at the metal surface enough energy to escape from the surface.

6 a i Light consists of photons of energy proportional to the light frequency. Each photon in the beam directed at the metal surface that is absorbed by a single electron at or near the metal surface gives all its energy to the electron. At this light frequency, the energy gained by such an electron is sufficient to enable it to leave the metal surface. The electrons that do leave the metal surface are attracted to the central anode and they then move round the circuit. The microammeter registers a current due to the flow of electrons from the anode round the circuit back to the metal.

    ii More photons are directed at the metal surface so more electrons are emitted from its surface. The current therefore increases.

  b i The light photons have less energy than in a) and each one does not have enough energy to enable an electron to escape from the metal. The energy of a light photon is less than the work function which is the minimum energy needed by an electron to escape from the metal surface.

    ii Making the incident light more intense increases the number of photons directed at the metal each second. However, no electrons can escape from the metal because no photon has enough energy to enable an electron to escape.

7 a i X to Z **ii** Y to Z **b** 3.4×10^{-22} J

8 a i 17.3 V

    ii Energy E of photon released = $(-14.2\,\text{eV}) - (-17.3\,\text{eV}) = 3.1\,\text{eV}$. Photon frequency = E/h = $(3.1\,\text{eV} \times 1.6 \times 10^{-19}\,\text{J/eV})/6.63 \times 10^{-34}$ J = 7.48×10^{14} Hz. Photon wavelength = c/f = $3.00 \times 10^{8}\,\text{m s}^{-1}/7.48 \times 10^{14}$ Hz = 4.0×10^{-7} m.

  b i B→E $6.8\,\text{eV}$ B→D $3.7\,\text{eV}$ B→C $1.1\,\text{eV}$ C→E $5.7\,\text{eV}$ C→D $2.6\,\text{eV}$ D→E $3.1\,\text{eV}$

    ii Energy of photon that causes excitation from E to D = $3.1\,\text{eV}$, so its wavelength = 4.0×10^{-7} m (as calculated in **aii**)

9 a i Kinetic energy of proton = $\frac{1}{2}mv^2$ = $0.5 \times 1.67 \times 10^{-27}$ kg $\times (1.7 \times 10^{7}\,\text{m s}^{-1})^2 = 2.41 \times 10^{-13}$ J

    ii Momentum, $p = 1.67 \times 10^{-27}$ kg $\times 1.7 \times 10^{7}\,\text{m s}^{-1}$ = $2.84 \times 10^{-20}\,\text{kg m s}^{-1}$; de Broglie wavelength = $h/p = 6.63 \times 10^{-34}$ J s/ $2.84 \times 10^{-20}\,\text{kg m s}^{-1} = 2.33 \times 10^{-14}$ m

b Momentum of photon = E/c = 2.41×10^{-13} J/$3.00 \times 10^{8}\,\text{m s}^{-1}$ = $8.03 \times 10^{-22}\,\text{kg m s}^{-1}$

10 a de Broglie wavelength = $h/mv = 6.63 \times 10^{-34}$ J s/ $(9.11 \times 10^{-31}\,\text{kg} \times 9.8 \times 10^{7}\,\text{m s}^{-1}) = 7.43 \times 10^{-12}$ m

  b 5° ring; $n = 1$ gives $2d \sin 5° = 7.43 \times 10^{-12}$ m, therefore $d = 7.43 \times 10^{-12}$ m/$2 \sin 5° = 4.26 \times 10^{-11}$ m 10° ring; $n = 2$ gives $2d \sin 10° = 2 \times 7.43 \times 10^{-12}$ m, therefore $d = 7.43 \times 10^{-12}$ m/$\sin 10° = 4.28 \times 10^{-11}$ m Therefore $d = 4.3 \times 10^{-11}$ m (Note: Data does not justify three significant figures in the answer.)

  c Increasing the speed of the electrons reduces their de Broglie wavelength and, since d does not change, the angle of diffraction of each ring decreases.

Chapter 21

1 a i $5.7 \times 10^{-8}\,\text{s}^{-1}$ **ii** 6.1×10^{-3} g

  b ii 16 W

2 a 2.7×10^{24} **b** 36 MBq

3 a Because radioactive decay is random, the probability of decay is the same for all the nuclei of a given isotope. Therefore, for a large number of nuclei, the number of nuclei that decay per second (i.e. the activity) depends only on the number of radioactive nuclei present. As a result, the number of radioactive nuclei decreases exponentially with time. The rate of decay is characterised by the half life of the isotope which is defined as the time taken for half the initial number of nuclei to decay. If the activity at a certain time is known, the number of atoms present at that time is given by the half life × the activity ÷ ln 2.

  b i 41 kBq **ii** 146 hours

4 a i 4.1×10^{14} Bq **ii** $86\,\text{J s}^{-1}$

  b 62 K

  c i Gamma radiation is weakly ionising and therefore can penetrate body tissue and kill cancer cells.

    ii The cobalt source is at the centre of a solid lead container which stops all the gamma radiation from the source except radiation that passes along a narrow channel from the source to the outside. A lead cover at the end of the channel used to stop the beam is removed only for a limited time when the beam is directed accurately at the part of the patient to be 'treated'. In this way, the patient is exposed to the beam for the least time necessary.

5 a The binding energy of a nucleus is the work that must be done to separate a nucleus into its constituent neutrons and protons.

  b i 28 MeV **ii** 2.2 MeV

  c The binding energy per nucleon of an alpha particle is about 7 MeV which is much greater than that of a deuterium nucleus which is about 1 MeV so the protons and neutrons in an alpha particle are bound much more strongly together than in a deuterium nucleus. In an alpha particle, the strong nuclear force holding the neutrons and protons together is much stronger than the electrostatic force of repulsion between the two protons. This force is independent of whether a nucleon is a neutron or a proton.

6 a $_3^6\text{Li} + {}_0^1\text{n} \rightarrow {}_1^3\text{H} + {}_2^4\alpha$
      X is an alpha particle
    b i 7.7×10^{-13} J   ii 4.00150 u
7 a i It has a range of $2-3 \times 10^{-15}$ m; it is an attractive force beyond 0.5×10^{-15} m and is repulsive at less than 0.5×10^{-15} m; it is charge-independent (i.e. the same between protons and neutrons).
      ii The binding energy per nucleon of the two fission nuclei is greater than that of the uranium −235 nucleus. The nucleons in the fission nuclei are more tightly bound than in the uranium nucleus. Therefore energy must be released when the uranium nucleus undergoes fission.
    b i $_{92}^{235}\text{U} + {}_0^1\text{n} \rightarrow {}_{54}^{140}\text{Xe} + {}_{38}^{93}\text{Sr} + 3{}_0^1\text{n}$
      ii 2.8×10^{-11} J
8 a Similarity: the binding energy per nucleon increases in both cases. Difference: a large nucleus splits into two smaller nuclei in a fission reaction; in a fusion reaction, two small nuclei fuse to form a larger nucleus.
    b i Two light nuclei need to approach each other within $2-3$ fm in order that a strong nuclear force can fuse them together. At 10^8 K, the nuclei have sufficient kinetic energy to overcome the electrostatic repulsion that would otherwise stop them fusing.
      ii $_1^2\text{H} + {}_1^2\text{H} \rightarrow {}_1^3\text{H} + {}_1^1\text{p}$ Energy released 4.1 MeV
9 a They contain many different fission products which are neutron-rich and therefore β⁻ and γ emitters which are highly radioactive. They also contain radioactive isotopes such as plutonium 239 which is highly radioactive.
    b i 5.2×10^9 Bq,   ii 4.4×10^8 Bq
10 a The binding energy of a nucleus is the work that must be done to separate a nucleus into its constituent neutrons and protons.
    b i 1 proton and 1 neutron   ii 1.118 MeV
    c i 3.8 ii B: For the nucleus $_Z^A\text{X}$, there are $A(A-1)/2$ force bonds which gives a BE ratio of $A(A-1)/2$ in relation to the $_1^2\text{H}$ nucleus. Applied to the helium $_2^4\text{He}$ nucleus, the ratio in relation to the $_1^2\text{H}$ nucleus is therefore 6, which differs greatly from the actual ratio which is about 12. Student A's suggestion doesn't work either as it predicts a ratio of 2 (= 4 for $_2^4\text{He}/2$ for $_1^2\text{H}$).
    d i Y is a proton: $_1^3\text{H} + {}_1^2\text{H} \rightarrow {}_2^4\text{He} + {}_1^1\text{p}$
      ii Energy released ≈ 17 MeV
11 a i Nuclear fission is the splitting of a large unstable nucleus into two smaller nuclei as a result of

the original nucleus being struck by an incident neutron.
      ii Uranium $_{92}^{235}\text{U}$
      iii When nuclear fission of a fissionable nucleus occurs, the nucleus splits in two and several neutrons referred to as fission neutrons are released. In a nuclear reactor, these fission neutrons collide with other fissionable nuclei causing a chain reaction in which further fission events occur and more neutrons are released which then cause more fission events and so on.
      iv In a nuclear reactor, the number of fission neutrons produced per second depends on the mass of the fissile material in the reactor core. However, the number of fission neutrons per second that are lost from the reactor core depends on the surface area of the fuel rods. For a sustainable chain reaction, the mass of fissile material must therefore be greater than a critical mass or else more neutrons per second are lost than are produced and chain reactions cease.
    b i In a thermal nuclear reactor, the fission neutrons are released with too much kinetic energy to induce fission events. The moderator is necessary in a thermal nuclear reactor because the moderator atoms reduce the kinetic energy of the fission neutrons that collide with them to a level at which further fission events occur.
      ii Boron or cadmium: The fission neutrons are absorbed by the control rod nuclei without causing fission. The rate of production of fission neutrons can therefore be reduced by inserting the control rods further into the reactor core or increased by partially withdrawing the control rods from the reactor core. Automatic sensors are used to monitor the rate of fission events in the reactor core and to control the position of the control rods in the core to maintain a constant fission rate in the core.
      iii The coolant must be fluid and non-corrosive.

Chapter 22

1 a 0.14 mm, ii 0.62 mm
    b i Ultrasonic waves of much lower frequency would diffract more due to their longer wavelength. Increased diffraction would reduce the resolution (i.e. detail) of the images.
      ii An ultrasonic probe contains a piezo-electric transducer in the shape of a disc is applied inside and at the end of a tube which contains a backing block that keeps the disc in place. When an alternating p.d. of frequency equal to the natural frequency of vibration of the disc is applied, the disc vibrates in resonance and creates ultrasound waves in the surrounding medium at the same frequency as the alternating p.d.

c i *Specific acoustic impedance* of a substance is the product of its density and the speed of ultrasound in it. **ii 1** 0.999 **2** 1.66 × 10⁻³

d The probe is applied to the body via a suitable gel so that most of the ultrasound energy enters the body. This is because an air-skin boundary reflects almost 100% of the incident ultrasonic energy, whereas a gel-skin boundary reflects very little of the incident ultrasonic energy.

2 a i $\lambda_{min} = \dfrac{hc}{eV} = \dfrac{6.63 \times 10^{-34}\,\text{Js} \times 3.00 \times 10^{8}\,\text{ms}^{-1}}{1.60 \times 10^{-19}\,\text{C} \times 40000\,\text{V}}$

$= 3.1 \times 10^{-11}\,\text{m}$

ii See Topic 22.2, Figure 2. The graph should show a continuous distribution between zero intensity at zero energy and at maximum energy with peak intensity between. X-ray spikes should be shown at specific energies.

b i A contrast medium is an X-ray absorbing substance that is used to fill a soft organ or blood vessel to make its image stand out from the surrounding tissues.

ii A barium meal containing barium sulphate is a contrast medium. It is given to a patient before an X-ray image of the stomach is taken.

c A CT scanner consists of an X-ray tube and a ring of thousands of small solid-state detectors linked to a computer. The patient lies stationary on a bed along the axis of the ring. The X-ray tube automatically moves round the inside of the ring, turning as it moves so the X-ray beam is always directed at the centre of the ring. The detector signals are simultaneously recorded by the computer each time the X-ray tube moves round the ring through a fraction of a degree until the tube has moved through 180°. For each position of the X-ray tube, the X-rays from the tube travel through the cross-section of the patient under investigation and reach the detectors.

The signal from each detector depends on the different types of tissue along the path and how far the X-ray beam passes through each type of tissue. By considering the patient as a collection of small volume elements (voxels), the signal from each detector is considered to have been attenuated (i.e. reduced in intensity) by each voxel along the path of the X-rays to the detector. The total attenuation along a given path is the sum of the attenuation by each voxel along the path. The computer is programmed to process all the detector signals to determine the attenuation due to each voxel, and then to display a digital image consisting of corresponding pixels, each of greyness or colour according to the attenuation caused by the voxel.

3 a i gamma radiation **ii** A positron-emitting substance is used. Nuclei of the substance are unstable and each one that decays emits a

positron which is the antiparticle of the electron. When the emitted positron collides with an electron, they annihilate each other and create two gamma ray photons which travel away from each other in opposite directions.

b i The two gamma ray photons are released simultaneously and because they travel in opposite directions, each one reaches the detector which is along its line of travel which is in the opposite direction to the other one.

ii The straight line XY is a chord of the circular ring which subtends an angle θ of 135° to the centre of the chord. The centre of the circle O, the midpoint M of the chord and either X or Y form a right-angle triangle with a hypotenuse of length 2.40 m. Since the angle MOX is 67.5° (= 0.5 × 135°), the distance MX = 2.40 m × sin 67.5° = 2.217 m. Therefore XY = 4.43 m.

iii The two gamma photons travel in opposite directions from a point at or near the line XY. The positron-emitting nucleus must therefore have been on or near the line XY because the positron emitted by it would have travelled no further than 2 or 3 mm from the nucleus.

iv More simultaneous or near-simultaneous detections of two gamma photons by separate detectors need to be recorded. For each such detection, straight line drawn on a scale diagram of the detector ring between the two relevant detectors would cross the XY line. Where all the lines or groups of lines cross at or near each other, the crossing points must correspond to where the positron-emitting substance is located.

OR The position of the positron-emitting nucleus along XY can be deduced if the time interval Δt between the arrival of the two gamma photons is measured. The time taken T for a photon to travel from X to Y can be calculated by dividing the length of XY by the speed of light. The distance along XY from the first detector triggered is equal to (Δt/T) × XY.

c 1. The PET scanner uses gamma radiation from a positron-emitting substance that has been taken in by the body. The CT scanner uses gamma radiation from an X-ray tube.

2. The PET scanner gives an image of the location of the positron-emitting substance in the body. The CT scanner gives an image of an internal organ in the body.

3. PET scanner detectors are designed to detect gamma photons in very short time intervals. CT scanner detectors measure the intensity of the gamma radiation from the X-ray tube after it has passed through the patient.

Chapter 23

1   a   i $\lambda_{max} T = 0.0029\,m\,K$ where T is the surface temperature of the star and λ_{max} is the wavelength at which the electromagnetic radiation emitted from the star is at maximum intensity.

   ii $T = 0.0029\,m\,K/\lambda_{max} = 0.0029\,m\,K/740 \times 10^{-9}\,m$
        $= 3900\,K$

   b $L = F \times 4\pi d^2 = 3.5 \times 10^{-8}\,W\,m^2 \times 4\pi\,(6.0 \times 10^{17}\,m)^2$
       $= 1.6 \times 10^{29}\,W$

   c   i   Stefan's law states that the luminosity of a star $L = \sigma AT^4$ where A is the surface area of the star and T is its surface temperature. Since luminosity is the radiant energy per second emitted from the surface, then L/A is the radiant energy per second emitted per unit surface area. Hence $L/A = \sigma T^4$.

   ii   Rearranging the above equation gives $A = L/\sigma T^4$
        $= 6.2 \times 10^{26}\,W/[5.67 \times 10^{-8}\,W\,m^{-2}\,K^{-4} \times (9700\,K)^4]$
        $= 1.2 \times 10^{22}\,m^2$. Since $A = 4\pi R^2$ where R is the surface radius, then $R = (A/4\pi)^{1/2} = 3.1 \times 10^{10}\,m$. Therefore its diameter $= 6.2 \times 10^{10}\,m$.

2   a   i   The surface temperature of a star is inversely proportional to its peak intensity wavelength. B is hottest because it has the shortest peak intensity wavelength and C is the coolest because it has the longest peak intensity wavelength.

   ii   The radiant flux intensity from a star is proportional to its luminosity $L \div$ its distance d squared. A calculation of L/d^2 using the above values gives $4.25\,L_{SUN}$ for A, $1.83\,L_{SUN}$ for B and $0.011\,L_{SUN}$ for C. Therefore, in terms of increasing radiant flux intensity, the order is C, B, A.

   b   i   For star A, $L = 10.6 \times 4.0 \times 10^{26}\,W = 4.2 \times 10^{27}\,W$

   ii   Surface temperature $T = 0.0029\,m\,K/380 \times 10^{-9}\,m = 7630\,K$ iii Rearranging $L = \sigma AT^4$ gives $A = L/\sigma T^4 = 4.2 \times 10^{27}\,W/(5.76 \times 10^{-8}\,W\,m^{-2}\,K^{-4} \times (7630\,K)^4 = 2.15 \times 10^{19}\,m$. Therefore $R = (A/4\pi)^{1/2} = 1.3 \times 10^9\,m$.

   c   i   Inserting $A = 4\pi R^2$ and $T = 0.0029\,m\,K/\lambda_{max}$ into $L = \sigma AT^4$ gives $L = \sigma(4\pi R^2) \times (0.0029\,m\,K)^4/\lambda_{max}{}^4$ which simplifies after rearrangement to $R^2 = z\lambda_{max}{}^4 L$ where $z = 4\pi\sigma\,(0.0029\,m\,K)^4$. Hence R^2 is proportional to $\lambda_{max}{}^4 L$.

   ii   Values of $L\lambda_{max}{}^4$ for stars B and C relative to A are $(130/10.6) \times (230\,nm/380\,nm)^4 = 1.6$ for B and $(0.0035/10.6) \times (940\,nm/380\,nm)^4 = 0.012$ for C. In terms of increasing radius, the order of the three stars is C, A, B.

3   a $L = 4\pi d^2 F = 4\pi\,(1.5 \times 10^{23}\,m)^2 \times 1.4 \times 10^{-13}\,W\,m^{-2}$
       $= 4.0 \times 10^{34}\,W$

   b   i $\lambda_{max} \approx 300\,nm$; $T = 0.0029\,m\,K/300 \times 10^{-9}\,m = 10000\,K$

   ii   Rearranging $L = \sigma AT^4$ gives $A = L/\sigma T^4 = 4.0 \times 10^{29}\,W/[5.67 \times 10^{-8}\,W\,m^{-2}\,K^{-4} \times (10000\,K)^4]$
        $= 7.1 \times 10^{10}\,m^2$. Since $A = 4\pi R^2$ where R is the surface radius, $R = (A/4\pi)^{1/2} = 7.5 \times 10^9\,m$. Therefore its diameter $= 15 \times 10^9\,m \approx 10 \times$ diameter of the Sun.

4   a   *Red shift* is the increase in the wavelength of the light from a light source that is moving away from the observer.

   b   The increase of wavelength $\Delta\lambda$ in the light from a receding star or galaxy is proportional to its speed of recession v in accordance with the Doppler shift equation $\Delta\lambda = (v/c)\,\lambda$. The increase of wavelength can be determined by comparing the wavelength of the lines of the line spectrum of the light from the star or galaxy with the wavelength of the same lines in the light from a laboratory source. The speed of recession is then calculated by dividing the increase of wavelength for each line by its wavelength and multiplying the result by the speed of light.

   c   i   See Figure 14.   ii $H =$ gradient $= 2.4 \times 10^{-18}\,s^{-1}$
        $(= 1000\,km\,s^{-1}/41 \times 10^{22}\,m)$

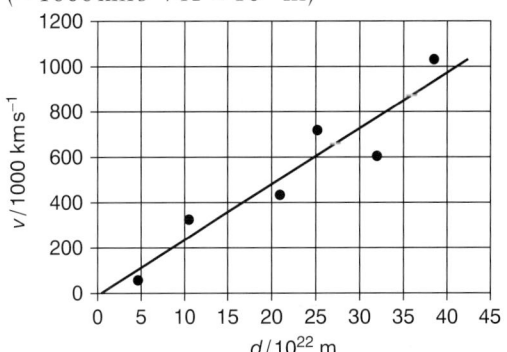

Figure 14

   iii   Uncertainty estimate $\pm 0.2 \times 10^{-18}\,s^{-1}$

   c   Hubble showed that the speed of recession is proportional to the distance from the Earth. All subsequent measurements confirm this relationship. The changed value of H does not affect the conclusion that the results confirm the proportionality relationship. The changed value of H is due to more accurate methods of measuring the distances.

Answers to questions for Key concepts: Forces (p.90–93)

		Answers	Marks	Comments
1 a i		to ensure its speed of projection was the same every time the test was carried out	1	
	ii	It might dislodge the bar on impact.	1	
		It might bend the tube as it rolled down inside it.	1	
b i		mean value of y/mm: 20.3, 71.7, 159, 278.3, 433, 637.7	2	2 marks for all correct
			2	1 mark for 3 correct
		x^2/m²: 0, 0.0420, 0.162, 0.355, 0.645, 0.990, 1.452		as above
	ii	$t = x/U$ (from $x = Ut$ rearranged)	1	
		substituting in $y = \frac{1}{2}gt^2$ gives		
		$y = \frac{1}{2}g(x/U)^2 = (\frac{1}{2}g/U^2)x^2 = kx^2$		
		where $k = \frac{1}{2}g/U^2$	1	
	iii	graph:		Three sets of measurements are given of the vertical distance fallen for different horizontal distances. In part b, the hypothesis is given that vertical distance is directly proportional to the horizontal distance squared. The mean value of each vertical distance has to be calculated then used to plot a graph of the vertical distance against the horizontal distance squared in order to test the hypothesis.

			Marks	Comments
		labelled correctly	1	
		correct units shown on axes	1	
		points plotted correctly	1	
		best-fit line drawn	1	
	iv	k = the gradient of the straight line	1	Accept answer for U in the range 3.1–3.5 m s⁻¹.
		$= 0.61/1.4 = 0.44$ m⁻¹	1	
		so $U = \sqrt{\dfrac{g}{2k}} = \sqrt{\dfrac{9.8}{2 \times 0.44}} = 3.3$ m s⁻¹	1	
c		The y values for each distance measurement have a range of more than 2 mm. (For example, for $x = 1.205$ m, the range of y is 12 mm.)	1	Part c asks about the precision and accuracy of the measurements. You need to realise here that the uncertainty of each mean value is not given by the precision of a mm scale, but by estimating the spread of each set of measurements (e.g. half the range of the vertical distance measurements).
		So, the y values have a measurement error of more than 2 mm.	1	
		For the x-values, although the readings using a mm rule can be made to within 1 mm, the exact position of impact on the bar is uncertain and introduces a further measurement error equal to the diameter of the bar.	1	
			1	
		Total mark	**20**	

Answers to questions for Key concepts: Electricity, waves and radioactivity (p.162–165)

	Answers	Marks	Comments
1 a i	four out of five components shown correctly	1	The question is about an investigation into the heating effects of an electric current. Part a tests your knowledge of electric circuits.
	for all components shown correctly	1	

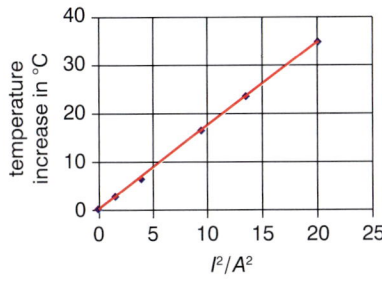

ii	by adjusting the variable resistor if the current (or the ammeter reading) changed to keep the ammeter reading the same in each test	1 1	Correct reference to ammeter and variable resistor needed for both marks.
b i	(Use the thermometer to) measure the water temperature at intervals after the current is switched off.	1	
	The insulation is effective if the water temperature does not decrease (or change) over at least 600 s.	1	Reference to time scale of at least 600 s necessary for the second mark.
ii	The temperature rise depends on the heating time and the volume (or mass of water).	1	
	If either of these quantities differs in a test, the temperature rise will be unreliable.	1	
c i	the energy supplied to the heater in the time $t = I^2 R t$	1	Here you have to use your knowledge of electrical power and resistance to justify a given prediction and then use your data analysis skills to plot a graph and use it to test the prediction.
	The temperature rise is proportional to the energy supplied.	1	
ii	I^2/A^2: 1.44, 4.00, 9.61, 13.7, 20.3	1	
iii	graph		

graph

	for correctly labelled axes	1	
	for suitable scales	1	
	for correctly plotted points	1	
	for best-fit line	1	
	The graph is a straight line through the origin, so the temperature increase is directly proportional to the current squared, as predicted.	1 1	
d	The resistance of the heater element increased as the water temperature increased.	1	You are expected to use your knowledge and understanding about the effect of temperature on the resistance of a metal to explain the observation.
	The battery provided a constant e.m.f.	1	
	So the resistance of the variable resistor had to be reduced to keep the total circuit resistance constant and hence keep the current constant.	1	
	Total mark	**18**	

Answers to questions for Key concepts: Fields (p.266–269)

	Answers	Marks	Comments
1 a i	With the object on the spring: mean value of x = 72 mm e = 70 mm	1	
ii	1.4%	1	Each reading was ±0.5 mm. As the extension was the subtraction of two readings, the absolute uncertainties are added to give an absolute uncertainty of ±1.0 mm and a percentage uncertainty of $\frac{1}{70} \times 100$ = ±1.4%.
b i	0.551 s	1	T_{av} = 11.02 s
ii	0.6%	1	The absolute uncertainty can be taken as half the range of the values, so the uncertainty in T_{av} is: (11.11 – 10.97)/2 = ±0.07 s
c i	$mg = ke$, therefore $e = mg/k$	1	
ii	Using the above equation gives $$\frac{m}{k} = \frac{e}{g}$$ Substituting this expression for $\frac{m}{k}$ into the mass–spring time period equation $$T = 2\pi\sqrt{\frac{m}{k}}$$ gives the required equation.	1 1	
d	Plot either T^2 against e or T against $\sqrt{e}$: correct labels and units suitable scales all points plotted correctly best-fit line According to the equation, the line should pass through the origin and the gradient is equal to $4\pi^2/g$ for the T^2 against e line or $2\pi/\sqrt{g}$ for the T against $\sqrt{e}$ line. To determine g, the gradient of the line should be measured. Gradient given the correct unit ($s^2\,m^{-1}$ or $s^2\,mm^{-1}$). A large triangle used correctly to determine the gradient and used with the appropriate gradient formula above to find g.	4 1 1 1 1 1	Draw a triangle or use points that cover over half of the line you have drawn. It improves accuracy. Make sure the points used are on your line and not just two of the plotted points. For example, the graph for T^2 against e is shown below. Its gradient = $4.074\,s^2\,m^{-1}$. Hence $g = 4\pi^2/4.074$ = $9.69\,m\,s^{-2}$ 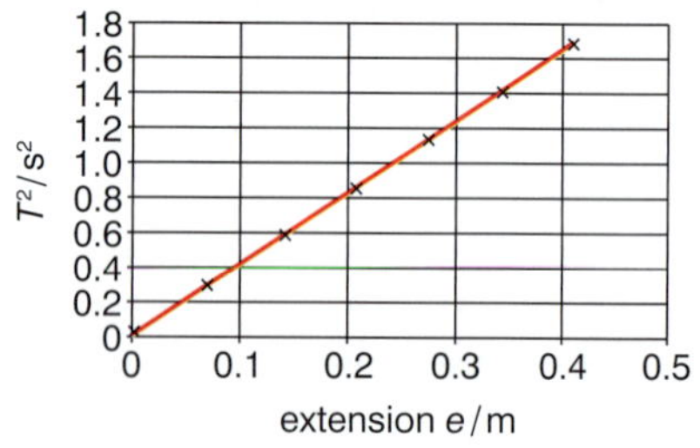
e	The line is straight through the origin. The uncertainty in the gradient can be estimated by considering the uncertainty in each measurement and put error bars on each plotted point, as explained on p.xxi. Draw best-fit lines through the error bars with maximum and minimum gradients, and then give an estimate of the uncertainty in the gradient value.		An alternative method is to estimate the uncertainty in the y-coordinate of the point where the extension is 0.400 m, which is between $1.62\,s^2$ and $1.65\,s^2$, giving a gradient of between $1.62\,s^2/0.400\,m$ (= $4.05\,s^2\,m^{-1}$) and $1.65\,s^2/0.400\,m$ (= $4.13\,s^2\,m^{-1}$) or $4.09 \pm 0.04\,s^2\,m^{-1}$. Prove for yourself that these estimates give $g = 9.65 \pm 0.09\,m\,s^{-2}$.
	Total mark	**20**	

Answers to questions for Key concepts: Thermal and nuclear physics, medical imaging and astrophysics and cosmology (p.351–355)

	Answers	Marks	Comments

1 a i The $_2^4$He nucleus has twice the charge of a $_1^1$H nucleus or proton but it cannot be just two protons because its mass is four times greater than the mass of proton. — **1**

- The extra mass of 2 units cannot be due to a single particle as mass numbers are integer units.

Alternative: — **1**

Rutherford used the neutron–proton model to explain why the mass number of any nucleus heavier than the $_1^1$H nucleus is greater than its atomic number.

The extra mass must be due to 'unit mass' particles, as mass numbers are integer units.

---

ii The $_1^1$H nucleus is the smallest known nucleus and has the least charge / is a single particle / is a proton. — **1**

---

b i $_4^9$Be, $_6^{12}$C — **1**

---

b ii any 3 from: — **3** — You may recognise that low-energy alpha particles are scattered elastically.

The α-particle has to have sufficient kinetic energy:

- to overcome the electrostatic repulsion of the nucleus
- to reach the nucleus closely enough
- to experience the strong nuclear force
- which pulls the α-particle into the nucleus
- as the strong nuclear force is stronger than the electrostatic force at close range.

---

c i $E_K = \frac{1}{2}mv^2 = \frac{p^2}{2m}$, where p = momentum = mv

From conservation of momentum, the momentum of the emitted α-particle $p_\alpha = -p_{nuc}$, where p_{nuc} is the momentum of the recoil nucleus. — **1** — Don't forget that the recoil nucleus has a mass of $206\,m_u$ because the a-particle has left the Po-210 nucleus.

Therefore $\dfrac{E_k \text{ for the recoil nucleus}}{E_k \text{ for the } \alpha \text{ particle}} = \dfrac{\left(\frac{p^2}{2m}\right)_{nuc}}{\left(\frac{p^2}{2m}\right)_\alpha} = \dfrac{m_\alpha}{m_{nuc}} = \dfrac{1}{206}$ — **1**

E_K for the recoil nucleus $= \dfrac{4}{206} \times 5.3\,\text{MeV} = 0.1\,\text{MeV}$ — **1**

So the total energy released = 5.3 + 0.1 = 5.4 MeV. — **1**

---

c ii If $R = kE^n$, a graph of $\ln R$ on the y-axis against $\ln E$ on the x-axis should give a straight line with a gradient equal to n. — **1** — This is an excellent example of the use of a log–log graph to determine the numerical power of an equation of the form $y = kx^n$. If this method does not generate a straight-line graph, the equation cannot be of the form above.

R / mm	39	48	53	57	66	78
E / MeV	5.3	6.0	6.5	6.8	7.4	8.3
ln R	3.66	3.87	3.97	4.04	4.19	4.36
ln E	1.67	1.79	1.87	1.92	2.00	2.12

Correct values to 2 or more significant figures for:

- ln R — **1**
- ln E — **1**

---

suitable scales — **1**

correctly labelled scales — **1**

all points plotted correctly — **1**

best-fit line — **1**

correct calculation of gradient — **1**

e.g. gradient $= \dfrac{4.34 - 3.72}{2.10 - 1.70} = 1.55$

$n = 1.55$

n in the range = 1.5 to 1.6 — **1**

Total mark — **20**

Index

Headings in **bold** indicate glossary terms. Page numbers in *italic* refer to figures; page numbers in **bold** refer to tables.